Autonomous Robot Vehicles

I.J. Cox G.T. Wilfong
Editors

Autonomous Robot Vehicles

Foreword by
T. Lozano-Pérez

With 326 Figures

Springer-Verlag
New York Berlin Heidelberg
London Paris Tokyo Hong Kong

Ingemar J. Cox
NEC Research Institute
Princeton, NJ 08540
USA

Gordon T. Wilfong
AT&T Bell Laboratories
Murray Hill, NJ 07974
USA

About the cover:
Two odometers, sketched by Leonardo da Vinci on the basis of a description written some 1,500 years earlier by the Roman engineer, Vitruvius, appear on the first page of the *Codex Atlanticus*. Photograph provided by the Biblioteca Ambrosiana, Italy.

Library of Congress Cataloging-in-Publication Data
Autonomous robot vehicles/Ingemar J. Cox, Gordon T. Wilfong, editors.
p. cm.
Includes bibliographical references.
ISBN 0-387-97240-4 (alk. paper)
1. Mobile robots. I. Cox, Ingemar J. II. Wilfong, Gordon Thomas, 1958–
TJ211.415.A88 1990
629.8'92—dc20 90-30961

Printed on acid-free paper.

Camera-ready copy supplied by the authors.
Printed and bound by Edwards Brothers, Inc., Ann Arbor, Michigan.
Printed in the United States of America.

9 8 7 6 5 4 3 2 1

ISBN 0-387-97240-4 Springer-Verlag New York Berlin Heidelberg
ISBN 3-540-97240-4 Springer-Verlag Berlin Heidelberg New York

To Susan, Zoe, and Vidal

and

To Lorna and Kevin

Foreword

This book presents a number of approaches to the myriad problems one faces in controlling autonomous mobile robots. These problems range from how to turn a robot's wheels to achieve a smooth turn, through how to obtain an optimal estimate of the robot's position based on multiple noisy measurements, to how to plan a collision-free trajectory in an obstacle-filled terrain. The papers in this volume are representative of the work in the field, although no single volume can survey the entire area. This book is timely, since there has been a great deal of progress recently in solving these problems. Furthermore, relevant articles are published in a bewildering array of journals and conferences; bringing them together here is a great service to workers in the field.

In this foreword I want to address briefly two questions about robotics in general and mobile robotics in particular. I hope that the examination of these questions will provide some added context to the material presented in this book:

- Where does the study of mobile robots fit in to the broader field of robotics?
- What are the key science and technology gaps in the field of mobile robots?

Mobile Robots and Robotics

Since its inception in the early 1960s, robotics has generally focused on the design and control of robot manipulators. To a large extent, this has been motivated by the needs of industry. Robot manipulators have found applications in factories (notably, automotive spot welding) at a rapid pace, in spite of the fact that the design, control, and programming of these robots are primitive. Much of the research work in robotics has had as its goal the manufacture of industrial robots that are faster, more accurate, and easier to program. Achievement of these goals can have a direct impact on the applicability of these robot manipulators to new industrial tasks.

The robotics research and development work in the past two decades has led to a variety of advances in industrial practice: Cartesian trajectories, robot programming languages, off-line programming systems, compliance devices for assembly, vision systems for parts location and inspection, seam tracking for arc welding, and direct-drive robots are among the most important examples. These technological advances have, in turn, made possible the automation of new applications: assembly, conveyor-belt following, and seam welding are some examples.

However, there has always been some tension between the constraints of traditional high-volume industrial applications and advances in the scientific and technological base in robotics. In my opinion, the key scientific and technological issue in robotics is that of coping with uncertainty (also known as error or noise) at all levels: modeling, actuating, and sensing. Solving robotics problems given perfect modeling, actuating, and sensing is drastically easier than solving them when substantial uncertainty is present. From the perspective of scientific research, this suggests a research focus on general techniques for a robot to cope with uncertainty, notably, sensing the state of the task, adapting to changes, and reasoning to select the actions needed to achieve a goal.

From the perspective of industrial application, it is equally valid (and usually cheaper in the short term) to structure the environment so as to reduce uncertainty to a negligible level. For example, specially designed parts feeders can be constructed to position precisely and orient the parts that the robot must grasp. Similarly, specially designed grippers can be constructed to ensure that the part is grasped securely and to provide the needed compliance to overcome residual positioning errors during assembly. Of course, the more structured the environment, the less call for robot sensing and reasoning. In fact, the vast majority of industrial robots work in environments that are so structured that they require no external sensing of the environment; the robots simply repeat a prespecified sequence of motions.

I am convinced that when we have robots that can efficiently interpret sensory data and plan a course of action to accomplish a task, they will make possible a host of new industrial and nonindustrial applications. But, how will we get there? Small steps in the direction of increased sensing and reasoning tend not to be cost-effective in large-volume industrial applications. Why? The reasons boil down to cost. Either new hardware is needed or there may be a slight slowdown in the operation; both cost money. And, since the robot will typically be doing the same task many times, it is more cost-effective to structure the environment to reduce the need for sensing and reasoning.

The above line of reasoning depends on the assumption that structuring the environment is in fact not too costly. Of course, the actual cost depends on two factors: the required precision and the size of the environment that needs structuring. The impact of precision is obvious, but one must also consider the size of the environment. To eliminate uncertainty effectively in the identity and position of objects within a well-defined workcell is feasible, whereas to eliminate uncertainty in a whole factory is almost certainly too costly, as is eliminating uncertainty in the great outdoors.

And, yet, there are a large variety of robot applications (industrial, military, scientific, household, and humanitarian) in which a robot will operate in large and unstructured domains, for example, delivering parts inside a large factory from an automated warehouse to an automated workcell, fire fighting, demolishing structures, cleaning industrial waste sites, maintaining nuclear plants, inspecting and repairing underwater structures, assembling structures in outer space, cleaning windows, and aiding the handicapped.

Mobile robots operating in such large, unstructured domains must be able to cope with significant uncertainty in the position and identity of objects. In fact, the uncertainty is such that one of the most challenging activities for a mobile robot is simply going from point A to point B, a relatively trivial activity for an industrial manipulator. As if in compensation for having to cope with more uncertain environments, we do not expect a mobile robot to follow paths or to reach destinations at the same level of precision that we would expect from an industrial manipulator, namely, on the order of a few thousandths of an inch.

The different domain of operation (as measured by uncertainty and required accuracy) of mobile robots leads to a different set of priorities in research, when compared to the research on industrial manipulators that has dominated robotics up to now. I believe that the priorities for mobile robots are squarely in the areas of sensing and reasoning. For this reason, I see research on mobile robots as a powerful driving force to obtain the scientific and technological advances in sensing and reasoning that will lead to a new generation of robots capable of coping with unstructured domains.

The Key Scientific and Technological Gaps

Starting from the premise that coping with uncertainty is the most crucial problem a mobile robot must face, we can conclude that the robot must have the following basic capabilities:

- **Sensory interpretation**: The robot must be able to determine its relationship to the environment by sensing. A wide variety of sensing technologies are available: odometry; ultrasonic, infrared and laser range sensing; and monocular, binocular, and trinocular vision have all been explored. The difficulty is in interpreting these data, that is, in deciding what the sensor signals tell us about the external world.
- **Reasoning**: The robot must be able to decide what actions are required to achieve its goal(s) in a given environment. This may involve decisions ranging from what paths to take to what sensors to use.

Industrial robots can get by with minimal environment sensing and no reasoning because they operate in essentially structured, static environments.

Sensory Interpretation

The key problem in sensing is in making the connection between the signal output by the sensor and the properties of the three-dimensional world. Note that for internal sensors, such as joint position encoders, this connection is fixed and known. For external sensors, the connection is tenuous at best.

Essentially all approaches to the interpretation of sensory information proceed by first identifying a set of features in the sensory data that potentially correspond to entities in the world. A widely used example drawn from image processing is the so-called "edge," which is some curve in an image in which a rapid intensity change occurs. Such a feature is hypothesized to correspond to some three-dimensional entity, either a change in reflectance, such as is due to paint, or a change in three-dimensional structure, such as a change in surface orientation. Many other types of features are also used.

Given these basic features, one can attempt to determine a more complete description of the world (such as building a three-dimensional description based on features in two images), one can try to identify objects or determine the displacement from some previous image, or one can try to combine features from different types of sensors. There is a great deal of research on how to combine features and how to match them to models. There is also a great deal written on how to detect features. In my opinion, the weak link is still feature detection.

Every feature detector I have ever seen in operation is extremely brittle. It may work well under some narrow set of circumstances, but it produces garbage under a slightly wider set of circumstances. By *garbage*, I mean that it will fail to detect the entities of interest in the world even when they are present and that it will detect what it thinks are these entities when they are not present. My contention is that the performance of any sensing system is largely determined after the feature detectors have finished their work. Subsequent processing may be very clever, but at best, it will avoid degrading output.

The trend is to attack this problem by so-called sensor fusion, that is, by combining the outputs of multiple feature detectors possibly operating on a variety of sensors or simply multiple observations of the same object. Much of this work has focused on applications of Kalman filtering, which essentially provides a mechanism for weighting the various pieces of data based on estimates of their reliability (covariances when we assume the noise has a Gaussian distribution). This is reasonable as far as it goes. But no amount of Kalman filtering will restore a missing feature or prevent a totally erroneous feature from having an impact on the estimate.

I believe that part of the problem is that models that characterize a sensor or feature detector by a covariance matrix are too weak; they do not capture the relevant physics. The fact is that, in some cases, the detector will provide very good estimates, whereas in other cases the estimates will be completely useless data. An extreme example is the output of an ultrasound sensor. If a surface is nearly normal to the beam, then a good distance measurement is obtained; otherwise, the beam is bounced and either not detected or, worse, detected after having bounced off several surfaces. In fact, examining the returned energy may allow one to differentiate these cases, but no single covariance matrix captures such extremes of behavior. To some extent, all sensor/feature detectors have this combination of low noise and drastic errors. Better sensor models and better processing based on those models are required to interpret the sensory data.

Furthermore, I believe systems make irreversible decisions (such as feature/no feature) too early in the processing of data. The motivation is usually to reduce the amount of data that need to be processed, but the results can be disastrous. Methods that postpone irreversible commitments should receive more attention.

Reasoning

In the robotics domain, the most prominent reasoning problems are primarily geometric in nature, such as picking safe paths or determining where to grasp an object. A great deal is known about these problems; many algorithms have been developed for cases in which complete information exists. Once again, one sees that the problems become dramatically more difficult when uncertainty is introduced. But, given that uncertainty is an unavoidable fact of life for a mobile robot, we must develop methods that can ensure that a task can be achieved in spite of the uncertainty.

Part of the solution is to exploit sensing to maintain as good a model of the current situation as possible. But, because of the traditional difficulty in interpreting sensory information, robot systems (both in the factory and in the laboratory) have been developed that minimize the amount of sensing they do. As a result, most robots operate open-loop relative to the task; that is, they perform a preprogrammed sequence of actions that are supposed to achieve their goals and seldom check to see that they, in fact, have achieved them. As a result, these programs are susceptible to even very small variations in the world.

But sensing, in itself, is not enough. We must be able to exploit sensed data in choosing the actions of the robot. The traditional encodings of robot action plans are either as a fixed sequence of actions or as an arbitrary computer program. Fixed sequences cannot cope with changes in the world, whether identified by sensors or not. Arbitrary programs can, in principle, cope with the output of sensors, but existing programming systems operate at too low a level (at the level of individual robot motions). It can take months or even years of programming to write a robust, sensor-based program for even a mildly complex task in an unstructured environment. This is unacceptable, since it effectively limits robots to repetitive operations in which the development time can be amortized over very large runs.

In many ways, the basic paradigm of robot programming through procedural languages is itself unsatisfactory. Procedural programs are intended for computation, not interaction; their structure is predicated on functional decomposition, which is predicated on having relatively few side effects. Even in traditional programming domains that emphasize interaction such as input/output, real-time operating systems, and graphics interface programming, special programming tools (such as interrupts and event-based programming) have been developed to supplement traditional procedural abstraction. The robot programming domain has even more severe requirements because of the prevalence of uncertainty (in computers, the state of the world is always knowable exactly, not so in robots) and the impact of three-dimensional geometry and robot kinematics (a small change in the position of a part is the difference between being safe and having a collision, between a part being reachable by the robot and being outside the workspace).

I believe a robot's action plan for unstructured environments must be encoded as a function of the state of the world. In this sense, a robot plan is like a feedback loop: a mapping from sensors to actuators. But, unlike a feedback loop, a robot plan should be a function from states of the world to goals for the robot (not robot actions directly). The goals can then be mapped into actions by geometric planners that can take into account the current state of the world and the capabilities of the robot. The reason why traditional feedback loops are not an adequate model for robot action plans is that feedback loops are applied to systems modeled by differential equations, usually linear differential equations. These systems are smooth: Small changes in the input produce small changes in the output. The three-dimensional interactions between objects do not have this property; small changes in the input must lead to large changes in the commanded output.

This approach presents a number of important questions, for example:

- What is a good choice for the structure of action plans?
- When can an action plan be guaranteed to achieve a goal?
- How can one evaluate the performance of an action plan?

All these questions have theoretical as well as experimental implications.

Summary

In summary, I believe that work in mobile robots represents an important development in robotics and is likely to be the cutting edge of robot research and applications in the coming years. I hope that this volume will encourage its readers to participate in these exciting developments.

Cambridge, Massachusetts Tomás Lozano-Pérez

Acknowledgments

The editors acknowledge the help and support of AT&T Bell Laboratories, in particular, G.L. Miller, Monnett Soldo, and members of the Robotics Principles Research Department. Special thanks are due J.H. Maddocks of the University of Maryland and W.L. Nelson of AT&T Bell Laboratories for many fruitful discussions regarding the chapter on guidance. We also thank Peter S. Maybeck of the Air Force Institute of Technology at Wright-Patterson Air Force Base, especially for his excellent course on Kalman filtering. We are indebted to Richard L. Greenspan of the Charles Stark Draper Labs who provided excellent advice and information for the chapter on navigation. Jim Nevins of Charles Stark Draper Labs, Norm Caplan of the National Science Foundation, and S. Yuta of Tsukuba University, Japan, provided information on international autonomous robot projects. Last, but not least, we thank the included authors for their invaluable support, assistance, and advice. The editors, assume full responsibility for any errors in or deficiencies of this book.

Contents

Introduction

There is significant interest in autonomous vehicles – vehicles that are capable of intelligent motion and action without requiring either a guide to follow or teleoperator control. Potential applications of autonomous vehicles are many and include reconnaissance/exploratory vehicles for space and undersea, land, and air environments; remote repair and maintenance; material handling systems for the office and the factory; and even intelligent wheelchairs for the handicapped. It is likely that autonomous vehicles will be commonplace in the not too distant future.

The papers included in this book are primarily concerned with land-based autonomous vehicles. However, the general principles remain the same whether the vehicle is intended to be driven, flown, floated, or submerged. All autonomous vehicles must navigate within an environment. In order to achieve this, the vehicle must be capable of (1) sensing its environment, (2) interpreting this sensor information to refine its knowledge of its position and the environment's structure, and (3) planning a route from an initial to a goal position in the presence of known or perhaps unknown obstacles. Having reached its destination, the autonomous vehicle may be required to perform some task such as manipulating objects, information gathering, or sample collecting. The tasks performed are very specific to the application in question. However, the fundamental principles involved in moving a vehicle from one place to another, for example. navigation, are common to all autonomous vehicles, and it is with these principles that this book is concerned. Some of these principles are unique to mobile robots; there are no analogies with robot manipulators. For example, navigational issues are unique to mobile robots: There is no corresponding, potentially unbounded uncertainty in the end effector position of a robot arm. Its position, to within a fixed tolerance, can always be determined by reading its joint encoder values. Consequently, this collection of papers addresses only issues of autonomous robot motion, with no coverage of the issues of manipulation, which it is believed are better dealt with under the topic of robot arms.

It is often hard to appreciate the difficulties inherent in the design of an intelligent autonomous vehicle. After all, people and animals have no trouble moving around in a complex environment – sensing, interpreting, and responding to new information. Why is it so difficult to construct an autonomous vehicle? The analogy with living animals is deceptive; no matter how "primitive" such systems are, they are still orders of magnitude more complex than the state-of-the-art autonomous vehicle systems being constructed today. At this juncture, a caveat is in order. Although the artificial intelligence community has long been a driving force for autonomous vehicle research, we believe that the field has matured into two distinct, though somewhat overlapping areas of research. This book is not about the design of "artificially intelligent" mobile machines, that is, machines that exhibit animal-like behavior. Although the behavior-oriented subdiscipline of the artificial intelligence community is an extremely interesting line of study, a second line of research is fundamentally task oriented: What are the fundamental problems in designing a vehicle to perform tasks such as material handling or planetary exploration? This book is concerned with the latter problem.

Even if an autonomous vehicle is given exact information about its environment, the problem remains complex. In reality, the problem is even further complicated by the fact that the information obtained by the vehicle's sensors is never free of noise. For example, small undulations in the floor cause errors to accumulate in odometric sensors. The long-term drift of an inertial guidance sensor also results in errors in the estimated position of the vehicle. If corrective action is not taken, the vehicle can easily become lost. Corrective action takes the form of additional sensing of the environment to reduce these position errors: Can we recognize a landmark, say? These other sensors are also noisy and may cause other problems. Ultrasonic sensors, for example, are plagued by problems of specular reflection. Television cameras are noisy, and low-level feature detection algorithms often perform poorly. The goal of autonomous robot vehicle research is to develop systems that are robust and reliable in the presence of many sources of noise. In addition, there is often an implicit compromise: cost! Ultrasonic range finders may have poor performance compared with optical range finders, but they are significantly cheaper.

Vehicular motion may be wheeled, tracked, or legged for land based systems, but irrespective of the mode of locomotion, many of the same problems persist. We believe that autonomous vehicle research can be categorized into six independent subdisciplines: guidance, sensors, navigation, cartography, sensing strategies, and motion planning.

This book begins by considering the *guidance* of various vehicle geometries. At this level, one is typically concerned with moving a vehicle along a desired trajectory (usually geometrically defined but possibly reflexive; e.g., follow the wall), when the vehicle's state, that is, position, heading, and velocity, is known exactly. Guidance is therefore solely concerned with trajectory generation and the kinematics, dynamics, and feedback control of the vehicle. In practice, neither the vehicle's nor the environment's state is known exactly, and sensing is therefore required. Chapter 2 describes some common and useful sensor systems and emphasizes accurate modeling of the sensor and its noise sources. *Navigation* is the science of position and course estimation. The algorithms developed assume no particular model of the sensors or vehicle kinematics. However, the performance of these navigation algorithms is significantly determined by the accuracy of the actual sensor and the associated sensor and vehicle kinematic models. Once more, the emphasis is on the need to quantify performance, this time the navigational estimate. Navigation almost always requires some form of map. Various methods for representing geometric and geographic information appear in Chapter 4. A map may not always be complete or remain current. Under these circumstances, it is necessary to update the map with new sensor information. Chapter 5 discusses several different approaches to minimizing the number of sensor readings necessary to perform these activities. At the highest level, motion planning is required to determine a collision-free motion between the initial state and the desired state of the vehicle. Several techniques for motion planning are discussed in Chapter 6. Autonomous vehicle research is multidisciplinary and requires expertise in many different areas. In addition, it must be remembered that autonomous vehicles are complex robot *systems*. Although this book does not offer any papers that explicitly deal with the construction of complex systems, Chapter 7 concludes with a description of four autonomous robot systems.

This book is intended to convey to the reader some of the successes and some of the hurdles associated with autonomous vehicle research. We have not attempted to refer to all the relevant publications. The omission of any particular reference should therefore not be interpreted as due to a lack of regard for its merit. However, we have included many representative papers from diverse references with the hope and intention of broadening the reader's perspective. We also hope the book appeals to a wide variety of disciplines, since mobile robot research is inherently multidisciplinary and there remain many open and interesting problems in all areas. We are excited by this expanding area of research and expect that autonomous vehicles will significantly affect the world in which we live.

Princeton, New Jersey
Murray Hill, New Jersey

Ingemar J. Cox
Gordon T. Wilfong

Permissions

On the Kinematics of Wheeled Mobile Robots
J.C. Alexander and J.H. Maddocks
© 1990 MIT Press. Reprinted with permission from *The International Journal of Robotics Research*. Not yet published.

Kinematic Modeling for Feedback Control of an Omnidirectional Wheeled Mobile Robot
P.F. Muir and C.P. Neuman
© 1987 IEEE. Reprinted with permission from *IEEE International Journal of Robotics and Automation,* 1772–1778.

An Automatic Guidance System of a Self-Controlled Vehicle
T. Hongo, H. Arakawa, G. Sugimoto, K. Tange, and Y. Yamamoto
© 1987 IEEE. Reprinted with permission from *IEEE Transactions on Industrial Electronics,* Vol. *IE-34,* No. 1, 5–10.

Local Path Control for an Autonomous Vehicle
W.L. Nelson and I.J. Cox
© 1988 IEEE. Reprinted with permission from *Proceedings of the 1988 IEEE International Conference on Robotics and Automation,* Philadelphia, Pennsylvania, April 24–29, 1988, 1504–1510.

Running with Symmetry
M.H. Raibert
© 1986 MIT Press. Reprinted with permission from *The International Journal of Robotics Research,* Vol. *5,* No. 4, 3–19.

Smooth Local Path Planning for Autonomous Vehicles
Y. Kanayama and B.I. Hartman
© 1989 IEEE. Reprinted with permission from *Proceedings of the 1989 IEEE International Conference on Robotics and Automation,* Scottsdale, Arizona, May 14–19, 1265–1270.

Sensor Models and Multisensor Integration
H.F. Durrant-Whyte
© 1988 MIT Press. Reprinted with permission from *The International Journal of Robotics Research,* Vol. *7,* No. 6, 97–113.

Location Estimation and Uncertainty Analysis for Mobile Robots
C. Ming Wang
© 1988 IEEE. Reprinted with permission from *Proceedings of the 1988 International Conference on Robotics and Automation,* Philadelphia, Pennsylvania, April 24–29, 1230–1235.

Inertial Navigation
M.M. Kuritsky and M.S. Goldstein
© 1983 IEEE. Reprinted with permission from *Proceedings of the IEEE,* Vol. *71,* No. 10, 1156–1176.

Continuous Transmission F.M. Sonar with One Octave Bandwidth and No Blind Time
P.T. Gough, A. de Roos, and M.J. Cusdin
© 1984 Institute of Electrical Engineers. Reprinted with permission from *IEE Proceedings,* Vol. *131,* Part F, No. 3, 270–274.

An Optical Rangefinder for Autonomous Robot Cart Navigation
G.L. Miller and E.R. Wagner
© 1987 Society of Photo-Optical Instrumental Engineers. Reprinted with permission from *Proceedings of the SPIE,* Vol. *852,* Mobile Robots II, 132–144.

Error Modeling in Stereo Navigation
L. Matthies and S.A. Shafer
© 1987 IEEE. Reprinted with permission from *IEEE Journal of Robotics and Automation,* Vol. *RA-3,* No. 3, 239–248.

Statistical Theory of Passive Location Systems
D.J. Torrieri
© 1984 IEEE. Reprinted with permission from *IEEE Transactions on Aerospace and Electronic Systems,* Vol. *AES-20,* No. 2, 183–198.

Estimating Uncertain Spatial Relationships in Robotics
R. Smith, M. Self, and P. Cheeseman
© 1988 Elsevier Science Publishers. Reprinted with permission from Lemmer/Kanal (Eds.), *Uncertainty in Artificial Intelligence,* Vol. 2, 435–461.

The Kalman Filter: An Introduction to Concepts
P.S. Maybeck
© 1979 Academic Press. Reprinted with permission from Maybeck, P.S., *Stochastic Models, Estimation and Control,* Vol. 1, 3–16.

Maintaining Representations of the Environment of a Mobile Robot
N. Ayache and O.D. Faugeras
© 1989 IEEE. Reprinted with permission from *IEEE Transactions in Robotics and Automation,* Vol. *5,* No. 6, 804–819.

Blanche: Position Estimation for an Autonomous Robot Vehicle
I.J. Cox
© 1989 IEEE. Reprinted with permission from *Proceedings of the IEEE/RSJ International Workshop on Robots and Systems* (IROS '89), Sept. 4–6, 1989, Tuksuba, Japan, 432–439.

Sonar-Based Real-World Mapping and Navigation
A. Elfes
© 1987 IEEE. Reprinted with permission from *IEEE Journal of Robotics and Automation,* Vol. *RA-3,* No. 3, 249–265.

Cartographic Databases
S.K. Honey and M.S. White
From CD ROM The New Papyrus. Reprinted by permission of Microsoft Press © 1986. All rights reserved.

Spatial Planning: A Configuration Space Approach
T. Lozano-Pérez
© 1983 IEEE. Reprinted with permission from *IEEE Transactions on Computers,* Vol. *C-32,* No. 2, 108–120.

Simplified Voronoi Diagrams
J. Canny and B. Donald

Solving the Find-Path Problem by Good Representation of Free Space
R.A. Brooks

Computational Complexity of Art Gallery Problems
D.T. Lee and A.K. Lin

Sensing Strategies for Disambiguating Among Multiple Objects in Known Poses
W.E.L. Grimson

Probing Convex Polytopes
D. Dobkin, H. Edelsbrunner, and C.K. Yap

An Automatic Motion Planning System for a Convex Polygonal Mobile Robot in 2-Dimensional Polygonal Space
K. Kedem and M. Sharir

Path-Planning Strategies for a Point Mobile Automaton Moving Amidst Unknown Obstacles of Arbitrary Shape
V.J. Lumelsky and A.A. Stepanov

Motion Planning for an Autonomous Vehicle
G.T. Wilfong

Real-Time Obstacle Avoidance for Manipulators and Mobile Robots
O. Khatib

The Stanford Cart and the CMU Rover
H.P. Moravec

An Integrated Navigation and Motion Control System for Autonomous Multisensory Mobile Robots
G. Giralt, R. Chatila, and M. Vaisset

A Mars Rover for the 1990's
B.H. Wilcox and D.B. Gennery

A Mobile Robot: Sensing, Planning, and Locomotion
D.J. Kriegman, E. Triendl, and T.O. Binford

1. Guidance: Kinematics, Control, and Trajectory Generation

The controlling intelligence understands its own nature, and what it does, and whereon it works.

Marcus Aurelius Antoninus, *Meditations VI* (AD 112–180)

We begin with a discussion of the lowest level of vehicular motion control, where we are concerned with moving a vehicle through a specified trajectory subject to kinematic and dynamic constraints.[1] The mechanical configuration of most wheeled land vehicles falls into one of two categories, steered-wheeled vehicles or differential-drive vehicles. The steered-wheeled configuration typically consists of a fixed rear axle and a single-steered (and often driven) front wheel. The tricycle configuration of many mobile robots is a common example, but general four-wheeled vehicles can be considered in this category if the two steered wheels are approximated by a single wheel at the center. The precise nature of the approximations necessary for a four- (or more) wheeled vehicle to be modeled by a tricycle is examined in Alexander and Maddocks [1]. The most common example of a differentially driven vehicle is the military tank, in which steering is effected by differences in the velocity of the two tracked wheels. An advantage of differential-drive vehicles is the ability to "turn on a dime," that is, they possess a zero turning radius. In comparison, the physical limits associated with the maximum steering angle can cause a steered-wheeled vehicle to have a large turning radius and thus make maneuvering more difficult in a cluttered environment.

The kinematics of a simple tricycle or a differential-drive vehicle is very straightforward and can usually be found as small sections within larger papers; for example, see the reprinted papers of Hongo et al., and of Nelson and Cox, respectively. However, both vehicle configurations have the disadvantage of not permitting arbitrary, rigid-body trajectories. For example, neither vehicle can move parallel to its fixed axle. Although this is a limitation of tricycle and differential-drive vehicles, it is not necessarily true of other mechanical configurations. We begin this chapter with a paper by Alexander and Maddocks that examines the question of what planar, rigid-body motions can be achieved for a given configuration with specified fixed and steered wheels and what steering and wheel velocities are required to attain these motions. Should incompatible steering and wheel velocities be applied to the vehicle, the subsequent motion will also include slippage. The paper also contains an analysis of slippage as well as many excellent references the reader is encouraged to pursue.

The invention of the Ilonator wheel [9] has allowed the construction of a further category of vehicle. The Ilonator wheel consists of a hub about which are mounted 12 rollers, each making a 45° angle to the wheel orientation. This arrangement allows the wheel to move in any direction, not just tangential to the wheel direction, without slipping. It has to been seen to be believed! A typical vehicle consists of a rectangular

[1] We are unaware of any introductory texts on kinematics and dynamics that are oriented toward mobile vehicles. The reader is therefore directed to the textbooks of Craig [4] and Paul [19] which discuss kinematics and trajectory generation for robot manipulators and to Dijksman [5], Martin [15], and Suh and Radcliffe [23] for more general texts on the kinematics and dynamics of mechanical systems.

frame on which are mounted four omnidirectional Ilonator wheels. Motion is accomplished by varying the relative velocity of the four wheels. The paper by Muir and Neuman examines the kinematics of an omnidirectional vehicle based on the use of Ilonator wheels. A very comprehensive treatment of the kinematics of steered-wheeled, differential-drive, omnidirectional, and ball-wheeled vehicles can be found in a paper by Muir and Neuman [16], but it is, unfortunately, too long to reprint here.

Of course, land vehicles need not be wheeled. Legged vehicles have the potential advantage of being capable of traveling over very rugged terrain that even tracked vehicles cannot traverse. Apart from obvious military applications, legged vehicles may have significant applications in mining, forestry, hazardous environment, and autonomous planetary exploration. Raibert's paper examines how symmetry properties can simplify both the kinematics and control of dynamic legged systems. This is one of many fascinating papers on a topic that has already been the subject of at least two books [20, 22]. We clearly cannot do this subject justice in this book, but we hope we have whetted the appetite of the reader.

The feedback control of an autonomous robot vehicle can sometimes present subtle and surprising problems. In particular, it is important to realize that a vehicle's position or pose is represented by three parameters (x,y,θ), two for translation and one for orientation, when we assume only planar motions. However, for common tricycle and differential-drive vehicle configurations, there are only two degrees of freedom for control, that is, steering angle and velocity or the independent velocity of the two wheels of a differential-drive vehicle. Such systems are referred to as nonholonomic [11-14, 24]. Although there has been a significant amount of work on the control of robot manipulators, there has been little theory on the control of autonomous mobile vehicles, presumably because of the newness of the field. We have included two papers on the control of mobile vehicles. The first, by Hongo et al., describes the control of a differential-drive vehicle. Note that the weighting of the control gains by the inverse of the velocity squared. This is intuitively reasonable, since only small turns of the steering wheel are required when traveling at high speed compared with much larger turns at low speeds. However, such a scheme may become unstable as the vehicle's speed slows to zero. The second paper, by Nelson and Cox, describes control of a tricycle. Here the velocity dependence of the gain parameters is ignored, although later work compensated for velocity. The interested reader is also directed to Graettinger and Krogh [7] for a discussion of time scaling of velocity to satisfy kinematic and dynamic constraints. Although both control schemes perform adequately, there appears to be more scope for significant work in this area to address issues of control for nonholonomic vehicles. Finally, we draw the reader's attention to interesting work on the human control of a nonholonomic skateboard [8], which also contains other fascinating references, for example, the human control of a bicycle.

Given initial and terminal poses, that is, position and orientation, it is necessary to generate a path linking the two points. Here we are not concerned with issues of obstacle avoidance and associated path planning (see Chapter 6), but rather the generation of curves possessing prescribed initial and terminal positions and tangents, when an obstacle-free environment is assumed. The earliest work on trajectory generation the editors are aware of is that of Dubins [6], who proved that, for a vehicle constrained to move only forward and having a finite turning radius, the shortest path from initial to terminal pose consists of at most three segments of the form of a circular arc, line, arc or arc, arc, arc. Reeds and Shepp [21] have recently extended this work to vehicles that are allowed to reverse. Cockayne and Hall [3] have also performed interesting work in this area. Given such a reference trajectory, it is necessary to select the guide point on the vehicle that is required to track this trajectory. For common steered-wheeled and differential-drive vehicles, it is usual to select the midpoint of the (rear) axle, since the direction of the vehicle is always tangent to the trajectory of this point. And herein lies a problem; for such a point to track a curve that has discontinuities in curvature, for example, the point where a circular arc meets a straight line, necessitates discontinuities in steering angle or wheel velocities. Three solutions exist. The first is to allow the vehicle to stop. Usually this is not desired, primarily because of time constraints. The second solution is to relocate the vehicle's guide point. For example, if, for the case of a tricycle, the guide point is moved to the point of contact between the steered wheel and the floor, then the vehicle is capable of following any curve for which the rate of change of the tangent direction does not exceed the maximum acceleration of the steering motor. However, a disadvantage of such an approach is that the orientation of the vehicle is no longer tracking the tangent direction of the curve. Consequently, after the front wheel has traversed a 90° arc segment, the vehicle's orientation has not changed by 90° [17].

The third solution lies in the development of trajectories that have continuous curvature. Kanayama and Miyake, in their early work on this subject, proposed the use of clothoid curves [10].[2] Nelson has recently shown [18] that such curves correspond to minimum-length curves under a peak-jerk, that is, the peak rate of change of centripetal acceleration constraint.[3] The last paper in this chapter describes more recent work by Kanayama and Hartman who derived curves that minimize the integral of the square of the derivative of curvature cost function.[4] Physically, these curves, called cubic spirals, minimize the integral-square-jerk. For "symmetric" postures, see the paper for definition, the work of Nelson [17] is of particular interest. Lastly, the computer graphics community has also been interested in geometric continuity of curves, for example, see Barsky and DeRose [2].

References

1. Alexander, J.C., and Maddocks, J.H. On the Maneuvering of Vehicles. *SIAM J. of Appl. Math. 48*, 1 (1988), 38-51.
2. Barsky, B.A., and DeRose, T.D. Parametric Curves. *IEEE Comput. Graphics Appl. 9*, 6 (1989), 60-68.
3. Cockayne, E.J., and Hall, G.W.C. Plane Motion of a Particle Subject to Curvature Constraints. *SIAM J. Control 13*, 1 (1975), 197-220.
4. Craig, J.J. *Introduction to Robotics Mechanics and Control.* Addison-Wesley, Reading, Mass. 1986.
5. Dijksman, E.A. *Motion Geometry of Mechanisms.* Cambridge University Press, Cambridge, England, 1976.
6. Dubins, L.E. On Curves of Minimal Length with a Constraint on Average Curvature, and with Prescribed Initial and Terminal Positions and Tangents. *Am. J. Math. 79* (1957), 497-516.
7. Graettinger, T.J., and Krogh, B.H. Evaluation and Time-Scaling of Trajectories for Wheeled Mobile Robots. In *Proceedings of the 1988 Americal Control Conference*, American Automatics Control Council, 1988, pp. 511-516.

[2]Recently, clothoid curves have been independently discovered by the designers of roller coaster amusement park rides. Most loops are now described by clothoids.

[3]Note that accelerations refer only to the guide point, not to, say, the center of mass of the vehicle.

[4]Whenever an axle is allowed to pivot, the center of curvature associated with the trajectory and the center of instantaneous rotation associated with the vehicle do not necessarily coincide [1].

8. Hubbard, M. Human Control of the Skateboard. *J. Biomechanics 13* (1980), 745-754.

9. Ilon, B.E. Wheels for a Course Stable Self-Propelling Vehicle Movable in any Desired Direction on the Ground or Some Other Base. U.S. Patent No. 3,876,255, 1975.

10. Kanayama, Y., and Miyake, N. Trajectory Generation for Mobile Robots. In *3rd International Symposium on Robotics Research*, O.D. Faugeras and G. Giralt, Eds., 1986, pp. 333-340.

11. Kane, R. Dynamics of Nonholonomic Systems. *J. Appl. Mechanics 28*, 4 (1961), 574-578.

12. Laumond, J. Feasible Trajectories for Mobile Robots with Kinematic and Environment Constraints. In *Proceedings on Intelligent Autonomous Systems*, 1986, pp. 346-354.

13. Lobas, L.G. Nonholonomic System Which Models Rolling of a Tricycle Along a Plane. *Soviet Appl. Mechanics 16* (1980), 346-352.

14. Lobas, L.G. Trajectories of a Two-Stage Mechanical System with Rolling. *Soviet Appl. Mechanics 16* (1981), 1084-1089.

15. Martin, G.H. *Kinematics and Dynamics of Machines.* McGraw-Hill, New York, 1982.

16. Muir, P.F., and Neuman, C.P. Kinematic Modeling of Wheeled Mobile Robots. *J. Robotic Systems 4*, 2 (1987), 281-340.

17. Nelson, W.L. Continuous Steering-Function Control of Robot Carts. *IEEE Trans. Indust. Electronics 36*, 3 (1989), 330-337.

18. Nelson, W.L. Continuous-Curvature Paths for Autonomous Vehicles. In *IEEE International Conference on Robotics and Automation*, IEEE, New York, 1989, pp. 1260-1264.

19. Paul, R.P. *Robot Manipulators: Mathematics, Programming and Control.* MIT Press, Cambridge, Mass., 1981.

20. Raibert, M.H. *Legged Robots that Balance.* MIT Press, Cambridge, Mass., 1986.

21. Reeds, J.A., and Shepp, L.A. Optimal Paths for a Car that Goes Both Forward and Backwards. *Pacific J. Math.* (to be published).

22. Song, S. and Waldron, K.J. *Machines that Walk: The Adaptive Suspension Vehicle.* MIT Press, Cambridge, Mass., 1988.

23. Suh, C.H. and Radcliffe, C.W. *Kinematics and Mechanisms Design.* Wiley, New York, 1978.

24. Whittaker, E.T. *A Treatise on the Analytic Dynamics of Particles and Rigid Bodies.* Cambridge University Press, London, England, 1964.

ON THE KINEMATICS OF WHEELED MOBILE ROBOTS

J. C. Alexander† and J. H. Maddocks‡

Department of Mathematics
University of Maryland
College Park, Maryland 20742

A wheeled mobile robot is here modelled as a planar rigid body that rides on an arbitrary number of wheels. The relationship between the rigid body motion of the robot and the steering and drive rates of wheels is developed. In particular, conditions are obtained that guarantee that rolling without skidding or sliding can occur. Explicit differential equations are derived to describe the rigid body motions that arise in such ideal rolling trajectories. The simplest wheel configuration that permits access of arbitrary rigid-body motions is determined. Then the question of slippage due to misalignment of the wheels is investigated by minimization of a nonsmooth convex dissipation functional that is derived from Coulomb's Law of friction. It is shown that this minimization principle is equivalent to the construction of quasi-static motions. Examples are presented to illustrate the models.

1. Introduction

In this paper, we analyze the kinematics of a *wheeled mobile robot*, or *WMR*. Such robots ride on a system of wheels and axles, some of which may be steerable or driven. There are many wheel and axle configurations that have been used for WMRs (Whitaker 1962; Lewis and Bejczy 1973; Smith and Coles 1973; Hollis 1977; Everett 1979; Giralt, Sobek and Chatila 1979; Moravec 1980; Iijima, Kanayama and Yuma 1981a; Iijima, Kanayama and Yuma 1981b; Balmer, Jr. 1982; Carlisle 1983; Helmers 1983b; Helmers 1983a; Ichikawa, Ozaki and Sadakane 1983; Moravec 1983; Johnson 1984; Nilsson 1984; Podnar, Dowling and Blackwell 1984; Rogers 1984; Helmers 1985; Holland 1985; Marrs 1985; Wallace et al. 1985; Wilson 1985; Moravec 1986). For other related work, see Kanayama and Miyake (1986), Fortune and Wilfong (1988), Wilfong (1988). The ultimate objective of our investigations is the complete description of the kinematics and inverse kinematics of such robots during low-speed maneuvering.

† The work of this author was partially supported by the National Science Foundation.

‡ The work of this author was supported by the U. S. Air Force Office of Scientific Research.

Much of the research cited above is described in a recent study of WMRs (Muir and Neuman 1987). This work develops a formalism that is used first to model the kinematics of each wheel, and second to amalgamate the information about individual wheels to describe the kinematics of the WMR regarded as a whole. A condition is developed that determines whether, given the configuration of the wheels in the WMR, ideal rolling is possible. If not, a least-squares fit to rolling is obtained. Some of the same questions are considered here. However, our approach is somewhat different, and, we believe, complementary. One major difference is that our analysis of cases in which ideal rolling fails is based on physical models of friction.

Attention is restricted to the problem of maneuvering a WMR on a horizontal plane. Precise and explicit connections between the steering and drive rates of the various wheels and the position and orientation of the robot are obtained. The inverse kinematic problem of determining the steering and drive rates that produce a prescribed robot trajectory is also resolved. We determine the simplest configuration of steerable and driven wheels that allows the robot to be maneuvered in arbitrary planar motions. Slippage resulting from wheel configurations that are incompatible with ideal rolling is also considered. Our description of slippage is based on the use of Coulomb's Law to model friction. It is shown that a quasi-static analysis, in which forces and moments are balanced, is equivalent to the minimization of a non-smooth, convex *dissipation* functional. This dissipation, or *friction*, functional is the weighted sum of the euclidian norms of certain vectors. For illustration, a complete analysis of the simplest application is presented.

The qualification of low speed arises in this work because our model of rolling does not consider inertial forces and accelerations. It will become apparent that this rolling model retains its validity until the inertial forces arising from accelerations are so large as to saturate the available frictional forces between the wheels and the surface. Thus the low-speed model may in fact be valid for relatively fast motions, provided either that the turns are not too tight or that the friction between the wheels and the surface is sufficiently large.

In Section 2, notation and definitions are introduced, and some basic results are presented. In the first instance, all wheels are assumed to roll; i.e., there is no slipping or sliding. This requirement places compatibility conditions on the motions of the wheels. These conditions and their consequences are discussed in Section 3, which is a reformulation and extension of results obtained in Alexander and Maddocks (1988). In this prior work, attention was focused on intrinsic properties of rigid-body trajectories, such as curvatures and centers of rotation. In the present development, more emphasis is given to formulations that would allow efficient numerical treatment. The conclusion of Section 3 is that the specification of steering and drive rates of two wheels are necessary and sufficient to specify arbitrary planar motions. The four rates under our control (two steering, two driving) must satisfy one (transcendental) compatibility condition. The explicit connections between the steering and drive rates and the rigid-body motion are described in Section 4, where a resolution of the inverse-kinematic problem is also given. A practical design is suggested in Section 5, and, in Section 6, the kinematics of three existing WMRs are compared with our proposed design.

Our kinematic development is then extended to consider failure of the rolling model due to slippage of the wheels. There are two distinct circumstances in which slippage will occur. The first has already been mentioned: the rolling model can fail to be a good approximation because of large inertial forces that saturate the available friction. This mode of failure of the rolling model is associated with high-speed maneuvering. The second mode of failure is that the steering and driving controls of the wheels are not compatible with ideal rolling. This lack of compatibility can arise at any speed. The first circumstance, which we call *skidding*, is not considered in this paper except to note that the friction functional described in Section 7 provides a technique to predict the onset of skidding. Quasi-static evolution equations for low speed maneuvering involving the second mode of failure, which we call *slippage* (sometimes called *scrubbing*), are developed in Section 7 and illustrated in Section 8.

2. Notation and Definitions, Elementary Results

A WMR is modelled as a planar rigid *robot body* that moves over a horizontal reference plane on wheels that are connected to the body by axles. For our purposes, the only role of the body of the WMR is to carry a moving coordinate system. Thus the *body coordinates* of a vector $\mathbf{x}$ are denoted ${}^{\mathrm{B}}\mathbf{x} = (x_1, x_2)$, while the underlying two-dimensional *space coordinates* are denoted ${}^{\mathrm{R}}\mathbf{x}$. The ith *axle*, $i = 1, \ldots, m$, is attached to the body at the *axle* or *constraint point* $\mathbf{x}_i$ with body coordinates ${}^{\mathrm{B}}\mathbf{x}_i$ and space coordinates ${}^{\mathrm{R}}\mathbf{x}_i$. The body coordinates are known constants; the space coordinates are unknown functions of time. The axle is supported by a single *simple wheel*, which is idealized as a disc, without thickness, of radius R_i that lies in a vertical plane through the axle point. A wheel can rotate in its vertical plane about its center point (which is attached to the axle). If it is *driven*, it will rotate at a prescribed angular speed. Otherwise it is *passive*, and its rotation rate is determined by the kinematics of the WMR. It may also be possible for the axle to rotate in the WMR about the vertical through the axle point. If the wheel (or axle) is *steered*, this rotation rate is prescribed. Otherwise the wheel is *unsteered*. A *fixed* wheel is one for which the axle cannot rotate. Technically, a fixed wheel is a special type of steered wheel, but it is useful to maintain a distinction.

There are a number of other types of wheels that are adopted in robot design, of which a simple wheel [called a *conventional wheel* in §3 of Muir and Neuman (1987)] is but the most straightforward. Our kinematic analysis can be expanded to include many of these more complicated wheels. Some remarks along this line are included in Sections §5 and 6, where practical wheel configurations for WMRs are considered.

Regarded as a rigid body in three-dimensional space, a simple wheel has a three-dimensional vector-valued angular velocity $\boldsymbol{\omega}_i$. Since the wheel remains in a vertical plane

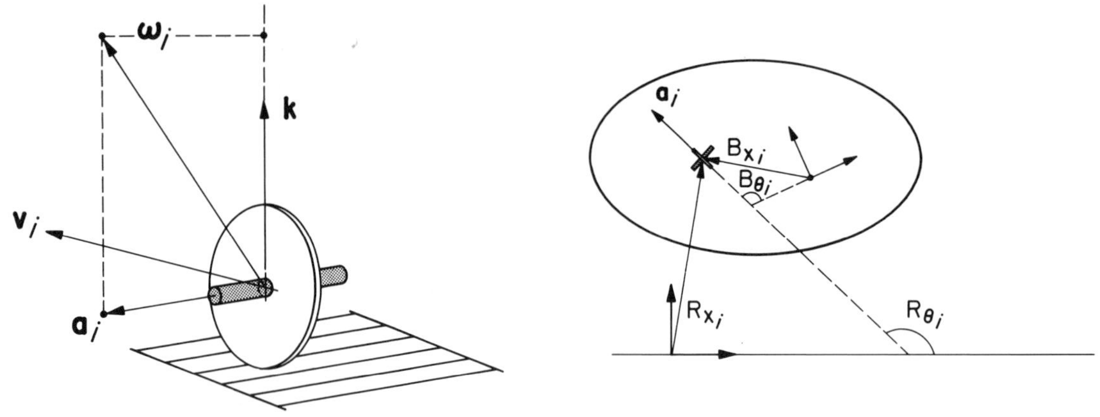

1. Schematic of vectors associated with wheel i. $\mathbf{a}_i$: axle vector along axle; $\mathbf{v}_i$: velocity of axle point $\mathbf{x}_i$; $\boldsymbol{\omega}_i$: angular velocity of wheel; $\mathbf{k}$: unit vertical vector. 2. Schematic of coordinate systems. ${}^{\mathrm{B}}\mathbf{x}_i$: body coordinates of axle point i; ${}^{\mathrm{R}}\mathbf{x}_i$: coordinates of axle point i with respect to frame fixed in space; ${}^{\mathrm{B}}\theta_i$: angle of axle vector $\mathbf{a}_i$ with respect to body frame; ${}^{\mathrm{R}}\theta_i$: angle of axle vector $\mathbf{a}_i$ with respect to fixed frame.

at all times, the 2-dimensional horizontal component of $\boldsymbol{\omega}_i$, denoted $\mathbf{a}_i$, is perpendicular to the plane of the wheel. The vector $\mathbf{a}_i$ is called the *axle vector.* The magnitude ρ_i of the axle vector is the *rotation rate* of the wheel. The axle vector $\mathbf{a}_i$ makes a *steering angle* ${}^{\mathrm{B}}\theta_i$ from a reference direction fixed in the body of the WMR, and an angle ${}^{\mathrm{R}}\theta_i$ from a reference direction fixed in space. The angles ${}^{\mathrm{B}}\theta_i$ and ${}^{\mathrm{R}}\theta_i$ are well-defined whenever $\rho_i \neq 0$. The conventions adopted here assume ${}^{\mathrm{B}}\theta_i$ to be interpreted modulo 2π and ρ_i to be nonnegative. In practical implementations, it may be more convenient to interpret ${}^{\mathrm{B}}\theta_i$ modulo π and to allow ρ_i to be negative (i.e., the wheel may rotate backward). We remark that the definitions of ${}^{\mathrm{B}}\theta_i$ and ${}^{\mathrm{R}}\theta_i$ differ by $\pi/2$ from those given in Alexander and Maddocks (1988). The scalar *pivot* or *steering speed* is defined to be ${}^{\mathrm{B}}\omega_i = {}^{\mathrm{B}}\theta_i'(t)$. The vertical component of the wheel angular velocity is ${}^{\mathrm{R}}\theta_i'(t) = {}^{\mathrm{B}}\omega_i + \Omega$, where Ω is the angular velocity of the robot body in the ambient space.

It is apparent that the motion of a simple wheel relative to the body of the WMR is mathematically characterized by the two scalars ρ_i and ${}^{\mathrm{B}}\theta_i$, or by ρ_i, ${}^{\mathrm{B}}\omega_i$, and an initial value ${}^{\mathrm{B}}\theta_i(0)$. Moreover, the steering angle and rotation rate are the physically relevant kinematic parameters. It should be stressed that the two scalars ρ_i and ${}^{\mathrm{B}}\theta_i$ uniquely characterize the components ${}^{\mathrm{B}}\mathbf{a}_i$ of the axle vector $\mathbf{a}_i$ *in the robot body.* The first major goal of this paper is to investigate how the assumption of ideal rolling determines the motion of the robot body in the ambient space in terms of the ${}^{\mathrm{B}} < a_i$.

The position (or configuration) of the robot body is determined by any two of the axle-point vectors ${}^{\mathrm{R}}\mathbf{x}_i(t)$. Alternatively the body configuration is totally specified given one vector ${}^{\mathrm{R}}\mathbf{x}_i(t)$ and the *body orientation angle* Θ between the reference directions fixed in the body and in space. The angular velocity of the body is $\Omega = \Theta'(t)$. It is apparent that

$$ {}^{\mathrm{R}}\theta_i(t) = \Theta(t) + {}^{\mathrm{B}}\theta_i(t), \tag{2.1} $$

and, by differentiation,

$$ {}^{\mathrm{R}}\theta_i'(t) = \Omega(t) + {}^{\mathrm{B}}\omega_i(t). \tag{2.2} $$

The fact that the robot is a rigid body is expressed by the *rigidity conditions*

$$ \frac{d}{dt}\left|{}^{\mathrm{R}}\mathbf{x}_i(t) - {}^{\mathrm{R}}\mathbf{x}_j(t)\right| = \frac{d}{dt}\left|{}^{\mathrm{B}}\mathbf{x}_i(t) - {}^{\mathrm{B}}\mathbf{x}_j(t)\right| = 0, \qquad i,j = 1,\ldots m. \tag{2.3} $$

If the *axle-point velocity* $\mathbf{v}_i(t) = {}^{\mathrm{R}}\mathbf{v}_i(t) = {}^{\mathrm{R}}\mathbf{x}_i'(t)$ is introduced, and the differentiation is performed, (2.3) can be expressed as

$$ \left(\mathbf{x}_i(t) - \mathbf{x}_j(t)\right) \cdot \left(\mathbf{v}_i(t) - \mathbf{v}_j(t)\right) = 0, \qquad i,j = 1,\ldots m. \tag{2.4} $$

This condition can be reformulated in another useful way. Let α_{ij} be the signed angle between the reference direction in the body and the line joining $\mathbf{x}_i$ to $\mathbf{x}_j$. By choosing basis vectors parallel and perpendicular to $\mathbf{x}_i - \mathbf{x}_j$ and considering components in those directions, (2.4) can be rewritten:

$$ |\mathbf{v}_i| \sin({}^{\mathrm{B}}\theta_i - \alpha_{ij}) - |\mathbf{v}_j| \sin({}^{\mathrm{B}}\theta_j - \alpha_{ij}) = 0. \tag{2.5} $$

One of the most elementary, yet elegant, results in kinematics (Chasles' theorem or Descartes' principle of instantaneous motion) asserts that the conditions in (2.4) imply that at each instant the motion of a planar rigid body coincides with either (1) a pure rotation about some point (the *instantaneous center of rotation* or *ICR*) or (2) a pure translation.

3. Rolling

The condition of rolling of the ith wheel relates the axle vector $\mathbf{a}_i$, which characterizes the wheel motion relative to the body, to the axle-point velocity $\mathbf{v}_i$, which partially describes the motion of the robot body in space. Explicitly, the *rolling conditions* are:

(1.) the directed angle from the axle-point velocity $\mathbf{v}_i$ to the axle vector $\mathbf{a}_i$ is $\pi/2$,

2. $|\mathbf{v}_i| = 2\pi R_i\,|\mathbf{a}_i| = 2\pi R_i \rho_i$.

In mathematical terms, $\mathbf{v}_i = 2\pi R_i \mathbf{a}_i \times \mathbf{k}$. Here, and throughout, $\mathbf{k}$ will denote the unit vertical vector, and any two-dimensional vectors, such as $\mathbf{a}_i$ and $\mathbf{v}_i$, are considered to be naturally embedded in three-dimensional space. Consequently the vector product is well-defined. Physically, $\mathbf{v}_i$ lies in the plane of the wheel, and its magnitude is determined by the rotation rate of the wheel.

The condition of rolling combines naturally with the notion of instantaneous center of rotation. If the instantaneous motion is a translation, then the axle vector of any rolling wheel must be orthogonal to the translation and be of given magnitude. More interestingly, if the body is rotating with instantaneous velocity $\Omega \neq 0$, and the ith wheel is rolling, then the ICR must be located at the point

$$\mathbf{x}_i + 2\pi R_i \Omega^{-1} \mathbf{a}_i. \tag{3.1}$$

Here we simplify notation by assuming, with no loss of generality, that $R_i = R_j = 1/2\pi$. This assumption merely represents a scaling of each axle vector $\mathbf{a}_i$. It then follows from (3.1) that any two rolling wheels, say the ith and the jth, must satisfy the *rolling compatibility conditions*:

$$\Omega(\mathbf{x}_i - \mathbf{x}_j) = (\mathbf{a}_j - \mathbf{a}_i). \tag{3.2}$$

This expression remains valid in the translational case $\Omega = 0$. It is apparent from (3.2) that once the body components of the axle vectors corresponding to two rolling wheels are given, then the angular velocity Ω, and therefore the axle vector of any other rolling wheel, is determined. On the other hand, even the first two axle vectors are not independent, for the vector equation (3.2) must be solvable for the scalar Ω. Note that the condition for (3.2) to be solvable is nothing other than (2.4), which, taken with the rolling conditions, states that $\mathbf{a}_j - \mathbf{a}_i$ must be parallel to $\mathbf{x}_i - \mathbf{x}_j$. Provided this solvability condition is satisfied, we find that

$$\Omega = -\frac{(\mathbf{a}_i - \mathbf{a}_j)\cdot(\mathbf{x}_i - \mathbf{x}_j)}{|\mathbf{x}_i - \mathbf{x}_j|^2}. \tag{3.3}$$

4. Rigid-Body Kinematics

In this section, we explicitly find the differential equations that determine the location of the robot body—in the reference frame—in terms of the axle-point locations and axle-vectors in the body frame, which are assumed to be known, and the initial location. Thus ${}^{\mathrm{B}}\mathbf{x}_i$, ${}^{\mathrm{B}}\mathbf{x}_j$, ${}^{\mathrm{B}}\mathbf{a}_i$, and ${}^{\mathrm{B}}\mathbf{a}_j$ are four known sets of coordinates that satisfy rolling compatibility condition (3.2). The variables that will be adopted are the space coordinates of one axle point ${}^{\mathrm{R}}\mathbf{x}_i$ and the body orientation Θ. There are four parameters associated with the two axle vectors, namely two steering angles and two rotation rates. These four scalars must satisfy the solvability condition (2.5). Although only three of the four scalars are independent, there is no convenient way to explicitly eliminate one of them.

The differential equations governing the motion are

$$ {}^{\mathrm{R}}\mathbf{x}_i' = {}^{\mathrm{R}}\mathbf{v}_i, \tag{4.1a} $$

$$ \Theta' = \Omega. \tag{4.1b} $$

However, with the assumption of rolling, the reference coordinates $\mathbf{v}_i$ can be expressed in terms of ρ_i, ${}^{\mathrm{B}}\theta_i$, and Θ, and (3.3) provides an expression for Ω. Thus (4.1) can be recast as a system of three first-order equations ((4.2a) is a vector equation written in terms of two coordinates):

$$ {}^{\mathrm{R}}\mathbf{x}_i' = \rho_i\big(\sin({}^{\mathrm{B}}\theta_i + \Theta), -\cos({}^{\mathrm{B}}\theta_i + \Theta)\big), \tag{4.2a} $$

$$ \Theta' = -\,x_{ij}^{-1}\left[\rho_i \cos({}^{\mathrm{B}}\theta_i - \alpha_{ij}) - \rho_j \cos({}^{\mathrm{B}}\theta_j - \alpha_{ij})\right], \tag{4.2b} $$

where $x_{ij} = \left|{}^{\mathrm{B}}\mathbf{x}_i - {}^{\mathrm{B}}\mathbf{x}_j\right|$. Here the reference direction in space is taken to lie along the positive ${}^{\mathrm{R}}x_1$-axis. Equations (4.2), taken with a set of initial conditions, determine the location and orientation of the robot, given a set of specified compatible parameters. The equation (4.2a), with i replaced by j, can be solved for ${}^{\mathrm{R}}\mathbf{x}_j$, and provided that (2.5) is satisfied, $\left|{}^{\mathrm{R}}\mathbf{x}_i(t) - {}^{\mathrm{R}}\mathbf{x}_j(t)\right|$ is automatically constant. Alternatively, ${}^{\mathrm{R}}\mathbf{x}_j$ can be constructed from the formula

$$ {}^{\mathrm{R}}\mathbf{x}_j = {}^{\mathrm{R}}\mathbf{x}_i + x_{ij}\big(\cos(\alpha_{ij} + \Theta), \sin(\alpha_{ij} + \Theta)\big). \tag{4.3} $$

The above analysis resolves the inverse kinematic problem *en passant.* For if ${}^{\mathrm{R}}\mathbf{x}_i$ and Θ are prescribed as functions of time, then (4.2a) can be regarded as two coupled transcendental equations that are uniquely solvable for the parameters $\rho_i \geq 0$, and ${}^{\mathrm{B}}\theta_i$ modulo 2π. Moreover, with given ${}^{\mathrm{R}}\mathbf{x}_i$ and Θ, (4.3) provides an expression for ${}^{\mathrm{R}}\mathbf{x}_j$. Consequently, (4.2a) provides a unique $\rho_j \geq 0$, and ${}^{\mathrm{B}}\theta_j$, modulo 2π. Similarly, if ${}^{\mathrm{R}}\mathbf{x}_i$ and ${}^{\mathrm{R}}\mathbf{x}_j$ satisfying (2.2) are prescribed as functions of time, then (4.3) provides Θ, and (4.2a) and (4.3) can be solved for the parameters as before. Any set of four parameters obtained in this way are automatically compatible.

More specifically, suppose that the position of a point $\mathbf{w}$ of the WMR and Θ are prescribed as functions of time. There is no loss of generality in assuming that $\mathbf{w}$ is actually an axle point ${}^{\mathrm{R}}\mathbf{x}_i$, because a "virtual" wheel can be imagined. From (4.2a), we obtain

$$\rho_i = |{}^{\mathrm{R}}\mathbf{x}_i'|, \tag{4.4a}$$

$${}^{\mathrm{B}}\theta_i = \beta_i - \Theta, \tag{4.4b}$$

where β_i is the known angle between ${}^{\mathrm{R}}\mathbf{x}_i'$ and the ${}^{\mathrm{R}}x_2$-axis of spatial coordinates. To obtain the parameters for the other wheels, differentiate (4.3) to obtain

$${}^{\mathrm{R}}\mathbf{x}_j' = {}^{\mathrm{R}}\mathbf{x}_i' + x_{ij}\Theta'\big(-\sin(\alpha_{ij} + \Theta), \cos(\alpha_{ij} + \Theta)\big),$$

and use (4.2a) and (4.3) to obtain

$$\rho_j = \sqrt{\rho_i^2 + x_{ij}^2(\Theta')^2 - 2\rho_i x_{ij}\Theta' \cos({}^{\mathrm{B}}\theta_i \alpha_{ij})}.$$

Thus the inverse kinematic equations are

$$\rho_j = \sqrt{|{}^{\mathrm{R}}\mathbf{x}_i'|^2 + x_{ij}^2(\Theta')^2 - 2|{}^{\mathrm{R}}\mathbf{x}_i'|x_{ij}\Theta' \cos(\beta_i - \alpha_{ij} - \Theta)} \tag{4.5a}$$

$${}^{\mathrm{B}}\theta_j = \tan^{-1}\left(\frac{|{}^{\mathrm{R}}\mathbf{x}_i'|\sin\beta_i - x_{ij}\Theta' \sin(\alpha_{ij} + \Theta)}{|{}^{\mathrm{R}}\mathbf{x}_i'|\cos\beta_i - x_{ij}\Theta' \cos(\alpha_{ij} + \Theta)}\right) - \Theta. \tag{4.5b}$$

Formulae (4.5) give closed-form expressions for the parameters ρ_j, ${}^{\mathrm{B}}\theta_j$ of an arbitrary wheel in terms of known quantities; there are no differential equations involved.

One of the simplest practical problems in the kinematics of a WMR is to move the robot body from one configuration, as represented by ${}^{\mathrm{R}}\mathbf{x}_i(0)$ and $\Theta(0)$ say, to a final configuration ${}^{\mathrm{R}}\mathbf{x}_i(T)$ and $\Theta(T)$. The above analysis demonstrates that to each rigid-body trajectory linking the initial and final configurations, or equivalently to each pair of functions ${}^{\mathrm{R}}\mathbf{x}_i(t)$ and $\Theta(t)$ with correct initial and final values, there corresponds a unique set of parameters. Consequently, many algorithms for the generation of controllable parameters can be developed — one for each rigid-body trajectory. For example, the mathematically simplest choice involves functions ${}^{\mathrm{R}}\mathbf{x}_i(t)$ and $\Theta(t)$ that are linear in time. In practice, other, non-kinematic, considerations enter into the question of which generating algorithm is, in some sense, best. For example, an optimal-control study could start from this analysis and would further determine which set of feasible control parameters actually minimized some functional, perhaps total work.*

The analysis of this section takes no account of constraints on the values that parameters may assume. For example, most practical designs will involve a maximum steering lock, and accordingly the parameter ${}^{\mathrm{B}}\theta_i$ must lie in some predetermined range. The development presented here certainly allows the admissibility of a given trajectory and its associated parameters to be checked *a posteriori*, but we do not attempt the much harder problem of the *a priori* characterization of trajectories that are attainable with a given restricted set of parameters.

* Possibly the total work of the designer.

5. Practical Designs

The analysis of the previous sections has demonstrated that in order to be able to maneuver a WMR in arbitrary planar rigid body motions, it is both necessary (in the considered class of designs) and sufficient that there be two driven and steered simple wheels. At first sight it appears that any two-wheeled robot must immediately fall, because it makes only two-point contact with the supporting plane. This argument is invalid, because the center of gravity of the robot body may lie below the axle attachment points. See, for example, Helmers (1983a), which discusses a robot with two non-simple wheels that are mounted in non-vertical planes. Nevertheless, the body of a robot with only two wheels will be susceptible to pendulum-like oscillations that may be excited during maneuvers. Consequently, we consider robots with wheel configurations that provide an intrinsically more stable, three-point contact with the supporting plane.

When the third support point is provided by a third simple wheel and rolling occurs, the analysis of Section 3 applies to completely determine the third axle vector in terms of the first two axle vectors. Consequently, of the six controllable parameters associated with three simple wheels, only three are independent. If there is an error in any one of these parameters, slippage must occur. Accordingly, it is apparent that there is considerable, essentially redundant, work required to calculate compatible wheel parameters. There are three plausible approaches to the removal of this redundancy.

First, the *mechanics* of the maneuvering robot may guarantee that some of the *kinematic* compatibility conditions are automatically satisfied when some of the wheels are not driven or steered. For example, if the third simple wheel is steered correctly and no control of its rotation rate is made, it will automatically rotate at a compatible speed in response to the frictional forces exerted by the surface. In contrast, there are no torques that would automatically orient the third wheel and allow it to be unsteered. Consequently, all redundancies cannot be eliminated merely by ignoring certain control parameters.

Second, compatibility problems are reduced if some mobility of the robot body is sacrificed by restriction of the range of some of the controls. For example in a three-wheeled robot, all compatibility conditions are removed if two steering angles ${}^{B}\theta_i$ are held constant and the associated drive rates are left passive. Having two fixed wheels limits possible rolling motions to the unacceptably small set of circular arcs with prescribed radius, *unless* the fixed axle vectors are parallel to each other *and* to the line between the two axle points. This last configuration is equivalent to having two simple wheels mounted on the same axle with independent rotation rates (such as the rear wheels of an automobile). This special three-wheel configuration with one steerable, two fixed, one driven, and two passive, wheels is a viable design that was adopted for the WMR Neptune (Podnar, Dowling and Blackwell 1984). Its kinematics are further discussed in Section 6.

Third, if restricted classes of planar rigid body motions are not acceptable, then a mechanically stable robot, with minimal control compatibility conditions, can be achieved by having two steered, driven wheels and a passive *castor*. For our purposes, a castor is a second (small) planar rigid body that has mounted in it a fixed simple wheel and that is attached to the robot body at a free pivot point. The addition of a castor to two simple wheels at an appropriate point certainly makes the robot body mechanically stable. The four kinematic parameters associated with the two simple wheels must satisfy one compatibility condition, but the castor can be left entirely passive with no rolling condition being violated. This last fact is physically obvious and is supported by practical experience. It can be derived from our mathematical analysis but will not be considered here.

6. Examples of Rolling

As an application of our theory, we consider the basic kinematic problem of determining the steering and drive rates that move a WMR from one prescribed rigid body configuration to another. Consider first a completely mobile robot, that is, a robot that can traverse arbitrary rigid-body motions with all wheels undergoing ideal rolling. Two such designs are the CMU Rover (Moravec 1983), which has three driven, steered simple wheels and the design suggested in the previous section, namely a WMR with two driven, steered simple wheels, and a supporting, passive castor. Because the WMR is assumed to be completely mobile, there is no question of which trajectories linking the given starting and ending configurations are actually accessible. We shall therefore pick a simple trajectory and determine the parameters that move one axle point $\mathbf{x}_i$ uniformly along the ${}^{\mathrm{R}}x_2$-axis from (0,0) to (0,h) in unit time, while simultaneously rotating the robot uniformly from $\Theta(0) = \Theta^0$ to $\Theta(1) = \Theta^1$. We set

$$ {}^{\mathrm{R}}\mathbf{x}_i(t) = (0,0) + t(0,h), \tag{6.1a} $$

$$ \Theta(t) = \Theta^0(1-t) + t\Theta^1. \tag{6.1b} $$

From (4.4b), ${}^{\mathrm{B}}\theta_i = \beta_i - \Theta = -\Theta^0(1-t) - t\Theta^1$. Moreover, from (4.4a), $\rho_i = |{}^{\mathrm{R}}\mathbf{x}_i'| = h$. The parameters at another simple wheel mounted at the axle point ${}^{\mathrm{R}}\mathbf{x}_j$ are obtained from equations (4.5), which reduce to

$$ \rho_j = \sqrt{h^2 + x_{ij}^2(\Theta^1 - \Theta^0)^2 - 2hx_{ij}(\Theta^1 - \Theta^0)\cos\big(\alpha_{ij} + \Theta(t)\big)}, \tag{6.2a} $$

$$ {}^{\mathrm{B}}\theta_j = -\tan^{-1}\left(\frac{x_{ij}(\Theta^1 - \Theta^0)\sin\big(\alpha_{ij} + \Theta(t)\big)}{h - x_{ij}(\Theta^1 - \Theta^0)\cos\big(\alpha_{ij} + \Theta(t)\big)}\right) - \Theta(t), \tag{6.2b} $$

and $\Theta(t)$ is given by (4.1b). Here x_{ij}, defined in Section 5, and α_{ij}, defined in Section 2, are constants determined by the geometry of the WMR. If any of the simple wheels of the robot are not driven exactly as is specified by (6.2), slippage of the wheels must occur, and the analysis of Sections 7 and 8 will be required.

Another practical WMR wheel configuration is kinematically equivalent to a child's tricycle. The kinematics of such vehicles are considered in Alexander and Maddocks (1988). The WMR Newt (Hollis 1977) is one tricycle-like WMR. It has two parallel driven wheels that are fixed in direction, along with a passive castor for stability. Another existing robot, Neptune (Podnar, Dowling and Blackwell 1984), has two passive parallel wheels fixed in direction and a driven, steered wheel. Because tricycle-like WMRs have a fixed axle, they are not completely mobile, and therefore arbitrary rigid-body trajectories cannot be specified. For example, such a WMR cannot be moved parallel to the fixed axle. However, the path of any given point not on the fixed axle can be prescribed arbitrarily (Alexander and Maddocks 1988). In particular, a WMR with the kinematics of a tricycle can be maneuvered with ideal rolling in any rigid-body trajectory for which the center of curvature always lies on the extension of the axle vector. Consequently the robot can be maneuvered with ideal rolling from one body configuration to another in a motion comprised of piecewise smooth arcs of pure rotations, and the prior analysis can be used to construct the appropriate steering and drive rates.

The Stanford Cart (Moravec 1980; Moravec 1983) is another WMR that is essentially kinematically equivalent to a tricycle. It has two parallel driven wheels fixed in direction and a pair of steerable wheels configured as a small automobile. For our purposes the only difference between the Stanford cart and a tricycle is that there are more wheels, and therefore more redundant controllable parameters that must be accurately activated to obviate slippage. For WMRs with four or more wheels, there is almost certain to be some slippage due to incompatibility of activated wheels.

7. Slippage

Sections 2–5 provide an analysis that uses a kinematic model of ideal rolling of simple wheels to predict the motion of a WMR in ambient space when the axle-vector body coordinates ${}^{B}\mathbf{a}_i$ are given as functions of time. The analysis does not involve a force balance and does not explicitly solve Newton's laws of motion. Instead, it is implicitly assumed that the undetermined static frictional forces, which act between the wheels and the supporting plane, are sufficiently large to balance the inertial forces that arise from accelerations. Naturally, the validity of this assumption depends on the relative sizes of the inertial and static frictional forces. Whenever the inertial forces dominate, the rolling model ceases to be valid. The subsequent motion involves a phenomenon we call skidding, and any analysis would have to be based on the full system of Newton's laws. On the other hand, the rolling model also fails, even if the inertial forces are negligible, whenever the wheel controls are incompatible [i.e., whenever rolling conditions (3.2) are not satisfied]. The subsequent motion involves the phenomenon we call *slippage*. The purpose of this section is to present a quasi-static model that can predict motions that include slippage due to wheel incompatibility. The model described is quasi-static because, in distinction from the preceding analysis, the inertial forces are assumed to be sufficiently small as to be totally negligible.

In the previously described model of rolling, we obtained closed-form expressions for the axle-point velocities $\mathbf{v}_i$ in terms of the controls, as represented by the axle vectors $\mathbf{a}_i$, and consequently obtained differential equations that describe the axle-point trajectories. Here we pursue a less ambitious goal; namely, to find an algorithm that determines the current axle-point velocities $\mathbf{v}_i$ given the current controls $\mathbf{a}_i$. We shall also consider the problem in which one or more drive controls are left passive. In this case, the direction of the axle vector is known, but its magnitude is undetermined.

Our first step is the construction of a *friction functional* that represents the power dissipated by Coulomb frictional forces during any planar rigid body motion of the WMR. The friction functional depends on the axle-point coordinates ${}^B\mathbf{x}_i$, the axle-vector coordinates ${}^B\mathbf{a}_i$, and on the planar rigid-body velocity of the robot, which is regarded as the unknown. Of course, a planar rigid-body velocity is a three-vector describing the planar linear velocity of some reference point and the angular velocity of the body. We will take the reference point to be the origin of the body coordinates ${}^B\mathbf{x}$, and $\mathbf{y}$ will denote the unique three-vector, with vertical component $\Omega = \mathbf{y} \cdot \mathbf{k}$, for which $\mathbf{y} \times \mathbf{k}$ is the linear velocity of the reference point. With this definition of $\mathbf{y}$, any axle-point velocity $\mathbf{v}_i$ can be written

$$\mathbf{v}_i = (\mathbf{y} - \Omega \mathbf{x}_i) \times \mathbf{k}. \tag{7.1}$$

Here, as before, planar vectors such as $\mathbf{x}_i$ are viewed as being embedded in three-dimensional space, so that the vector product is well defined. We shall show that the friction functional is a convex function of $\mathbf{y}$. Consequently the minimization problem is well posed. It is, however, mathematically delicate, because the friction functional is neither strictly convex nor everywhere differentiable. Thus a minimizer need not be unique, and the standard conditions applying at a minimizer(namely, the vanishing of the derivative) must be interpreted in the sense of the subdifferential from convex analysis.

We obtain an evolutionary system for the WMR from the *principle of least dissipation*, which states that the actual motion of the WMR is the one that minimizes the friction functional at all times. It is then shown that motions that satisfy minimal dissipation are quasi-static in the sense that the forces and moments that arise are in equilibrium. Accordingly, the principle of least dissipation leads to approximate solutions of Newton's laws of motion, provided that inertial terms are negligible.

Consider a simple wheel that is undergoing a combination of rolling and slippage. Define the *rolling velocity* $\mathbf{r}_i$ by $\mathbf{r}_i = \mathbf{a}_i \times \mathbf{k}$ where, as before, the $\mathbf{a}_i$ are the axle vectors (scaled with a factor $2\pi R_i$). In the case of ideal rolling, $\mathbf{r}_i$ coincides with the velocity $\mathbf{v}_i$ of the ith axle point, but in the case of combined rolling and sliding, there can be a discrepancy. We define the *slippage velocity* $\mathbf{s}_i$ to be this discrepancy. Explicitly, for all i,

$$\mathbf{s}_i = \mathbf{v}_i - \mathbf{r}_i = \{\mathbf{y} - \Omega \mathbf{x}_i - \mathbf{a}_i\} \times \mathbf{k}. \tag{7.2}$$

It should also be remarked that the wheel rotation rate ρ_i satisfies

$$\rho_i = |\mathbf{r}_i| = |\mathbf{a}_i|. \tag{7.3}$$

According to Coulomb's law, the sliding friction $\mathbf{f}_i$ between a particle and a surface is given by the expression

$$\mathbf{f}_i = -\frac{\alpha_i \mathbf{s}_i}{|\mathbf{s}_i|}, \tag{7.4}$$

where $\mathbf{s}_i \neq \mathbf{0}$ is the relative velocity of the particle and the surface, and $\alpha_i \geq 0$ is the product of the normal load and the coefficient of sliding friction. If the coefficients of sliding and static friction are assumed equal, (7.4) also describes Coulomb's law of static friction, provided that in the limit $\mathbf{s}_i \to \mathbf{0}$, $\mathbf{s}_i/|\mathbf{s}_i|$ is interpreted as an arbitrary vector with magnitude less than or equal to one.

The total instantaneous power P being dissipated by sliding friction is the sum of the scalar products

$$P = -\sum_i \mathbf{f}_i \cdot \mathbf{s}_i, \tag{7.5}$$

or, by (7.4),

$$P = \sum_i \alpha_i |\mathbf{s}_i|. \tag{7.6}$$

Of course, the construction of (7.6) applies to the contact points between the wheels of the WMR and the supporting plane, with the relative velocity being the slippage velocity $\mathbf{s}_i$. Consequently, (7.2) can be used to rewrite (7.6) as

$$P = P(\mathbf{y}; {}^{\mathrm{B}}\mathbf{x}_i, {}^{\mathrm{B}}\mathbf{a}_i, \alpha_i) = \sum_i \alpha_i |(\mathbf{A}_i \mathbf{y} - {}^{\mathrm{R}}\mathbf{a}_i) \times \mathbf{k}|. \tag{7.7}$$

Here A_i denotes the 2×3 matrix defined by

$$\mathbf{A}_i = \begin{pmatrix} 1 & 0 & -{}^{\mathrm{B}}x_1^i \\ 0 & 1 & -{}^{\mathrm{B}}x_2^i \end{pmatrix}.$$

We call (7.7) the *dissipation* or *friction functional.* The variable $\mathbf{y}$ is unknown, but ${}^{\mathrm{B}}\mathbf{x}_i$, ${}^{\mathrm{B}}\mathbf{a}_i(t)$ and α_i are viewed as known parameters. Because $(\mathbf{A}_i\mathbf{y} - {}^{\mathrm{R}}\mathbf{a}_i)$ lies in the horizontal plane and $\mathbf{k}$ is the vertical unit vector, the dissipation functional can be rewritten as:

$$P = P(\mathbf{y}; {}^{\mathrm{B}}\mathbf{x}_i, {}^{\mathrm{B}}\mathbf{a}_i, \alpha_i) = \sum_i \alpha_i |\mathbf{A}_i \mathbf{y} - {}^{\mathrm{R}}\mathbf{a}_i|. \tag{7.8}$$

It is straightforward to show that the dissipation functional is convex in $\mathbf{y}$. In particular, by the triangle inequality, we have that for $0 \leq \tau \leq 1$,

$$\sum_i \alpha_i |\mathbf{A}_i(\tau\mathbf{y}_1 + (1-\tau)\mathbf{y}_2) - {}^{\mathrm{B}}\mathbf{a}_i| \leq \tau \sum_i \alpha_i |\mathbf{A}_i\mathbf{y}_1 - {}^{\mathrm{B}}\mathbf{a}_i| + (1-\tau) \sum_i \alpha_i |\mathbf{A}_i\mathbf{y}_2 - {}^{\mathrm{B}}\mathbf{a}_i|. \tag{7.9}$$

Consequently the first variation of the dissipation functional provides both necessary and sufficient conditions to determine minima.

Provided that no term $|\mathbf{A}_i\mathbf{y} - {}^B\mathbf{a}_i|$ vanishes, the dissipation function is a differentiable function of $\mathbf{y}$, and the first-order conditions are obtained by differentiation with respect to $\mathbf{y}$:

$$\sum_i \alpha_i \mathbf{A}_i^T \frac{\mathbf{A}_i\mathbf{y} - {}^B\mathbf{a}_i}{|\mathbf{A}_i\mathbf{y} - {}^B\mathbf{a}_i|} = \mathbf{0}. \tag{7.10}$$

Moreover, the theory of convex functionals—see for example Rockafellar (1970)—can be applied to treat the singular cases. In particular, provided that any singular terms

$$\frac{\mathbf{A}_i\mathbf{y} - {}^B\mathbf{a}_i}{|\mathbf{A}_i\mathbf{y} - {}^B\mathbf{a}_i|} \tag{7.11}$$

are interpreted as arbitrary vectors $\mathbf{z}_i$ with $|\mathbf{z}_i| \leq 1$, the left side of (7.10) always represents the subdifferential of the dissipation functional. Consequently, any minimizing vector $\mathbf{y}$ must satisfy equation (7.10) either in the classical sense or, when there are singular terms, in a generalized sense involving vectors $\mathbf{z}_i$.

Equations (7.10) can be expanded to:

$$\sum_i \alpha_i \frac{\mathbf{A}_i\mathbf{y} - {}^B\mathbf{a}_i}{|\mathbf{A}_i\mathbf{y} - {}^B\mathbf{a}_i|} = \mathbf{0}, \tag{7.12}$$

and

$$\sum_i \alpha_i {}^B\mathbf{x}_i \cdot \frac{\mathbf{A}_i\mathbf{y} - {}^B\mathbf{a}_i}{|\mathbf{A}_i\mathbf{y} - {}^B\mathbf{a}_i|} = \mathbf{0}. \tag{7.13}$$

The definition of $\mathbf{A}_i$, taken with equations (7.2) and (7.4), implies that the cross-product with $\mathbf{k}$ of equations (7.12) and (7.13) can be written in the form

$$\sum_i \mathbf{f}_i = \mathbf{0}, \tag{7.14}$$

and

$$\sum_i (\mathbf{x}_i \times \mathbf{f}_i) \cdot \mathbf{k} = 0 \qquad (7.15).$$

Consequently, minimization of the friction functional with respect to the velocity $\mathbf{y}$ generates *quasi-static* motions, in which the frictional forces and moments acting on the robot are in equilibrium. This conclusion is valid whether (7.10) represents a classic derivative or a subdifferential. The non-smooth case arises precisely in the absence of slippage at one or more wheels, and the associated indeterminacy in the subdifferential is exactly matched by the indeterminacy in Coulomb's law of static friction.

It should be observed that the dissipation functional is always nonnegative and is zero only if each term in the sum (7.7) vanishes. Provided that there are at least two wheels, this requirement leads to an overdetermined system of equations involving the three-vector $\mathbf{y}$. The solvability conditions for this overdetermined system are precisely the rolling compatibility conditions (3.2) derived previously. Accordingly the principle of minimal dissipation predicts ideal rolling of all wheels whenever ideal rolling is possible.

We have shown that the principle of minimal dissipation provides a quasi-static dynamic system that models motions involving a combination of rolling and slipping. The principle characterizes instantaneous rigid-body motions from minimization over the variable $\mathbf{y}$ while the (known) parameters ${}^{\mathrm{B}}\mathbf{x}_i$, ${}^{\mathrm{B}}\mathbf{a}_i$, and α_i are held fixed. A minimizing vector $\mathbf{y}$ determines the linear velocity (in body coordinates) of the origin of the body-coordinate system and the angular velocity of the WMR. For most parameter values the minimal $\mathbf{y}$ will be unique, but the example of the next section shows that nonuniqueness is possible.

In the case of a passive wheel the interpretation is slightly different. Then the axle vector ${}^{\mathrm{B}}\mathbf{a}_i$ is of the form $\rho_i{}^{\mathrm{B}}\mathbf{u}_i$, where ${}^{\mathrm{B}}\mathbf{u}_i$ is a given unit vector, but the rotation rate ρ_i is *a priori* unknown. Any unknown rotation rates are treated as additional variables over which the slippage functional must be minimized, and further first-order conditions are obtained. More precisely, partial differentiation with respect to each passive rotation rate ρ_i implies

$$(\mathbf{A}_i\mathbf{y} - \rho_i{}^{\mathrm{B}}\mathbf{u}_i) \cdot {}^{\mathrm{B}}\mathbf{u}_i = 0. \tag{7.16}$$

However this equation is equivalent to the geometric condition that the slippage velocity is orthogonal to the rolling velocity, or $\mathbf{s}_i \cdot \mathbf{r}_i = 0$. Hence there is a minimization principle at each passive wheel: for given $\mathbf{v}_i$ and ${}^{\mathrm{B}}\mathbf{u}_i$, the rotation rate ρ_i is such that the slippage vector $\mathbf{s}_i$ has minimal length. Alternatively, equation (7.16) characterizes the rotation rate for which the frictional force is parallel to the axle vector. Consequently, the frictional force exerts no torque about the axle vector, which is in accord with quasi-static motion.

8. Examples Involving Slippage

Before a general analysis of the kinematics of WMRs with slippage can be attempted, the dissipation functional must be better understood. We present, as an illustrative example, an analysis of the case of two wheels.

As was noted in Section 5, when a WMR has two wheels both of which are driven and steered, then there is one constraint (3.2) on the parameters that must be satisfied in order for ideal rolling to be possible; if this constraint is violated, slippage must occur. We shall analyze the motion, including slippage, for such two-heeled robots. The remarks concerning castors made in Section 5 apply with equal force here and imply that castors have no effect on the kinematics of a WMR. Thus it is apparent that the subsequent analysis also applies to robots with two simple wheels and passive stabilizing castors.

Denote the wheels by $i = 1, 2$. Recall that the $\alpha_i \geq 0$ are the product of the normal load on the ith wheel and the coefficient of friction. There are two cases to consider: (1) the symmetric case $\alpha_1 = \alpha_2$ and (2) the asymmetric case $\alpha_1 \neq \alpha_2$.

In the case of two wheels with asymmetric weights, a minimal dissipation solution must have one wheel undergoing ideal rolling. This follows from the following reasoning. In the smooth case (7.12) reduces to the weighted sum of two unit vectors, and accordingly has a solution if and only if the two unit vectors are anti-parallel and $\alpha_1 = \alpha_2$. Therefore, when $\alpha_1 \neq \alpha_2$, the minimum must occur at a nondifferentiable point of the dissipation functional. In this case the simplest approach is to consider the functional (7.8) directly. Suppose that $\alpha_1 < \alpha_2$. It is then apparent that (7.8) is minimized when it is the second wheel that undergoes ideal rolling. The ideal rolling condition

$$\mathbf{A}_2\mathbf{y} = \mathbf{a}_2, \tag{8.1}$$

provides an underdetermined set of equations for $\mathbf{y}$ that reduce to the condition

$$\mathbf{y} - \Omega\mathbf{k} = \Omega\mathbf{x}_2 + \mathbf{a}_2,$$

where the scalar angular velocity Ω is arbitrary. Then the dissipation functional (7.8) reduces to

$$\alpha_1|\Omega(\mathbf{x}_1 - \mathbf{x}_2) + (\mathbf{a}_1 - \mathbf{a}_2)|, \tag{8.2}$$

which is minimized when

$$\Omega = -\frac{(\mathbf{a}_1 - \mathbf{a}_2)\cdot(\mathbf{x}_1 - \mathbf{x}_2)}{|\mathbf{x}_1 - \mathbf{x}_2|^2}. \tag{8.3}$$

Remarkably, the angular velocity Ω in (8.3) that is predicted by the slippage model coincides with the angular velocity in (3.3) that was predicted by the ideal rolling model. Notice that the second wheel undergoes ideal rolling in such a manner that the slippage of the first wheel is minimized. Furthermore, the minimal dissipation is strictly positive unless $\mathbf{a}_1 - \mathbf{a}_2$ is parallel to $\mathbf{x}_1 - \mathbf{x}_2$. In this last case the compatibility condition for ideal rolling of both wheels is satisfied, $P = 0$, and no slippage occurs.

Consider now the symmetric case. Here there is no difference between the smooth and non-smooth cases. Equation (7.12) always implies that the horizontal component of $\mathbf{y}$ lies on the straight line segment between $\Omega^R\mathbf{x}_1 + {}^R\mathbf{a}_1$ and $\Omega^R\mathbf{x}_2 + {}^R\mathbf{a}_2$. That is, for $0 \leq \gamma \leq 1$,

$$\mathbf{y} - \Omega\mathbf{k} = (1-\gamma)(\Omega^R\mathbf{x}_1 + {}^R\mathbf{a}_1) + \gamma(\Omega^R\mathbf{x}_2 + {}^R\mathbf{a}_2). \tag{8.4}$$

Then (7.13) further implies that

$$\Omega = -\frac{(\mathbf{a}_1 - \mathbf{a}_2)\cdot(\mathbf{x}_1 - \mathbf{x}_2)}{|\mathbf{x}_1 - \mathbf{x}_2|^2}, \tag{8.5}$$

as before. Notice that although the angular velocity Ω of the WMR is uniquely determined, condition (8.4) does *not* provide a unique solution for the horizontal component of $\mathbf{y}$. Consequently the WMR has several possible linear velocities, all corresponding to quasi-static motion. In practice, it might be expected that the load α_i on one wheel or the other will dominate at any particular time, and the non-uniqueness will be resolved so that the dominant wheel will experience ideal rolling. Small perturbations could cause this dominance to jump intermittently from one wheel to the other, and the motion of the WMR will vary discontinuously.*

9. Discussion

This paper concerns the forward and inverse kinematics of a WMR with simple wheels that is maneuvering over a horizontal plane. The investigation has two main parts. In the first instance the wheels are assumed to be undergoing ideal rolling with static contact between the wheels and supporting plane. We describe the explicit relations between the motion of the wheels in the robot body—as characterized by the steering and driving rates—and the motion of the robot body in the plane, and *vice versa.* The analysis allows us to ascertain the mobility of any given robot design. That is, for a given wheel configuration, we can determine those rigid-body trajectories that are accessible with ideal rolling of all wheels, and we can construct the steering and drive rates that access the trajectory.

The central result in our kinematic investigations is the ideal rolling compatibility conditions, which imply that only three steering and drive rates are independent. The second part of the paper concerns motions that ensue when these rolling compatibility conditions fail. Our model for combined rolling and slipping is formulated and analyzed as a minimization principle involving a friction functional that measures the dissipation due to Coulomb friction. It is shown that this principle of minimal dissipation is equivalent to the characterization of quasi-static motions for which frictional forces and torques are in equilibrium. It should be stressed that the unknowns in our analysis are linear and angular velocities rather than forces and torques. This remark may be of practical significance, because the servo-control of velocities is typically easier than that of forces. We also mention that the theory for numerical minimization of non-smooth convex functionals of a class encompassing the dissipation functional is well developed (Overton 1983).

The dissipation functional constructed here has connections with Rayleigh's dissipation function (Lord Rayleigh 1896; Whittaker 1961, p. 230; Pars 1965, §10.11). The classic form of Rayleigh's dissipation function arises in the Lagrangian description of a finite-dimensional dynamic system that has dissipative forces that are linearly dependent on the speeds and directly opposed to the velocities. For such systems the Rayleigh dissipation function has a smooth dependence on the coordinates and is quadratic in the speeds. Of course, Coulomb's model of sliding friction leads to dissipative forces that

---

* J. C. Alexander has experienced this effect in a very badly aligned automobile. As a WMR, an automobile can be modelled with two steerable wheels and one driven wheel (the rear axle). The effect is the same as in the text. Usually one wheel will dominate; however, on a wet road, the coefficients of friction can vary so that the other wheel will dominate for a second or two causing the automobile to suddenly lurch sideways. The author remembers the sensation well.

are directly opposed to the velocity but are independent of speed. Consequently we are led to the more complicated non-smooth dissipation functional described above. The principle of minimal dissipation is also closely related to ideas introduced by Peshkin and Sanderson (1986) to describe quasi-static sliding of systems of particles, but rolling was not considered there. We also mention the work of Moreau (1979), who exploits methods of convex analysis, and of Goyal and Ruina (1988). Both of these works construct models for the forces that arise at wheels in terms of anisotropic friction laws. However, the thrust of their research is considerably different from our development, which is focused on a simultaneous analysis of several wheels with the unknown being the rigid-body velocity of the robot.

References

Alexander, J. C. and Maddocks, J. H. 1988. On the maneuvering of vehicles. *SIAM J. Appl. Math.* 48:38–51.

Balmer, Jr., C. 1982. Avatar: A home built robot. *Robotics Age* 4(1):20–27.

Carlisle, B. 1983. An OMNI-directional mobile robot. *Developments in Robotics*. ed. B. Rocks. Kempston, England: IFS Publishing Ltd.. pp. 79–88.

Everett, H. R. 1979. A second-generation autonomous sentry robot. *Robotics Age* 7(4):29–32.

Fortune, F. and Wilfong, G. 1988. *Planning constrained motion*. Murray Hill, New Jersey: AT&T Bell Laboratories. preprint.

Giralt, G., Sobek, R. and Chatila, R. 1979. A multi-level planning and navigation system for a mobile robot; A first approach to Hilare. *Proc. of the IJCAI*. Tokyo, Japan. pp. 335–338.

Goyal, S. and Ruina, A. 1988. Relation between load and motion for a rigid body sliding on a planar surface with dry friction: limit surfaces, incipient and asymptotic motion. submitted to *Wear*.

Helmers, C. 1983a. The wandering roboteers. *Robotics Age* 5(3):6–10.

Helmers, C. 1983b. Ein Heldenleben (or, A Hero's life, with apologies to R. Strauss). *Robotics Age* 5(2):7–16.

Helmers, C. 1985. Photo essay: A first glimpse at Gemini. *Robotics Age* 7(2):12–13.

Holland, J. M. 1985. Rethinking robot mobility. *Robotics Age* 7(1):26–30.

Hollis, R. 1977. Newt: A mobile, cognitive robot. *Byte* 2(6):30–45.

Ichikawa, Y., Ozaki, N. and Sadakane, K. 1983. A hybrid locomotion vehicle for nuclear power plants. *IEEE Trans. Systems Man Cybernet.* SMC-13(6):1089–1093.

Iijima, J., Kanayama, Y. and Yuma, S. 1981a. Elementary functions of a self-contained robot 'Yamabiko 3.1'. *Proc. of the 11th ISIR.* Tokyo: Society of Biomechanisms Japan, Japan Industrial Robot Association. pp. 211–218.

Iijima, J., Kanayama, Y. and Yuma, S. 1981b. A mobile robot with sonic sensors and its understanding of a simple world. Technical Report No. ISE-TR-81-22. Tsukuba, Japan: University of Tsukuba Institute of Information Science and Electronics.

Johnson, R. 1984. Part of the beginning. *Robotics Age* 6(4):35–37.

Kanayama, Y. and Miyake, N. 1986. Trajectory generation for mobile robots. *Proc. Third International Symposium of Robotics Research.* Gouvieux, France. pp. 333–340.

Lewis, R. A. and Bejczy, A. K. 1973. Planning considerations for a roving robot with arms. *Proc. of the 3rd IJCAI.* Stanford, California. pp. 308–316.

Lord Rayleigh 1894-1896. *The Theory of Sound* 2nd edition, two volumes. London: McMillan. reprinted by Dover, 1945, New York .

Marrs, T. 1985. *The Personal Robot Book.* Blue Ridge Summit, Pennsylvania: Tab Books, Inc..

Moravec, H. P. 1980. Obstacle avoidance and navigation in the real world by a seeing robot rover. Ph. D. thesis. Stanford University Department of Computer Science.

Moravec, H. P. 1983. The Stanford Cart and the CMU Rover. *Proc. IEEE* 71(7):872–884.

Moravec, H. P., ed.1986. Autonomous mobile robots annual report. Technical Report CMU-RI-MRL-86-1. Pittsburgh, Pennsylvania: Carnegie-Mellon University Robotics Institute.

Moreau, J. J. 1979. Application of convex analysis to solve problems of dry friction. *Trends in Applications of Pure Mathematics to Mechanics,* vol. 2. ed. H. Zorski. Marshfield, Masachusetts: Pitman. pp. 263–280.

Muir, P. F. and Neuman, C. P. 1987. Kinematic modeling of wheeled mobile robots. *J. Robotic Systems* 4(2):281–333.

Nilsson, N. J. 1984. Shakey the robot. Technical Report 323. Menlo Park, California: SRI International Artificial Intelligence Center, Computer Science and Technology Division.

Overton, M. L. 1983. A quadratically convergent method for minimizing a sum of Euclidean norms. *Math. Programming* 27:34–63.

Pars, L. A. 1965. *A Treatise on Analytical Dynamics.* New York, NY: John Wiley & Sons. reprinted by Oxbow Press, Woodbridge, Connecticut, 1979.

Peshkin, M. A. and Sanderson, A. C. 1986. A variational principle for quasistatic mechanics. Technical Report CMU-RI-TR-86-16. Pittsburgh, Pennsylvania: Carnegie-Mellon University.

Podnar, G., Dowling, K. and Blackwell, M. 1984. A functional vehicle for autonomous mobile robot research. Technical Report No. CMU-RI-TR-84-28. Pittsburgh, Pennsylvania: Carnegie-Mellon University Robotics Institute.

Rockafellar, R. T. 1970. *Convex Analysis.* Princeton Math. Series. Vol. 28. Princeton, New Jersey: Princeton University Press.

Rogers, M. 1984. Birth of the killer robots. *Newsweek* CIII(26):51.

Smith, M. H. and Coles, L. S. 1973. Design of a low cost, general purpose robot. *Proc. of 3rd IJCAI.* Stanford, California. pp. 324–335.

Wallace, R., Stentz, A., Thorpe, C., Moravec, H., Whittaker, W. and Kanade, T. 1985. First results in robot road-following. *Proc. of the 9th IJCAI.* Los Angeles, California. pp. 1089–1095.

Whitaker, L. E. 1962. Stair climbing device. U. S. Patent No. 3,058,754.

Whittaker, E. T. 1961. *A Treatise on the Analytical Dynamics of Particles and Rigid Bodies* 4th edition. New York: Cambridge University Press.

Wilfong, G. T. 1988. Motion planning for an autonomous vehicle. *Proc. 1988 IEEE International Conference on Robotics and Automation.* Philadelphia, Pennsylvania. pp. 529–533.

Wilson, E. 1985. Denning mobile robotics: Robots guard the pen. *High Technology* 5(6):15–16.

KINEMATIC MODELING FOR FEEDBACK CONTROL OF AN OMNIDIRECTIONAL WHEELED MOBILE ROBOT

Patrick F. Muir and Charles P. Neuman

Department of Electrical and Computer Engineering
The Robotics Institute
Carnegie Mellon University
Pittsburgh, PA 15213

Abstract

We have introduced a methodology for the kinematic modeling of wheeled mobile robots[1-2]. In this paper, we apply our methodology to *Uranus*[3], an omnidirectional wheeled mobile robot which is being developed in the Robotics Institute of Carnegie Mellon University. We assign coordinate systems to specify the transformation matrices and write the kinematic equations-of-motion. We illustrate the ***actuated inverse*** and ***sensed forward*** solutions; i.e., the calculation of actuator velocities from robot velocities and robot velocities from sensed wheel velocities. We apply the actuated inverse and sensed forward solutions to the kinematic control of Uranus by: calculating in *real-time* the robot position from shaft encoder readings (i.e., dead reckoning); formulating an algorithm to detect wheel slippage; and developing an algorithm for feedback control.

1. Introduction

We have formulated a methodology for the kinematic modeling[1-2] of wheeled mobile robots (WMRs). In this paper, we apply our methodology to the kinematic modeling and control of *Uranus*[3], a WMR which is being constructed in the Robotics Institute of Carnegie Mellon University for research and development in control, computer vision, path planning and sonar sensing. Uranus provides omnidirectional mobility (i.e., independent and simultaneous X and Y translations and Z rotation) with a stable, rectangular wheelbase. Wheelchairs[4] which utilize this kinematic structure have already shown advantages over conventional designs. For example, the sideways X translational degree-of-freedom allows more maneuverability between closely spaced obstacles than turning the wheelchair in the direction of travel before translating.

In Section 2, we describe the kinematic structure of Uranus, omnidirectional wheels, and our control system hardware. Then, in Section 3, we develop the kinematic equations-of-motion by assigning coordinate systems, writing the transformation matrices and formulating the wheel Jacobian matrices. The Jacobian matrix of each of the four wheels resolves the robot body velocities as linear combinations of the wheel velocities. The four wheel equations-of-motion are then combined to form the composite robot equation-of-motion. We solve the composite robot equation-of-motion in Section 4. The actuated inverse solution (in Section 4.1) calculates the velocities of the actuators (i.e., wheel drive motors) from robot body velocities. The sensed forward solution (in Section 4.2) calculates the robot body velocities from the sensed wheel velocities (i.e., wheel shaft encoder measurements). In Section 5, we turn to kinematic control of Uranus. Dead-reckoning (discussed in Section 5.1), which is the real-time calculation of the robot position from the wheel shaft encoder measurements, plays the role of feedback in our kinematic control algorithm. If undesired wheel slip occurs, the dead reckoned robot position is erroneous. We then formulate in Section 5.2 an algorithm to detect wheel slippage. In Section 5.3, we design a kinematics-based feedback control algorithm for real-time implementation on Uranus. Finally, in Section 6, we highlight our kinematic controller design.

* This research has been supported by the Office of Naval Research under Contract N00014-81-K-0503 and the Department of Electrical and Computer Engineering, Carnegie Mellon University. The authors acknowledge Hans Moravec, Head of the Mobile Robot Laboratory, for his encouragement and support of this research.

2. Uranus

Uranus[3] (pictured in Figure 1) has the kinematic structure of the Wheelon wheelchair[4]: four omnidirectional wheels mounted at the corners of a rectangular frame. For kinematic modeling, we sketch the wheelbase in Figure 2. Each omnidirectional wheel[5] (shown in Figure 3) consists of a wheel hub about which 12 rollers are mounted at the angle $\eta = 45°$ to the wheel orientation. Each omnidirectional wheel has three degrees-of-freedom (DOFs). One DOF is in the direction of the wheel orientation; the second DOF is in the direction of the roller rotation; and the third DOF is the rotation about the point-of-contact between the wheel and the floor. In Figures 2 and 3, the rollers on the underside of the wheel (i.e., the rollers touching the floor), rather than the rollers which are actually visible from a top view, are shown to facilitate kinematic analysis.

Figure 1: Uranus

The control system hardware consists of an onboard Motorola 68000 control processor, motors, shaft encoders and inter-

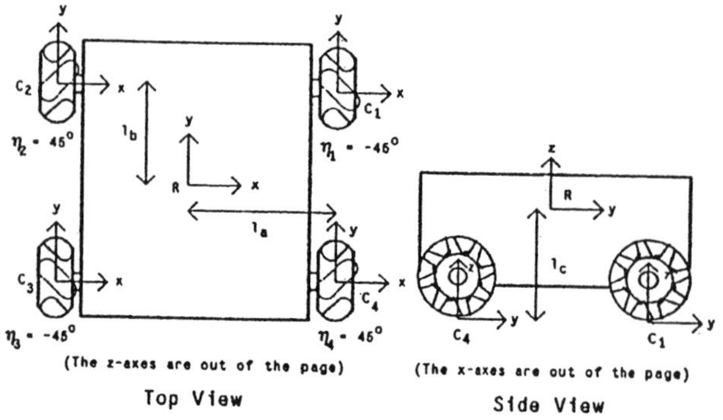

Figure 2: Coordinate System Assignments for Uranus

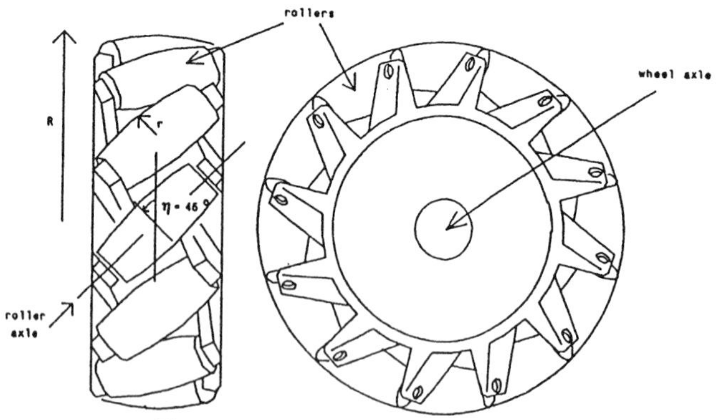

Figure 3: Omnidirectional Wheel

face electronics. The inner hub of each omnidirectional wheel is driven by a brushless DC motor and its position is sensed by an optical shaft encoder. The position and velocity of each wheel are thus available to the control processor through the interface electronics. The interface electronics also handle electronic commutation and pulse-width modulation for the brushless DC motors. Since the dynamic model of the motor velocity is *linear* in the pulse-width[6], the control processor need only calculate the wheel motor velocities. The desired robot position trajectory is communicated to the control processor from an independent offboard processor. A stiff suspension mechanism is built into each wheel assembly. The hardware construction of Uranus is nearing completion.

Three practical assumptions make the kinematic modeling of Uranus tractable. First, the motion of the suspension and compliance of the rollers are negligible. The assumption permits us to model the kinematics of the robot independently of the dynamics of these flexible components. Second, Uranus moves on a planar surface. We thus neglect irregularities in the floor surface. Even though this assumption restricts the range of practical applications, environments which do not satisfy this assumption (e.g., rough, bumpy or rocky surfaces) do not lend themselves to energy-efficient wheeled-vehicle travel[7]. Third, the rotational friction at the point-of-contact between a wheel and the surface is small enough to permit rotational slip. We make this assumption to treat the rotation about the point-of-contact (i.e., rotational wheel slippage) as a wheel DOF. Automobiles illustrate the practicality of this assumption since the wheels must rotate about their points-of-contact to navigate a turn.

3. Kinematic Modeling

3.1. Coordinate System Assignments

We apply the Sheth-Uicker convention[8] to assign coordinate systems. The Sheth-Uicker convention differs from the Denavit-Hartenburg[9] convention, which is applied to model stationary manipulators, because two coordinate systems are assigned at each joint (one in each link) instead of one. The displacement between the pair of coordinate systems at a joint is specified by the joint parameters. The Sheth-Uicker convention thereby allows the modeling of the higher-pair, multidimensional wheel motion and eliminates ambiguities in coordinate transformation matrices of closed-link chains.

Coordinate system assignment is illustrated in Figure 2 for Uranus. The stationary *floor* coordinate system F serves as the absolute reference coordinate frame for robot motions. The floor coordinate system is assigned with the z-axis orthogonal to the surface of travel. The *robot* coordinate system R is assigned to the robot body with the z-axis orthogonal to the surface of travel. The motion of the robot coordinate system is interpreted as the motion of the robot. We have assigned the robot coordinate system at the center of the robot body (l_a and l_b are one-half of the robot width and length, respectively), and at a height l_c above the floor. We assign a *contact point* coordinate system C_i (for $i = 1,2,3,4$) at the point-of-contact between each wheel and the floor with the y-axis parallel to the wheel orientation, and the x-y plane tangent to the surface of travel.

For stationary serial link manipulators, all joints are one-dimensional lower-pairs: prismatic joints allow translational Z motion and revolute joints allow rotational θ motion. The velocity of a manipulator joint can be specified independently of its absolute position by referencing the velocity of the coordinate system in the link on one side of the joint to the coordinate system in the adjacent link on the other side of the joint. In contrast, WMRs have three-dimensional higher-pair wheel-to-floor and robot-to-floor joints allowing simultaneous X, Y and θ motions. Such multi-dimensional joints require the assignment of an *instantaneously coincident coordinate system*[1-2] to specify the velocity of the joint independently of its position. An instantaneously coincident coordinate system $\overline{A}$ for the coordinate system A is defined to be at the same position and orientation in space as the A coordinate system, but stationary relative to the absolute F coordinate system. For example, an *instantaneously coincident robot* coordinate system $\overline{R}$ is defined at the same position and orientation in space as the robot coordinate system R, but stationary relative to the F coordinate system. We may thus specify the velocities of the robot coordinate system R independently of the robot position by referencing the velocities of the R coordinate system to the $\overline{R}$ coordinate system (as in Section 3.3). When the robot coordinate system moves relative to the floor coordinate system, we assign a different instantaneously coincident robot coordinate system at each time instant. Similarly, the *instantaneously coincident contact point* coordinate system $\overline{C}_i$ (for $i = 1,2,3,4$) coincides with the contact point coordinate system C_i and is stationary relative to the floor coordinate system. We require the instantaneously coincident contact point coordinate system to specify the wheel velocities independently of the wheel positions.

3.2. Transformation Matrices

Homogeneous (4×4) transformation matrices express the relative positions and orientations of coordinate systems[10]. The homogeneous transformation matrix ${}^{A}\Pi_{B}$ transforms the coordinates of the point $^B\mathbf{r}$ in the coordinate frame B to its corresponding coordinates $^A\mathbf{r}$ in the coordinate frame A:

$$^A\mathbf{r} = {}^A\mathbf{\Pi}_B \; {}^B\mathbf{r} \,. \tag{1}$$

We adopt the following notation. Scalar quantities are denoted by lower case letters (e.g., w). Vectors are denoted by lower case boldface letters (e.g., $\mathbf{r}$). Matrices are denoted by upper case boldface letters (e.g., $\mathbf{\Pi}$). Pre-superscripts denote reference coordinate systems. For example, $^A\mathbf{r}$ is the vector $\mathbf{r}$ in the A coordinate frame. Post-subscripts denote coordinate systems or components of a vector or matrix. For example, the transformation matrix $^A\mathbf{\Pi}_B$ defines the position and orientation of coordinate system B relative to coordinate frame A; and r_x is the x-component of the vector $\mathbf{r}$.

Our nomenclature for scalar rotational and translational displacements and velocities is:

$^A\theta_B$: The rotational displacement about the z-axis of the A coordinate system between the x-axis of the A coordinate system and the x-axis of the B coordinate system (counterclockwise by convention); and $^A\dot{\theta}_B = {}^A\omega_B$.

$\theta_{w_i x}$: The rotational displacement of wheel i (for $i = 1,2,3,4$) about an axis centered in the wheel and parallel to the x-axis of C_i; and $\dot{\theta}_{w_i x} = \omega_{w_i x}$.

$\theta_{w_i r}$: The rotational displacement of the roller of wheel i (for $i = 1,2,3,4$) which is in contact with the floor about an axis aligned with the axle of the roller; and $\dot{\theta}_{w_i r} = \omega_{w_i r}$.

$\theta_{w_i z}$: The rotational displacement of wheel i (for $i = 1,2,3,4$) about an axis centered in the wheel and parallel to the z-axis of C_i; and $\dot{\theta}_{w_i z} = \omega_{w_i z}$.

$^Ad_{Bj}$: The translational displacement along the j-axis (for $j \in [x,y,z]$) of the A coordinate system between the origins of the A and B coordinate systems; and $^A\dot{d}_{Bj} = {}^Av_{Bj}$.

Since any two coordinate systems A and B in our WMR model are located at non-zero x, y and z-coordinates relative to each other, each transformation matrix must contain the translations $^Ad_{Bx}$, $^Ad_{By}$ and $^Ad_{Bz}$. We have assigned all coordinate systems with the z-axes perpendicular to the surface of travel so that all rotations between coordinate systems are about the z-axis. The transformation matrix $^A\mathbf{\Pi}_B$ thus embodies a rotation $^A\theta_B$ about the z-axis of coordinate system A, and the translations $^Ad_{Bx}$, $^Ad_{By}$ and $^Ad_{Bz}$ along the respective coordinate axes:

$$^A\mathbf{\Pi}_B = \begin{pmatrix} \cos{}^A\theta_B & -\sin{}^A\theta_B & 0 & ^Ad_{Bx} \\ \sin{}^A\theta_B & \cos{}^A\theta_B & 0 & ^Ad_{By} \\ 0 & 0 & 1 & ^Ad_{Bz} \\ 0 & 0 & 0 & 1 \end{pmatrix} \,. \tag{2}$$

The pertinent coordinate transformation matrices are written for Uranus directly from Figure 2:

$$^R\mathbf{\Pi}_{C_1} = \begin{pmatrix} 1 & 0 & 0 & l_a \\ 0 & 1 & 0 & l_b \\ 0 & 0 & 1 & -l_c \\ 0 & 0 & 0 & 1 \end{pmatrix} , \quad ^R\mathbf{\Pi}_{C_2} = \begin{pmatrix} 1 & 0 & 0 & -l_a \\ 0 & 1 & 0 & l_b \\ 0 & 0 & 1 & -l_c \\ 0 & 0 & 0 & 1 \end{pmatrix}$$

$$^R\mathbf{\Pi}_{C_3} = \begin{pmatrix} 1 & 0 & 0 & -l_a \\ 0 & 1 & 0 & -l_b \\ 0 & 0 & 1 & -l_c \\ 0 & 0 & 0 & 1 \end{pmatrix} , \quad ^R\mathbf{\Pi}_{C_4} = \begin{pmatrix} 1 & 0 & 0 & l_a \\ 0 & 1 & 0 & -l_b \\ 0 & 0 & 1 & -l_c \\ 0 & 0 & 0 & 1 \end{pmatrix} .$$

Only one coordinate transformation matrix is written for each wheel of Uranus. More generally, wheels with steering links require the assignment of multiple coordinate systems[1-2]. The *sparse* transformation matrices result from assigning the robot coordinate system parallel to the wheel coordinate systems. Our robot coordinate system assignment thus facilitates kinematic modeling and control.

3.3. Wheel Jacobian Matrices

The Jacobian matrix $\mathbf{J}_i$ for wheel i resolves the robot velocities $^{\overline{R}}\dot{\mathbf{p}}_R$ as linear combinations of the wheel velocities $\dot{\mathbf{q}}_i$:

$$^{\overline{R}}\dot{\mathbf{p}}_R = \mathbf{J}_i\dot{\mathbf{q}}_i \qquad for\ i = 1,2,3,4 \,. \tag{3}$$

The robot velocity vector $^{\overline{R}}\dot{\mathbf{p}}_R$ consists of three components: the translational robot velocities $^{\overline{R}}v_{Rx}$ and $^{\overline{R}}v_{Ry}$, and the rotational robot velocity $^{\overline{R}}\omega_{Rz}$. The three DOFs of omnidirectional wheel i, wheel hub rotation, roller rotation, and rotation about the point-of-contact, correspond respectively to the three components $\omega_{w_i x}$, $\omega_{w_i r}$, and $\omega_{w_i z}$, of the wheel velocity vector $\dot{\mathbf{q}}_i$. The omnidirectional wheel Jacobian matrix $\mathbf{J}_i$ is constructed from the transformation matrix $^R\mathbf{\Pi}_{C_i}$, the wheel and roller radii R_i and r_i, and the roller angle η_i [1-2]:

$$\mathbf{J}_i = \begin{pmatrix} -R_i\sin{}^R\theta_{C_i} & r_i\sin\left({}^R\theta_{C_i}+\eta_i\right) & ^Rd_{C_iy} \\ R_i\cos{}^R\theta_{C_i} & -r_i\cos\left({}^R\theta_{C_i}+\eta_i\right) & -^Rd_{C_ix} \\ 0 & 0 & 1 \end{pmatrix} \,. \tag{4}$$

Since the determinant of $\mathbf{J}_i$ is $(-r_iR_i\sin\eta_i)$, the omnidirectional wheel Jacobian matrix in (4) is non-singular when $\eta_i \neq 0$ *or* $180°$; i.e., when the rollers are not aligned with the wheel orientation. Since the roller angles on Uranus are $\pm 45°$, the rank of each wheel Jacobian matrix is three, indicating that each wheel has three DOFs (as stated in Section 2).

To simplify fabrication, all wheels are identical. Consequently, the radii of all wheels and rollers are equal (i.e., $R_1 = R_2 = R_3 = R_4 = R$, and $r_1 = r_2 = r_3 = r_4 = r$). The magnitude of the roller angles on all four wheels are also equal; however, wheels 1 and 3 are mounted so that their roller angles are reversed from the roller angles of wheels 2 and 4 (i.e., $\eta_1 = \eta_3 = -45°$ and $\eta_2 = \eta_4 = 45°$). From the Jacobian matrix for omnidirectional wheels in (4), we write the equation-of-motion for each wheel:

$$Wheel\ 1: \quad ^{\overline{R}}\dot{\mathbf{p}}_R = \mathbf{J}_1\dot{\mathbf{q}}_1 \tag{5}$$

$$\begin{pmatrix} ^{\overline{R}}v_{Rx} \\ ^{\overline{R}}v_{Ry} \\ ^{\overline{R}}\omega_R \end{pmatrix} = \begin{pmatrix} 0 & -r\sqrt{2}/2 & l_b \\ R & -r\sqrt{2}/2 & -l_a \\ 0 & 0 & 1 \end{pmatrix} \begin{pmatrix} \omega_{w_1x} \\ \omega_{w_1r} \\ \omega_{w_1z} \end{pmatrix}$$

$$Wheel\ 2: \quad ^{\overline{R}}\dot{\mathbf{p}}_R = \mathbf{J}_2\dot{\mathbf{q}}_2 \tag{6}$$

$$\begin{pmatrix} ^{\overline{R}}v_{Rx} \\ ^{\overline{R}}v_{Ry} \\ ^{\overline{R}}\omega_R \end{pmatrix} = \begin{pmatrix} 0 & r\sqrt{2}/2 & l_b \\ R & -r\sqrt{2}/2 & l_a \\ 0 & 0 & 1 \end{pmatrix} \begin{pmatrix} \omega_{w_2x} \\ \omega_{w_2r} \\ \omega_{w_2z} \end{pmatrix}$$

$$Wheel\ 3: \quad {}^{\overline{R}}\dot{\mathbf{p}}_R = \mathbf{J}_3\dot{\mathbf{q}}_3 \tag{7}$$

$$\begin{pmatrix} {}^{\overline{R}}v_{Rx} \\ {}^{\overline{R}}v_{Ry} \\ {}^{\overline{R}}\omega_R \end{pmatrix} = \begin{pmatrix} 0 & -r\sqrt{2}/2 & -l_b \\ R & -r\sqrt{2}/2 & l_a \\ 0 & 0 & 1 \end{pmatrix} \begin{pmatrix} \omega_{w_3x} \\ \omega_{w_3r} \\ \omega_{w_3z} \end{pmatrix}$$

$$Wheel\ 4: \quad {}^{\overline{R}}\dot{\mathbf{p}}_R = \mathbf{J}_4\dot{\mathbf{q}}_4 \tag{8}$$

$$\begin{pmatrix} {}^{\overline{R}}v_{Rx} \\ {}^{\overline{R}}v_{Ry} \\ {}^{\overline{R}}\omega_R \end{pmatrix} = \begin{pmatrix} 0 & r\sqrt{2}/2 & -l_b \\ R & -r\sqrt{2}/2 & -l_a \\ 0 & 0 & 1 \end{pmatrix} \begin{pmatrix} \omega_{w_4x} \\ \omega_{w_4r} \\ \omega_{w_4z} \end{pmatrix}.$$

3.4. Composite Robot Equation

The motion of Uranus results from the simultaneous motions of the wheels. To model the motion, we combine the equations-of-motion of the four wheels in (5)-(8) to form the ***composite robot equation-of-motion***:

$$\begin{pmatrix} \mathbf{I} \\ \mathbf{I} \\ \mathbf{I} \\ \mathbf{I} \end{pmatrix} {}^{\overline{R}}\dot{\mathbf{p}}_R = \begin{pmatrix} \mathbf{J}_1 & 0 & \dots & 0 \\ 0 & \mathbf{J}_2 & \ddots & \vdots \\ \vdots & \ddots & \mathbf{J}_3 & 0 \\ 0 & \dots & 0 & \mathbf{J}_4 \end{pmatrix} \begin{pmatrix} \dot{\mathbf{q}}_1 \\ \dot{\mathbf{q}}_2 \\ \dot{\mathbf{q}}_3 \\ \dot{\mathbf{q}}_4 \end{pmatrix} \tag{9}$$

$$\text{or} \quad \mathbf{A}_o {}^{\overline{R}}\dot{\mathbf{p}}_R = \mathbf{B}_o\dot{\mathbf{q}} \tag{10}$$

in which $\mathbf{I}$ is a (3×3) identity matrix, $\mathbf{A}_0$ is a (12×3) matrix, $\mathbf{B}_0$ is a (12×12) block diagonal matrix, and $\dot{\mathbf{q}}$ is the composite wheel velocity vector. The 12 wheel equations in (10) must be solved simultaneously to characterize the WMR motion.

4. Kinematic Solutions

4.1. Actuated Inverse Solution

We compute the actuated inverse solution by calculating the actuated wheel velocities in (10) from the robot velocities. Because of the closed-link chains in WMRs, we need not actuate all of the wheel variables θ_{w_ix}, θ_{w_ir}, and θ_{w_iz}, for $i = 1,2,3,4$. To separate the actuated "a" and unactuated "u" wheel variables, we partition the wheel equation in (3) into two components:

$${}^{\overline{R}}\dot{\mathbf{p}}_R = \mathbf{J}_{ia}\dot{\mathbf{q}}_{ia} + \mathbf{J}_{iu}\dot{\mathbf{q}}_{iu} \qquad for\ i = 1,2,3,4\,. \tag{11}$$

We combine the four partitioned wheel equations in (11) to rewrite the composite robot equation in (9) and (10) as

$$(\mathbf{I}\ \ \mathbf{I}\ \ \mathbf{I}\ \ \mathbf{I})^T\ {}^{\overline{R}}\dot{\mathbf{p}}_R$$

$$= \begin{pmatrix} \mathbf{J}_{1a} & 0 & \dots & 0 & \mathbf{J}_{1u} & 0 & \dots & 0 \\ 0 & \mathbf{J}_{2a} & \ddots & \vdots & 0 & \mathbf{J}_{2u} & \ddots & \vdots \\ \vdots & \ddots & \mathbf{J}_{3a} & 0 & \vdots & \ddots & \mathbf{J}_{3u} & 0 \\ 0 & \dots & 0 & \mathbf{J}_{4a} & 0 & \dots & 0 & \mathbf{J}_{4u} \end{pmatrix} \begin{pmatrix} \dot{\mathbf{q}}_{1a} \\ \dot{\mathbf{q}}_{2a} \\ \dot{\mathbf{q}}_{3a} \\ \dot{\mathbf{q}}_{4a} \\ \dot{\mathbf{q}}_{1u} \\ \dot{\mathbf{q}}_{2u} \\ \dot{\mathbf{q}}_{3u} \\ \dot{\mathbf{q}}_{4u} \end{pmatrix} \tag{12}$$

$$\text{or} \quad \mathbf{A}_0\ {}^{\overline{R}}\dot{\mathbf{p}}_R = \mathbf{B}_{0p}\begin{pmatrix} \dot{\mathbf{q}}_a \\ \dot{\mathbf{q}}_u \end{pmatrix}. \tag{13}$$

In the absence of wheel slippage, the wheel velocities are coupled and hence the problem of calculating the wheel velocities from the robot velocities is *overdetermined*[1-2]; i.e., there are more independent equations than unknowns (wheel velocities). We compute the least-squares solution

$$\begin{pmatrix} \dot{\mathbf{q}}_a \\ \dot{\mathbf{q}}_u \end{pmatrix} = (\mathbf{B}_{0p}^T\mathbf{B}_{0p})^{-1}\ \mathbf{B}_{0p}^T\ \mathbf{A}_0\ {}^{\overline{R}}\dot{\mathbf{p}}_R \tag{14}$$

to obtain the actuated wheel velocities[1-2]:

$$\dot{\mathbf{q}}_a = \begin{pmatrix} [\mathbf{J}_{1a}^T\boldsymbol{\Delta}(\mathbf{J}_{1u})\mathbf{J}_{1a}]^{-1}\mathbf{J}_{1a}^T\boldsymbol{\Delta}(\mathbf{J}_{1u}) \\ [\mathbf{J}_{2a}^T\boldsymbol{\Delta}(\mathbf{J}_{2u})\mathbf{J}_{2a}]^{-1}\mathbf{J}_{2a}^T\boldsymbol{\Delta}(\mathbf{J}_{2u}) \\ [\mathbf{J}_{3a}^T\boldsymbol{\Delta}(\mathbf{J}_{3u})\mathbf{J}_{3a}]^{-1}\mathbf{J}_{3a}^T\boldsymbol{\Delta}(\mathbf{J}_{3u}) \\ [\mathbf{J}_{4a}^T\boldsymbol{\Delta}(\mathbf{J}_{4u})\mathbf{J}_{4a}]^{-1}\mathbf{J}_{4a}^T\boldsymbol{\Delta}(\mathbf{J}_{4u}) \end{pmatrix} {}^{\overline{R}}\dot{\mathbf{p}}_R = \mathbf{J}_a\ {}^{\overline{R}}\dot{\mathbf{p}}_R \tag{15}$$

in which we define the ***Delta*** matrix function $\boldsymbol{\Delta}(\bullet)$ as:

$$\boldsymbol{\Delta}(\mathbf{U}) = \begin{cases} -\mathbf{I} & for\ \ \mathbf{U} = null \\ \mathbf{U}(\mathbf{U}^T\ \mathbf{U})^{-1}\mathbf{U}^T - \mathbf{I} & Otherwise \end{cases} \tag{16}$$

where the argument $\mathbf{U}$ is a $(c \times d)$ matrix of rank d. In (15), $\mathbf{J}_a$ is the [number of actuators (four for Uranus) $\times$ 3] ***actuated*** wheel Jacobian matrix.

For Uranus, we identify the actuated wheel velocities as ω_{w_ix}, for $i = 1,2,3,4$, and the unactuated wheel velocities as ω_{w_ir} and ω_{w_iz}, for $i = 1,2,3,4$, and apply the actuated inverse solution in (15) to obtain:

$$\begin{pmatrix} \omega_{w_1x} \\ \omega_{w_2x} \\ \omega_{w_3x} \\ \omega_{w_4x} \end{pmatrix} = \frac{1}{R}\begin{pmatrix} -1 & 1 & (l_a + l_b) \\ 1 & 1 & -(l_a + l_b) \\ -1 & 1 & -(l_a + l_b) \\ 1 & 1 & (l_a + l_b) \end{pmatrix} \begin{pmatrix} {}^{\overline{R}}v_{Rx} \\ {}^{\overline{R}}v_{Ry} \\ {}^{\overline{R}}\omega_R \end{pmatrix}. \tag{17}$$

Since all of the wheel Jacobian matrices for Uranus are nonsingular, we compute the actuated inverse solution in (17) by solving wheel equations (5)-(8) independently for each of the actuated wheel velocities. This alternate, straightforward approach reduces the computational requirements of our least-squares solution in (11)-(17) by a factor of thirteen.

4.2. Sensed Forward Solution

The sensed forward solution computes the robot velocity vector ${}^{\overline{R}}\dot{\mathbf{p}}_R$ from the sensed wheel positions and velocities $\mathbf{q}_s$ and $\dot{\mathbf{q}}_s$. Development of the sensed forward solution parallels our development of the actuated inverse solution in Section 4.1. The first step is to separate the sensed "s" and not-sensed "n" wheel velocities and rewrite the wheel equations-of-motion in (3) as:

$${}^{\overline{R}}\dot{\mathbf{p}}_R = \mathbf{J}_{is}\dot{\mathbf{q}}_{is} + \mathbf{J}_{in}\dot{\mathbf{q}}_{in} \qquad for\ i = 1,2,3,4\ . \tag{18}$$

We combine the four wheel equations in (18) with the unknown robot and wheel velocities on the left-hand side:

$$\begin{pmatrix} \mathbf{I} & -\mathbf{J}_{1n} & 0 & \dots & 0 \\ \mathbf{I} & 0 & -\mathbf{J}_{2n} & \ddots & \vdots \\ \mathbf{I} & \vdots & \ddots & -\mathbf{J}_{3n} & 0 \\ \mathbf{I} & 0 & \dots & 0 & -\mathbf{J}_{4n} \end{pmatrix} \begin{pmatrix} {}^{\overline{R}}\dot{\mathbf{p}}_R \\ \dot{\mathbf{q}}_{1n} \\ \dot{\mathbf{q}}_{2n} \\ \dot{\mathbf{q}}_{3n} \\ \dot{\mathbf{q}}_{4n} \end{pmatrix}$$

$$= \begin{pmatrix} \mathbf{J}_{1s} & 0 & \dots & 0 \\ 0 & \mathbf{J}_{2s} & \ddots & \vdots \\ \vdots & \ddots & \mathbf{J}_{3s} & 0 \\ 0 & \dots & 0 & \mathbf{J}_{4s} \end{pmatrix} \begin{pmatrix} \dot{\mathbf{q}}_{1s} \\ \dot{\mathbf{q}}_{2s} \\ \dot{\mathbf{q}}_{3s} \\ \dot{\mathbf{q}}_{4s} \end{pmatrix} \tag{19}$$

$$\text{or} \quad \mathbf{A}_n\begin{pmatrix} {}^{\overline{R}}\dot{\mathbf{p}}_R \\ \dot{\mathbf{q}}_n \end{pmatrix} = \mathbf{B}_s\dot{\mathbf{q}}_s\,. \tag{20}$$

Since there are more independent equations (three for each wheel) than unknowns (two not-sensed wheel velocities for each wheel plus three robot velocities), the problem of computing the unknown wheel and robot velocities from the sensed wheel velocities is overdetermined. We thus apply the least-squares solution[1-2]:

$$\begin{pmatrix} {}^{R}\dot{\mathbf{p}}_R \\ \dot{\mathbf{q}}_n \end{pmatrix} = (\mathbf{A}_n^T \mathbf{A}_n)^{-1} \mathbf{A}_n^T \mathbf{B}_s \dot{\mathbf{q}}_s \,, \tag{21}$$

to calculate the robot velocities[1-2]:

$$ {}^{R}\dot{\mathbf{p}}_R = \mathbf{\Psi}^{-1}\, \mathbf{S}\, \dot{\mathbf{q}}_s \tag{22}$$

where $\mathbf{\Psi} = [\Delta(\mathbf{J}_{1n}) + \Delta(\mathbf{J}_{2n}) + \Delta(\mathbf{J}_{3n}) + \Delta(\mathbf{J}_{4n})]$ is a (3×3) matrix and

$$\mathbf{S} = [\Delta(\mathbf{J}_{1n})\mathbf{J}_{1s} \quad \Delta(\mathbf{J}_{2n})\mathbf{J}_{2s} \quad \Delta(\mathbf{J}_{3n})\mathbf{J}_{3s} \quad \Delta(\mathbf{J}_{4n})\mathbf{J}_{4s}] \,.$$

Hence, solving the system of linear algebraic equations in (22) is *not* a computational burden.

For Uranus, we identify the sensed wheel velocities as $\omega_{w_i x}$, for $i = 1,2,3,4$, and the not-sensed wheel velocities as $\omega_{w_i r}$ and $\omega_{w_i z}$, for $i = 1,2,3,4$, and apply the least-squares sensed forward solution in (22) to obtain:

$$\begin{pmatrix} {}^{R}v_{Rx} \\ {}^{R}v_{Ry} \\ {}^{R}\omega_R \end{pmatrix} = \frac{R}{4\, l_{ab}} \begin{pmatrix} -l_{ab} & l_{ab} & -l_{ab} & l_{ab} \\ l_{ab} & l_{ab} & l_{ab} & l_{ab} \\ 1 & -1 & -1 & 1 \end{pmatrix} \begin{pmatrix} \omega_{w_1 x} \\ \omega_{w_2 x} \\ \omega_{w_3 x} \\ \omega_{w_4 x} \end{pmatrix} , \tag{23}$$

where $l_{ab} = l_a + l_b$. The least-squares forward solution in (23) need not produce a zero error because of sensor noise, wheel slippage, and non-planar surfaces. In the presense of these error sources, we cannot calculate the robot velocity exactly. Our least-squares solution provides an optimal solution by minimizing the sum of the squared errors of the velocity components.

5. Applications

5.1. Dead Reckoning

Dead reckoning[13] is the real-time calculation of the WMR position from wheel sensor measurements. To determine the robot position in real-time, we integrate the robot velocity over each sampling period. The integration begins when the robot is at rest (i.e., ${}^{F}\dot{\mathbf{p}}_R(0) = \mathbf{0}$). The initial robot position ${}^{F}\mathbf{p}_R(0)$ is either specified or determined by computer vision or sonar ranging. We assume that the robot motion is adequately modeled by piecewise constant accelerations since the robot is being actuated by constant force/torque generators in each control sampling period (the control sampling period matches the dead reckoning sampling period). The robot velocity ${}^{R}\dot{\mathbf{p}}_R$ in the sampling period from time $t = (n-1)T$ to time $t = nT$ is

$$\begin{aligned} {}^{R}\dot{\mathbf{p}}_R(t) &= {}^{R}\dot{\mathbf{p}}_R[(n-1)T] \\ &+ \frac{{}^{R}\dot{\mathbf{p}}_R(nT) - {}^{R}\dot{\mathbf{p}}_R[(n-1)T]}{T}(t - [(n-1)T]), \end{aligned} \tag{24}$$

where the robot velocity ${}^{R}\dot{\mathbf{p}}_R(nT)$ at each sampling instant is calculated according to the sensed forward solution in (23). We apply the orthogonal velocity transformation matrix $\mathbf{V}$ to transform the robot velocity to the floor coordinate system:

$$\begin{aligned} {}^{F}\dot{\mathbf{p}}_R &= \mathbf{V}\, {}^{R}\dot{\mathbf{p}}_R \\ &= \begin{pmatrix} \cos {}^{F}\theta_R & -\sin {}^{F}\theta_R & 0 \\ \sin {}^{F}\theta_R & \cos {}^{F}\theta_R & 0 \\ 0 & 0 & 1 \end{pmatrix} {}^{R}\dot{\mathbf{p}}_R \,. \end{aligned} \tag{25}$$

We calculate the robot position at the current sampling instant $t = nT$ by integrating the velocity over the sampling period and adding the result to the robot position at the previous sampling instant $t = (n-1)T$:

$$ {}^{F}\mathbf{p}_R(nT) = {}^{F}\mathbf{p}_R[(n-1)T] + \int_{(n-1)T}^{nT} {}^{F}\dot{\mathbf{p}}_R(t)dt \,. \tag{26}$$

By subtituting (24) and (25) into (26), we express the current robot position in terms of the position at the previous sampling instant and the robot velocity at the current and previous sampling instants:

$$\begin{aligned} {}^{F}\mathbf{p}_R(nT) &= {}^{F}\mathbf{p}_R[(n-1)T] \\ &+ \frac{T}{2}\, \mathbf{V}[(n-1)T]\left\{ {}^{R}\dot{\mathbf{p}}_R[(n-1)T] + {}^{R}\dot{\mathbf{p}}_R(nT) \right\}. \end{aligned} \tag{27}$$

Upon substituting the sensed forward solution in (23) into (27), expanding, and simplifying, we obtain algorithms for the *direct* calculation of the current orientation ${}^{F}\theta_R(nT)$, and the *recursive* calculation of the current translations ${}^{F}d_{Rx}(nT)$ and ${}^{F}d_{Ry}(nT)$ of Uranus:

$$ {}^{F}\theta_R(nT) = \frac{R}{4(l_a + l_b)}\theta(nT) + {}^{F}\theta_R(0) - \frac{R}{4(l_a + l_b)}\theta(0) \tag{28}$$

where $\theta(\bullet) = [\theta_{w_1 x}(\bullet) - \theta_{w_2 x}(\bullet) - \theta_{w_3 x}(\bullet) + \theta_{w_4 x}(\bullet)]$, and

$$\begin{aligned} \begin{pmatrix} {}^{F}d_{Rx}(nT) \\ {}^{F}d_{Ry}(nT) \end{pmatrix} &= \begin{pmatrix} {}^{F}d_{Rx}[(n-1)T] \\ {}^{F}d_{Ry}[(n-1)T] \end{pmatrix} \\ &+ \frac{TR}{8} \begin{pmatrix} -a & b & -a & b \\ b & a & b & a \end{pmatrix} [\dot{\mathbf{q}}_s[(n-1)T] + \dot{\mathbf{q}}_s(nT)] \end{aligned} \tag{29}$$

where $a = [\cos {}^{F}\theta_R(nT) + \sin {}^{F}\theta_R(nT)]$ and $b = [\cos {}^{F}\theta_R(nT) - \sin {}^{F}\theta_R(nT)]$. The direct calculation of the robot orientation indicates that (28) is a *holonomic* constraint[14]. Since the robot translations cannot be calculated directly, (29) are non-holonomic constraints. Errors in the recursively calculated robot translations due to finite precision, sensor noise, wheel slip, and non-planar surfaces will accumulate; whereas, the direct calculation of the holonomic orientation constraint in (28) will not accumulate these errors.

5.2. Wheel Slip Detection

Since the sensed forward solution for Uranus in (23) is overdetermined, the kinematic equations-of-motion are inconsistent in the presence of wheel slip. The error in the least-squares forward solution is then non-zero. We thus detect wheel slip by calculating the error in the least-squares forward solution. If the error in the forward solution exceeds a threshold, we conclude that the forward solution, and hence the dead reckoned robot position, are not sufficiently accurate for feedback control calculations. In the improbable case that all four wheels slip simultaneously in such a manner that the equations-of-motion remain consistent, our approach will fail to detect wheel slip.

In practice, error sources such as finite precision, sensor noise and non-planar surfaces also render the kinematic equations-of-motion inconsistent. We expect that the least-squares error induced by these sources will be small in comparison with the error caused by wheel slippage. We thus compare the least-squares error with an error threshold. If the least-squares error in the forward solution exceeds the threshold, we conclude that wheel slip has occurred. When we detect that wheel slip has occurred, we resort to absolute methods (e.g., computer vision, ultrasonic ranging sensors, and laser range finders) to determine the robot position. Since current absolute locating methods are computationally slow relative to the robot motion, the WMR should halt motion until the dead reckoning algorithm is updated by an absolute locating method.

To calculate the error in the least-squares forward solution, we relate the robot velocity vector (computed by the sensed forward solution) to the sensed wheel velocities by eliminating the not-sensed wheel velocities from the composite robot equation in (10). We express the not-sensed wheel velocities as linear combinations of the robot velocities by applying the actuated inverse solution in (15) with the not-sensed ("n" subscripts) and sensed ("s" subscripts) wheel velocities playing the roles of the actuated ("a" subscripts) and unactuated ("u" subscripts) wheel velocities, respectively:

$$\dot{\mathbf{q}}_n = \begin{pmatrix} [\mathbf{J}_{1n}^T\Delta(\mathbf{J}_{1s})\mathbf{J}_{1n}]^{-1}\mathbf{J}_{1n}^T\Delta(\mathbf{J}_{1s}) \\ [\mathbf{J}_{2n}^T\Delta(\mathbf{J}_{2s})\mathbf{J}_{2n}]^{-1}\mathbf{J}_{2n}^T\Delta(\mathbf{J}_{2s}) \\ [\mathbf{J}_{3n}^T\Delta(\mathbf{J}_{3s})\mathbf{J}_{3n}]^{-1}\mathbf{J}_{3n}^T\Delta(\mathbf{J}_{3s}) \\ [\mathbf{J}_{4n}^T\Delta(\mathbf{J}_{4n})\mathbf{J}_{4n}]^{-1}\mathbf{J}_{4n}^T\Delta(\mathbf{J}_{4n}) \end{pmatrix} {}^{\bar{R}}\dot{\mathbf{p}}_R \,. \tag{30}$$

We partition the sensed and not-sensed wheel velocities in the composite robot equation in (9) and substitute (30) for the not-sensed wheel velocities to obtain the robot *sensing* equation:

$$\begin{pmatrix} \mathbf{I}-\mathbf{J}_{1n}[\mathbf{J}_{1n}^T\Delta(\mathbf{J}_{1s})\mathbf{J}_{1n}]^{-1}\mathbf{J}_{1n}^T\Delta(\mathbf{J}_{1s}) \\ \mathbf{I}-\mathbf{J}_{2n}[\mathbf{J}_{2n}^T\Delta(\mathbf{J}_{2s})\mathbf{J}_{2n}]^{-1}\mathbf{J}_{2n}^T\Delta(\mathbf{J}_{2s}) \\ \mathbf{I}-\mathbf{J}_{3n}[\mathbf{J}_{3n}^T\Delta(\mathbf{J}_{3s})\mathbf{J}_{3n}]^{-1}\mathbf{J}_{3n}^T\Delta(\mathbf{J}_{3s}) \\ \mathbf{I}-\mathbf{J}_{4n}[\mathbf{J}_{4n}^T\Delta(\mathbf{J}_{4n})\mathbf{J}_{4n}]^{-1}\mathbf{J}_{4n}^T\Delta(\mathbf{J}_{4n}) \end{pmatrix} {}^{\bar{R}}\dot{\mathbf{p}}_R = \begin{pmatrix} \mathbf{J}_{1s} & 0 & \dots & 0 \\ 0 & \mathbf{J}_{2s} & \ddots & 0 \\ \vdots & \ddots & \mathbf{J}_{3s} & 0 \\ 0 & \dots & 0 & \mathbf{J}_{4s} \end{pmatrix} \dot{\mathbf{q}}_s \,, \tag{31}$$

$$\text{or} \qquad \mathbf{A}_s {}^{\bar{R}}\dot{\mathbf{p}}_R = \mathbf{B}_s\dot{\mathbf{q}}_s \,. \tag{32}$$

Calculation of the sensed forward solution in (22) is the first step in determining the least-squares error. We substitute the computed robot velocity vector ${}^{\bar{R}}\dot{\mathbf{p}}_R$ for the actual robot velocity vector in (32). The least-squares error vector $\mathbf{e}$ is then calculated by subtracting the right-hand side of (32) from the left-hand side:

$$\mathbf{e} \;=\; \mathbf{A}_s \, {}^{\bar{R}}\dot{\mathbf{p}}_R \;-\; \mathbf{B}_s \, \dot{\mathbf{q}}_s \,. \tag{33}$$

We compare the two-norm of the least-squares error $\sqrt{\mathbf{e}^T\mathbf{e}}$ with the scalar threshold e_t. If the norm exceeds the threshold, we conclude that wheel slip has occurred. For Uranus, this algorithm is:

$$If \quad \frac{R}{2}(\omega_{w_1x}+\omega_{w_2x}-\omega_{w_3x}-\omega_{w_4x}) \;>\; e_t \,, \qquad \textit{wheel slip has occurred} \,. \tag{34}$$

5.3. Kinematic Feedback Control

The documented WMR control systems are kinematically based[11–12]; i.e., they do not incorporate a dynamic model of the robot motion. A reference robot trajectory is provided by a trajectory planner and the task of the control system is to produce signals for the wheel actuators so that the WMR tracks the reference trajectory.

A kinematics-based control system for Uranus is diagrammed in Figure 4. Directed arrows indicate the flow of information. The number of scalar variables represented by each arrow is indicated within the flow lines. At time nT, we sample the wheel positions $\mathbf{q}_s(nT)$ and velocities $\dot{\mathbf{q}}_s(nT)$ from the wheel shaft encoders and interface logic, and receive the desired robot position vector ${}^F\mathbf{p}_d(nT)$ from an offboard trajectory planning processor. We apply the dead reckoning algorithm in (28)-(29) to compute the robot position ${}^F\mathbf{p}_R(nT)$. We compare the desired robot position with the actual robot position to calculate the robot position error:

$${}^F\mathbf{e}_R(nT) \;=\; {}^F\mathbf{p}_d(nT) - {}^F\mathbf{p}_R(nT). \tag{35}$$

The position error is multiplied by the (3×3) diagonal feedforward gain vector $\mathbf{K}$ and is then transformed to the $\bar{r}$obot coordinate frame by applying the inverse motion matrix $\mathbf{V}^{-1}(nT) = \mathbf{V}^T(nT)$ in (25). Under the assumption that the robot tracking error remains small, we treat the robot position error ${}^{\bar{R}}\mathbf{e}_R$ as the differential displacement ${}^{\bar{R}}\delta\mathbf{p}_R$. We apply the actuated inverse solution in (15) to transform this robot differential displacement (as velocities are transformed) into actuator displacements $\delta\mathbf{q}_a$:

$$\delta\mathbf{q}_a = \mathbf{J}_a \, {}^{\bar{R}}\delta\mathbf{p}_R \,. \tag{36}$$

The resulting actuator velocities are then communicated to the pulse-width modulators.

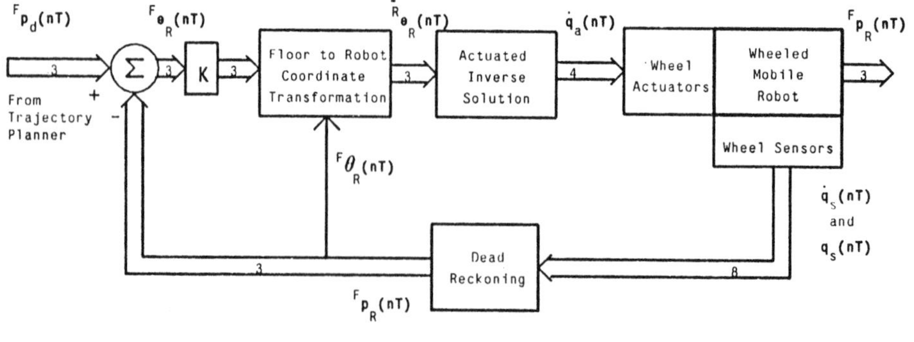

Figure 4: Kinematics-Based WMR Control System

The actuator velocities are calculated by matrix-vector multiplication according to our kinematic feedback control algorithm as:

$$\dot{\mathbf{q}}_a(nT) = \mathbf{J}_a \mathbf{V}^T(nT) \mathbf{K} \, {}^F\mathbf{e}_R(nT) \quad (37)$$

which, for Uranus, is:

$$\begin{pmatrix} \omega_{w_1 x}(nT) \\ \omega_{w_2 x}(nT) \\ \omega_{w_3 x}(nT) \\ \omega_{w_4 x}(nT) \end{pmatrix} = \frac{1}{R} \begin{pmatrix} -ak_x & -bk_y & lk_z \\ bk_x & ak_y & -lk_z \\ -ak_x & -bk_y & -lk_z \\ bk_x & ak_y & lk_z \end{pmatrix} {}^F\mathbf{e}_R(nT) . \quad (38)$$

The elements k_x, k_y, and k_z of the diagonal feedforward gain matrix $\mathbf{K}$ are adjusted experimentally to provide a fast robot tracking response without excessive overshoot or oscillations about the reference trajectory.

6. Summary

We have illustrated our methodology for the kinematic modeling of WMRs[1-2] through the omnidirectional WMR Uranus. We have solved the kinematic equations-of-motion to calculate the actuated inverse and sensed forward solutions. Finally, we applied our kinematic methodology to dead reckoning, wheel slip detection and feedback control algorithm design. These three applications are the fundamental components of the kinematics-based feedback control algorithm (in Figure 4) for Uranus which is executed every sampling period:

(1) Read the present desired robot position vector from the off-board path-planning processor, and sample the positions and velocities of the wheels.

(2) Evaluate (34) to determine whether wheel slip has occurred. If wheel slippage is detected, halt the robot and request the present position of the robot from a vision, sonar, or laser-rangefinding processor.

(3) Execute the dead reckoning algorithm in (28)-(29) to compute the present robot position.

(4) Calculate the robot position error in (35).

(5) Execute the control algorithm in (38) to compute the actuator velocities.

(6) Communicate the resulting actuator velocities to the wheel motors.

Over the past twenty years, manipulator control systems have progressed from independent joint-space control[15], to kinematics-based Cartesian-space control[16], to dynamics-based Cartesian-space feedback control[17], to robust dynamics-based feedback control[18-19] and adaptive control algorithms[20]. Because we anticipate that future WMR control systems will incorporate dynamic models, we are continuing our research by formulating methodologies for the *dynamic modeling* and *dynamics-based feedback control* of WMRs.

7. References

[1] P. F. Muir and C. P. Neuman, "Kinematic Modeling of Wheeled Mobile Robots," Technical Report No. CMU-RI-TR-86-12, The Robotics Institute, Carnegie Mellon University, Pittsburgh, PA, 15213, July 1986.

[2] P. F. Muir and C. P. Neuman, "Kinematic Modeling of Wheeled Mobile Robots," *Journal of Robotic Systems*, Vol. 4, No. 2, Spring 1987.

[3] H. P. Moravec (editor), "Autonomous Mobile Robots Annual Report - 1985," Robotics Institute Technical Report No. CMU-RI-MRL-86-1, Carnegie Mellon University, Pittsburgh, PA, January 1986.

[4] Alvema Rehab, "Wheelon - the New Movement," (advertisement), P.O. Box 17017,S-16117 Bromma, Sweden.

[5] B. E. Ilon, "Wheels for a Course Stable Selfpropelling Vehicle Movable in any Desired Direction on the Ground or Some Other Base," U.S. Patent No. 3,876,255 , 1975.

[6] P. F. Muir and C. P. Neuman, "Pulsewidth Modulation Control of Brushless DC Motors for Robotic Applications," *IEEE Transactions on Industrial Electronics*, Vol. IE-32, No. 3, August 1985, pp. 222-229.

[7] M. G. Bekker, *Introduction to Terrain-Vehicle Systems*, The University of Michigan Press, Ann Arbor, MI, 1969.

[8] P. N. Sheth and J. J. Uicker, Jr., "A Generalized Symbolic Notation for Mechanisms," *Journal of Engineering for Industry*, Series B, Vol. 93, No. 70-Mech-19, 102-112 (1971).

[9] J. Denavit and R. S. Hartenberg, "A Kinematic Notation for Lower-Pair Mechanisms Based on Matrices," *Journal of Applied Mechanics*, Vol. 77, No. 2, 215-221 (1955).

[10] R. P. Paul, *Robot Manipulators: Mathematics, Programming and Control*, The MIT Press, Cambridge, MA, 1981.

[11] T. Hongo, et al., "An Automatic Guidance System of a Self-Controlled Vehicle: The Command System and the Control Algorithm," IEEE Proceedings of the IECON, San Francisco, CA, November 1985, pp. 18-22.

[12] D. J. Daniel, "Analysis, Design, and Implementation of Microprocessor Control for a Mobile Platform," Master's Project Report, Department of Electrical and Computer Engineering, Carnegie Mellon University, Pittsburgh, PA, 15213, August 1984.

[13] *McGraw-Hill Encyclopedia of Science & Technology*, 5th Edition, McGraw-Hill Book Company, New York, Vol. 4, pp. 36-38, 1982.

[14] L. D. Landau and E. M. Lifshitz, *Mechanics*, Third Edition, Pergamon Press, New York, 1976.

[15] R. Paul, et al., "Advanced Industrial Robot Control Systems," Technical Report No. TR-EE79-35, School of Electrical Engineering, Purdue University, West Lafayette, IN, 47907, July 1979.

[16] D. E. Whitney, "Resolved Motion Rate Control of Manipulators and Human Protheses," *IEEE Transactions on Man-Machine Systems*, Vol. MMS-10, No. 2, June 1969, pp.47-53.

[17] J. Y. S. Luh, M. W. Walker, and R. P. C. Paul, "Resolved-Acceleration Control of Mechanical Manipulators," *IEEE Transactions on Automatic Control*, Vol. AC-25, No. 3, June 1980, pp. 468-474.

[18] V. D. Tourassis and C. P. Neuman, "Robust Nonlinear Feedback Control for Robotic Manipulators," *IEE Proceedings - D: Control Theory and Applications*, Special Issue on Robotics, Vol. 132, No. 4, July 1985, pp. 134-143.

[19] C. P. Neuman and V. D. Tourassis, "Robust Discrete Nonlinear Feedback Control for Robotic Manipulators," *Journal of Robotic Systems*, Vol. 4, No. 1, 1987.

[20] Nicosia, S. and Tomei, P., "Model Reference Adaptive Control Algorithms for Industrial Robots," *Automatica*, Vol. 20, No. 5, September 1984, pp. 635-644.

An Automatic Guidance System of a Self-Controlled Vehicle

TAKERO HONGO, HIDEO ARAKAWA, GUNJI SUGIMOTO, KOICHI TANGE, AND YUZO YAMAMOTO

Abstract—**An automatically guided vehicle, traveling without fixed guide ways, has been developed. In this paper, the construction of the vehicle, the control algorithm, and its general performance are described.**

The vehicle measures its own position by using measuring wheels and corrects the accumulated error at some intervals while it travels. Its measurement is so accurate that the required correction interval could be more than 50 m.

A sequence of simple commands is used to show the vehicle the way to the destination. A programmed course is composed of some straight and curved lines. According to the command sequence the vehicle leads itself by determining a current command, which represents the currently required movement. Referring to it, a microcomputer always controls its movement to diminish the errors in the speed, in the position, and in the direction generally. Therefore, like some kinds of mobile robots, the vehicle can travel along a programmed course.

The experimental results indicate that the vehicle is able to travel a long distance continuously. The movement is smooth and its control is very reliable. These high performances confirmed that the vehicle could be put into practical applications and we are now testing it for transportation of documents in our laboratory.

I. Introduction

In accordance with the recent development of flexible automations, an automatically guided vehicle traveling without a fixed guide way has been required, since the track for the vehicle can be changed easily and quickly. But this type of vehicle is now in the improvement stage for practical applications, and several problems remain unsolved. They can be described as follows.

1) It must be accurate, easy, and immediate for the vehicle to measure its own position and heading wherever it goes.

2) Its traveling course should be instructed easily and able to be changed quickly.

3) It must be controlled precisely to travel along the desired track with high reliability.

In order to achieve these requirements, there have been several proposals, especially in the field of robotics, such as introducing the Dead Reckoning Navigation System [1] and taking a simple command system [2].

The Dead Reckoning Navigation System, based on the measurement of the rotations of the wheels and/or some directional information, is more preferable in satisfying requirement 1), because this system requires no equipment in the environment. This indicates that the track can be changed quickly. But this system has a defect in that its measurement error would increase cumulatively. Therefore, in order to accomplish any long-distance traveling, the measurement errors must be corrected periodically. It is required to diminish the frequency of these corrections. In order to accomplish this requirement, it is necessary that both the Dead Reckoning Navigation and the correction of the accumulated error be fairly accurate. At this point we have not achieved the navigation accuracy satisfactory for practical applications.

Some simple commands are suitable for an operator or a supervisory computer to instruct the vehicle about its traveling course. In order to make it simple to represent the traveling course, the command should correspond to a unit path rather than discrete points. If the command specifications are presented by the coordinate system fixed to the ground rather than the vehicle, it would be easier to supervise the vehicle. In order to satisfy these requirements, it is necessary to establish the movement control method according to the command system.

Some control algorithms of such vehicles have been proposed [3]. They were divided into two types, namely the continuous path (CP) method and the point-to-point (PTP) method. The CP method is used to control the vehicle to move on a continuous line. The PTP method is used to control the vehicle to follow a transient target point. Though the CP method requires more time for calculation, there is no need to generate the series of transient target points. From this standpoint, the CP method is more preferable, because its movement would be smoother and it would be easier to change the traveling course during real-time execution. But only a few vehicles adopted the CP method, and its performance has never been known clearly. For instance, it has never been discussed how far the vehicle's speed influenced the control ability. In addition, it has been necessary to establish the new algorithm, which makes it possible to control the vehicle to move on a curved line. Usually most of the vehicles were steered only by referring each one's current position and heading to the determined course, irrespective of each vehicle's speed. It was feared that the speed change made the movement unstable, because the effect of the steering on the position is evidently influenced by the speed. To obtain a stable control ability, it is necessary to comprehend the response of the vehicle to the values of the control variables.

The authors have adopted the following method to satisfy the above requirements and have developed an automatically guided vehicle without fixed guide ways.

1) This vehicle calculates its position and heading from the rotations of the measuring wheels, not the driving wheels, and

Manuscript received March 30, 1985; revised July 8, 1986.

The authors are with Toyota Central Research and Development Laboratories, Inc., 41-1, Aza Yokomichi, Oaza Nagakute, Nagakute-cho, Aichi-gun, Aichi-ken, 480-11, Japan.

IEEE Log Number 8611671.

corrects its measurement errors at a predetermined position by use of an optical method.

2) The instructions for its traveling course are represented by a sequence of simple commands, each of which represents an operation such as traveling along a path or stopping at a point.

3) The rotational speeds of the driving wheels are determined from their calculated effects on the control error, comprising velocity error, position error, and direction error.

In order to prove the performance of the developed guidance system, we have constructed an experimental vehicle and carried out some experimental travelings. We obtained good results and were convinced that its performance was more than satisfactory for practical applications. In this paper the construction of the experimental vehicle is presented. Then the command system and the new movement control algorithm contrived for the experimental vehicle is described. Lastly the excellent performance of the experimental vehicle is presented.

II. CONSTRUCTION OF THE VEHICLE

The block diagram of the guidance system of the experimental vehicle is shown in Fig. 1. This system consists of a leading unit, a movement control unit, a location unit, a calibration unit, and a driving unit. The leading unit receives the instructions for the traveling course and instructs the movement control unit to execute the required movement by a current command. The movement control unit gives the value of the desired rotational speed of each driving wheel to the driving unit. The driving unit turns the driving wheels at the determined rotational speeds to move the vehicle. Fig. 2 shows the structure of the vehicle and the arrangement of the two driving wheels driven by two dc servo motors and the two measuring wheels that rotate independently of the driving wheels. The servo amplifier controls the electric current to the motors by comparing the actual rotational speed and the determined one, computed by the microcomputer. The location unit is always calculating the current position and heading of the vehicle. The displacement of the vehicle for a short period is calculated from the rotational displacements of the measuring wheels, which are detected by rotary encoders. Then the current position and heading of the vehicle are obtained by adding its displacement to the last calculations. The principle of this calculation is presented in Fig. 3. The current position and heading are used to determine the current command and the rotational speeds of the driving wheels.

The calibration unit corrects the accumulated error of the current position. It calculates the position and the direction of the vehicle relative to the mark board (Fig. 4). Then referring to the location data of the mark board, stored in a memory, it calculates the current position and heading to correct the accumulated error of the location unit. The mark board is composed of the head and tail units, and each unit is divided into three sections. Two optical range finders are attached to the front and the rear of the left side of the vehicle. They recognize the unit when they detect extreme changes in reflectivity, from high to low, corresponding to the boundary line between the left and the center sections, and they measure

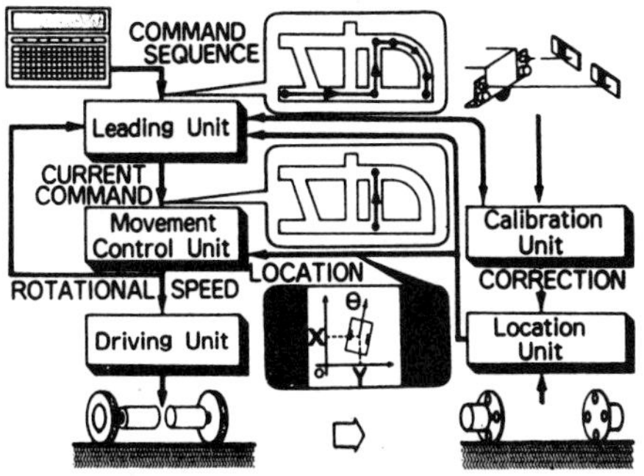

Fig. 1. Block diagram of the guidance system.

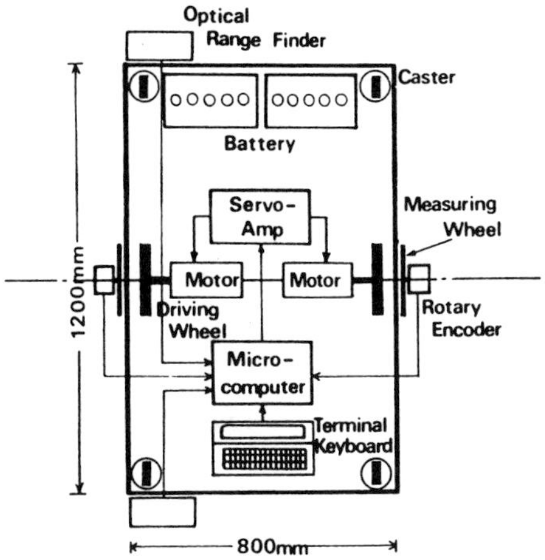

Fig. 2. Schematic diagram of the experimental vehicle.

$$\theta_1 = \theta_0 + \frac{N_L - N_R}{Tread}$$
$$X_1 = X_0 + \frac{N_L + N_R}{2} \cos \frac{\theta_0 + \theta_1}{2}$$
$$Y_1 = Y_0 + \frac{N_L + N_R}{2} \sin \frac{\theta_0 + \theta_1}{2}$$

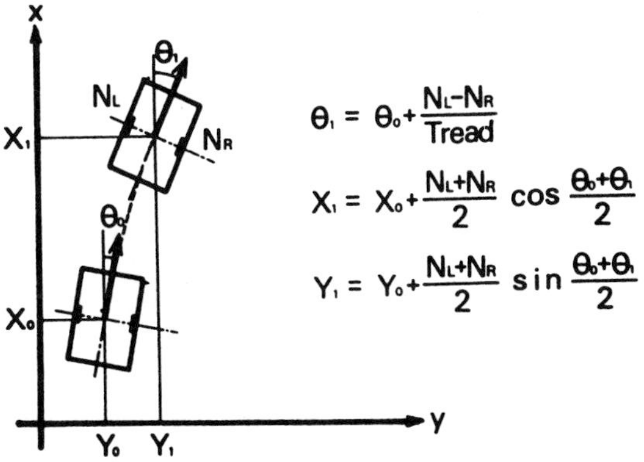

Fig. 3. Principle of the position and heading measurement.

$$\theta = \arctan \frac{l_1 - l_2}{W}$$
$$x_1 = x_0 - (C + \frac{l_1 + l_2}{2}) \sin \theta$$
$$y_1 = y_0 + (C + \frac{l_1 + l_2}{2}) \cos \theta$$

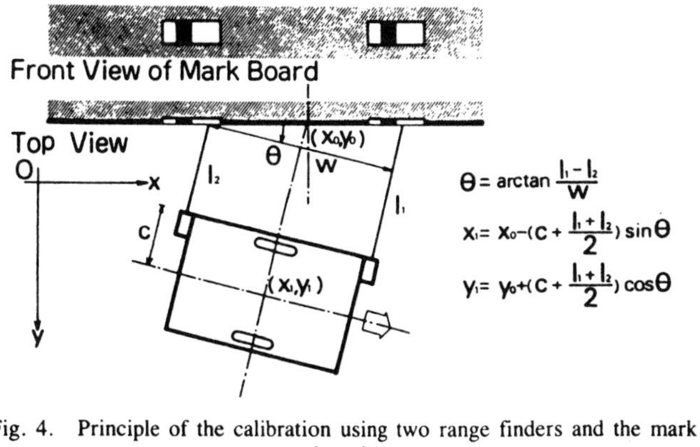

Fig. 4. Principle of the calibration using two range finders and the mark board.

the distance to the measured section, which is the right section of the unit. The actual calibration procedure is executed as follows.

1) When the front detects the tail unit, the vehicle decreases its velocity.

2) When the rear detects the tail unit, the vehicle is controlled to move straight at a constant speed.

3) When the front detects the head unit, the location unit is suspended.

4) While both range finders detect the measured sections, they repeat the measurement of the distance.

Then the calculations are executed in the following order:

a) the position and the direction relative to the mark board at instant 4);
b) the position and the heading at instant 4);
c) the position and the heading at instant 3); and
d) the current position and heading.

Then the current position and heading are corrected and the location unit will resume its calculations.

The vehicle uses a microcomputer (MC68000 with a clock rate of 8 MHz) as a calculator and a controller to execute the operations of these units. It is designed to receive instructions through a terminal keyboard and signals from the range finders and the rotary encoders. This vehicle is powered by four batteries with a total capacity of 130 A·h at 24 V. Each measuring wheel is 20 cm in diameter and its shaft is connected to a 1800-count/revolution optical shaft encoder. It is free to rotate touching the ground at a constant pressure. The driving wheel is 20 cm in diameter and its shaft is connected to a motor through gears, a timing belt, and a clutch. The vehicle, as in Fig. 5, is 80 cm in height, 80 cm in width, 120 cm in length, and about 350 kg in weight. The maximum acceleration is more than 0.6 m/s^2, and the top speed is 10 km/h. The optical range finders are contrived for this vehicle. According to the optical triangulation, each of them throws an infrared light beam to the measured spot, and detects the position of its image by a lateral effect photodiode. Its total photocurrent gives the information about reflectivity. The test results showed that they could recognize the mark board and measure the distance from 350 to 1250 mm with an error of less than 1 mm within 13 ms.

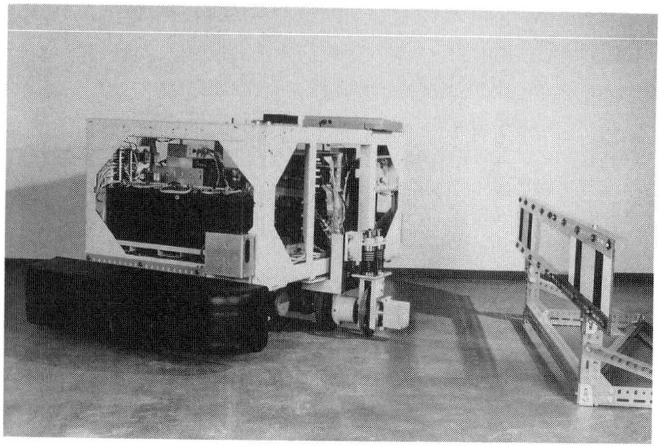

Fig. 5. Picture of the experimental vehicle and the mark board.

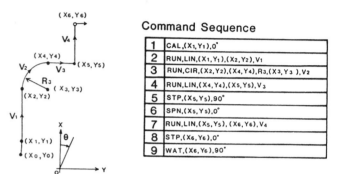

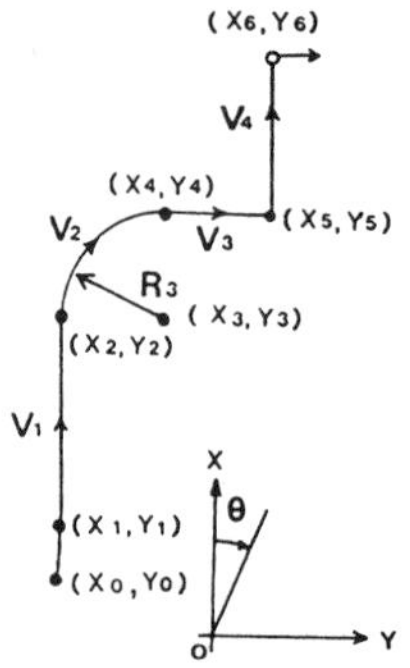

Command Sequence

1	CAL,(X1,Y1),0°
2	RUN,LIN,(X1,Y1),(X2,Y2),V1
3	RUN,CIR,(X2,Y2),(X4,Y4),R3,(X3,Y3),V2
4	RUN,LIN,(X4,Y4),(X5,Y5),V3
5	STP,(X5,Y5),90°
6	SPN,(X5,Y5),0°
7	RUN,LIN,(X5,Y5),(X6,Y6),V4
8	STP,(X6,Y6),0°
9	WAT,(X6,Y6),90°

Fig. 6. Example of traveling course and its command sequence.

III. Command System

In this guidance system, the instructions given to the vehicle are a sequence of elementary movements at a point or along a path, which represent the traveling along a course. In order to express the elementary movements, a new command system is introduced. Each command corresponds to an elementary movement and is composed of some of the following specifications: an action, a path specification, a speed specification, and a heading specification. A path specification consists of some of the following: path type, start and/or end position of the path, and the radius and the center position of a curved path. The positions and headings in these specifications are represented by the coordinates or the angle from the X axis as shown in Fig. 6. The symbols used for these specifications are listed below:

X_s, Y_s	coordinates of the start position of the path;
X_e, Y_e	coordinates of the end position of the path;
Θ_e	angle of the heading;
$\bar{V}$	velocity;
R_o	radius of the curved path; and
X_o, Y_o	coordinates of the center position of the curved path.

The instructions for the vehicle, described as a sequence of commands, are stored in the memory of the microcomputer. An example of a command sequence is shown in Fig. 6. The leading unit selects the current command, which is executed by the movement control unit, and leads the vehicle along the determined course. The leading unit is always monitoring the vehicle's state, such as its current position or current velocity, to recognize the instant when the terminating conditions of the current command are satisfied. Then it selects the next command according to the command sequence. The terminating conditions are determined to make it easy to proceed to the next movement. For example when the current command is "run" and the next is "stop," the stopping distance is always compared with the remaining distance to the end position. Therefore the vehicle is guided to travel smoothly along the determined course.

IV. Movement Control

The movement control unit determines the exact rotational speeds of the driving wheels in accordance with the current command. In order to avoid sudden changes in velocity, the movement control unit determines the accelerations of the two driving wheels, represented by A_l and A_r, rather than the speeds as control variables. It then determines the rotational speeds by integrating them (Fig. 7). The principle of the movement control will be described below.

First, in order to express the vehicle's state, the state variables are introduced: X and Y are coordinates of a position, Θ heading angle, V_l and V_r velocities of the left and right wheels, respectively, and V_x and V_y the velocity components. The relationships between these state variables and the control variables can be expressed as follows:

$$V_l(t+T)=V_l(t)+\int_t^{t+T} A_l \cdot d\tau \tag{1}$$

$$V_r(t+T)=V_r(t)+\int_t^{t+T} A_r \cdot d\tau \tag{2}$$

$$\Theta(t+T)=\theta(t)+\int_t^{t+T} \frac{V_l(\tau)-V_r(\tau)}{T_{read}} \cdot d\tau \tag{3}$$

$$V_x(t+T)=\cos\ (\Theta(t+T)) \cdot \frac{V_l(t+T)+Vr(t+T)}{2} \tag{4}$$

$$V_y(t+T)=\sin\ (\Theta(t+T)) \cdot \frac{V_l(t+T)+V_r(t+T)}{2} \tag{5}$$

$$X(t+T)=X(t)+\int_t^{t+T} V_x(\tau) \cdot d\tau \tag{6}$$

$$Y(t+T)=Y(t)+\int_t^{t+T} V_y(\tau) \cdot d\tau \tag{7}$$

where T_{read} is the length between the two wheels.

Because the current values $X(t)$, $Y(t)$, $\Theta(t)$, $V_l(t)$, and $V_r(t)$ are known as measured or determined values at any time t, each value of the state variables at any given time $t+T$ should be obtained as function of the control variables A_l and A_r. Therefore we can adjust the control variables A_l and A_r so that the vehicle's state in a period will agree with the current command.

In order to evaluate the predicted vehicle's state, an evaluation function E of the state variables is defined in accordance with the current command. This function represents the control error and is expressed as follows:

$$E=k_0 \cdot \epsilon_0+k_1 \cdot \epsilon_1+k_2 \cdot \epsilon_2 \tag{8}$$

where ϵ_0 velocity error, ϵ_1 position errors, ϵ_2 direction error, and k_0, k_1, and k_2 weighting factors. Each error is defined as follows.

1) Velocity error is the square of the deviation from the desired rotational speeds $\bar{V}_l$ and $\bar{V}_r$, calculated from the specifications of the command:

$$\epsilon_0=(V_l-\bar{V}_l)^2+(V_r-\bar{V}_r)^2 \tag{9}$$

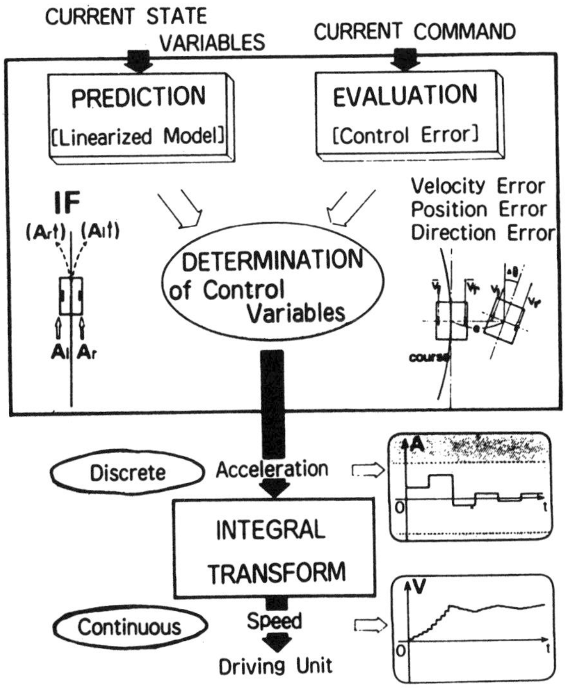

Fig. 7. Block diagram of the movement control method.

where

$\bar{V}_l=\bar{V}, \quad \bar{V}_r=\bar{V}$ run straight line

$\bar{V}_l=\bar{V} \cdot \left(1+\frac{T_{read}}{R_o}\right), \quad \bar{V}_r=\bar{V} \cdot \left(1-\frac{T_{read}}{R_o}\right)$

run curved line

$\bar{V}_l=0, \quad \bar{V}_r=0$ stop, spin, or wait.

2) Position error is the square of the deviation from the desired track specified by the path specification of the command:

$$\epsilon_1=\left[\frac{(X-X_s)\cdot(Y_e-Y_s)+(Y-Y_s)\cdot(X_e-X_s)}{(X_e-X_s)^2+(Y_e-Y_s)^2}\right]^2 \cdot \frac{1}{V^2}$$

run straight line (10)

$$=[\sqrt{(X-X_o)^2+(Y-Y_o)^2}-R_o]^2 \cdot \frac{1}{V^2}$$

run curved line (11)

$$=(X-X_e)^2+(Y-Y_e)^2 \quad \text{stop, spin, or wait} \tag{12}$$

where

$$V=\frac{V_l+V_r}{2}.$$

3) Direction error is the square of the rate of the deviation

from the desired track:

$$\epsilon_2 = \left[\frac{V_x \cdot (Y_e - Y_s) + V_y \cdot (X_e - X_s)}{(X_e - X_s)^2 + (Y_e - Y_s)^2}\right]^2 \cdot \frac{1}{V^2}$$

run straight line (13)

$$= \left[\frac{V_x \cdot (X - X_o) + V_y \cdot (Y - Y_o)}{\sqrt{(X - X_o)^2 + (Y - Y_o)^2}}\right]^2 \cdot \frac{1}{V^2}$$

run curved line (14)

$$= (\Theta - \Theta_e)^2 \qquad \text{stop, spin, or wait.} \tag{15}$$

The evaluation function $\bar{E}$ at a given time $t+T$ can be expressed as function of the control variables A_l and A_r, because the values of the parameters X_s, X_e, Y_s, Y_e, X_o, Y_o, R_o, and Θ_e are given as specifications of the current command and the values of the state variables at given time $t+T$ are expressed as functions of the control variables A_l and A_r, as shown in (1)–(7). Therefore the effects of variation in accelerations on $\bar{E}$ can be calculated to determine the accelerations as follows:

$$A_l = -\frac{d\bar{E}}{dA_l} \tag{16}$$

$$A_r = -\frac{d\bar{E}}{dA_r}. \tag{17}$$

Thus the movement control unit determines the values of A_l and A_r from the current values of the state variables and the specifications of the current command by the equations derived so far. And then the values of the accelerations are limited to the predetermined value $A_{\max}$, because of the limited power of the drive unit. The desired rotational speeds of the driving wheels are determined by integrating the calculated accelerations.

V. Experiments

The actual traveling experiments were carried out to prove the performance of the guidance system.

First, the actual program on the microcomputer was tested. The microprocessor (MC68000) was used with a clock rate of 8 MHz. Most of the programs were written in Pascal and the rest in machine language. It took about 100 ms to calculate the accelerations, and about 3 ms to calculate the position. These programs were executed under a multitasking operating system. The microcomputer determined the accelerations at an interval of 300 ms and the rotational speeds at an interval of 100 ms, and calculated the position at an interval of 10 ms. The total program size was about 48 kbytes.

Next, the performance of the movement control was tested. The control parameters of k_0, k_1, k_2, and T were roughly determined by referring to the computer simulation results, and were finely tuned by the experimental movements. This experimental vehicle could record its position and velocity in the memory at regular intervals to obtain the control performance. Some of the experimental results are shown in Figs. 8 and 9. The purpose of the first experiment was to test the

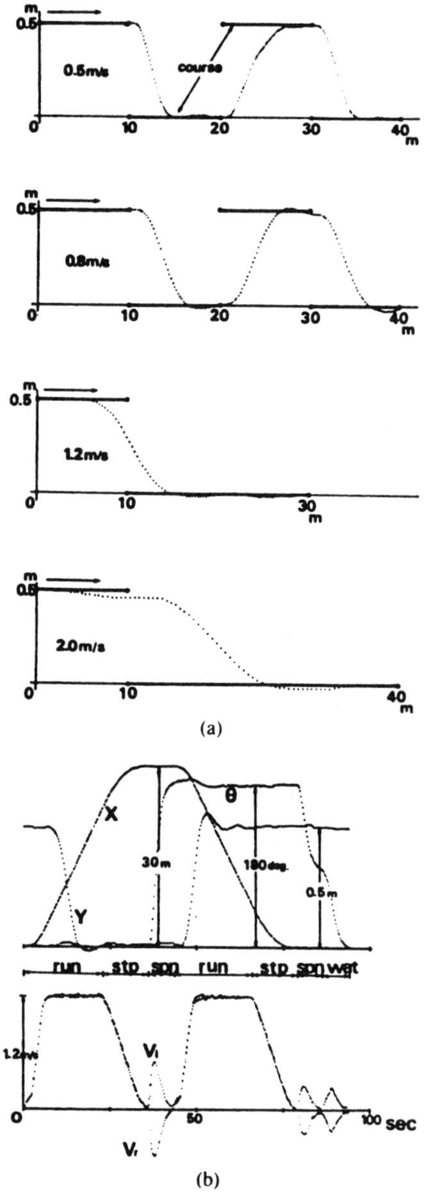

Fig. 8. Experimental traveling along straight lines: (a) programmed course and its trajectory (every dot indicates the position at an interval of 250 ms); and (b) changes of state variables.

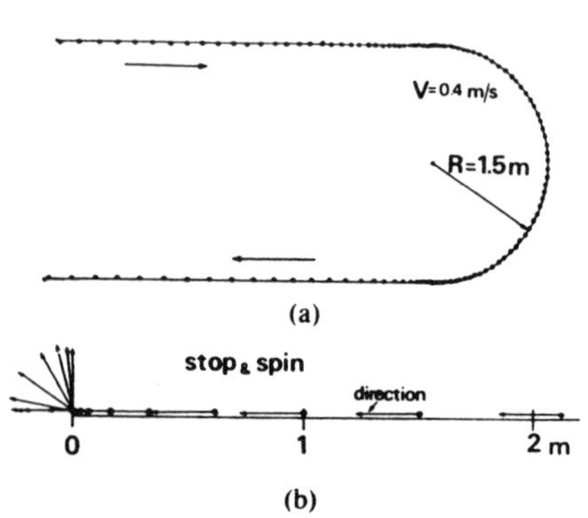

Fig. 9. Experimental results of operations: (a) running along a curved line; and (b) stop and spin.

response and the stability of the vehicle. The traveling course was composed of parallel lines, separated at a distance of 0.5 m, and the vehicle followed them one after the other. The vehicle precisely followed the course without winding out at each speed range and its velocity was smoothly changed as in Fig. 8. Then its ability to travel along a curved path was examined. Fig. 9 shows that the vehicle was precisely controlled to follow its determined course. The test results of its stop and spin operations showed, as in Fig. 9, that it was precisely controlled to stop at the determined position and that it could turn to the determined direction smoothly. Other experimental travelings along a straight or curved path showed that its movement was smooth and stable over a wide range of velocities, up to 2.7 m/s; and the deviation from the desired track was a few centimeters at low speeds of about 1 m/s, and did not exceed 20 cm at any speed. The stop position could be controlled within 2 cm from the target in the direction of the movement.

Then the location unit and the calibration unit were tested. The actual diameters of the measuring wheels were determined from the relationships between their rotations and the traveling distance on a drum. The experimental vehicle executed its calibration process successfully. The test results showed that the measurements of its position and heading were corrected within errors of 20 mm and 0.2°, respectively, while the vehicle was moving at a constant speed, up to 0.5 m/s. In order to estimate the total error, several experimental travelings were carried out. The vehicle executed the calibration process at the beginning, and traveled ahead along a straight path to the destination point which was 50 m away from the start point. The results showed that the traveling of 50 m on an asphalt road produced a position error of about 200 mm. Fig. 10 shows the typical results of the distribution of its arrival points after the traveling of 50 m. The deviation from the destination point contains the position error and the control error, but the latter was smaller. These results indicate that the calibration process at an interval of about 50 m enables the vehicle to travel a long distance without a fatal deviation from the determined course.

Finally its long-distance traveling was tested to prove its general performance. Several mark boards were set up at an interval of about 50 m. Its traveling course was cyclic and its circumference was more than 200 m. The vehicle succeeded in continuing its long-distance travel along the cyclic course repeatedly.

VI. Conclusions

The performances of the experimental vehicle can be summarized as follows.

1) It is easy to set or change the traveling course.

2) Because of the error correcting ability, the vehicle can be precisely controlled.

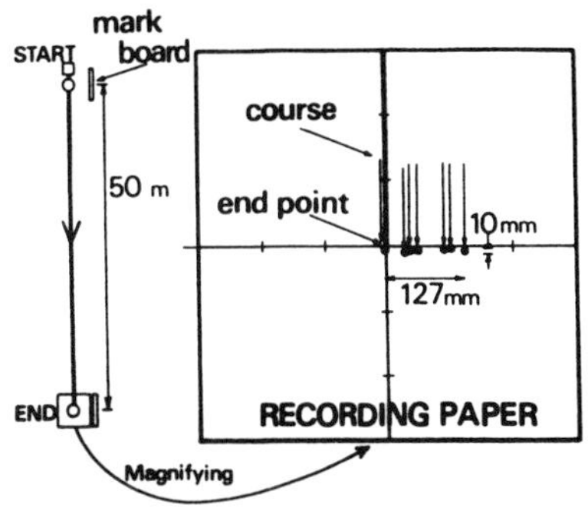

Fig. 10. Typical distribution of arrival points after the traveling of 50 m.

3) This system enables the vehicle to travel a long distance along a given course with little deviation.

These good results indicate that the measuring method, depending on the rotations of the measuring wheels, is accurate enough to lead the vehicle for more than 50 m on a well-paved road. It is also proved that it is possible to correct the accumulated errors with the necessary accuracy. That is, our measuring method is one of the simplest ways to obtain the high accuracy needed for the automatically guided vehicle.

Long-distance traveling is possible only when the vehicle can be precisely controlled to move as instructed. At this point, the control method is proved to be good. The control method, depending on the evaluation of control errors and the prediction of control effects, is reliable and makes it possible for the vehicle to travel smoothly; and its calculation program can be executed in real time by the microcomputer. Experimental travelings indicate that the requirement on the control, needed for the automatically guided vehicle, is satisfied by this method.

These experiments confirm that the vehicle is reliable enough for practical use, therefore, we are going to put the vehicle into trial use for transportation of documents and books in our laboratory. We are convinced that this automatically guided vehicle has sufficient mobility for research into autonomous mobile robots.

References

[1] T. Tsumura *et al.*, "An experimental system for automatic guidance of roboted vehicle following the route stored in memory," in *Proc. 11th ISIR,* 1981, pp. 187-194.

[2] J. Iijima *et al.*, "A locomotion control system for mobile robots," in *Proc. 7th IJCAI,* 1981, pp. 779-784.

[3] S. Uta *et al.*, "An implementation of mitchi—a locomotion command system for intelligent mobile robots," in *Proc. 15th ICAR,* 1985, pp. 127-134.

[4] S. Tachi *et al.*, "Guide dog robot—feasibility experiments with MELDOG MARK III," in *Proc. 11th ISIR,* 1981, pp. 95-102.

[5] H. P. Moravec, "Rover visual obstacle avoidance," in *Proc. 7th IJCAI,* 1981, pp. 785-790.

[6] H. P. Moravec, "The stanford cart and the CMU rover," *Proc. IEEE,* vol. 71, no. 7, pp. 872-884, 1983.

Local Path Control for an Autonomous Vehicle

Winston L. Nelson
Ingemar J. Cox

AT&T Bell Laboratories
Murray Hill, NJ 07974, USA

ABSTRACT

This paper describes a control system for an autonomous robot cart designed to operate in well-structured environments such as offices and factories. The onboard navigation system comprises a reference-state generator, an error-feedback controller, plus cart location sensing using odometry. There is a convenient separation between the path guidance and control logic. Under normal operating conditions, the controller ensures that the errors between the measured and reference states are small. These errors only exceed set limits if the cart is malfunctioning. Major hardware failures can be detected in this way and failsafe procedures invoked. Results on the control system performance derived from a computer simulation of the cart and its operating environment, and from an experimental cart, indicate that it can provide reliable, accurate, and safe operation of autonomous robot carts.

1. INTRODUCTION

Robot carts are coming into increased use in automated factories and other such reasonably well-structured environments.[1] Most of the automated carts in present use, however, are not really autonomous (i.e., self-guided) but rely on "tracks" imbedded in, or painted on, the factory or office floor to guide them from one station to another.[2]

This paper describes some preliminary results on a local path control system for an experimental cart, "Blanche",[3] shown in Figure 1. It is a tricycle configuration with a single front wheel which serves both for steering and driving the cart and two passive load-bearing rear wheels. The superstructure in the front supports an optical rangefinder unit.[4]

The cart is capable of navigating in structured environments such as offices and factories without external guide tracks. The onboard control system uses odometry to provide the position and heading information needed to guide the cart along specified paths between stations. The overall path control system for the cart is diagrammed in Figure 2. The external, or off-board, part of the system consists of a global path planner which stores the structured environment data and produces the path plans from one station to the next.[5] It also includes a communication link between this supervisory controller and the cart.

The path plan directing the cart from one station to the next comprises a list of path segments of three basic types: lines, arcs, and splines. Although line and arc segments provide optimal (minimum distance) paths between any initial and final positions and headings,[6] spline segments have been included because they reduce the number of segments needed to specify paths for cases such as simple lane-change maneuvers.

The local path planner takes the path segment data and passes each segment in turn to the reference state generator, which serves the function of converting the sparse *spatial* data of the path plan into detailed *temporal* path data produced as the cart moves along the path. Having the RPG on board the cart facilitates autonomous operation of the cart with only an occasional low-data rate link to the supervisory controller. As indicated in Figure 2, this reference state is input to the cart controller, which compares it with the measured state of the cart (derived from the odometry sensors) and outputs control torque signals to the cart steering and drive motors.

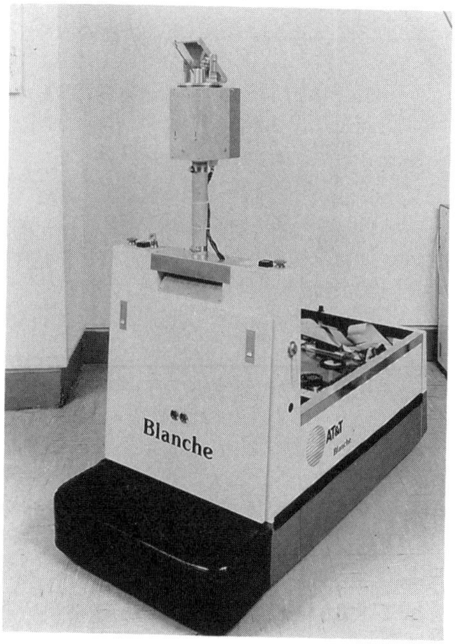

Figure 1. Experimental robot cart.

Algorithms for the autonomous cart navigation system are being developed and tested both on a computer-aided design workstation and on the experimental cart. The simulation permits the testing of path-planning, path-generation, and path-control algorithms in a safe and readily re-structurable operating environment before porting them to the computer on board the cart. The computer system and sensors on the cart, and the path planning system system in the supervisory controller, are discussed in the companion papers.[3]-[5]

2. REFERENCE PATH GENERATOR

2.1 Rationale

The reference path generator (RPG) functions as an onboard navigator that continually provides an appropriate reference state for the cart controller to track. The reference state is determined by three factors: (1) the path plan, (2) the acceleration/deceleration capabilities of the cart, and (3) the current operating status of the cart.

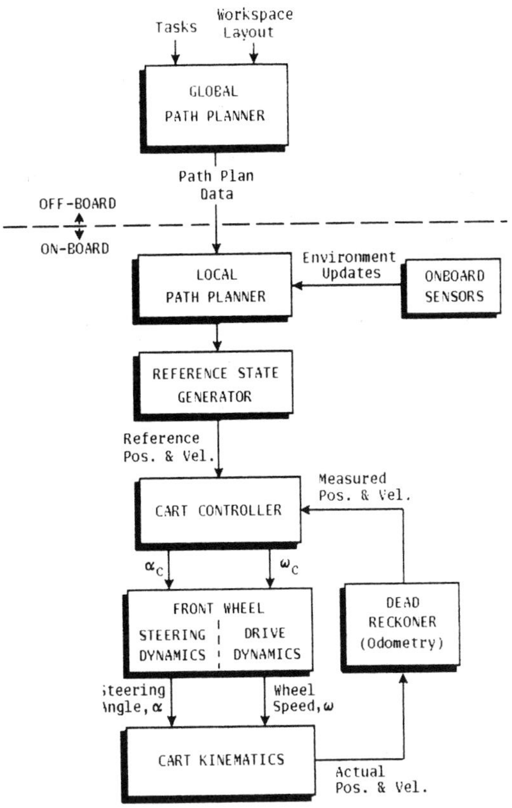

Figure 2. Block diagram of path control system.

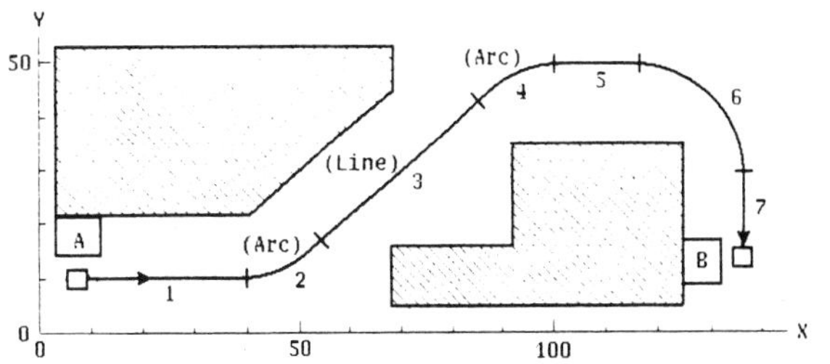

(a) PATH SEGMENTS

Segment Number	Segment Type*	x_d	y_d	θ_d	v_d
		Desired End State			
1	1	40.00	10.00	0.00	0.50
2	2	56.97	17.03	45.00	0.50
3	1	83.03	42.97	45.00	0.50
4	2	100.00	50.00	0.00	0.50
5	1	116.20	50.00	0.00	0.50
6	2	136.20	30.00	-90.00	0.40
7	1	136.20	14.22	-90.00	0.00

* 1 = line, 2 = arc

(b) PATH PLAN DATA

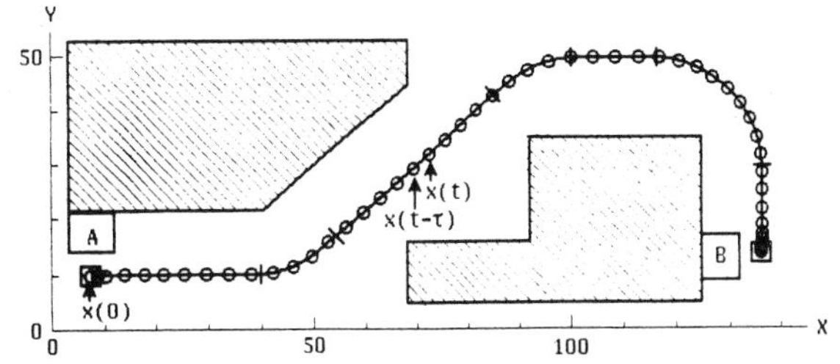

(c) REFERENCE POINTS

Figure 3. Example of a path plan, data, and reference path.

The path plan received from the global path planner consists of a list, each line of which specifies a segment of the path. The segment specification includes the segment type and the desired state at the end of the segment. Figure 3(a) shows a possible path plan between two cart stations (A and B). This sample path plan contains seven segments, four line segments separated by three arc segments. The corresponding path data used by the RPG to evolve the reference path are the seven lines of numbers listed in Figure 3(b).

Although the sample plan in Figure 3(a) consists only of lines and arcs, the RPG, as presently implemented, can also handle spline segments. Splines are especially convenient for "lane-change" maneuvers, because one spline segment can replace an arc-line-arc sequence. For example, segments 2, 3, and 4 of the path shown in Figure 3(a) could be represented by one spline segment.

The RPG computes at each control update cycle the reference state on the path, the reference steering angle and drive speed, and the distance to the end of the current path segment. When the distance to the end-point of the segment becomes less than the distance between the preceding reference points, the RPG requests the next segment in the path plan and begins generating reference states along the new segment. An illustration of the reference points generated for the path shown in Figure 3(a) is given in Figure 3(c).*

The main advantage of the reference path guidance scheme over point-to-point guidance schemes[7], [8] is that the cart's onboard system is always providing a current point on the path where the cart should be, with smooth transitions across path-segment boundaries. A point-to-point control scheme, because of measurement errors, may not detect the point at which the control algorithm should switch to a new target point, and hence may fail to make the proper transitions from one path segment to the next.

The RPG scheme also provides a means for onboard monitoring of the operating status of the cart. Under normal operating conditions, the difference between the measured state of the cart and the current reference state should be small. If this state error ever exceeds some established limits, then the cart must be functioning abnormally. The cart controller can test for abnormal errors and signal the local path planner to take remedial action. By this means, system malfunctions, such as failure of motors or odometry sensors, can be detected in time for the cart to be safely stopped.

A third advantage of the RPG scheme is that it provides a convenient separation between the path-guidance logic and the controller logic, which must be tuned to the characteristics of the closed-loop dynamics of the steering and drive systems of the cart. Once properly adjusted, the controller design remains fixed throughout all the different operating situations that may arise along the path; all the controller needs to know is what the current reference and measured states are. The RPG takes care of dealing with the geometric computations necessary to evolve the path plan as points in time, and also with anomalies which may require local path modifications, either spatially, temporally, or both.

2.2 Reference Speed

In order for the RPG to be a "considerate" guide to the cart controller, it should use speed profiles compatible with the recommended acceleration and speed limits of the drive system. For start-up from a stop, and for speed changes across path segments, a ramp change in reference speed is programmed using the recommended acceleration. For the terminal segment of the path, or for emergency

* The actual reference points form a very dense set of points along the path; a sparse subset of points is shown in Figure 3(c) for clarity.

stops along the path, a ramp-down of speed, based on a gain G_r, multiplying the remaining distance to the stop point, is programmed to provide smooth stops without overshoot. The value of G_r is chosen to produce an appropriate compromise between the competing requirements for docking. The value should be large enough so that the cart remains maneuverable up to a point near the stop point, but not so large that the decrease in the reference speed exceeds the deceleration capability of the cart. Maneuverability is needed near the docking point to correct for any heading error or position error normal to the path, while reasonable deceleration is needed to correct for position error tangential to the path.

2.3 Reference Position and Heading

The reference position and heading on the desired path at the current update time, t, depends on the position and heading at $t-\tau$, the current path-segment type, and the reference speed, $v_r(t)$, discussed above. The geometric computations needed to generate the new reference position and heading depend on the segment type. Before describing these, it is useful to introduce some notation: The *desired* position-heading vector at the end point of the current segment is denoted by $\mathbf{z}_d \equiv [x_d, y_d, \theta_d]^T$, where the superscript T denotes transpose. The *beginning* position-heading vector for the current segment (which is the $\mathbf{z}_d$ of the previous segment) is denoted by $\mathbf{z}_b \equiv [x_b, y_b, \theta_b]^T$. The *reference* position-heading vector at the update time t is denoted by $\mathbf{z}_r(t) \equiv [x_r(t), y_r(t), \theta_r(t)]^T$. We define $\mathbf{z}_f \equiv [x_f, y_f, \theta_f]^T$ to be the *former* reference position-heading vector, transformed to a coordinate system whose origin is at (x_b, y_b) and whose X-axis is aligned with θ_b, i. e.,

$$\mathbf{z}_f = \mathbf{B}\,(\mathbf{z}_r(t-\tau) - \mathbf{z}_b) \qquad (2.1)$$

where the base transformation matrix, $\mathbf{B}$, is

$$\mathbf{B} = \begin{bmatrix} \cos\theta_b & \sin\theta_b & 0 \\ -sin\theta_b & \cos\theta_b & 0 \\ 0 & 0 & 1 \end{bmatrix} \qquad (2.2)$$

This transformation of reference positions and headings with respect to the beginning of the current path segment simplifies the update computations within the different segment types. We also define $\mathbf{z}_p \equiv [x_p, y_p, \theta_p]^T$ to be the *projected* position-heading vector in this transformed coordinate space. Finally, we denote by α_r the reference steering angle and by ω_r the reference drive rotational speed needed to move the cart along the current path segment when the error between the reference and the actual state of the cart is zero.

2.3.1 Line segments. If the segment type is a line, the incremental path motions are a sequence of $v_r\tau$ steps along the transformed X-axis. Hence, the projected position-heading vector along a line segment is

$$\mathbf{z}_p = [x_f + v_r(t)\tau \;,\; 0 \;,\; 0]^T \qquad (2.3)$$

The reference steering angle and drive speed for straight line segments are obviously

$$\alpha_r = 0 \;,\quad \omega_r = v_r/R \qquad (2.4)$$

where R is the radius of the front wheel of the cart.

2.3.2 Arc segments. When the path seqment is a circular arc, a radius of curvature parameter is computed as

$$r_c = y_e/(1-cos\theta_e) \qquad (2.5)$$

where y_e and θ_e are the second and third elements of the transformed desired end-point vector,

$$\mathbf{z}_e = \mathbf{B}\,(\mathbf{z}_d - \mathbf{z}_b) \qquad (2.6)$$

The parameter r_c has a magnitude equal to the radius of the arc and a sign which is positive for a counter-clockwise turn and negative for a clockwise turn.

The projected position-heading vector along an arc is

$$\mathbf{z}_p = [r_c \sin\theta_p \;,\; r_c(1-cos\theta_p) \;,\; \theta_p]^T \qquad (2.7)$$

where $\theta_p = \theta_f + v_r(t)\tau / r_c$. The reference steering angle and drive speed for arc segments are constants, given by

$$\alpha_r = \tan^{-1}(b/r_c) \;,\quad \omega_r = v_r/(R\,\cos\alpha_r) \qquad (2.8)$$

where b is the wheelbase of the cart (see Figure 4).

2.3.3 Spline segments. A cubic spline is useful for connecting two line segments with a smooth curve that matches both the position and slope conditions of the line segments at the junction points.* Its general form is a cubic polynomial having four coefficients to match the four end-point conditions:

$$y = Kx^3 + Lx^2 + Mx + N$$

In the base-transformed space in which the RPG computes the reference path, however, only the first two coefficients are non-zero. These coefficients are determined from the elements of the vector $\mathbf{z}_e$ in (2.6) as follows:

$$\begin{aligned} K &= (tan\theta_e - 2y_e/x_e)/x_e^2 \;, \\ L &= (3y_e/x_e - \tan\theta_e)/x_e \end{aligned} \qquad (2.9)$$

The elements of the projected position-heading vector, $\mathbf{z}_p \equiv [x_p, y_p, \theta_p]^T$, along the spline segment are given by

$$\begin{aligned} x_p &= x_f + v_r(t)\tau\cos\theta_f \;, \\ y_p &= Kx_p^3 + Lx_p^2 \;, \\ \theta_p &= \tan^{-1}[(y_p - y_f)/(x_p - x_f)] \end{aligned} \qquad (2.10)$$

The reference steering angle and drive speed for spline segments are

$$\alpha_r = \tan^{-1}(b\,c_p) \;,\quad \omega_r = v_r/(R\,\cos\alpha_r) \qquad (2.11)$$

where c_p is the spline curvature at x_p, given by

$$c_p = (6Kx_p + 2L)/(1 + (3Kx_p^2 + 2Lx_p)^2)^{3/2} \qquad (2.12)$$

The position-heading vector for the reference state at time t along the line, arc, and spline segments is obtained by the transformation of the position-heading vector, $\mathbf{z}_p$, back to the original XY-space:

$$\mathbf{z}_r(t) = \mathbf{B}^{-1}(\mathbf{z}_p) + \mathbf{z}_b \qquad (2.13)$$

where $\mathbf{B}^{-1}$ is obtained simply by interchanging the off-diagonal terms in (2.2). The position-heading vector given by (2.13), and the programmed speed, $v_r(t)$, described in Section 2.2, form the new reference state, which with the new reference steering angle, α_r, and reference drive speed, ω_r, comprise the data output by the RPG to the cart controller in each control update cycle.

3. CART CONTROLLER

3.1 Cart Coordinates and Kinematics

The coordinates by which the cart is controlled are the front wheel steering angle, α, and the front wheel drive speed, ω. If the location of the cart is measured with respect to the the center of rotation (CR) of the cart, located at the mid-point between the rear odometry wheels, (see Figure 4), the speed, v, heading, θ, and position (x,y) of the cart are related to the steering angle and drive speed coordinates (α,ω) by the kinematic equations

$$\begin{aligned} v &= R\omega\cos\alpha \;, & \dot{\theta} &= (R/b)\omega\sin\alpha \\ \dot{x} &= v\cos\theta \;, & \dot{y} &= v\sin\theta \end{aligned} \qquad (3.1)$$

where R is the radius of the drive wheel, and b is the wheelbase, as shown in Figure 4.*

The task for the cart controller is to steer and drive the front wheel of the cart so that the CR point accurately tracks the reference points "laid down" by the reference path generator. The cart position and velocity measurements necessary for performing this task are obtained from the odometry system indicated in the cart control block diagram shown in Figure 5. The odometry for this system is provided by optical

* If a continuity of curvature at the junction points is also desired a quintic spline may be used.

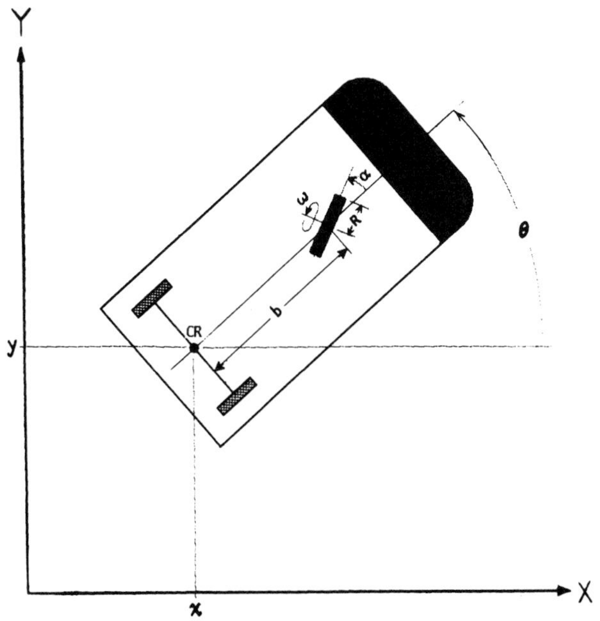

Figure 4. Workspace and cart coordinates.

encoders mounted on two non-load-bearing, knife-edge wheels located near the rear wheels. Since any odometry system has cumulative errors, provision for nullifying these errors at regular calibratrion points in the workspace is needed. The optical ranger can be used to obtain these position fixes.[3]

The cart controller, shown as a shaded block in Figure 5, consists of a path controller and two motor-control units. The motor-control units provide inner control loops for the steering and drive motors, while the path controller provides the outer control loop for path tracking.

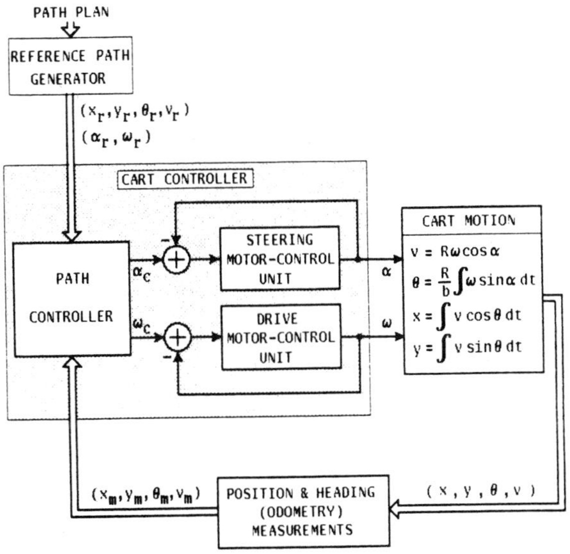

Figure 5. Block diagram of the cart control structure.

3.2 Motor Error Control

The front wheel steering angle and the front wheel drive speed depend on the motor-load dynamics in response to the control signals to the motors. The purpose of the motor-control units is to compensate for the dynamics in such a way that the motor outputs (α,ω) respond rapidly and smoothly to accurately follow the *commanded* values (α_c,ω_c) output from the path controller. This is done by conventional feedback compensation, based on error signals between the commanded values and the motor outputs values, measured by angle encoders on the motor shafts. To achieve reasonably high-gain error control without causing undesirable, or even unstable response, controller chips containing digital pole-zero compensating filters are used.

The dynamics of the motor-load combination for the two units are represented to a close approximation by the linear model:

$$\ddot{\alpha} = a_1 - H_1\dot{\alpha} \qquad (3.2)$$
$$\dot{\omega} = a_2 - H_2\omega$$

where a_1 and a_2 are the control accelerations (torque per unit moment of inertia) developed by the steering and drive motors, respectively, and H_1 and H_2 are the friction coefficients of the two motor-load systems. The digital error signals driving the control units are

$$e_\alpha(k) = C_r(\alpha_c(k)-\alpha(k)) \;,\quad e_\omega(k) = C_r(\omega_c(k)-\omega(k)) \qquad (3.3)$$

where k denotes the k-th sampling period, and C_r is the angle encoder gain (counts/radian). The difference between the two control units is that one is controlling the angle while the other is controlling the angular velocity. Because of the extra integration in the angle control dynamics, the steering angle is the more difficult one to compensate properly.

3.3 Path error control

The path controller generates the steering and drive command signals (α_c,ω_c), using the reference values (α_r,ω_r) from the RPG, and path-errors derived from the measured state of the cart relative to the reference state, as shown in Figure 5. The path error is resolved into four components. As diagrammed in Figure 6, the distance error is resolved into *tangential* error, e_t, and *normal* error, e_n:

$$e_t = (x_r-x_m)cos\theta_r + (y_r-y_m)sin\theta_r \qquad (3.4)$$
$$e_n = -(x_r-x_m)sin\theta_r + (y_r-y_m)cos\theta_r$$

The velocity error is resolved into *heading* error, e_h, and *speed* error, e_v:

$$e_h = \theta_r - \theta_m \;,\quad e_v = v_r - v_m \qquad (3.5)$$

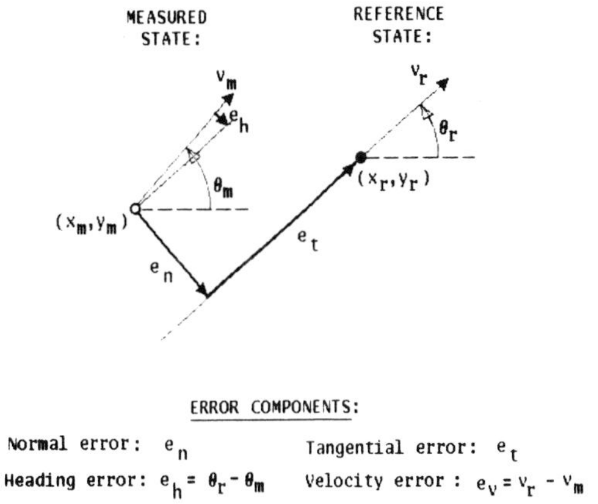

Figure 6. Error components used for path control.

* The choice of x,y position determines what point on the cart tracks the reference path, as well as the form of the kinematic equations. The choice to have the CR point track the path, rather than, say, the front wheel, was made for reasons related to path planning and cart maneuverability.

Given the reference values α_r, ω_r and these error components, the command steering and drive signals generated by the path controller in each update cycle are:

$$\alpha_c = \alpha_r + C_1 e_n + C_2 e_h \ , \qquad (3.6)$$
$$\omega_c = \omega_r + C_3 e_t + C_4 e_v$$

Note that when the cart is following the reference path without error, the command steering and speed are equal to their reference values.

The weighting constants, $C_1,...,C_4$ in (3.6), are chosen to optimize the overall path-tracking performance of the cart. As with the motor control units, it was the steering control part of the outer loop design which was the most sensitive to the choice of the control weights in (3.6). Because the cart speed affects the optimum steering gains through the kinematic relations (3.1), work is currently under way to find an appropriate functional dependence on speed for these weights, such that path control is maintained at near optimum during the critical docking segments, when the speed is being reduced to zero.

4. RESULTS

4.1 Implementation

The reference path generator and cart controller procedures outlined in Sections 2 and 3 have been implemented both in a simulation and on a real vehicle. The simulation program is written in the C language and runs on a VAX11/750 and Silicon Graphics IRIS workstation. The current version of the simulation program requires as input a file containing the path plan in a format like that shown in Figure 3(b), and a file containing system and run-control parameters. It produces as output a run parameter file, two files of system variables as a function of time, and a file containing reference path and cart motion data for the cart animation program. The animation program shows the cart moving in response to the steering and drive commands, along with the reference path and the layout of the workspace obstacles and corridors.

The C++ language[9] is used to program the experimental cart. Consequently, only minor changes in the simulation code are required for it to execute directly on the vehicle. This language compatability has facilitated rapid transfer of program developments between the simulation environment and the actual robot cart. The simulation also permits the testing of the cart's performance in workspace conditions that cannot be duplicated in the laboratory environment.

Program development for the vehicle is performed under UNIX.* The compiled code is then downloaded via ethernet to the onboard real-time system consisting of a Multibus-based MC68020 microprocessor running a real-time UNIX-derived executive called NRTX[10].

During program execution, the cart is completely autonomous. Any interaction is performed using a portable battery-powered Data General One PC acting as a terminal emulator. Diagnostic data from the cart may be temporarily saved in the PC and later downloaded to the UNIX host for analysis.

4.2 Motor Control Units

The mathematical model of the cart dynamics used in the simulation is a simple approximation of the actual cart dynamics. Nevertheless, it was found to be an adequate representation for obtaining good agreement of the simulated response with the observed response of the cart steering angle and drive speed to changes in their command values, once the model parameters were properly adjusted.

The model for the steering motor-control unit involves three unknown parameters: the friction coefficient H_1, the motor torque/inertia gain factor G_l, and the motor acceleration limit A_1. The other parameter values needed for both the motor control units on the cart and in the simulation program are the digital filter gain, C, and zero and pole values, A and B. To obtain appropriate parameter values, an iterative design procedure was used, in which initial estimates of the unknown parameters were made. The small-error (linear) closed-loop system model was then analyzed to determine loop gain and pole-zero values that provided good transient response. These values were used in the cart controller chip[11] to get experimental data on step response to different command steering angles. The friction and motor torque parameters in the simulation model were then adjusted to get similar step responses. After a second iteration of this procedure, a good agreement between the simulation model and the actual cart response was obtained. Figure 7 shows the steering angle response times, for positive and negative steering commands in the cart motor control unit, compared with the response times obtained from simulation runs. The response times were the times at which the wheel angle, α, reached 90% of the command angle, α_c. Also listed in Figure 7 are the design parameters obtained from this procedure.

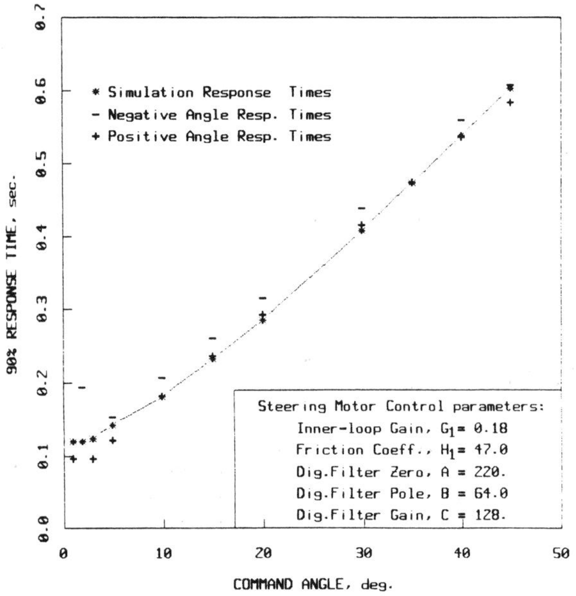

Figure 7. Steering response times for a range of commanded steering angles.

The response time data in Figure 7 indicate a good agreement between the cart and the simulation over the full ±45 degree range of possible steering angles, in spite of the simple model (3rd-order linear system dynamics with hard acceleration limits) used in the simulation. At large steering angles, the motor control is at the acceleration limit during most of the response. At small steering angles, the acceleration limit is a less significant factor in the response, but nonlinear friction effects may become more significant, which probably accounts for why the positive and negative angle response times for the cart diverge somewhat. Note that the simulation data for angles below about 4 degrees lies between the positive and negative experimental data, at a relatively constant response time of about 0.12 s.

The drive motor control unit has essentially the same elements as the steering motor unit, the main difference being that velocity-error rather than angle-error is used to control the drive motor. This reduces the control loop to a 2nd-order system, which is simpler to stabilize. As a result, the digital filter gain for the drive unit could be increased to four times that used in the steering unit (with approximately the same pole-zero values) without causing oscillations in the speed response. The present cart drive control yields a speed response rate of about 10 in/s/s. This is adequate for the cart to track changes in the path reference speed, which are currently programmed at 6 in/s/s.

4.3 Path Controller

4.3.1 Simulation Results. The reference path procedures described in Section 2 are essentially independent of the cart control, once the

* UNIX is a trademark of AT&T Bell Laboratories

physical constraints, such as the angle, velocity, and acceleration limits discussed above have been properly taken into account. The path plans represent feasible, collision-free paths in terms of such constraints as the minimum turning-radius of the cart (about two feet) and the physical area swept out by the cart as it moves along the prescribed path.

Given a feasible path to track, then, it is the task of the path controller, described in Section 3.3, to keep the tracking errors small enough to ensure smooth, collision-free motion along the paths. This task is complicated by the nonlinear kinematics (3.1) linking the cart position and velocity to the steering angle and drive speed and by the nonlinear transformations in the computation of the error components (3.4). At this stage of the study, a reasonably good design for the path controller has been achieved through a combination of small-error analysis and stabilization methods on the overall closed-loop system, followed by verification and fine-tuning of the control algorithm, using the full nonlinear system simulation program.

When the control weights in the control law (3.6) are properly adjusted, a satisfactory tracking of the reference path is achieved, as indicated by the simulation run shown in Figure 8. Because the scale of the overall trajectory is large compared to the path tracking errors, it is difficult to distinguish the actual cart path from the reference path in Figure 8. The cart speed in this run was 4 in/s. The errors in the position measurement system were set to zero in this simulation run, so any path errors were due to control errors, rather than measurement errors. The position control errors for the run shown in Figure 8 were less than 1/4 inch normal to the path, and less than 3/4 inch tangential to the path, while the heading error did not exceed 3 degrees. At the three stop points (A, B, and C), the normal and tangential errors were less than .06 and .53 inch, respectively, while the heading errors were less than half a degree.

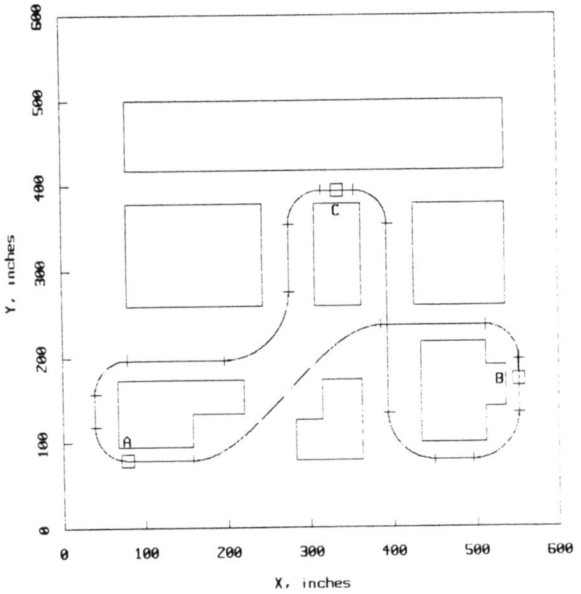

Figure 8. Example of a simulation run.

4.3.2 Experimental Results. The test runs for the experimental cart have so far been limited to a 14 by 12 foot "workspace" in the robot laboratory. Hence the test paths have been mostly figure-eight patterns, with tight turns in which the steering angle is at or near the hard limit of 45 degrees. Figure 9 illustrates the performance of the cart over such a test path. The reference trajectory is the solid line, while the cart's measured trajectories are shown for three operating speeds: 2 in/sec (dotted), 4 in/sec (dashed) and 6 in/sec (large dashed). The control algorithm clearly performs better at the slower speeds as the gains have been optimized for the slower speeds. The performance degradation at

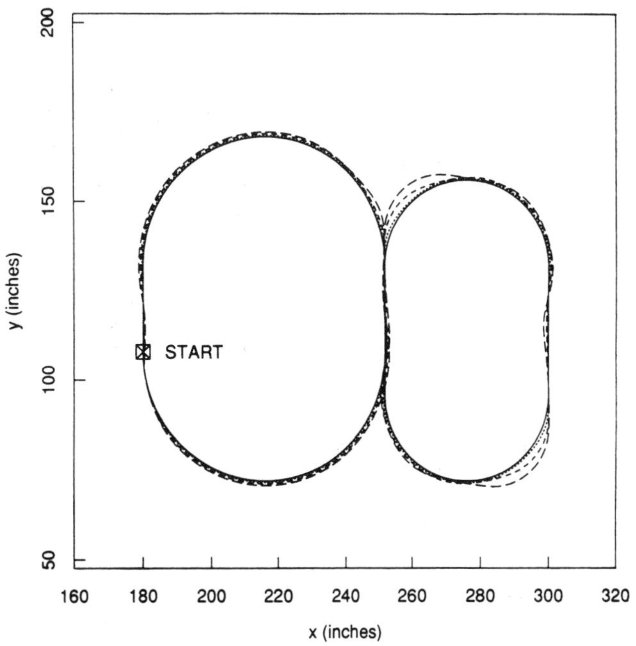

Figure 9. Measured path trajectories for the experimental cart.

higher speeds is usually acceptable since high speed transit maneuvers typically do not require as much precision as low speed docking maneuvers. Some overshoot at the beginning of the arc segments is unavoidable, since the control cannot make the steering wheel follow the angle discontinuity at the transition point between line and arc segments. Furthermore, as the vehicle enters the tighter loop on the right, the overshoot at high speeds is exaggerated due to the steering wheel hitting the hard limit stop.

The discontinuity in steering angle could be avoided if we were to control the point directly under the drive wheel rather than the midpoint between the rear wheels. However, controlling the orientation of the vehicle becomes more complex, as does path planning. A further alternative would be to replace the arc segments with segments of curves which are continuous in curvature[12].

5. CONCLUSION

Guidance and control procedures for enabling a robot cart to accurately and autonomously navigate around a structured environment, given a set of path data from a global path planner, have been described. These procedures have been implemented in an experimental robot cart and in a computer simulation that closely models the physical characteristics of the experimental cart. The use of an onboard reference path generator (RPG) has a number of advantages over generation of the path by the global path planner: (1) it reduces greatly the amount of data needed from the path planner, so even a very low data-rate, intermittant link to the cart is sufficient; (2) it provides a convenient separation between the path guidance and the cart control logic; (3) it supports fail-safe operation through continuous monitoring of the cart's control status; and, (4) it allows the cart relative independence in evolving the detailed path in time, once it is commanded to start. For example, unexpected obstacles detected by the onboard sensors can immediately trigger the local path planner to insert a stop segment at the current path location, and then generate a startup segment to continue the cart along the original path. This logic has been tested in the simulator and implemented on the cart, so that if a person is in the cart's path, for example, it stops until the person moves out of the way, then starts up automatically and continues along the path. Fixed obstacles within the structured environment are known to the path planner. If an unknown obstacle were encountered by the cart, the vehicle would stop, relay the obstacle data back to the global path planner and await an updated path plan.

As mentioned in Sec. 3.3, work is under way to find an appropriate functional dependence of the control weights in (3.11) on speed, in order to optimize the positioning accuracy of the cart during docking maneuvers, where the cart's position and heading as the speed reaches zero are more critical than while the cart is moving. The current performance with fixed weights gives docking accuracies of less than an inch in position and a few degrees in heading, provided the final path segment is a line several cart lengths long. The aim of the parametric control design is to reduce docking errors to less than about 1/10 inch and 1/10 degree, for arcs as well as line approach-segments.

The experience gained so far in this study indicates that the basic approach described above for guidance and control of a robot cart is sound. The separation of navigation and control functions in the RPG and the cart controller units provides a robust method for maintaining accurate tracking of any feasible path, once the controller unit has been properly adjusted to stabilize the closed-loop dynamics of the path control loop. Because the computer language of the simulation is the same as that used by the cart's onboard computer, program improvements developed in the simulation environment can be readily transferred to the onboard computer for experimental verification. The simulation environment includes a graphics workstation on which a wide range of factory and office layouts can be set up. A path-planning program[5] generates minimum-turn, collision-free paths, allowing the cart to navigate between any selected points in the layout. These can then be input to the cart simulation program for verification of cart performance over the prescribed paths.

The work reported in this and the companion papers[3]-[5] is continuing with the goal of achieving a versatile, autonomous robot cart that will be useful for transport of materials in reasonably well-structured environments, such as factories, warehouses, and offices.

5.1 Acknowledgements

The authors wish to thank W. J. Kropfl, G. L. Miller, E. R. Wagner, and G. T. Wilfong, for their help and useful discussions during this work.

REFERENCES

1. *Proc. of the Second International Conf. on Automated Guided Vehicle Systems*, Stuttgart, W. Germany, 7-9 June 1983, H. J. Warnicke, Ed., North Holland Publ. Co., 1983.

2. T. Tsumura, "Survey of Automated Guided Vehicle in Japanese Factory," *Proc. 1986 IEEE International Conf. on Robotics and Automation*, San Francisco, CA, Vol. II, p. 1329.

3. I. J. Cox, "Blanche: An autonomous robot vehicle for structured environments," *Proc. 1988 IEEE International Conf. on Robotics and Automation*, Philadelphia, PA.

4. G. L. Miller and E. R. Wagner, "An Optical Rangefinder for Autonomous Robot Cart Navigation," *Proc. IECON/SPIE '87*, Cambridge, MA, Nov. 1-6, 1987.

5. G. T. Wilfong, "Motion Planning for an Autonomous Vehicle," *Proc. 1988 IEEE International Conf. on Robotics and Automation*, Philadelphia, PA.

6. L. E. Dubins, "On Curves of Minimal Length With a Constraint on Average Curvature, and With Prescribed Initial and Terminal Positions and Tangents", *American Journal of Mathematics* **79**, 1957 pp. 497-516.

7. T. Hongo *et al.*, "An automatic guidance system of a self-controlled vehicle," *IEEE Trans. on Ind. Electronics*, Vol. IE-34, No. 1, 1987, pp. 5-10.

8. S. Uta *et al.*, "An implementation of michi-a locomotion command system for intelligent mobile robots," *Proc. 15th ICAR*, 1985, pp. 127-134.

9. B. Stroustrup, *The C++ Language*, Addison Wesley, New York(1986).

10. D. A. Kapilow, "Real-Time Programming in a UNIX Environment," *1985 Symposium on Factory Automation and Robotics*, Courant Institute of Mathematical Sciences, New York University, p. 28.

11. "General Purpose Motion Control IC HCTL-1000" Hewlett Packard Technical Data Sheet, 1985.

12. W. L. Nelson, "Continuous Steering-Function Control of Robot Carts", Bell Labs Internal Document, Oct. 1987, submitted for publication, 1988.

Marc H. Raibert
Department of Computer Science and
The Robotics Institute
Carnegie-Mellon University
Pittsburgh, Pennsylvania 15213

Running With Symmetry

Abstract

Symmetry can simplify the control of dynamic legged systems. In this paper, the symmetries studied describe motion of the body and legs in terms of even and odd functions of time. A single set of equations describes symmetric running for systems with any number of legs and for a wide range of gaits. Techniques based on symmetry have been used in laboratory experiments to control machines that run on one, two, and four legs. In addition to simplifying the control of legged machines, symmetry may help us to understand legged locomotion in animals. Data from a cat trotting and galloping on a treadmill and from a human running on a track conform reasonably well to the predicted symmetries.

1. Introduction

Running is a series of bouncing and ballistic motions that exert forces on the body during every stride. The bouncing motions are caused by the vertical rebound of the body when the legs push on the ground, and the ballistic motions occur between bounces when the body is airborne. If a legged system is to keep its forward running speed fixed and its body in a stable upright posture despite these motions, then the net acceleration of the body must be zero over each entire stride. This requires that the torques and horizontal forces exerted on the body by the legs must integrate to zero over each stride and that the vertical forces must integrate to the body's weight times the duration of the stride. This is equally true for running machines and for running animals.

Although there are many patterns of body and leg motion that can satisfy these requirements, a particularly simple solution arises when each variable has an even or odd symmetry during the time a foot is in contact with the ground.

$$\text{Body Symmetry} \begin{cases} x(t) = -x(-t), \\ z(t) = z(-t), \\ \phi(t) = -\phi(-t), \end{cases} \tag{1}$$

$$\text{Leg Symmetry} \begin{cases} \theta(t) = -\theta(-t), \\ r(t) = r(-t). \end{cases} \tag{2}$$

x, z, and ϕ are the forward position, vertical position, and pitch angle of the body, and θ and r are the angle and length of the leg, all measured in the *sagittal plane** (see Fig. 1). For simplicity, t and x are defined so that $t = 0$ halfway through the stance phase, and $x(0) = 0$. These symmetry equations specify that forward body position, body pitch angle, and leg angle are each odd functions of time throughout the stance phase, and that body elevation and axial leg length are even functions of time. The symmetry also requires that the actuators operate with even and odd symmetry:

$$\text{Actuator Symmetry} \begin{cases} f(t) = f(-t), \\ \tau(t) = -\tau(-t), \end{cases} \tag{3}$$

where τ is the torque exerted about the hip and f is the force exerted along the leg axis.

These symmetries are significant because they result in accelerations of the body that are odd functions of time throughout a stride. Odd functions integrate to zero over symmetric limits, leaving the forward running speed, body elevation, and body pitch angle unchanged from one stride to the next.

We first recognized the value of symmetry when exploring control for machines that balance as they hop on one leg (Raibert and Brown 1984; Raibert,

This research was supported by a grant from the System Development Foundation and by a contract from the Engineering Applications Office of the Defense Advanced Research Projects Agency.

The International Journal of Robotics Research,
Vol. 5, No. 4, Winter 1986,

* The sagittal plane for animals is defined by the fore–aft and up–down directions.

Fig. 1. Definition of variables used in symmetry equations. Positive τ acts about the hip to accelerate the body in the positive φ direction. Positive f acts along axis of the leg and pushes the body away from the ground.

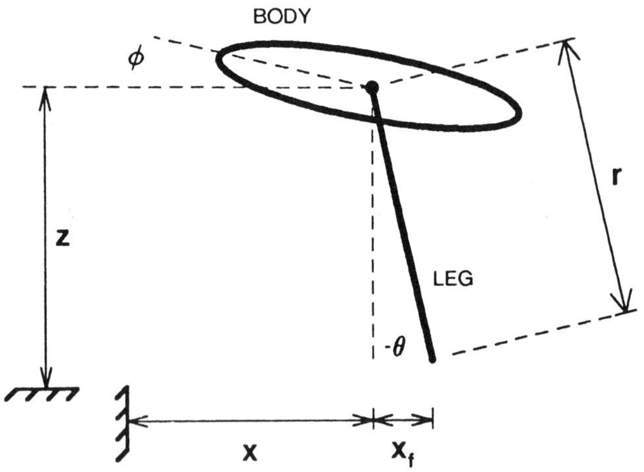

Fig. 2. When the foot is placed on the neutral point, there is a symmetric motion of the body. The figure depicts running from left to right. The left-most drawing shows the configuration just before the foot touches the ground, the center drawing shows the configuration halfway through stance when the leg is maximally compressed and vertical, and the right-most drawing shows the configuration just after the foot loses contact with the ground.

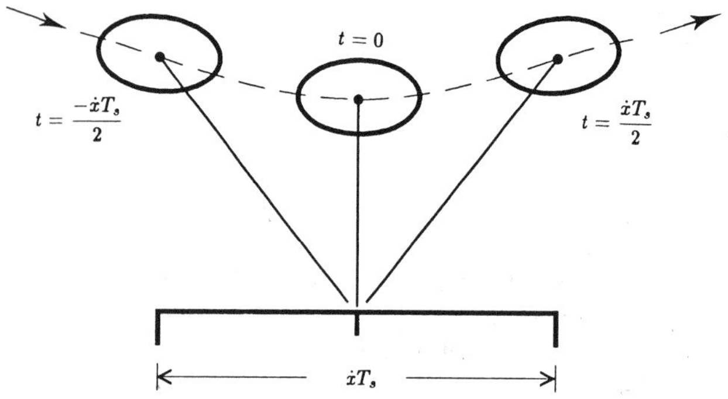

Brown, and Chepponis 1984). Like an inverted pendulum, a one-legged system tips and accelerates when its point of support is not located directly below the center of mass. Symmetric motion ensures that tipping and acceleration in one direction is balanced by equal tipping and acceleration in the opposite direction (see Fig. 2). The control system we implemented for the one-legged machines uses a knowledge of the dynamics to manipulate the machine's initial configuration on each step to produce a symmetric motion during the ensuing support phase. Symmetry simplifies the control because it frees the control system from regulating the details of the trajectory—the details are determined passively by the mechanical system. One-legged hopping machines that run and balance using these techniques are described in Raibert and Brown (1984) and Raibert, Brown, and Chepponis (1984). The approach was recently extended to the control of a running biped (Hodgins, Koechling, and Raibert, 1986) and to a trotting quadruped (Raibert, Chepponis, and Brown 1986).

This paper introduces motion symmetry in the context of one-legged systems and then generalizes to more complicated cases. Symmetry is particularly simple for one-legged machines because only one leg provides support at a time, each support interval is isolated in time by periods of ballistic flight, and the hip is located at the center of mass. After reviewing the one-legged case, we consider motions that span several support intervals and the use of several legs for support during a single support interval. One set of symmetry equations applies for one, two, and four legs, and for gaits that use legs singly and in combination.

In addition to suggesting simple control for legged robots, the symmetries developed in this paper may help us to understand the control mechanisms at work in running animals. Hildebrand (1965, 1966, 1968, 1976) established the importance of symmetry in animal locomotion when he observed that the left half of a horse often uses the same pattern of footfalls as the right half, but is 180 degrees out of phase. He devised a simple and elegant characterization of the symmetric walking and running gaits for a variety of quadrupeds, using just two parameters: the phase angle between the front and rear legs and the duty cycle of the legs. By mapping each observation of symmetric behavior into a point in phase/duty-cycle space, Hildebrand was able to classify systematically gaits for over 150 quadruped genera.

Rather than look at relationships between the footfalls of the left and right legs as Hildebrand did, I measured the trajectories of the feet with respect to the body and the trajectory of the body through space in the sagittal plane. Data for the trotting and galloping cat and for the running human show that they sometimes move as the symmetries predict.

2. Mechanics of Symmetry

A number of simplifications ease the analysis of symmetry. The analysis is based on a model that is restricted to move in the plane, with massless legs and no losses anywhere in the system. The body is a rigid object that moves fore and aft and up and down and that pitches in the plane, with position and orientation given by $[x\ z\ \phi]$. Each leg is a single member that pivots about its hip on a hinge-type joint and that lengthens and shortens by telescoping. The length of a leg and its angle with respect to the vertical are given by $[r\ \theta]$. A foot at the end of each leg provides a single point of support. Friction between a foot and the ground prevents the foot from sliding when there is contact. A foot in contact with the ground acts mechanically like a hinge joint.

Each leg actuator exerts a force f along the leg's axis between the body and the ground. Positive f accelerates the body away from the ground, and, because the feet are not sticky, $f \geqslant 0$. This force is zero when there is no contact between the foot and the ground. Normally the leg is springy in the axial direction, in which case f is a function of leg length. A second actuator acts at the hip, generating a torque τ between the leg and the body. Positive τ accelerates the body in the positive ϕ direction. Equations of motion for this sort of model are given in Appendix A.

In normal operation, the models follow a regular pattern of activity, alternating between periods of support and periods of flight. The transition from flight to support is called *touchdown* and the transition from support to flight is called *lift-off*. During support, a foot remains stationary and the leg exerts a combination of vertical and horizontal forces on the body. Because legs are springy, the body's vertical motion is an elastic rebound that returns the system to the flight phase after each collision with the ground. Once airborne, the body follows a ballistic trajectory. The body may derive support from one or more legs during a single support interval depending on the number of legs in the system and the gait. Because the legs have no mass and the entire system is lossless, bouncing motions continue undiminished without an external source of energy.

2.1. Symmetric Motion with One Leg

Imagine that at time $t = 0$ the foot of a one-legged system is located directly below the center of mass, the body is upright, and the velocity of the body is purely horizontal: $\theta = 0$, $\phi = 0$, and $\dot{z} = 0$. Figure 3 shows this configuration. Because it has left-right symmetry and there are no losses, the system's expected behavior proceeding forward in time is precisely the same as its past behavior receding backward in time, but with a reflection about the line $x = 0$. This behavior is described by the body symmetry equations, which state that $x(t)$ and $\phi(t)$ are odd functions of time and $z(t)$ is an even function. Because the body moves along a symmetric trajectory with respect to the origin and because the foot is located at the origin during support, the body symmetry equations imply that the foot's motion is symmetric with respect to the body, which gives the leg symmetry equations, Eq. (2).

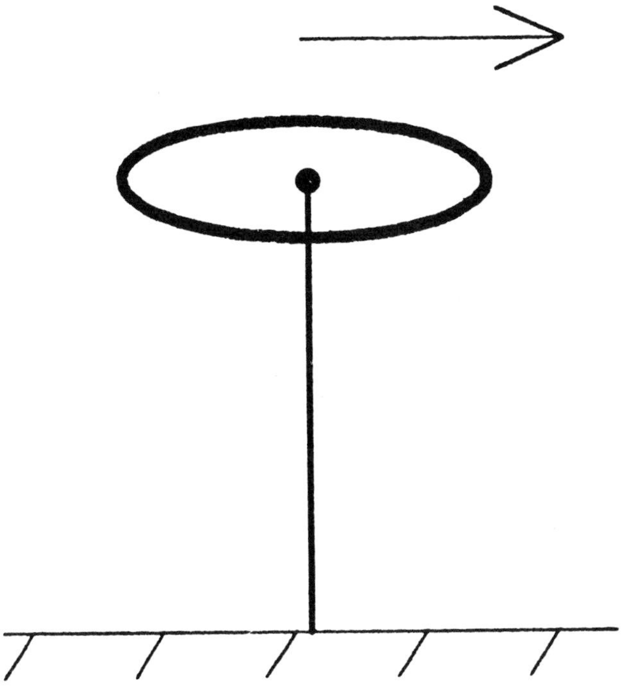

Fig. 3. Symmetric configuration of a one-legged system halfway through stance, when it has fore-aft symmetry (left-right as shown in diagram) as well as symmetry moving forward and backward in time. The vertical velocity is zero, the support point is located directly under the center of mass, and the body is upright: $\dot{\theta}(0) = x_f(0) = \phi(0) = 0$.

Symmetric motion of the body and legs requires symmetric actuation, as given in Eq. (3). From the equations of motion (see Appendix A), we see that hip torque is the only influence on body pitch angle, so odd ϕ implies odd τ. With the evenness and oddness of the other variables specified, f must be even to satisfy the equations of motion.

A locomotion system operates in steady state when the state variables, measured at the same time during each stride cycle, do not vary from stride to stride. The state variables of interest are the body's forward velocity, vertical position, vertical velocity, pitch angle, and pitch rate. With the state vector **S** representing these variables, $\mathbf{S} = [\dot{x}\ z\ \dot{z}\ \phi\ \dot{\phi}]$, steady state is defined by

$$\mathbf{S}(t) = \mathbf{S}(t + T), \tag{4}$$

where T is the duration of one stride.

Symmetric body and leg motion results in *steady-state locomotion.** For the forward speed to remain unchanged from stride to stride, the horizontal force f_x acting on the body must integrate to zero over a stride:

$$\int_{stride} f_x \, dt = 0. \tag{5}$$

Assume that $f_x = 0$ during flight and that the forward speed does not change. From the equations of motion we get $f_x = f \sin\theta - \left(\frac{\tau}{r}\right)\cos\theta$ during stance, which is an odd function because f and r are even and τ and θ are odd. Therefore

$$\dot{x}(t_{lo}) - \dot{x}(t_{td}) = \int_{t_{td}}^{t_{lo}} f_x \, dt = 0. \tag{6}$$

This confirms that symmetric motion provides no net horizontal force on the body, and running speed proceeds in steady state from stride to stride.

The vertical position and velocity also proceed in steady state for a symmetric motion. The elevation of the body is an even function of time during stance, so $z(t_{lo,i}) = z(t_{td,i})$ and $\dot{z}(t_{lo,i}) = -\dot{z}(t_{td,i})$. During flight, the body travels a parabolic trajectory that is also even, if we specify $t = 0$ halfway through flight: $z(t_{td,i+1}) = z(t_{lo,i})$ and $\dot{z}(t_{td,i+1}) = -\dot{z}(t_{lo,i})$. Consequently, $z(t_{td,i}) = z(t_{td,i+1})$ or $z(t_{td}) = z(t_{td+T})$, which is the steady-state condition on z, and $\dot{z}(t_{td}) = \dot{z}(t_{td+T})$, which is the steady-state condition on $\dot{z}$.

The torque acting on the body is zero during flight and an odd function during stance, so the body pitch rate undergoes zero net acceleration during stance, $\dot{\phi}(t_{lo}) = \dot{\phi}(t_{td})$. This satisfies the steady-state condition on $\dot{\phi}$. For the pitch angle of the body to proceed in steady state, its value at the end of flight must be equal and opposite to its value at the beginning of the flight phase. Assuming that symmetry holds during stance so that $\phi_{lo} = -\phi_{td}$ and that no torques act on the body during flight, a repeating pattern requires that

$$\frac{\dot{z}(t)}{-g} = \frac{\phi(t)}{\dot{\phi}(t)}, \tag{7}$$

where g is the acceleration of gravity. This constraint prescribes the relationship among pitch angle, pitch rate, and vertical velocity needed for steady-state running. It is trivially satisfied when there is no pitching motion, $\phi(t) = 0$ and $\dot{\phi}(t) = 0$. Equation (7) results in a second symmetric configuration that occurs during flight. This configuration, given by $f = 0$, $\dot{z} = 0$, and $\phi = 0$, ensures that the body's behavior is symmetric during flight.

For a one-legged system, symmetric body motion can be obtained *only* from symmetric leg motion. That is, if x, z, and ϕ obey the symmetries of Eq. (1), then r must be even and θ must be odd. The proof is given in Appendix B.

The symmetry equations given in Eqs. (1–3) are consistent with the equations of motion given in Appendix A. This is shown by labeling the symmetry for each term in the equations of motion according to the behavior of the body, the behavior of the legs, and the required actuation. Each variable is labeled with a preceding superscript indicating that it is either even 'e' or odd 'o'. Substituting these labeled variables in the equations of motion and further labeling each term, we get:

$$\overbrace{m\,{}^{o}\ddot{x}}^{\text{odd}} = \overbrace{{}^{e}\!f \sin {}^{o}\theta}^{\text{odd}} - \overbrace{\frac{{}^{o}\tau}{{}^{e}r}\cos {}^{o}\theta}^{\text{odd}}, \tag{8}$$

$$\overbrace{m\,{}^{e}\ddot{y}}^{\text{even}} = \overbrace{{}^{e}\!f \cos {}^{o}\theta}^{\text{even}} + \overbrace{\frac{{}^{o}\tau}{{}^{e}r}\sin {}^{o}\theta}^{\text{even}} - \overbrace{mg}^{\text{even}}, \tag{9}$$

$$\overbrace{I\,{}^{o}\ddot{\phi}}^{\text{odd}} = \overbrace{{}^{o}\tau}^{\text{odd}}. \tag{10}$$

---

* A trajectory that provides steady-state locomotion is one that provides a nominal motion that would repeat from cycle to cycle if there were no disturbances. It does not mean that there are restoring forces that will return the system to the trajectory if it deviates as the result of a disturbance. Restoring forces are also required for stability once the nominal trajectory has been determined. Asymmetry in the motion is a source of such restoring forces.

Fig. 4. Asymmetric trajectories. Displacement of the foot from the neutral point accelerates the body by skewing its trajectory. When the foot is placed behind the neutral point, the body accelerates forward during stance (left). When the foot is placed forward of the neutral point, the body accelerates backward during stance (right). Dashed lines indicate the path of the body and solid horizontal lines under each figure indicate the CG-print.

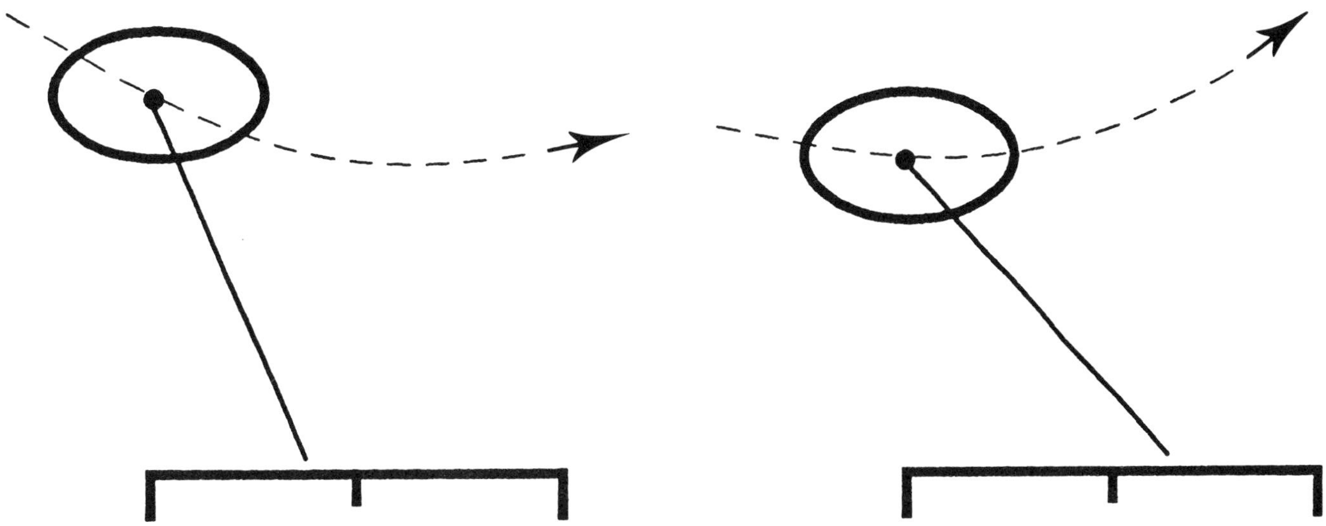

A similar procedure shows that the symmetries are consistent with the equations of motion for multilegged systems as well.

3. Generating Symmetric Motions

The discussion has focused on the nature and value of symmetric motion without addressing the generation of symmetric motion. What action must a control system take to produce symmetric behavior? Recall that a legged system moves with symmetry if θ, $\dot{z}$, and ϕ all equal zero at the same time during support, but the control system must commit the foot to a position on the ground before touchdown when neither $\dot{z}$ nor ϕ is zero. The task of orchestrating such a rendezvous is to predict where the center of mass will be when the body's vertical velocity and pitch angle are both zero.

General solutions to this problem are not known. The difficulty in accomplishing this task is that placement of the foot influences the path of the body. If we had an expression for the body's trajectory during support as a function of $\dot{x}(t_{td})$, $\dot{z}(t_{td})$, and $\theta(t_{td})$—a solution to the equations of motion—then we could solve for the desired foot placement. We have not found a closed-form expression for the path of the body during support, even for simple models.

Despite the lack of a general solution, approximate solutions exist for gaits that use just one leg for support at a time, the one-foot gaits. The simplest approximate solution assumes that forward speed is constant during support and that the period of support T_s is constant, depending on only the spring mass characteristics of the leg and body. These approximations estimate the length of the *CG-print,* the forward distance traveled by the center of mass during support, as $\dot{x}T_s$:

$$x_f = \frac{\dot{x}T_s}{2} - k_{\dot{x}}(\dot{x} - \dot{x}_d), \tag{11}$$

where

x_f	is the forward displacement of the foot with respect to the projection of the center of mass,
$\dot{x}$	is the forward velocity of the body,
$\dot{x}_d$	is the desired forward velocities of the body,
T_s	is the duration of a support period, and
$k_{\dot{x}}$	is a gain.

The first term in Eq. (11) is the neutral foot position that provides symmetry. The second term introduces asymmetry that accelerates the system to correct for errors in running speed, as shown in Fig. 4. It displaces the foot from the neutral point to skew the pattern of body motion. A set of systematically skewed motions is shown in Fig. 5. These displacements accelerate the body to stabilize its motion against disturbances and to change running speed. Control systems for one-,

Fig. 5. Path of the body during stance for several forward foot positions. Only the neutral foot position results in a symmetric body trajectory (bold), whereas those to either side are skewed, either forward or backward. The initial forward speed is the same for each trajectory. The circles indicate the location of the body at touchdown, and the origin is the foot position. These data are from simulations of a model with a linear leg spring. Adapted from Stentz (1983).

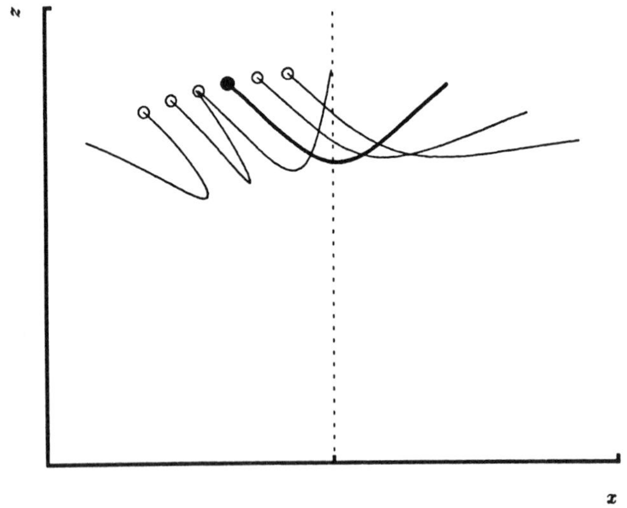

Fig. 6. Symmetry data recorded from a physical, 3-D, one-legged, hopping machine. The behavior of the machine obeys the symmetry equations when the foot is placed on the neutral point. Data for three consecutive support intervals are superimposed. The leg is longer at lift-off than at touchdown because it lengthens during support to provide thrust that compensates for various mechanical losses in the system. The time axes were adjusted so that t = 0 halfway through the support interval, and the x-origin was adjusted so that x(t = 0) = 0. Running speed is about 1.6 m/s. Dashed vertical lines indicate touchdown and lift-off.

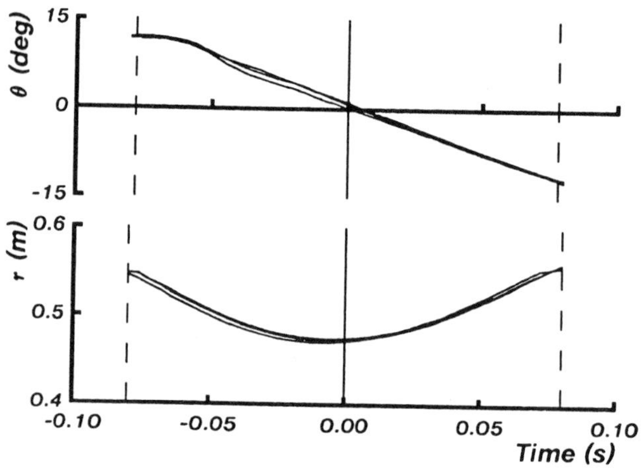

two-, and four-legged machines use this approximation to choose a forward displacement for the foot during flight (Raibert 1986*b*).

Our experience with this approximation is that it provides good symmetry at low and moderate running speeds. The data shown in Fig. 6 were recorded from a three-dimensional one-legged machine as it traveled at constant speed (Raibert, Brown, and Chepponis, 1984). The leg and body moved with good symmetry.

Another approach to the problem of predicting behavior during support might be to use a tabulated solution to the equations of motion. A table could be indexed by forward, vertical, and desired velocities, and it would provide the necessary leg touchdown angle. Depending on the size of the table, this approach could provide any desired degree of accuracy. Tabular solutions to this problem are discussed in Raibert (1986*b*).

3.1. Pairs of Antisymmetric Steps

Motion symmetry need not be confined to just one step. Although we have concentrated on symmetry that applies on a step-by-step basis, the symmetries apply equally well when pairs of steps produce complementary accelerations, with the symmetry distributed over more than one support interval. This case is discussed next.

Suppose that single support periods deviate from symmetry but that two sequential support periods each deviate from symmetry in a complementary fashion. Figure 7 shows a sequence of such antisymmetric steps. The trajectory of the body during each step is asymmetric, and the system accelerates because the foot is displaced from the neutral point. If the foot position on the next step compensates, however, then the body motions on successive steps balance with equal and opposite accelerations. Equations (1 – 3) still describe the behavior of the body and leg, provided that we define $t = 0$ at the point halfway between the two steps.

So far we have assumed that the forward running speed is nonzero, but it need not be. Antisymmetric pairs of steps can apply to running in place with no forward speed. For instance, if the foot were placed so that the horizontal component of the body velocity is just reversed during support, and this were done on each step, then the average forward running speed would be zero and the system would bounce back and forth on each step. This is just the sort of behavior observed in the frontal plane of the human and the pacing quadruped.

Figure 8 presents data from a physical demonstration of symmetry distributed over a pair of steps for which the forward running speed is zero. To generate

Fig. 7. Pairs of antisymmetric steps. If the foot is positioned behind the neutral point on one step, and in front of it on the next step, then the pair of steps may have symmetry that stabilizes the forward running speed, even though the motion during each step is no longer symmetric. We redefine the stride to include the symmetric pair of steps. The body and leg are drawn once for each touchdown and lift-off. The vertical dashed lines indicate the planes of symmetry, which occur halfway through the strides and between the strides.

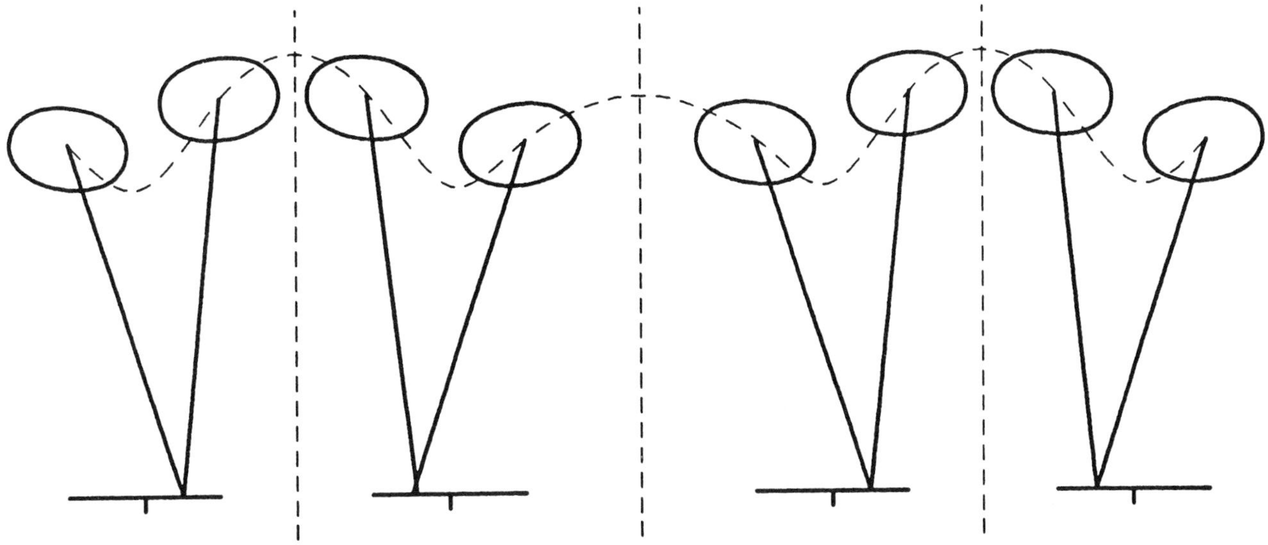

these data, we modified the control algorithm for a physical one-legged hopping machine to add an offset Δx to the desired foot placement on every even-numbered hop and to subtract Δx on every odd-numbered hop. For small values of Δx, the system hopped from side to side with no net forward acceleration. The system maintained its balance, provided that the offset of the foot was small enough so that the system did not tip over entirely before the next step.

3.2. Symmetry with Several Legs

A system with two legs can run with a variety of gaits. The two legs can operate precisely in phase, precisely out of phase, or with intermediate phase. Figure 9 shows several examples that differ with regard to the amount of body pitching, the variation in forward running speed within a stride, and the degree of temporal overlap in the support provided by the two legs. In each case, however, symmetric body and leg motion results in steady-state locomotion.

The body symmetries for a system with several legs are the same as for one leg, but the leg and actuator symmetries are modified slightly. Each leg and actuator variable, θ, r, τ, and f, have the same meanings as before but with subscripts to distinguish among the individual legs:

$$\theta_j(t) = -\theta_k(-t), \tag{12}$$

$$r_j(t) = r_k(-t), \tag{13}$$

$$\tau_j(t) = -\tau_k(-t), \tag{14}$$

$$f_j(t) = f_k(-t). \tag{15}$$

For a system with two legs, $j = 1$ and $k = 2$. For four legs, two pairings are possible: $j = [1, 4]$ and $k = [2, 3]$ or $j = [1, 4]$ and $k = [3, 2]$, depending on the gait, where 1 is left front, 2 is left rear, 3 is right rear, and 4 is right front.

Symmetric body motion no longer requires an individual leg to move with a symmetry of its own. Instead, the behavior of one leg is linked to the behavior of another leg, so that they operate with reciprocating symmetry. This frees the variables describing any one leg to take on arbitrary functions of time while preserving the symmetric forces and moments impinging on the body during support. These motion symmetries apply when legs operate in unison, when legs have different but overlapping support periods, and when the legs provide support separately. As before, the equations that describe leg motion apply only when $f > 0$, so it does not matter how the legs move when

Fig. 8. Symmetric pairs of steps. The curves show the recorded path of the body for a physical, one-legged hopping machine hopping in place. The control algorithms were those described by Raibert, Brown, and Chepponis (1984), but an offset, Δx, was added to the foot position on even-numbered hops and subtracted on odd-numbered hops. The magnitude of Δx was set to two different values, shown separately in the two curves. The plot shows the motion of the body in a vertical plane that contains the foot offset.

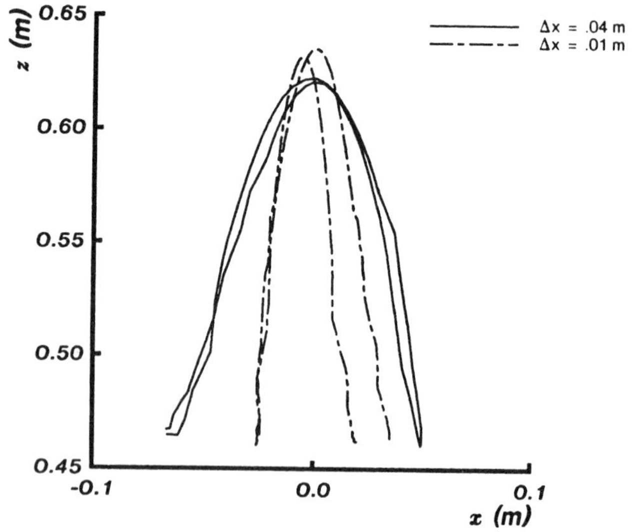

Fig. 9. Running with two legs separated by a long body. Symmetry can be achieved when both feet provide support simultaneously, when there is partial overlap in the support periods, and when the legs provide support in sequence. These three cases are distinguished by the phasing of the legs. It may be difficult to place the feet on the neutral points when the hips are widely separated. Displacements of the feet from the neutral points influence pitching of the body and the duration of each flight phase.

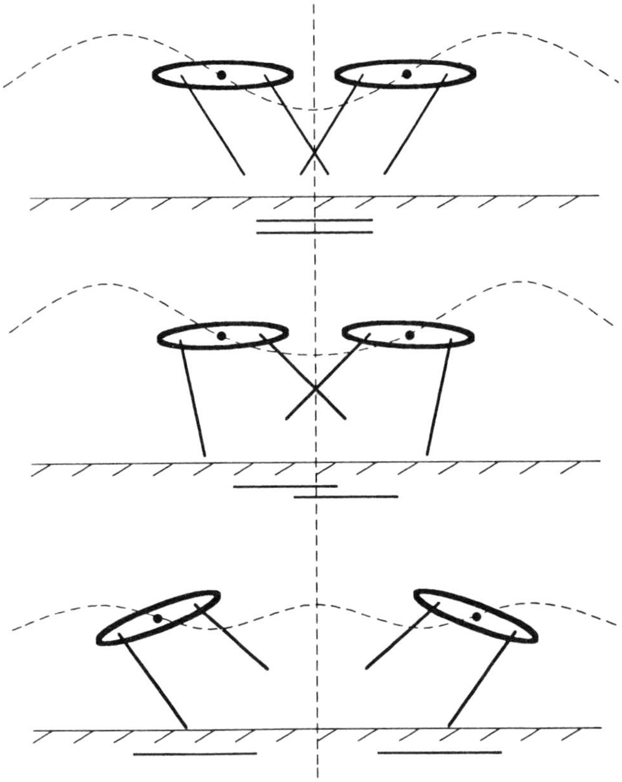

they are not touching the ground. Equations (12–15) reduce to the one-legged case when $j = k = 1$.

For a system with two legs, the conditions for symmetric behavior are nearly the same as for one leg, but it is no longer required that an individual foot be located under the center of mass. For instance, Fig. 10 shows two symmetric configurations that place no feet directly under the center of mass. In both cases, the *center of support* is located under the center of mass. It is also possible to have no support in the symmetric configuration, as the bottom gait in Fig. 9 suggests. The antisymmetric activity of the two legs operating as a pair produces symmetric motion of the body when measured over the stride. This is very much like the behavior of the one-legged system when it uses pairs of steps to achieve symmetry.

A characteristic of locomotion when pairs of legs work in reciprocation is that the individual feet need not be placed on the neutral point to achieve steady-state behavior. This is important because it may be difficult for a legged system to reach far enough under the center of mass when the hips and shoulders are located at the extremes of a long body. This situation arises in the sagittal plane for the quadruped bound and gallop and to a lesser extent in the frontal plane for the quadruped pace.

4. Symmetry in Animal Running

The importance of symmetry in the control of legged robots raises the question of what role it might play in the behavior of running animals. Can the symmetries developed for legged robots help us to describe and understand the running behavior of legged animals?

I analyzed data for a cat trotting and galloping on a treadmill and for a human running on an outdoor cinder track. The cat data were obtained by digitizing 16 mm, 100 fps film provided by Wetzel, Atwater, and Stuart (1976). Each frame showed a side view of the cat on the treadmill and a 1-ms counter used to calibrate the film speed. Treadmill markers spaced at 0.25-m intervals provided a scale of reference and permitted registration of each frame. Small circular markers attached to the cat's skin made the digitizing easier. Running speeds with respect to the treadmill

Fig. 10. Symmetric configuration during support for two legs. Configuration of two-legged systems halfway through the support interval. The center of support *is located under the center of mass, vertical velocity is zero, and the body is upright:* $\theta_i + \theta_j = \dot{z} = \phi = 0$.

Fig. 11. Body motion of the galloping cat. Data are shown for one stride of a cat running on a treadmill with a rotary gallop. According to symmetry theory, forward body position x *and body pitch angle* ϕ *should each have odd symmetry, and body height* z *should have even symmetry. The symmetry displayed in these plots is good. Dashed vertical lines indicate the beginning and end of the stance phase. Solid vertical line indicates the symmetry point, when* t = 0.

Fig. 12. Leg motion for the galloping cat. Leg angle θ *should have odd symmetry and leg length* r *should have even symmetry. Symmetry in behavior of the legs is found when they are considered in reciprocating pairs, e.g.,* $\theta_{RR}(t) = -\theta_{RF}(-t)$ *and* $r_{RR}(t) = r_{RF}(-t)$. *Symbols indicate pairs of points that should have symmetric positions with respect to the origin (for odd symmetry), or the z-axis (for even symmetry). Both leg angle and leg length show very good symmetry. Data for each leg are shown only when its foot touches the support surface. Dashed vertical lines indicate the beginning and end of the stance phase. Solid vertical line indicates the symmetry point, when* t = 0. *The data are from the same stride as in Fig. 11.*

Fig. 13. Data for the cat trotting on a treadmill. The left front and hind legs form one pair of legs that operate in reciprocating symmetry, and the right front and rear legs form the other reciprocating pair. Running speed was 2.2 m/s.

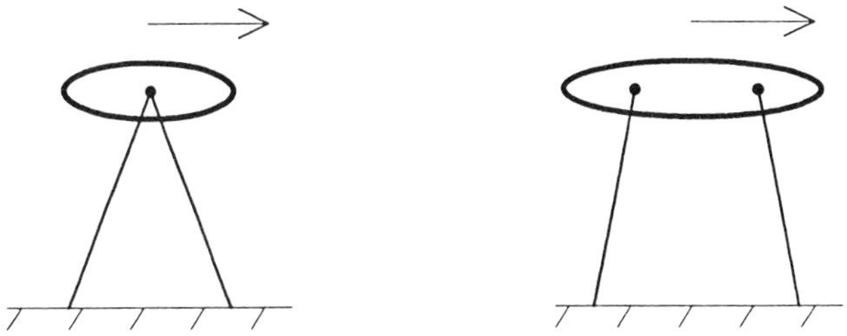

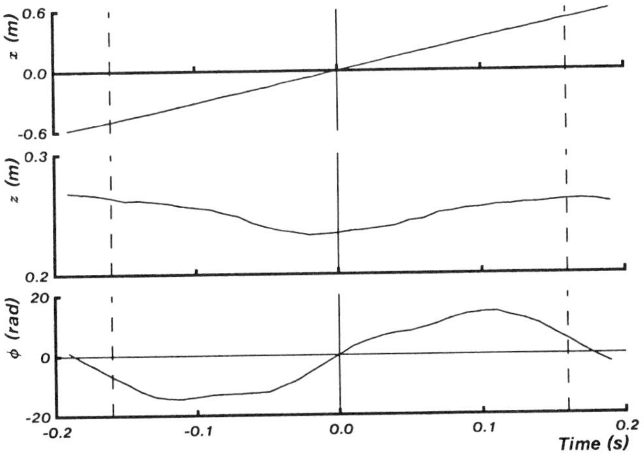

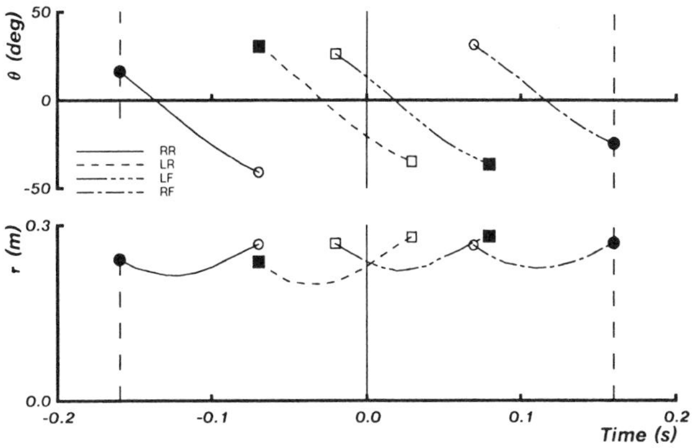

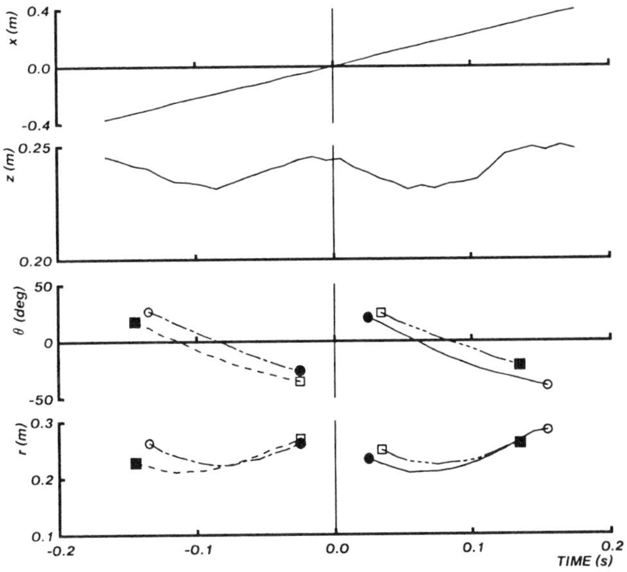

surface were about 2.2 m/s for trotting and 3.1 m/s for galloping.

The human measurements were made by digitizing 16 mm film of a runner on the semicircular section of an outdoor cinder track. The camera was mounted on a tripod located at the center of the semicircle and panned to track the runner. Ground markers spaced at 1.0-m intervals provided scale and registration as before. Running speed was about 3.8 m/s.

In digitizing both the cat and the human data, the point of support provided by each foot was estimated visually. A straight line from this point to the hip, or shoulder for the cat's front legs, was used to find the leg length r and the leg angle θ. The center of mass of the cat was taken as the midpoint between the shoulder and the hip. The pitch angle of the body was the angle between the horizontal and the line connecting shoulder to hip, offset so that $\phi(0) = 0$. These mea-

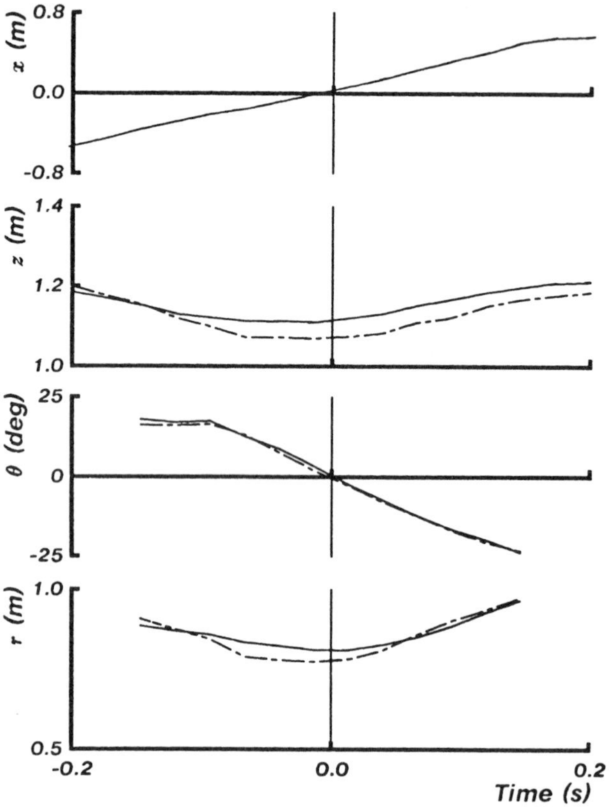

Fig. 14. Data for one stride of a human running on an outdoor cinder track. Data for the right (stippled) and left legs are superimposed. Running speed was 3.8 m/s. Subject MHR.

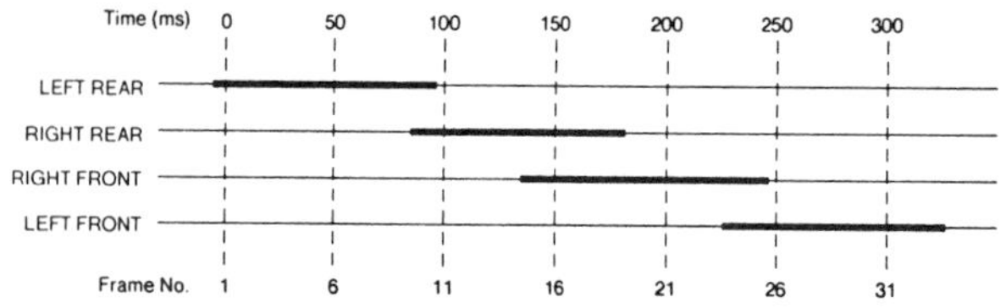

Fig. 15. Pattern of foot contacts in the rotary gallop of a cat. Horizontal bars indicate that the foot is in contact with the support surface. The duration of an entire stride is 350 ms. Vertical dotted lines indicate the seven frames used in Fig. 16.

surements provided three parameters of the body's motion: its forward position, vertical position, and pitch angle $[x\ z\ \phi]$. The measurements also gave two parameters of each leg's motion: its length and angle with respect to the vertical $[r\ \theta]$. In addition to information about the timing of footfalls, these measurements provided information about where on the ground the feet were placed with respect to the body and how the body itself moved.

Data for one stride of the cat gallop, trot, and human run are plotted in Figs. 11–14. In each case, the data are in agreement with the even and odd symmetries predicted by the symmetry equations. The cat data for galloping show a remarkable degree of symmetry.

These running symmetries can be visualized graphically. The symmetry equations imply that if we reverse both the direction of forward travel and the direction of time, $x = -x$, $t = -t$, then the pattern of forward body movement and of footfalls should not be affected: $x(t) = -x(-t)$. This invariance is illustrated in Fig. 16. Of particular interest is the precise overlap of the footfalls for the forward and reverse running sequences. This overlap was predicted by the symmetry equations.

Some of the data examined reveal a bias in foot position toward the rear of the animal. The timing diagram of Fig. 15 illustrates such a bias. For each leg, $|\theta(t_{lo})| > |\theta(t_{td})|$, and the last leg providing support to the body stayed in contact with the ground longer than the first leg providing support. According to the principles outlined in this paper, such bias or skew might mean that a net forward force was generated on the body. Such a force could accelerate the system forward, compensate for an external disturbance, or compensate for losses occurring elsewhere in the system.

Another explanation, however, might be that the axial leg force does not obey Eq. (15). For instance, because the legs are not massless, their collisions with the ground on each step result in asymmetric forces. One might also expect the legs to deliver thrust actively during support, in order to make up for losses and to maintain the vertical bouncing motion. This active thrusting would result in a violation of Eq. (15). Without knowing the actual force that each leg exerts on the ground, it is difficult to draw definite conclusions regarding the implication of the observed asymmetry in foot position.

Animals do not run with a pattern of motion that is precisely repeatable from one stride to the next, even for a single gait. This variability has been reported in studies of interlimb coordination in the cat: for example see (Stuart et al. 1973; Miller, van der Burg, and van der Meché 1975; English 1979; and Vilensky and Patrick 1984). In principle, variability need not influence symmetry, and the two may be orthogonal. A legged system can switch from one symmetric pattern of motion on one stride to a different symmetric pattern on the next stride. A legged system can maintain

Fig. 16. Graphical interpretation of symmetry in the galloping cat. (left) *Photographs of a galloping cat taken at 50-ms intervals.* (right) *The shaded figures show the forward translation and the configuration of the cat during normal running, and the outlines show reverse running. The outlines were made from the same photographs as the shaded figures, but were reflected about the vertical axis and are presented in reverse sequential order,* x(t) = −x(−t). *Therefore the outline at the top was made from the photograph at the bottom after reversing its orientation. The positions of supporting feet and the rightward motion of the body correspond quite well in the two sequences, as predicted by symmetry.* (diagram construction) *The relative placement of the figures for each sequence—the photographs, the shaded silhouettes, and the outlines—accurately reflects the forward progress of the cat with respect to the surface of the treadmill. After each set of figures was assembled according to the forward travel, the three sets were positioned relative to one another. Photographs are from film provided by Wetzel, Atwater, and Stuart (1976).*

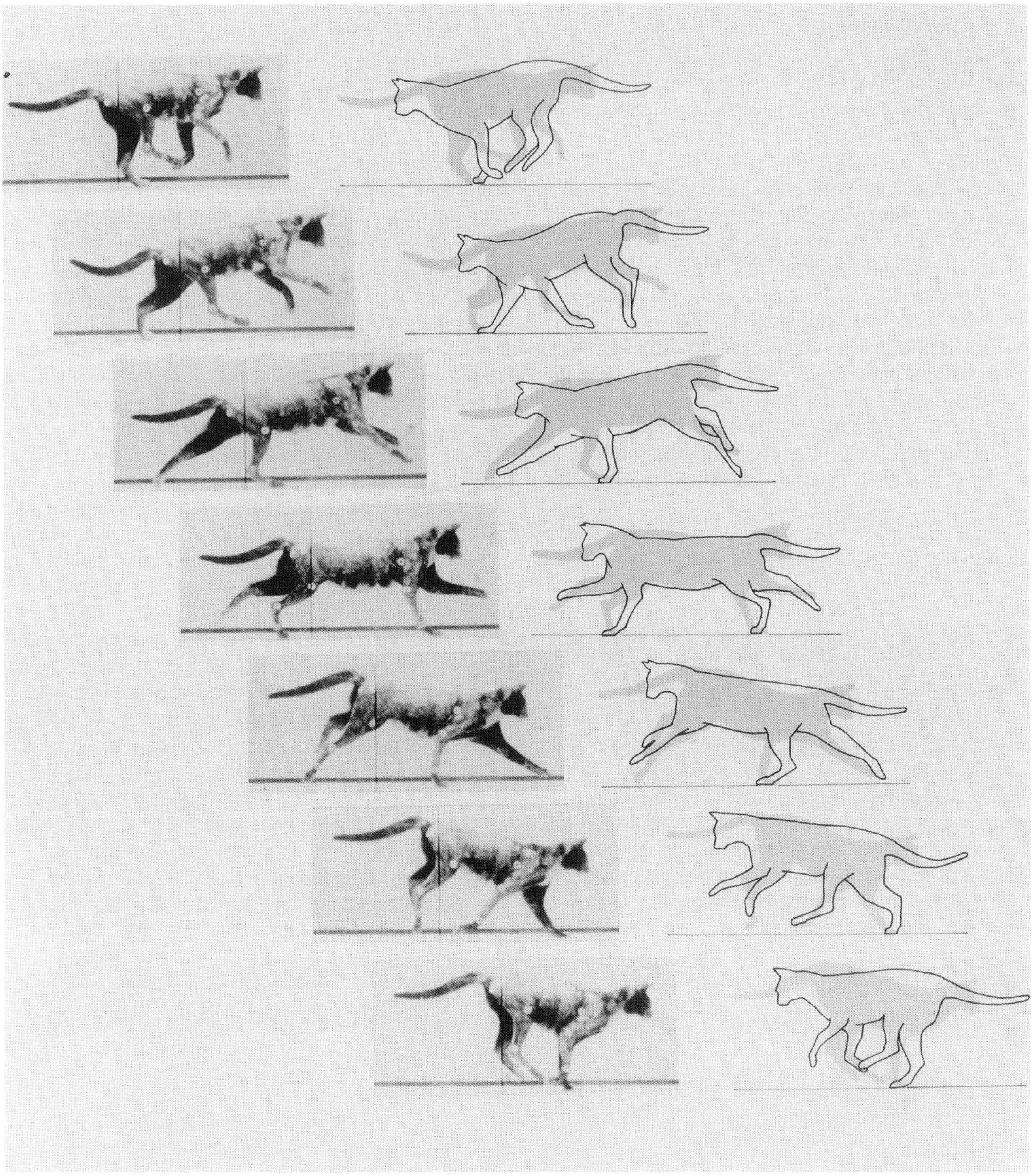

symmetry despite variations, because the symmetry equations describe a *class* of body and leg motions rather than a particular motion. On the other hand, variations may have asymmetric components. Such asymmetric components are expected when a system accelerates as described earlier, but asymmetric steady-state variation is also possible. The variability reported in the literature has not been analyzed to reveal the relative contributions of symmetric and asymmetric components.

5. Scissor Symmetry

When a human runs, the two legs form roughly symmetric angles with respect to a vertical axis passing through the hip. The angle formed between the hip and foot of the forward leg and the vertical axis is about equal and opposite to the corresponding angle for the rearward leg. This symmetry is largely independent of the speed, bounce, stride, and other parameters of the gait. This behavior reminds one of the way one orients the blades of a scissors to the paper they cut.

A consequence of scissor symmetry is that the angle of the leg to be placed is about equal to the angle of the leg that was just lifted. Can this symmetry be used to formulate an algorithm that correctly places the foot on each step, eliminating the need for a calculation that depends on forward running speed and the duration of support? A scissor algorithm would specify that

$$\theta(t_{td,i+1}) = -\theta(t_{lo,i}), \tag{16}$$

where

$\theta(t_{lo,i})$ is the angle of the lift-off leg on step i, and
$\theta(t_{td,i+1})$ is the angle of the landing leg on step $(i+1)$.

The scissor algorithm of Eq. (16) could be used to specify foot placement for systems with any number of legs, provided that the gait used only one leg for support at a time. What sort of behavior would result?

When running with constant forward speed $\dot{x}$ and uniform stance duration T_s, the foot moves a distance $\dot{x}T_s$ backward with respect to the hip during the support interval. For a given landing angle of the leg $\theta(t_{td,i})$, the lift-off angle is

$$\theta(t_{lo,i}) = \arcsin\left(\frac{\dot{x}T_s + r\sin\theta_{td,i}}{r}\right). \tag{17}$$

Combining Eqs. (16) and (17), we obtain

$$\theta(t_{td,i+2}) = \arcsin \left(\frac{\dot{x}T_s + r\sin\left(-\arcsin\left(\frac{\dot{x}T_s + r\sin\theta(t_{td,i})}{r}\right)\right)}{r}\right) = \theta(t_{td,i}). \tag{18}$$

During running at constant speed, the algorithm generates pairs of steps that have symmetry, like those discussed earlier. A pattern of paired antisymmetric steps gives balance, provided that the degree of asymmetry is relatively small and the step rate is large. When $\theta(t_{lo,i}) = \arcsin(\dot{x}T_s/2)$, the scissor algorithm generates the same foot placement on every step, and the placements are the same as those produced by using the CG-print calculation for the neutral point.

The scissor algorithm can also work properly during forward accelerations. Suppose that during the support interval an external disturbance accelerates the system forward. The result is that the stance leg sweeps farther back and the lift-off angle of the stance leg is larger than it would have been without the disturbance. The other leg is placed correspondingly further forward, compensating for the increased velocity. A decelerating disturbance works in a corresponding manner. The acceleration need not be due to an external disturbance but could be caused by actions of the hip actuator that are intended to stabilize the body attitude. They might be caused by the driving or swinging actions of other legs in a more complicated system.

One way to look at the scissor algorithm is that it provides an alternative method for estimating the length of the CG-print. The lift-off angle of the leg serves to indicate both the forward velocity and ground time. The faster the body moves forward relative to the ground, the further backward the foot moves during stance. The foot also moves backward further when the system spends more time on the ground. Therefore the angle of the leg at lift-off is determined by the product of the average forward velocity and the duration of stance. The scissor algorithm is attractive because it is difficult to estimate the length of the CG-print accurately. It avoids the need to measure explicitly the forward velocity of the body and the duration of stance.

There are several difficulties with the scissor algorithm. First, the leg angle at touchdown is also influenced by the leg angle at lift-off. The product of average velocity and ground time relates only the change in leg angle, so the starting angle of the leg at touchdown determines where it is at lift-off. In principle, the algorithm can generate a sequence of uniform symmetric steps. In practice, there is no mechanism to keep from drifting to antisymmetric pairs of skewed steps with diverging skew.

This problem might be overcome by somehow damping the foot placement excursions or by using information from previous steps to filter the two-step oscillations. Another alternative might be to take both the touchdown and lift-off angles into account when calculating the next foot placement:

$$\theta(t_{td,i+1}) = \frac{\theta(t_{td,i}) - \theta(t_{lo,i})}{2}. \qquad (19)$$

Another problem with the scissor algorithm is that it may not be responsive to sudden changes in the forward speed of the body. The forward speed that determines the next foot placement is the average from the entire previous support interval. The latency inherent in this indirect measurement could result in sluggish response to disturbances.

5.1. Asymmetry in Running

Despite the value of symmetry, there are several reasons why one should not expect to see perfect symmetry in the behavior of legged machines and animals. One reason for asymmetry is that legs are not lossless. The arguments used to motivate the relationship between symmetric motion and steady-state behavior do not apply in the presence of friction. In particular, the behavior of the system moving forward in time is no longer symmetric to its behavior moving backward in time. The details of the discrepancy depend on the details of the losses and on the geometry of the system. Another energy loss contributing to asymmetric motion is due to unsprung mass in the legs. Each time a foot strikes or leaves the ground, the system loses a fraction of its kinetic energy. In order to maintain stable locomotion, the control system must resupply energy on each cycle to compensate for these losses. For instance, the leg lengthens during the support interval and shortens during flight to maintain a stable hopping height. This can be done only by delivering asymmetric forces and torques through the actuators.

Another reason for asymmetric behavior is asymmetry in the mechanical system. Most animals have large heavy heads at one end of their bodies that are not counterbalanced by large heavy tails at the other end. Front and rear legs often vary in size, and the hips and shoulders may not be equally spaced about the center of mass. Each of these factors may induce asymmetry in the motions that can provide balanced, steady-state behavior. This is less of a problem for laboratory machines because they can be designed to conform to whatever mechanical symmetry is required.

Naturally, we shouldn't expect to see symmetric motion when the control system purposely skews the motion to change running speed. In this case, asymmetry in the motion provides the forces that accelerate the body. An external load, such as that produced by wind resistance or a draw-bar load, would also require a component of asymmetry in the motion of the body and legs. A runner at the start of a footrace and the driver of a jinrikisha demonstrate these sorts of asymmetric behavior.

Perhaps a better view is to think of locomotion in terms of the sum of a symmetric part and an asymmetric part. The symmetric part of the motion during each stride maintains steady-state behavior. Deviations from symmetry compensate for losses and provide acceleration.

The symmetry discussed in this paper postulates that each body variable, each leg variable, and each actuator variable has an even or odd symmetry. The net result of their interaction is to constrain the forces acting on the body throughout a stride so that they preserve the body's forward speed, elevation, and pitch angle. One might imagine a less complete symmetry that does not require symmetry of the basic variables individually but requires symmetry only in the net

Fig. 17. Two functions that integrate to zero. One is symmetric and one is not. Symmetry provides a sufficient condition but not a necessary condition for zero net forward acceleration.

forces and torques acting on the body:

$$\begin{aligned} f_x(t) &= -f_x(-t), \\ f_y(t) &= f_y(-t), \\ \tau_\phi(t) &= -\tau_\phi(-t). \end{aligned} \tag{20}$$

Stated alternatively, the body moves with symmetry while the legs do not. We have proved that this cannot be the case when only one leg is used for support at a time. The proof is given in Appendix B. However such solutions may be workable with additional legs.

6. What Does Symmetry Mean?

We can interpret symmetry in several ways. First, it is useful in the control of legged machines. The strategy used to control running machines was built around symmetry, and symmetry may play a role in achieving more complicated running behavior in the future. For instance, reciprocating leg symmetry is important in making a quadruped gallop.

Symmetry also helps us to characterize and understand the behavior we observe in animals. The analysis of symmetry in the cat and human shows that it describes how animals move when they trot, gallop, and run, and we expect to find that the same symmetries describe the motions of other animals running with other gaits. Perhaps most important is the idea that symmetry and balance give us tools for dealing with a dynamic system without requiring detailed solutions to complex formulations. Symmetry implies that each motion has two parts with opposing effects, just as balance requires equal and compensating forces and torques.

In certain respects, these symmetries are limited. They do not specify the details of a particular body motion that provides locomotion but merely give a broad classification that embodies several interesting features of the motion. The symmetries provide only sufficient conditions for successful locomotion, not necessary conditions (see Fig. 17). As far as we have been able to determine, the behavior of a legged system may violate the motion symmetries we have described with impunity, without limiting its ability to run and balance. Finally, these symmetries do not yield a specific prescription for control. They suggest only how the system should ultimately move and hint at possible avenues of attack.

In other respects, the symmetries described here are quite powerful. Three simple equations outline plausible body motions for systems with any number of legs engaged in a wide variety of gaits. Another small set of equations describes how the legs move. Although the symmetries do not specify individual motions or how to produce them, they provide rules that govern a large class of successful motions and suggest a wide variety of experiments.

This work on symmetry falls into a broader context that splits responsibility for control between the control system and the mechanical system being controlled. In this context, the control of locomotion is a low bandwidth activity that takes advantage of the intrinsic properties of the mechanical system. Rather than use a high bandwidth servo to move each joint of the legged system along a prescribed trajectory at high rate, the control system makes adjustments just once per stride. Once the foot has been positioned on each step, the mechanical system passively determines the details of the motion for the remainder of the stride. This approach depends on having a passive nominal motion that is close to the desired behavior. In the present context, symmetry is the means of achieving the nominal motion. This sort of approach may have value only for systems that perform repetitive behaviors. For instance, aside from juggling and handwriting (Hollerbach 1980), robot manipulation may be unsuited to this approach.

7. Summary

Symmetric motions of the body in space and of the feet with respect to the body provide nominal motions for steady-state locomotion. A control system for run-

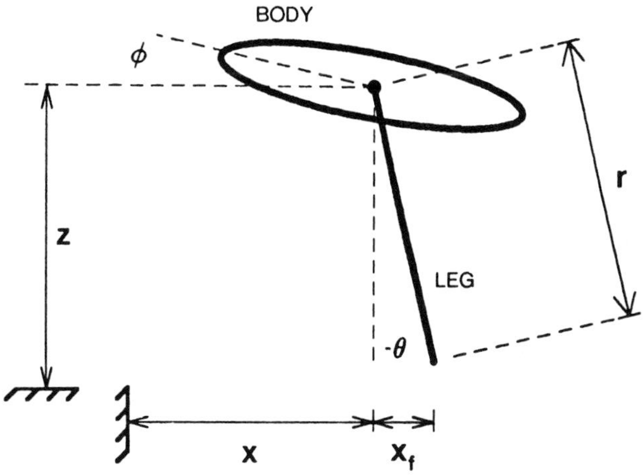

Fig. A1. *Model of planar, one-legged system with a massless leg.*

ning can produce steady-state behavior by choosing motions of the legs that give $x(t)$ and $\phi(t)$ odd symmetry and $z(t)$ even symmetry. The leg motions chosen are themselves described by odd and even symmetries. This method applies to a number of legged configurations and helps to describe the behavior of running animals.

The significance of these symmetric motions is that they permit a control system to manipulate the symmetry and skewness of the motion, rather than the detailed shape of the motion. When the system's behavior conforms to Eqs (1–3), all forces acting on the body integrate to zero throughout one stride, so the body experiences no net acceleration. When behavior deviates from symmetry, the net acceleration of the system deviates from zero in a manageable way. The control task becomes one of manipulating these deviations.

The conditions for symmetric body motion can be stated simply: at a single point in time during the support period, the center of support must be located under the center of mass, the pitch angle of the body must be zero, and the vertical velocity of the body must be zero, i.e., $\theta_j(0) + \theta_k(0) = 0$, $\phi(0) = 0$, and $\dot{z}(0) = 0$. The body follows a symmetric trajectory during stance when these conditions are satisfied.

Symmetric running motions may have great generality. In principle, a wide variety of natural running gaits can be achieved using body and leg motions that exhibit the symmetries described. These include the trot, the pace, the canter, the gallop, the bound, and the pronk, as well as the intermediate forms of these gaits. Although we have plotted symmetry data only for the cat and the human, we expect to find a wide variety of natural legged systems using nearly symmetric motions when they run.

Acknowledgments

I am grateful to Mary Wetzel for providing me with the cat film, to Jessica Hodgins and Jeff Koechling for helping film the human runners, and to Jeff Koechling for writing the computer programs used to digitize and process data.

Appendix A. Equations of Motion for Planar Systems

A.1. Equations of Motion for a Planar, One-Legged System

The equations of motion for a planar one-legged model, with a massless leg and the hip located at the center of mass (as shown in Fig. A1), are:

$$m\ddot{x} = f \sin\theta - \frac{\tau}{r}\cos\theta, \tag{A1}$$

$$m\ddot{z} = f \cos\theta + \frac{\tau}{r}\sin\theta - mg, \tag{A2}$$

$$J\ddot{\phi} = \tau, \tag{A3}$$

where

x, z, ϕ	are the horizontal, vertical, and angular positions of the body,
r, θ	are the length and orientation of the leg,
τ	is the hip torque (positive τ accelerates body in the positive ϕ direction),
f	is the axial leg force (positive f accelerates the body away from the ground),
m	is the body mass,
J	is the body moment of inertia, and
g	is the acceleration of gravity.

Fig. A2. Model of planar, two-legged system with separated hips. It can represent the lateral half of a quadruped projected onto the sagittal plane or a biped projected onto the frontal plane.

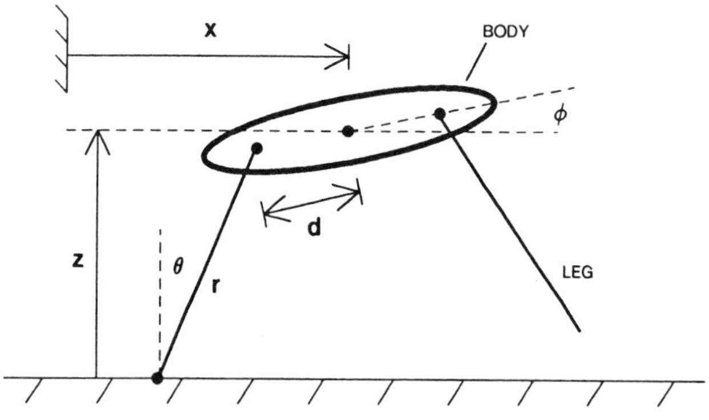

A.2. Equations of Motion for a Planar, Two-Legged System

The equations of motion for planar model with two massless legs and hips located a distance d from the body's center of mass (as shown in Fig. A2) are:

$$m\ddot{x} = f_1 \sin\theta_1 + f_2 \sin\theta_2 - \frac{\tau_1}{r_1}\cos\theta_1 - \frac{\tau_2}{r_2}\cos\theta_2, \quad \text{(A4)}$$

$$m\ddot{z} = f_1 \cos\theta_1 + f_2 \cos\theta_2 + \frac{\tau_1}{r_1}\sin\theta_1 + \frac{\tau_2}{r_2}\sin\theta_2 - mg, \quad \text{(A5)}$$

$$J\ddot{\phi} = f_1 d\cos(\theta_1 - \phi) - \frac{\tau_1 d}{r_1}\sin(\phi - \theta_1) + \tau_1 - f_2 d\cos(\theta_2 - \phi) + \frac{\tau_2 d}{r_2}\sin(\phi - \theta_2) + \tau_2. \quad \text{(A6)}$$

Appendix B. Proof of Symmetric Leg Motion

In this appendix we prove that symmetric body motion requires symmetric leg motion for the one-legged case. Body and leg motions are symmetric when $x(t)$, $\phi(t)$, and $\theta(t)$ are odd and $z(t)$ and $f(t)$ are even.

The equations of motion can be rewritten expressing each element of the leg motion as the sum of an even and odd part. For instance, the angle of the leg with respect to the vertical is $\theta = {}^e\theta + {}^o\theta$, where ${}^e\theta$ represents the even part and ${}^o\theta$ represents the odd part. Also, r can be replaced with $1/({}^ez + {}^oz)$:

$$m\ddot{x} = ({}^ef + {}^of)\sin({}^e\theta + {}^o\theta) - \tau({}^ez + {}^oz)\cos({}^e\theta + {}^o\theta), \quad \text{(B1)}$$

$$m\ddot{z} = ({}^ef + {}^of)\cos({}^e\theta + {}^o\theta) + \tau({}^ez + {}^oz)\sin({}^e\theta + {}^o\theta) - mg, \quad \text{(B2)}$$

$$J\ddot{\phi} = {}^o\tau + {}^e\tau. \quad \text{(B3)}$$

From Eq. (B3), τ must be odd, and because ϕ is odd, we assume that $\ddot{\phi}$ is odd. To specify that the body moves with the desired symmetry, the even part of the right-hand side of Eq. (B1) is set to zero, and the odd part of the right-hand side of Eq. (B2) is also set to zero:

$$0 = {}^ef\sin{}^e\theta\cos{}^o\theta + {}^of\cos{}^e\theta\sin{}^o\theta + \tau\,{}^ez\sin{}^e\theta\sin{}^o\theta - \tau\,{}^oz\cos{}^e\theta\cos{}^o\theta. \quad \text{(B4)}$$

$$0 = -{}^ef\sin{}^e\theta\sin{}^o\theta + {}^of\cos{}^e\theta\cos{}^o\theta, + \tau\,{}^ez\sin{}^e\theta\cos{}^o\theta + \tau\,{}^oz\cos{}^e\theta\sin{}^o\theta. \quad \text{(B5)}$$

Solutions to Eqs. (B4) and (B5) require that

$$\tan{}^e\theta = \frac{\tau\,{}^oz}{{}^ef} \quad \text{and} \quad \tan{}^e\theta = -\frac{{}^of}{\tau\,{}^ez}. \quad \text{(B6)}$$

During the support interval, the foot remains stationary with respect to the ground, so motion of the body with respect to the ground determines motion of the foot with respect to the body. Therefore the symmetries of Eq. (1) and the solutions to Eq. (B6) also govern the trajectory of the foot with respect to the body. They require that $x_f(t) - x_f(0) = -x_f(-t) + x_f(0)$ and $z_f(t) = z_f(-t)$. The leg motion is symmetric if $x_f(0) = 0$.

Because odd functions equal zero when $t = 0$, Eq. (B6) requires that ${}^e\theta(t = 0) = 0$, implying that $x_f(0) = 0$. Hence ${}^e\theta = {}^or = {}^of = 0$, leaving θ odd and r and f even. They obey the leg symmetries given by Eq. (2).

Goldberg (1985) simplified the proof as follows. The foot remains stationary with respect to the ground during the support interval, so motion of the body with respect to the ground determines the motion of the foot with respect to the body. Therefore the symmetries of Eq. (2) govern the trajectory of the foot with respect to the body:

$$x_f(t) - x_f(0) = -x_f(-t) + x_f(0), \tag{B7}$$

$$z_f(t) = z_f(-t). \tag{B8}$$

The leg motion is symmetric if $x_f(0) = 0$.

Let f_x and f_z be the horizontal and vertical forces between the foot and the ground. The torque at the hip can be written as

$$\tau = -f_x z_f + f_z x_f. \tag{B9}$$

From the equations of motion, we know that τ and f_x are odd and that f_z is even, so Eq. (B9) requires that x_f be odd. Therefore leg angle $\theta = \arctan(x_f/z_f)$ is odd and leg length $r = \sqrt{x_f^2 + z_f^2}$ is even. Axial leg force $f = (f_x x_f + f_z z_f)/r$, which is even.

References

English, A. W. 1979. Interlimb coordination during stepping in the cat: an electromyographic analysis. *J. Neurophysiol.* 42:229–243.

Goldberg, K., 1985. Private communication.

Hildebrand, M. 1965. Symmetrical gaits of horses. *Science* 150:701–708.

Hildebrand, M. 1966. Analysis of the symmetrical gaits of tetrapods. *Folia Biotheoretica* 4:9–22.

Hildebrand, M. 1968. Symmetrical gaits of dogs in relation to body build. *J. Morphology* 124:353–359.

Hildebrand, M. 1976. Analysis of tetrapod gaits: general considerations and symmetrical gaits. In *Neural control of locomotion,* eds. R. N. Herman et al. New York: Plenum Press, pp. 203–236.

Hodgins, J., Koechling, J., and Raibert, M. H. 1986. Running experiments with a planar biped. *3rd Int. Symp. Robotics Res.* Cambridge: MIT Press.

Hollerbach, J. M., 1980. An oscillation theory of handwriting. Ph.D. thesis. Cambridge: Massachusetts Institute of Technology. Artificial Intelligence Laboratory.

Miller, S., van der Burg, J., and van der Meché, F. G. A. 1975. Coordination of movements of the hindlimbs and forelimbs in different forms of locomotion in normal and decerebrate cats. *Brain Research* 91:217–237.

Raibert, M. H. 1986*a*. Symmetry in running. *Science* 231:1292–1294.

Raibert, M. H. 1986*b*. *Legged robots that balance.* Cambridge: MIT Press.

Raibert, M. H., and Brown, H. B., Jr. 1984. Experiments in balance with a 2D one-legged hopping machine. *ASME J. Dyn. Sys. Measurement, Cont.* 106:75–81.

Raibert, M. H., Brown, H. B., Jr., and Chepponis, M. 1984. Experiments in balance with a 3D one-legged hopping machine. *Int. J. Robotics Res.* 3:75–92.

Raibert, M. H., Chepponis, M., and Brown, H. B. Jr. 1986. Running on four legs as though they were one. *IEEE J. Robotics and Automation* 2:70–82.

Stentz, A. 1983. Behavior during stance. In *Dynamically stable legged locomotion— Third Annual Report,* eds. M. H. Raibert et al. Pittsburgh: Robotics Institute, Carnegie-Mellon University CMU-R1-TR-83-20, pp. 106–110.

Stuart, D. G., Withey, T. P., Wetzel, M. C., and Goslow, G. E., Jr. 1973. Time constraints for inter-limb coordination in the cat during unrestrained locomotion. In *Control of posture and locomotion,* eds. R. B. Stein et al. New York: Plenum Press, 537–560.

Vilensky, J. A., and Patrick, M. C. 1984. Inter- and intratrial variation in cat locomotor behavior. *Physiology & Behavior* 33:733–743.

Wetzel, M. C., Atwater, A. E., and Stuart, D. G. 1976. Movements of the hindlimb during locomotion of the cat. In *Neural Control of Locomotion,* eds. R. N. Herman et al. New York: Plenum Press, 99–136.

Smooth Local Path Planning for Autonomous Vehicles

Yutaka Kanayama and Bruce I. Hartman

Community and Organization Research Institute
Department of Computer Science
University of California
Santa Barbara, CA 93106

Abstract

Two cost functions of paths for smoothness are defined; Path curvature and the derivative of path curvature. Through these definitions, two classes of simple paths are obtained; the set of circular arcs and the set of cubic spirals. A cubic spiral is a curve whose tangent direction is described by a cubic function of path distance s. These sets of simple paths are used for solving path planning problems of symmetric posture (position and orientation) pairs. For a non-symmetric posture pair, we use two simple paths as a solution. In order to find those paths, we use the fact that the locus of split postures is a circle or a straight line. A posture q is said to be a split posture of a pair (p_1, p_2) of postures, if p_1 and q are symmetric and so are q and p_2. The resultant solutions are smoother than those obtained by one of the authors using clothoid curves. This algorithm has been successfully implemented on the autonomous mobile robot Yamabico-11 at UCSB.

1. Introduction

We proposed a robot's path description method in which a path is described by a sequence of postures (a posture is a position with an orientation) instead of a sequence of curve segments [1]. For instance, a path shown in Figure 1 is described by a sequence $(p_1, p_2, \cdots, p_8)$ of eight postures. In this method, we are to find each path segment which is determined solely by two endpostures. Thus, in Figure 1, we are to solve a local path-finding problem for pairs of postures (p_1, p_2), ... , (p_7, p_8) independently. This paper describes a method of finding the "smoothest" local path joining a given pair of postures.

Smoothness of paths is essential for mobile robot navigation, because unsmooth motions may cause slippage of wheels which degrades the robot's deadreckoning ability. In order to control "smoothness" of paths, we propose to define the cost of a path for smoothness. A unit cost for smoothness at a point on a path is proposed as (1) square of its curvature, or (2) square of the derivative of its curvature in this paper. The total cost of a path is taken as the integration of either of these incremental costs. Using calculus of variations, we obtained circles, or a new class of curves which we call *cubic spirals* for solutions. The class of cubic spirals has an advantage that the curvature is continuous at each specified posture, because the curvature is null there. The curvature continuity is essential for vehicle path design.

When navigating itself, an autonomous vehicle has only two degrees of freedom; velocity and curvature. Therefore, if we focus only on the static features of robot's paths, a path can be naturally described by its curvature $\kappa(s)$ of the traveling distance s. (The traveling distance is the robot's only coordinate.) This is the reason why we do not adopt spline curves in formalizing this smooth path planning theory. It is not easy for the robot to control (x, y) in a Cartesian coordinate system.

Komoriya and Hongo independently adopted a description method using a sequence of straight lines and circular arcs [2][3] and so do many other researchers. This paper gives a mathematical interpretation to their work; if we take curvature itself as the cost of a path, the solutions are circles and straight lines. However, one major problem in the method is that the curvature of a generated path is not necessarily continuous and thus, strictly speaking, a vehicle is not able to follow them smoothly [4]. The use of clothoid curves, or Cornu spirals, was proposed and implemented on Stanford "mobi" [5]. This class of curves has linear curvature change and has been used in highway and railway curve design. The paths have curvature continuity. Our results show that cubic spirals are better in terms of the maximum curvature.

This algorithm has been successfully implemented on an autonomous mobile robot Yamabico-11 which has been developed at the University of California at Santa Barbara. This local path planning function is expressed in the form of elementary *move* and *stop* functions integrated into a higher-

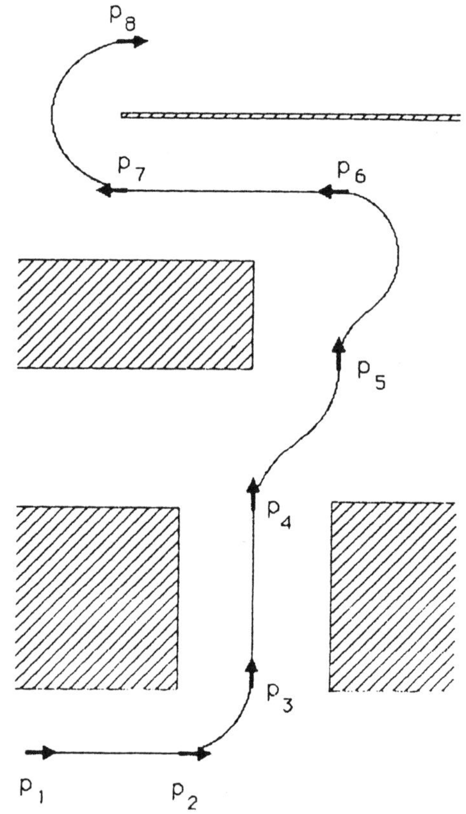

Fig. 1 Path Specification by Postures

level mobile robot language, MML, a dialect of the C language [6][7][8][9]. For complete proofs for some Propositions, refer to [10].

2. Overview of the Method

We define a 2D global Cartesian coordinate system in a world for a mobile robot. Let us call a combination of position and orientation (x, y, θ) a *posture*. The direction θ is taken counter-clockwise from the X-axis. Any vehicle possesses this three degrees of freedom in its positioning. If $p = (x, y, \theta)$ is a posture, "point p" stands for the point (x, y). Let $p_1 = (x_1, y_1, \theta_1)$ and $p_2 = (x_2, y_2, \theta_2)$ be two postures with $(x_1, y_1) \neq (x_2, y_2)$. The orientation β from the point (x_1, y_1) to (x_2, y_2) is (Fig. 2),

$$\beta = \tan^{-1}\left(\frac{y_2 - y_1}{x_2 - x_1}\right) \tag{1}$$

The pair of two postures p_1 and p_2 are said to be *symmetric* if and only if the the following relation is satisfied:

$$\theta_1 - \beta = -(\theta_2 - \beta) \tag{2}$$

We write $sym(p_1, p_2)$ if p_1 and p_2 are symmetric. For instance, in Figure 1, $sym(p_1, p_2)$, $sym(p_2, p_3)$, $sym(p_3, p_4)$, $sym(p_6, p_7)$, and $sym(p_7, p_8)$. But ˉ$sym(p_4, p_5)$ and ˉ$sym(p_5, p_6)$, where a ˉ symbol means negation. If $sym(p_1, p_2)$ and there is an intersection p_0 of the "forward ray" of p_1 and the "backward ray" of p_2 (or if there is an intersection p_0 of the "backward ray" of p_1 and "forward ray" of p_2), the triangle $p_0p_1p_2$ is isosceles (Fig. 2). However, in some cases, there is no such an intersection associated with a symmetric posture pair, for instance, (p_1, p_2) and (p_7, p_8) in Figure 1.

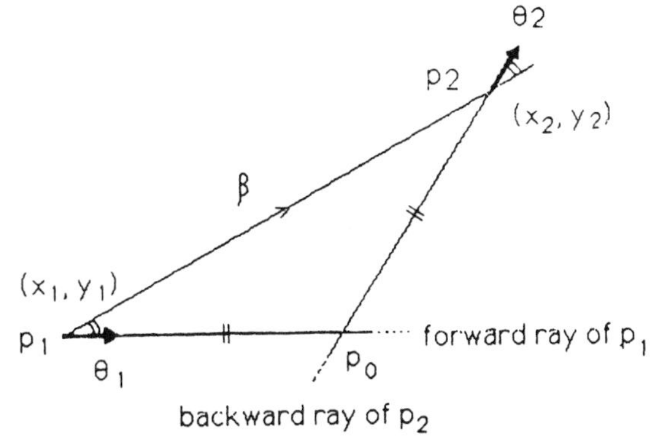

Fig. 2 Symmetric Postures

A *size* d and an *deflection* α of a pair (p_1, p_2) of a symmetric posture pair are defined as the Euclidean distance of the two points and the angle between the directions of its postures respectively:

$$d = ((x_2 - x_1)^2 + (y_2 - y_1)^2)^{1/2} \tag{3}$$

$$\alpha = \theta_2 - \theta_1 \tag{4}$$

A posture q is said to be a *split posture* of a pair of postures (p_1, p_2) if $sym(p_1, q)$ and $sym(q, p_2)$.

We assume that the tangent direction of a path Π (with a finite length) is defined everywhere on the path. For each point (x, y) on a path Π, we associate a posture (x, y, θ), where θ is a tangent direction of Π at the point. As a special case, for an endpoint of the path, we associate an endposture. A path (with a finite length) is said to be *simple* if its endpostures are symmetric. For instance, every circular arc is simple (cf. Proposition 1).

In our method of finding solutions, the following preparation steps are required: Define a cost function for smoothness of paths. Thus, for each path Π, its cost $cost(\Pi)$ is explicitly determined. Through this definition of a cost, we define a set S of simple paths, in which there is one and only one simple path with a given size d and a deflection α. After this preparation is done, we are ready to describe the local path planning method when being given two postures $p_1 = (x_1, y_1, \theta_1)$ and $p_2 = (x_2, y_2, \theta_2)$, where $(x_1, y_1) \neq (x_2, y_2)$.

The Algorithm

(Case I) If $sym(p_1, p_2)$, find a path in S with the same size and deflection of (p_1, p_2) and make appropriate translation and rotation to fit the pair of end-postures. This resultant simple path is expressed as $\Pi(p_1, p_2)$. (Notice that the size and deflection completely characterize a simple path.)

(Case II) If ˉ$sym(p_1, p_2)$, first, find the locus of split postures of (p_1, p_2). Next we find a split posture q which makes the sum of costs, $\mathrm{cost}(\Pi(p_1, q)) + \mathrm{cost}(\Pi(q, p_2))$, minimum.

3. Path Representation

Let us define the representation of closed paths with finite length. A path Π is a pair (l, κ) where l is a positive *length*, and κ is a continuous curvature function: $[-l/2, l/2] \rightarrow R$ of Π, where R is the set of real numbers. We stipulate that $x(0) = y(0) = \theta(0) = 0$. That is, we are adopting a standard positioning convention of paths. The tangent direction function θ is given by,

$$\theta(s) = \int_0^s \kappa(t)dt \quad -l/2 \le s \le l/2 \tag{5}$$

Therefore, a point $(x, y) = (x(s), y(s))$ on Π is given by,

$$\begin{cases} x = x(s) = \int_0^s \cos\theta(t)\,dt \\ y = y(s) = \int_0^s \sin\theta(t)\,dt \end{cases} \tag{6}$$

The *endpostures* of the path Π is $(x(-l/2), y(-l/2), \theta(-l/2))$ and $(x(l/2), y(l/2), \theta(l/2))$. The following Proposition relates the "symmetric" property of the curvature of a path, and the symmetric property of endpostures of the path.

Proposition 1. If a curvature function κ is "symmetric", namely, if $\kappa(-s) = \kappa(s)$ holds, so do the following relations,

$$\begin{cases} \theta(s) = -\theta(s) \\ x(s) = -x(-s) \\ y(s) = y(-s) \end{cases} \tag{7}$$

for all s. Therefore, in a special case of $s = l/2$, its endpostures are symmetric.

4. Cost Functions and Complete Set of Simple Paths

Let us define two cost functions of paths for smoothness:

$$\mathrm{cost}_1(\Pi) = \int_{-l/2}^{l/2} F_1 ds = \int_{-l/2}^{l/2} \kappa^2 ds = \int_{-l/2}^{l/2} (\dot{\theta})^2 ds \tag{8}$$

$$\mathrm{cost}_2(\Pi) = \int_{-l/2}^{l/2} F_2 ds = \int_{-l/2}^{l/2} (\dot{\kappa})^2 ds = \int_{-l/2}^{l/2} (\ddot{\theta})^2 ds \tag{9}$$

Although we do not consider any dynamic properties of robot motion control, we can interpret these definitions from the viewpoint of dynamics. Assuming constant-velocity navigation of a vehicle along the path Π, the instantaneous centripetal acceleration of the vehicle is proportional to its curvature $\kappa(s)$. Notice that the radius of the osculating circle at $(x(s), y(s))$ is $r(s) = 1/\kappa(s)$. Therefore, in the first definition of Equation (8), we are taking the integration of (square of) acceleration as its path cost. In the definition of $cost_2$ of Equation (9), we take the integration of square of the *variation of acceleration* (or *jerk*) as a path cost. Since jerk is considered to be minimized for comfortable vehicle control, it would be rather reasonable to adopt this definition. Let us characterize classes of paths which make the path costs of Equations (8) and (9) minimum:

Proposition 2. For a fixed length l, the following path minimizes $\text{cost}_1(\Pi)$ of Equation (8):

$$\begin{cases} \dot{\theta}(s) = A \equiv \kappa(s) \\ \theta(s) = As + B \end{cases} \tag{10}$$

where A and B are integral constants.

Proof. $F_1 = (\dot{\theta})^2$ by Equation (8). By a textbook on calculus of variations [11], we have

$$\frac{\partial F_1}{\partial \theta} - \frac{d}{ds}\left(\frac{\partial F_1}{\partial \dot{\theta}}\right) = 0$$

Therefore,

$$-\frac{d}{ds}(2\dot{\theta}) = -2\frac{d^2\theta}{ds^2} = 0$$

Equation (10) is obtained, by simply integrating this two times. □

Paths given by Equation (10) are merely circular arcs or straight lines. Since the curvature function is constant and is "symmetric", each circular arc is simple by Proposition 1. In this case, the set S of simple paths are the set of all circular arcs with various sizes and deflections.

Proposition 3. For a fixed length l, the following path minimizes $\text{cost}_2(\Pi)$ of Equation (9):

$$\begin{cases} \dot{\theta}(s) = \frac{1}{2}As^2 + Bs + C \equiv \kappa(s) \\ \theta(s) = \frac{1}{6}As^3 + \frac{1}{2}Bs^2 + Cs + D \end{cases} \tag{11}$$

where A, B, C, and D are integral constants.

Proof. $F_2 = \ddot{\theta}^2$ by Equation (9). Similarly, we have

$$\frac{\partial F_2}{\partial \theta} - \frac{d}{ds}\left(\frac{\partial F_2}{\partial \dot{\theta}}\right) + \frac{d^2}{ds^2}\left(\frac{\partial F_2}{\partial \ddot{\theta}}\right) = 0$$

Therefore,

$$2\frac{d^2}{ds^2}\ddot{\theta} = 2\frac{d^4\theta}{ds^4} = 0$$

Equation (11) is obtained, by integrating this four times. □

These curves are more complex than circles. Let us call this class of curves *cubic spirals*, since the direction function θ is a cubic function. Figure 3 shows an example of curvature and tangent direction functions of a cubic spiral. Figure 4 shows an example of a whole cubic spiral. We will use only the part of a cubic spiral $[s_1, s_2]$ in Figure 3, between the two points where curvature is null. Since the curvature of this part is clearly "symmetric", the path is also simple. Hereafter, a *cubic spiral* refers to that finite simple path out of a whole spiral. Figure 5 shows a class of cubic spirals with unit length. These are said to be *standard* cubic spirals.

Once, the use of clothoid curves was proposed by the authors [5]. (Precisely speaking, *clothoid pairs* were used.) This class of curves also possesses the advantage of having curvature continuity (See Conclusion).

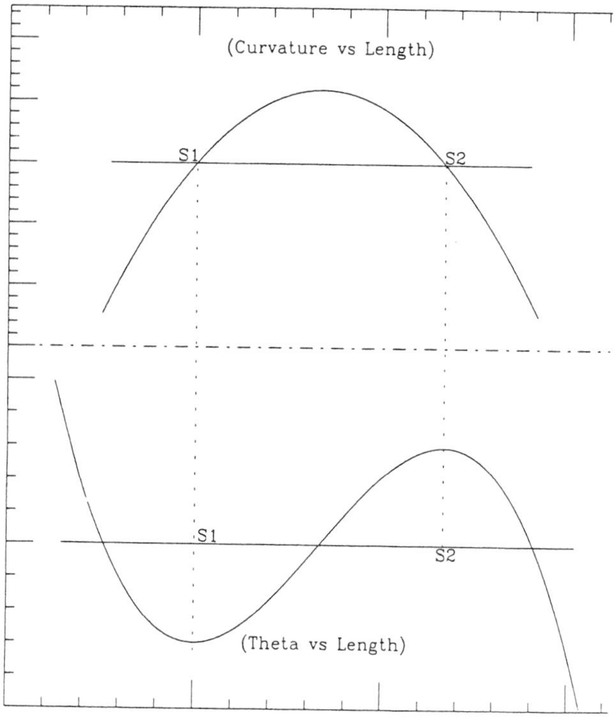

Fig. 3 Curvature and Tangent Direction of Cubic Spiral

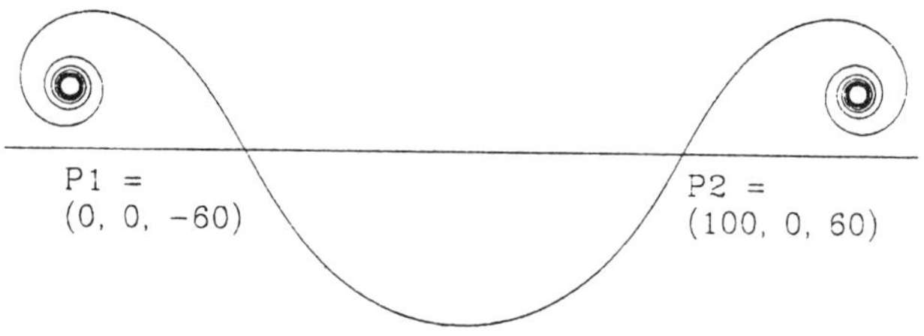

Fig. 4 A Whole Cubic Spiral

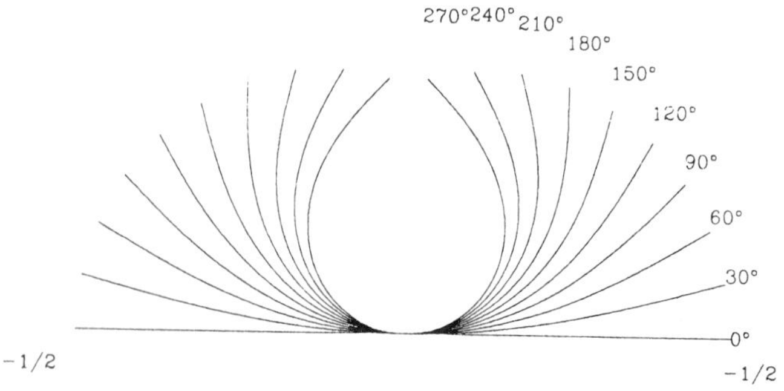

Fig. 5 Standard Cubic Spirals

5. Solving Symmetric Cases

In our smooth path planning problem, finding a simple path for a given symmetric posture pair (p_1, p_2) is the most elementary part (Section 2). In this case, the size d and the deflection α of (p_1, p_2) are given. The following Proposition gives a circle solution:

Proposition 4. Given the distance d and deflection α of a circular arc Π, its length, curvature and cost are:

$$l = \frac{\alpha/2}{\sin(\alpha/2)} d \tag{12}$$

$$\kappa(s) = \frac{2\sin(\alpha/2)}{d} \tag{13}$$

$$\mathrm{cost}_1(\Pi) = \frac{2\alpha\sin(\alpha/2)}{d} \tag{14}$$

In the cubic spiral case, we need to characterize the set of cubic spirals. By the definition of the cubic spiral, its curvature is,

$$\kappa(s) = c\left(\frac{l^2}{4} - s^2\right), \quad \text{for } s \in [-l/2, l/2] \tag{15}$$

where c is a constant. A cubic spiral with a unit length is called *standard*. Let $D(\alpha)$ denote the size of a standard cubic spiral with a deflection α. For instance, $D(0) = 1$, $D(\pi/2) = 0.8558$, and $D(\pi) = 0.4861$.

Proposition 5.

(I) All cubic spirals with the same deflection α are similar to each other.

(II) If the size d and deflection α of a cubic spiral Π is given, its length, curvature, and cost are:

$$l = \frac{d}{D(\alpha)} \tag{16}$$

$$\kappa(s) = \frac{6\alpha D(\alpha)^3}{d^3}\left(\frac{d^2}{4D(\alpha)^2} - s^2\right) \tag{17}$$

$$\mathrm{cost}_2(\Pi) = \frac{12\alpha^2 D(\alpha)^3}{d^3} \tag{18}$$

(III)

$$D(\alpha) = 2\int_0^{1/2} \cos\left(\alpha(3/2 - 2s^2)s\right) ds \tag{19}$$

Figures 6 and 7 show symmetric cubic spirals for a simple left-turn and u-turn, respectively. They also clarify the differences among solutions using circles, cubic spirals, and clothoids. The length of the circle is minimum while that of the clothoid is maximum. Maximum curvature of the cubic spiral is less than that of the clothoid curve. This is one of the advantages of cubic spirals. To compensate for this, however, we need to have a bigger κ at the start and end of the path.

6. Solving Non-Symmetric Cases

The next question is how to solve the problem if the input pair of postures $p_1 = (x_1, y_1, \theta_1)$ and $p_2 = (x_2, y_2, \theta_2)$ with $(x_1, y_1) \neq (x_2, y_2)$ is non-symmetric. The pair (p_1, p_2) is said to be *parallel* if and only if $\theta_1 = \theta_2$.

Proposition 6. The locus of points of split postures (x, y, θ) of an input posture pair $p_1 = (x_1, y_1, \theta_1)$ and $p_2 = (x_2, y_2, \theta_2)$ is:

(I) If (p_1, p_2) is parallel,

The line determined by the points p_1 and p_2:

$$(x - x_1)(y - y_2) = (x - x_2)(y - y_1) \tag{23}$$

(II) If (p_1, p_2) is not parallel,

The following circle which goes through the points p_1 and p_2:

$$((x - x_1)(x - x_2) + (y - y_1)(y - y_2))\tan\left(\frac{\theta_2 - \theta_1}{2}\right)$$
$$= (x - x_1)(y - y_2) - (x - x_2)(y - y_1) \tag{24}$$

The center p_c of the circle (24) is:

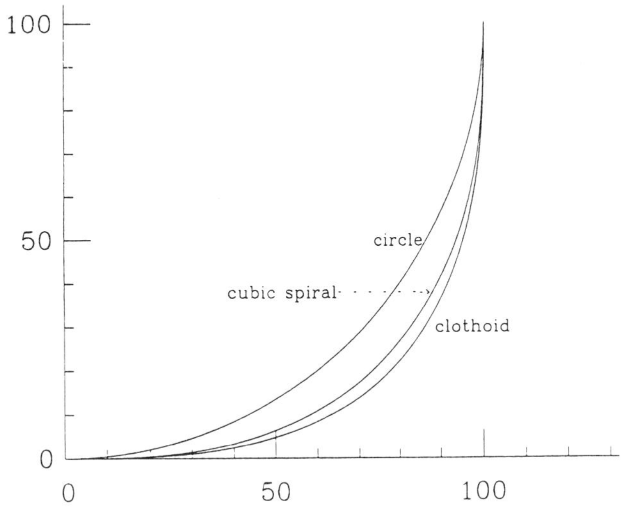

Fig. 6 Simple Curves with $\alpha = \pi/2$

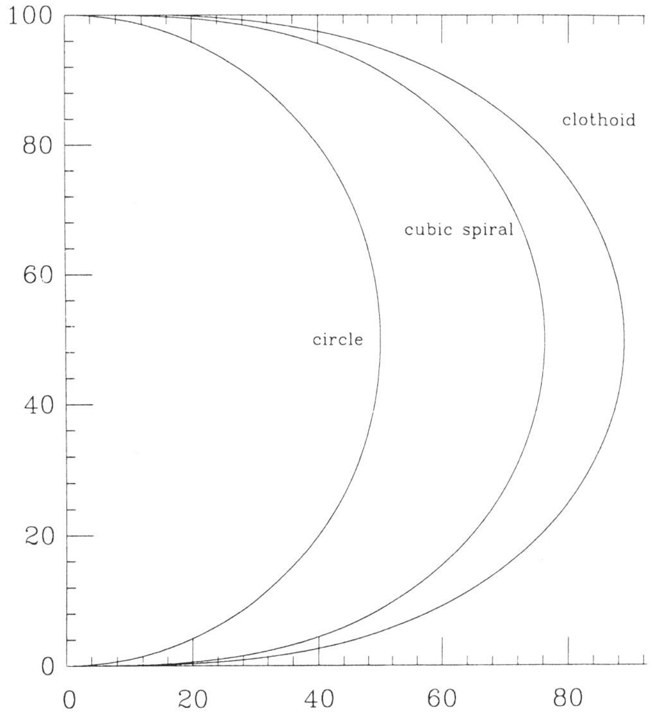

Fig. 7 Simple Curves with $\alpha = \pi$

$$p_c = (x_c, y_c) = (\frac{x_1 + x_2 + c(y_1 - y_2)}{2}, \frac{y_1 + y_2 + c(x_2 - x_1)}{2}), \quad (25)$$

where $c \equiv cot((\theta_2 - \theta_1)/2)$.

Proof. By Equations (2),

$$\begin{cases} \dfrac{\theta_1 + \theta}{2} = \tan^{-1}(\dfrac{y - y_1}{x - x_1}) \\ \dfrac{\theta + \theta_2}{2} = \tan^{-1}(\dfrac{y_2 - y}{x_2 - x}) \end{cases}$$

By taking the difference of both sides to cancel θ,

$$\frac{\theta_2 - \theta_1}{2} = \tan^{-1}(\frac{y_2 - y}{x_2 - x}) - \tan^{-1}(\frac{y - y_1}{x - x_1})$$

By applying the tan function to the both sides,

$$\tan(\frac{\theta_2 - \theta_1}{2}) = \tan(\tan^{-1}(\frac{y_2 - y}{x_2 - x}) - \tan^{-1}(\frac{y - y_1}{x - x_1}))$$

$$= \frac{\dfrac{y - y_2}{x - x_2} - \dfrac{y - y_1}{x - x_1}}{1 + (\dfrac{y - y_2}{x - x_2})(\dfrac{y - y_1}{x - x_1})} = \frac{(x - x_1)(y - y_2) - (x - x_2)(y - y_1)}{(x - x_1)(x - x_2) + (y - y_1)(y - y_2)}$$

This is equal to Equation (24). In the parallel case, its left hand side is zero and Equation (23) is obtained. □

Figures 8 and 9 shows examples of the sets of split postures for a parallel and non-parallel cases respectively. In non-parallel cases, if $\alpha > 0$, three points p_c, p_1 and p_2 are placed counterclockwise (p_c is the center of the locus, see Figure 12); if $\alpha < 0$, three points p_c, p_1 and p_2 are placed clockwise.

In selecting a split point, we further stipulate the following in order to avoid impractical solutions:

(I) In parallel cases ($\alpha = 0$), we select a split point on the segment $\overline{p_1 p_2}$.

(II) In case of $\alpha > 0$, we select a split point on the arc which is taken between p_1 and p_2 in counterclockwise order.

(III) In case of $\alpha < 0$, we select a split point on the arc which is taken between p_1 and p_2 in clockwise order.

Under these restrictions, we are to find a split posture q which makes the total of $\text{cost}(\Pi(p_1, q)) + \text{cost}(\Pi(q, p_2))$ minimum.

6.1. Parallel Cases

Let $\theta \equiv \theta_1 = \theta_2$.

Proposition 7. For either cost function $cost_1(\Pi)$ or $cost_2(\Pi)$, the least cost split point of $p_1 = (x_1, y_1, \theta)$ and $p_2 = (x_2, y_2, \theta)$ is

$$q = (\frac{x_1 + x_2}{2}, \frac{y_1 + y_2}{2}, \beta - (\theta - \beta)) \quad (26)$$

where $\beta = \tan^{-1}(\dfrac{y_2 - y_1}{x_2 - x_1})$

This solution leads to a combination of two identical simple paths which are mirror images of each other. Figure 10 shows several resultant paths of parallel cases.

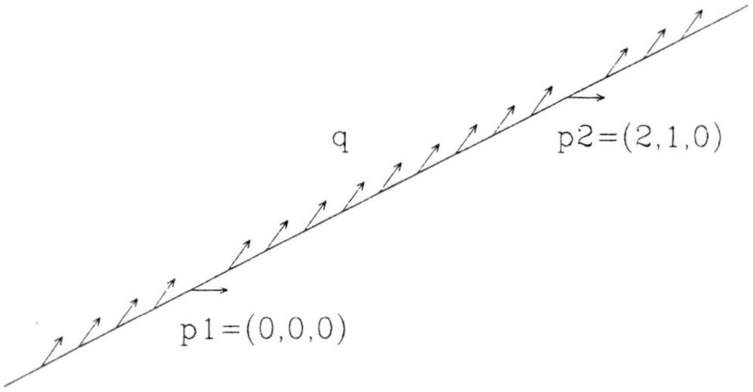

Fig. 8 Split Postures in Parallel Case

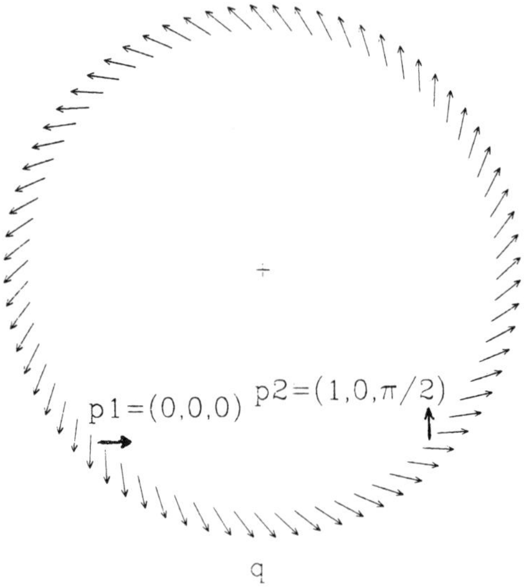

Fig. 9 Split Postures in Non-Parallel Case

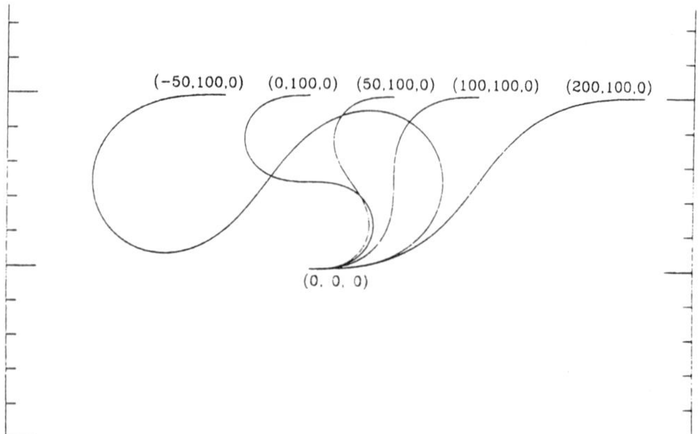

Fig. 10 Solutions to Parallel Cases

6.2. Non-Parallel Cases

In this case, the locus of split points is a circle of Equation (23).

Proposition 8. In a non-parallel case, there is a split point on the permissible arc of Equation (24) such that the sum of the costs of two cubic spirals is minimum.

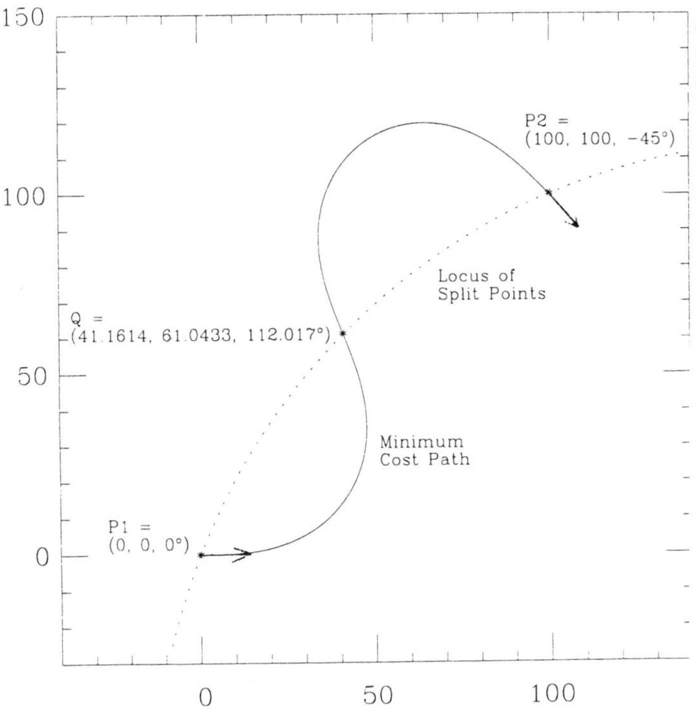

Fig. 11 Example of Non-Parallel Case

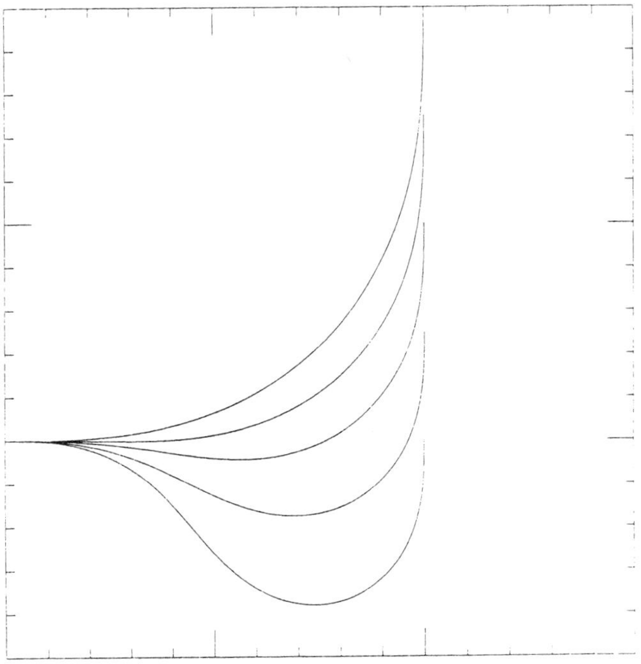

Fig. 12 Series of Non-Parallel Case Solutions

Proof. Since the total cost function is positive and continuous. □

Because no analytical method for finding the minimum solution has been found yet, we adopt an iterative method based on the Newton-Raphson algorithm. In Figure 11, the input postures are $p_1 = (0, 0, 0)$ and $p_2 = (100, 100, -\pi/4)$. This posture pair is neither symmetric nor parallel. The locus of possible split points is indicated by the broken line whose center is (136.603, --36.603), which is out of bounds of this figure. The cost is minimum at the split point Q shown in the figure. Figure 12 shows a sequence of non-parallel case solutions.

7. Conclusion

In this paper, two definitions of smoothness of vehicle paths is proposed; the cost of a path is its curvature or the derivative of its curvature. After implementing on our mobile robot Yamabico-11, we confirmed the vehicle moves far smoother than when the clothoid method is used. Let us compare the maximum curvature $\kappa(\alpha)$ at the middle point of a standard cubic spiral, and the maximun curvature $\kappa'(\alpha)$ at the middle point of a standard cubic spiral pair, with the same deflection α. Let R be their ratio:

$$R(\alpha) = \frac{\kappa(\alpha)}{\kappa'(\alpha)} = \frac{3}{4}\,\frac{D(\alpha)}{D'(\alpha)} \tag{27}$$

The functions $D(\alpha)$ is defined by Equation (19). $D'(\alpha)$ is the size of the standard clothoid pair with a deflection α. Typical values of R are: $R(\alpha) \approx 0.75$ for small α, $R(\pi/4) = 0.7528$, $R(\pi/2) = 0.7624$, $R(3\pi/4) = 0.7832$, and $R(\pi) = 0.8309$. Thus, curvatures of cubic spirals are always smaller for normal deflections, and the smoothness of cubic spirals are numerically certified.

A cost function for path planning is said to be *size-free* if the problem is enlarged by n, the solution is also enlarged by n. Both cost functions proposed here are clearly size-free. Also, a cost function for path planning is said to be *continuous* if a small change in the problem causes a small change in the solution. Again, both cost functions are continuous. These are among some essential meta-properties in path planning theories.

References

[1] Y. Kanayama, "A Path/Motion Description Method and Its Application to Mobile Robot Language Design", Technical Report of Department of Computer Science at University of California at Santa Barbara, February 1989

[2] Hongo T., Arakawa, H., Sugimoto, G., Tange, K. and Yamamoto, Y. 1985. "An Automatic Guidance System of a Self-Controlled Vehicle - The Command System and the Control Algorithm ---," Proc. IECON.

[3] Komoriya, K., Tachi, S. and Tanie, K. 1986. "A Method of Autonomous Locomotion for Mobile Robots," Advanced Robotics, vol. 1, no. 1, pp. 3-19

[4] I. J. Cox, "Branche: An Autonomous Robot Vehicle for Structured Environments", Proc. IEEE International Conference on Robotics and Automation, pp. 978-982, 1988

[5] Y. Kanayama and Miyake N., "Trajectory Generation for Mobile Robots", Robotics Research, vol. 3, MIT Press, pp. 333-340, 1986

[6] Y. Kanayama, A. Nilipour and C. A. Lelm, "A Locomotion Control Method for Autonomous Vehicles", Proc. IEEE Conference on Robotics and Automation, pp. 1315-1317, 1988

[7] B. I. Hartman, Y. Kanayama, and T. Smith, "Model and Sensor Vased Navigation by an Autonomous Mobile Robot", Proc. of International Conference on Advanced Robotics, June 1989, to appear.

[8] Y. Kanayama, "A Path/Motion Description Method and Its Application to Mobile Robot Language Design", Technical Report of the Department of Computer Science at University of California at Santa Barbara, TRCS89-05, February 1989

[9] Y. Kanayama and T. Noguchi, "Spatial Learning by an Autonomous Mobile Robot with Ultrasonic Sensors", Technical Report of the Department of Computer Science at University of California at Santa Barbara, TRCS89-06, February 1989

[10] Y. Kanayama and B. I. Hartman, "Smooth Local Path Planning for Autonomous Vehicles", Technical Report of the Department of Computer Science at University of California at Santa Barbara, TRCS88-15, June 1988

[11] R. Weinstock, "Calculus of Variations", New York, Dover Publishing Inc., 1974

2. Sensors

Nothing reaches the intellect before making its appearance in the senses.

Latin Proverb

Chapter 1 developed the principles of controlling a vehicle when it was implicitly assumed that the vehicle's position and its environment were completely known. Neither of these conditions exist in practice, and sensing is therefore a prerequisite to the navigation of a mobile vehicle. The primary reasons for sensing are to provide information for (1) navigation or position estimation, (2) obstacle detection and/or avoidance, and (3) exploration of an unknown environment, perhaps for the purpose of constructing a map. The choice of sensor technology for a mobile vehicle is governed by many constraints, including basic physics, available technology, preference/prejudice, environmental considerations, and cost. Some of these constraints vary from application to application.

It is important to realize the significance of a precise characterization of sensors. An accurate model of the sensor's precision and noise characteristics should be developed and any systematic errors (biases) removed. Only after this is done can sensor data be meaningfully interpreted. Further, most, if not all autonomous vehicles will be equipped with two or more complementary sensors, the data from which must integrated (combined). Although there has been work in qualitative integration [1, 11] we agree with Durrant-Whyte, see reprinted paper, that the principal objection to these heuristic approaches is their inability *to quantify* performance. Good sensor models allow that integration of sensory information be performed with quantifiable accuracy. After all, what good are two independent position estimates if their accuracy is unknown? Conversely, only when measurement accuracies are known can conventional statistical theory be used to integrate them. Sensor integration, in the context of navigation, is dealt with in Chapter 3. The reader is also directed to a special issue of the *International Journal of Robotics Research* [7] on sensor data fusion and to a National Science Foundation Workshop [12] on multisensor integration.

Navigation sensors can be broadly thought to belong to one of two categories: "dead reckoning" or external/environmental sensors. Dead reckoning sensors allow a vehicle's position to be estimated by integrating sensor information over time. External/environmental sensors provide information on the surrounding environment.

Dead reckoning is usually performed by odometry or inertial guidance sensors or both. Odometry is the most common form of sensor available on a mobile vehicle. It is also one of the oldest, dating back to perhaps the time of Archimedes, approximately 250 BC [21]. Using dead reckoning, position errors grow without bounds unless an independent position fix is used to reduce these errors. It is worth noting, however, that odometry requires only a single integration,[1] instead of the double integration of accelerometer-based devices. Consequently, the uncertainty grows only as a function of time rather than time squared. Odometry should be very practical in structured environments such as the office or the factory, but it may be less appropriate under

[1]Thus, if the noise is assumed to be white noise, the error in position is the integral of white noise, which is Brownian motion!

more natural circumstances such as roads or undeveloped terrain [18, 23]. Ming's paper provides a theoretical treatment of the errors associated with odometry. However, the editors are unaware of any comprehensive experimental paper that discusses the expected size of these errors as a function of, say, road surface. This is an omission that needs attention. Further, the performance of odometry sensors can vary significantly, depending on their position and method of mounting. Ideally, for a tricycle configuration, nonload-bearing, knife edge odometry wheels should be mounted in line with the rear axle. Such a configuration minimizes errors due to slippage and avoids changes in the wheel's circumference due to variations in vehicle load.

Odometry may not always be practical, especially in situations in which wheel slippage or very rough terrain is common. In these situations, conventional inertial guidance sensors based on the use of accelerometers can be utilized. These sensors, although currently expensive – one-axis gas rate gyros cost approximately $2000 in 1989 – are expected to decline in price, while performance is expected to improve steadly, especially for ring laser gyros. If such projections materialize, we can expect conventional inertial guidance sensors to be increasingly used. Kuritsky and Goldstein's paper discusses several common types of inertial guidance sensors. An important element of this paper is that the sensors are very well modeled with respect to accuracy and noise. We emphasize that this type of characterization is imperative if integration of multiple sensory information is to be accomplished. The last section of this paper is also noteworthy in its discussion of a stellar-inertial mechanism, whereby star fixes are used to significantly reduce accumulated position error. The integration of the inertial guidance estimate with the independent reference fix is performed by a Kalman filter, as discussed more thoroughly in Chapter 3.

For many autonomous vehicles, a star fix is not practical, particularly in office or factory environments! However, some form of sensing of the surroundings is required so that known reference points within the environment can be recognized. Range information is especially important for mobile vehicles. From it, information can be easily derived concerning the vehicle's position, areas of free space, and the presence of obstacles. Ultrasonic range finders are a very common form of sensor for this purpose and they are also useful for detecting unexpected obstacles in the path of the vehicle. Gough et al. provide a clear exposition of the principle of operation of ultrasonic range finders. A popular brand of ultrasonic range finder can be purchased from the Polaroid Corporation [10]. This sensor has been improved [19] to provide better performance at short distances. Most recently, Leonard and Durrant-Whyte [17] have had good success at modeling such sensors based on the concept of "regions of common depth."

Unfortunately, specular reflection is a major problem with ultrasonic range finders. When a wave is incident on the surface of an object, two forms of reflection commonly take place, diffuse and specular [6, 24]. *Specular reflection* is the reflection predicted by geometric optics when a beam is incident on a perfectly reflecting mirror. It is the primary reflectance mode when the surface of an object is smooth with respect to the illuminating wavelength, which is commonly the case for ultrasonic wavelengths (approximately 3 mm at 100 kHz if the velocity of sound in air is assumed to be approximately 330 m/s). The practical effect of specular reflection is twofold. First, a "true" signal is returned to the transmitter/receiver only if the illuminated surface is approximately normal to the incident beam. Second, a "spurious" signal may also be returned, due to multiple reflections of the beam. These signals always increase the apparent distance of a surface to the receiver. In situations in which the environment is initially unknown, interpretation of ultrasonic information is particularly difficult [4]. The use of ultrasonic range finders has been described as analogous to a man finding his way through a completely dark maze of mirrors with the aid of a flashlight strapped to his forehead!

Lambert's law, which holds for much diffuse reflectance, states that, for isotropic radiation (independent of direction) and a plane surface, the reflected intensity in any direction varies as the cosine of the angle between the direction and the normal to the surface. Diffuse reflectance allows a much larger angle of incidence between the illuminating beam and the surface normal before insufficient signal is returned. Diffuse reflectance usually occurs when the illuminated surface is rough compared with the illuminating wavelength. This is almost always true at optical frequencies where the wavelength is between 0.3 and 1 micron. Optical range finders therefore have significantly better operating characteristics, but they cost significantly more than ultrasonic range finders. For example, the Odetics laser range finder, capable of scanning a 128x128 field of view in 835 ms and ranging to 30′ with a resolution of 0.7″ cost in the region of $125,000 in 1989. The major difficulty with designing an optical range finder is associated with the fact that light travels approximately 1′ every nanosecond. The associated timing or phase measurement electronics can be difficult to design, particularly if cost is an important consideration. The paper by Miller and Wagner describes a low-cost optical range finder designed explicitly for mobile robot applications. Further information on this sensor can be found in Cox [8].

Ultrasonic and optical range finders measure the distance from the sensor to an object directly. Besl [5] provides an excellent review of active range measurement sensors. However, range does not have to be measured directly. An alternative approach is to infer range indirectly. The computer vision community has extensively studied the problem of depth from stereo, optical flow, shading etc. [3, 14, 15, 20]. Here, the physical sensor is a TV camera followed by an algorithm to extract depth information. The concept of a logical sensor [13] is of interest here, since it abstracts the physical measuring process from the information provided by the sensor. That is, a logical range finder provides range information and, it is hoped, an associated measurement accuracy, but this information can be acquired by any number of alternative range sensors and, perhaps, associated processing. Logical sensors are strongly related to data abstraction principles of computer science [9] and are particularly relevant when systems are implemented. Other interesting references to the use of vision for localization include [16, 22]. The final paper by Matthies and Shafer discusses how depth information can be extracted from stereo cameras. The principal merit of this paper is the modeling of the noise and uncertainty in the depth estimates. A further excellent paper on this topic is Ayache and Faugeras' [2] paper, reprinted in Chapter 3.

References

1. Allen, P.K. Integrating Vision and Touch for Object Recognition Tasks. *Intl. J. Robotics Res. 7*, 6 (1988), 15-33.

2. Ayache, N., and Faugeras, O.D. Maintaining Representations of the Environment of a Mobile Robot. *IEEE Trans. Robotics Automation RA-5*, 6 (1989), 804-819.

3. Ballard, D.H., and Brown, C.M. *Computer Vision.* Prentice-Hall, Englewood Cliffs, N.J., 1982.

4. Barshan, B., and Kuc, R. Differentiating Sonar Reflections from Corners and Planes by Employing an Intelligent Sensor. *IEEE Trans. Pattern Analysis Machine Intelligence* (to be published).

5. Besl, P.J. Active Optical Range Imaging Sensors. In *Advances in Machine Vision*, J.L.C. Sanz, Ed. Springer-Verlag, New York, 1989.

6. Born, M., and Wolf, E. *Principles of Optics.* Pergamon, Elmsford, N.Y., 1980.

7. M. Brady, (Ed.) *Intl. J. Robotics Res.*, 7, 6, (1988).

8. Cox, I.J. Blanche: An Autonomous Robot Vehicle for Structured Environments. In *IEEE International Conference on Robotics and Automation*, IEEE, New York, 1988, pp. 978-982.

9. Cox, I.J. C++ Language Support for Guaranteed Initialization, Safe Termination and Error Recovery in Robotics. In *IEEE International Conference on Robotics and Automation*, IEEE, New York, 1988, pp. 641-643.

10. Division, C.B. *Ultrasonic Ranging System.* Polaroid Corporation, 1984.

11. Flynn, A.M. Combining Sonar and Infrared Sensors for Mobile Robot Navigation. *Intl. J. Robotics Res. 7*, 6 (1988), 5-14.

12. Henderson, T. Workshop on Multi-Sensor Integration. UUCS-87-006, University of Utah Computer Science, Salt Lake City, 1987.

13. Henderson, T., and Shilcrat, E. Logical Sensor Systems. *J. Robotic Systems 1*, 2 (1984), 169-193.

14. Horn, B.K. *Robot Vision.* MIT Press, Cambridge, Mass., 1986.

15. Kanade, T., (Ed.) *Three-Dimensional Machine Vision.* Kluwer Academic, Boston, Mass., 1987.

16. Krotkov, E. Mobile Robot Localization Using a Single Image. In *Proceedings of the IEEE International Conference on Robotics and Automation*, Vol. 3, IEEE, New York, 1989, pp. 978-983.

17. Leonard, J.J., and Durrant-Whyte, H.F. A Unified Approach to Mobile-Robot Navigation, (to be published).

18. McMillan, J.C. An Integrated System for Land Navigation. *J. Instit. Navigation 34*, 1 (1987), 43-63.

19. Miller, G.L., Boie, R.A., and Sibilia, M.J. Active Damping of Ultrasonic Transducers for Robotic Applications. In *IEEE International Conference on Robotics and Automation*, IEEE, New York, 1984, pp. 379-384.

20. Shirai, Y. *Three-dimensional computer vision.* Springer-Verlag, New York, 1986.

21. Sleeswyk, A.W. Vitruvius' Odometer. *Sci. Am.* (1981), 188-198.

22. Sugihara, K. Some Location Problems for Robot Navigation Using a Single Camera. *Comput. Vision, Graph. Image Process. 42*, 1 (1988), 112-129.

23. Tsumura, T., Fujiwara, N., and Hashimoto, M. An Experimental System for Self-Contained Position and Heading Measurement of Ground Vehicle. In *International Conference on Advanced Robotics*, 1983, pp. 269-276.

24. Walsh, J.W.T. *Photometry.* Constable and Company, London, 1958.

Hugh F. Durrant-Whyte
Department of Engineering Science
University of Oxford, Oxford, U.K.

Sensor Models and Multisensor Integration

Abstract

We maintain that the key to intelligent fusion of disparate sensory information is to provide an effective model of sensor capabilities. A sensor model is an abstraction of the actual sensing process. It describes the information a sensor is able to provide, how this information is limited by the environment, how it can be enhanced by information obtained from other sensors, and how it may be improved by active use of the physical sensing device. The importance of having a model of sensor performance is that capabilities can be estimated a priori *and, thus, sensor strategies developed in line with information requirements.*

We describe a technique for modeling sensors and the information they provide. This model treats each sensor as an individual decision maker, acting as a member of a team with common goals. Each sensor is considered as a source of uncertain geometric information, able to communicate to, and coordinate its activities with, other members of the sensing team. We treat three components of this sensor model: the observation model, which describes a sensor's measurement characteristics; the dependency model, which describes a sensor's dependence on information from other sources; and the state model, which describes how a sensor's observations are affected by its location and internal state. We show how this mechanism can be used to manipulate, communicate, and integrate uncertain sensor observations. We show that these sensor models can deal effectively with cooperative, competitive, and complementary interactions between different disparate information sources.

1. Introduction

Recently there has been an increasing interest in the development of robot systems capable of using many different sources of sensory information (Henderson 1987). This interest arises from a realization that there are *fundamental* limitations on any attempt at building descriptions of the environment based on a single source of information: A single source of sensory information can only provide partial information about an environment, information that is insufficient to constrain possible interpretations and that is limited in resolving ambiguity. Diverse information from many different sources can be used to overcome the limitations inherent in the use of single sensors through coordinated constraint of partial interpretations and by cooperative resolution of ambiguity.

The sensors of a multisensor system are characteristically diverse and necessarily logically distinct. The information obtained by the different sensors of a multisensor system is always uncertain, usually partial, occasionally spurious or incorrect, and often geographically or geometrically incomparable with other sensor views. The motivating goal behind all sensor integration techniques is to actively *utilize* this diversity of information in overcoming the limitations of any one sensor system. We will maintain that the only way to understand and utilize the disparity between different sensor views is to explicitly model the sensor and the information it provides in a manner that can describe both individual sensor abilities and the interactions that must take place between sensors. A sensor model is an abstraction of the physical sensing process whose purpose is to describe the ability of a sensor to extract descriptions of the environment in terms of the information available to the sensor itself. Sensor models should provide a quantitative ability to analyze sensor performance, allowing capabilities to be estimated *a priori* and decision procedures developed in line with information requirements.

We will develop a probabilistic model of individual sensor performance, embedding these descriptions in a team-theoretic framework to describe interactions between different sensors. The information obtained by each sensor is represented in terms of the geometric elements it is able to extract. This geometry is described probabilistically, enabling different sensors to communicate to each other in a common dimensionless language. This model of individual sensor performance is then embedded in a team structure which describes when information is communicated, and how this information can be used to improve the performance of the overall sensing system. Within this framework, we develop three important models: the observation model, describing the observations and decisions made by individual sensors; the dependency model, describing the communication of information between different sensors; and the state model, describing how a sensor's observations are affected by its location and internal state. By describing the resulting structure as a multi-Bayesian team, we develop some simple fusion procedures that demonstrate how these sensor models can be used to improve communication and cooperation between diverse sensing systems, how individual sensors can be controlled to acquire useful information, and how the description of competitive and complementary interactions between sensors can be used to improve the reliability and usefulness of multisensor robot systems. In conclusion, we argue for the development of other multisensor models and show how the team model described might fit into these.

2. Sensor Models

The realization that a multisensor system must be modeled to be understood is not a new idea. However, this need has often been subsumed into an implicit recognition of the limitations of a sensor with respect to some prescribed task. This approach to modeling leads to impossibly complex integration procedures and an inability to understand or utilize the potential power available from a multisensor system.

The most comprehensive rationalization of sensor abilities in the context of a multisensor robot system is probably Henderson's "logical" sensor descriptions (Henderson and Hansen 1985). These logical sensor models provide a description of sensor capabilities in terms of a set of rules specifying input-output characteristics of a device or algorithm. Logical models are defined in a modular manner, allowing different functions to be specified in terms of "characteristic building blocks," which can be used to construct a complete system. One advantage of this approach is that it concentrates on the information provided by different sensors and algorithms rather than on the physical sensing process itself. This lends itself to ideas of modularity and extensibility.

Flynn (1985) has also proposed a rule-based sensor model for integrating sonar and infrared sensor information as applied to mobile robot map building. Three rules were proposed for deciding when data from one or another sensor should be considered valid. Although this is a rather limited model, the way in which the rules were derived from direct observation of sensor data has a certain appeal. Allen (1987) has also used an implicit rule-based model of sensor performance in describing the ability of a tactile finger to explore geometries occluded during visual inspection.

Our principal objection to these heuristic, rule-based approaches to sensor modeling is their inability to *quantify* sensor performance. A quantitative model of sensor information is essential in providing a means of analyzing performance, in estimating system abilities, and in developing sensing strategies in line with task requirements. The importance of providing quantitative models of sensor performance is well established in conventional sensing domains. Powerful techniques exist for developing probabilistic noise models and for studying modes of sensor failure; Gelb (1974) describes a number of these models and the way they are used. The utility of existing probabilistic models of sensor information has been realized by a number of researchers in robotics. Most notably, Faugeras and Ayache (Faugeras and Ayache 1986; Ayache and Faugeras 1987) have employed a probabilistic model to fuse noisy line segment descriptions extracted from stereo visual images. Porril et al. (1987) and Bolle and Cooper (1986) have employed similar statistical models to estimate object locations from visual images. Durrant-Whyte (1987a) has developed probabilistic models of both visual and tactile information.

However, even in more mature sensing research areas, only single sensor models are well understood. The integration of information from many diverse, geographically disparate sensor systems is still an open research issue; see for example the discussion in Athans (1987). In robotics, multisensor integration is made more difficult because of often indivisible relationships between different sensing modalities (in manipulation or in hand-eye coordination, for example).

Terzopoulos (1986) has considered this problem, in describing a cooperative integration of sparse depth information with local surface constraints obtained from visual intensity arrays. A further exploration of the dynamics involved in cooperative multisensor interactions is introduced in Hager (1987) and Hager and Durrant-Whyte (1988). This study describes a number of mechanisms for communicating and coordinating information flow between sensors, and develops a simulated scenario, involving active sensors, to explore these ideas.

In general, a sensor model has two important components: an individual model of performance, and a group model of available information. The individual model describes to the sensor itself the observations that an independent sensor can obtain. The group model describes to a single sensor the observations made by the other sensors in the system. The purpose of these models is to provide each sensor with an ability to make decisions, based on its own observations and in the light of information provided by other sensors.

2.1. Characterizing Sensors

In modeling a sensor's ability to make observations of the environment, we should be concerned with a number of important sensing characteristics.

Device Complexity

When a sensor is composed of many physical devices, each making contributions to the observation extracted by the sensor, the exact distribution of uncertainty and character of the sensor is complex and difficult to describe as an exact model. For example, a camera system comprises a charge coupled device (CCD) array, lenses, digitizers, etc; their cumulative effect on the sensor character is difficult to predict. This problem is encountered in other sensor-based research areas as well as robotics.

Observation Error

The feature observations reconstructed from sensor data have more uncertainty associated with them than just sensor noise (Bajcsy, Krotkov, and Mintz 1986). Error may occur due to inaccuracies in the placement of devices, incorrect interpretation of measurements, or device failure, for example. Uncertainty also arises when a feature observation is incomplete or partial, when information is unavailable in one or more degrees of freedom, or all two-dimensional visual cues, for example.

Observation Disparity

Robot sensors are characterized by the diversity of observations that can be obtained: edges, normals, locations, or texture for example. If sensor information from many disparate sources is to be combined, we must have an ability to transform one type of uncertain geometric feature to another (edges to surface normals for example). This ability allows observations from different cues to be compared or used to complement each other in deriving a consensus description of the environment.

Multiple Viewpoints

When we have two or more sensors which are geographically separated, we must be able to transform their observations in to a common coordinate system so that they can be compared. In vision this is called the viewpoint problem—how to combine observations taken from a number of different viewpoints into a consensus description. The central issue here is to describe the ability of the sensor model to transform and manipulate uncertain descriptions of the environment. This is an important consideration in active sensor descriptions.

To provide a homogeneous method for describing these diverse characteristics, we will develop a model of sensor observations in terms of uncertain geometry; see Durrant-Whyte (1988) for a complete description of this subject. All geometric objects (features, locations and relations) can be described by a parameter vector $\mathbf{p}$ and a vector function

$$\mathbf{g}(\mathbf{x}, \mathbf{p}) = \mathbf{0}, \quad \mathbf{x} \subseteq \Re^n, \quad \mathbf{p} \in \Re^m. \tag{1}$$

This function can be interpreted as a model of the physical geometric object that maps a (compact) *region* $x \subseteq \Re^n$ in Euclidean n-space to a *point* $\mathbf{p} \in \Re^m$ in the parameter space. Each function $\mathbf{g}$ describes a particular type of geometric object: all straight lines, or the family of quadratic surfaces, for example. Each value of $\mathbf{p}$ specifies a particular instance of a geometric object molded by $\mathbf{g}$. For example, *all* plane surfaces can be represented by the equation

$$g(\mathbf{x}, \mathbf{p}) = p_x x + p_y y + p_z z + 1 = 0, \qquad (2)$$

with $\mathbf{x} = [x, y, z]^T$ and $\mathbf{p} = [p_x, p_y, p_z]^T$. A specific plane can be represented as a point $\mathbf{p} \in \Re^3$ in the parameter space. A sensor can be considered as observing values of $\mathbf{p}$. The uncertain *event* that the sensor "sees" a specific instance of this geometric object can be described by taking the parameter vector $\mathbf{p}$ to be a random variable. The likelihood that a particular instance of a geometric feature is observed can now be described by a probability density function (p.d.f.) $f_g(\mathbf{p})$. For a specific type of feature represented by Eq. (1), this p.d.f. describes the probability or likelihood of observing a particular instance of the associated geometric object. It should be noted that this is *not* just a noise model.

The advantage of describing sensor information in terms of an uncertain parameterized function is that geometric description itself can easily be transformed between different coordinate systems and different object representations, providing a simple but effective means of communicating information between different sensors. We will consider every sensor as a "geometry extractor," able to extract uncertain geometric information about the environment and communicate this information to the system in terms of a p.d.f. $f(\mathbf{p})$ on the observed geometric parameter vector $\mathbf{p}$. The communication of information then reduced to the transformation of stochastic geometric functions of the form of Eq. (1).

This geometric description of sensor information provides two essential components: first, a natural way to embed noise models into the structure, and second a mechanism to communicate information between different sensors in the common language of geometry.

2.2. Sensors as Members of a Team

We will develop a description of a multisensor system as a *team* of decision makers. Each sensor of this team will be considered as an individual decision maker taking observations, making local decisions, and implementing its own actions. Together, the sensors must coordinate their activities, guide each other to view areas of interest, and ultimately come to some team-consensus view of the environment.

Team decision theory was originally developed to provide a quantitative mechanism for describing economic organizations. The basis for the analysis of cooperation among structures with differing opinions or interest was formulated by Nash (1950) in the well-known bargaining problem. In their book, Marshak and Radnor (1972) describe the extension of the bargaining problem to multiperson organizations. The primary emphasis of this work is to provide a means of coordinating a distributed decision making process. Each decision maker is only allowed access to partial state information, from which local decisions must be made, reflecting some global optimality criteria. This team structure has also been developed as a means of describing problems in multiperson control. In a series of papers, Ho (Ho and Chu 1972; Ho 1980) has investigated the distributed linear quadratic Gaussian control problem, as applied to a group of sequential decision makers.

The characteristics of a team structure capture many of the desirable properties of a multisensor system: Team members can make local decisions based on a team goal, providing a means of delegating observation tasks and actions. Team members can help each other by exchanging information unattainable to individual members, thus providing the team with a more complete description of events. Team members can disagree with each other as to what is being observed; resolving differences of opinion can supply a mechanism for validating each other's operation. The most important observation about a team is that the coordinated activity of the members is more effective than the sum of their individual actions.

The idea of a team is essentially simple. It provides a mechanism for describing the observations made by a number of individual decision makers, how this information is communicated to other members of a team, and how team decisions and actions can be arrived at. In general, the group decision problem can be very complex. Our development of multisensor teams eliminates much of this complexity by reducing the more general group decision problem to one of team estimation. Further simplifications arise from providing a common, geometric, communication language, and restricting the basis of local decision making to likelihood comparisons rather than, more generally, utility.

We propose to describe a multisensor system as a team in the following sense: The sensors are considered as members of the team, each observing the environment and making local decisions based on the information available to them. The observations $\mathbf{z}$ made by a sensor are described by an information structure η. Each sensor can make a decision δ, based on these observations, resulting in an action $\mathbf{a}$, usually an estimate of a feature description, describing the opinion of the sensor. The opinions of each sensor are integrated to provide a team decision and action.

Our interest in this team structure centers on the model of individual decision makers as described by each team member's information structure.

2.3. Information Structure

The ith team member is described by an *information structure* η_i. This structure is a model of sensor capabilities. Formally an information structure is defined by the relation between observations, state and decisions.

DEFINITION The *information structure* of the ith sensor or team member ($i = 1, \ldots, n$) is a function η_i which describes the observations $\mathbf{z}_i$ made by a sensor in terms of its physical state $\mathbf{x}_i$, available prior information about the state of the environment $\mathbf{p}_i \in \mathcal{P}_i$, and the other sensors' or team members' actions $\mathbf{a}_j \in \mathcal{A}_j, j = 1, \ldots, n$. Thus $\mathbf{z}_i = \eta_i(\mathbf{x}_i, \mathbf{p}_i, \mathbf{a}_1, \ldots, \mathbf{a}_{i-1}, \mathbf{a}_{i+1}, \ldots, \mathbf{a}_n)$.

Collectively the n-tuple $\boldsymbol{\eta} = (\eta_1, \ldots, \eta_n)$ is called the information structure of the team. The action $\mathbf{a}_i$ of the ith team member is related to its information $\mathbf{z}_i$ by a decision function $\delta \in \mathcal{D}_i$ as $\mathbf{a}_i = \delta_i(\mathbf{z}_i)$. Collectively the n-tuple $\boldsymbol{\delta} = (\delta_1, \ldots, \delta_n)$ is called the team decision function.

We will be concerned with the model of sensor performance provided by the information structure. We consider three types of sensor model: an *observation* model, a *dependence* model, and *state* model. The observation model η_i^p describes the character of measurements given the state of the sensor and all other sensor actions. The dependence model η_i^δ describes the effect of other sensor actions (observations) on the sensor measurements. The state model η_i^x describes the observation dependence on the internal state and location of the sensor. We will embed all these models in a common information structure η_i.

Within this structure, we will describe a multisensor system as a team of observers, taking observations of geometric features, and making decisions which are estimates of geometric descriptions $\delta_i(\mathbf{z}_i) \mapsto \mathbf{p}_i \in \mathcal{P}$. With this interpretation, the formal definition of the information structure can be simplified and considered as a function transforming uncertain geometric feature estimates from one sensor into prior information or predictive hypotheses for use by another sensor.

$$\mathbf{z}_i = \eta_i(\mathbf{x}_i, \mathbf{p}_i, \delta_1(\mathbf{z}_1), \ldots, \delta_{i-1}(\mathbf{z}_{i-1}), \delta_{i+1}(\mathbf{z}_{i+1}), \ldots, \delta_n(\mathbf{z}_n)), \tag{3}$$

where the observations $\mathbf{z}_i$ are random vectors and η_i is a stochastic function transforming the decisions $\delta_j(\cdot) \in \mathcal{P}_j$, the prior information $\mathbf{p}_i \in \mathcal{P}_i$ and the sensor state $\mathbf{x}_i$ into elements of $\mathcal{P}$.

We will model the observations $\mathbf{z}_i$ by a p.d.f. $f_i(\mathbf{z}_i | \eta_i(\cdot))$. This density function serves as the natural realization of the information structure; it describes the observations as a probabilistic function of the information provided to the sensor. The description of observations is now dimensionless, so we can manipulate this function using standard probabilistic techniques and communicate information between different sensory sources by transformations of uncertain geometric feature descriptions. The decisions $\delta_i(\cdot)$ can be related through f_i to any common decision philosophies—maximum likelihood, for example. This will enable us to develop decision procedures capable of using many disparate sources of information. Another important advantage of incorporating the information structure into a probability distribution function is that the variables (prior information, state, and other sensor decisions) can be separated from each other by expanding f_i as a series of conditional probability distributions. This in turn allows us to decouple the three types of sensor model (η^p, η^δ, and η^x) from each other. Let

$$\bar{\delta}_i = (\delta_1, \ldots, \delta_{i-1}, \delta_{i+1}, \ldots, \delta_n)$$

be the information communicated to the ith sensor by all other sensors. Assuming that the observation's dependence on state, prior information, and other sensor decisions is separable, we can write

$$\begin{aligned} f(\mathbf{z}_i|\eta_i(\mathbf{x}_i, \mathbf{p}_i, \bar{\delta}_i)) &= f_{\eta i}(\mathbf{z}_i|\mathbf{x}_i, \mathbf{p}_i, \bar{\delta}_i) \\ &= f_x(\mathbf{z}_i|\mathbf{x}_i) f_p(\mathbf{z}_i|\mathbf{p}_i) f_\delta(\mathbf{z}_i|\bar{\delta}_i) \\ &= f_x(\eta_i^x) f_p(\eta_i^p) f_\delta(\eta_i^\delta). \end{aligned} \tag{4}$$

The state model $f_x(\eta_i^x)$ now describes the dependence of a sensor's observations on its location and internal state *given* any prior information and all other sensor opinions. The observation model $f_p(\eta_i^p)$ describes the dependence of sensor measurements on the state of the environment *given* all other sensor decisions. The dependence model $f_\delta(\eta_i^\delta)$ describes the prior information supplied by the other sensors in the system. The product of these three models describes the observations made by the sensor.

This ability to decouple sensor models enables the effect of different types of information to be analyzed independently. This in turn provides a powerful framework in which to develop descriptions of sensor performance. In the remainder of this article, we will develop these three different types of sensor models.

2.4. Observation Models

An observation model describes the measurements made by a sensor, given all prior information, other sensors' information, and the state of the sensor itself. It is essentially a model of sensor noise and error. Observation models have been studied extensively in related sensor research areas but are still relatively underused in robotics. We will briefly discuss some more popular models and introduce our own favorite.

Consider a sensor taking observations of a geometric feature, an instance of a given family described by $\mathbf{g}(\mathbf{x}, \mathbf{p}) = \mathbf{0}$ and parameterized by the vector $\mathbf{p}$. The observation model of the ith sensor is represented as a conditional probability distribution function $f_i(\mathbf{z}_i|\mathbf{p}_i)$, describing the likelihood of feature observation given all prior information about $\mathbf{p}_i$. The exact form of $f(\cdot|\mathbf{p})$ will depend on many physical factors. It is unlikely that we can obtain an exact description of the probabilistic character of observations in all but the simplest cases. It may in fact be undesirable to use an exact model even if it were available because of its likely computational complexity and its inability to model non-noise errors such as software failures or algorithmic misclassifications. It is usual to assume a Gaussian or uniform distribution model for the conditional observations (Faugeras and Ayache 1986). These models allow the development of computationally simple decision procedures. However, they are not able to represent poor information and can fail with catastrophic results even when the observations deviate only a small amount from the assumed model. In the absence of an exact observation description, it is possible to build robust approximations to the true character of sensor measurements. There are two different ways to approach this approximation problem: We can model the observation noise as a *class* of possible distributions, or as some nominal distribution *together* with an additional unknown likelihood of errors and mistakes. McKendall and Mintz (1987) have conducted experiments on a stereo camera system and have proposed modeling visual noise as a class of probability distributions. Applying a minimax philosophy to this class results in a soft quantizer as a decision procedure. In the long term these models show great promise, but they currently have two major problems: the relative computational complexity of the decision procedures, and their inability to describe non-noise errors. An alternative to this approach is to use an approximation that describes the observations by some nominal distribution together with an unknown likelihood of errors or mistakes. Then we shall require that our decision procedures be robust to the possibility of error, but otherwise assume a nominal noise model. These distributions are termed *gross error models* (Huber 1981) and have the general form

$$\mathcal{P}_\epsilon(F_0) = \{F | F = (1 - \epsilon)F_0 + \epsilon H, H \in \mathcal{M}\}. \quad (5)$$

These models are described by a set of distributions F which is composed of some nominal distribution F_0 together with a small fraction ϵ of a second probability measure H. This second measure is often assumed unknown and acts to contaminate the nominal distribution with unexpected observations. This model results in decision procedures which cluster observations and trim outliers from consideration in the integration process. If the nominal model has finite moments, then this clustering process converges to a Gaussian observation model. For this reason, we consider a particular case of the gross error model called the contaminated Gaussian distribution, which has the general form

$$\begin{aligned} f(\mathbf{z}|\mathbf{p}) = & \frac{1-\epsilon}{(2\pi)m/2|\Lambda_1|^{1/2}} \\ & \times \exp\left[-\frac{1}{2}(\mathbf{z}-\mathbf{p})^T\Lambda_1^{-1}(\mathbf{z}-\mathbf{p})\right] \\ & + \frac{\epsilon}{(2\pi)m/2|\Lambda_2|^{1/2}} \\ & \times \exp\left[-\frac{1}{2}(\mathbf{z}-\mathbf{p})^T\Lambda_2^{-1}(\mathbf{z}-\mathbf{p})\right], \end{aligned} \quad (6)$$

with $0.01 < \epsilon < 0.05$ and $|\Lambda_1| \ll |\Lambda_2|$. The spirit of this model is that the sensor behaves as $N(\mathbf{p}, \Lambda_1)$ most of the time but submits occasional spurious measurements from $N(\mathbf{p}, \Lambda_2)$. With the type of sensors that we are considering (vision, tactile, range, etc.), the contaminated model is intended to represent the fact that we would normally expect quite accurate observations within a specific range, but must be robust to problems like miscalibration, spurious matching, and software failures. We assume that we have knowledge of the value of Λ_1 from the character of the sensor, but we do not explicitly assume anything other than bounds on the values of Λ_2 and ϵ. This has the property of forcing any reasonable integration policy developed to be robust to a wide variety of sensor observations and malfunctions. This model is thought to be a sufficiently conservative estimate of sensor behavior, in that by choosing Λ_1, Λ_2, and ϵ sufficiently large, we can encompass all the possible uncertainty characteristics we may expect to encounter. The intention of this model is to approximate a sensor's true characteristics without having to analyze possible sensor responses. This forces us to develop decision procedures which are robust to the exact specification of observation model, which provide efficient results in the light of our poor knowledge of sensor characteristics, and that will be robust to spurious or gross contaminations.

The observation model developed here describes the distribution of a single feature measurement. We must also be concerned with multiple-sample characteristics, the statistical correlation between observations, the likely density of observations, and the effect of algorithms that aggregate observations into higher-level environment descriptions. We maintain that these issues should be considered in terms of decision procedures rather than in the sensor model *per se*. The ϵ-contamination model has been used to develop decision rules which are sympathetic to the *actual* character of sensor observations and which are robust with respect to spurious information and model specification (Durrant-Whyte 1987b).

2.5. Dependency Models

To understand and utilize diverse sensor capabilities, it is important that we be able to model the interactions and exchange of information between different sensors. Unlike the development of observation models, there has been essentially no research in dynamic sensor models in robotics or any other sensor-based research areas. We will develop a model of dependence between sensory systems by considering the dependence model $f_\delta(\mathbf{z}_i | \cdot)$ in terms of a set of conditional probabilities or functions describing the information provided by other sensors and cues.

Consider the dependence of the *i*th sensor or team members' observations on other sensor information as described by the probabilistic dependence model

$$f_\delta(\mathbf{z}_i|\bar{\delta}_i) = f_\delta(\mathbf{z}_i|\delta_1(\mathbf{z}_1), \ldots, \delta_{i-1}(\mathbf{z}_{i-1}), \ldots, \delta_{i+1}(\mathbf{z}_{i+1}), \delta_n(\mathbf{z}_n)). \quad (7)$$

The decisions made by other sensors, communicated to the ith sensor can be described probabilistically by the distribution function $f_\delta(\bar{\delta}_i)$. With this definition, we can interpret the dependence model as a posterior probability of observations, and the information communicated to the sensor as a prior probability. This interpretation makes statistical sense as the joint feature density is found by multiplying the conditional observation model by the prior information, so by following Bayes' rule, we have $f_\delta(\mathbf{z}, \bar{\delta}_i) = f_\delta(\mathbf{z}_i|\bar{\delta}_i)f_\delta(\bar{\delta}_i)$. This interpretation is also intuitively appealing because it is the observations made by other sensors, communicated to the ith sensor, that provide the initial prior information.

Physically, the information provided by other sensors, as described by the distribution $f_\delta(\bar{\delta}_i)$, is geometric: The decisions $\delta_j(\cdot) \in \mathcal{P}_j$ communicated to the ith sensor are parameter vectors of the geometric feature observed by the jth sensor. The distribution function $f_\delta(\cdot)$ describes the uncertainty in these communicated decisions. This then leads us to interpret the dependency model $f_\delta(\mathbf{z}_i|\cdot)$ as a geometric function, describing the transformation of geometric features observed by the jth sensor into geometric features observed by the ith sensor. This transformation is stochastic and must make use of the tools developed in the field of uncertain geometry (Durrant-Whyte 1988; Faugeras and Ayache 1986). Our interest now centers on the communication of uncertain geometric information between sensor systems.

The internal structure of the distribution $f_\delta(\cdot)$ must, in general, describe the interdependence of all sensor information contributing to the ith sensor's information structure $f_\delta(\mathbf{z}_i|\cdot)$. These dependencies should represent the sequence in which information is passed between sensors and describe the way in which observations combine in different ways to provide a complete environment description. Each observation $\mathbf{z}_i$ made by a sensor is a measurement of a specific type of feature $\mathbf{p}_i \in \mathcal{P}_i$; the decision $\delta_i \in \mathcal{P}_i$ is an estimate of this feature based on the observation, such as the parameter vector of an edge or surface equation. The information structure is an interpretation of this estimate in terms of a distribution function defined on the parameter vector. The dependency model $f_\delta(\mathbf{z}_i| \ldots, \delta_j, \ldots)$ is therefore a *transformation* of feature descriptions obtained by the jth sensor into feature descriptions required by the ith sensor. This transformation represents a change in stochastic feature descriptions.

The interpretation of the dependency model as a prior probability distribution allow us to expand $f_\delta(\cdot)$ as a series of conditional distributions describing the effect of each individual sensor on the ith sensor's prior information. For example, if the numeric order of decision making is also the natural precedence, then

$$f_\delta(\bar{\delta}_i) = f^i_\delta(\delta_1, \ldots, \delta_n) = f^i_\delta(\delta_1|\delta_2, \ldots, \delta_n) f^i_\delta(\delta_2|\delta_3, \ldots, \delta_n) \ldots f^i_\delta(\delta_n).$$

Each term $f^i_\delta(\delta_j|\delta_k)$ describes the information contributed by the jth sensor to the ith sensor's prior information, given that the information provided by the kth sensor is already known. The transformation affected by $\eta^i_j(\delta_j)$ takes the jth sensor observation and interprets it in terms of observations made by the ith sensor: $\eta^i_j(\cdot) \in \mathcal{P}_i$. The term $\eta^\delta_j(\delta_k)$ is the dependence model describing the kth sensor's contribution to the ith sensor's prior information. This in turn can be described by the probability distribution $f^j_\delta(\delta_k)$.

This decomposition by conditionals can, in general, be written in any appropriate order. If there is a natural precedence order in which sensors take observations, then an expansion in that order is appropriate. For example, if the decision δ_j only depends on information provided by the $(j-1)$th sensor, then the ith dependence model can be represented by

$$f^i_\delta(\delta_1, \ldots, \delta_{i-1}) = f^i_\delta(\delta_{i-1}|\delta_{i-2}) \ldots f^i_\delta(\delta_2|\delta_1)f^i_\delta(\delta_1), \quad (8)$$

describing a Markovian chain of decision makers, as shown in Fig. 1.

The use of conditional probability distributions in this way induces a *network* structure of relations between different sensors and cues. This network is a constraint exposing description of sensor capabilities: Each arc in the network describes a dependence be-

Fig. 1. The Markovian team decision network.

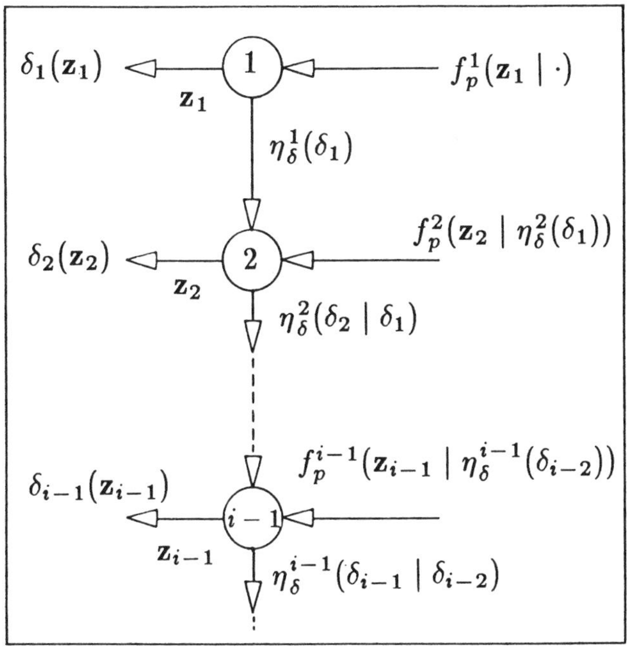

tween sensor observations, represented by a constraining transformation and implemented by the propagation of prior information between sensors. Each decision-making node or sensor uses this prior information to guide or constrain the acquisition of new observations. There are two distinct processes going on during this interaction: the local decision made by each sensor, and the transformation of this information between decision-making nodes. Each sensor takes observations described by the observation model $f_p(\mathbf{z}_i|\mathbf{p})$ and is provided with prior information from other sensors through the dependence model $f_\delta(\mathbf{z}_i|\cdot)$. The sensor makes a decision $\delta_i(\cdot)$ based on the product of observed and prior information. This is passed on to the next sensor and interpreted in terms of its dependence model $f_\delta(\mathbf{z}_{i+1}|\delta_i(\cdot), \ldots)$, which transforms the information provided by δ_i into the feature description understood by δ_{i+1}. This transformation and combination of information from disparate sources demonstrates the clear advantage of describing information in terms of a dimensionless probability distribution function.

These dependency models and opinion networks provide a powerful means of describing the dynamic exchange of information between different sensor systems. These models will be developed further in the context of the decision problem in the next section.

2.6. State Models

Many robot sensors are able to change their observation characteristics by either relocating themselves or altering their internal state. For example, a camera with controllable zoom and focus (Krotkov 1986), a mobile ranging device with variable viewing angle, or a tactile probe taking observations by touching objects in the environment (Cameron, Daniel, and Durrant-Whyte 1988). To be able to use these *active* sensor capabilities in an intelligent manner, we must be able to model the dependency of observations on sensor state or location. A model of this dependency will allow us to develop sensor control strategies and provide a mechanism with which to actively observe and localize objects in the environment.

We will describe the dependence of sensor observations on location and state by a state model $f_x(\mathbf{z}_i|\mathbf{x}_i)$, which describes the posterior likelihood of feature observation in terms of a state vector $\mathbf{x}_i$. This model acts as a "view-modifier" on the other sources of information supplied to the sensor, a function that transforms the prior information $f_\delta(\mathbf{z}_i|\cdot)$ and observation model $f_p(\mathbf{z}_i|\cdot)$ to the current viewpoint of the sensor. This transformation is just a product of distribution functions:

$$f_i(\mathbf{z}_i|\mathbf{x}_i, \mathbf{p}_i, \bar{\delta}_i) = f_x(\mathbf{z}_i|\mathbf{x}_i)[f_p(\mathbf{z}_i|\mathbf{p}_i)f_\delta(\mathbf{z}_i|\bar{\delta}_i)]. \quad (9)$$

There are two related parts to this description of a state model: the dependency of observation uncertainty on sensor state, and the transformation of prior world information to the current sensor location to determine if a feature is in view. Both of these considerations involve a process of transforming feature descriptions between coordinate systems.

Consider a mobile sensor located in space by the vector $\mathbf{x} = [x, y, z, \phi, \theta, \psi]^T$, observing a given feature $\mathbf{g}(\mathbf{x}, \mathbf{p}) = \mathbf{0}$ parameterized by the vector $\mathbf{p}$ as shown in Fig. 2. Suppose that the observations made by this sensor when in a fixed location can be described by a

Fig. 2. A mobile sensor observing a single geometric feature.

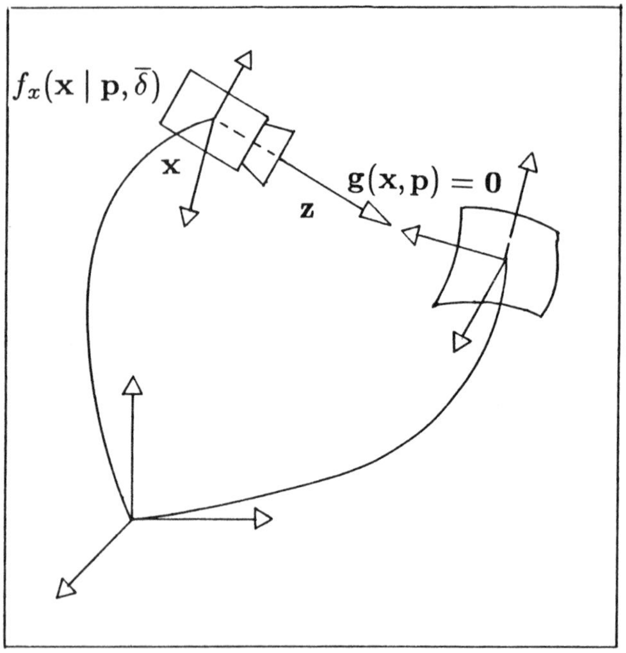

Gaussian distribution $\mathbf{z}_i \sim N(\hat{\mathbf{p}}, \Lambda_p)$. This is our observation model. When a Gaussian feature is transformed between coordinate systems as $\mathbf{p}_j = {}^j\mathbf{h}_i(\mathbf{p}_i)$, the mean vector follows the usual laws of geometry (Durrant-Whyte 1988), $\hat{\mathbf{p}}_j = {}^j\mathbf{h}_i(\hat{\mathbf{p}}_i)$, and with ${}^j\mathbf{J}_i = \partial^j\mathbf{h}_i/\partial\mathbf{p}_i$ the variance matrix is transformed by

$$^j\Lambda_p = {}^j\mathbf{J}_i(\mathbf{x})\,{}^i\Lambda_p\,{}^j_i\mathbf{J}^T_i(\mathbf{x}).$$

This transform can be interpreted as a change in sensor location resulting in a change in feature observation. In this case, the observation model will be state dependent:

$$\mathbf{z}_i \sim N({}^j\mathbf{h}_i(\hat{\mathbf{p}}_i), \mathbf{J}\Lambda_p\mathbf{J}^T) \sim N(\hat{\mathbf{p}}_j, \Lambda_p(\mathbf{x})),$$

describing how the uncertainty and perspective in feature observation are affected by the location of the sensor. Now suppose that we have some prior information about a feature to be observed, either from a prior world model or from observations made by other sensors, described again as $\mathbf{g}(\mathbf{x}, \mathbf{p}) = \mathbf{0}$, with $\mathbf{p} \sim N(\hat{\mathbf{p}}, \Lambda_p)$. Neglecting the possibility of occlusion, this information can be transformed into the coordinate system of the sensor through the relation $\mathbf{p}' = \mathbf{h}(\mathbf{p})$. This transformation is described by the state model $f_x(\cdot\,|\mathbf{x})$ and implemented by equations relating the mean and variance in different coordinate systems. This allows us to describe the prior information available to a sensor in terms of the features it may actually view.

The development of sensor strategies follows naturally from this description of observation dependence on sensor state. The sensor observation model described in terms of the mean and variance of sensor observations is now a function of the state and location of the sensor. If, for example, information about some particular feature is required, then this dependence of feature description on state can be used to determine the value of $\mathbf{x}$ which would give the measurements made by the sensor some appropriate characteristics. For example, knowing the function $\Lambda_\mathbf{p}(\mathbf{x})$, we can proceed to find the sensor state $\mathbf{x}$ that minimizes (in some sense) the uncertainty in taking observations of $\mathbf{p}$.

A second important problem in modeling state dependencies is describing the effect on a sensor's observations of *other* sensor states. This most obviously occurs when two or more sensors obtain information from the same physical device. A harder problem arises when observations made by two different physical devices are dependent on each other's state, such as the problem of hand-eye coordination. A simple example of this is discussed in Hager and Durrant-Whyte 1988).

3. Integrating Sensor Information

We are primarily interested in teams of observers, sensors making observations of the state of the environment. In this case individual team members can be considered as Bayesian estimators. The team decision is to come to a consensus view of the observed state of nature. The static team of estimators is often called a multi-Bayesian system (Weerahandi and Zidek 1981; 1983). Each individual sensor or team member communicates information in the common language of uncertain geometry. This allows different sensors to exchange information and utilize diverse opinions from other, different, sources. Each sensor must also communicate a preference order on its decisions to provide a means of resolving any differences of opinion among sensors. In a team of Bayesians, this preference order is described by a likelihood function. The description of preferences in terms of a probability distribution provides a dimensionless means of comparing estimates of disparate geometric features in a common framework. The communication of observations and preferences forms the basis for coordinating the acquisition of information and provides a mechanism for the generation and verification of team hypotheses.

The description of a multisensor system as a team of decision makers contains a number of important elements:

How individual sensors make local decisions based on their own observations
How the capabilities and information provided by each sensor are described to other members of the team
How information obtained by one sensor is described and communicated to other sensors
How local decisions made by individual sensors can be integrated to provide a team decision and action.

We will develop a description of a sensor as a Bayesian observer of geometric states in the environment. The capabilities of each sensor are represented by the probabilistic information structure, describing the observation of geometric features with respect to the state of the sensor and the availability of prior information. The communication of information will be described by the transformation and interpretation of uncertain geometric features. Decisions will be made by testing of dimensionless likelihoods on different geometric observations and consequent integration using Bayesian decision procedures.

3.1. The Team Structure

The observations $\mathbf{z}_i$ made by different sensors can be described, in terms of uncertain geometry, as a probability distribution function. Each sensor must make a decision δ_i, based on these observations to estimate some geometric feature in the environment $\delta_i(\mathbf{z}_i) \in \mathcal{P}_i$.

The decisions made by each sensor must be evaluated with respect to some measure of preference. This preference ordering, or utility function $u_i(\cdot, \delta_i) \in \Re$, allows different observations to be compared in a common framework. Consider for example two sensors taking observations $\mathbf{z}_1$ and $\mathbf{z}_2$ of two physically disparate geometric features $\mathbf{p}_1$ and $\mathbf{p}_2$, respectively. Suppose that the observed features are related in some way so that the estimate $\delta_1(\mathbf{z}_1)$ is constrained by the estimate $\delta_2(\mathbf{z}_2)$. In general, these two estimates cannot be compared directly and so the constraints between them cannot be made explicit. However, we can compare the contribution or utility they each provide to some joint consensus. Thus by first choosing some consensus estimate, we can compare diverse opinions by evaluating $u_i(\cdot, \delta_i(\mathbf{z}_i))$. The local decision is to maximize u_i, the group decision is to incorporate geometric constraint between sensors in this maximization.

A sensor or team member will be considered *rational* if for each observation $\mathbf{z}_i$ of some prior feature $\mathbf{p}_i \in \mathcal{P}_i$ it makes the estimate $\delta_i(\mathbf{z}_i) \in \mathcal{P}_i$, which maximizes its individual utility $u_i(\mathbf{p}_i, \delta_i(\mathbf{z})_i) \in \Re$. In this sense, utility is just a metric for constructing a complete lattice of decisions, thus allowing any two decisions to be compared in a common framework.

Individual rationality alone imposes insufficient structure on the organization of team members: If each sensor is allowed to make its own decisions regardless of other sensor opinions, then a group consensus will never be reached. We must provide a structure which allows for sensor opinions to be constrained or biased toward a common team objective. To do this, we will define a *team* utility function defined on the state of the environment $\mathbf{p}$ and the team decision function $\boldsymbol{\delta}$:

$$U = U(\mathbf{p}, \delta_1, \delta_2, \ldots, \delta_n). \tag{10}$$

This is a real-valued function which for every team decision $\boldsymbol{\delta}$ assigns a team utility, admitting a preference (in some sense) on group decisions. The fundamental problem in team decision theory is based on finding this preference ordering on $\boldsymbol{\delta}$. This problem, the formulation of group rationality axioms, can be quite complex and is still an open research issue; see, for example, Harsanyi (1977) and Bachrach (1975).

We can considerably simplify the team decision problem by restricting the type of group preference orderings that can be considered "rational." Consider a multisensor team taking observations of a number of different features in the environment. Each sensor has a preference ordering on the decisions that it can make; if it estimates the location of a feature, then it would prefer to have this estimate verified rather than disproved or relocated by the group decision. Clearly it must allow its decisions to be changed by at least a small amount so that it can be brought into line with a consistent interpretation of all sensors' observations. However, it should not allow its opinion to be changed so much that the group opinion now bears no resemblance to the sensors' original observations. This dichotomy between consensus and disagreement is fundamental to the multisensor decision problem.

If the sensor system suggests an explanation for all the different sensor observations, then each sensor must interpret this decision in terms of an explanation for its own measurements. If the sensor cannot reconcile its views with those of the team, then it must disagree and not support the team decision. If, however, it can support the team interpretation of sensor observations, the sensor should reflect this in its individual preference ordering. In this situation, the sensor system will try to find decisions that explain as many different sensor observations as possible, suggesting consistent interpretations for the different opinions, but allowing sensors to disagree. If sensors were not allowed to disagree, then they may be required to support hypotheses that are actually incorrect. The role of the sensors and sensor system in this case is to dynamically find a consistent, consensus estimate of the state of the environment by alternately suggesting and trying to verify possible environment hypotheses. This process of dynamic interaction, agreement, and disagreement is called the *bargaining* problem (Nash 1950).

3.2. The Multi-Bayesian Team

We will consider a team structure composed of individual Bayesian observers, each making observations of uncertain geometric features in the environment, communicating prior information to each other through geometric transforms, and making decisions based on likelihood.

A multi-Bayesian team works by considering the likelihood function $f_i(\cdot|\mathbf{p})$ of each team member as the (normalized) utility of individual observers. Then the team utility is considered to be the joint posterior distribution function $F(\mathbf{p}|\mathbf{z}_1, \ldots, \mathbf{z}_n)$ after each sensor i has made the observation $\mathbf{z}_i$. The advantage of considering the team problem in this framework is that both individual and team utilities are normalized so that interperson comparison of decisions can be performed easily, supplying a simple and transparent interpretation to the group rationality problem.

Before developing the general multi-Bayesian system, it is helpful to study the simpler case of two scalar homogeneous observers. Consider two team members, each observing the same scalar variable $p \in \mathcal{P}$, with observation density $f_i(\cdot|p)$, $i = 1, 2$. Suppose each observer takes a single observation z_i, considered independent and derived from a Gaussian distribution with mean $\hat{p}$ and variance σ_i^2. The goal of this team is to come to some consensus estimate of state $\bar{p} \in \mathcal{P}$, based on the two observations z_1 and z_2. In the multi-Bayesian system, each team member's individual utility function is given by the posterior likelihood $f(p|z_i) \sim N(\hat{p}, \sigma_i^2)$. The team utility function is given by the joint posterior likelihood $F(p|z_1, z_2) = f_1(p|z_1)f_2(p|z_2)$. A team member will be considered *individually* rational if it chooses the estimate $\bar{p} \in \mathcal{P}$ which maximizes its local posterior density:

$$\bar{p} = \arg\max_{p\in\mathcal{P}} f_i(p|z_i), \qquad i = 1, 2. \tag{11}$$

The team itself will be considered group rational if together the team members choose the estimate $\bar{p} \in \mathcal{P}$ which maximizes the joint posterior density:

$$\begin{aligned} \bar{p} &= \arg\max_{p\in\mathcal{P}} F(p|z_1, z_2) \\ &= \arg\max_{p\in\mathcal{P}} f_1(p|z_1)f_2(p|z_2). \end{aligned} \tag{12}$$

Whether or not the individual team members will arrive at a consensus team estimate will depend on some measure of how much they disagree, $|z_1 - z_2|$. If z_1 and z_2 are "close enough," then the two Bayesians should agree to use the posterior density $F(p|z_1, z_2)$ as their joint utility function and provide a joint estimate that satisfies Eq. (12). As $|z_1 - z_2|$ increases, they should "agree to disagree" and use their individual posterior densities as utility functions, providing individual estimates satisfying Eq. (11).

In a multi-Bayesian system, the point of disagreement occurs when the likelihood of individual estimates exceeds that of the group estimate. The best way to visualize this is by considering the *set*

$$\mathbf{f}(p) \times [f_1(p|z_1), f_2(p|z_2)]^T \in \Re^2$$

describing the preference space of the two observers. When $|z_1 - z_2|$ is small, this space is convex, and a joint utility vector can be found which is larger than any individual utility vector. As $|z_1 - z_2|$ increases, this space becomes concave, so that individual utility vectors become larger than any joint consensus (Weerahandi and Zidek 1983).

To find the point at which this space is no longer convex and disagreement occurs, all we need ensure is that the second derivative of the function $F(p|\cdot)$ is positive. Differentiating, we obtain

$$\frac{\partial^2 F}{\partial p^2} = \frac{1}{f_1}\frac{d^2 f_1}{dp^2} + \frac{1}{f_2}\frac{d^2 f_2}{dp^2} + \frac{2}{f_1 f_2}\frac{df_1}{dp}\frac{df_2}{dp}$$
$$= (\sigma_1^{-2} + \sigma_2^{-2}) - [\sigma_1^{-2}(p - z_1) + \sigma_2^{-2}(p - z_2)]^2.$$

For this to be positive and hence $\mathbf{f}(p)$ to be convex, we are required to find a consensus p which satisfies

$$[\sigma_1^{-2}(p - z_1) + \sigma_2^{-2}(p - z_2)]^2[\sigma_1^{-2} + \sigma_2^{-2}]^{-1} \leq 1. \quad (13)$$

Notice that Eq. (13) is no more than a normalized weighted sum, a scalar equivalent to the Kalman gain matrix. It follows, therefore, that if a p can be found which satisifes Eq. (13), then the consensus $\bar{p}$ which maximizes F will be given by

$$\bar{p} = (\sigma_1^{-2} z_1 + \sigma_2^{-2} z_2)/(\sigma_1^{-2} + \sigma_2^{-2}). \quad (14)$$

Substituting Eq. (14) into Eq. (13), we can write

$$(z_1 - z_2)(\sigma_1^2 + \sigma_2^2)^{-1}(z_1 - z_2) = D_{12}(z_1, z_2),$$

where $D_{12} \leq 1$ in Eq. (13). Figure 3 shows a plot of the

Fig. 3. The space of preferences for a two-Bayesian team.

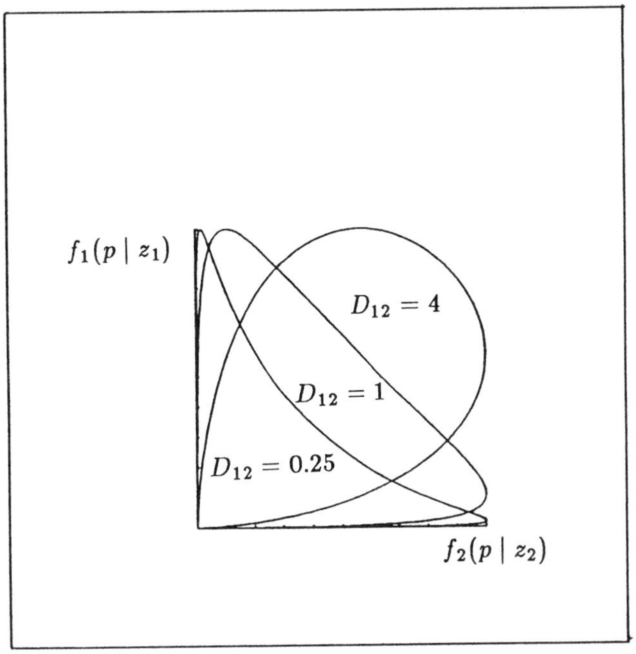

set $\mathbf{f}(p)$ for two Bayesians. The disagreement measure $D_{12} = D_{12}(z_1, z_2)$ is called the Mahalanobis distance. Figure 3 shows that for $D_{12} \leq 1$, the set $\mathbf{f}(p)$ is convex and a consensus, given by Eq. (14), exists. As the difference $|z_1 - z_2|$ increases, D_{12} becomes larger and eventually the set $\mathbf{f}(p)$ becomes concave, indicating disagreement. The consensus value of p describes an envelope of extrema of $\mathbf{f}(p)$. This provides an important visualization of agreement or disagreement between a set of Bayesians.

We will now generalize this two-Bayesian system to a team of n Bayesians all taking different observations of geometric features in the environment. Consider n sensors each taking observations $\mathbf{z}_i \in \mathcal{P}_i$ of different geometric features $\mathbf{p}_i \in \mathcal{P}_i$, $\mathcal{P}_i \neq \mathcal{P}_j$. To compare geometrically disparate observations, we must hypothesize the existence of some common geometric object $\mathbf{p} \in \mathcal{P}$ to which these observations are related. For example, this object could be the location of a centroid or the description of a common surface. The observations made by a sensor can be related to this common object through the decision function $\delta_i(\mathbf{z}_i) \in \mathcal{P}$. Although $\delta_i^{-1}(\mathbf{p}) \in \mathcal{P}_i$ is usually well defined, $\delta_i(\mathbf{z}_i) \in \mathcal{P}$ is often only a partial mapping. For example, given the centroid of a known polyhedra, we can uniquely determine the location of all its edges and surfaces. Conversely, given an observed edge of this polyhedra, it is not possible to uniquely determine the object's centroid without recourse to other constraining information. However, although the decision $\delta_i(\mathbf{z}_i)$ contains only partial information about $\mathbf{p}$, we can consider the observation in terms of uncertain geometry and thus manipulate or transform it using the techniques described in the previous section.

Following our development of the two-Bayesian team, we will consider an individual sensor's preference ordering on consensus descriptions of a common geometric object $\mathbf{p}$ as the posterior density $f_i(\mathbf{p}|\delta_i(\mathbf{z}_i))$, and the joint preference ordering as the joint posterior density:

$$F(\mathbf{p}|\delta_1(\mathbf{z}_1), \ldots, \delta_n(\mathbf{z}_n)) = \prod_{i=1}^{n} f_i(\mathbf{p}|\delta_i(\mathbf{z}_i)). \quad (15)$$

We will call the set

$$\mathbf{f}(\mathbf{p}) = [f_1(\mathbf{p}|\delta_1(\mathbf{z}_1)), \ldots, f_2(\mathbf{p}|\delta_1(\mathbf{z}_n))]^{\mathrm{T}} \subseteq \Re^n$$

the opinion space of the n-sensor team. This is an n-dimensional space with basis axis corresponding to individual sensor preferences f_i. The set $\mathbf{f}(\mathbf{p})$ describes an $(n-1)$-dimensional surface in this space, parameterized by $\mathbf{p}$ and enclosing a volume $v(\mathbf{z}_1, \ldots, \mathbf{z}_n) \in \Re^n$. A consensus of *all* sensors requires that this volume be convex along each axis. The consensus value lies on the surface enclosing this volume. If $v \in \Re^n$ is concave along the jth axis, then the jth sensor observation will be unable to agree with other sensor opinions; its individual maximum likelihood estimate is preferred to the team decision. Thus if the volume v is concave in one or more directions f_i, then coalitions supporting different decisions will be formed. This occurrence is quite natural; individual sensor systems may indeed be viewing different features and so should not agree with a single consensus decision. Further, a partitioning of the f_i into coalitions should tell the sensor system that its suggested consensus is only partially correct and should be altered to bring the different coalitions into a single, convex opinion.

Following similar arguments to those of the two-Bayesian system, the observations $\mathbf{z}_i$ will only provide a consensus estimate for $\mathbf{p}$ when the subspaces of the opinion space are convex. To determine convexity, we need only ensure that the hessian of F is positive semidefinite. It can be shown (Durrant-Whyte 1987b) that a consensus value of $\mathbf{p}$ which satisfies this also satisfies

$$\left[\sum_{i=1}^{n} \Lambda_i^{-1}(\mathbf{p} - \delta_i(\mathbf{z}_i))\right]^{\mathrm{T}} \left[\sum_{i=1}^{n} \Lambda_i^{-1}\right]^{-1} \left[\sum_{i=1}^{n} \Lambda_i^{-1}(\mathbf{p} - \delta_i(\mathbf{z}_i))\right] \leq 1. \quad (16)$$

If it exists, the consensus parameter estimate $\hat{\mathbf{p}}$ that maximizes Eq. (15) and minimizes the left side of Eq. (16) is given by the modified Kalman minimum-variance estimate

$$\hat{\mathbf{p}} = \left[\sum_{i=1}^{n} \Lambda_i^{-1}\right]^{-1} \left[\sum_{i=1}^{n} \Lambda_i^{-1} \delta_i(\mathbf{z}_i)\right]. \quad (17)$$

To compare and integrate different sensor information, we must first allow each sensor to make its best geometric decision $\delta_i(\cdot)$, based on its own observations, and transform this decision geometrically into a description understood by other sensors. In this common framework, different decisions can then be compared using Eq. (16) and, if there are grounds for agreement, integrated using Eq. (17).

3.3. Integrating Sensor Information

The hypothesis $\{\mathbf{p}_i\}$ generated by each sensor from its observations can be interpreted in terms of partial hypothesis $\{\mathbf{p}\}$ of some underlying global geometric environment description $\mathbf{g}(\mathbf{x}, \mathbf{p}) = \mathbf{0}$, $\mathbf{p} \in \mathcal{P}$. The fact that individual sensor hypothesis $\mathbf{p}_i$ can only ever supply partial estimates of the global geometry $\mathbf{p}$ is a primary motivation for the use of many sources of sensory information. The partial hypothesis on global geometry provided by the sensors provides a means of comparing and combining different sensor views.

Consider a single sensor taking observations $\{\mathbf{z}_i\}$ of a particular type of feature in the environment $\mathbf{g}_i(\mathbf{x}, \mathbf{p}_i) = \mathbf{0}$. Each sensor observation is described through the observation model $f_i^p = f_i(\mathbf{z}_i|\mathbf{p}_i)$. The features described by $\{\mathbf{p}_i\}$ are considered to be related to some underlying global geometric description $\mathbf{g}(\mathbf{x}, \mathbf{p}) = \mathbf{0}$ through the transformation $\{\mathbf{p}_i\} = \mathbf{h}_i(\mathbf{p})$, $\mathbf{p}_i \in \mathscr{P}_i$, $\mathbf{p} \in \mathscr{P}$. This is a stochastic transformation of one type of geometry to another. The forward transformation $\mathbf{h}_i(\cdot)$ is usually well defined, and if we have some prior information about $\mathbf{p}$, $\mathbf{h}_i$ can be used to obtain prior feature information. However, in general no prior information will be available. In this case we must use the inverse transform to generate hypotheses about the underlying geometry: $\mathbf{p} = \mathbf{h}_i^{-1}(\mathbf{p}_i)$. The inverse transform is usually indeterminant, a single estimate $\mathbf{p}_i$ insufficient to generate a complete hypothesis $\mathbf{p}$.

Consider now a multisensor system comprising n sensors $S_1, \ldots, S_n$, each taking a set of geometric observations $\{\mathbf{z}_i\}$ of geometric features $\{\mathbf{p}_i\}$ in the environment. From these observations, each sensor can make individual estimates $\delta_i(\mathbf{z}_i) \in \mathscr{P}_i$ of possible features of a particular type, resulting in a set of feature hypotheses $\{\mathbf{p}_i'\} = \{\delta_i(\mathbf{z}_{i,1}), \ldots, \delta_i(\mathbf{z}_{i,1})\} = \{\mathbf{p}_{i,1}', \ldots, \mathbf{p}_{i,1}'\}$. We will consider the posterior distribution $f(\mathbf{p}_i|\mathbf{z}_i)$ as the preference order induced by the observations $\mathbf{z}_i$ on possible hypothesis $\mathbf{p}_i$. With this identification, we can consider the estimates $\mathbf{p}_i'$ as modeled by a Gaussian $\mathbf{p}_{i,j}' \sim N(\mathbf{p}_{i,j}, \mathbf{\Lambda}_{i,j})$. To transform these feature hypotheses on some underlying geometry, we need to apply the transformation $\mathbf{h}^{-1}(\cdot)$ to the set $\{\mathbf{p}_i'\}$. If the elements of this hypothesis set are considered Gaussian, we can approximate this transform by a transformation of mean and variance, so that for each $\delta_i(\mathbf{z}_i)$, we have

$$\hat{\mathbf{p}}' = \mathbf{h}_i^{-1}[\mathbf{p}_i'] = \mathbf{h}_i^{-1}[\delta(\mathbf{z}_i)],$$
$$\mathbf{\Lambda}_p^{-1} = \left(\frac{\partial \mathbf{h}_i^{-1}}{\partial \mathbf{p}_i}\right) \mathbf{\Lambda}_i^{-1} \left(\frac{\partial \mathbf{h}_i^{-1}}{\partial \mathbf{p}_i}\right)^{\mathrm{T}}. \tag{18}$$

Thus by transforming each feature hypothesis, we can obtain from each sensor S_i a set of hypotheses $\{\mathbf{p}\}_i$ on the underlying environment geometry, each of which can be described by a mean and variance. The information matrix $\mathbf{\Lambda}_p^{-1}$ will be singular, no information in one or more degrees of freedom, because the transform $\mathbf{h}_i^{-1}(\cdot)$ is indeterminant. This provides for the feature hypotheses $\mathbf{p}_i'$ to generate only partial estimates of the underlying geometry.

We now have a means of generating partial hypotheses of the environment geometry, regardless of the existence of prior information, in a language common to all sensors. It remains now to compare, verify, and combine these partial estimates to provide a complete description of the underlying environment geometry. We will formulate this problem in terms of a multi-Bayesian team, verifying individual sensor hypotheses by comparing their contribution to some global environment description.

Consider the preference placed on environment descriptions $\mathbf{p}$ by each sensor's partial estimates $\mathbf{p}'$ as the posterior intensity $f_i(\mathbf{p}|\mathbf{p}') = f_i(\mathbf{p}|\mathbf{h}_i^{-1}[\delta_i(\mathbf{z}_i)])$, $i = 1, \ldots, n$. This preference ordering represents an *individual* sensor's preferred contribution to a team consensus. We will define the *team* preference ordering as the joint posterior density on all contributions:

$$F(\mathbf{p}|\mathbf{h}_1^{-1}[\delta_1(\mathbf{z}_1)], \ldots, \mathbf{h}_n^{-1}[\delta_n(\mathbf{z}_n)]) = \prod_{i=1}^{n} f_i(\mathbf{p}|\mathbf{h}_i^{-1}[\delta_i(\mathbf{z}_i)]). \tag{19}$$

We will seek consensus values of $\mathbf{p}$ which make the opinion space convex. The hypotheses generated from different sensor observations will be associated with many different values of $\mathbf{p}$. Rather than comparing all of these hypotheses in one opinion pool, it makes sense to compare different estimates pairwise, recursively clustering hypotheses into groups, one associated with each value of $\mathbf{p}$.

From Eq. (16), the pairwise comparison of two hypotheses $\mathbf{h}_i^{-1}[\delta_i(\mathbf{z}_i)]$ and $\mathbf{h}_j^{-1}[\delta_j(\mathbf{z}_j)]$ requires that we find a consensus hypothesis $\mathbf{p}$ which satisifes

$$\bar{\mathbf{p}} = [\mathbf{\Lambda}_i^{-1} + \mathbf{\Lambda}_j^{-1}]^{-1} \times [\mathbf{\Lambda}_i^{-1}\mathbf{h}_i^{-1}[\delta_i(\mathbf{z}_i)] + \mathbf{\Lambda}_j^{-1}\mathbf{h}_j^{-1}[\delta_j(\mathbf{z}_j)]]. \tag{20}$$

Thus for two sensors to agree on a hypothesis, their interpretations must satisfy

$$(\mathbf{h}_i^{-1}[\delta_i(\mathbf{z}_i)] - \mathbf{h}_j^{-1}[\delta_j(\mathbf{z}_j)])(\mathbf{\Lambda}_i + \mathbf{\Lambda}_j)^{-1} \times (\mathbf{h}_i^{-1}[\delta_i(\mathbf{z}_i)] - \mathbf{h}_j^{-1}[\delta_j(\mathbf{z}_j)])^{\mathrm{T}} \leq 1. \tag{21}$$

Note that the transformation of observations is often only partial, so the sensor may have invariant (or indifferent) opinion preference to certain degrees of freedom of the team hypothesis.

3.4. Coordinating Information Flow

Our development of multisensor system models has relied on the probabilistic description and transformation of uncertain geometric objects. We have outlined how these objects can be communicated to other sensors, compared, and integrated to provide a group estimate of geometric elements describing the environment. An important aspect of this development is the dynamic exchange of information between different sensors. The dynamic use of information may occur in a number of ways, all of which can be classified in terms of either competitive, complementary, or cooperative information intercation. The dependence model can be used to describe all three of these dynamic effects.

1. *Competitive information* interaction occurs whenever two (or more) sensors supply information in the same location and degrees of freedom. For a consensus decision to be made, the opinion space formed by $f_\delta^i(\mathbf{p} \mid \cdot)$ must be convex; otherwise differences of opinion must be resolved by resorting to further information. Note that sensors can describe quite different geometric features but, when transformed by f_δ^i to a common description, can still be competitive. Examples of competitive information interaction include feature clustering algorithms or any homogeneous observation integration problem.
2. *Complementary information* interaction occurs whenever two (or more) information sources supply different information about the same geometric feature, in different degrees of freedom. In this case, the joint information is composed of different nonoverlapping information sources. It follows that each dependence model f_δ^i must only supply partial information about the ith sensor's observations and, consequently, expresses no preference ordering on the information obtained. A simple example of this is the observation of different beacons of a triangulation system. Complementary information can often be considered in terms of one sensor "filling in the gaps" in another sensor's observations.
3. *Cooperative information* interaction occurs whenever one sensor relies on another for information, prior to observation. Of particular importance is the use of one sensor's information to guide the search for new observations; the guiding of a tactile sensor by initial visual inspection is an example of this.

These different mechanisms for information exchange provide a basis for the coordination of information flow between sensors. They provide a mechanism to resolve conflicts or disagreements and to allow the development of dynamic sensing strategies.

4. Summary and Conclusions

We have argued that the development of use of sensor models is essential in providing a basis for understanding and utilizing multisensor systems. We have described a method of modeling sensors which considers the sensors of a multisensor system as members of a team communicating, cooperating, and coordinating their actions toward the solution of a common goal. The sensors of this team communicate information to each other in terms of a probabilistic description of observed geometric features. The description of sensors and sensor information in geometric terms allows this communication of information to be performed using well-understood tools from geometry and probability theory.

The sensor models developed are based on the idea of a team information structure, describing a sensor's observations as a probabilistic function of state and decisions communicated from other information sources. We have described three models based on this structure: the observation model, the dependence model, and the state model. These models, when used in conjunction with our description of sensor information as uncertain geometry, provide a probabilistic description of sensor capabilities which allow the development of general methods for the integration, coordination, and control of multisensor robot systems.

We have presented a general outline of how these models can be used to provide sensing strategies in multisensor systems. These techniques are far short of full development, and much could be done to improve their utility in more general settings. In particular, we would encourage the development of better methods to manipulate and interpret geometric information. Further, it is clear that these techniques would benefit from being developed in conjunction with other types of sensor models, able to describe aspects of sensing which are not geometric.

References

Allen, P. 1987. *Object recognition using vision and touch.* Boston: Kluwer Academic.

Athans, M. 1987. Command and control theory: a challenge to controlscience. *IEEE Trans. Automatic Control* 32(4):286–293.

Ayache, N., and Faugeras, O. 1987 (London). Building, registrating, and fusing noisy visual maps. *Int. Conf. Computer Vision.*

Bachrach, M. 1975. Group decisions in the face of differences of opinion. *Management Sci.* 22:182.

Bajcsy, R., Krotkov, E., and Mintz, M. 1986. *Models of errors and mistakes in machine perception.* Technical Report MS-CIS-86-26, U. Pennsylvania, Dept. Computer Science.

Bolle, R. M., and Cooper, D. B. 1986. On optimally combining pieces of information, with application to estimating 3-d complex-object position from range data. *IEEE Trans. Pattern Analysis and Machine Intelligence* 8:619.

Cameron, A. R., Daniel, R., and Durrant-Whyte, H. 1988 (April 25–29, Philadelphia, Pa.). Touch and motion. *Proc. IEEE Int. Conf. Robotics and Automation.*

Durrant-Whyte, H. F. 1987a. Consistent integration and propagation of disparate sensor information. *Int. J. Robotics Research* 6(3):3–24.

Durrant-Whyte, H. F. 1987b. *Integration, coordination, and control of multi-sensor robot systems.* Boston: Kluwer Academic.

Durrant-Whyte, H. F. 1988. Uncertain geometry in robotics. *IEEE J. Robotics and Automation* 4(1):23–31.

Faugeras, O., and Ayache, N. 1986 (San Francisco, Calif.). Building visual maps by combining noisy stereo measurements. *Proc. IEEE Conf. Robotics and Automation.*

Flynn, A. M. 1985. *Redundant sensors for mobile robot navigation.* Technical Report M.Sc. Thesis, MIT.

Gelb, M. 1974. *Applied optimal estimation.* Cambridge, Mass.: MIT Press.

Hager, G. 1987. *Information maps for active sensor control.* Technical Report MS-CIS-87-07, U. Pennsylvania, Dept. Computer Science.

Hager, G., and Durrant-Whyte, H. F. 1988. Information and multi-sensor coordination. In *Uncertainty in Artificial Intelligence 2,* North-Holland.

Harsanyi, S. 1977. *Rational behavior and bargaining.* New York: Cambridge University Press.

Henderson, T. 1987. *Workshop on multi-sensor integration.* Technical Report UUCS-87-006, U. Utah Computer Science.

Henderson, T., and Hansen, C. 1985. The specification of distributed sensing and control. *J. Robotic Systems* 2:387–396.

Ho, Y. C. 1980. Team decision theory and information structures. *Proc. IEEE* 68:644.

Ho, Y. C., and Chu, K. C. 1972. Team decision theory and information structures in optimal control. *IEEE Trans. Automatic Control* 17:15.

Huber, P. J. 1981. *Robust statistics.* New York: Wiley.

Krotkov, E. 1986. *Focusing.* Technical Report MS-CIS-86-22, U. Pennsylvania Dept. Computer Science.

Marshak, J., and Radnor, R. 1972. *The economic theory of teams.* New Haven, Conn.: Yale University Press.

McKendall, R., and Mintz, M. 1987. *Models of sensor noise and optimal algorithms for estimation and quantization in vision systems.* Technical Report MC-CIS-87, U. Pennsylvania Dept. Computer Science.

Nash, J. F. 1950. The bargaining problem. *Econometrica* 155.

Porril, J., Pollard, S. B., and Mayhew, J. E. W. 1987. Optimal combination of multiple sensors including stereo vision. *Image and Vision Computing* 5:174–180.

Terzopoulos, D. 1986. Integrating visual information from multiple sources. In *From pixels to predicates,* ed. A. P. Pentland. Ablex Press.

Weerahandi, S., and Zidek, J. V. 1981. Multi-bayesian statistical decision theory. *J. Royal Statistical Society* 44:85.

Weerahandi, S., and Zidek, J. V. 1983. Elements of multi-bayesian decision theory. *The Annals of Statistics* 11:1032.

LOCATION ESTIMATION AND UNCERTAINTY ANALYSIS FOR MOBILE ROBOTS

C. Ming Wang

Computer Science Department
General Motors Research Laboratories
Warren, Michigan 48090

Abstract

A motion controller for the autonomous mobile vehicle commands the robot's drive mechanism to keep the robot near its desired path at all times. In order for the controller to behave properly, the controller must know the robot's position at any given time. The controller uses the information provided by the optical encoders attached to the wheels to determine vehicle position. This paper analyzes the effect of measurement errors, wheel slippage, and noise on the accuracy of the estimated vehicle position obtained in this manner. Specifically, the location estimator and its uncertainty covariance matrix are derived.

1. Introduction

Determining the location of a robot is an important problem in navigating an autonomous vehicle in an unstructured environment. In a two dimensional space, the location of a mobile robot can be represented by a triplet (x, y, θ) ([7]) where x, y and θ are, respectively, the position and orientation of the robot. A problem arises from the fact that there is always error associated with the robot's motion. For example, in a two-wheeled drive system, the robot controller uses the information provided by the optical encoders attached to the wheels to command the robot's drive mechanism to keep the robot near its desired path at all time. However, the robot drive mechanism and controller may or may not follow commands very well and the measurements from the optical encoders are not error-free. These imprecisions are assumed random and can be modeled by a parametric distribution. The parameters of the error distribution can be determined experimentally.

Several methods for quantifying and dealing with this uncertainty have been proposed in the literature. Chatila and Laumond[3] used a scalar error estimate as the uncertainty measure for position and were not concerned with angular error. The uncertainty measure is used as the weight in combining the redundant measurements of the same entity. Brooks[1], employing a min/max error bounds approach, developed the uncertainty manifold for each location estimator. Smith and Cheeseman[8] (also see Smith, Self and Cheeseman[9]) used the covariance matrix as the uncertainty measure for the location estimator. They also introduced two operations which can be used to manipulate the relationship between any coordinate frame, given the chain of uncertain relative transformations linking them. The first operation is called compounding and can be better explained by Figure 1 (taken from Smith and Cheeseman[8]).

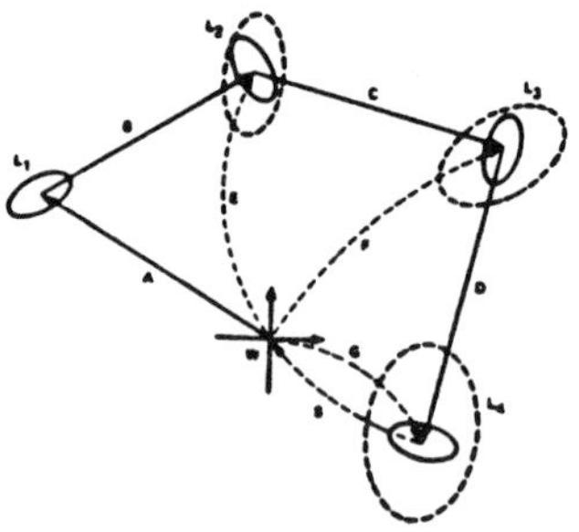

Figure 1. A Sequence of Transformations

In this example, the robot makes a number of moves and ends up near its initial position W. The solid ellipses represent the relative uncertainty of the robot with respect to its last position, while the dashed ellipses are the uncertainty of the robot with respect to W. The compounding process calculates the nominal location and associated error (dashed ellipse) of any object relative to any other object linked through a chain of transformations. The resulting compounded transformation has greater uncertainty than its components. That is, as the robot moves from one place to another, the uncertainty about its location with respect to the initial location grows.

The uncertainty could grow to a point that it becomes impossible for a robot to make any meaningful inference about its whereabouts. By using some auxiliary information, such as data from sensing landmarks, the uncertainty can be reduced. The reduction is done by employing a technique such as the Kalman filter to combine the existing uncertainty with new information. This is the merging process used by Smith and Cheeseman[8].

The calculation of the covariance matrix, in most cases, is not straightforward because of the nonlinear nature of the location estimator. Approximations are often needed. For example, Smith and Cheeseman[8] used the first-order Taylor expansion to evaluate the uncertainty covariance matrix. Wang[11] analytically verified the appropriateness of the first-order approximation, and described the limits of applicability.

In this article, the two-wheeled robot drive system based on the information from the optical encoders is examined. The estimation of robot states and the uncertainty analysis for such a system are presented. Specifically, the location estimator and its covariance matrix are derived. A comparison among existing methods for calculating the covariance matrix of the location estimator is also studied by simulation.

2. The Two-Wheeled Robot Drive System

A two-wheeled robot has two opposed drive wheels, mounted on the left and right sides of the robot, with their common axis passing through the center of the robot. The movement of the robot as a whole is indicated by the motion of the midpoint of the axis. In Figure 2 (also discussed by Tsumuru and Fujiwara[10]), the left and right wheel positions are denoted by A and C, and the midpoint of the axis, B, is the robot position reference point.

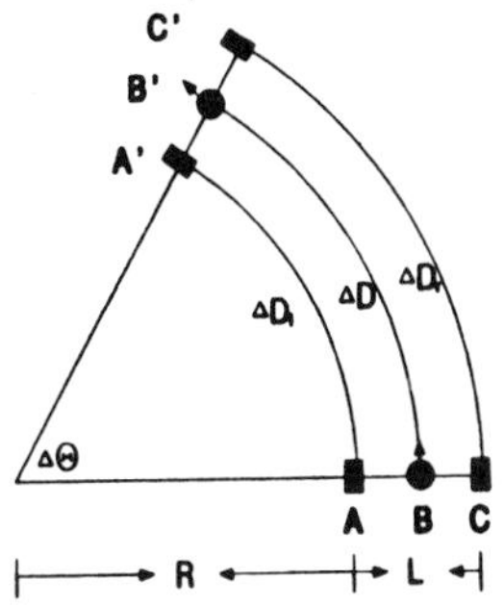

Figure 2. Illustration of the Drive System

In this example, the robot moves from B to B'. The length of the axis is $L = AC = A'C'$. The distance travelled, ΔD, and the angle changed, $\Delta\theta$, resulting from the movement can be calculated in terms of the incremental changes of the odometric measurements of the right and left wheel motions. Let ΔD_r and ΔD_l denote the covered distances of the right and left wheels respectively, then

$$\Delta D_r = (L+R)\,\Delta\theta, \quad \Delta D_l = R\,\Delta\theta.$$

Thus, we have

$$\begin{aligned}\Delta D &= (\Delta D_r + \Delta D_l)/2,\\ \Delta\theta &= (\Delta D_r - \Delta D_l)/L.\end{aligned} \tag{2.1}$$

That is, ΔD is the average of the outer and inner arcs and $\Delta\theta$ is proportional to the difference of the outer and inner arcs.

3. The Location of the Robot

In a two-dimensional space, the location (or state) of the robot at step (or time) n can be represented by

$$\mathbf{U}_n = (X_n\; Y_n\; \theta_n)^t.$$

Suppose we know that at time $n-1$, the robot is located at $O = (X_{n-1}, Y_{n-1})$ and is oriented in the direction of the point A as shown in Figure 3, i.e. $\theta_{n-1} = \angle AOF$. We wish to determine the location and orientation of the robot at time n, given that we know ΔD_n and $\Delta\theta_n$.

The new orientation, θ_n, is given by $\theta_{n-1} + \Delta\theta_n$. But the position of the robot is unknown. The robot can take any path that starts at O, has total arc length ΔD_n and turns by a total of $\Delta\theta_n$. To determine the position of the robot, we need to make some assumptions about the type of path it follows. If we assume a circular path, then (see Figure 3) $\Delta D_n = arc\ OP$ and $\Delta\theta_n = \angle ABP$. The location of the robot at time n, P, can be calculated as follows.

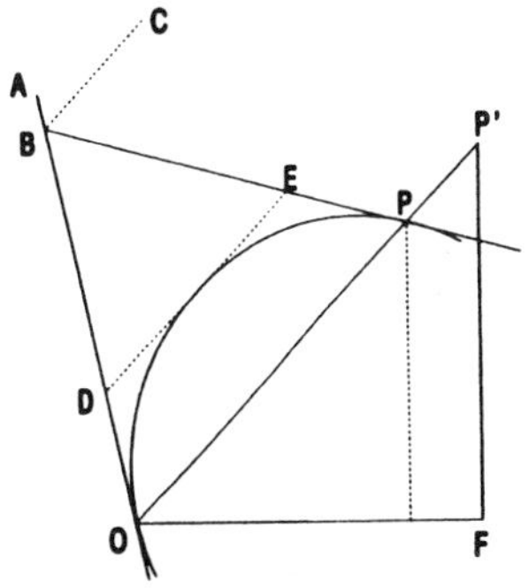

Figure 3. Illustration of Relationship of Positions

It can be shown that BC, which is parallel to DE and to OP, bisects the angle $\angle ABP$. That is,

$$\angle ABC = \angle ADE = \angle AOP = \Delta\theta_n/2.$$

Therefore, we have

$$\angle POF = \angle AOF - \angle AOP = \theta_{n-1} + \Delta\theta_n/2.$$

For the general case of arbitrary path, the length of OP is unknown, one commonly used method (e.g. Tsumura and Fujiwara[10], Julliere et al. [6], Iijima et al. [4]) is to approximate P by P' with the length of $OP' = arc\ OP = \Delta D_n$. With this approximation, we have

$$\begin{aligned}\Delta X_n &\approx \Delta D_n \cos(\theta_{n-1} + \Delta\theta_n/2)\\ \Delta Y_n &\approx \Delta D_n \sin(\theta_{n-1} + \Delta\theta_n/2),\end{aligned} \tag{3.1}$$

and the location of the robot at time n is approximated by

$$\begin{aligned}X_n &= X_{n-1} + \Delta X_n \approx X_{n-1} + \Delta D_n \cos(\theta_{n-1} + \frac{\Delta\theta_n}{2})\\ Y_n &= Y_{n-1} + \Delta Y_n \approx Y_{n-1} + \Delta D_n \sin(\theta_{n-1} + \frac{\Delta\theta_n}{2})\\ \theta_n &= \theta_{n-1} + \Delta\theta_n.\end{aligned} \tag{3.2}$$

However, with the added assumption of circular path, the relationship between the length of OP and arc OP can be obtained as (also see notations in Figure 2)

$$\frac{OP}{arc\,OP} = \frac{2(L/2+R)\sin(\Delta\theta_n/2)}{(L/2+R)\Delta\theta_n} = \frac{\sin(\Delta\theta_n/2)}{\Delta\theta_n/2}.$$

That is,

$$OP = \frac{\sin(\Delta\theta_n/2)}{\Delta\theta_n/2}\Delta D_n.$$

With this result, we have

$$\begin{aligned}\Delta X_n &= \frac{\sin(\Delta\theta_n/2)}{\Delta\theta_n/2}\Delta D_n \cos(\theta_{n-1} + \frac{\Delta\theta_n}{2})\\ \Delta Y_n &= \frac{\sin(\Delta\theta_n/2)}{\Delta\theta_n/2}\Delta D_n \sin(\theta_{n-1} + \frac{\Delta\theta_n}{2}),\end{aligned} \tag{3.3}$$

and the location of the robot at time n is given by

$$\begin{aligned}X_n &= X_{n-1} + \frac{\sin(\Delta\theta_n/2)}{\Delta\theta_n/2}\Delta D_n \cos(\theta_{n-1} + \frac{\Delta\theta_n}{2})\\ Y_n &= Y_{n-1} + \frac{\sin(\Delta\theta_n/2)}{\Delta\theta_n/2}\Delta D_n \sin(\theta_{n-1} + \frac{\Delta\theta_n}{2})\\ \theta_n &= \theta_{n-1} + \Delta\theta_n.\end{aligned}$$

or in matrix form

$$\mathbf{U}_n = \mathbf{U}_{n-1} + \Delta\mathbf{U}_n. \tag{3.4}$$

Note that $\sin\phi/\phi \to 1$ as $\phi \to 0$, the term $\sin(\Delta\theta_n/2)/(\Delta\theta_n/2)$ can be viewed as an adjustment factor for the robot location due to the circular movement.

4. The Error of the Location Estimation

There is error in calculating the location of the robot at each time step. The error arises from the fact that the odometric measurements from both wheels are imperfect due to sensor noise and wheel slippage. These imprecisions are assumed random and can be modeled by a parametric distribution. In other words, we have

$$\Delta\hat{D}_r = \Delta D_r + \epsilon_r$$
$$\Delta\hat{D}_l = \Delta D_l + \epsilon_l,$$

where $\Delta\hat{D}_r$ and $\Delta\hat{D}_l$, respectively, are the observed distance that right and left wheels have moved; and ϵ_r and ϵ_l are random error. Furthermore, if we can assume that ϵ_r and ϵ_l are independently normally distributed with means 0 and variances σ_r^2 and σ_l^2, then

$$\Delta\hat{D}_r \sim N(\Delta D_r, \sigma_r^2)$$
$$\Delta\hat{D}_l \sim N(\Delta D_l, \sigma_l^2),$$

and the estimated incremental displacement $\Delta\hat{D} = (\Delta\hat{D}_r + \Delta\hat{D}_l)/2$ is distributed as normal with mean ΔD and variance $\sigma^2_{\Delta\hat{D}} = (\sigma_r^2 + \sigma_l^2)/4$. Similarly, the distribution of the estimated incremental orientation $\Delta\hat{\theta} = (\Delta\hat{D}_r - \Delta\hat{D}_l)/L$ is normal with mean $\Delta\theta$ and variance $\sigma^2_{\Delta\hat{\theta}} = (\sigma_r^2 + \sigma_l^2)/L^2$.

The covariance between $\Delta\hat{D}$ and $\Delta\hat{\theta}$ is given by

$$\begin{aligned} cov[\Delta\hat{D}, \Delta\hat{\theta}] &= cov[(\Delta\hat{D}_r + \Delta\hat{D}_l)/2, (\Delta\hat{D}_r - \Delta\hat{D}_l)/L] \\ &= (\sigma_r^2 - \sigma_l^2)/2L. \end{aligned}$$

If $\sigma_r^2 = \sigma_l^2$; that is, the error variance of the right wheel is the same as that of the left wheel, which is a reasonable assumption in practice, then the covariance between $\Delta\hat{D}$ and $\Delta\hat{\theta}$ vanishes.

Under normality, $cov[X,Y] = 0$ implies that X and Y are independent. Thus, if the measurement errors of the optical decoders at both wheels are independently and identically normally distributed, then $\Delta\hat{D}$ and $\Delta\hat{\theta}$ are independent normal random variables. This result greatly simplifies the derivation of the covariance matrix of the location estimator of (3.4).

5. The Location Estimator

The trajectory-following controller commands right and left wheel velocities based on the information about the current desired location and current estimated location of the robot. The location of the robot at time n is estimated by

$$\hat{X}_n = \hat{X}_{n-1} + \frac{\sin(\Delta\hat{\theta}_n/2)}{\Delta\hat{\theta}_n/2}\Delta\hat{D}_n \cos(\hat{\theta}_{n-1} + \frac{\Delta\hat{\theta}_n}{2})$$
$$\hat{Y}_n = \hat{Y}_{n-1} + \frac{\sin(\Delta\hat{\theta}_n/2)}{\Delta\hat{\theta}_n/2}\Delta\hat{D}_n \sin(\hat{\theta}_{n-1} + \frac{\Delta\hat{\theta}_n}{2})$$
$$\hat{\theta}_n = \hat{\theta}_{n-1} + \Delta\hat{\theta}_n,$$

or

$$\hat{\mathbf{U}}_n = \hat{\mathbf{U}}_{n-1} + \Delta\hat{\mathbf{U}}_n. \tag{5.1}$$

Table 1. Biases of Adjustment Factor.

$\sigma_{\Delta\hat{\theta}_n}$ \ $\Delta\theta_n$	0°	10°	30°	90°	180°
0°	.0	.0	.0	.0	.0
1°	.000012	.000012	.000012	.000010	.000004
2°	.000050	.000049	.000049	.000041	.000018
3°	.000114	.000112	.000111	.000093	.000041
4°	.000203	.000201	.000198	.000166	.000073
5°	.000317	.000315	.000310	.000260	.000114
6°	.000456	.000454	.000447	.000375	.000165
7°	.000621	.000619	.000608	.000510	.000224
8°	.000811	.000808	.000794	.000666	.000293
9°	.001027	.001023	.001005	.000843	.000371
10°	.001267	.001263	.001241	.001041	.000458
11°	.001533	.001529	.001501	.001260	.000554
12°	.001824	.001819	.001787	.001499	.000660

Knowledge of the statistical behavior of (5.1) is necessary in order to determine when the landmarks should be sensed and when such information is to be used to keep a guard on the robot's location information. We use the covariance matrix of (5.1) as the uncertainty measure for the location estimator.

Because of the presence of the adjustment factor, there is no exact method for calculating the covariance matrix of the location estimator in (5.1). Approximation methods are available. Commonly used approaches range from linearization to dropping off $\sin(\Delta\hat{\theta}_n/2)/(\Delta\hat{\theta}_n/2)$ in the calculation of the covariance matrix. Before we assess the virtues of various approximations, we shall examine this adjustment factor more closely (under the normality assumption).

Since $\Delta\theta_n$ lies between $-\pi$ and π, and $\sin\phi/\phi$ is an even function of ϕ, we will only study the behavior of $\sin\hat{\phi}/\hat{\phi}$ with $\phi \in [0, \pi/2]$ and $\hat{\phi}$ the estimate of ϕ.

First, we investigate the bias of the adjustment factor, i.e. see how close on average $\sin\hat{\phi}/\hat{\phi}$ is to $\sin\phi/\phi$. It is shown (Wang[12]) that

$$E\left[\frac{\sin\hat{\phi}}{\hat{\phi}}\right] - \frac{\sin\phi}{\phi} = -s_1 + s_2 - s_3 + \cdots + (-1)^n s_n + \cdots,$$

where

$$s_n = \frac{\sigma^{2n}}{n!\,2^n}\left(\frac{1}{2n+1} - \frac{1}{2n+3}\frac{\phi^2}{2!} + \cdots + \frac{(-1)^k}{2n+2k+1}\frac{\phi^{2k}}{(2k)!} + \cdots\right)$$

with σ the error variance of $\hat{\phi}$, or $\sigma = \sigma_{\Delta\hat{\theta}_n}/2$.

It can be shown that $s_{2n-1} - s_{2n} \geq 0$ and is a decreasing function of ϕ. Thus, for fixed σ^2, the bias

$$\left|E\left[\frac{\sin\hat{\phi}}{\hat{\phi}}\right] - \frac{\sin\phi}{\phi}\right| = (s_1 - s_2) + (s_3 - s_4) + \cdots + (s_{2n-1} - s_{2n}) + \cdots$$

attains its maximum at $\phi = 0$. Furthermore, it can be shown that

$$\left|E\left[\frac{\sin\hat{\phi}}{\hat{\phi}}\right] - \frac{\sin\phi}{\phi}\right| = \frac{\sin\phi}{\phi} - \int_0^1 e^{-\sigma^2 x^2/2}\cos(\phi x)dx. \tag{5.2}$$

The values of (5.2) for various combinations of $\sigma_{\Delta\hat{\theta}_n}$ and $\Delta\theta_n$ are given in Table 1. It is seen that the biases are quite small even for moderate values of angular error.

Next, we evaluate the variance of the adjustment factor, i.e. to see the spread of the distribution of $\sin\hat{\phi}/\hat{\phi}$. It is shown that

$$E\left[\frac{\sin^2\hat{\phi}}{\hat{\phi}^2}\right] = \int_0^1 \int_0^{2x} e^{-\sigma^2 y^2/2} \cos(\phi y) dy\, dx.$$

Thus, we have

$$var\left[\frac{\sin\hat{\phi}}{\hat{\phi}}\right] = \int_0^1 \int_0^{2x} e^{-\sigma^2 y^2/2} \cos(\phi y) dy\, dx - \left[\int_0^1 e^{-\sigma^2 y^2/2} \cos(\phi y) dy\right]^2. \tag{5.3}$$

Although we can not show, analytically, any monotonicity properties of (5.3) as a function of ϕ (for fixed σ), the numerical results (from numerical integrations by routines DBLIN and DCADRE of IMSL[5]) do indicate that the variance of $\sin\hat{\phi}/\hat{\phi}$ is an increasing function of ϕ. Table 2 reports the variances of $\sin\hat{\phi}/\hat{\phi}$ for some selected values of $\sigma_{\Delta\hat{\theta}_n}$ and $\Delta\theta_n$.

Table 2. Variances of Adjustment Factor.

$\Delta\theta_n$	0°	10°	30°	90°	180°
$\sigma_{\Delta\hat{\theta}_n}$					
0°	.0	.0	.0	.0	.0
1°	.000000	.000000	.000000	.000004	.000012
2°	.000000	.000000	.000002	.000018	.000050
3°	.000000	.000000	.000005	.000041	.000112
4°	.000000	.000001	.000009	.000073	.000200
5°	.000000	.000001	.000014	.000115	.000312
6°	.000000	.000002	.000020	.000165	.000449
7°	.000000	.000003	.000028	.000225	.000611
8°	.000001	.000005	.000037	.000294	.000798
9°	.000002	.000007	.000048	.000373	.001010
10°	.000003	.000009	.000059	.000460	.001245
11°	.000004	.000012	.000073	.000557	.001506
12°	.000006	.000015	.000088	.000663	.001791

The above results (bias and variance) indicate that the distribution of the adjustment factor $\sin(\Delta\hat{\theta}_n/2)/(\Delta\hat{\theta}_n/2)$ is highly concentrated around $\sin(\Delta\theta_n/2)/(\Delta\theta_n/2)$ for a wide range of angular errors and incremental orientations. This simple fact enables us to find a good approximate covariance matrix for the location estimator in (5.1)

6. The Covariance Matrix of the Location Estimator

The covariance matrix of $\hat{\mathbf{U}}_n$ of (5.1) is given by

$$cov[\hat{\mathbf{U}}_n] = cov[\hat{\mathbf{U}}_{n-1}] + cov[\Delta\hat{\mathbf{U}}_n] + cov[\hat{\mathbf{U}}_{n-1},\ \Delta\hat{\mathbf{U}}_n] + cov[\Delta\hat{\mathbf{U}}_n,\ \hat{\mathbf{U}}_{n-1}],$$

where $cov[\mathbf{Z}]$ represents the *variance/covariance* matrix of $\mathbf{Z}$, while $cov[\mathbf{V},\ \mathbf{W}]$ is the *cross-covariance* matrix between $\mathbf{V}$ and $\mathbf{W}$. As we mentioned in the previous section, the presence of the adjustment factor makes the calculation of the covariance matrix of (5.1) difficult. To circumvent this problem and since for small $\Delta\hat{\theta}_n$, the adjustment factor is close to 1, an approximate method one would consider first is to discard $\sin(\Delta\hat{\theta}_n/2)/(\Delta\hat{\theta}_n/2)$ in the calculation of the covariance matrix. With this approximation, the derivation of $cov[\hat{\mathbf{U}}_n]$ is straightforward (see Wang[12]). For example, the $(1,1)$ diagonal element of matrix $cov[\Delta\hat{\mathbf{U}}_n]$ is given by

$$var_1(\Delta\hat{X}_n) = (1-a_n^2)^2\Delta X_n^2/2 + (1-a_n^4)\Delta Y_n^2/2 + (1+a_n^4\cos 2\phi_n)\sigma^2_{\Delta\hat{D}_n}/2 \tag{6.1}$$

and the $(1,2)$ off-diagonal element is

$$cov_1(\Delta\hat{X}_n,\ \Delta\hat{Y}_n) = a_n^2(a_n^2-1)\Delta X_n\Delta Y_n + a_n^4 \sin\phi_n \cos\phi_n \sigma^2_{\Delta\hat{D}_n} \tag{6.2}$$

where

$$\phi_n = \theta_{n-1} + \Delta\theta_n/2$$
$$\hat{\phi}_n = \hat{\theta}_{n-1} + \Delta\hat{\theta}_n/2$$
$$a_n = exp[-var(\hat{\phi}_n)/2] = exp[-(4\sigma^2_{\hat{\theta}_{n-1}} + \sigma^2_{\Delta\hat{\theta}_n})/8].$$

The $(1,1)$ element of matrix $cov[\hat{\mathbf{U}}_{n-1},\ \Delta\hat{\mathbf{U}}_n]$ is given by

$$cov_1(\hat{X}_{n-1},\ \Delta\hat{X}_n) = b_n(1-c_n)^2\Delta X_{n-1}\Delta X_n/2 + b_n(1-c_n^2)\Delta Y_{n-1}\Delta Y_n/2$$

where

$$b_n = exp[-var(\hat{\phi}_{n-1} - \hat{\phi}_n)/2] = exp[-(\sigma^2_{\Delta\hat{\theta}_{n-1}} + \sigma^2_{\Delta\hat{\theta}_n})/8]$$
$$c_n = exp[-cov(\hat{\phi}_{n-1},\ \hat{\phi}_n)] = exp[-(2\sigma^2_{\hat{\theta}_{n-1}} - \sigma^2_{\Delta\hat{\theta}_{n-1}})/2].$$

The second approach uses the Taylor series to approximate the adjustment factor. Specifically, since

$$\sin(\hat{\theta}_{n-1} + \Delta\hat{\theta}_n) \approx \sin\hat{\theta}_{n-1} + \Delta\hat{\theta}_n \cos\hat{\theta}_{n-1} - \frac{(\Delta\hat{\theta}_n)^2}{2}\sin\hat{\theta}_{n-1},$$

thus,

$$\Delta\hat{X}_n = \frac{\sin(\Delta\hat{\theta}_n/2)}{\Delta\hat{\theta}_n/2}\Delta\hat{D}_n \cos(\hat{\theta}_{n-1} + \frac{\Delta\hat{\theta}_n}{2})$$
$$= \Delta\hat{D}_n \frac{\sin(\hat{\theta}_{n-1} + \Delta\hat{\theta}_n) - \sin\hat{\theta}_{n-1}}{\Delta\hat{\theta}_n}$$
$$\approx \Delta\hat{D}_n(\cos\hat{\theta}_{n-1} - \frac{\Delta\hat{\theta}_n}{2}\sin\hat{\theta}_{n-1}).$$

Similarly,

$$\Delta\hat{Y}_n \approx \Delta\hat{D}_n(\sin\hat{\theta}_{n-1} + \frac{\Delta\hat{\theta}_n}{2}\cos\hat{\theta}_{n-1}).$$

These results are then used in the calculation of $cov[\hat{\mathbf{U}}_n]$.

Both approaches discussed so far require that $\Delta\hat{\theta}_n$, the incremental orientation at time n, be small. However, the observer of the robot controller will update the location of the robot, but not calculate the covariance matrix, at each step because of the real-time consideration. Thus, by the time the observer evaluates the covariance matrix of the location estimate, the orientation change accumulated through steps may not be small enough to make the above two approximations valid. A better approximation which can tolerate a wider range of $\Delta\hat{\theta}_n$ is needed.

Instead of treating $\sin(\Delta\hat{\theta}_n/2)/(\Delta\hat{\theta}_n/2)$ as unity and dropping it from the calculation of the covariance matrix, the results in the previous section suggest that we can treat the adjustment factor as a constant in the calculation of the covariance matrix. That is, the diagonal elements of $cov[\Delta\hat{\mathbf{U}}_n]$ are given by

$$\begin{aligned}
var(\Delta\hat{X}_n) &= var\left[\frac{\sin(\Delta\hat{\theta}_n/2)}{\Delta\hat{\theta}_n/2}\Delta\hat{D}_n\cos(\hat{\theta}_{n-1}+\frac{\Delta\hat{\theta}_n}{2})\right] \\
&\approx \frac{\sin^2(\Delta\hat{\theta}_n/2)}{(\Delta\hat{\theta}_n/2)^2} var\left[\Delta\hat{D}_n\cos(\hat{\theta}_{n-1}+\frac{\Delta\hat{\theta}_n}{2})\right] \\
&= \frac{\sin^2(\Delta\hat{\theta}_n/2)}{(\Delta\hat{\theta}_n/2)^2} var_1(\Delta\hat{X}_n) \qquad (6.3)\\
var(\Delta\hat{Y}_n) &\approx \frac{\sin^2(\Delta\hat{\theta}_n/2)}{(\Delta\hat{\theta}_n/2)^2} var_1(\Delta\hat{Y}_n) \\
var(\Delta\hat{\theta}_n) &= var_1(\Delta\hat{\theta}_n),
\end{aligned}$$

and the off-diagonal elements are

$$\begin{aligned}
cov(\Delta\hat{X}_n,\, \Delta\hat{Y}_n) &\approx \frac{\sin^2(\Delta\hat{\theta}_n/2)}{(\Delta\hat{\theta}_n/2)^2} cov_1(\Delta\hat{X}_n,\, \Delta\hat{Y}_n) \\
cov(\Delta\hat{X}_n,\, \Delta\hat{\theta}_n) &\approx \frac{\sin(\Delta\hat{\theta}_n/2)}{\Delta\hat{\theta}_n/2} cov_1(\Delta\hat{X}_n,\, \Delta\hat{\theta}_n) \qquad (6.4)\\
cov(\Delta\hat{Y}_n,\, \Delta\hat{\theta}_n) &\approx \frac{\sin(\Delta\hat{\theta}_n/2)}{\Delta\hat{\theta}_n/2} cov_1(\Delta\hat{Y}_n,\, \Delta\hat{\theta}_n),
\end{aligned}$$

where $var_1(\cdot)$ and $cov_1(\cdot)$ are given in (6.1) and (6.2). For the elements of the cross-covariance matrix $cov[\hat{\mathbf{U}}_{n-1},\, \Delta\hat{\mathbf{U}}_n]$, we have

$$\begin{aligned}
cov(\hat{X}_{n-1},\, \Delta\hat{X}_n) &= cov(\Delta\hat{X}_{n-1},\, \Delta\hat{X}_n) \\
&\approx \frac{\sin(\Delta\hat{\theta}_{n-1}/2)}{\Delta\hat{\theta}_{n-1}/2}\,\frac{\sin(\Delta\hat{\theta}_n/2)}{\Delta\hat{\theta}_n/2} cov_1(\hat{X}_{n-1},\, \Delta\hat{X}_n) \\
cov(\hat{\theta}_{n-1},\, \Delta\hat{X}_n) &\approx \frac{\sin(\Delta\hat{\theta}_n/2)}{\Delta\hat{\theta}_n/2} cov_1(\hat{\theta}_{n-1},\, \Delta\hat{X}_n).
\end{aligned}$$

Since $0 \le \sin\phi/\phi \le 1$, the elements of $cov[\Delta\hat{\mathbf{U}}_n]$ in (6.3) and (6.4) are smaller than the elements in (6.1) and (6.2). When $\Delta\hat{\theta}_n$ is small, the differences are negligible. However, for larger $\Delta\hat{\theta}_n$, the differences can be significant. For example, $\sin^2(\Delta\hat{\theta}_n/2)/(\Delta\hat{\theta}_n/2)^2 = 0.912$ when $\Delta\hat{\theta}_n = 60°$, i.e. $var(\Delta\hat{X}_n)$ and $var(\Delta\hat{Y}_n)$ in (6.3) are almost 10% smaller than their counterparts in (6.1). For $\Delta\hat{\theta}_n = 90°$, it would result in a nearly 20% reduction. Thus, the question of interest is whether (6.1) and (6.2) overestimate or (6.3) and (6.4) underestimate the covariance matrix $cov[\Delta\hat{\mathbf{U}}_n]$. We will try to answer this question in the next section.

7. Comparisons of Different Approximations

We conducted simulation studies to compare the three approximations for various combinations of parameters and angular errors. One thousand independent random deviates $\Delta\hat{D}_n$, $\hat{\theta}_{n-1}$ and $\Delta\hat{\theta}_n$ were generated from normal with means ΔD_n, θ_{n-1} and $\Delta\theta_n$, and variances $\sigma^2_{\Delta\hat{D}_n} = (a + b\,\Delta D_n)^2$, $\sigma^2_{\hat{\theta}_{n-1}}$ and $\sigma^2_{\Delta\hat{\theta}_n}$, respectively. Covariance matrices of $(\Delta\hat{X}_n,\, \Delta\hat{Y}_n,\, \Delta\hat{\theta}_n)^t$ for different approximations are then constructed to compare with the nominal one. Since there is no approximation involved in the calculation of $var(\Delta\hat{\theta}_n)$, we compare only the covariance matrices of $(\Delta\hat{X}_n,\, \Delta\hat{Y}_n)^t$, $cov(\Delta\hat{X}_n,\, \Delta\hat{\theta}_n)$ and $cov(\Delta\hat{Y}_n,\, \Delta\hat{\theta}_n)$.

A total of 84 cases were run, but only 2 cases will be reported and discussed here. For detailed results, see Wang[12].

Figures 4 and 5 display the 95% confidence regions corresponding to the covariance matrices of $(\Delta\hat{X}_n,\, \Delta\hat{Y}_n)^t$. The parameters used were $\Delta D_n = 50$, $\sigma_{\Delta\hat{D}_n} = 4.25$ for both figures, $\theta_{n-1} = -45°$, $\sigma_{\hat{\theta}_{n-1}} = 10°$, $\Delta\theta_n = -10°$, $\sigma_{\Delta\hat{\theta}_n} = 4°$, for Figure 4, and $\theta_{n-1} = -90°$, $\sigma_{\hat{\theta}_{n-1}} = 15°$, $\Delta\theta_n = 120°$, $\sigma_{\Delta\hat{\theta}_n} = 5°$, for Figure 5. Ellipses labeled 0, 1, 2 and 3 correspond to nominal, method 1, method 2 and method 3 respectively. Method 1 evaluates the covariance matrix by using the Taylor expansion approximation; method 2 drops the adjustment factor from (5.1); and method 3 treats the adjustment factor as a constant in the calculation of the covariance matrix. It is seen that for small $\Delta\theta_n$ (Figure 4), methods 1, 2 and 3 give almost identical results which agree well to the nominal covariance matrix. Specifically, in Figure 4, the covariance matrices of $(\Delta\hat{X}_n,\, \Delta\hat{Y}_n)^t$ for nominal and methods 1 to 3 are, respectively,

$$\begin{pmatrix} 86.6017 & \\ -5.2264 & 18.2567 \end{pmatrix}, \begin{pmatrix} 87.6494 & \\ -4.9043 & 18.3491 \end{pmatrix},$$
$$\begin{pmatrix} 86.8769 & \\ -5.1544 & 18.2732 \end{pmatrix}, \text{ and } \begin{pmatrix} 86.6564 & \\ -5.1413 & 18.2268 \end{pmatrix}.$$

It is evident that method 3 is by far the best approximation method when $\Delta\theta_n$ is large (Figure 5). The other 2 methods, especially method 1, are too conservative. The covariance matrices in Figure 5 are

$$\begin{pmatrix} 40.745 & \\ 47.032 & 102.119 \end{pmatrix}, \begin{pmatrix} 205.030 & \\ 167.263 & 228.001 \end{pmatrix},$$
$$\begin{pmatrix} 60.668 & \\ 69.740 & 147.878 \end{pmatrix}, \text{ and } \begin{pmatrix} 41.492 & \\ 47.696 & 101.137 \end{pmatrix}.$$

Similar results are also obtained when comparing the covariance of $\Delta\hat{Y}_n$, $\Delta\hat{\theta}_n$ and the covariance of $\Delta\hat{X}_n$, $\Delta\hat{\theta}_n$.

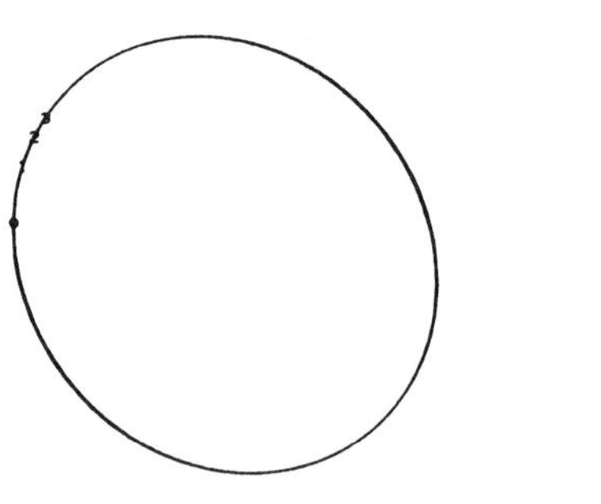

Figure 4. Comparisons with small $\Delta\theta_n$

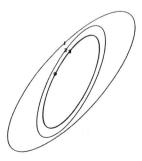

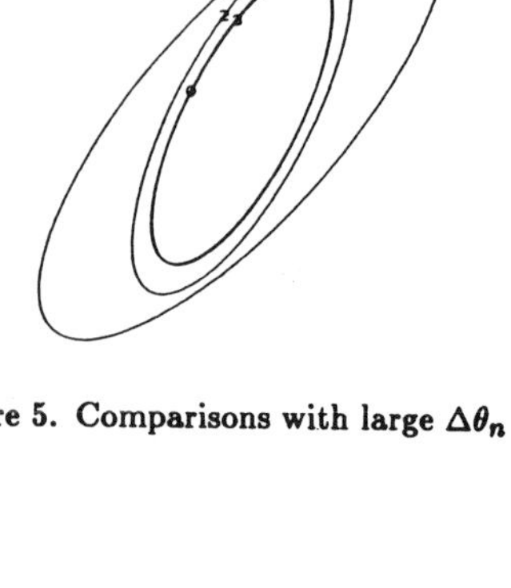

Figure 5. Comparisons with large $\Delta\theta_n$

From the above discussions, we can conclude that method 3 is a good approximation method for calculating the covariance matrix of the location estimator in (5.1). It is simpler than method 2, yet produces much more adequate results. It can tolerate a much wider range of $\Delta\theta_n$ than can methods 1 and 2.

8. Concluding Remarks

The role of the observer of a mobile robot controller is to maintain and update the information about the vehicle position and associated uncertainties, so the vehicle can behave properly. In this paper, we have presented a method for estimating the robot's location for the two-wheeled robot drive system based on the information from the optical encoders. This result differs from the earlier work by introducing a circular adjustment factor into the estimator. Since the variance of the adjustment factor is rather small under regular conditions, the proposed method gives a more accurate estimate of the robot's location without increasing the uncertainty of the estimate.

We have also proposed an approximate method for calculating the covariance matrix of the location estimator. It is shown that the covariance matrix so obtained approximates extremely well to the nominal one. The covariance matrix enables us to continuously maintain an approximation to uncertainty in location estimation and helps us to make decisions about the docking strategies and the need of sensing.

The real-time implementation of the controller makes the calculation of the covariance matrix an important issue. The first-order approximation used by Smith and Cheeseman[8] is very easy to compute and generalize. In addition, by precomputing and storing a set of Jacobian matrices that represent useful and frequently encountered spatial transformations, not only can an even greater computational efficiency be achieved, but also various types of transformations can be handled by a *common* general procedure. Although the appropriateness of the first-order approximation has been analytically verified by Wang[11] for uncertain transformations described by Smith and Cheeseman[8], the goodness of the first-order approximation in calculating the covariance matrix for (5.1) is generally unknown. Further study is planned to investigate the behavior of the observer obtained by the first-order approximation and compare it with that of the observer proposed in this article. The motivation is that if the first-order approximation can compare favorably enough, we would prefer the first-order approximation because of its simplicity, efficiency and generality.

References

1. Brooks, R.A. (1985), "Visual Map Making for a Mobile Robot", *Proc. IEEE Int. Conf. on Robotics and Automation*, 824-829, St. Louis, Missouri.
2. Chapman, J.W. (1970), "Moments, Variances, and Covariances of Sines and Cosines of Arguments which are Subject to Random Error", *Technometrics*, 12, 693-694.
3. Chatila, R. and Laumond, L-P. (1985), "Position Referencing and Consistent World Modeling for Mobile Robots", *Proc. IEEE Int. Conf. on Robotics and Automation*, 138-145, St. Louis, Missouri.
4. Iijima, J., Yuta, S. and Kanayama, Y. (1981), "Elementary Functions of A Self-Contained Robot - YAMABICO 3.1", *Proceedings of the 11th International Symposium on Industrial Robots*, 211-218, Tokyo.
5. IMSL Inc. (1980), *The IMSL Library*, Houston, Texas.
6. Julliere, M., Marce, L. and Place, H. (1983), "A Guidance System for a Mobile Robot", *Proceedings of the 13th International Symposium on Industrial Robots*, 58-68, Chicago, Illinois.
7. Lozano-Perez, T. (1983), "Spatial Planning: A Configuration Space Approach", *IEEE Transactions on Computers*, C-32, 108-120.
8. Smith, R.C. and Cheeseman, P. (1987), "On the Representation and Estimation of Spatial Uncertainty", *Int. Journal of Robotics Research*, 5, 56-68.
9. Smith, R.C., Self, M. and Cheeseman, P. (1986), "Estimating Uncertain Spatial Relationships in Robotics", *Proc. Second Workshop on Uncertainty in Artificial intell.*, Philadelphia, AAAI.
10. Tsumura, T. and Fujiwara, N. (1978), "An Experimental System for Processing Movement Information of Vehicle", *28th IEEE Vehicle Technology Conference Proceedings*, 163-168, Denver, Colorado.
11. Wang, C.M. (1986), "Error Analysis of Spatial Representation and Estimation of Mobile Robots", *GM Research Publication GMR-5573*, General Motors Research Laboratories, Warren, Michigan.
12. Wang, C.M. (1987), "Location Estimation and Error Analysis for an Autonomous Mobile Robot", *GM Research Publication GMR-5897*, General Motors Research Laboratories, Warren, Michigan.

Inertial Navigation

Edited by
MORRIS M. KURITSKY, MEMBER, IEEE, AND MURRAY S. GOLDSTEIN, SENIOR MEMBER, IEEE

Invited Paper

Abstract—**Inertial Navigation Systems have found universal application both militarily and commercially. They are self-contained, nonradiating, nonjammable, and sufficiently accurate to meet the requirements of users in a most satisfactory manner. An overview of inertial navigation is provided, followed by several sections detailing a specific, but different mechanization approach. A Ring Laser Gyro (RLG) based navigation system design is reviewed with special emphasis directed at requirements for navigation accuracy and alignment time. Along with discussions of the RLG unit, an introduction to a novel accelerometer approach, the Vibration Beam Accelerometer (VBA), is provided. A gimballed, self-contained High Accuracy Inertial Navigation System, denoted HAINS, represents one approach toward achieving navigation capability of 0.2 nmi / h and an rms velocity of 1.5 ft / s per axis while retaining the form and fit and affordability of standard inertial tactical flight navigators. The Stellar–Inertial Navigation section illustrates the bounding of position and verticality errors thus achieving exceptional accuracies. Two gyroscopic approaches, presently in development are finally discussed. The Fiber Optic Gyroscope (FOG) and Magnetic Resonance Gyroscopes (MRG's) are of interest for navigation because of their potential for low cost and excellent reliability.**

Authors

The authors are with the Kearfott Division, The Singer Company, Wayne, NJ 07470.

Murray S. Goldstein
Ivan A. Greenwood
Morris M. Kuritsky
Harold Lerman
James E. McCarthy
Thomas Shanahan
Marvin Silver
James H. Simpson

I
Introduction and Overview of Inertial Navigation

MORRIS M. KURITSKY, MEMBER, IEEE, AND MURRAY S. GOLDSTEIN, SENIOR MEMBER, IEEE

I. Introduction

Navigation has many facets, many definitions, and many subsets. One begins by simply wanting to know "where am I?" and then expands upon that simple statement with information desired on how to get from "where I am" to "where I want to be." The question then arises "referenced to what?" and the problem becomes more complex, involving coordinate systems, spheroids, ellipsoids, space, etc. The subset of navigation to be discussed herein is that of inertial navigation. It is a class of implementation which addresses Newton's laws, the constancy of momentum, of light, of stars, of gravity, etc. Those physical laws of nature dealing with acceleration ($F = ma$), gravitational attraction ($F = G(M_1 M_2/R^2)$), momentum ($H = k$), velocity of light ($c = f\lambda$), and the inertial sphere represented by the stars are sensed and utilized to provide the means by which the simple "where am I" question is answered.

There have been several excellent papers on the evolution of navigation [1], [2]; hence this paper will not treat that aspect. This paper will 1) present a cursory explanation of the theory of inertial navigation; 2) illustrate implementation approaches by several application examples; and 3) include some mathematical treatises and examples of inertial system technology so the reader can fully appreciate this very important aspect of navigation. The paper will cover a) Introduction and Overview of Inertial Navigation; b) Strapdown Systems; c) High-Accuracy Inertial Navigation; d) Stellar Inertial; and e) New Evolving Instrument Technologies. Apologies are offered to those agencies, companies, or other groups, or applications which are not sufficiently credited, either due to limitation in space or by inadvertent omissions. Also, we have attempted to write the group of papers in a way to be meaningful to a broad spectrum of readers.

II. Overview

One way of providing an overview to self-contained inertial navigation is to design such an inertial system and illustrate pitfalls, requirements, and some fundamental issues. We start with the basic requirement of navigation on our terrestrial body —Earth. We could have selected space or any other consistent reference since the fundamental issues are similar. The general problem which must be solved is that of three-dimensional navigation in an appropriate reference coordinate system. The reference coordinate system selected for illustration is Earth's North, West, Up triad.

Self-contained inertial navigation starts with the double integration of acceleration sensed in the Newtonian (Inertial) space frame. Several additional key physical laws must also be properly utilized. Depending upon the mechanization approach, these laws involve the constancy of momentum or the constancy of the speed of light, the existence of gravity, and the accurate measurement of time or its equivalent.

The outputs of this system are to be a set of position coordinates for any time (t); usually velocity and attitude are also provided. There are two coordinate systems which are most often used. The first is Earth-referenced as selected for illustration and provides position in terms of latitude, longitude, and altitude.

Manuscript received April 20, 1983; revised May 1983.
M. M. Kuritsky and M. S. Goldstein are with the Kearfott Division, The Singer Company, Wayne, NJ 07470.

The second is Newtonian space or the stars to define an astronomical reference. When stellar observations are used, both frames must be integrated consistently. Some instrument references prefer inertial space operation because of performance criteria independent of star utilization.

III. Sensor Reference

The primary sensor in an inertial navigation system is the accelerometer. This instrument produces a precise output, in either analog or digital form, which is proportional to acceleration applied along the input axis of the sensor. Although we will limit our discussion to single-axis accelerometers, two-axis accelerometers have been successfully built and used. If three single-axis accelerometers are mounted so their input axes form an orthogonal triad, any acceleration of this assembly will be resolved to define an acceleration vector. It is necessary, however, that the accelerometer assembly be referenced to a coordinate system which can be maintained or defined in a precise manner. It does not matter how that accelerometer triad moves, as long as we keep track of its precise position, angularly and linearly. The nature of this reference coordinate system depends on the nature of the vehicle and its mission. A manned aircraft or submarine generally uses an Earth referenced coordinate system. A space vehicle is probably more concerned with a space-fixed or inertial reference. Regardless of the coordinate system, it must be handled consistently and information can be provided as needed. To accomplish the orientation control of the accelerometers, an inertial sensor, the gyroscope, is used. The gyroscope has the required characteristic of being able to prescribe a reference in inertial space. The three-axis reference may be obtained by the use of either three single-degree-of-freedom gyros or two two-degree-of-freedom gyros, or combinations thereof. Mounting the accelerometers to the gyro reference package in turn provides a defined reference for the acceleration vector.

The gyroscopes and accelerometers are generally mounted in a cluster arrangement which is then gimballed or strapped down to measure vehicle motions about and along three orthogonal axes. Gyros and gimbals are used in conjunction with electronics and gimbal torquers to create null-seeking servo loops for the gimballed case. Any angular motion about the axes is sensed by the corresponding gyro and via appropriate gimbal control maintains the cluster fixed within the reference frame. Output transducers on the gimbals provide attitude output. Or, the mechanical assembly of inertial sensors can be "strapped down" and computationally a reference attitude matrix is determined, which is effectively the stabilized reference system. The two approaches produce similar results.

IV. Basic Truths

Before embarking upon the mechanization of an inertial system design, there are several tenets which should be examined and remembered. These include the following:

1) Acceleration is an inertially derived vector. (Keep track of the different rotational coordinate frames.)

2) The constancy of momentum or of the speed of light again is inertially derived.

3) The accelerometer, not the gyroscope, ties an object to Earth via gravitational mass attraction.

4) The gyroscope (no matter the type) measures angle (or angular rate) of the device upon which it is mounted relative to inertial space. Applying a torque to the gyro causes a controlling rate reaction if the gyro uses a torquer. Although a Ring Laser Gyro (RLG) may not fit the classical gyroscopic definition, we continue to use that term.

5) Star sightings provide directional information.

6) Strapped down systems and gimballed systems operate similarly with appropriate coordinate system definition, either electromechanically or computationally developed. (Instrument error sources will propagate differently between the two approaches and different error sources exist. These are covered in some detail in subsequent sections.)

V. Mechanization

Given that the initial position of a "plate" on the rotating Earth is latitude, λ_0, and longitude, ϕ_0, we can effectively stop the rotation of that "plate" or measure and control its rate in inertial space with angle rate measuring devices called gyroscopes. A controlled or known horizontal rate of $\omega_e \cos \lambda$ and a vertical rate of $\omega_e \sin \lambda$ is imposed upon the plate. This is accomplished with three axes of gyroscopic control via a supporting gimbal arrangement or measured and computationally defined. Other rotations with respect to the reference frame due to motion or disturbances are either decoupled or measured and compensation introduced. Upon this same plate, accelerometers are mounted and the two horizontal sensing units are positioned to yield a null output, which effectively causes the gravity vector to be perpendicular to these horizontal accelerometer sensing axes. The plate is now tangent to the Earth's ellipsoid. (The undulative geoid due to gravity anomalies and some discussion of exact gravity representation is treated in the high-accuracy navigation and stellar–inertial sections.) Now, if the plate moves linearly, relative to space, the accelerometers sense the acceleration and when properly integrated, yields the velocity and distance traversed. The Earth-induced acceleration components at the point of interest must also be treated properly (the greater the accuracy the greater the complexity in treating the actual geoid versus a simplified ellipsoid).

Since we are dealing with navigation about a (more or less) spherical Earth, linear motion is very simply related to angular motion by means of the Earth's radius, R, i.e., change in platform attitude $\theta = D/R$ where D is the linear distance from the departure point as derived from accelerometer information and measured along the corresponding great-circle course; similarly, $\Omega(\text{rad/s}) = V(\text{ft})/(\text{s})/R(\text{ft})$, where $\Omega = d\theta/dt$.

Thus the reference sensors are effectively precessed at a rate corresponding to the linear velocity of the aircraft.

Because the mechanization involves the double integration of acceleration to produce distance or angular traversal coupled with feedback from gravity to the accelerometer, a quasi-oscillator is effectively produced. Since Earth's radius is used, errors in the loop (accelerometer biases, gyro drifts, initial tilt, velocity errors, etc.) oscillate with the effective period $T = 2\pi\sqrt{R_e/g}$ or 84.4 min. In fact, this is the famous Schuler period characteristically encountered when an inertial system navigates relative to Earth. A considerable amount of intuition and mathematical proof formed the foundation of the above simple statement before it was reduced to practice [3]. Refer to Table I for a brief listing of error propagation characteristics. Note position error increases with time, but many other errors are bounded in growth and oscillate with the periodicity shown.

Some reference has been made relative to error propagation differences between a gimballed and a strapdown system. Two examples of that difference are illustrated by the way the accelerometer bias error and the East gyro bias drift error propagates.

TABLE I
PROPAGATION OF ERRORS

Error Source	Position Error	Source
Accelerometer bias ϵ_a	$\frac{\epsilon_a}{\omega_0^2}(1 - \cos \omega_0 t)$	instrument
Initial velocity error ϵ_v	$\epsilon_v \frac{\sin \omega_0 t}{\omega_0}$	initial condition
Vertical gyro drift ω_v	$R\omega_v\left(t - \frac{\sin \omega_0 t}{\omega_0}\right)$	instrument
Initial vertical alignment θ_0	$R\theta_0(1 - \cos \omega_0 t)$	initial condition, calibration, instrument
Initial azimuth alignment ψ_0	$\psi_0 VI_c t$	initial condition calibration, instrument
Azimuth gyro drift ω_z	$\omega_z VI_c\left[\frac{t^2}{2} + \frac{\cos \omega_0 t - 1}{\omega_0^2}\right]$	instrument

where
ω_0 Schuler frequency (ω_s, alternately used)
g gravity
R Earth's radius
VI_c inertial velocity, cross direction
λ present latitude
t time

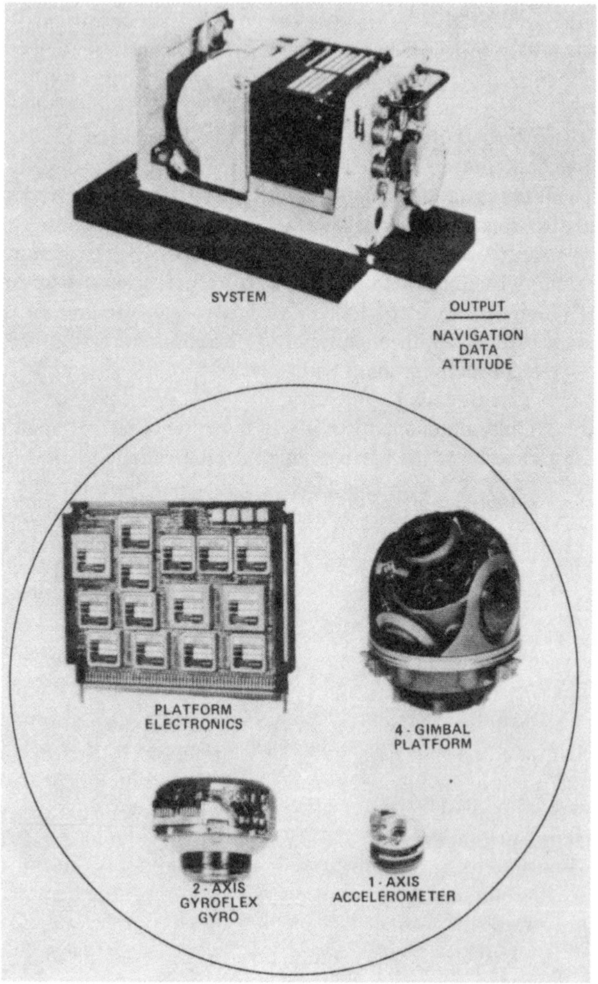

Fig. 1. Typical aircraft gimballed inertial system.

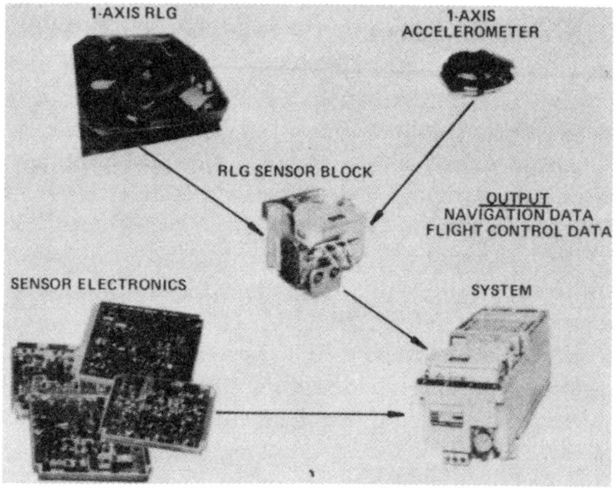

Fig. 2. Typical strapdown inertial system.

On the gimballed system, both errors cause correlated errors. Accelerometer bias causes platform tilt with a corresponding null output and gyro bias drift causes an azimuthal error which then couples Earth rate into the system again to null the gyro bias. If the tilt and azimuthal error remain fixed in the sensor coordinate system, these particular errors do not propagate significantly. In the strapped down system, the errors decorrelate due to sensor motion coincident with the vehicle, thus decoupling the sensor reference and initial navigation coordinate reference.

A typical gimballed system and a strapdown inertial system as well as their major component parts are shown in Figs. 1 and 2.

Note that we have encountered two methods that Newtonian forces, or the mechanization as utilized, reduces or increases error growth: 1) the Schuler oscillation; and 2) the correlation/decorrelation of errors in a manner to self-cancel or to add dependent upon trajectory and mechanization. Another important error compensation effect is that caused by latitude–azimuthal cross coupling of errors.

VI. SHIP'S INERTIAL SYSTEM

The Ship's Inertial Navigation System operates purely inertially or in an aided mode, and is discussed in many papers [4]. It has a long history of development and significant successes. Although its implementation is accomplished in a much expanded and needed complex manner, the basic portion of its instrumentation includes the inertial reference cluster previously discussed. An understanding of the mentioned cross-coupling error is required to understand the error propagation of such a system or any inertial system over many hours.

The latitude, longitude, and azimuth errors are affected by the gyro drift in each channel by way of cross coupling. One of the most significant cross-coupling effects propagates as follows:

A North velocity error generates latitude error. As the latitude error increases, the resolution of the inertial rate into the instrument reference coordinates becomes incorrect. The azimuth gyro torquing error $\epsilon(W_U)$ becomes

$$\epsilon(W_U) = \left[W_e \cos\lambda + \frac{V_{GE}}{R}\sec^2\lambda \right]\epsilon\lambda + \epsilon\frac{(V_{GE})}{R}\tan\lambda$$

(ω_e = Earth rate, V_{GE} = East ground velocity, ϵ = error, λ = latitude) and it generates azimuth error. The effect of azimuth error δ_u on the North velocity error is the same as that of a hypothetical East gyro whose drift is $-\delta_U(\omega_e + \dot{\phi})\cos\lambda$. A North velocity error will thus be generated opposing the original velocity error. The long-term average azimuth error adjusts itself to drive the long-term average North velocity error to zero. It follows that the latitude and azimuth errors remain bounded over the long term.

VII. Instruments

The final requirement in this simplified criterion for system design is the selection of inertial components and their installation. A wide array of components and installations is available as illustrated in the following:

A. Gyro Types

Single-degree-of-freedom, two-degree-of-freedom, free rotor, and solid-state including:

1) Rotating Wheel:

a) Wheel within float in buoyant fluid

1) jewelled bearing support
2) magnetically supported
3) other exotic support.

b) Wheel supported by universal joint (hinge)

1) torsional hinges
2) flex hinges.

c) Wheel or rotor electrostatically supported.

d) Momentum element support

1) ball bearing
2) gas bearing
3) fluid bearing
4) electrostatic bearing.

2) Optical Gyro (Solid-State Gyro):

a) Ring Laser Gyro (Passive and Active Resonator Cavity)

b) Fiber Optic Gyro (Phase or Frequency Detection).

3) Nuclear Magnetic Resonant Gyro.

4) Multisensor—Combined Gyro and Accelerometer.

B. Accelerometer Types

These include single, dual axis, "freely supported," and the following:

1) Proofmass Supported by Pendulum:

a) Electromagnetic restoring loop (analog or digital).

b) Proofmass supported by beam or string or tuning fork in oscillator configuration:
Oscillator system's frequency changed by acceleration's effect on mass. (Further explanation of the vibrating beam accelerometer is found in the strapdown section.)

c) Proofmass (effectively) supported by gyroscope:

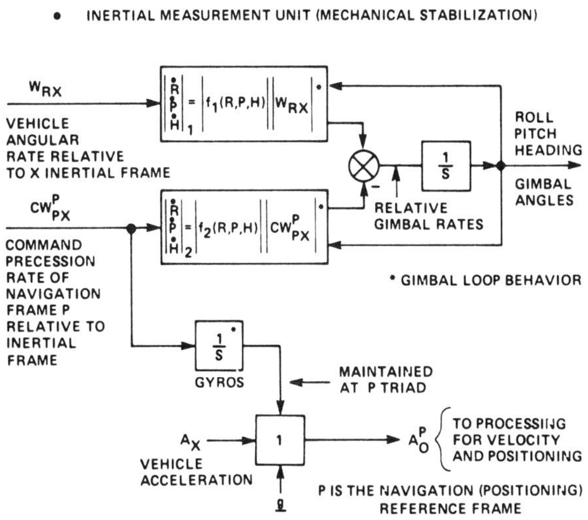

Fig. 3. Use of gimballed gyros.

Acceleration causes mass unbalanced torque and gyro yields on output as function of acceleration.

d) Proofmass supported electrostatically with appropriate readout capability.

C. Reference Platform Mechanization Approaches

a) Four gimbals utilized to support a cluster of gyros and accelerometers.

b) Three gimbals similarly used with attendant loss of gimbal freedom.

c) Strapdown—the computer utilizes gyroscopic outputs to computationally establish the reference desired coordinate frame.

d) Hybrid strapdown—gimballed; a combination of gimballed electro-servo isolation and strapdown reference action. (One or two gimbals.)

e) Multifunction platform assembly—combination of inertial and aided sensor operation (i.e., stellar–inertial).

The actual design of the inertial system obviously depends upon accuracy requirements, mission environments, usage time, reaction time, reliability, cost, and many other factors. Figs. 3 and 4 describe some of the coordinate system processing differences between strapdown and gimballed systems.

VIII. Self-Contained Stellar–Inertial System

Our overview continues into the self-contained realm of stellar-aided inertial navigation. Celestial bodies will remain one of the most accurate means of determining position. This fact is being recognized in the development of hybrid stellar–inertial navigation systems. In a manner completely analogous to a human navigator updating his position by taking star fixes, inertially determined position can be refined by auxiliary stellar information.

A one-star fix permits correction of heading misalignment, whereas a two-star fix will permit determination of position. An excellent way of illustrating how a single star provides azimuthal information is to consider the following: If a known vertical exists at a reference point then a vertical plane determined by two lines: 1) that vertical and 2) the line-of-sight (LOS) to a star is established. That plane's intersect with Earth in a great circle determines azimuth. Two-star sightings yield positional information. The elevation angle to the first star is satisfied (assuming in this case that a reference vertical exists) by any reference tracker

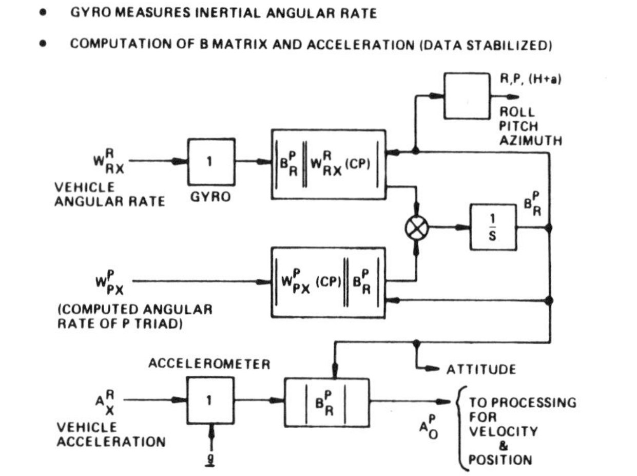

A_O^P – TOTAL ACCELERATION INCLUDING GRAVITY EXPRESSED IN P COORDINATES

ω_{RX}^R – RATE OF R COORDINATE FRAME RELATIVE TO X INERTIAL COORDINATE FRAME, EXPRESSED IN R COORDINATE

B_R^P – TRANSFORMATION OF R FRAME TO P FRAME (LOCAL VERTICAL FRAME)

(CP) – CROSS PRODUCT EXPRESSED AS A SKEW SYMMETRIC MATRIX

ω_{PX}^P – RATE OF P (NAVIGATION FRAME) RELATIVE TO X COORDINATE FRAME, EXPRESSED IN P COORDINATE.

Fig. 4. Gyroscopes—strapdown application.

on a circular locus on Earth whose tracker elevation angle is the same. A second star provides a second locus circle on Earth and the two intersect. The ambiguity of two intersections is easily handled. The reference point established by the vertical is at the location of one intersection. Fig. 17 (in Section IV) depicts the angles and directions discussed.

The stellar–inertial system is completely self-contained and is capable of high-accuracy long-range (long flight time) navigation. This system can also be ideal for tactical uses, when mobility is paramount and accurate initial alignment conditions cannot be established prior to flight. Using a stellar monitor, it is possible to determine azimuth (one-star fix) alignment by measuring the orientation of the platform with respect to the stars in flight (assuming vertical and initial position are known).

As indicated in the Introduction, the following sections will address in somewhat greater detail specific application of inertial navigation as well as the use of potentially more accurate and/or more reliable inertial sensors.

II
RLG Strapdown System Navigation: A System Design Viewpoint

MARVIN SILVER

I. Introduction

A Ring Laser Gyro (RLG) based inertial navigation system design is reviewed, continuing the theme of this paper group. Special emphasis is directed at requirements for navigation accuracy, alignment time, ambient temperature range, shock, vibration environments, and recalibration interval. These are quantitatively evaluated with respect to their system impact.

The ramifications of the listed basic system requirements on RLG quality are shown by detailing the components error budget magnitude and sensitivity to each system requirement. RLG lock-in circumvention, path length control, current control in addition to calibration accuracy and recalibration intervals are also treated.

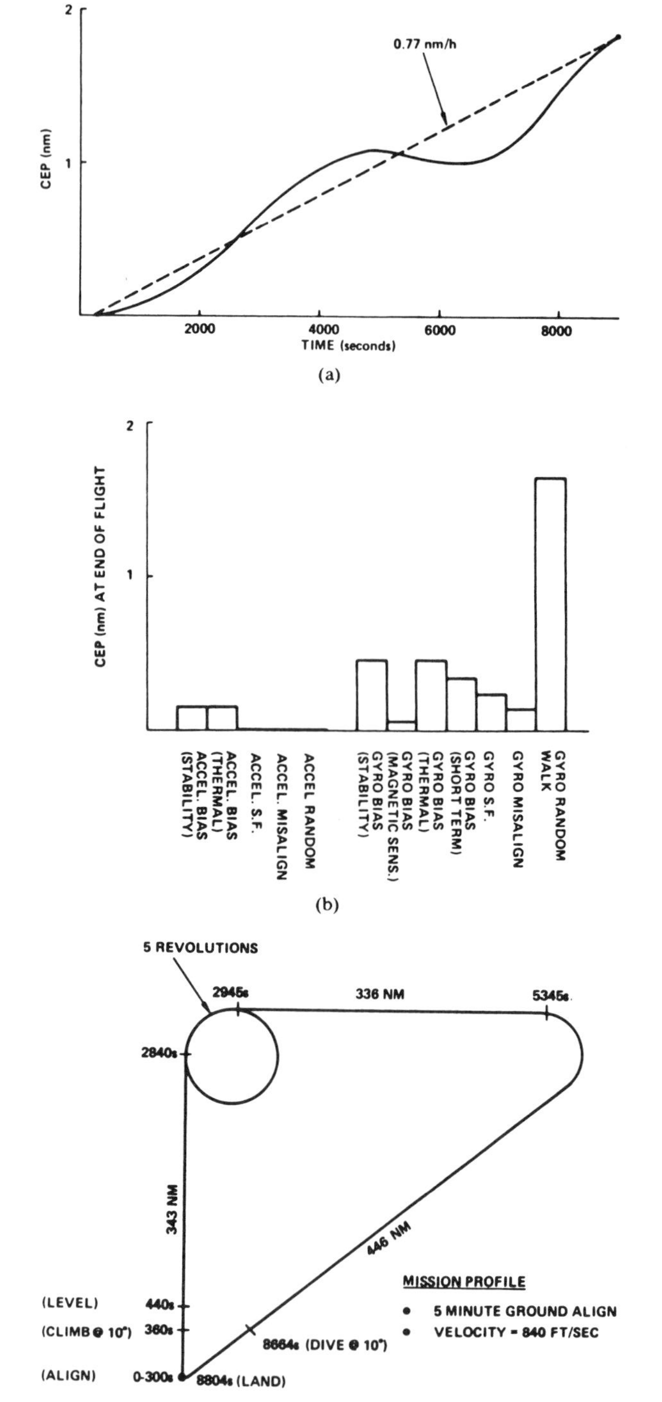

Fig. 5. Propagation of position error for given mission. Profile and budget of Table II. (a) Performance. (b) Individual error contributors.

II. Basic Inertial Navigation System Error Allocation

The quickest way for the system designer to gain a quantitative understanding of the RLG performance requirements is to review the RLG error sources in relation to the total system error budget. Table II presents a typical error budget for the 0.5–1.0-nmi/h class RLG navigation system. Fig. 5 details the contribution of each error source to system navigation error for the given vehicle trajectory. While error propagation is greatly affected by trajectory, Fig. 5 still gives a quick view as to the relative impact of various error sources. Reference [5] provides the interested reader with a listing of many strapdown papers.

TABLE II
TYPICAL ERROR BUDGET FOR RLG

Gyro Errors (Includes Electronics)	
Random walk	$0.003°/\sqrt{h}$
Bias stability error	0.004°/h
Bias thermal error over operating temperature range	0.004°/h
Short-term bias stability	0.003°/h
Bias magnetic sensitivity	0.001°/h/G
Scale factor error	5 ppm
Scale factor error thermal effect over operating range	5 ppm
Scale factor asymmetry	1 ppm
Scale factor nonlinearity over operating range	5 ppm
Axis to axis orthogonality	3 arc-seconds
Accelerometer Errors (Includes Electronics)	
Bias stability error	20 μg
Bias thermal error over operating range	20 μg
Short-term bias stability	5 μg
Scale factor error	50 ppm
Scale factor thermal error	40 ppm
Scale factor asymmetry	20 ppm
Misalignment with respect to gyro axis	3 arc-seconds
Magnetic field sensitivity	10 μg/G

III. RANDOM-WALK ERROR

The basic operation of the RLG is to excite and sustain two oppositely directed traveling waves that can oscillate with different magnitudes and frequencies. The frequency difference between the two traveling waves under ideal conditions is a direct function of the inertial rate perpendicular to the plane of the traveling waves. The frequency difference, or beat frequency expression is

$$\Delta f = \frac{4A\Omega_{in}}{L\lambda} \tag{1}$$

where

- A area enclosed by the laser cavity
- L perimeter of cavity for the traveling wave
- λ wavelength of emitted radiation
- Δf beat frequency (Hz)
- Ω_{in} inertial rate perpendicular to cavity plane (input axis rate).

This frequency difference is measured optically via the two lightwave interference patterns. As in any mechanical system that sustains two modes of oscillation, problems occur when the two frequencies approach each other. Energy is traded between the two modes and the frequencies tend to lock and become one and (1) is violated at lock-in. This trading of energy or coupling is in large part caused by back-scattered radiation from imperfect mirrors. Loss producing mechanisms within the cavity such as outgassing of epoxies also contribute to lock-in. Fig. 6 shows the relationship of gyro beat frequency versus input rotational rate, and shows the lock-in zone magnitude definition.

Current mirror technology produces lock-in magnitudes in the vicinity of 100°/h, a far cry from the requirement of less than 0.01°/h for the 1-nmi/h class system. The major technique utilized, at this point in time to circumvent this lock-in has been dubbed "the dither" technique. The technique consists of mechanically rocking the gyro through a stiff dither flexure suspension, which acts as a rotary spring, built into the gyro assembly, which

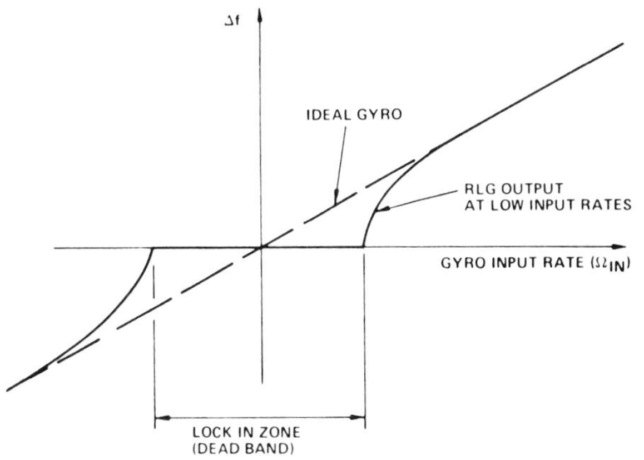

Fig. 6. Gyro output versus input rate—illustration lock-in zone.

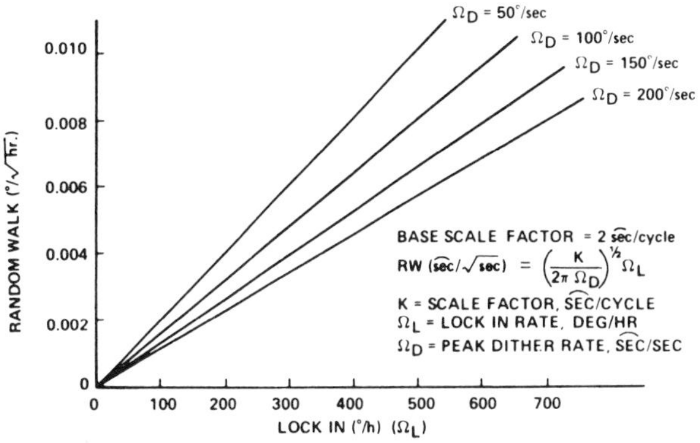

Fig. 7. Random walk versus lock-in and dither amplitude.

produces a rate about the gyro input axis that causes the gyro to rapidly enter and leave the lock-in zone producing a random drift error. Piezoelectric electric transducers provide the force to rotate the laser cavity block.

The output of the fringe detection scheme over a time ΔT is equal to a pulse of N pulses proportional to $\Delta\theta$

$$\Delta\theta \approx \int_t^{t+\Delta t} \Omega_{in}\, dt + \int_t^{t+\Delta t} \text{random noise}\, dt. \tag{2}$$

The random-noise term in the integral sum is the error in the beat frequency caused by the dither rate going into and out of lock-in with random phase. The integral of this random noise has the form of an angle error that increases as a function of $\sqrt{\text{time}}$ which is the classical random-walk function. The magnitude of the random-walk error is related to the dither amplitude and the lock-in rate per Fig. 7. Residual lock-in with randomized dither is at a level of under 0.002°/h and random-walk error of $0.001°/\sqrt{h}$ to $0.003°/\sqrt{h}$ are being routinely achieved. A price has been paid, however, with respect to additional complexity and errors caused by the dither spring mechanism. These additional errors and their impact on system design will be addressed in later paragraphs.

IV. IMPACT OF RANDOM-WALK ERROR ON GYROCOMPASS ALIGNMENT AND NAVIGATION

Random walk continues to be, even at present achievable levels of error, the major limitation in reducing the required time of ground gyrocompassing prior to system flight (unaided systems). As previously described, the random-walk error produces an attitude error that builds up as a function of $\sqrt{\text{time}}$. This

TABLE III
GROUND ALIGNMENT HEADING ERRORS VERSUS
RANDOM WALK AND ALIGNMENT TIME

Random	Alignment Time (Minutes)										
Walk $°/\sqrt{h}$	1	2	3	4	5	6	7	8	9	10	
0.001	2.5	1.8	1.4	1.3	1.1	1.0	0.95	0.89	0.84	0.79	Heading
0.003	7.5	5.3	4.3	3.7	3.4	3.1	2.8	2.7	2.5	2.4	Error
0.007	17.5	12.4	10.1	8.8	7.8	7.2	6.6	6.2	5.8	5.5	arc-min

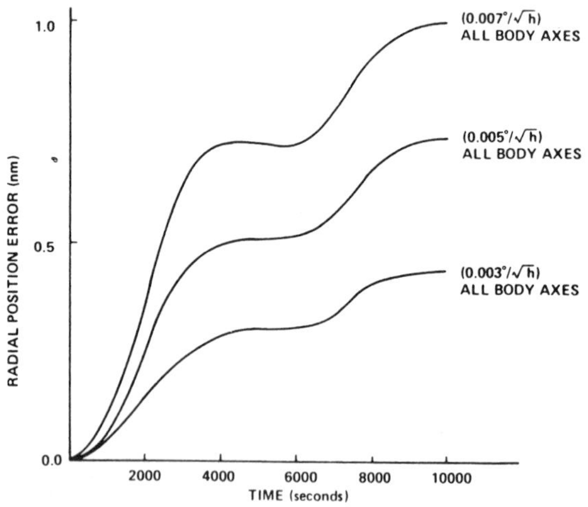

Fig. 8. Navigation error versus random-walk magnitude. (Zero vehicle velocity, alignment error due to random walk not included in navigation error.)

attitude error divided by the alignment time can be viewed as an erroneous drift rate causing a heading error equal to

$$\epsilon\psi_{\mathrm{RW}} = \left\{\begin{matrix}\text{heading error (radians) due to} \\ \text{random walk during ground} \\ \text{gyrocompass}\end{matrix}\right\} \frac{(\mathrm{RW})}{\sqrt{T}\, W_e \cos\lambda} \tag{3}$$

where RW is the random-walk coefficient in ($°/\sqrt{h}$) along East axis; T is alignment time (hours); and $W_e \cos\lambda$ the horizontal components of Earth's rate ($°$/h).

Table III shows the heading error as a function of the random-walk coefficient and alignment time at 45° latitude.

Once in flight, the heading error caused by random walk during ground alignment produces a velocity error per (4) and (5).

$$\epsilon V_n = \left\{\begin{matrix}\text{North velocity error due to} \\ \text{ground alignment caused by} \\ \text{random walk}\end{matrix}\right\}$$

$$\approx V_{\mathrm{East}} * \epsilon\psi_{\mathrm{RW}} + (\mathrm{RW}_e \cos\lambda)\epsilon\psi_{\mathrm{RW}}(1 - \cos\omega_s t) \tag{4}$$

$$\epsilon V_E = \left\{\begin{matrix}\text{East velocity error due to} \\ \text{ground alignment error caused} \\ \text{by random walk}\end{matrix}\right\} \approx V_n * \epsilon\psi_{\mathrm{RW}}. \tag{5}$$

Equations (4) and (5) show the impact on velocity error due to heading error caused by random walk for short flight times. In addition, random-walk error causes a velocity error during flight solely due to the random drift integration into attitude error during the flight. Fig. 8 shows the magnitude versus flight time for this error as a function of various random-walk magnitudes (does not include velocity error due to alignment error). As evidenced by Fig. 8 and (4) and (5), the random-walk error is a critical error source for a system that requires rapid ground alignment.

V. GYRO BIAS ERROR

Gyro bias error is defined as the difference between the true low-frequency gyro bias (period greater than mission time) and the calibrated gyro bias value loaded in the computer to compensate for this error. As long as this term remains stable and the calibrated value is subtracted from the gyro output, the compensated gyro output will indicate zero rate for the zero-rate input condition. One year stability of this error to better than 0.004°/h is achievable in production instruments to date.

Instrument factors affecting gyro bias stability are as follows:

1) Stability of the mirror's optical axis, and mirror surface erosion.
2) Outgassing of epoxy materials within the laser cavity.
3) Precise path length control to correct for changes due to expansion, contraction, and bending of the gyro block material. Equation (1) shows the path length–beat frequency relationship.
4) Control of the current required to sustain lasing of each beam with current differences to less than 50 nA.
5) Sufficient control of the dither amplitude to maintain any errors induced by dither to be constant. The forces acting to change dither amplitude are changes in the piezoelectric element's scale factor over temperature; dither change due to external vibration and dither cross coupling.

VI. IMPACT OF GYRO BIAS ERROR ON GYROCOMPASS ALIGNMENT AND NAVIGATION

Heading error during ground gyrocompass alignment is given by

$$\epsilon\psi_B \approx \frac{\epsilon D_{EB}}{W_E \cos\lambda} + \frac{D_{\mathrm{up}} T}{2} \tag{6}$$

where ϵD_{EB} is the East component of the total gyro bias error vector during alignment (total drift vector formed by the vector addition of the three gyro's drift errors reflected in the East axis); D_{up} is the component of the total gyro bias error vector along the Up axis; and T is the alignment time. The velocity error is approximated by (7)

$$\epsilon V_n \approx V_E \epsilon\psi_B + D_{\mathrm{up}} \mathrm{RW}_e \cos\lambda \left[t^2/2 - W_s^2(1 - \cos W_s t)\right] + \frac{1}{2} D_{\mathrm{up}} V_E t$$

$$\epsilon V_E \approx V_n \epsilon\psi_B + R(\epsilon D_n)(1 - \cos W_s t) + \frac{1}{2} D_{\mathrm{up}} V_N t \tag{7}$$

where ϵD_n is the error in North gyro bias and t is flight time: t_i = time after maneuver.

As soon as the vehicle changes its orientation causing the gyro axes to change from the orientation at ground alignment an additional velocity error in the form of $A(1 - \cos W_s t_i)$ occurs. This is called the "decorrelation" effect of strapdown systems caused by the gyro drift vector changing with respect to the local level axes as discussed in the Overview (Section I). Fig. 9 shows

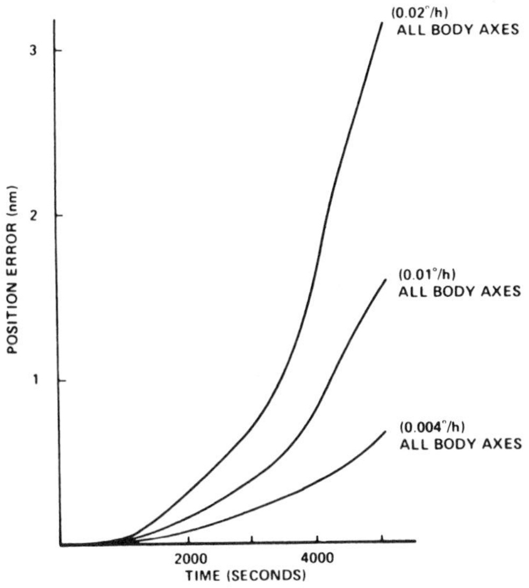

Fig. 9. Gyro drift error navigation impact. (Zero vehicle velocity; 5-min ground align, enter nav mode, turn 180° in azimuth after 20 min of nav.)

the navigation error for various magnitudes of gyro drift error after a 5-min ground alignment, navigation for 20 min at zero vehicle velocity, and then a 180° turn, and another 1 h of navigation.

VII. Gyro Bias Thermal Error

RLG inertial navigation systems are usually designed without heating or temperature control of the gyro and accelerometer. The low thermal conductivity of the RLG and its sensitivity to thermal gradients makes heating of the inertial instruments counterproductive. A typical form for the thermal compensation model is shown in (8).

Thermal Gyro Bias Compensation

$$= B_1 e^{t/\lambda} + B_2(T_p - T_r) + B_3(T_p - T_r)^2 \qquad (8)$$

where

- λ self-heat time constant
- T gyro temperature; p—present, r—reference
- B_1, B_2, B_3 calibrated thermal coefficients (B_1 is a function of T_i; i—initial)
- t time from turn-on.

Factors affecting thermal drift and stability are similar to those listed in the gyro bias error paragraph, but now include thermal effects. Current systems are achieving thermal modeling (difference between true thermal drift and model compensated thermal drift) to under 0.004°/h over the thermal range of −40 to 190°F.

VIII. Impact of Gyro Bias Thermal Error on Alignment and Navigation

For thermal drift errors that are essentially constant during alignment and navigation, error behavior can be approximated by lumping this error with the gyro bias error previously discussed. For environments that are harsher, the warmup error may follow an exponential. Even for this extreme case, the thermal time constant of the RLG is sufficiently large so as to allow approximation of the effect as a bias error for short flights.

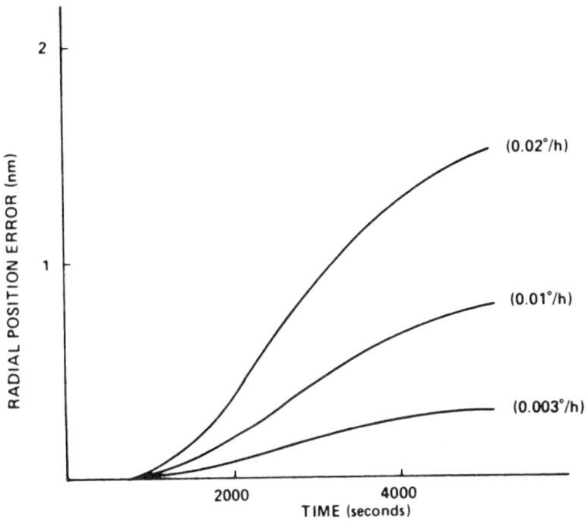

Fig. 10. Impact of gyro randomness, first-order Markovian, τ = 300 s (5-min align, 20-min nav, 180° turn-in azimuth, continue nav).

IX. Short-Term Gyro Bias Stability

This error source is a catch-all that includes all effects of gyro drift error that have a period substantially less than the mission time. Jags and level shifts of various causes fall into this category. An approximation for the alignment error causes by this source is shown in (9).

$$\epsilon\psi_{ST} \approx \frac{D_{E\text{AVG}}}{W_e \cos\lambda} + \frac{D_{U\text{AVG}}T_0}{2} \qquad (9)$$

where $D_{E\text{AVG}}$, $D_{U\text{AVG}}$ are the average value of random drift vector in the east and up directions over the alignment interval. Fig. 10 shows the navigation impact of short-term bias stability.

X. Gyro Drift Error Magnetic Sensitivity

The Earth's magnetic field is approximately 1 G total, with horizontal and vertical components latitude dependent. In addition, magnetic fields can be created by other instrumentation in proximity to the system. One of the prime mechanisms for RLG magnetic sensitivity is the property of a magnetic field to change the property of light that is not linearly polarized in such a way that a change in gyro drift occurs. All elements of the optical path are optimized to reduce this distortion of the desired linear polarized light. Unshielded gyros can have errors up to 0.04°/h/G. Reduction of magnetic sensitivity of 60:1 is achievable with proper shielding, and values of 0.001°/h/G sensitivity are achievable. Error propagation is similar to that due to a short-term drift change.

XI. Gyro Scale Factor Error

Gyro scale factor is defined as the conversion of gyro output pulse obtained by the fringe motion detector circuits into indicated angle. The basic gyro scale factor is dictated by the inertial angle change to produce one cycle of motion of the beat frequency. From (1), θ_{rad}/cycle $= L\lambda/4A$, where θ_{rad} is the inertial angle change about the axis perpendicular to the cavity plane to produce one cycle of beat frequency and one cycle of fringe motion. For a 32-cm path equilateral triangle gyro with a light wavelength of 0.63 μm the nominal scale factor is 2.11 arc-seconds/cycle. Utilization of two photodiodes spaced 1/4 of a fringe apart to detect the fringe motion allows a basic scale factor resolution improvement of four times yielding an effective scale factor of 0.5275 arc-seconds/pulse for the example shown.

A number of factors effect the gyro scale factor magnitude and stability, however, the major influences are the frequency-dependent index of refraction of the gas, and the pulling of the cavity resonance toward the center of the atomic gain curve. Both effects are triggered by fluctuations in laser gain or oscillation frequency. Basic scale factor stability of 5 ppm one year after calibration and less than 15 ppm after five years are achievable with current systems.

XII. Gyro Scale Factor Thermal Error

RLG technology has advanced to a point where scale factor thermal stability is good enough over a temperature range as large as −40 to 190 °F to make the requirement for thermal modeling unnecessary to meet thermal scale factor errors of 5 ppm. However, if thermal models are employed to shave off a few parts per million of error or a desire to achieve 10 ppm over several years they take the simplified form of

$$SF = S_0 + S_1(T_p - T_r) \tag{10}$$

where $S_{0,1}$ is the scale factor at reference temperature and linear thermal coefficient and $T_{p,r}$ is the gyro temperature; present and reference.

XIII. Impact of Gyro Scale Factor on Alignment and Navigation

RLG scale factor error, because of its small values, leads to negligible error during ground alignment. Error occurs during vehicle attitude change. The mechanism for error can be quickly approximated for simple maneuvers as

$$\epsilon\theta_{SF} \approx (\epsilon K)(\Delta\phi) \tag{11}$$

where $\epsilon\theta_{SF}$ is the attitude error due to scale factor after vehicle maneuver and ϵK is the scale factor error.

A 360 ° roll maneuver will produce an attitude error about the roll axis of 6.5 arc-seconds for a 5-ppm scale factor error. For a roll axis in the horizontal plane during the maneuver, the attitude error would be a tilt error with respect to local vertical (horizontal plane). Velocity error would propagate immediately after the maneuver in the form of

$$\epsilon V \approx \frac{g\epsilon\theta}{W_s} \sin W_s t \tag{12}$$

(terms are defined in Table I).

The 6.5-arc-second attitude error of the example given would yield a peak velocity error of 0.81 ft/s. A second roll of 360 ° in the same direction would double the velocity error.

XIV. Gyro Scale Factor Asymmetry Error/Vehicle Vibration

This is another critical error source that is dramatically reduced using RLG technology over mechanical gyros. The error is defined as the unknown difference between the scale factor for positive rates and that for negative rates.

$$\epsilon K_{\mathrm{ASYM}} = \frac{(K_+) - (K_-)}{2} \tag{13}$$

where K_+ is the scale factor error for positive rates and K_- is the scale factor error for negative rates.

A true input sinusoidal rate with zero mean value will be transformed through this error with a nonzero mean or a net drift error (rectification effect) as long as the motion continues. The form of this error for a sinusoidal rate input is

$$\epsilon D_{\mathrm{ASYM}} = \frac{2\dot{\theta}\epsilon K_{\mathrm{ASYM}}}{\pi} \tag{14}$$

where $\epsilon D_{\mathrm{ASYM}}$ is the effective drift error caused by rectification effect of gyro scale factor asymmetry, $\dot{\theta}$ is the peak magnitude of sinusoidal rate input, and $\epsilon K_{\mathrm{ASYM}}$ is the scale factor asymmetry error.

Sensor block rotation is caused by vehicle linear vibration acting through unbalances in the isolation system. The linear-to-rotational transfer function is quadratic in nature with a resonant peak. Sensor block rocking amplitudes can be as high as 75 arc-seconds under severe vibration environments (MIL-E-5400 vibration). For an isolator resonance at 80 Hz, a peak rocking rate of 10 °/s would be induced. Per (14), a 1-ppm asymmetry would produce a 0.01 °/h effective drift. More realistic vibration environments produce rocking at magnitudes of under 1 °/s, and asymmetry caused drift of 0.001 °/h acting as a gyro bias. Measured scale factor asymmetry for production RLG systems are being held to well under 1 ppm.

XV. Gyro-to-Gyro Orthogonality/Gyro to Accelerometer Orthogonality

The stability and knowledge of the gyro input axes relative to each other and relative to the accelerometer axes is critical to strapdown navigation systems. This puts stringent requirements on: system design and system calibration; system mounting rigidity; matched thermal coefficients of expansion between sensors and sensor block material; no heat application to avoid rapid thermal shock; isolators to avoid stresses due to body vibration and shock; and instruments whose internal axes stability is consistent with the required overall stability. The present state of the art for RLG navigation systems in this accuracy class is to have a combined calibration and stability error in the better than 3-arc-second class.

XVI. Alignment and Navigation Error Due to Sensor Axes Misalignment Error

The basic data flow for strapdown navigation is the transfer of sensed body specific force acceleration measurements into a computational local level coordinate frame; compensate for local gravity acceleration to obtain total acceleration; and integrate to get velocity and position. Equation (15) shows this flow

$$(A)_L = (B)_B^L (A)_B + (g) \tag{15}$$

where

- $(A)_L$ total acceleration vector in local level coordinates;
- $(B)_B^L$ direction cosine matrix that transforms body coordinate vectors to local-level coordinate vectors;
- $(A)_B$ specific force as sensed by a triad of body mounted accelerometers corrected for calibrated misalignment errors;
- (g) local level gravity vector.

The B matrix is initialized during the alignment phase (ground alignment for unaided systems). During this ground alignment, accelerometer misalignment errors contribute to errors in the initialized (B) matrix. After initialization, the (B) matrix is computed by integrating the gyro outputs corrected for known misalignment errors. Errors in gyro misalignments will cause a given vehicle rotation to be calculated by the system computer to

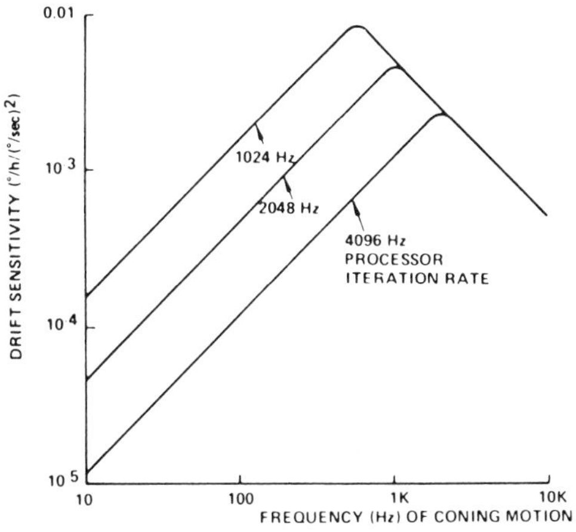

Fig. 11. Attitude matrix drift sensitivity to coning motion.

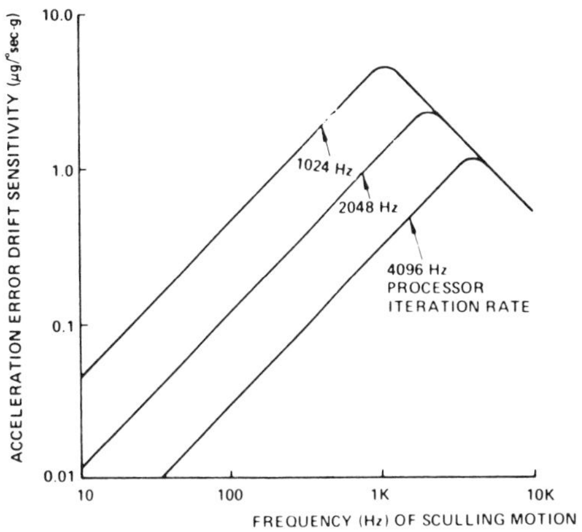

Fig. 12. Acceleration error sensitivity to sculling motion.

appear rotated in space by the nonorthogonality error of the gyro-to-gyro axes. For example, a Z-axis rate will cause an erroneous rate in the Y gyro of $(\psi_{YZ})\dot{\theta}_Z$ where ψ_{YZ} is the Y gyro orthogonality error in the YZ plane. Thus the computer instead of seeing the rate vector $\dot{\theta}_Z\bar{Z}$ sees the rotation vector $\dot{\theta}_Z\bar{Z} + (\dot{\theta}_Z)(\psi_{YZ})\bar{Y}$ causing a net error rotation of the rate vector of ψ_{YZ}. An error in the (B) matrix will result in attitude error causing incorrect transformation of body accelerations and velocity error. For the example given and for a 180° rotation about the Z body axis will cause the gyro-to-gyro orthogonality error to propagate as an error velocity in the horizontal plane of

$$\epsilon V = \left(\frac{2g\psi_{YZ}}{W_S}\right)\sin W_S t. \tag{16}$$

A 5-arc-second nonorthogonality error for this case would cause a peak velocity error of 1.2 ft/s. Accelerometer misalignment errors in addition to causing alignment errors will cause improper transformation of sensed acceleration into the local level system.

XVII. Vehicle Vibration, Coning, and Sculling Errors

A sensor block motion or rocking that causes, with respect to the reference frame, a sinusoidal Euler angle about one axis, and a cosinusoidal Euler angle about an orthogonal axis will have the third axis transcribe a cone in space. A gyro along this third axis will sense a net bias rate. Gyros along the other two will sense sinusoidal rates 90° out of phase. When the gyro outputs are transformed from body space to the reference frame without distortion, the two sinusoidal outputs are synthesized to buck out (cancel) the net bias rate and the system properly shows that the device is not spining in space. All is well. However, if the sinusoidal gyro outputs are improperly transformed into the reference frame because the computer bandwidth (computation rate) is insufficient to reconstruct the sinewaves in the reference frame, the bias coning rate will not be totally canceled out and the computer will erroneously compute a spinning or drift about the coning axis. This is coning error. Fig. 11 shows the relationship between sensor block rocking amplitude, coning drift error, and computer processing speed for the coning correction of gyro outputs.

A sensor block rocking about one axis with simultaneous linear acceleration, in quadrature, about a second orthogonal axis will cause an accelerometer sensing along the third orthogonal axis to sense a net bias acceleration. When the measurements of three accelerometers are transformed, without distortion, into the reference frame the net bias acceleration is canceled and, as with coning errors, do not accrue. However, if the sinusoidal accelerometer outputs are not transformed with fidelity into the reference frame, the bias error is not canceled. This is the sculling error. Fig. 12 shows the relationship between sensor block rocking amplitude, linear acceleration, sculling error, and computer processing speed required in transforming accelerometer outputs through a sculling correction algorithm. This algorithm in conjunction with a coning correction algorithm, increases the effective computation bandwidth of the computer to the processing speed of the algorithms.

Vehicle linear vibrations, acting through unbalances in the isolator structure, set up conditions for both coning and sculling motion. An important part of the design of the isolation system is to minimize vibration-induced rocking.

Another cause of sensor block rocking is the self-induced rocking caused by the dither mechanism.

A good portion of this self-induced sensor block rocking due to dither may cause a coning of the block and a potential coning problem if the computation speed of the coning algorithm is not fast enough. Coning and sculling processing speeds of 2 to 6 kHz will drop coning and sculling errors to less than 0.001°/h and $5\mu g$, respectively, for most flight vibration environments.

XVIII. Isolation System Considerations

The isolation system for a dithered RLG navigation system is critical. The isolators must protect the instruments from shock by attenuating the shock spectrum to below the component maximum acceleration rating. High damping (low Q), and low resonant frequency have minimum transmissibility to the higher frequencies of the usual shock spectrum. However, the isolator system must be stiff enough to prevent violation of the sway space allocation; excessive attitude error with respect to vehicle axes, and excessive rocking. The maximum rocking is in most cases at the isolator systems rotational resonance. Care must be taken that this resonance is not too low so as to be in the autopilot's passband. Another critical consideration is to minimize the dither torque required from the piezoelectric devices which have limited torque capability. Required torque to produce a given dither amplitude and rate increases dramatically as the block natural frequency approaches the gyro's natural frequency. If the block frequency is made greater than the dither frequency to obtain separation, namely, a stiff mount, excitation of box resonance is a design danger plus the greater transmissibility of shock acceleration. If the block frequency is made lower than the dither frequency to achieve the separation, block rocking may be in the passband of the autopilot. The usual choice is to select the dither frequency for best lock-in circumvention, with the block frequency significantly lower than the dither frequency to a point that balances the considerations of shock attenuation and maximum torque efficiency to provide the required dither rate. Another dither-related problem is the fact that there are three dithering gyros on the same sensor block. A number of mechanisms have been developed to neutralize cross-talk problems, the simplest of which is to have the gyros dithering at different frequencies to avoid cross talk. Other techniques solve this problem without this required sexing of the gyros.

XIX. Alternate Accelerometer Consideration

Recent trends in inertial navigation, especially in strapdown, have led to the need for new accelerometers that can operate in more demanding environments, yield improved performance and reliability at a lower cost, and have a digital output. Quartz-crystal resonators have long demonstrated an inherent accuracy and stability in frequency control and time-keeping applications. Performance requirements are typically: range $\pm 20g$; bias stability (one year one sigma) 20 μg; scale factor stability (one year, one sigma) 10 ppm; and, operating temperature range -40 to $+70\,^\circ$C.

An approach in the application of quartz crystals to accelerometers has been to use flexure-mode resonators in a beam configuration to restrain the proof mass, hence a Vibrating Beam Accelerometer (VBA). The behavior of a vibrating beam in tension is somewhat like a string in tension whereby an increase in tension will cause the resonant frequency to increase. The beam, however, has the following advantages over the string: first, a beam requires no bias tension which is the major cause of bias instability in vibrating string instruments; second, a beam also responds to compression. (Fig. 13 represents a VBA schematic.) References [6] and [7] provide further details of the VBA.

The described accelerometer approach is only one of several that are usable in today's strapdown systems. Others are listed in the Overview section. Significantly, more money has gone toward the gyroscope over the years. However, the accelerometer is now demanding its day in court. The accuracy requirements enumerated are being met in the laboratory by the VBA and will soon undergo field testing.

XX. The Challenge

The major system considerations of the RLG with respect to performance have been addressed; however, the most important "error" facing RLG technology is cost. The cost is in error, it is too high. The instrument is a natural for applications not only in the high- and medium-accuracy class of navigation, but except for cost, for attitude and reference systems (AHRS), and for hybrid navigation systems utilizing aids (2–5-nmi/h class). Industry is actively working on ways and means of dropping this cost. Areas of significant cost reduction activity are in: automated (robotic) manufacturing processes (RLG is ideally suited for this automation); mirror cost (dropping rapidly in cost from its present 25 percent of instrument cost); and more leverage in reducing cost by reducing accuracy requirements.

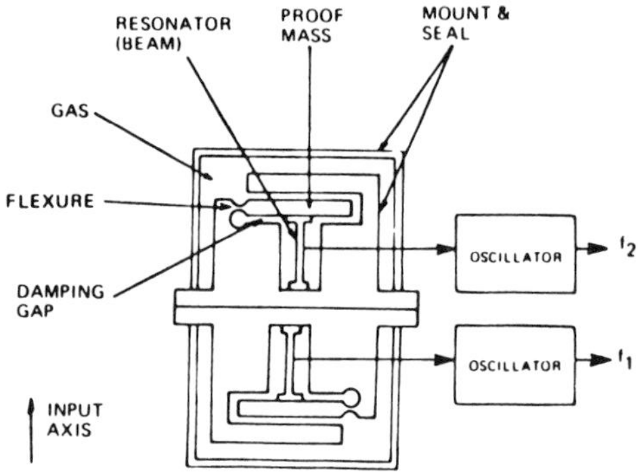

Fig. 13. Vibrating beam accelerometer (VBA) schematic.

In summary, RLG performance is well under control, and cost must be reduced to broaden the instruments acceptance. This will broaden the production base and further lower the cost.

III
High-Accuracy Inertial Navigation

THOMAS SHANAHAN and
JAMES E. MC CARTHY

I. Introduction

The last two decades have seen the increased application, both military and commercial, of inertial navigation equipment (commonly referred to as an Inertial Navigation System or INS). The marketplace is a crowded one characterized by acute competition in terms of product cost, performance, and reliability.

In characterizing the performance for aircraft applications, INS's are generally sorted into two categories:

medium accuracy	position error 1 nmi/h (CEP)
	velocity error 2.5 ft/s (rms)
high accuracy	position error 0.25 nmi/h (CEP)
	velocity error 1.5 ft/s (rms).

Gimballed platforms have been the norm for medium-accuracy applications for many years, although a growing trend for strapdown is seen in this latter accuracy category. The higher accuracy bracket has until recently been a thin market populated by exotic and relatively expensive equipment. One of the more accurate long-term inertial navigators in existence today is found in Fleet Ballistics Missile submarines [4]. The instruments utilized are of a higher quality than those discussed herein, but they also contribute to a much larger, heavier, and more expensive system. The gyroscopic instruments for this system range from the single-degree-of-freedom floated units to electrostatic-supported gyro rotor two-degree-of-freedom units.

Recent trends in the aircraft market, however, driven by military imperatives, disclose a significant growth in requirements for a high-accuracy aircraft compatible system. Concerns for precise weapon delivery, reconnaisance, mapping, and survivability, to mention a few, have motivated a series of design initiatives aimed at upgrading the gimballed INS's to meet these requirements. The following discussion is a case study of the analysis and key design improvements employed to effect the transition from a Medium-Accuracy INS to a High-Accuracy INS (HAINS).

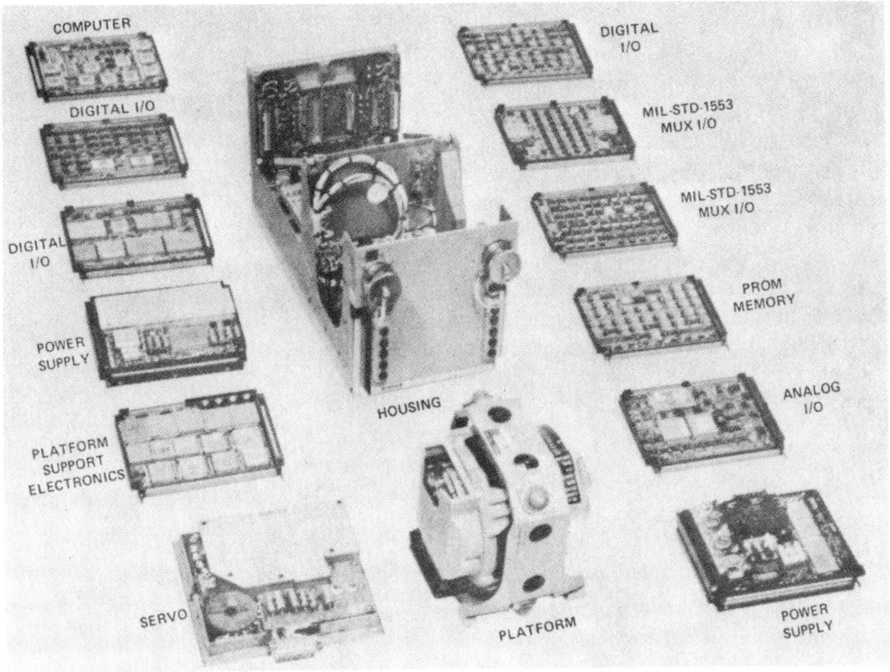

Fig. 14. Singer Kearfott SKN-2400 inertial navigation system.

II. Parameters of the Problem

In embarking on the HAINS design, which is essentially a performance-improvement program, several of the following important factors must be considered:

1) *The Need for HAINS is Immediate*. This eliminates consideration of esoteric design approaches. The design approaches must be conservative with inherent low risk.

2) *The Environment is Highly Competitive*. The objective is to produce HAINS for a cost penalty of no more than 10 percent over a conventional medium-accuracy INS, and to further enhance reliability and maintainability features to maintain a low life-cycle cost.

3) *A Broad Range of Applications Must be Met*. Potential aircraft requirements include both manned strategic and tactical missions, as well as cruise missiles. It is imperative that the HAINS supply the requisite accuracy without the volumetric and weight penalty of predecessor high-accuracy systems.

With these general guidelines, a HAINS design approach was staked out which embraced the following features:

Maintain the basic architecture and internal partitioning of the very mature and proven medium-accuracy INS (see Fig. 14).

Incorporate essential performance improvements in the inertial sensor to ensure a produceable design to the more stringent requirements of HAINS.

Exploit advances that have materialized in airborne computer technology to introduce improved software.

Configure the HAINS along the lines of the USAF Standard INS Specification (ref USAF Standard ENAC 77-1) to ensure a well-accepted form factor and interface capability.

III. Mission/Performance Analysis

The first step in this enterprise is to perform a series of comprehensive mission simulations and to identify the individual performance parameters which drive the total position and velocity performance of the INS. Fortunately, the error models and error propagation for conventional gimballed navigators are quite well understood and one can follow a top-down analytical approach to assessing system performance. Of interest here are mission scenarios, which include cases where the HAINS is aligned on the ground prior to flight as well as those where alignment is performed in air using external references such as acceleration matching, velocity aiding, and position fixing.

The performance capability of the Inertial Navigation Unit (INU) is evaluated utilizing an Error Analysis Simulator which provides a statistical error analysis of aided (hybrid) and unaided INU performance. The simulator contains math models of the INU and sensor aid error sources. It also contains the dynamics required to simulate flight trajectories. Error sources considered include:

individual inertial instrument errors
system level sources
reference data errors (Doppler and position update)
environmental induced errors (warmup and gravity anomalies).

Mission scenarios, of course, vary with application, but several benchmark trajectories have been established for evaluating performance for Tactical/Strike, Strategic/Standoff, and Penetration Missions. It is illuminating to look in detail at the analysis of a typical strategic mission employing in-air alignment to identify the pressure points for HAINS performance. The aircraft trajectory for this case is shown in Fig. 15. Table IV summarizes the major error sources of the INS and the resultant position error effects for this scenario and immediately points out the terms of concern for the desired HAINS performance.

After several iterations, including consideration of alternate mission scenarios, a revised error budget is established which can supply HAINS accuracy with reasonable margin. This error budget is also shown in Table IV.

In summary, the key design requirements for the HAINS focus

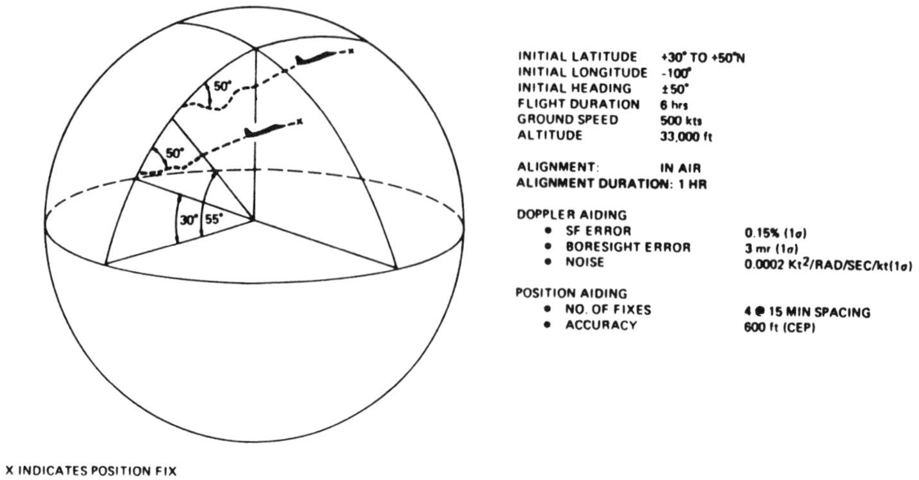

Fig. 15. Typical strategic mission trajectory.

TABLE IV
MAJOR ERROR SOURCES OF INS (6-h FLIGHT)

Error Source	Medium Accuracy Error Value (1 sigma)	Medium Accuracy CEP Rate (nmi/h)	High Accuracy Error Value (1 sigma)	High Accuracy CEP Rate (nmi/h)
Gyroscope				
Run-To-Run Drift				
Vertical Gyro	0.01 °/h	0.098	0.005 °/h	0.081
Azimuth Gyro	0.015 °/h	0.268	0.006 °/h	0.107
Random Drift				
Vertical Gyro	0.005 °/h	0.257	0.001 °/h	0.051
Azimuth Gyro	0.01 °/h	0.091	0.003 °/h	0.027
Heading Dependent Drift				
Vertical Gyro	0.007 °/h	0.360	0.001 °/h	0.051
Azimuth Gyro	0.015 °/h	0.137	0.003 °/h	0.027
Torquer Scale Factor	0.05%	0.127	0.015%	0.051
Accelerometer				
Scale Factor	0.05 °/h	0.050	0.02%	0.021
Bias	50 μg	0.035	50 μg	0.035
Asymmetry	150 PPM	0.042	NEG	–
Misalignments	1 min	0.149	10 s	0.025
Other Errors*		0.234	–	0.078
Position Fix, Doppler, Gravity Anomaly Errors**		0.110	–	0.110
CEP Rate (Total)		0.65	–	0.21

*Inertial sensor nonlinear anisoelastic, mass unbalance effects.
**Position fix, 600′; Doppler, 0.1 percent; gravity anomaly, 25 μg.
Geoid model/gravitational errors

The HAINS software employs the WGS Geoid Model which represents the components of the gravity vector as a function of altitude and latitude. After implementation of this model, the remaining gravitational influences on INS performance derive from anomalies and deflections of the vertical. Correction of these phenomena requires the use of gravity survey data which must be stored in the INS nonvolatile memory. Since the net effects of these phenomena, if not compensated, are normally less than 0.1 nmi/h, mapping or modeling of these terms is not presently incorporated for HAINS, but would obviously be needed when one approaches this latter requirement.

in on the task of improving the inertial sensor in several significant areas:

Gyroscope— restraint (long-term), random and heading sensitive drifts
— scale factor and mechanical stability
Accelerometer— scale factor, asymmetry, and mechanical stability
Component Alignments.

The following paragraphs elaborate on the approaches employed to realize the desired improvements and include a summary of test results.

IV. MAJOR PERFORMANCE IMPROVEMENTS

In incorporating design improvements for gyro drift, which in its various manifestations is the most important contributor to navigation system errors, several enhancements are required. A key consideration in current low-cost tuned rotor gyroscopes is the support flexure or hinge which connects the gyroscopic ele-

TABLE V
GROUND TEST SUMMARY

Test	RPER (nmi/h)	V_n(KTS—rms)	V_e(KTS—rms)
Six 10-h static navigation runs	0.113 (4-h avg) 0.126 (6-h avg) 0.140 (10-h avg)	0.166 (10-h avg)	0.140 (10-h avg)
Two 3-h Scorsby tests	0.27, avg	0.048, avg	0.038, avg
Four 84-min heading sens. tests	0.069, avg	0.147, avg	0.076, avg
One 4.5-h simulated flight test	0.120	0.134	0.128
One 84-min sine vibration test at 0.5 *g*	0.096	0.105	0.085

ment to the gyro motor. Factors such as residual hinge spring rate, mechanical/kinematic phenomena which produce rectified torques on the gyroscopic mass as a result of vibration, and mechanical instabilities which result in mass shifts about the hinge axis are of paramount concern. Another significant contributor to gyro drift is magnetic coupling between the gyro motor and the gyroscopic mass. Thus gyro improvements for HAINS have concentrated on the following:

A reconfigured support flexure providing:
- mechanical trimming to reduce anisoelasticity
- improved flexure axis orthogonalities
- reduced sensitivity to acceleration
- improved flywheel to hinge stability.

An alternate motor control circuit and induction motor.

A programmable speed controller.

A. Heading Sensitive Drifts

This error manifests itself as a change in gimbal drift as a function of platform case-to-cluster azimuth angle and is due to magnetic, thermal, and vibrational nonsymmetries acting on the inertial components. In the HAINS platform, these effects have been reduced by a factor of more than two from standard medium accuracy navigators and have demonstrated high repeatability to permit further compensation by software.

B. Component Misalignment

Misalignments between the sensitive axes of the inertial components are a significant contributor to navigation error. In the standard INU these misalignments are sufficiently small and controlled only by specified machining tolerances to keep navigation error within allowable limits. These misalignments are stable, however, and they can be accurately measured during calibration of the inertial sensors. The measured misalignments are stored in the flight computer program.

C. Electronics Thermal Errors

Component tests demonstrate that the electronics modules which supply torquing pulses to the gyros exhibit scale factor temperature sensitivity. The effect (in the order of 10 to 20 ppm/°C) is small for standard INU requirements, but becomes significant for the HAINS. This has been overcome by mounting the sensitive gyro pulse torquing bridge circuits and current source within the temperature-controlled platform.

D. Accelerometer Errors

In gimballed inertial navigators, the position error contribution from the accelerometers is generally of secondary importance. Improved velocity error, however, is also a concern for HAINS applications. Of some importance then is the elimination of asymmetry errors (i.e., scale factors differing between positive and negative acceleration inputs). Again, the availability of flight computers with additional throughput permit the effective elimination of this error. The computer I/O and software are organized to process the positive and negative acceleration outputs independently.

Strapdown navigation system applications are inherently more dependent on accelerometer performance and pressure from this sector of the industry has resulted in development of instrument enhancements which supply improved bias and scale factor stability. For these applications, Kearfott has developed improved magnet materials and stiffer pendulum assemblies. The resulting accelerometers utilizing these features have been incorporated into HAINS.

V. RESULTS

A number of prototype HAINS systems incorporating the improvements noted have been assembled and subjected to extensive laboratory and flight testing, including tests at government facilities. The evaluation consisted of a ground test phase and a flight test phase. The ground test results are summarized in Table V. The HAINS was subjected to a program of twenty flights in C-130 and C-141 cargo aircraft. The flight trajectories were especially designed to give results which emphasize the errors inherent in the system. There were six East/West trajectories, six West/East, and six North/South. Two of each of the six flights were for 3 h and the others for 6 h. In addition, there were two 10-h flights. The accuracy was determined by comparison with an accurate reference system, whose position and velocity accuracies were 12 ft (1 sigma) and 0.1 ft/s (1 sigma), respectively. The position error CEP for the ensemble of flights was 0.18 nmi/h (see Fig. 16). The rms velocity errors were 1.05 ft/s for V_n and 1.25 ft/s for V_e.

VI. SUMMARY

A high-accuracy development program has been successfully completed which has resulted in introduction of improved versions of the GYROFLEX[R] Gyro and Accelerometers. These improved inertial instruments have been combined with several higher level enhancements in platform assembly and shielding and software-based compensations to produce an Inertial Navigator with significantly better error characteristics than predecessor systems.

[R]Registered in U.S. Patent Office.

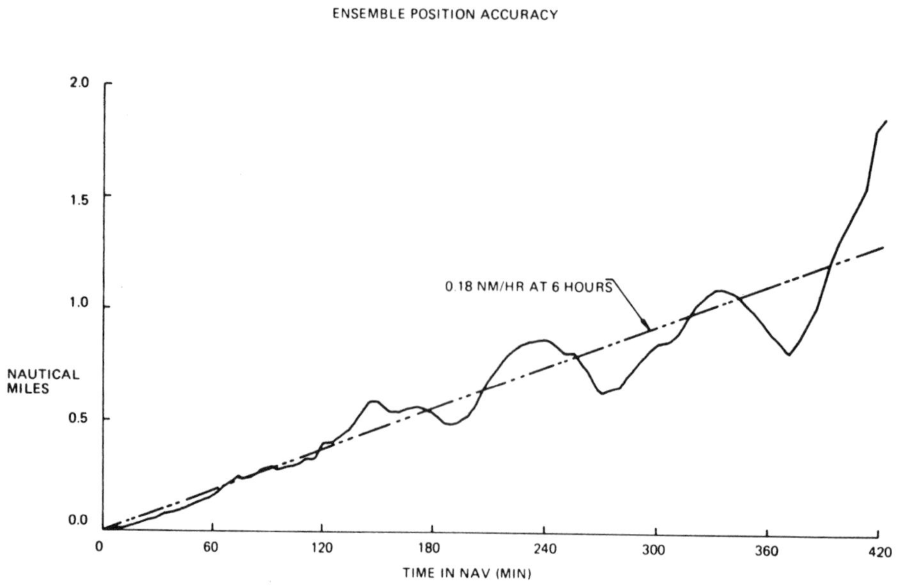

Fig. 16. Flight test.

IV
Terrestrial Stellar–Inertial Navigation Systems

HAROLD LERMAN, MEMBER, IEEE

I. Introduction

Inertial Navigation Systems are self-contained, nonradiating, nonjammable dead-reckoning navigation systems. The position error propagation of an unaided inertial navigator unfortunately grows with time and distance.

Radio position aids (e.g., GPS, Loran, TACAN, etc.) can be used to bound the position error of inertial navigators. However, the integrated hybrid radio-inertial navigation system is no longer self-contained, nonradiating, nonjammable. For those situations, where the advantages of a self-contained highly accurate navigation system are critical, a stellar–inertial navigation system satisfies the requirement.

The vertical error of an unaided inertial navigator is basically bounded. This was previously discussed in the Overview Section. In a stellar–inertial system, the position error growth is limited to the *bounded* vertical error. (Position and/or velocity aids can be used to further reduce this error.) This bounding of error and additional (if any) improvements are maintained by the stellar–inertial system during self-contained operation. The integration of the various sensors and the stellar observations are accomplished by an optimal filter operating within the airborne digital processor.

This scope of this paper is limited to terrestrial stellar–inertial navigation systems for aircraft and cruise-vehicle applications. These systems are also used on various missiles [8] for position fixes and/or heading fixes and for space applications like Space Shuttle.

This section covers the following:

brief description of celestial navigation basics;
stellar–inertial configuration, platforms, and system configurations;
performance, error discussions, and simulations.

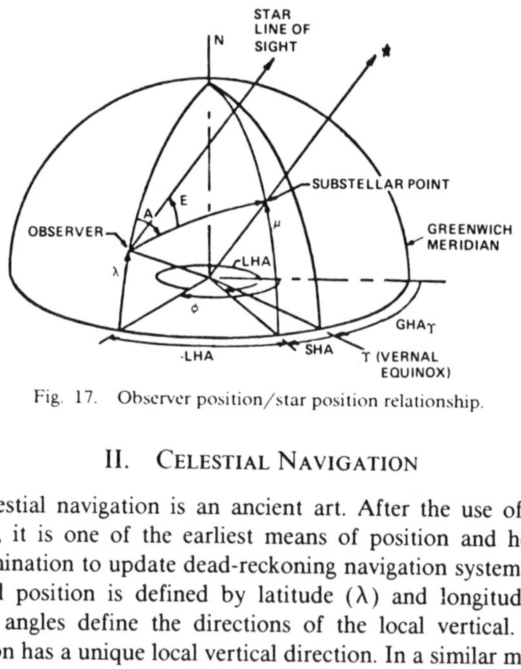

Fig. 17. Observer position/star position relationship.

II. Celestial Navigation

Celestial navigation is an ancient art. After the use of landmarks, it is one of the earliest means of position and heading determination to update dead-reckoning navigation systems. Terrestrial position is defined by latitude (λ) and longitude (ϕ). These angles define the directions of the local vertical. Every position has a unique local vertical direction. In a similar manner, star position is defined by declination (μ) and Sidereal Hour Angle (SHA). These angles define the direction of the star line of sight.

The relationship between the rotating Earth prime meridan and the celestial space reference plane (vernal equinox) is called the Greenwich Hour Angle (GHA ♈) and is a function of Greenwich Mean Time. It is interesting to note that it was the need for accurate celestial navigation which lead to the development of the chronometer. One second time error causes 1500-ft position error at the equator.

Fig. 17 shows the relationship between observer's terrestrial position and star position. Previously introduced terms: λ, ϕ, μ, SHA, and GHA ♈ are shown. New terms: LHA, A, and E are introduced. Local Hour Angle (LHA) is the hour angle difference between the observer present position and the star position. Elevation (E) and azimuth (A) represent the star line of sight at the observer's present position.

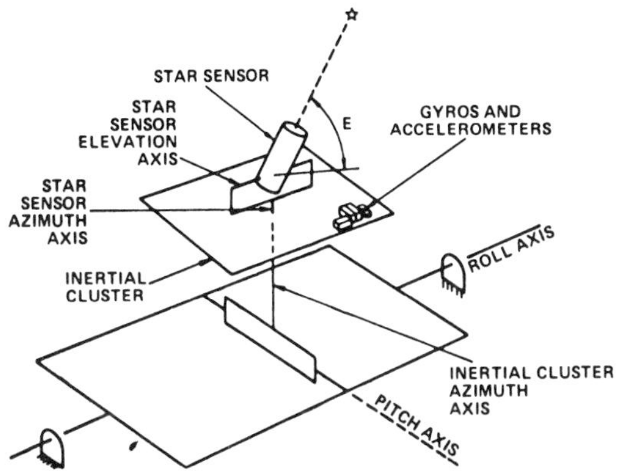

Fig. 18. Conventional stellar–inertial measurement unit.

A star sensor is used to measure the elevation and azimuth angles defined. The star sensor must be referenced to a geographic coordinate system at the observer's position. The equations for the elevation and azimuth angles as a function of observer position (λ, ϕ) and star position (μ, SHA) and GHA ♈ are

$$\sin A \cos E = -\cos \mu \sin(\mathrm{LHA}) \tag{17}$$

$$\cos A \cos E = \sin \mu \cos \lambda - \cos \mu \sin \lambda \cos(\mathrm{LHA}) \tag{18}$$

$$\sin E = \sin \mu \sin \lambda + \cos \mu \cos \lambda \cos(\mathrm{LHA}) \tag{19}$$

where

$$\mathrm{LHA} = \mathrm{SHA} + \mathrm{GHA}\,♈ - \phi. \tag{20}$$

Equations (17) to (20) can be used to directly solve for latitude and longitude. However, latitude and longitude are known to the accuracy of the dead-reckoning system; thus small angle approximations can be used resulting in the following:

$$\Delta E = -(\epsilon\lambda - \theta_W)\cos A - (\epsilon\phi \cos\lambda - \theta_N)\sin A \tag{21}$$

where

$\epsilon\lambda, \epsilon\phi$ error in latitude and longitude of the dead-reckoning system;

θ_W, θ_N vertical error about West and North axes of star-sensor platform;

ΔE difference between expected and measured elevation angle (residual).

Equation (21) shows not only the impact of position error on the elevation residual, but also the impact of vertical error on the elevation residual. Vertical error directly limits the accuracy of the celestial position fix. This should come as no surprise, since a celestial position fix is accomplished by measuring the angular direction of the observer vertical. Note that (21) is a line of positions in the vicinity of the observer, instead of the circle of positions described in the Overview. Star azimuth is not usually used to determine position. The star azimuth residual has a low sensitivity to position error and a high sensitivity to attitude error. A minimum of two star observations are required to determine a position fix. For zero vertical error and two star observations, (21) is used to solve for the position error of the dead-reckoning system

$$\epsilon\lambda = \frac{\Delta E_1 \sin A_2 - \Delta E_2 \sin A_1}{\sin(A_2 - A_1)} \tag{22}$$

$$\epsilon\phi = \frac{\Delta E_1 \cos A_2 - \Delta E_2 \cos A_1}{\sin(A_2 - A_1)\cdot\cos\lambda} \tag{23}$$

It is important to select two stars roughly 90° apart in azimuth to obtain optimal position fix performance. The procedure for the celestial position fix is 1) star selection; 2) star pointing per (17)–(20); 3) star observation (measure residuals); and 4) position update per (22) and (23).

In sophisticated stellar–inertial systems, both the star elevation and azimuth residuals are processed by a real-time Kalman filter to both update and calibrate the inertial system. The stellar–inertial system is configured to continuously observe multiple stars in sequence to obtain optimal performance.

III. Stellar–Inertial Mechanizations

A. *Stellar–Inertial Platform*

The star sensor of a stellar–inertial system must be mounted on a gimballed structure with known attitude (azimuth and vertical). This permits aiming of the star sensor optical axis with respect to North and the horizontal at the estimated star line of sight, without restricting vehicle operation. If this were not a constraint, the strapped-down mounting of a star sensor would be permissible. Inertial systems, gimballed or strapped down, must know the attitude of their inertial sensors and thus serve as an ideal reference for celestial navigation star sensors.

State-of-the-art star sensors are generally solid state. They have an array of pixel elements to sense and locate the star with respect to the telescope optical axis. Daylight sensing of stars is a requirement for aircraft stellar–inertial systems. To minimize the effect of sky background and maintain a reasonable number of pixel elements, a small star sensor field of view is used. For example, a 0.25° field of view permits star acquisition without going through a search pattern for a maximum system error of ±7.5 mi.

The telescope is of the folded or Cassagrain design to achieve a reasonable form factor for gimbal mounting. The overall platform assembly requires an optical window and gimbal freedom to permit a large sky field of view for adequate star selection/availability. A 95° field of view is adequate and achievable.

A gimballed inertial platform requires at least three axes (roll, pitch, and heading) for operation utilizing isolation gyros. The inner cluster containing the inertial sensors (gyros and accelerometers) is normally maintained vertical and at a known angle with respect to North. An obvious approach is to mount a two-gimbal assembly containing a star sensor on the inertial cluster. Fig. 18 is a mechanical schematic of such a device. It is sophisticated and somewhat complex. The fact that the star sensor must have optical access over a large field of view contributers to this complexity. The inertial system operates as a conventional gimballed system neglecting star sensor updates.

Only two gimbals are required to position a star sensor to a commanded star line of sight. By mounting a strapdown inertial system on a two-gimbal telescope cluster the azimuth and elevation of the telescope can be computed and controlled. A similar two-gimbal stellar–inertial system was successfully demonstrated on the T-16 missile used during test phases of the Assault Breaker missile program. This concept as well as the described conventional approach can be considered for aircraft applications. The inertial system, Fig. 19, essentially behaves as a strapdown inertial system in computing position and velocity. Vehicle attitude is determined by combining data from the inertial system and the

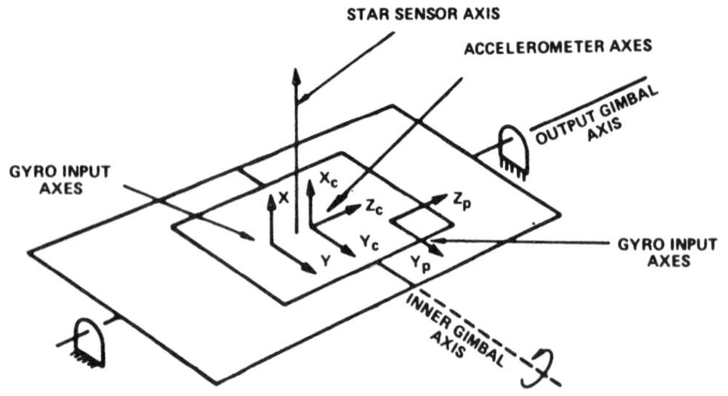

Fig. 19. Two-gimbal stellar–inertial measurement unit.

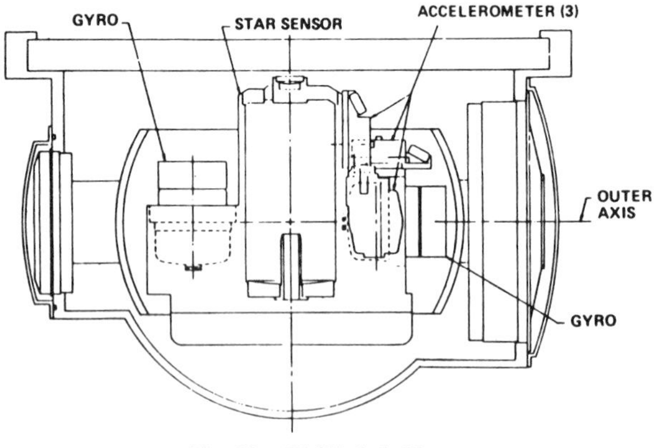

Fig. 20. SIMU gimbal layout.

gimbal transducers. The strapdown gyros permit rapid repositioning of the star sensor for multiple star operation. Fig. 20 is a layout of this assembly.

The five-gimbal and two-gimbal stellar–inertial platforms are functionally equivalent. The two-gimbal stellar–inertial platform has the advantage of simplicity, smaller size, and weight. The isolation gyros of the five-gimbal platform generally have better performance than the strapdown gyros of the two-gimbal platform (given that the same kind of gyroscopic design is utilized).

B. Stellar–Inertial Navigation Systems

Modern stellar–inertial navigation systems have a configuration similar to all hybrid–inertial navigation systems; they are integrated by a real-time Kalman filter. In fact, redundant position and velocity observation can and should (when possible) be processed by the same Kalman filter. Fig. 21 is a simplified mechanization block diagram of a generic stellar–inertial navigation system. The inertial system continuously processes the inertial sensor information. When stellar observations are present, the Kalman filter uses the data to update and calibrate the integrated system. Based on a system error model using sensor *a priori* error statistics and real-time knowledge of vehicle motion, the Kalman filter determines an optimal gain matrix. The product of the error residuals (difference between integrated system outputs and observations) and the Kalman fitler gain is used to update and calibrate the integrated system. The properties of an integrated stellar–inertial system are 1) effects of gyro drift bias are eliminated; 2) the position error is bounded to the vertical error; and 3) initial conditions of velocity error and accelerometer bias are not observable, and thus not correctable by a stellar–inertial system unless other aids are utilized.

In-air alignment of a stellar–inertial system is not possible (see item 3, above) unless position and/or velocity observations are

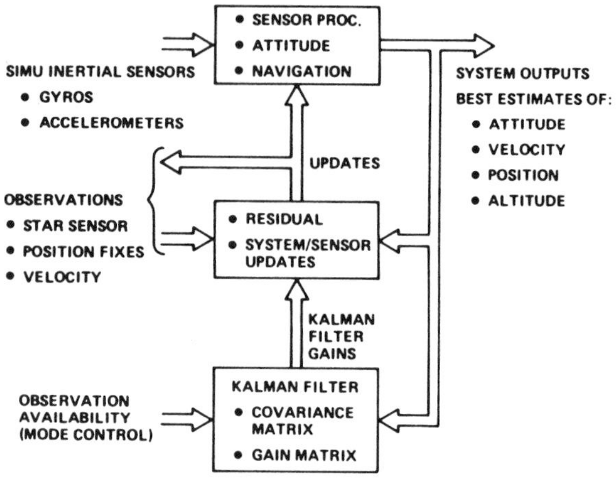

Fig. 21. Stellar–inertial mechanization.

available until alignment is completed. The continuous availability of velocity data (even of low quality) is very beneficial to a stellar–inertial system. Velocity damping eliminates the Schuler oscillation and prevents growth of the vertical error due to gravity anomaly. The reduced vertical errors improves the position performance of the stellar–inertial system (refer to item 2, above).

Relatively frequent star observations virtually eliminate the effect of gyro drift bias and substantially reduce the effect of low-frequency random gyro drift. During periods of unaided inertial operation, the random gyro drift is a predominant error source. Gravity anomalies (affecting the vertical, refer to item 2, above) is another major error source.

To eliminate the effect of the larger random gyro drift of the strapdown in the two-gimbal stellar–inertial platform, a unique system approach has been configured. This approach is utilized when redundant inertial systems are used. The integrated system uses both a two-gimbal stellar–inertial system and a high-accuracy inertial navigation system, of the type previously described in Section III. Both systems are modeled in a Kalman filter. This configuration has the equivalent effect of physically mounting the star sensor on the inertial cluster of the high-accuracy inertial navigation system. It is equivalent or better in function and performance to the five-gimbal stellar–inertial approach.

It is important to point out that this approach does not depend on any critical mechanical alignment between the two navigation systems. Velocity matching between the two systems via the Kalman filter achieves the desired results. It is to be noted that gravity is always an observable to both subsystems, permitting the all important vertical to be identical in the subsystems.

IV. Performance

The behavior of a stellar–inertial system can best be understood by reviewing simulation results. A simple North–East flight was selected using a 0.2-mi/h (CEP) inertial navigator and a stellar subsystem star sensor with 2-arc-second (1 sigma) observation noise.

Six runs are shown on Fig. 22 for different modes and conditions. Runs identified with 1, 2, or 3 refer to the systems modes of 1) unaided inertial, 2) stellar–inertial, and 3) stellar–inertial with velocity damping. The velocity data errors are 2-ft/s (1 sigma) bias and 0.5-ft/s (1 sigma) noise averaged over 100-s period. Runs identified with the letter *A* or *B* refer to the modeling of gravity anomaly errors. Runs 1*A*, 2*A*, and 3*A* contain a gravity anomaly random error source of 15μg (1 sigma) with an autocorrelation time constant of 120 s. Vehicle speed is

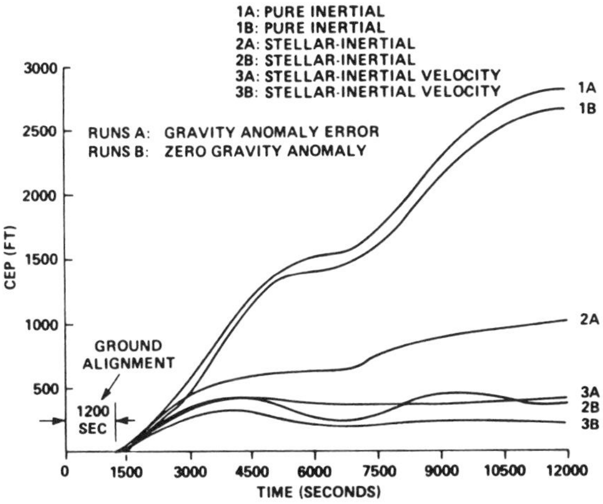

Fig. 22. Flight-mode error comparison.

1000 ft/s. Runs $1B$, $2B$, and $3B$ assume perfect knowledge of gravity. The stellar observation interval is 100 s alternating between two stars. First star observation occurred 300 s after take-off.

Refer to Fig. 22. Runs $1A$ and $1B$ serve as the reference for unaided inertial performance. The impact of the gravity anomaly error source is not significant for unaided inertial operation. Runs $2A$ and $2B$ illustrate the improvement caused by stellar–inertial operation. (100 ft is equivalent to 1 arc-second on the Earth's surface.) The unaided inertial position error is reduced to the vertical error of the stellar–inertial system. There is significant difference caused by the gravity anomaly error. As can be seen, the introduction of random gravity anomaly noise into an inertial or stellar–inertial system causes the vertical errors to grow. Run $2B$ shows the bounded position error (and, therefore, vertical error) for perfect modeling of gravity. The need for gravity modeling to optimize stellar–inertial performance is obvious.

Runs $3A$ and $3B$ show the advantage of velocity damping in a stellar–inertial system. Both the Schuler oscillation and the unbounded vertical error growth are eliminated. The significant improvement is from Run $2A$ to $3A$ where the effect of the gravity anomaly error is greatly reduced by velocity damping. The velocity source can be Doppler radar, multimode radar, or true air speed (for low wind noise conditions).

V. Summary

The stellar–inertial navigation system offers the user self-contained operation and excellent performance. Modular building block techniques exist for the simple installation and incorporation of stellar equipment with existing inertial navigation systems.

Va
Fiber-Optic Gyroscopes

IVAN A. GREENWOOD, MEMBER, IEEE

I. Introduction

Fiber-Optic Gyros (FOG's) have the potential of being simpler, more reliable, and less costly to build than Ring Laser Gyros (RLG's) for many reasons, including: little or no lock-in phenomena; no plasma flow problems; no complicated base block fabrication; and no critical mirror fabrication or aging problems.

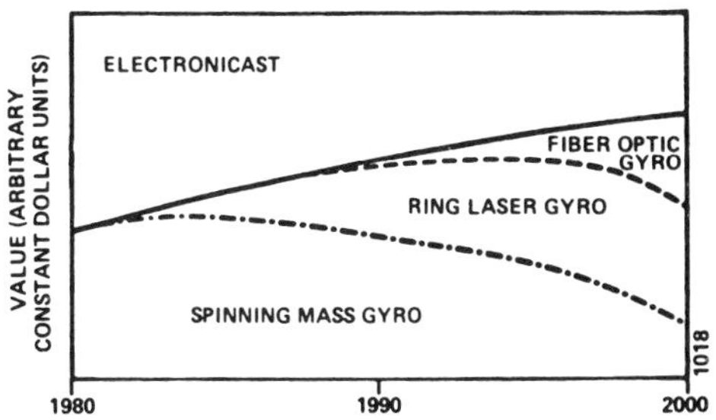

Fig. 23. High-performance gyroscope technology trends, as forecast by Jeff Montgomery of Electronicast, Redwood City, CA.

There is a widespread belief that FOG's will begin to impact the RLG high-performance market late in the 1980's and take over important segments of this market in the 1990's. Fig. 23 shows the FOG forecast of J. Montgomery of Electronicast. This enthusiasm for the future of high-performance FOG's has developed since several research groups demonstrated FOG random-walk coefficients of only a few millidegrees per root hour. Rate bias stability has been a major problem in FOG research. Early in 1983, inertial grade rate bias stability under laboratory conditions was demonstrated, reinforcing belief in the future of FOG's for high-performance applications.

The most popular FOG is a Sagnac interferometer. The basic idea is that light traveling in CW and CCW directions through a fiber-optic coil emerges with a slight time and hence optical phase difference between the two paths when the coil is rotating about the axis of the coil. The Sagnac phase shift, $\Delta\phi$, which is the measure of rotation rate is

$$\Delta\phi = 2\pi L D \Omega / \lambda_0 c_0 \tag{24}$$

where L is the fiber length, D is the coil diameter, Ω is the rotation rate, λ_0 is the vacuum length of the light, and c_0 is the velocity of light in vacuum. Ezekiel and Ardity [9] provide a tutorial review to which the reader is referred for basic information on FOG's.

II. Major Problems

Sagnac's interferometer experiment in 1913 produced a sensitivity of 2 r/s. In 1976, Vaili and Shorthill [10] demonstrated a fiber-optic Sagnac interferometer. Between that date and the present, a number of major sources of both rate noise and bias instability were identified and ingenious approaches developed which have now brought laboratory FOG performance close to that predicted by theory.

A. Rayleigh Backscatter

In 1980, a major source of rate noise was identified by Cutler, Newton, and Shaw [11]. This is the Rayleigh backscatter noise from the fiber itself. Cures for this problem were found to be optical source short correlation times (wide spectrum widths) and proper phase modulation within the fiber loop.

B. Magnetic Effects

An important contribution to bias drift, the effect of magnetic fields acting on imperfect fibers, was described by Bohm, Petermann, and Weidel [12]. With improved fiber design and simple shielding, magnetic effects are now not a major problem. With perfect polarization-maintaining fiber, there would be no magnetic sensitivity.

C. Dynamic Thermal Gradients

Another significant contribution to bias drift was identified in 1980 by Shupe [13]. This is the effect of changing thermal gradients along the length of the fiber coil. Once recognized, this problem has been addressed by symmetrical coil design and improved thermal gradient control.

D. Insertion Loss, Stability

Most early FOG's used discrete components. In 1981, Bergh, Lefevre, and Shaw [14] reported the first all single-mode fiber-optic gyroscope. This was a major advance in two respects. Insertion loss from source to detector was very small, leading to outstanding signal-to-noise ratio. Instability due to relative component motion was eliminated. In addition, this approach may lead to lower production costs, although this is not yet established. All-fiber gyros have so far been reported only in connection with zero or small rate inputs.

E. Rate Measurement

For strapdown applications, very large rate dynamic ranges, good scale factor accuracies, and digital outputs usually are required. These requirements led to a search for means other than analog phase measurement for converting the Sagnac phase shift into a digital output representing input rotation. Ezekiel and Balsamo [15] in 1977 described a resonant ring gyro in which differential frequency represented input rate. Pool and Sellers [16] and Cahill and Udd [17] in 1978 filed patents on frequency shift phase-nulled Sagnac gyros. In this approach, the output of the gyro becomes identical to a ring laser gyro of the same enclosed area, independent of the number of turns on the fiber coil. Although they have many technical problems, acoustooptical frequency shifters are normally used in such gyros. Fiber-optic frequency shifters for all-fiber gyros have not yet been reported. A frequency shifter described by Heismann and Ulrich [18] may be applicable to this pressing need.

F. Kerr Effect

Early in 1982, Ezekiel, Davis, and Hellwarth [19] reported differential-intensity-induced nonreciprocal phase-shift errors in a narrow spectrum laser fiber-optic gyro. In only a few months, two solutions [10], [21] to this problem were found. Square-wave modulation of the laser source in a Sagnac interferometer was shown to overcome the effect, and more generally, a broad spectrum source with appropriate statistics was shown to do as well. Because of the Kerr effect, most fiber-optic Sagnac interferometer gyros now use broad spectrum laser diode or superluminescent diode light sources.

G. Polarization, Polarization Control

Until recently, fiber-optic gyros used a combination of a polarizer between the two couplers, and a polarization controller (or a depolarizer) in the coil loop path [22]–[24]. Most residual rate bias drifts were ascribed to lack of polarization control due to changes in fiber-optics properties over time and environmental conditions. In March 1983, Burns *et al.* [25] reported experiments (at zero input rate) with a gyro having a fiber-optic coil and its associated coupler fabricated from polarization-maintaining fiber. The optical source was a superluminescent diode. Over a 24-h period they obtained a rate sigma of less than 0.01°/h with a filter time constant of 125 s.

III. The Present State of the Art

Several groups had demonstrated FOG short-term random walk substantially as predicted by theory, about equal to RLG performance, and adequate for strapdown inertial navigation. Inertial grade long-term rate bias stability had not been reported by these groups as of March, 1983. It was widely believed that polarization instability was the remaining major problem which caused actual performance to differ from theoretical predictions over long periods of observation.

The results obtained by Burns *et al.* [25] confirmed this view, and showed further that polarization-maintaining optical fiber is a viable solution to the problem at or very close to the level required for navigational quality strapdown inertial systems, at least under benign conditions. These observations close a period of intense research productivity during which the major and mysterious problems besetting the original FOG experiments were identified and understood.

IV. The Future

The FOG is now entering its development and engineering phase. For applications not requiring high performance, the emphasis will be on finding adequate components of very low cost and assembly techniques which can be semi-automated.

For navigational applications, much quality component development with reasonable cost objectives still lie ahead. The two most pressing needs are low insertion-loss frequency shifters for all-fiber gyros, and inertial sensor grade polarization-maintaining optical fibers and couplers. For high-performance inertial sensors, the field will very likely shift to the 1.3–1.6-μm wavelength region, since the available lower insertion losses are important and component research for this wavelength region is progressing rapidly due to the demands of the telecommunications industries.

Vb
Magnetic Resonance Gyroscope

JAMES H. SIMPSON, SENIOR MEMBER, IEEE,
AND IVAN A. GREENWOOD, MEMBER, IEEE

Most fundamental particles, electrons, protons, neutrons, as well as nuclei, possess an intrinsic angular momentum. The use of this angular momentum for gyroscopic measurements has long been an intriguing idea [26]. Several approaches have been suggested for the design of instruments using this intrinsic angular momentum and performance approaching navigation requirements has recently been realized in selected laboratory experiments.

In all cases, the intrinsic angular momentum of the nucleus has been used in some form of Nuclear Magnetic Resonance (NMR). As in all NMR, a net nuclear magnetization is established, in most cases by optical pumping, the exchange of angular momentum between atoms in the vapor state, and circularly polarized resonant radiation [27].

The basic rotation information is obtained from observations on the precessional motion of the net nuclear angular momentum in an applied magnetic field H_0. The Larmor frequency, the frequency of precession about H_0, is proportional to the magnitude of H_0, $\omega_0 = \gamma H_0$, where the constant of proportionality γ the gyromagnetic ratio is a characteristic of the particular nuclide.

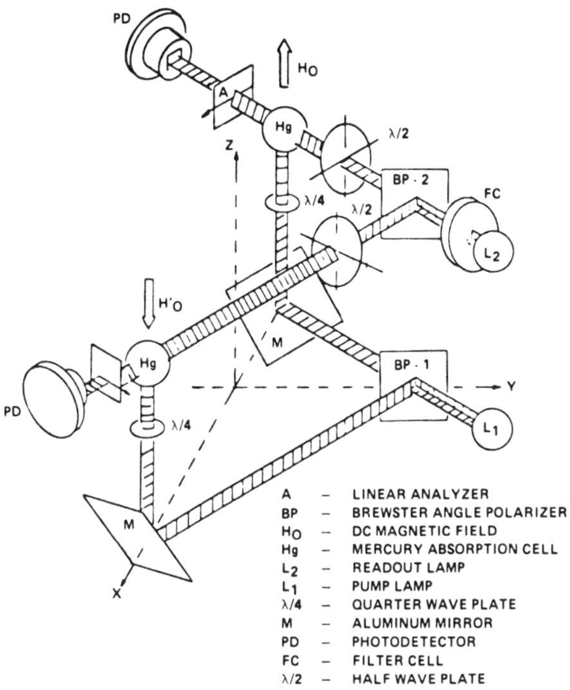

Fig. 24. Arrangement of optical components.

However, if the precession is observed from a coordinate frame rotating at an angular rate ω_r about the direction of H_0, the observed frequency will be shifted by ω_r, $\omega = \gamma H_0 - \omega_r$. For ω_r in the range required for a practical navigational gyro, measurement using this equation would require impractically precise knowledge of H_0. This problem is overcome by using two different types of spin particles in the same magnetic field, resulting in two observables, Larmor frequencies, and two unknowns, the magnetic field and the rate of rotation.

$$\omega_1 = \gamma_1 H_0 - \omega_r$$

and

$$\omega_2 = \gamma_2 H_0 - \omega_r.$$

One approach based on the implementation of these equations employs noble gases as the resonant nuclei [28]. The net nuclear magnetic moments are produced by spin exchange between optically oriented rubidium atoms and the noble gases. The magnetic resonances are also detected through the rubidium magnetic resonance. One way of looking at this phenomenon is to regard the resonance in rubidium as a sensitive magnetometer that detects the nuclear magnetizations. Any method based on the pair of equations above relies upon the constancy and knowledge of the ratio of gyromagnetic ratios γ_1/γ_2.

The requirement to know or control this ratio can be eliminated by use of two magnetic fields in opposite directions and a resonance cell containing both nuclei in each magnetic field. This second approach is shown schematically in Fig. 24. In this implementation, the resonant nuclei are the two odd stable isotopes of mercury, Hg^{199} and Hg^{201}. The resonance radiation for optically pumping the samples has a wavelength of 253.7 nm. Both pumping beams are derived from the same light source by means of a polarizing beam splitter. The hyperfine structure of the 253.7 nm line is such that the component from Hg^{204} is coincident with one of the components of each of the odd isotopes.

The magnetic resonance is driven by an ac magnetic field generated by a small coil near the resonant sample. The mercury is in the form of a low-pressure vapor, approximately 0.1 mtorr, and is contained in a spherical cell about 1 cm in diameter that is made from high-purity fused silica, as shown in Fig. 25. The magnetic resonance is detected optically through the transverse rf Faraday effect. The readout beams for both cells are also derived from a common source.

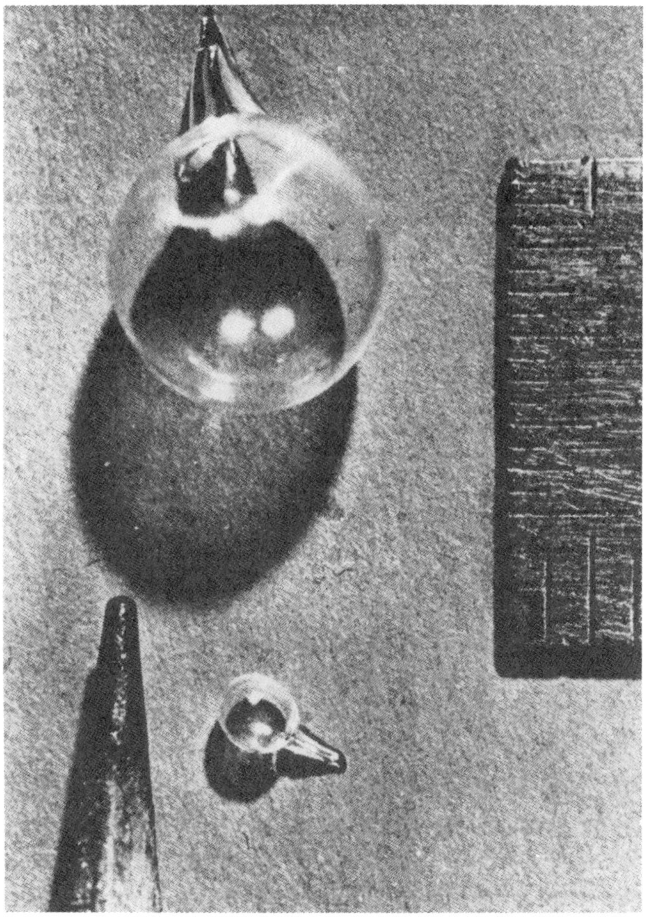

Fig. 25. Fused-silica mercury-vapor resonance cells.

The ac magnetic fields for driving the magnetic resonances are obtained by amplifying the outputs of the photodetectors in a broad-band amplifier and feeding the output of the amplifier to the drive coils to form a self-oscillating circuit. With a magnetic field of 1.3 G, the resonant frequencies of Hg^{199} and Hg^{201} are 1000 and 369 Hz, respectively.

The output of the gyro is obtained by forming the phase difference between the 1000-Hz signals and the phase difference between the 369-Hz signals from the two cells. An error signal for control of one of the H_0 magnetic fields is formed by adding the two phase differences. The gyro output is the difference of the two phase differences and is simply four times the angle of rotation. The gyro scale factor does not depend upon the geometry or size of the instrument.

The performance of an MRG is characterized by an angle output noise, which depends on the signal-to-noise spectral density ratio, an angle random walk, which depends on the angle noise and the relaxation times of resonance, and a variety of bias effects [29]. Recent laboratory tests have demonstrated performance in the range of a few hundredths of a degree per hour. Tests on experimental models and developments of theory and new approaches continue.

References

[1] C. S. Draper, "Origins of inertial navigation," *J. Guidance Contr.*, Sept., Oct. 1981.

[2] H. Marc, "Navigation, the government and industry: An ancient partnership," *J. Ion*, Spring 1979.

[3] G. Pitman, *Inertial Guidance*. New York: Wiley, 1962.

[4] B. Mc Kelvie and H. Galt, Jr., "The evolution of the ships, inertial navigation system for the FBM program," *J. Ion*, Fall 1978.

[5] S. C. Garag, L. D. Morrow, and R. Mamen, "Strapdown navigation technology: A literature survey," *J. Guidance Contr.*, May, June 1978.

[6] M. S. Goldstein *et al.*, "Stellar guidance (stellar acquisition feasibility flight)," Kearfott Tech. News Bull. 1966.

[7] W. C. Albert and R. E. Weber, "Vibrating beam accelerometer for strapdown applications," presented at the IEEE Plans 82 Position Location Navigation Symp., Dec. 1982.

[8] W. C. Albert, "Vibrating quartz crystal beam accelerometer," *Instr. Aerospace Industry*, vol. 28, May 1982.

[9] S. Ezekiel and H. J. Arditty, *Fiber Rotation Sensors and Related Technologies*. New York: Springer, 1982.

[10] V. Vali and R. W. Shorthill, "Fiber ring interferometer," *Appl. Opt.*, vol. 15, pp. 1099–1100, 1976.

[11] C. C. Cutler, S. A. Newton, and H. J. Shaw, "Limitation of rotation sensing by scattering," *Opt. Lett.*, vol. 5, pp. 488–490, Nov. 1980.

[12] K. Bohm, K. Petermann, and E. Weidel, "Sensitivity of a fiber-optic gyroscope to environmental magnetic fields," *Opt. Lett.*, vol. 7, pp. 180–182, Apr. 1982.

[13] D. M. Shupe, "Thermally induced nonreciprocity with fiber-optic interferometer," *Appl. Opt.*, vol. 19, pp. 654–655, Mar. 1, 1980.

[14] R. A. Bergh, H. C. Lefevre, and H. J. Shaw, "All single-mode fiberoptic gyroscope," *Opt. Lett.*, vol. 6, pp. 198–200, Apr. 1981.

[15] S. Ezekiel and S. R. Balsamo, *Appl. Phys. Lett.*, vol. 30, pp. 478–480, May 1, 1977.

[16] R. H. Pool and G. W. Sellers, "Interferometer gyroscope having relaxed detector linearity requirements," U.S. Patent 4 273 444, June 16, 1981.

[17] R. F. Cahill and E. Udd, "Phase nulling optical gyro," U.S. Patent 4 299 490, Nov. 10, 1981.

[18] F. Heismann and R. Ulrich, "Integrated-optical single sideband modulator and phase shifter," *IEEE J. Quantum Electron.*, vol. QE-18, pp. 767–771, Apr. 1982.

[19] S. Ezekiel, J. L. Davis, and R. W. Hellwarth, "Observation of intensity-induced nonreciprocity in a fiber-optic gyroscope," *Opt. Lett.*, vol. 7, pp. 457–459, Sept. 1982.

[20] R. A. Bergh, H. C. Lefevre, and H. J. Shaw, "Compensation of the optical Kerr effect in fiber-optic gyroscopes," *Opt. Lett.*, vol. 7, pp. 282–284, June 1982.

[21] R. A. Bergh, B. Culshaw, C. C. Culver, H. C. Lefevre, and H. J. Shaw, "Source statistics and Kerr effect in fiber-optic gyroscopes," *Opt. Lett.*, vol. 17, pp. 563–565, Nov. 1982.

[22] E. C. Kintner, "Polarization control in optical-fiber gyroscopes," *Opt. Lett.*, vol. 6, pp. 154–156, Mar. 1981.

[23] G. A. Pavlath and H. J. Shaw, "Birefringence and polarization effects in fiber gyroscopes," *Appl. Opt.*, vol. 21, pp. 1752–1757, May 15, 1982.

[24] R. Ulrich and M. Johnson, "Fiber-ring interferometer: Polarization analyses," *Opt. Lett.*, vol. 4, pp. 152–154, May 1979.

[25] W. K. Burns, R. P. Moeller, C. A. Villarruel, and M. Abebe, "Fiber optic gyroscope with polarization holding fiber," paper PD#2, presented at the Topical Meet. on Optical Fiber Communication, New Orleans, LA, Feb. 28–Mar. 2, 1983, OSA/IEEE.

[26] J. H. Simpson, *Astron. Aeron.*, vol. 2, p. 42, Oct. 1964.

[27] B. Cagnac, *Ann. Phys.* (Paris), vol. 6, p. 467, 1961.

[28] E. Kanegsberg, in *Proc. SPIE 157*, p. 73, Aug. 1978.

[29] I. A. Greenwood and J. H. Simpson, in *Proc. NAECON '77*, p. 1246, 1977.

Continuous transmission FM sonar with one octave bandwidth and no blind time

P.T. Gough, B.E., Ph.D., A. de Roos, B.E., and M.J. Cusdin, N.Z.C.S.

Indexing terms: *Signal processing, Sonar*

Abstract: Despite the advantage of high average power output, traditional continuous transmission FM (CTFM) sonars suffer from two major defects: separate transmit and receive transducers are required, and the demodulated signal is unavailable for a proportion of the sweep period. The paper shows how, by using a dual-demodulation system, the demodulated signal can be made continuous, resulting in the complete elimination of the blind time and in a range resolution as good as two wavelengths of the mean frequency. Experimental results using a CTFM air sonar justify both these claims.

1 Introduction

The first sonars radiated a repetitive train of impulsive 'pings' and used the time delay between the transmission of an impulse and the detection of an echo as a measure of the target's range. In an effort to increase the signal/noise ratio, the impulse was replaced by a modulated pulse, and the received echo(es) passed through a narrowband filter centred at the modulation frequency. As the filter rejected the out-of-band noise, the signal/total-noise ratio improved. However, the range resolution, i.e. the ability to distinguish two targets not well separated in range, is improved by transmitting as short a pulse envelope as possible. Typically, this pulse envelope may be several tens of wavelengths in length, but a lower practical limit of 8 or 10 wavelengths is often seen as a compromise between the desire for good range resolution and the bandpass filter's ability to reject out-of-band noise.

In this simple system, the greatest limiting factor is that the average power radiated is proportional to the pulse length. As the pulse length is decreased to improve the range resolution, the average power is decreased, and, as well, the total noise is now increased since the bandpass filter must be wider to pass the shorter pulse. Consequently, short-pulse sonars have a poor signal/noise ratio and a short range of operation.

In an effort to increase the average power radiated and yet retain the range-resolution capability of a short pulse, the transmitted pulse was 'chirped'. The chirped pulse is many wavelengths long with the onset frequency either higher or lower than the terminal frequency: usually this change in frequency within the pulse envelope is linear.

A receiver that produces an optimum singal/noise ratio incorporates a matched filter. The matched filter for a chirped pulse train includes a delay line which has a frequency sensitive delay and delays the onset frequency by the length of the chirp and the terminal frequency not at all. This process is intended to compress the pulse into an impulse.

Mathematically, if the transmitted signal is:

$$S(t)_{chirp} = \exp j2\pi f_H t \exp -j2\pi bt^2 \operatorname{rect}\left[\frac{t - \frac{T_c}{2}}{T_c}\right] \qquad (1)$$

for the period $0 < t < T$ and repeated every T s, where

f_H = highest frequency transmitted
$(f_H - 2bT_c)$ = lowest frequency transmitted
T = pulse repetition period
T_c = pulse length ($T \gg T_c$)
b = constant

Paper 2960F (E15), first received 1st September 1983 and in revised form

The authors are with the Department of Electrical & Electronic Engineering, University of Canterbury, Christchurch 1, New Zealand

the received signal (echo) from a single target at a range R is

$$E(t)_{chirp} = AS\left(t - \frac{2R}{c}\right) \qquad (2)$$

repeated every T s, where A is a constant and c is the speed of sound in the medium. The delay line now cross-correlates $E(t)$ with $S(t)$ to find $2R/c$.

When the length of the chirp T_c equals the pulse repetition period T, the chirped pulse transforms into a CTFM (or CWFM) waveform. (In actual fact, CTFM preceded chirped pulses by several years). Since the duty cycle of the transmitter is now 100%, a CTFM sonar both transmits and receives continuously, and so two separate transducers are required. This is a major limitation of all CTFM sonars.

The delay line required to compress such a long 'pulse' is quite different to any other system since the differential time delay may be of the order of seconds. In fact, the CTFM 'matched filter' comprises a demodulator with the local-oscillator input a replica of the transmitted waveform, followed by a spectrum analyser.

2 Traditional CTFM system

If the signal transmitted by a CTFM system is

$$S(t) = \exp j2\pi f_H t \exp -j2\pi bt^2 \qquad 0 \leqslant t < T \qquad (3)$$

and repeated every T s to $\pm\infty$, the received echo from a single target at range R is

$$E(t) \doteqdot AS\left(t - \frac{2R}{c}\right) + AS\left(t + T - \frac{2R}{c}\right) \qquad 0 \leqslant t < T \qquad (4)$$

From here, repetition every T s is assumed.

Fig. 1 shows the frequency of the transmitted signal over time and received echos.

Let us now demodulate $E(t)$ by multiplying it with a local oscillator which is a replica of $S(t)$. Then lowpass filtering the resultant signal gives:

$$D(t) = A \exp -j2\pi\left(T - \frac{2R}{c}\right)bt \qquad 0 \leqslant t < \frac{2R}{c} \qquad (5a)$$

$$= A \exp -j2\pi \frac{2Rb}{c} t \qquad \frac{2R}{c} \leqslant t < T \qquad (5b)$$

Note that in eqn. 5*b* a single target at a range of R produces a single tone whose frequency $2Rb/c$ is directly proportional to the range.

To illustrate this demodulation process further, consider a collection of M targets at M different and contiguous ranges. $D(t)$ over the full sweep period T is quite complicated, but within a portion of the period T:

$$D(t) = \sum_{i=1}^{M} A_i \exp -j2\pi(2R_i b/c)t \qquad \frac{2R_m}{c} \leqslant t < T \tag{6}$$

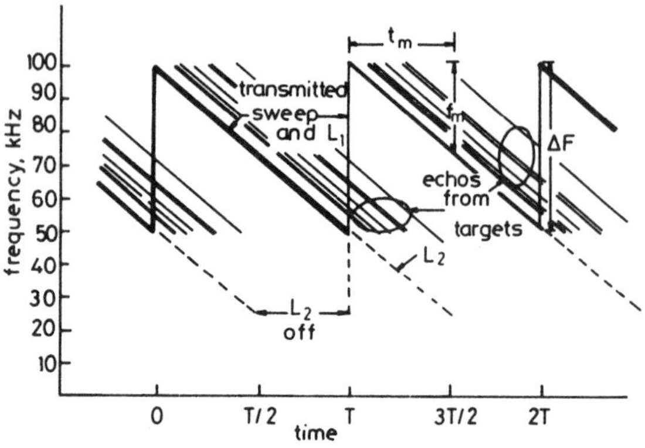

Fig. 1 *Time dependence of transmitter frequency, returned echos and the two local-oscillator frequencies when $k = 0$*

Also, $R_m = 0.5\ (cT/2)$ so that $L_2(t)$ may be gated off for a significant portion of the sweep period T

where

A_i = echo strength of ith target
R_i = range of ith target
R_m = maximum range

Passing this signal through a spectrum analyser, the spectrum of $D(t)$ is:

$$\tilde{D}(f) \doteqdot \sum_{i=1}^{M} A_i \delta\left(f - \frac{2R_i b}{c}\right) \tag{7}$$

for large T, where

$$\tilde{D}(f) = \int_{2R_m/c}^{T} D(t) \exp -j2\pi ft\ dt \tag{8}$$

Note that in the portion of the period not covered by eqn. 6, i.e.

$$0 \leqslant t < \frac{2R_m}{c}$$

the demodulated output is of no use since the direct relationship between range and frequency is not valid for all ranges. The normal procedure is to ignore $D(t)$ during this portion of the period, and thus the sonar is 'blind' for a time 0 to $2R_m/c$.

For small maximum ranges when R_m is much less than $cT/2$, this blind time is no real problem, but for large maximum ranges (i.e. when R_m becomes a noticeable fraction of $cT/2$) the sonar may be blind for a large portion of its sweep repetition period. An additional complication occurs with the effect of the sudden application of the signal $D(t)$ to the spectrum analyser. Many analysers comprise banks of high Q-factor bandpass filters which are somewhat underdamped to achieve maximum selectivity. The sudden application of the signal $D(t)$ may produce unwanted transients in all of the bandpass filters [1], and these transients can mask a weak signal.

The difficulties, introduced by blind time and the on-off nature of the useful part of the demodulated signal, have so far limited CTFM sonars to narrowband systems where the demodulated bandwidth is typically 10% of the transmitted bandwidth, consequently

$$R_m \ll cT/2 \tag{9}$$

However, there are still real benefits in using a simple CTFM system despite limitations imposed by using a narrow bandwidth/short range.

The first and most obvious is that since the duty cycle is 100%, the average power can be very high with no difficulties caused by massive peak powers or subsequent cavitation. The second advantage is more subtle. The response time of the spectrum analyser is proportional to the bandwidth of the individual bandpass filters, and by making these filters less selective they respond more rapidly. So now we have the choice of covering the spectral range of $D(t)$ with a few wideband filters having a rapid response time or many narrowband filters having a slow response time. Thus we can trade-off range resolution for speed of response; something no pulsed sonar can do once the maximum range is fixed. Consequently, using a CTFM sonar, the number of independent looks at the full range is now no longer limited by the sweep repetition period T, but only on the response time of the spectrum analyser. (In comparison, a pulsed sonar gets only one independent look every T s. This response time cannot be altered and is independent of the range accuracy and resolution).

In the CTFM sonar, the signal $D(t)$ is fed to the spectrum analyser capable of resolving K frequencies to a resolution Δf. Thus the response time of the spectrum analyser is $1/\Delta f$ and so the number of range cells is K, where

$$K = (2R_m/c)\ \Delta f \tag{10}$$

and we can perform this analysis T/f times per sweep repetition period T. However, despite these advantages, we are still limited to narrow demodulated bandwidth sonars, where

$$R_m \ll cT/2$$

in order to keep the blind time an insignificant fraction of T.

2.1 *CTFM system using an interlaced double demodulator*

Consider now a sonar system using the same transmitted signal $S(t)$ and the same received signal $E(t)$. However, instead of demodulating $E(t)$ by $S(t)$, let us use two demodulators with two local oscillators having outputs $L_1(t)$ and $L_2(t)$, where

$$L_1(t) = \exp j2\pi(f_H + kbT)t \exp -j2\pi bt^2 \tag{11}$$

and

$$L_2(t) = \exp j2\pi(f_H + (k-1)bT)t \exp -j2\pi bt^2 \tag{12}$$

where $0 < t < T$ and k is any positive integer including zero. Let us now look at the resultant process if, for example, k is unity.

If $D_1(t)$ is the output from the first demodulator and $D_2(t)$ the output from the second demodulator, then, for $k = 1$:

$$D_1(t) + D_2(t) = \sum_{i=1}^{M} A_i \exp -j2\pi\left(f_H - \frac{2R_m}{c}\right)t + G(t) \tag{13}$$

for all t, where $G(t)$ comprises all signals of frequency greater than f_H or less than $(f_H - bT)$ and is usually filtered out. Note that now there is no blind time and R_m is only limited by the sweep-repetition period T, so that

$$R_m \leqslant \frac{cT}{2} \tag{14}$$

Of course, there may be good reasons for limiting the useful range to much less than $cT/2$, such as range ambiguities, spectrum analyser capabilities, etc.

Also note that, depending on k, the range may now be reversed with respect to demodulated frequency, with zero range at a frequency of f_H and maximum range R_m at $(f_H - bT)$ as it is when k is unity. Signals in this band can be brought down to baseband (DC to $+bT$) by a third demodulator with a fixed local oscillator at f_H, or the spectrum analyser can cover the band f_H to $(f_H - bT)$, whichever is the more convenient.

2.2 Range ambiguities

In pulsed sonars, range ambiguities occur at multiples of the maximum range. Thus, the ith target displayed as being at a range of R_i m may, in fact, be at $lcT/2 + R_i$ m, where l is a positive integer.

In CTFM sonars, range ambiguities can occur at ranges of $lcT/2 + R_i$ and $lcT/2 - R_i$, as the system demodulates echos which are both higher and lower in frequency than the local-oscillator signal(s) $L(t)$. In traditional CTFM sonars, where $R_m \ll cT/2$ (to keep the blind time small), the range corresponding to the first ambiguity $(cT/2 - R_i)$ is much greater than the displayed range R_i, and the extra transmission loss (mainly due to spherical spreading) ensures that the phantom target is well attenuated.

However, in dual-demodulation CTFM sonars, where R_m is a significant portion of $cT/2$, the signal strengths of phantom targets can be comparable to those of real targets.

It is possible to eliminate range ambiguities at ranges of $cT/2 - R_i$. For example, using narrow bandpass filters, cascaded mixers and single sideband techniques, circuits can be designed to force the system to use only echos above (or below) the local oscillator. These techniques usually reduce the range ambiguities to those of the pulse sonar, with the disadvantage of increased complexity.

3 CTFM sonar using dual demodulation

CTFM sonar systems have been produced and used at the University of Canterbury, New Zealand for many years. Until recently, they were all analogue systems using a single demodulator. Blind times were kept relatively short by keeping R_m a small portion of $cT/2$ (typically 10%). Operators listened to the demodulated outputs to locate and identify targets, since the ear is a most effective spectrum analyser in the range 200 Hz to 12 kHz. Devices based on these principles found application as experimental sonars, heart monitors, divers' navigational aids and mobility aids for the partially sighted [2, 3, 4].

Although these devices are still useful, recent advances in digital frequency synthesis now make interlaced or multiple local-oscillator techniques feasible. More specifically, the advent of repeatable, linear and phase-controllable swept frequency generators make the interlaced dual-demodulator system quite practical. In the sonar described here, the transmitter sweeps down from 100 kHz to 50 kHz (one octave) in a sweep period T which is user-selectable. This transmit sweep also doubles as the first local oscillator L_1; i.e. $k = 0$ in eqn. 11 and 12. The second local oscillator, L_2, is synthesised to join onto the end of L_1 and sweeps from 50 kHz down to 25 kHz (Fig. 2).

The outputs of the two demodulators ($D_1(t)$ and $D_2(t)$) are summed together (Fig. 2) to produce continuous signals for all targets up to the maximum range R_m, now limited to $0.5cT/2$. The received echos are then demodulated directly into baseband (DC to 25 kHz) and the signals are fed to the spectrum analyser. This R_m is less than the maximum permitted $(cT/2)$ in order to reduce range ambiguities.

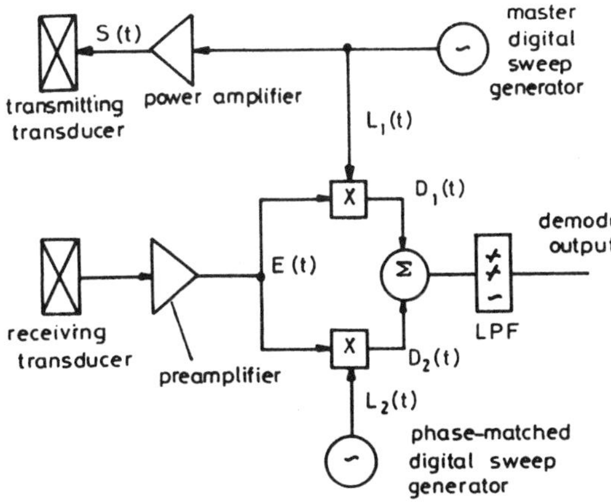

Fig. 2 *Block diagram of double-demodulator CTFM sonar*

4 Test results using an air sonar

The absorption of sound in air is much greater than it is in water (2.9 dB/m compared with 0.02 dB/m at 80 kHz). Thus, a suitable testing area for an air model of an underwater sonar is a large room with sound-absorbing material on the walls, floor and ceiling.

The transmitter and receiver electronics of our sonar were connected to a pair of solid dielectric transducers [5] comprising a grooved brass backplate and a sheet of gold-coated mylar. These transducers have been developed at the University of Canterbury to provide a useful sensitivity over the wide frequency bands [6].

The demodulated output of our sonar was fed into a Nicolet Scientific Corporation model 444 Mini-Uniquitous 1024-point spectrum analyser, which displays the lowest 400 spectral lines.

The target strength (TS) is determined by the ratio of the intensity of sound, returned by the target at a distance of 1 m from its acoustic centre, to the incident intensity from a distant source, i.e.

$$\mathrm{TS} \equiv 10 \log_{10}\left(\frac{I_r}{I_i}\right)_{r=1\,\mathrm{m}} \tag{15}$$

Target strengths for commonly occurring targets have been derived in various texts. In particular, the target strength of a sphere is given by Reference 7:

$$\mathrm{TS} = 10 \log (a^2/4) \tag{16}$$

where a is the radius of the sphere.

We used a polished aluminium sphere with a radius of 73.5 mm as our reference target. According to eqn. 16, it has a target strength of -28.69 dB in the physical-optics limit.

This sphere was used to obtain Figs. 3*a* and *b*. Fig. 3*a* shows the time waveform of the demodulated output, while Fig. 3*b* shows the spectrum of this particular waveform.

For comparison, a 'rough' target has the demodulated time waveform shown in Fig. 4*a* and the spectrum shown in Fig. 4*b*.

4.1 Dynamic range

We define dynamic range as the difference in demodulator output signal strength when we place, at a given range, a target of size such that overload is just avoided, compared with the signal strength when a sufficiently small target is placed at that same range so that a signal/noise ratio of 10 dB results. On this basis, a dynamic range of 65 dB was achieved.

This dynamic range is restricted by two factors: system noise limits the detection of small targets, while the mixers used for the demodulation process limit the size of the large targets that can be detected without introducing distortion.

At a range of 1 m we can detect, without distortion, targets from a TS of +10 dB to −55 dB. The signal/noise ratio for the smaller target was +10 dB for a single sweep, producing a probability of false alarm equal to 10^{-4} and a probability of detection equal to 0.6 [8].

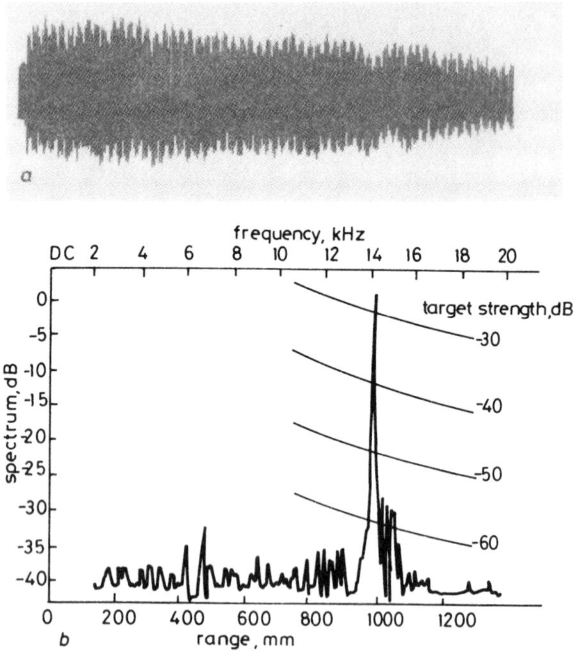

Fig. 3 *Using a −28 dB sphere as a target at a range of 1 m*
a Demodulated time waveform $D(t)$
b Frequency spectrum of Fig. 3*a*

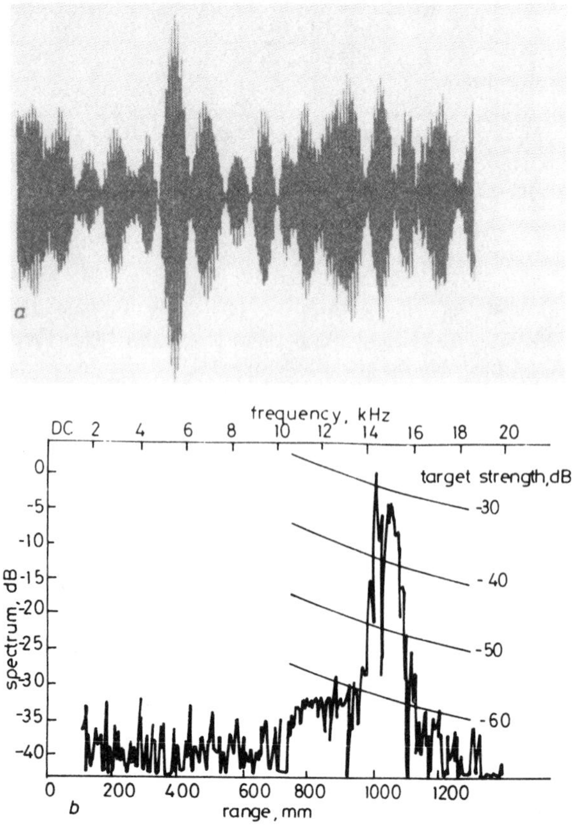

Fig. 4 *Using 'standard' rough object at a range of 1 m for comparison with Fig. 3*
a Demodulated waveform $D(t)$
b Frequency spectrum of Fig. 4*a*

4.2 Range accuracy and range resolution

We define range accuracy as the precision with which we can determine the range of a single target. Range resolution is defined as the separation in range between two targets that can be simultaneously detected by the sonar (for which at least a 3 dB dip is required between the returns from two targets).

Now, referring to Fig. 1, the time taken for sound waves to travel to a target at the maximum range and back is:

$$t_m = 2R_m/c \tag{17}$$

where R_m is the distance to the maximum range; or

$$R_m = ct_m/2 \tag{18}$$

Referring to Fig. 1 and using similar triangles:

$$t_m/f_m = T/\Delta F \tag{19}$$

Rearranging eqn. 19 gives:

$$t_m = Tf_m/\Delta F \tag{20}$$

where $\Delta F \equiv f_H - (f_H - bT)$. Substituting eqn. 20 into eqn. 18 gives:

$$R_m = Tf_m c/(2\Delta F) \tag{21}$$

With a sweep period T, we can resolve the demodulated output into spectral lines spaced $1/T$ Hz apart. To convert the frequency spacing to an equivalent range spacing, we note that the number of resolvable range cells is K, as defined in eqn. 10, and

$$K = f_m T \tag{22}$$

so that the spacing between the range cells is:

$$\Delta R = Tf_m c/2\Delta F f_m T$$
$$= c/(2\Delta F) \tag{23}$$

In systems with very large signal/noise ratios, we can determine the range of a known single target to distances much smaller than ΔR by comparing the response of adjacent range cells; i.e. there is no lower limit to the range accuracy. However, practical considerations, such as poor signal/noise ratios and multipath propagation, ensure that range accuracies of less than ΔR are seldom reached.

Range resolution usually requires that there be a 3 dB dip between the returns from two targets. This dip must be centred on one of the spectral lines (range cells) as otherwise we have no way of distinguishing two separate targets positioned on adjacent spectral lines from a single target centred between two spectral lines. Thus the frequency spacing between two resolvable targets is at least $2/T$, giving a range resolution of:

$$\text{range resolution} = c/\Delta F \tag{24}$$

Using this criteria for resolution, the range resolution for the air sonar using in these experiments is (332 m/s)/50 kHz or 6.6 mm.

Given that c is constant, we note that both the range accuracy and range resolution are functions of the transmitted bandwidth ΔF only.

To measure the range resolution, we needed very small

targets; so small in fact that their echos were of the same order of magnitude as the noise. To lower the noise, the spectrum was time averaged. Fig. 5 is the time-averaged

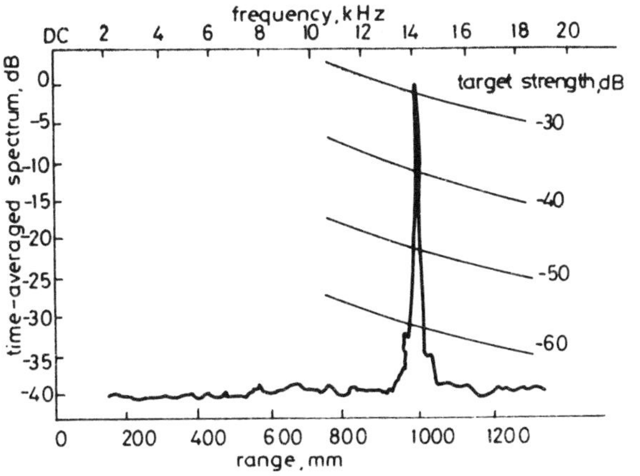

Fig. 5 *Spectrum of echos from the − 28 dB sphere at a range of 1 m and averaged over 32 sweep repetition periods*
Note the lower noise fluctuations when compared with Fig. 3*b*

spectrum of the echos from the sphere (an integration over 32 sweep periods). Notice the much reduced noise fluctuations compared with Fig. 3*b*

The range resolution was measured using a target comprising two strands of 0.2 mm wire. With one strand fixed in space, the range of the second was varied until a 3 dB dip resulted. The sonar measured the separation to be 6.0 mm ± 1.4 mm. Measurement of the range resolution was limited by the spacing of the spectral lines of the spectrum analyser. The time-averaged spectrum of the two strands is shown in Fig. 6, where $\bar{\lambda}$ is the mean wavelength of the transmitted waveform.

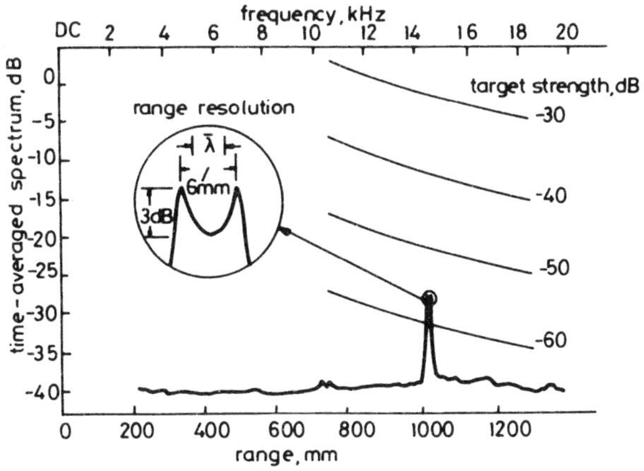

Fig. 6 *Spectrum of echos from a double strand of 0.2 mm wire at a range of 1 m and averaged over 32 sweep periods*

Since the centre frequency of the transmission was 75 kHz (corresponding to a wavelength of 4.4. mm), the range resolution achieved is close to 1.5 wavelengths. To achieve the same resolution using a pulsed sonar, 1.5 cycles of a nominal 75 kHz signal would have to be detected with a system having a bandwidth of 50 kHz. With such a wide bandwidth, the wanted echos would be difficult to extract from the noise.

5 Comparison of range resolution attainable by pulsed and CTFM sonars

The number of range cells theoretically attainable with pulsed and CTFM sonars (using identical bandwidths, analysis times and ranges) is the same. However, the number of range cells that can be achieved in practice is limited by different factors for the two sonars.

In pulsed sonars, this limit is imposed by the pulse length. The shorter the pulse (to increase the number of range cells), the wider the system bandwidth required to pass the pulse, and, hence, the more noise that can enter the system. A minimum signal/noise ratio usually puts an upper limit on the number of resolvable range cells in pulsed sonars.

This contrasts with CTFM sonars, where the sweep period T determines the smallest filter bandwidth ($1/T$) attainable. Note that as the filter bandwidth decreases (to give more resolvable range cells) so too does the noise bandwidth intercepted by the filter.

At present, the number of range cells in CTFM sonars is limited by practical spectrum analysers. However, recent years have seen significant improvements in spectrum analyser capabilities (computational speed, number of spectral lines, cost). So long as such improvements continue, the CTFM option for achieving large numbers of range cells is worth constant evaluation.

6 Conclusions

Although we do not envisage the CTFM system replacing pulsed sonars, there are features of CTFM sonars that make them attractive alternatives under some circumstances. For example, traditional CTFM sonars can have average transmitted powers comparable to pulsed sonars with much lower instantaneous transmitted powers. Furthermore, because of the continuous nature of the transmission, echos from targets are continuously received and almost continuously demodulated.

An important feature of the new sonar system described herein is that, by using dual demodulation, we have managed to eliminate the blind time associated with conventional CTFM sonars, thus making the demodulated output truly continuous: a real benefit for targets with a fluctuating return or for a moving side-scan sonar.

Finally, we note that both the time waveform and spectrum of the smooth sphere are considerably different to those of the 'rough' target. This suggests that there is scope for using these features for target classification. Work is currently progressing in this area.

7 Acknowledgements

We acknowledge Prof. L. Kay for the use of his air transducers and his laboratory and also his comments. We also acknowledge the comments and help of our colleague J. Sinton.

8 References

1 HAYKIN, S.: 'Communication systems' (John Wiley, 1978), pp. 101–103
2 KAY, L.: 'A sonar aid to enhance spatial perception of the blind engineering design and evaluation', *Radio & Electron. Eng.*, 1974, **44**, pp. 605–627
3 KAY, L., BOYS, J.T., CLARK, G., and MASON, J.: 'The echocardiophone: a new means for observing spatial movement of the heart', *Ultrasonics*, 1977, pp. 136–141
4 DE ROOS, A., KAY, L., CUSDIN, M.J., and VERNON. A.: 'A diver's sonar with binaural display'. Ultrasonics International '81 Conf. Proc. (IPC Science and Technology Press, 1981), pp. 38–42.
5 KUHL, W. et al.: 'Condenser transmitters and microphones with solid dielectric for airborne ultrasonics', *Acoustica*, 1954, **14**, pp. 519–32
6 BOYS, J.T., STRELOW, E.R., and CLARK, G.R.S.: 'A prosthetic aid for a developing blind child', *Ultrasonics*, 1979, **17**, pp. 37–42
7 ULRICK, R.J.: 'Principles of underwater sound' (McGraw-Hill, 2nd edn., 1975), p. 274
8 SKOLNIK, M.I.: 'Introduction to radar systems' (McGraw-Hill, 2nd edn., 1980)

An Optical Rangefinder for Autonomous Robot Cart Navigation

Gabriel L. Miller and Eric R. Wagner

AT&T Bell Laboratories
600 Mountain Avenue, Murray Hill, New Jersey 07974

ABSTRACT

A simple, low cost, infra-red rangefinder has been developed for investigation of autonomous robot cart navigation in factories and similar environments. A 2mW 0.82μ LED source (not a laser) is 100% amplitude modulated at 5MHz and used to form a collimated 1" diameter transmit beam that is unconditionally eye-safe. Returning scattered radiation is focussed by a 4" diameter coaxial Fresnel lens onto a p-i-n silicon photodiode. Range is determined from the phase shift between the 5MHz modulation on the transmitted and received signals. Use of a rotating mirror provides 360° polar coordinate coverage of both distance and reflectance out to ∿20 ft. around the vehicle. Both radial and angular resolution correspond to ∿1 inch at a ranging distance of 5 ft., with an overall bandwidth of ∿1KHz. The ranging resolution at 20 ft. is ∿2.5 inches, which is close to the theoretical limit possible for the radiated power, bandwidth, optics and receiver employed. The system is capable of reading wall bar codes "on the fly" and is in addition capable of simultaneously ranging and acting as a wideband optical communication receiver. Total parts cost for the optical transmitter, Fresnel lens, receiver and all the electronics is <$200. The remaining major parts, consisting of the rotating mirror, ring mounting, motor and incremental encoder, cost <$500.

1. INTRODUCTION

The use in factories of robot carts or "Autonomously Guided Vehicles" (AGV's) is becoming quite widespread.(1) There are thousands of AGV's in use world-wide at the time of writing, and this number is increasing steadily. The reason has little to do with direct labor cost, but more accurately reflects the continuously increasing need to impose greater discipline on factory material flow. As such this trend can be expected to continue unabated for some time, making the investigation and development of such systems a worthwhile long-term endeavour.

The vast majority of existing factory AGV's are simple predesignated path followers, guided by cables buried in the factory floor. All sorts of other schemes have been investigated,(2) paint stripes on the floor, various kinds of beacons, metal tapes, optical guiding paths, etc. but so far only the buried cable has gained widespread acceptance by virtue of navigational reliability, physical durability, noise immunity, and communication capability. (Office mail cart AGV's, on the other hand, operate in a much more benign environment and in that domain both fluorescent dye paint stripes and metal tape navigational guide paths have proved satisfactory.)

Of course AGV's for factories are but a small part of the overall effort on robot vehicles. Significant US research efforts are underway (with major funding by DARPA) oriented towards the development of robot vehicles for military applications. Obviously such tasks are vastly different from the factory AGV problem, are clearly incomparably more difficult, and of course reflect a very different cost structure. Similar remarks hold for "Robots for Hazard" programs such as the Japanese JUPITER project. Such programs are aimed at the development of very flexible, agile, teleoperator-telepresence vehicles to assist in a wide variety of serious accident situations.

By contrast the factory AGV problem involves the need for the development of relatively simple vehicles, to operate in a structured environment, but with exceedingly reliable navigational and error-recovery capabilities together with reasonable overall construction costs. This immediately imposes numbers of constraints on system designs.

2. SYSTEM DESIGN

The navigational reliability of existing factory AGV systems stems directly from the use of a buried cable. This can be thought of as a limiting case of imposing structure on an environment. Here one intentionally deploys physical hardware (i.e. buried cables) throughout the environment to serve no other purpose but that of guidance (and to some extent communication). Clearly this is an expensive and inflexible solution, albeit an extraordinarily robust one.

However the typical factory is already dense with structure, taking the form of all of its existing machinery, aisles, pillars, walls, doorways, etc. Most of these features are substantially invariant with time and in principle can be used as extremely useful aides for navigation, although obviously that was not their initial primary intent. It is precisely the existence of this highly structured and relatively stable environment that simplifies and separates the factory AGV navigation problem from the military vehicle navigation problem.

Having once decided to investigate the use of existing features for navigational aides the remaining design issues are relatively straightforward. Ultrasonic sensing is one obvious approach whereby a vehicle can sense its environment.(3,4) However one is limited to the use of acoustic wavelengths of ∿1mm or more by virtue of rapidly increasing atmospheric absorption at higher frequencies. On a scale of millimetres most man-made objects (walls etc.) are quite smooth, leading to essentially specular reflections of a probing ultrasonic beam. This fact creates substantial and inherent difficulties in the use of ultrasonics to explore wide regions of man-made environments.

Given that the problem is that of wavelength there is really only one other serious candidate, namely light. Assuming appropriate sources and detectors to exist, then anywhere in the range of infrared to optical frequencies would be satisfactory from the point of view of wavelength. The first obvious choice would simply be to use two conventional CCD cameras and associated machine vision hardware to obtain stereo information, as is being attempted in numbers of other cart projects. However this turns out to be complicated, computationally very demanding, slow and exceedingly expensive, rendering it unattractive for factory AGV application.

Alternatively, enormous simplification in 3-D vision can be achieved by using just one TV camera and illuminating the scene of interest with one or more <u>planes</u> of light, as initiated by the GM CONSIGHT structured light system(5) and variants thereof.(6,7,8) However the expense and complexity of a complete machine vision system is unfortunately still required.

An even simpler alternative is that of optical triangulation ranging, using a finely collimated probing laser beam and imaging the returning scattered light onto a linear CCD. The position of the received focussed light spot translates into the distance at which the laser beam was intercepted. This scheme has been applied to robot carts(9) and has the added advantage that it can read wall mounted bar codes on the fly as the robot cart moves. However there is some difficulty in making such a system provide 360° coverage over a wide range of distances at reasonably high speed.

A different approach has been investigated in the current work, namely that of obtaining polar-coordinate plots all around the vehicle by direct time-of-flight optical ranging with a rotating infra-red beam. The factors that led to the investigation of this method were the realization that it was potentially very simple and could in principle also provide readily interpretable data on both range and reflectivity. The major difficulty was obvious, namely that of achieving sufficiently good time resolution to make the system viable. Before describing how this was actually achieved it may be worth noting that before embarking on this course <u>all</u> of the currently available commercial optical autofocus camera systems were also studied as potential candidates. If suitably augmented some of these appeared usable for cart application, but none ultimately looked as attractive as the straightforward time-of-flight system that was ultimately implemented. (A particular difficulty with the camera systems was in speed of response, since all of the accurate ones use some form of electromechanical linkage in addition to their optical and electronic components.)

Once having decided to use an optical time-of-flight system the next choice became that of either using essentially δ function light flashes with a high speed amplifier and time-pickoff at the receiver,(10) or of using a CW phase-shift method. Both approaches have been employed for optical ranging in the past. Nitzan et al(11) were apparently the first to report on the latter method for 3D machine vision in the open literature, making use of a mechanically scanned RF modulated He-Ne laser beam and a photomultiplier receiver. (Conceptually very similar complete high resolution 3D imaging systems have also been custom fabricated for DARPA, in the form of the ERIM scanner,(12) at a price of ∿$125,000). It is known that the information that can be obtained from either the δ function or CW approach is in principle identical, provided only that each run at the same average optical power (i.e. equal numbers of photons per second) and use the same optics. For the present purposes it turned out to be particularly advantageous to use the CW approach since this can lead to very simple low-cost designs. A schematic diagram of the system actually employed is shown in Figure 1.

Here an ∿2mW LED (it is noteworthy that this is <u>not</u> a laser) emits 0.82μ infra-red radiation that is formed into an ∿1" diameter collimated beam, thereby providing unconditional eye safety, (a <u>very</u> important issue for factory applications). The LED is 100%

amplitude modulated at 5MHz, resulting in an average radiated optical power of ∿1mW.

Returned scattered radiation is focussed by a 4" diameter coaxial Fresnel lens onto a totally depleted p-i-n silicon photodiode. The diode photocurrent in turn drives a high speed low noise, JFET-input, integrator. All of the front-end noise sources are shown in Figure 2. Here the JFET series noise spectral density e_n corresponds to an equivalent series noise resistance R_s of no more than ∿100Ω. The total parallel noise is dominated by two contributions, namely the photodiode leakage current I_L and the noise from the feedback resistor R_F. The totally depleted photodiode detector capacitance C_D is ∿10pf, which is also close to both the value of the feedback capacitance C_F and the JFET gate-channel capacitance C_{IN}. These values, together with a total stray capacitance C_s of ∿10pf yield a total effective (cold) input capacitance C_T of ∿40pf. This, together with a feedback resistor R_F of 100K, leads to a noise corner time constant of ∿120ns, indicating that the system is operating not far from the series noise limit. Indeed the low value of R_F was chosen precisely with this goal in mind. This is due to the fact that the system needs to operate in situations with high ambient background illumination and of course all of the "background" photocurrent must perforce also flow through R_F. This indicates that R_F should be chosen to be as small as possible, without introducing an inordinate amount of noise. (If R_F is not small then changing illumination changes the preamplifier DC operating point and hence can affect its phase shift.)

Of course the other method of minimizing background light effects is to shield the detector with a narrow-band multi-layer optical filter precisely tuned to the LED wavelength. This is not satisfactory for the present design, however, because of the wide range of incident angles of the light arriving from the short focal length Fresnel lens. As a compromise, therefore, a multi-layer "cold mirror" filter was mounted over the detector. This effectively removes all radiation with a wavelength shorter than 0.75μ, while the silicon band-gap energy of Eg=1.2ev automatically cuts out everything longer than 1.0μ. This scheme thereby brackets the 0.82μ LED wavelength of interest.

Following the preamplifier the signal enters a tuned amplifier, ultimately driving a high quality limiter and balanced-modulator phase detector. The time delay between the transmitted and received 5MHz signal, corresponding to ∿2ns per foot of range, is thereby determined and constitutes the final output signal.

However, to be able to obtain the required data in this way a significant problem has to be solved, namely that of handling the very wide dynamic range of optical intensity expected in the received signal. This dynamic range arises as the product of the inverse square law effect with distance multiplied by the differing effective infra-red reflectivity of different target surfaces. Overall this can span a range of several thousand to one. The difficulty here is not the wide range per se, but the fact that it needs to be handled without introducing any change in phase greater than $\sim 3 \times 10^{-3}$ radians at a frequency of 5MHz. (This is a very stringent condition indeed and one that is ordinarily of no concern for conventional automatic gain control (AGC) systems.) The solution to this problem represented the most difficult task in the design. It was solved by factoring it into a product of two parts, a passive optical part providing a factor of ∿20 and an active electronic AGC part, providing a factor of ∿50.

However, before discussing the solution it must be noted that the problem cannot be well handled by instead servoing the LED intensity. This arises for two reasons. First, an electrical reference phase signal is of course needed that is accurately synchronized with the transmitted light. For a range of the order of a thousand to one in optical intensity one has therefore merely transferred the phase shift problem from the receiver to the transmitter. Even more tellingly, however, by servoing the light intensity one would not be making use of the improved distance resolution that is inherently available at short ranges. This clearly indicates that the problem is best handled in the receiver.

The optical part of the dynamic range situation is shown in simplified form in Figure 3. Here light from infinity is brought to the Fresnel lens focal point at A, which is where the photodetector is actually mounted. However for a (hypothetical) point source scattering centre at B none of the scattered light can reach the detector since all the extreme rays focus behind the detector at C, while all the paraxial rays are automatically blocked by the presence of the transmitter optics. The actual situation is somewhat more complicated than this because the transmit beam typically illuminates a disc, not a point, but the final result is qualitatively similar. The end result is that the optical system automatically and increasingly discriminates against light that is scattered from targets that are closer than 5 feet from the ranger.

The low-phase shift AGC stage that was finally developed is shown in simplified form in Figure 4. Here the input signal drives an amplifying element Q_0 (a bipolar transistor of JFET), with some known g_m, to provide an output signal current I_s at the common Q_1, Q_2 emitters. This current I_s is small compared to the standing current I_0 flowing through Q_0.

The base of Q_2 is held at a convenient fixed potential while the base of Q_1 is voltage driven from the AGC loop. Depending on the Q_1, Q_2 base voltage difference the current I_0 splits into I_1 and I_2, while of course $I_1+I_2=I_0$. Now clearly the dynamic impedance looking into the emitter of Q_1 is given by $Z_1=(KT/qI_1)$ while that at Q_2 is $Z_2=(KT/qI_2)$. Here K, T and q have the usual significance of Boltzmann's constant, absolute temperature and the electronic charge, respectively. The total dynamic impedance Z_T to ground at point A is essentially just that of Z_1 and Z_2 in parallel, namely $Z_T=(KT/qI_0)$ which is a constant independent of how I_1 and I_2 share the current. The signal voltage at A is therefore given by $V_S=I_SZ_T$, and consequently the signal current flowing through Q_2 is just $V_S/Z_2=(I_SI_2/I_0)$, i.e. the "gain" is exactly linear in I_2. Furthermore Q_2 is run grounded base, and even at low values of I_2 (high attenuation) it remains fast since its emitter is voltage driven by the much lower emitter impedance of Q_1, while its base transit time and collector depletion layer sweepout time remain essentially constant.

This scheme by itself is almost, but not quite, good enough, due to the inevitable existence of small capacitances from point A to the Q_2 collector.

This problem can be handled in turn by noting the fact that essentially all the current I_0 at point A must also flow at point B. As a consequence the dynamic impedance of diode D1 must also equal KT/qI_0, and therefore also the signal voltage at B must always be essentially equal and opposite to that at A. (Capacitor C1 is large, holding the anode of D1 at AC ground.) As a result the trimmer C2 provides a bridge admustment to cancel the stray C across Q_2, thereby finally allowing a wide range of gain control with exceedingly low overall phase shift.

The physical construction of the system is shown in Figure 5. The optical ranger is mounted vertically under a rotating mirror to provide full 360° coverage. The drive motor is fitted with an integral incremental encoder, while the mirror mounting contains an optical sensor to provide $\Theta=0°$ indication on each full revolution. (This is the simplest possible mirror driving scheme, future plans may well include putting the mirror itself under full computer control.)

3. EXPERIMENTAL RESULTS

The ranger provides three outputs, target distance, target angle and AGC gain. The onboard robot cart ADC's span ±1.5 volts at their inputs, so the range and gain output signals are also arranged to lie in that range. A measured range calibration curve is shown in Figure 6. It will be observed that the output is linear from 1′ to ∿17′ and quite usable out to ∿20′. Also shown on the same graph is the amplifier gain (AGC) signal. (It will be recalled from Figure 4 that this is simply proportional to the collector current I_2 of Q_2.) This signal spans 0 to 1.25 volts, with the gain itself being exceedingly accurately proportional to the observed number.

It is evident that the amplifier gain drops monotonically, as expected, as the target is moved in from 20′ to ∿4′. The measured gain actually dropped from 0.977 to 0.047, i.e. by ∿21:1, over this distance. On moving closer, however, from 4′ to 1′, the required amplifier gain increases markedly from 0.047 to 0.540. This is a consequence of the geometrical paraxial ray shielding effect described in connection with Figure 3. It will be noted that the overall dynamic range that had to be handled in this case was therefore only ∿20:1, whereas the full inverse square law effect from 1′ to 20′ would have been expected to be ∿ 400:1.

Also apparent from this figure is the lack of phase shift in the AGC system. For, in moving the target from 4′ to 1′ the signal shifted 6ns forward in time (i.e. 3′ at 2ns per foot), while the gain increased from 0.047 to 0.540. However on moving the target from 4′ to 14′ the signal shifted 20ns backward in time, while the gain again increased by just about the same factor. Over this whole range, however, the calibration curve is exceedingly linear. (This cannot be the case if the AGC exhibits any gain-dependent phase shift effects.)

Apart from linearity and dynamic range the other quantity of interest is the system ranging resolution. This is readily measured by connecting a wideband (10Hz to 10MHz) true RMS voltmeter directly to the ranger output. The noise bandwidth that is actually being employed is then being set by the ∿1KHz bandwidth limiting filter on the final amplifier following the ranger phase detector. Calibration for this measurement was provided from the results of Figure 6 which indicates that 150mV corresponds to 1′. The observed RMS noise at 4′, for example, was 7mV which therefore represents an RMS ranging error σ_R of 0.56 inches. It is to be noted that this is in a 1 KHz bandwidth. These results are plotted in Figure 7, and are in reasonable agreement with theoretical analysis which indicated that time resolution of <100 picoseconds should be achievable, as is indeed observed in the noise minimum at 4′.

It is to be stressed, of course, that the ranging resolution is a function of the actual signal to noise ratio. This means, for example, that a "darker" object at 4′ would also indicate a larger value of σ_R. Similarly a more highly reflecting object at 20′ would still indicate 20′ but could well give σ_R of only half an inch, and so on.

There is a limit to this behaviour however. The system is not intended to operate with gains of less than $\sim$0.020. For larger input signals the AGC stage starts to overload, leading to erroneous readings. By the same token the system is not intended to run at the maximum gain of 1.250 either, since if this is the case it means that there is not enough signal for the AGC to turn on and properly stabilize the amplitude at the input to the limiter. Usable range results are therefore only produced if the corresponding gain signal is less than 1.250 and greater than 0.020. This is one of three reasons why the gain signal is provided. Another reason is that the availability of this signal allows detailed reflectance data to be obtained for the observed scene. This is very useful additional signature and recognition information, as stressed in reference (11). And finally the magnitude of the gain signal itself is proportional to the observed amplifier noise, thereby allowing the ranging resolution to be known at all times.

4. APPLICATIONS

The optical ranger has been mounted on an existing research tricycle cart (named "Blanche") having one steered and powered wheel and two trailing idling wheels. Initially, cart navigation was by odometry (using incremental angle encoders on each trailing wheel), augmented by the use of a gas-jet gyroscope for additional rate-of-turn information. The added capability of the optical ranger, mounted above the cart as indicated in Figures 8 and 9, allows the vehicle to sense its surroundings. A typical sample of the ranger output is shown in Figure 10. The expected operating environment will usually be an aisle in a factory (or office), having a total width of $\sim$7 to $\sim$10 feet and essentially indefinite length. The vehicle will therefore typically be continuously interacting with a roughly linear length of artifacts (machines, walls, cross aisles, pillars, etc.) perhaps $\sim$3 to $\sim$5 or so feet away on either side, and "observed" for maybe up to $\sim$10′ ahead and behind on either side. The ranger was designed with this environment in mind. As the vehicle moves forward (factory experience shows $\sim$1.5 f.p.s. to be a reasonable speed) the data in the slice of the world seen by the ranger "flows" by on either side. New features are continuously added at the front while old ones drop out at the back. Most of the data between these two vision extremes is of course both highly correlated and redundant from view-to-view since it is essentially the same information continuously translated in space and seen from a slowly changing angular viewpoint. It is worth recalling in this regard that of course the odometry is simultaneously providing both distance moved and heading information at the same time. This odometry information is therefore available to continuously cross check and augment that from the ranger, and vice versa. (Odometry is of course also crucial in the case of open areas where the ranger provides little or no useful output.) It is conjectured that just these two low-cost complementary capabilities, namely odometry and optical ranging, will be capable of providing reliable factory cart navigation when suitably combined with stored map information and appropriate heuristics. The present hardware was developed with the goal of investigating this issue. Should it prove necessary, substantial added capability can be made available by providing on-board computer control of both angular degrees of freedom of the rangers scanning mirror.

As a final point, it is of course recognized that cart signalling and communication is also an important issue. In this regard it is worth noting that the ranger can read $\sim$1" wide striped bar codes "on the fly". Indeed the bandwidth of $\sim$1KHz was set with this in mind since such stripes at 10′, together with a mirror spin rate of $\sim$one rotation per second, give rise to a $\sim$720Hz signal. The bar code signal is available from the ranger AGC monitor output. (The $\sim$1" wide tape sold to enhance joggers visibility at night provides exceptionally strong IR reflection, and can be used to construct ranger-readable stripes. Totally "black" stripes can be fabricated by using the black conducting sponge plastic sheet that is used to package MOS integrated circuits.)

Also relating to communication, it will be noted from Figure 1 that the 5MHz tuning of the optical receiver takes place <u>after</u> the wideband low noise integrator. This means that a signal can be tapped off at the preamplifier output, operating essentially anywhere in frequency up to its measured limit of $\sim$50MHz. This opens the way to the use of small LED transmitters to be used as wideband communication links to the cart. Such transmitters can either run continuously, or be turned on by detecting the carts optical scanner probing beam itself. Such links have the obvious advantage of being essentially interference free.

5. CONCLUSION

A simple eye-safe polar-coordinate time-of-flight optical ranger has been developed specifically to investigate the potentialities of such an instrument as an aid in AGC navigation. The system is inexpensive (total parts cost <\$700) and capable of providing

∿1% target range, angle and reflectance information out to a distance of ∿20′, while operating with a ∿1KHz bandwidth. It also has the capability of reading wall-mounted bar codes and acting as a wideband optical communication receiver.

The physical scaling in size of the ranger has also been studied. It appears to be quite feasible to employ the same low-cost approach to range over a distance of ∿2 feet with ∿1" diameter optics while maintaining a resolution of ∿0.1". Such a capability could be useful for a number of other applications.

6. ACKNOWLEDGMENTS

It is a pleasure to acknowledge the help of I. J. Cox both in connection with initial investigations on the optical transmitter and also for providing plots of the results shown in Figure 10.

7. REFERENCES

1. Proceedings of the Second International Conference on Automated Guided Vehicle Systems, Stuttgart, W. Germany 7-9 June 1983. Ed. Prof. Dr. Ing. H. J. Warnecke. North Holland Publishing Co. 1983.
2. T. Tsumura, Proceedings of the 1986 IEEE International Conference on Robotics and Automation, San Francisco, CA, USA, April 7-10, 1986, Vol. II, p. 1329.
3. H. P. Moravec and A. Elfes, 1985 IEEE International Conference on Robotics and Automation, St. Louis, MO, March 25-28, 1986, p. 116.
4. A. Elfes, Proceedings of the 1986 IEEE International Conference on Robotics and Automation, San Francisco, CA, USA, April 7-10, 1986, Vol. II, p. 1151.
5. S. Holland, L. Rossol, M. Ward, "Computer Vision and Sensor Based Robots", Plenum Press, New York, NY, 1979.
6. J. S. Albus, E. Kent, M. Nashman, P. Mansbach, L. Palombo and M. Shneier, Proceedings of the SPIE Technical Symposium, Crystal City, VA, May 1982, p. 142.
7. J. Le Moigne and A. M. Waxman, Proceedings of the Seventh CIPPRS/IAPR International Conference on Pattern Recognition, Montreal, Canada, July 30 - August 2, 1984, p. 203.
8. G. C. Morgan, Proceedings of the Third International Conference on Robot Vision and Sensory Control, ROVISEC 3, Cambridge, MA, USA, November 1983, p. 615.
9. T. Hongo, H. Arakawa, G. Sugimoto, K. Tange and Y. Yamamoto, Proceedings of the IEEE/SICE International Conference on Industrial Electronics and Control Instrumentation IECON '85, San Francisco, CA, USA, November 18-22, 1985, p. 535.
10. R. Ahola and R. Myllyla, Proceedings of the IEEE/SICE International Conference on Industrial Electronics and Control Instrumentation IECON '84, Tokyo, Japan, October 22-26, 1984, p. 812.
11. D. Nitzan, A. E. Brain and R. O. Duda, Proceedings of IEEE, Vol. 65 (1977), p. 206.
12. Information supplied by the ERIM organization.

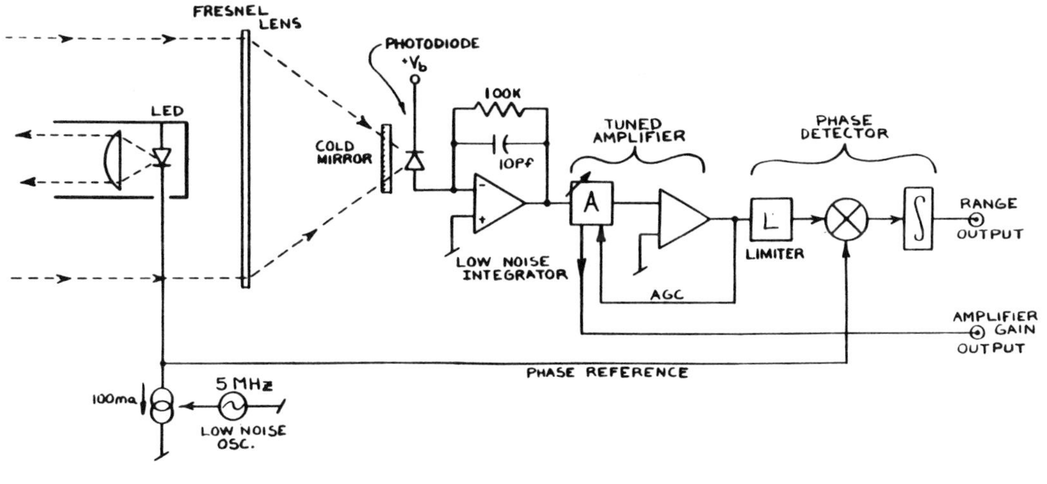

FIGURE 1

Schematic block diagram of the optical ranger. A linear rectifier at the output of the tuned amplifier drives the variable gain element A in such a way that essentially constant RF amplitude is always applied to the input of the limiter L, thereby providing a cascaded limiting action at the input to the phase detector.

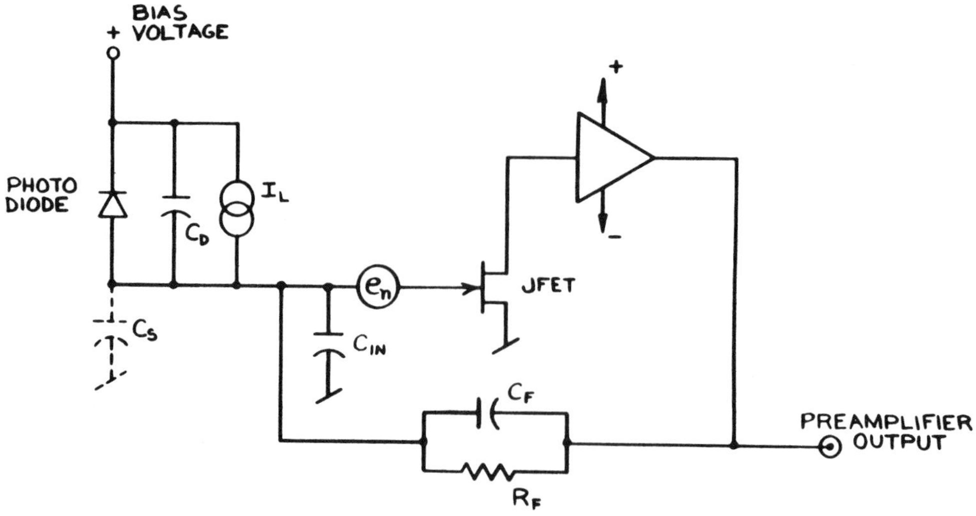

FIGURE 2

Simplified schematic of the low noise integrating preamplifier, showing its dominant capacitance and series and parallel noise sources.

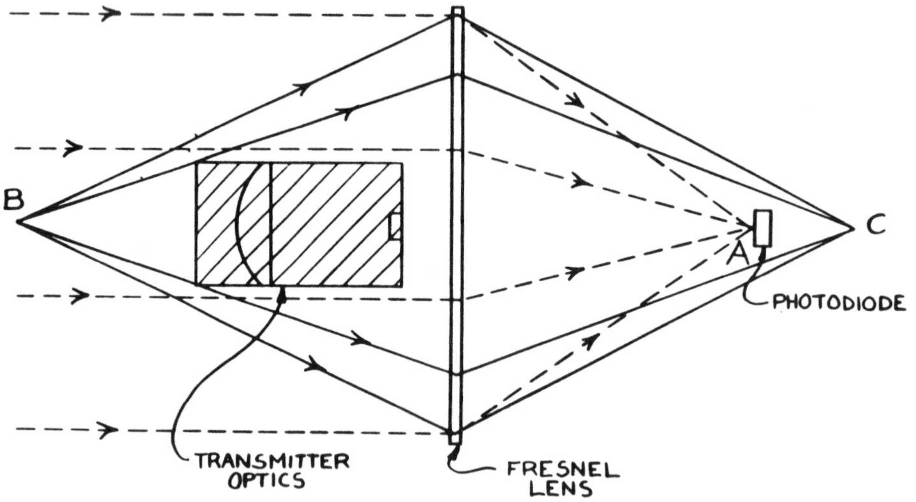

FIGURE 3

The elements of the coaxial optical system, showing the shielding effect for closely spaced scattering centers.

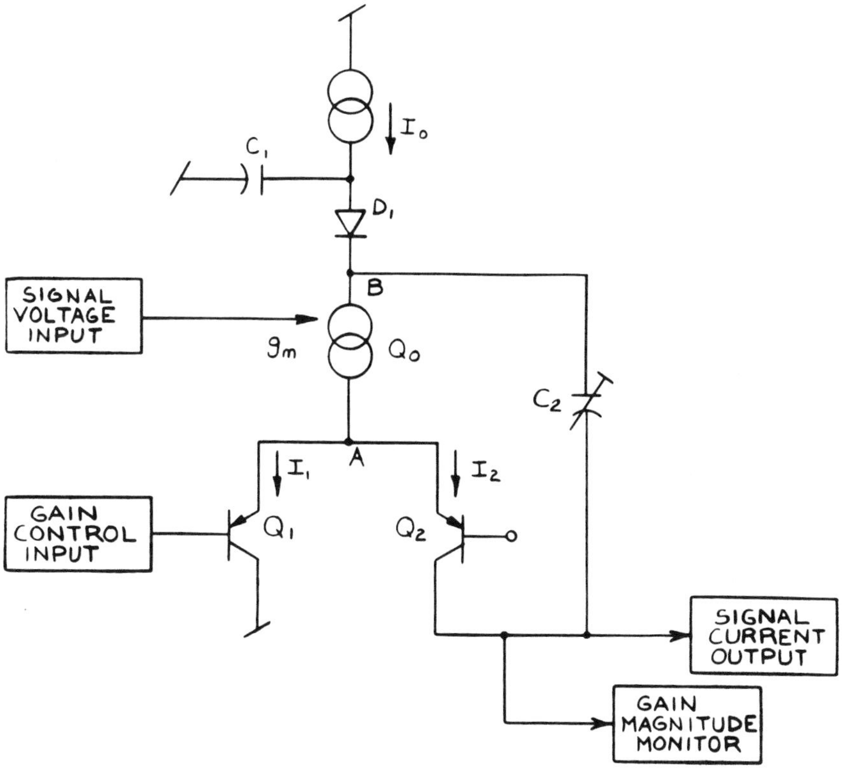

FIGURE 4

Schematic of the low phase shift variable gain stage.

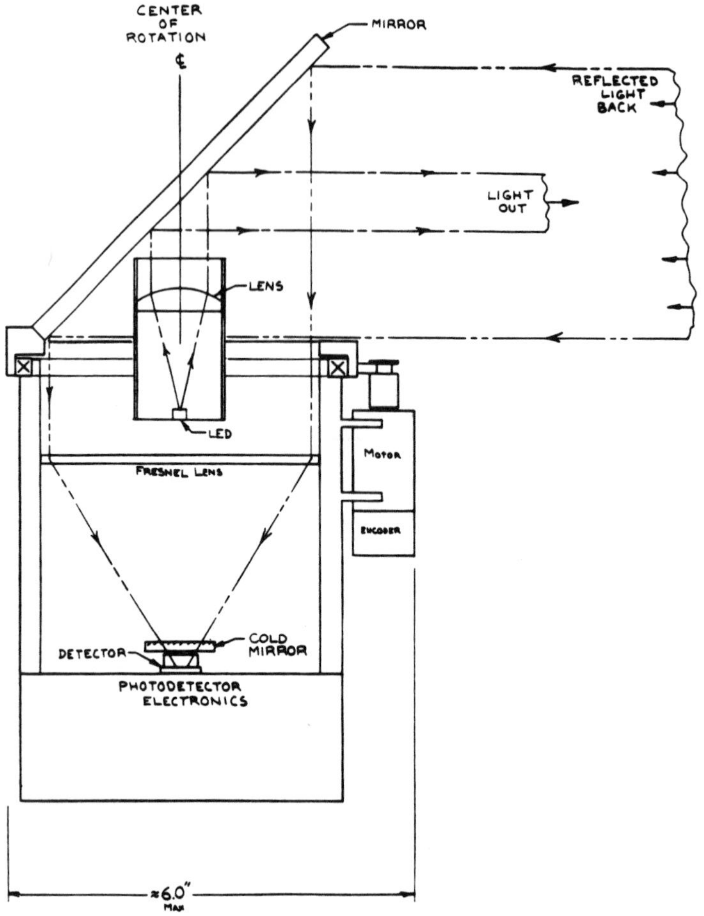

FIGURE 5

Physical construction of the optical ranger. The receiver electronics are mounted on one 4"x4" board directly below the photodetector.

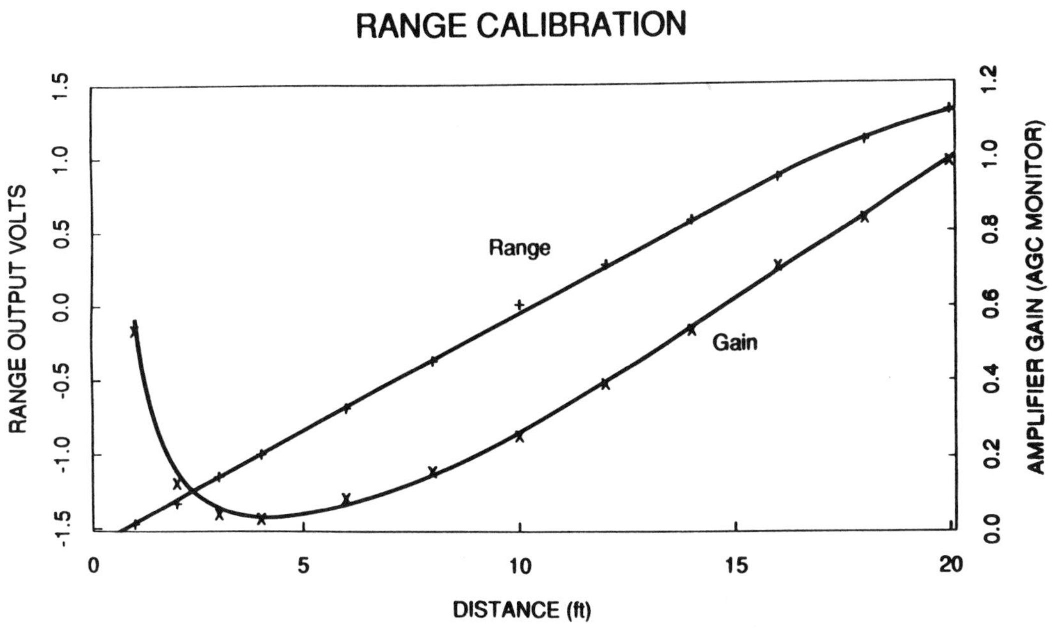

FIGURE 6

Range calibration, and corresponding receiver gain, for a white (diffuse) surface, normal to the IR beam, as a function of distance from the ranger.

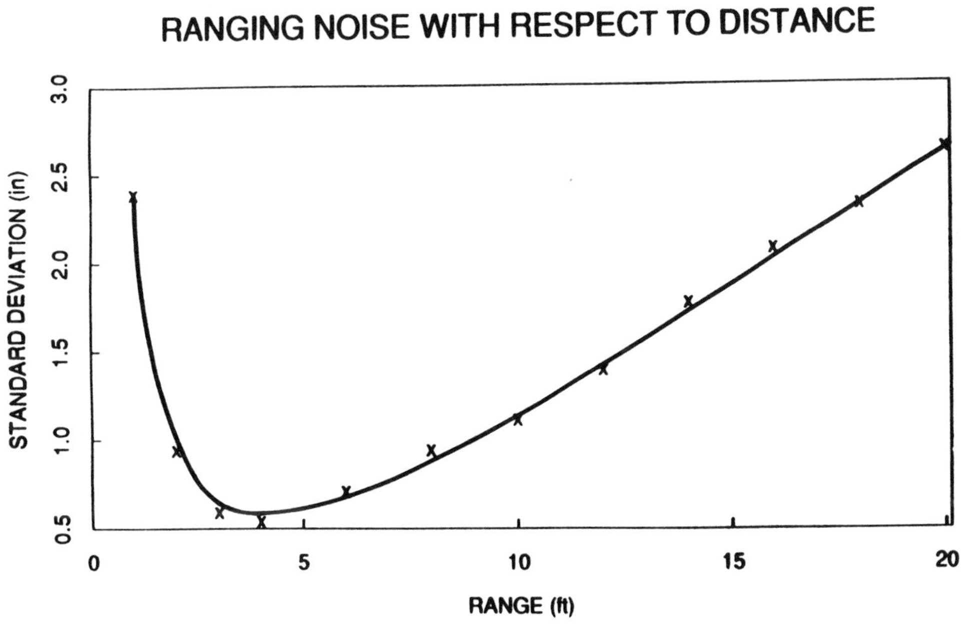

FIGURE 7

Ranging noise as a function of distance, (for the same target and conditions employed in obtaining the data of Figure 6).

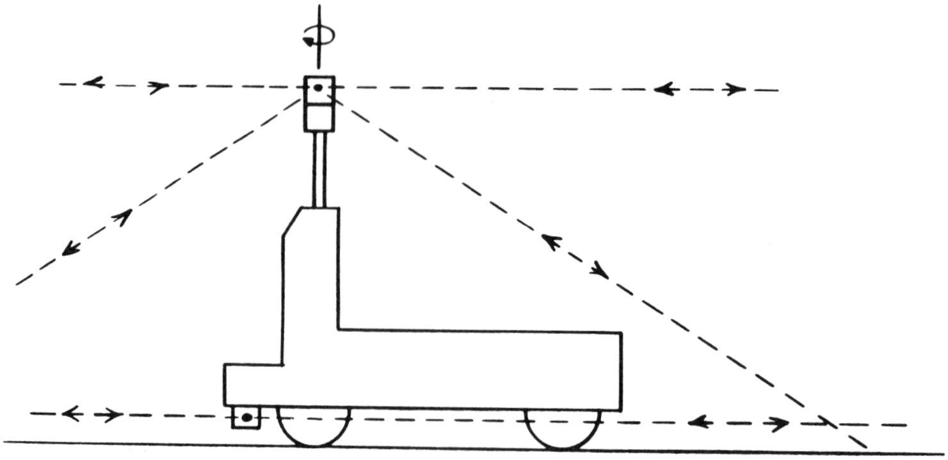

FIGURE 8

The ranger is mounted directly above the cart in such a way that it can either provide a radial scan or, by tilting the mirror, a conical scan to detect any object in its path. An alternative mounting that has also been considered (but not implemented) is in the space between the bottom of the cart and the floor.

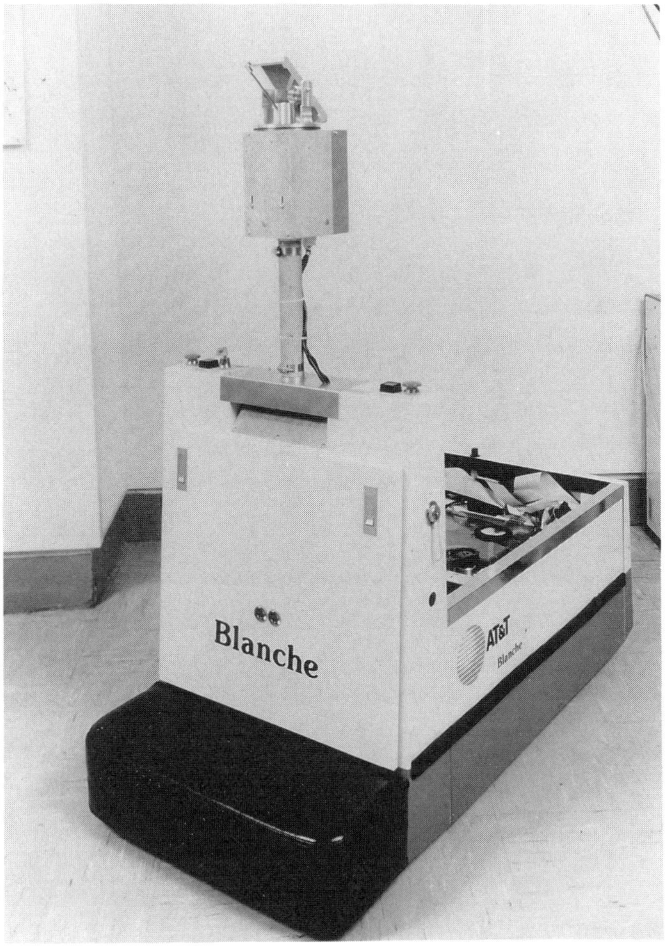

FIGURE 9

The Murray Hill Robot Cart ("Blanche"), showing the mounting of the optical ranger. The mirror orientation is such that the ranger is looking directly out of the page, making the coaxial optical system clearly visible.

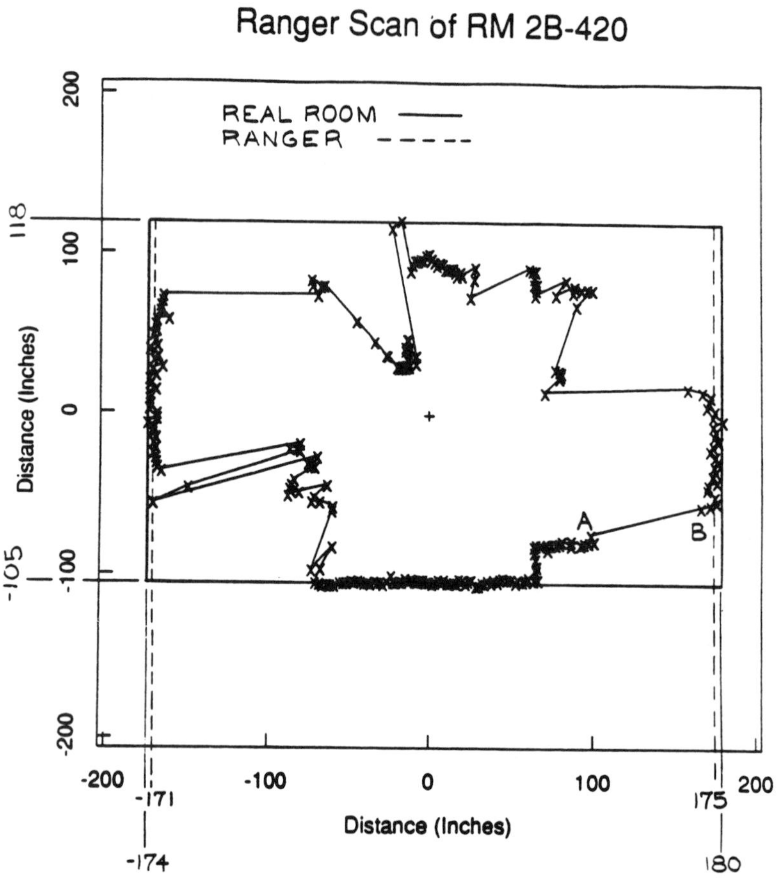

FIGURE 10

A representative ranger scan taken in a laboratory environment. The ranger is mounted on the robot cart and located at the (0,0) position, marked by a cross, in the centre of the plot. The data collection corresponded to a single mirror rotation taking ∿1 second. . The display computer program employed connects adjacent data points with straight line segments. The reason for the absence of any data points in regions such as that labelled AB in the figure, is that in such ranges the signal is outside the AGC window and therefore no data are accumulated. All of the objects in the room were found to correspond accurately to their position as indicated by the ranger.

Error Modeling in Stereo Navigation

LARRY MATTHIES AND STEVEN A. SHAFER, MEMBER, IEEE

Abstract—**In stereo navigation, a mobile robot estimates its position by tracking landmarks with on-board cameras. Previous systems for stereo navigation have suffered from poor accuracy, in part because they relied on scalar models of measurement error in triangulation. Using three-dimensional (3D) Gaussian distributions to model triangulation error is shown to lead to much better performance. How to compute the error model from image correspondences, estimate robot motion between frames, and update the global positions of the robot and the landmarks over time are discussed. Simulations show that, compared to scalar error models, the 3D Gaussian reduces the variance in robot position estimates and better distinguishes rotational from translational motion. A short indoor run with real images supported these conclusions and computed the final robot position to within two percent of distance and one degree of orientation. These results illustrate the importance of error modeling in stereo vision for this and other applications.**

I. Introduction

CONSIDER a robot given the task of going from A to B. At a coarse level its route is planned from a prestored map, while at a fine level the route is determined by sensor information gathered along the way. Incremental motion estimates are integrated to keep track of the robot's position in the map, which in turn is used to predict upcoming landmarks, hazards, or arrival at the destination.

To realize this scenario, a robot needs sensors that can measure its position and detect the presence of three-dimensional (3D) objects nearby. Stereo vision can provide both kinds of information. Stereo matching at one point in time provides a local 3D model for route planning and obstacle avoidance. Selected points in this model become landmarks that are tracked by the stereo system to monitor the robot's progress. Using stereo in this way, to detect nearby objects and to estimate the motion of the robot, is what we refer to as stereo navigation.

We are interested in stereo in this scenario for a number of reasons. First, other motion sensors can be in error, such as shaft encoders when wheels slip or lose contact with the ground. Second, other sensors, such as sonar and radar, can be inappropriate for reasons of concealment, possible confusion with the broadcasts of other robots nearby, or because color and reflectivity information are important. Lastly, we are interested in stereo *per se* and believe that methods developed for this domain can be transferred to other applications.

Manuscript received July 25, 1986; revised December 1, 1986. This work was sponsored in part by the Office of Naval Research under Contract N00014-81-K-0503 and in part by the Defense Advanced Research Projects Agency under Contracts DACA 76-85-C-0003 and F33615-84-K-1520. This work was presented at the ACM/IEEE Fall Joint Computer Conference, Dallas, TX, November 5, 1986.

The authors are with the Computer Science Department, Carnegie-Mellon University, Pittsburgh, PA 15213, USA.

IEEE Log Number 8714999.

Methods for extracting shape and motion information from image sequences can be classified as correspondence-based or flow-based. Correspondence methods [7], [11], [18], [24] track distinct features such as corners and lines through the image sequence and compute 3D structure by triangulation. Flow-based methods [1], [25] treat the image sequence as function $I(x, y, t)$ of row, column, and time, restrict the motion between frames to be small, and compute shape and motion in terms of differential changes in I. This paper deals with error modeling issues in the correspondence paradigm.

One of the first systems for correspondence-based stereo navigation was that built by Moravec [18]. This system moved a robot in a stop–go–stop fashion, digitizing and analyzing images at every stop. Features were matched in stereo images to build a world model consisting of 3D points. After moving and acquiring more images, the points in the world model were matched in the new images to find their coordinates relative to the new robot location. A least squares procedure was applied to the differences between the new and old point locations to infer the actual motion of the robot. The contribution of each landmark point to this motion estimate was multiplied by a scalar weight that varied inversely with the distance to the point.

In earlier work with Moravec [17], we found the motion solving part of this system to be somewhat inaccurate and unstable. This has been a common experience with visual motion solving algorithms in general. In the case of correspondence-based algorithms, this can partly be attributed to inadequate modeling of measurement error in triangulation. In triangulation, 3D coordinates are computed by intersecting rays projected through corresponding points in two images. Errors in locating the image points induce errors in the 3D coordinates, which in turn cause errors in motion estimates based on the 3D information. Modeling the measurement errors can reduce their effect on motion estimates. However, we will demonstrate that using scalar weights to model uncertainty in 3D coordinates leads to poor performance.

More sophisticated methods have been used in a number of places. In photogrammetry [20], two-dimensional (2D) and 3D normal distributions are used to model error in image coordinates and 3D point locations, respectively. Gennery [11] has used 2D normal distributions of image coordinates in camera calibration for computer vision. Hallam [15] used normal error models in conjunction with Kalman filters to track points and estimate robot motion from sonar data. Broida and Challeppa [5] used similar methods to track a known object in monocular image sequences, and Faugeras [9] has discussed the application of these methods to stereo.

This paper shows how these methods can be applied to

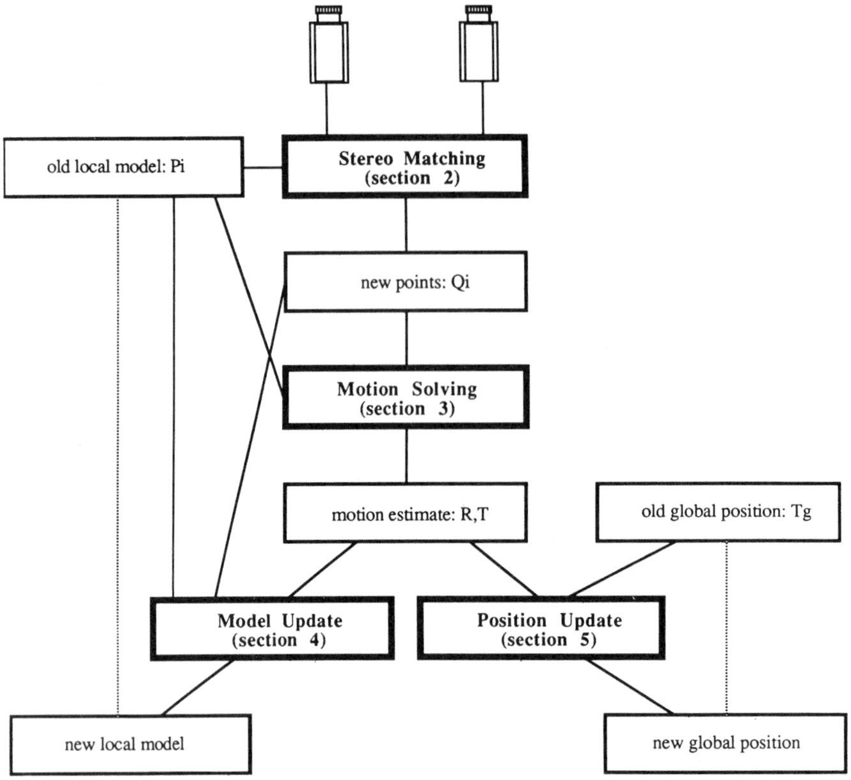

Fig. 1. System block diagram.

stereo navigation and demonstrates by results with real images that they lead to markedly better performance. The system we will describe has evolved from Moravec's [18] and is shown in Fig. 1. The main data structures are a set of 3D points P_i, called the local model and described in robot-centered coordinates, and the robot's current estimate of its position in some fixed global reference frame. The points in the local model are obtained by stereo matching and are used as landmarks. When a new stereo pair is digitized, points from the local model are matched in the images to determine their current locations Q_i relative to the robot. A motion solving algorithm estimates the rotation and translation (R and T) relating the new and old coordinates. The model updating system transforms the old local model into the current coordinate frame and combines it with the new points to create a new local model. Finally, the motion estimate is used to update the robot's global position. The cycle then repeats with the acquisition of a new pair of images.

Section II shows how to model triangulation error in the stereo matcher with 3D normal distributions. In Section III this is incorporated in an algorithm for finding the rotation and translation between successive stereo pairs. The covariance matrix of this transformation is used in Section IV to update the local model with Kalman filters and in Section V to estimate the robot's global position uncertainty. Simulations described in Section VI show that compared to scalar error models this system reduces the variance of position estimates and better distinguishes rotational motion from translation. An experiment with real images, using 54 stereo pairs covering 5.4 m and fully automatic feature tracking, supported these conclusions and computed the final robot position to within two percent of distance and one degree of orientation. Conclusions are summarized in Section VII.

II. Modeling Stereo Triangulation Error

The geometry of stereo triangulation is shown schematically in Fig. 2 for the case of 2D points projecting onto one-dimensional (1D) images. The tick marks on the image planes denote pixel boundaries, and the radiating lines extend these boundaries into space. Suppose point P projects onto the left image at x_l and the right image at x_r. Because of errors in measurement, the stereo system will determine x_l and x_r with some error, which in turn causes error in the estimated location of P. Fig. 2 illustrates this for errors caused by image quantization; because of resolution limits, the estimated location of P can lie anywhere in the shaded region surrounding the true location [22]. Random contributions to measurement error will blur the boundaries of this region, but the qualitative shape will be similar. We want to take this uncertainty into account in any reasoning based on measurements of P.

Three approaches to modeling such uncertainty are discrete tolerance limits, scalar weights, and multidimensional probability distributions. Tolerance regions have been used in object recognition to test candidate model to image matches [14] and to constrain three-dimensional relationships between objects [4], [6]. For example, Baird [4] used tolerance regions in finding the transformation between a two-dimensional set of model points and their measured image positions. Uncertainty was represented with convex polygons surrounding the measured point locations, and the transformed model points were required to lie within these polygons. Acceptable transforma-

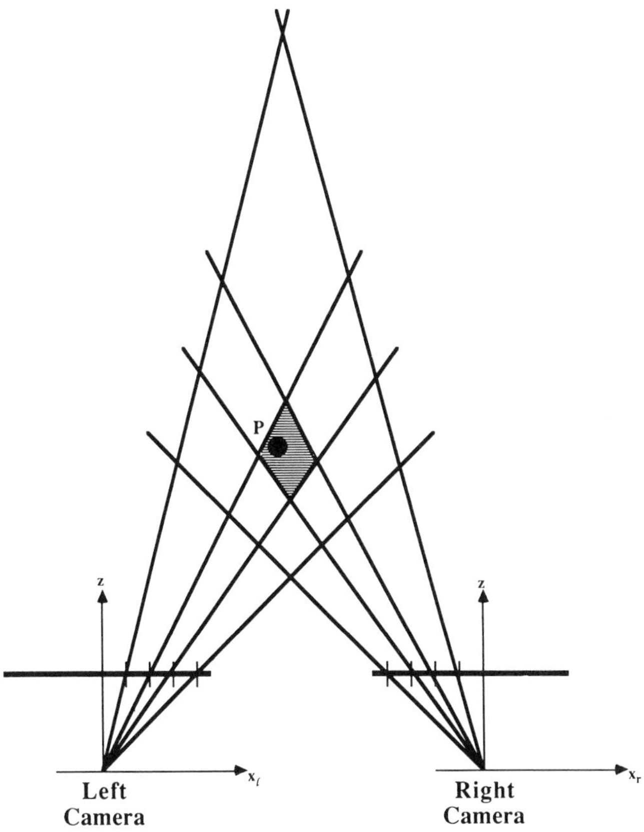

Fig. 2. Stereo geometry showing triangulation uncertainty.

tions were found by linear programming. In our application, statistical minimization and methods are more appropriate because of the stochastic nature of measurement errors and the need to filter time sequences of measurements.

The motivation for using scalar weights is that uncertainty grows with distance, so it can be modeled by weighting points inversely with distance [18]. However, as Fig. 2 shows, the uncertainty induced by triangulation is not a simple scalar function of distance to the point; it is also skewed and oriented. Nearby points have a fairly compact uncertainty, whereas distant points have a more elongated uncertainty that is roughly aligned with the line of sight to the point. Scalar error measures do not capture these distinctions in shape.

These distinctions can be captured by using 3D probability distributions to characterize the uncertainty in point locations. Our approach is to assume 2D, normally distributed (i.e., Gaussian) error in the measured image coordinates and to derive 3D Gaussian distributions describing the error in the inferred 3D coordinates. Similar approaches have been used in photogrammetry [20] and elsewhere in computer vision [11], [5], [12], [15], [9]. The use of Gaussian distributions to model image coordinate error is a common [11], [5], convenient approximation that gives adequate performance, as will be seen in Section VI. For the 3D coordinates, the true distribution *will* be non-Gaussian because triangulation is a nonlinear operation; we approximate this as Gaussian for simplicity and because it gives an adequate approximation when the distance to points is not extreme. We will discuss shortly the cases where this breaks down.

We will now show the details of the triangulation and error model calculation for the general case of 3D points projecting onto 2D images. We assume a camera geometry with parallel image planes, aligned epipolar lines, and image coordinate systems centered at the piercing point of each camera. Let the image coordinates be given by $l = [x_l, y_l]$ and $r = [x_r, y_r]$ in the left and right image, respectively. Consider these as normally distributed random vectors with means μ_l and μ_r and covariance matrices V_l and V_r. From l and r we need to estimate the coordinates $[X, Y, Z]^T$ of the 3D point P. We take the simple approach of using the ideal noise-free triangulation equations $P = [X, Y, Z]^T = f(l, r)$, or

$$X = b(x_l + x_r)/(x_l - x_r)$$

$$Y = b(y_l + y_r)/(x_l - x_r)$$

$$Z = 2b/(x_l - x_r) \tag{1}$$

(assuming a unit focal length and a baseline of $2b$) and inferring the distributions of X, Y, and Z as functions of random vectors l and r. If (1) was linear, P would be normal [8] with mean $\mu_p = f(\mu_l, \mu_r)$ and covariance

$$V_p = J \begin{bmatrix} V_l & 0 \\ 0 & V_r \end{bmatrix} J^T \tag{2}$$

where J is the matrix of first partial derivatives of f or the Jacobian. Since f is nonlinear these expressions do not hold exactly, but we use them as satisfactory approximations.

The true values of the means and covariances of the image coordinates needed to plug into (1) and (2) are unknown. We approximate the means with the coordinates returned by the stereo matcher and the covariances with identity matrices. This is equivalent to treating the image coordinates as uncorrelated with variances of one pixel. Better covariance approximations can be obtained by several methods [2], [11].

What does this error model mean geometrically? Constant probability contours of the distribution of P describe ellipsoids about the nominal mean that approximate the true error distribution. This is illustrated in Fig. 3 where the ellipse represents the contour of the error model and the diamond represents quantization error of Fig. 2. For nearby points the contours will be close to spherical; the farther the points, the more eccentric they become. A covariance matrix with structure $V = wI$, equal to a scalar times the identity matrix, describes only spherical contours. This is the difference between attaching scalar weights to 3D coordinate vectors and using the full 3D distribution; that is, scalar weights are equivalent to spherical covariances whereas the full distribution permits ellipsoidal covariances. In the balance of the paper we will often refer to scalar weights as a spherical error model and the full distribution as an ellipsoidal error model.

Where the Gaussian approximation breaks down is in failing to represent the longer tails of the true error distribution. The true distribution is skewed not unlike the diamond in Fig. 3, whereas normal distributions are symmetric. The skew is not significant when points are close, but becomes more pronounced the more distant the points. A possible consequence is

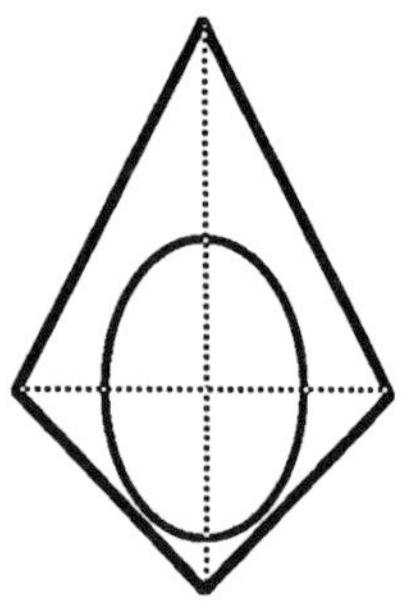

Fig. 3. Quantization error with normal approximation.

biased estimation of point locations, which may lead to biased motion estimates. We will return to these issues in Section VI.

III. Solving for Robot Motion

The previous section showed how to model measurement error in stereo triangulation. In this section we show how to incorporate the error model into an algorithm for estimating the motion between successive stereo pairs. We will begin by showing how motion is computed with scalar weights, then derive an algorithm based on the 3D Gaussian error model, and finally give this algorithm a geometric interpretation.

Referring back to Fig. 1, at this stage in the cycle the robot has two sets of 3D points that have been obtained by stereo matching: a local model of points P_i defined relative to its previous position and the coordinates Q_i of these points relative to its current position. The correspondences between P_i and Q_i are known, but the motion between them is not. Thus we have a set of equations

$$Q_i = RP_i + T$$

in which P_i and Q_i are known point vectors, R is the matrix of the unknown rotation, and T is the unknown translation.

Using scalar weights, one finds R and T by expressing the errors of fit by

$$\epsilon_i = Q_i - RP_i - T$$

and minimizing the weighted sum of squares

$$\sum_{i=1}^{n} w_i \epsilon_i^T \epsilon_i \tag{3}$$

where w_i are the weights. Although the rotation makes this optimization problem nonlinear, two methods are known that give the solution essentially in closed form. The method we have used is due to Schonemann [19]. It treats the nine elements of R as unknowns and applies Lagrange multipliers to force R to be orthogonal. The only iterative part of the algorithm involves taking the singular value decomposition of a 3×3 matrix. Readers are referred to [19] for details. The alternate method, described in [16] and [26, p. 426], parameterizes the rotation as a quaternion and obtains the quaternion elements as the eigenvector corresponding to the largest eigenvalue of a 4×4 matrix.

As will be shown in Section VI, the scalar model of uncertainty embodied in (3) leads to poor performance. Using the 3D Gaussian error model the solution takes a similar, but more complicated form. For simplicity we begin with the case of translational motion. In this case the motion equation is

$$Q_i = P_i + T$$

which we may rewrite as

$$Q_i - P_i = M_i = T$$

to emphasize the role of $M_i = Q_i - P_i$ as measurements of T. From Section II, P_i and Q_i are modeled as normally distributed, uncorrelated random vectors with covariances U_i and V_i, respectively. Therefore, M_i will also be normally distributed with covariance $U_i + V_i$. Now if we consider M_i to be a sequence of noisy measurements of T, each corrupted by noise with zero mean and covariance $U_i + V_i$, application of the maximum likelihood method leads to minimizing the following expression over possible values of T [8]:

$$\sum_{i=1}^{n} \epsilon_i^T W_i \epsilon_i \tag{4}$$

where $\epsilon_i = M_i - T$ and $W_i = (U_i + V_i)^{-1}$. The solution to this is

$$T = \left(\sum_{i=1}^{n} W_i \right)^{-1} \sum_{i=1}^{n} W_i M_i$$

and the covariance matrix of the estimation errors is

$$V_T = \left(\sum_{i=1}^{n} W_i \right)^{-1}.$$

The covariance matrix can be analyzed to assess the quality of the motion estimate. It is also used later in modeling the uncertainty of the robot's global position estimate.

An intuitive interpretation of (4) is shown in Fig. 4. The weight matrices W_i function as norms that measure distance differently for each point. Error vectors making equal contributions to the total error of fit lie on ellipsoidal contours. For example, in Fig. 4, residuals ϵ_a and ϵ_b contribute equally to the total error, but ϵ_c contributes more because $\epsilon_a^T W \epsilon_a = \epsilon_b^T W \epsilon_b < \epsilon_c^T W \epsilon_c$. This effectively gives more weight to errors perpendicular to the line of sight than parallel to it, which, given the nature of stereo, is what we would like to do. The "spherical" error model obtained by using the scalar weights of (3) has the obvious mnemonic meaning that residual vectors making equal contributions to the total error lie on spherical contours. This distinction is what gives the ellipsoid model its power.

Generalizing this method to handle rotation is complicated by the fact that the equations become nonlinear. The function to be optimized takes the form

$$\sum_{i=1}^{n} \epsilon_i^T W_i \epsilon_i \tag{5}$$

with $\epsilon_i = Q_i - RP_i - T$ and $W_i = (RU_iR^T + V_i)^{-1}$.

We have not been able to find direct solutions to this problem or even to approximations in which W_i is not a

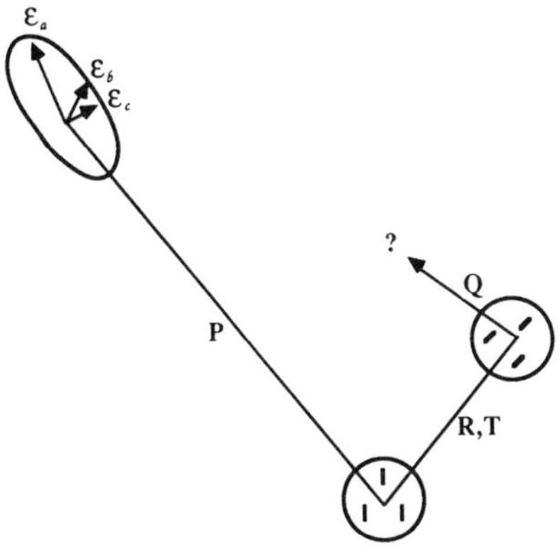

Fig. 4. Interpretation of (4): W_i scales residual vectors, lengthening them parallel to line of sight and shortening them perpendicularly to it.

function of R. Our approach has been to use the direct solution of Schonemann [19] for scalar weights to get an initial estimate of the transformation and to apply the Gauss–Newton method [13, p. 134] to (5) to refine iteratively the estimate. Convergence behavior is good unless all points are very distant; for example, in the experiments with real data described later, the final estimates were obtained after four to eight iterations.

To recap, this section incorporated the error model of Section II in an algorithm for finding the rotation and translation between two 3D points sets. The algorithm replaces the scalar weights of (3) with weight matrices based on the covariances of corresponding points. When the motion is purely translational, the problem is linear and has a direct solution, but when the motion involves rotation we resort to an iterative solution. The error covariance of the motion solution will be used in the following two sections in updating the robot's local model and global position estimate.

IV. Updating the Local Model

So far we have described how to model error in triangulation and how to solve for the motion between two successive stereo pairs. This section deals with how to process a long sequence of stereo pairs. At issue is how to average information from successive images to achieve more accurate landmark localization and consequently more accurate estimates of robot position.

An appropriate tool for this is the Kalman filter [10]. In filtering terminology the quantity to be estimated is called the "state," and when a measurement is taken the filter updates the current estimate of the state. Kalman filters incorporate known statistical properties of the measurements into the update process and produce error covariances for the state estimate. They are widely used in terrestrial and aerospace navigation and guidance applications [10], [26]. In computer vision they have been used in object recognition [3], tracking of known objects with monocular image sequences [5], [12], and for robot navigation and object tracking with sonar data [15].

In our application, the state consists of the locations of the landmark points in the local model. A question arises as to whether the landmarks should be represented in a global stationary frame of reference or in a local moving robot-centered frame. In either case, the update involves transforming coordinates from one frame to the other and applying the filter. If a fixed number of landmarks are being tracked, there is no difference in cost between the two. There will be a difference in the uncertainty of the resulting model; this difference depends on the relative uncertainties of the old model, the new measurements, and the intervening motion. We have not completed an analysis of this situation, but are currently keeping the landmark model in robot-centered coordinates.

The update involves transforming the old local model to the current coordinate frame, inflating its uncertainty to account for the uncertainty of the transformation, and filtering the old model with the new measurements to create the updated model. Let P_{t-1} be the coordinate vector of a single point in the old local model at time $(t - 1)$, and let V_{t-1} be its covariance. For purely translational motion, P_{t-1} is transformed to the current frame by

$$\mathcal{P}_{t-1} = P_{t-1} + T \tag{6}$$

where T is the translation from time $(t - 1)$ to time t. The translation has an error covariance matrix V_T so the transformed point has covariance

$$\mathcal{V}_{t-1} = V_{t-1} + V_T. \tag{7}$$

Equation (6) introduces some correlation between points that is not accounted for in (7), but we assume this is small enough to ignore. To extend this to rotation, we rewrite (6) as

$$\mathcal{P}_{t-1} = RP_{t-1} + T. \tag{8}$$

This is nonlinear, so to compute $\mathcal{V}_{t-1}$ we proceed by analogy to (2); that is, we premultiply the covariance of R, T, and P_{t-1} by the Jacobian of the transformation and postmultiply by the Jacobian transposed. Since we treat P_{t-1} as uncorrelated with R and T, this leads to

$$\mathcal{V}_{t-1} = J_m V_m J_m^T + RV_{t-1}R^T$$

where J_m contains the derivatives of (8) with respect to the motion parameters and V_m is the covariance of the motion parameters.

Now let Q_t be the measurement of the same point at time t, and let U_t be the covariance of this measurement. Some manipulation of the basic Kalman filter equations leads to the following estimates of the updated point location and covariance:

$$V_t = (\mathcal{V}_{t-1}^{-1} + U_t^{-1})^{-1} \tag{9}$$

$$P_t = \mathcal{P}_{t-1} + V_t U_t^{-1}(Q_t - \mathcal{P}_{t-1}). \tag{10}$$

The intuition behind (10) is as follows. The second term takes the difference $(Q_t - \mathcal{P}_{t-1})$ of the new measurement from the old estimate, weights the difference by $V_t U_t^{-1}$, and applies the result as an update to the old estimate $\mathcal{P}_{t-1}$. Matrix U_t^{-1} will be "larger" the more precise the new measurement, giving it more weight in the update, and smaller the less precise the measurement, giving it less weight. Conversely, V_t will be small if the old estimate is precise and large otherwise. Hence if the old estimate is already good, the new measurement receives little weight; if it is poor, the new measurement receives more weight.

The procedure we have described assumes that the error in the motion estimate is uncorrelated with the error in the landmark points. When the motion estimate is obtained by using the methods of the previous section this will not be true, although if other sensors are also contributing to the motion estimate, it will be approximately true. This is an issue we are investigating.

V. Updating the Global Robot Position

By using the modules discussed in the previous sections, the robot computes estimates of its motion between successive stereo pairs. Combining these to estimate its global position is a simple matter of concatenating the transformation matrices. It may also be desirable to estimate the uncertainty of the global position, which can be done by propagating the covariance matrices of the incremental motions into a covariance of the global position. For translation this is also very simple. If the global position at time $(t - 1)$ is $T_{g_{t-1}}$ and the next incremental translation is T_t, then the next global position is

$$T_{g_t} = T_{g_{t-1}} + T_t. \tag{11}$$

Since this is linear, if the incremental translation estimates have uncorrelated zero-mean Gaussian errors, then T_{g_t} will also have zero-mean, Gaussian error with covariance given by

$$V_{g_t} = V_{g_{t-1}} + U_t$$

where $V_{g_{t-1}}$ and U_t are the covariances of $T_{g_{t-1}}$ and T_t, respectively. The case of motion in the plane, where there are two parameters for translation and one for rotation, has been dealt with by Smith and Cheeseman [21]. In summary, one obtains an equation analogous to (11) in which the three parameters of the global position are expressed as functions of the previous position and the incremental motion. These are nonlinear and error propagation is done by linearization. For general motion in three dimensions, this is not straightforward with the Euler angle representation of rotation we have used to date. In this case other parameterizations of rotation, such as quaternions, may be preferable [9], [26]. We are exploring this further.

VI. Performance

Our evaluation to date has concentrated on comparing the use of the spherical and ellipsoidal error models in the motion solving methods of Section III. Results of tests with simulated and real data are described below.

A. Simulations

Three sets of simulation data will be presented. The first is a base case that compares the standard deviations of position estimates obtained with each error model for a single step of vehicle motion. That is, it considers motion between only two consecutive stereo pairs. It illustrates the difference in the variability of position estimates with each model and reveals the effects on the motion estimates of coupling between the translational and rotational degrees of freedom. The second set also considers only two consecutive stereo pairs and tests limiting performance by tracking progressively more distant points. The last set examines both long-range performance over many images and the effect on performance of different stereo baselines.

The simulations were generated as follows. The "scene" consisted of random points uniformly distributed in a 3D volume in front of the simulated cameras. For the first set of simulations, this volume extended 5 m to either side of the cameras, 5 m above and below the cameras, and from 2 to 10 m in front of the cameras. The cameras themselves were simulated as having 512 × 512 pixels and a field of view of 53°. The stereo baseline was 0.5 m. Image coordinates were obtained by projecting the points onto the images, adding Gaussian noise to the floating point image coordinates, and rounding to the nearest pixel. These coordinates were input to the triangulation and motion solving algorithms. For the ellipsoidal error model, covariance matrices were computed as described in Section II. In the scalar case, weights were derived by taking the Z variance from the covariance matrix. Scalars obtained by several other methods were tried and found to give very similar results. These include the volume and length of the major axis of the standard error ellipsoid and Moravec's half-pixel shift rule [18].

The first set of simulations determined the standard deviation of the estimated motion between two consecutive stereo pairs when the true motion was 1 m. The results are given in Figs. 5 and 6, plotted against the number of points used to compute the motion estimate. For any given number of points tracked, the standard deviations are taken over 5000 random trials with entirely new points generated for each trial. In both figures, the top three curves were obtained with spherical modeling and the bottom three with ellipsoidal. Tilt implies rotation of the camera up or down, pan is the rotation about the vertical axis, and roll the rotation about the camera axis. The most significant thing to note is that the standard deviations obtained with the ellipsoidal model are a factor of 5–10 less than those of the spherical model. The size of the difference will vary with the distance to the points; for example, when they are within 1–2 m of the cameras the factor is 2–4, and when they are within 2–5 m it is 3–6. The case shown in the figures (points from 2–10 m away) approximates the conditions of the indoor run with real data described later. Another point to note is that with the spherical model the estimates of roll and forward translation show less variation than the remaining parameters. This is because lateral translations and panning rotations have coupled effects on the errors of fit, as do vertical translations and tilting rotations. This shows up in the covariance matrix of the computed motion parameters as larger correlations between these pairs of parameters than other pairs. These correlations are present with both error models, but the effects on the variance of the individual parameters are greater in the spherical case. Lastly, note that for a given level of performance fewer points are needed with the ellipsoidal model than the spherical, offsetting the greater expense of the iterative motion solution needed in the

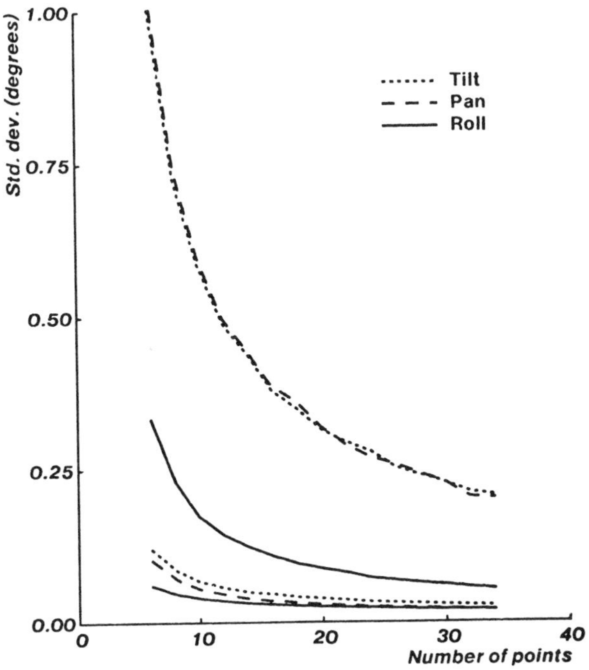

Fig. 5. Standard deviation versus number of points for rotations. Top three curves are for spherical model, bottom three are for ellipsoidal model. Use of ellipsoidal model gave significantly lower variance in estimates.

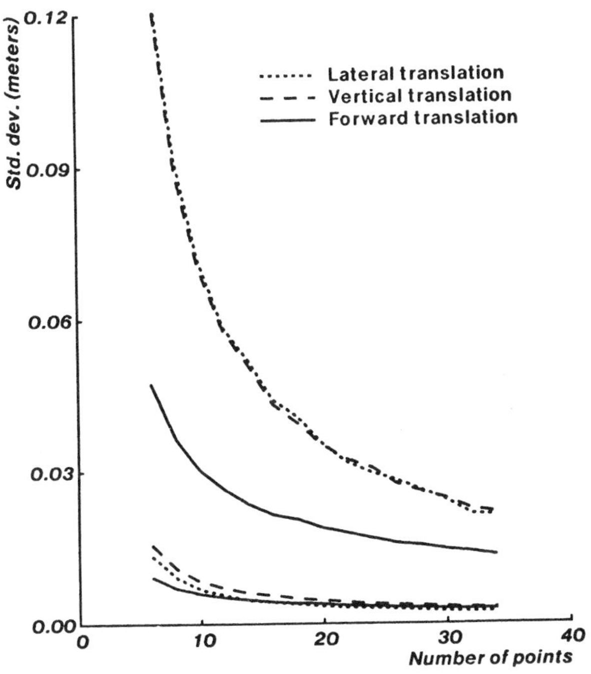

Fig. 6. Standard deviation versus number of points for translations. Top three curves are for spherical model, bottom three, are for ellipsoidal model.

ellipsoidal case. The exact relationship will depend on the camera configuration.

The second set of simulations illustrates the dependence of the standard deviation on the distance to the points in the scene. The initial volume for generating points was 2–4 m away; this was expanded by moving the far limit back in stages until the final volume was 2–40 m. As with the previous experiment, for each volume 5000 random trials were performed with different points generated for each trial. Fig. 7

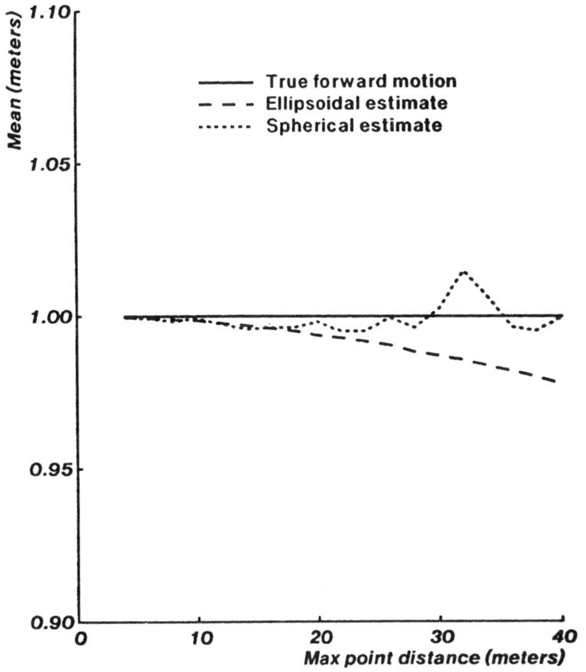

Fig. 7. Mean of estimated forward distance traveled versus maximum distance to points.

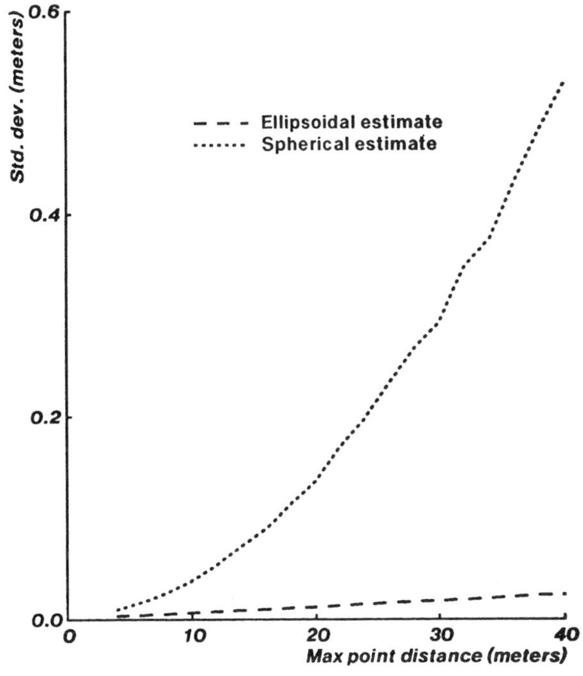

Fig. 8. Standard deviation of estimated forward distance traveled versus maximum distance to points.

shows the mean of the forward translation estimates as a function of the maximum distance to the points, and Fig. 8 shows the standard deviation. The true forward motion was one meter. The standard deviation tells most of the story. With the ellipsoidal model, the standard deviation remains modest throughout the range of the experiment, reaching a maximum of about three percent of the actual motion. On the other hand, with the spherical model the standard deviation is initially modest but grows rapidly to the point that the estimates are unusable. The other motion parameters, though not shown, behave similarly. Looking at the means, with the ellipsoidal model there is negligible bias when points are nearby, with a growing tendency to underestimate the distance traveled as the points themselves become more distant. For the spherical model there also appears to be some underestimation when points are nearby, but the rapid growth of the standard deviation makes further interpretation of little value. Thus this experiment illustrates the strong contrast between the algorithms that develops with increasing distance to points.

The last simulation looked at motion over a long sequence of images, both to confirm the above results and to test a hypothesis suggested by the previous simulation: that for equivalent performance, the ellipsoidal model may permit the use of a shorter stereo baseline than the spherical. This is an important consideration, because length of the baseline directly affects the difficulty of stereo matching. Each trial in this experiment involved tracking points 2–10 m from the cameras, with new points added when existing ones passed out of view. Fig. 9 shows the standard deviation of the estimated distance as a function of the true distance. The travel between images was 0.64 m, so the figure represents about 90 images. It shows curves for a 0.5-m baseline with the spherical model and 0.125-, 0.25-, and 0.5-m baselines for the ellipsoidal model. Comparing the curves for 0.5-m baselines, the ellipsoidal model does outperform the spherical. It appears that the curves may eventually run parallel, so that the difference between the methods would be an additive constant rather than multiplicative. Looking at the effects of different baselines, results with the ellipsoidal model are still better than the spherical model with a 0.25-m baseline, though not with 0.125-m. Based on standard deviations of position, it does appear possible to use a shorter baseline. However, another factor involved is bias of the motion estimates. In general, we have found that the narrower the baseline, the more motion is underestimated. The same occurs when we increase the variance of the noise in the image coordinates. This requires further investigation. For the moment we just note that bias can be a problem with short baselines or nontrivial noise levels.

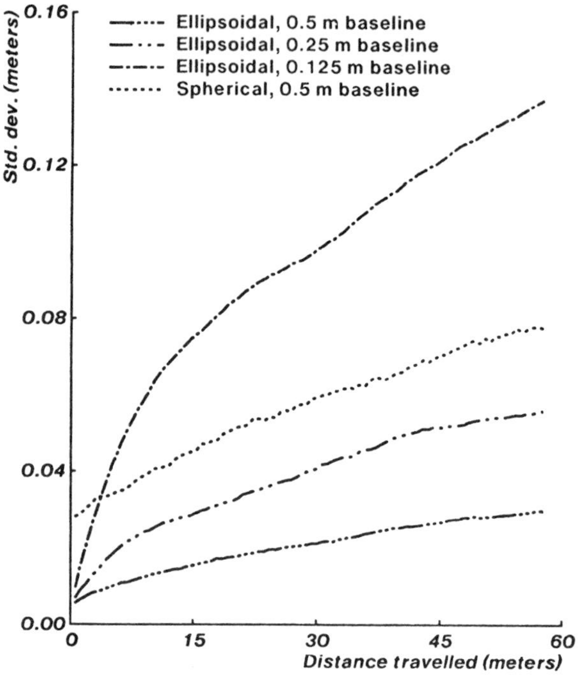

Fig. 9. Standard deviation of estimated forward distance traveled versu true distance.

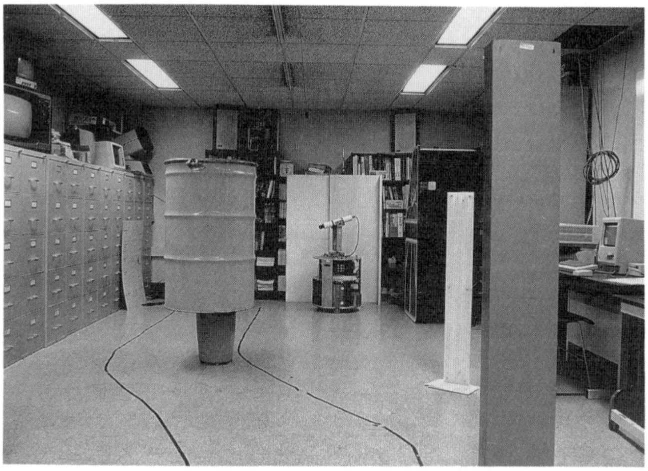

Fig. 10. One image from lab sequence.

B. Real Images

To verify the simulations on real images, we used both error models to estimate the position of a stereo-equipped robot traveling across the floor of our lab. The scene is pictured in Fig. 10. The robot was driven straight forward in 54 steps of slightly less than 10 cm each. The cameras were on a 20-cm baseline and had a 36° field of view. The FIDO feature-tracking system [23] was used to track points through the image sequence, and the resulting set of matched image coordinates were input to the algorithms described earlier to estimate the robot's position at each step. We will briefly describe the operation of FIDO before discussing the results of the experiment.

FIDO uses the Moravec interest operator and coarse-to-fine correlation algorithm to pick and match point features in stereo pairs. The interest operator is applied to one image of a stereo pair to pick points where intensity varies in all directions; typically these are sharp corners or intersections of lines. The correlator finds these points in the other image of the stereo pair. To find the same points in subsequent stereo pairs, an *a priori* motion estimate is used to predict the location of the point in the new images, a constraint window is defined around the predicted location based on the uncertainty of the motion estimate, and the correlator is applied to find the position of best match within the constraint window. Incorrect matches are culled with a threshold on the correlation coefficient and with a 3D error heuristic called the "3D prune" stage. This heuristic uses the fact that under rigid motion the distance between two 3D points does not change over time. Points which appear to violate this condition are discarded. The advantage of this test is that it does not require knowledge of the motion between stereo pairs. Points that survive this test become input to the motion solving al-

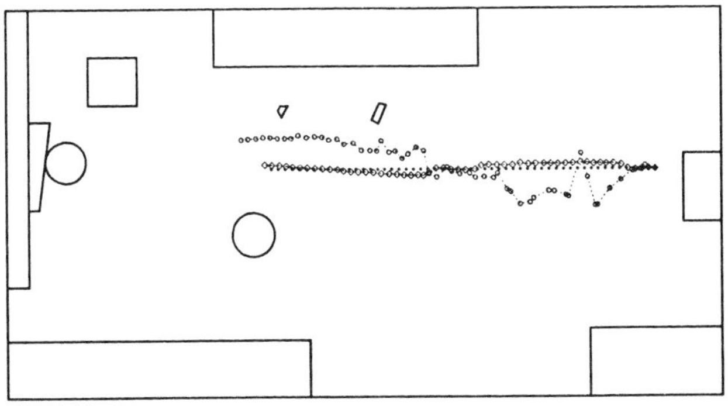

Fig. 11. Position estimates obtained with 3 DOF algorithm and clean data. Dots show actual vehicle positions, diamonds show positions estimated with ellipsoidal model, and circles show positions estimated with spherical model.

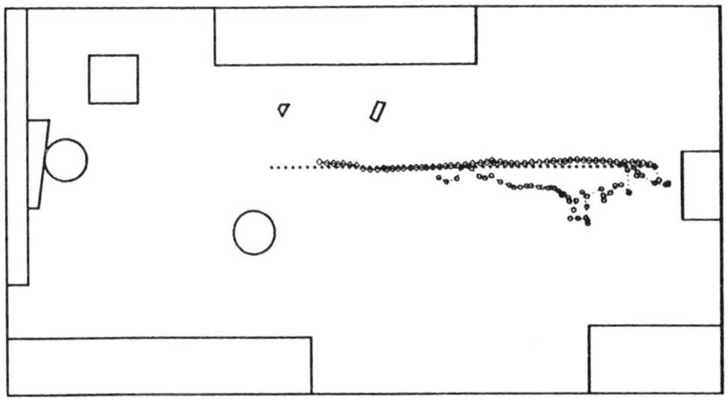

Fig. 12. Results with noisy data. As in Fig. 11, dots show actual vehicle positions, diamonds show ellipsoidal estimates, and circles show spherical estimates.

gorithms. In the experiments to follow, between 30 and 40 points usually remained.

Fig. 11 compares the true motion to the position estimates obtained with the spherical and ellipsoidal error models. For this figure a "planar" motion solver was used that solved only for the parameters of motion in the plane, that is two degrees of translation and one of rotation. The line of heavy dots shows the true position at every step, the path marked with circles shows the positions estimated with the spherical model, and the path marked with diamonds shows the same for the ellipsoidal model. The final position estimated with the ellipsoidal model was correct to within two percent of the distance and 1° of orientation. With the spherical model the corresponding figures were eight percent and 7°.

To gauge the effect of noisier image matches, we adjusted the threshold of the prune stage so that progressively fewer points were discarded. The general effect was to increasingly underestimate the distance traveled. Fig. 12 shows what happened when the prune stage was entirely disabled, leaving only the correlation threshold to detect matching errors. Estimates with the spherical model were initially very bad. We attribute this to matching errors caused by large depth discontinuities around the foreground objects. When these objects fell out of view, the estimates were better behaved. The behavior with the ellipsoidal model was much less erratic.

Finally, we repeated the first experiment (i.e., clean data) with the algorithm that computes all six degrees of freedom (DOF) of motion. The results were in accord with the planar case, with roughly the same levels of error in the final position estimate. It was notable that with the spherical model the error in roll was less than a degree, while in the other rotations it was between 5° and 12°. This is consistent with the observation made from the first simulation about coupled rotation and translation.

VII. Conclusion

Comparing motion estimates obtained with the spherical (scalar) and ellipsoidal (3D Gaussian) error models, there is no question that the ellipsoidal model is preferred. Simulations showed that position estimates with the ellipsoidal model had less variance and live trials confirmed that they were more accurate and less influenced by matching errors. The contrast between algorithms is strongly influenced by the distance to the points being tracked; with nearby points, the difference will be moderate, but it grows very rapidly with increasing distance.

The possibility of bias arose with very large distances to objects and high noise levels. We attribute this to the non-Gaussian nature of the true error distribution in these situations. Under these conditions, better error modeling is an area for further research. The question of whether the ellipsoidal method permits a shorter baseline has only been tested in simulation; based on the variance of the estimates it appears feasible, but the bias issue is unresolved.

Perhaps the most valuable result is demonstrating that accurate position estimates can be achieved in a fully automatic system when an adequate error model is used. The true motion in the examples we showed was pure translation, but we believe that the results will hold for general motion and preliminary simulations bear this out. With matching to subpixel resolution, matching of extended features instead of points, and more sophisticated error detection, it may be possible to obtain much better performance than that quoted here. Another interpretation of our results is that they show the importance of error modeling in stereo and probably other aspects of vision. One area we plan to explore this is in shape from stereo, beginning with the local update paradigm of Section V.

Acknowledgment

We are indebted to Hans Moravec for making us aware of Kalman filters, Peter Highnam for introducing us to Schonemann's algorithm, and Takeo Kanade for pointing out the possibility of using a shorter baseline.

References

[1] G. Adiv, "Determining three-dimensional motion and structure from optical flow generated by several moving objects," *IEEE Trans. Pattern Anal. Machine Intell.*, vol. PAMI-7, pp. 384–401, July 1985.
[2] P. Anandan and R. Weiss, "Introducing a smoothness constraint in a matching approach for the computation of displacement fields," in *Proc. ARPA IUS Workshop*, SAIC, Dec. 1985, pp. 186–197.
[3] N. Ayache and O. D. Faugeras, "HYPER: A new approach for the recognition and positioning of two-dimensional objects," *IEEE Trans. Pattern Anal. Machine Intell.*, vol. PAMI-8, pp. 44–54, Jan. 1986.

[4] H. S. Baird, *Model-Based Image Matching Using Location*. Cambridge, MA: MIT Press, 1985.

[5] T. J. Broida and R. Chellappa, "Estimation of motion parameters from noisy images," *IEEE Trans. Pattern Anal. Machine Intell.*, vol. PAMI-6, pp. 90–99, Jan. 1986.

[6] R. A. Brooks, "Symbolic reasoning among 3-D models and 2-D images," *Artificial Intell.*, vol. 17, pp. 285–348, 1981.

[7] L. Dreschler and H.-H. Nagel, "Volumetric model and 3D trajectory of a moving car derived from monocular TV frame sequences of a street scene," *Comput. Graph. Image Processing*, vol. 20, pp. 199–228, 1982.

[8] T. F. Elbert, *Estimation and Control of Systems*. New York: Van Nostrand Reinhold, 1984.

[9] O. D. Faugeras, N. Ayache, B. Faverjon, and F. Lustman, "Building visual maps by combining noisy stereo measurements," in *Proc. IEEE Int. Conf. Robotics and Automation*, Apr. 1986, pp. 1433–1438.

[10] A. Gelb, Ed., *Applied Optimal Estimation*. Cambridge, MA: MIT Press, 1974.

[11] D. B. Gennery, "Modelling the environment of an exploring vehicle by means of stereo vision," Ph.D. dissertation, Stanford Univ., Stanford, CA, June 1980.

[12] ——, "Tracking known three-dimensional objects," in *Proc. AAAI*, 1982, pp. 13–17.

[13] P. E. Gill, W. Murray, and M. H. Wright, *Practical Optimization*. New York: Academic, 1981.

[14] W. E. L. Grimson and T. Lozano-Perez, "Model-based recognition and localization from sparse range or tactile data," *Int. J. Robotics Res.*, vol. 3, pp. 3–35, Fall 1984.

[15] J. Hallam, "Resolving observer motion by object tracking," in *Proc. Int. Joint Conf. Artificial Intelligence*, 1983.

[16] M. Hebert, "Reconnaissance de formes tridimensionelles," Ph.D. dissertation, L'Universite de Paris-Sud, Centre d'Orsay, Sept. 1983.

[17] L. H. Matthies and C. E. Thorpe, "Experience with visual robot navigation," in *Proc. IEEE Oceans'84 Conf.*, Washington, DC, Aug. 1984.

[18] H. P. Moravec, "Obstacle avoidance and navigation in the real world by a seeing robot rover," Ph.D. dissertation, Stanford Univ., Stanford, CA, Sept. 1980.

[19] P. H. Schonemann and R. M. Carroll, "Fitting one matrix to another under choice of a central dilation and a rigid motion," *Psychometrika*, vol. 35, pp. 245–255, June 1970.

[20] C. C. Slama, Ed., *Manual of Photogrammetry*. Falls Church, VA: American Society of Photogrammetry, 1980.

[21] R. C. Smith and P. Cheeseman, "On the representation and estimation of spatial uncertainty," Tech. Rep. (draft), SRI International, 1985.

[22] F. Solina, "Errors in stereo due to quantization," Tech. Rep. MS-CIS-85-34, Univ. Pennsylvania, Sept. 1985.

[23] C. E. Thorpe, "Vision and navigation for a robot rover," Ph.D. dissertation, Carnegie-Mellon Univ., Dec. 1984.

[24] R. Y. Tsai and T. S. Huang, "Uniqueness and estimation of three-dimensional motion parameters of rigid objects with curved surfaces," *IEEE Trans. Pattern Anal. Machine Intell.*, vol. PAMI-6, pp. 13–26, Jan. 1984.

[25] A. M. Waxman and J. J. Duncan, "Binocular image flows," in *Proc. Workshop on Motion: Representation and Analysis*, May 1986, pp. 31–38.

[26] J. R. Wertz, Ed., *Spacecraft Attitude Determination and Control*. D. Reidel Publishing, 1978.

3. Navigation: Position and Course Estimation

The least initial deviation from the truth is multiplied later a thousandfold.

Aristotle, *On the Heavens*, Bk. 1, Ch. 5 (384–322 BC)

Navigation is a fundamental requirement of autonomous mobile robots. It is defined as ''the science of getting ships, aircraft, or spacecraft from place to place; *esp*: the method of determining position, course, and distance traveled'' [16] . It is essential to know the whereabouts of the vehicle, although the accuracy of the estimate may vary according to the requirements of the vehicle's path [18]. Position estimation has been extensively studied in aircraft, missile, space, and marine areas. There are several excellent journals and conferences specializing in this area, including *Navigation* (the Journal of the Institute of Navigation), *The Journal of Navigation*, (the Journal of the Royal Institute of Navigation), *IEEE Transactions on Aerospace and Electronic Systems*, the annual IEEE PLANS symposiums (Position, Location and Navigation Symposium), and the annual NAV conferences of the Royal Institute of Navigation. In addition, there are at least two institutes – the Institute of Navigation in Washington DC, USA. and The Royal Institute of Navigation in London, England – as well as organizations dedicated to such specific navigational aids as the International Omega Association (Omega), and the Wild Goose Association (Loran).

It is surprising to find that position estimation for mobile robots is still problematic, given the available knowledge and experience in the land, marine and aerospace communities. The reason for this is more than a lack of knowledge regarding the navigation literature from other disciplines, although this may be a contributing factor. A more important reason is that autonomous vehicle research has, until quite recently, avoided any use of artificial beacons, that is, either active or passive beacons that are placed in the vehicle's environment for navigational purposes. This is because the primary interest in autonomous vehicles has been from research into artificial intelligence with the goal of developing ''intelligent'' mobile machines the behavior of which closely resembles that of animals. As a consequence, vehicle navigation systems have had to identify naturally occurring landmarks in the environment. The recognition of naturally occurring reference points within a robot's environment is not always easy, due to noise and/or difficulties in interpreting the sensory information. This has certainly been true of machine vision and ultrasonic sensors. The choice of these sensors has been based primarily on technological and cost considerations and, to a lesser degree, on a desire to mimic the human visual system. The use of radar, so prevalent in aerospace applications, is not practical in structured office environments because, with a wavelength of the order of centimeters, radar has worse specular reflection characteristics and is significantly more expensive than ultrasonics.[1] Upgrading the sensor may not be possible because of, say, technological or economic constraints.

Placing easy to recognize beacons in the robot's workspace is an effective way of avoiding many of the problems associated with recognizing naturally occurring landmarks. Artificial beacons may be either active or passive. In the active case, the beacons may transmit periodic waveforms that are received by the (passive) receiver aboard the vehicle. Active beacons are a very common method of navigation for both aircraft and shipping [22], and many marine and aerospace navigation systems have been developed, including LORAN and GPS [19], and have been used for vehicle localization [4]. Passive beacons, for example, bar codes or retroreflectors [26], are more commonly used for mobile vehicle applications [7, 21, 27], in which case the vehicle is usually equipped with an active transmitter such as a laser. Passive beacons have the advantages of robustness and reliability – since they require no power and are completely inert – and economy – since a single active transmitter on the vehicle can be serviced by many passive beacons. Beacons can be uniquely identified using, for example, bar codes. In other cases, all beacons may look identical and determination of the correspondence is required. This problem is examined in the paper by Sugihara [25]. Autonomous mobile vehicles that navigate with the aid of beacons are now available commercially. However, insufficient experience with these technologies prevents any authoritative comparison of current approaches at this time.

Navigation can be broadly separated into two distinct approaches: reference and dead reckoning. Reference guidance refers to navigation with respect to a coordinate frame based on visible external landmarks. Dead reckoning refers to navigation based on odometry, inertial guidance, or some other "self-contained" sensing. Dead reckoning usually provides the vehicle with an estimate of its position.[2] Its disadvantage is that the position error grows without bound unless an independent reference is used periodically to reduce the error. Reference guidance has the advantage that position errors are bounded, but detection of external references or landmarks and real-time position fixing may not always be possible. Clearly, dead reckoning and reference navigation are complementary and combinations of the two approaches can provide very accurate positioning systems. For example, military aircraft have long combined inertial navigation systems (INS) with Tacan, a radar-based position update, to bound INS error growth [13]. More recently, GPS and INS have been combined in systems that also estimate such INS sensor errors as misalignment and drift rates, to lead to the slowest possible growth of INS navigational error in between GPS updates [23].

Before the basic concepts of navigation are introduced, one further issue should be addressed. Navigation based on directions of the form "follow the road for about five miles until you come to a traffic light, then take a left," may appear, at first sight, to be decidedly different from either dead reckoning or reference-based navigation. It is not. Instructions of the form "follow the road" or "turn left at the traffic light"

[1] Interestingly, marine navigation techniques developed by the early 1900s were found to be inappropriate for aircraft navigation. For example, marine sextants measure altitude of celestial bodies above the horizon, which can generally be seen only by low-flying aircraft. It remains to be seen whether autonomous land vehicles will develop alternative navigational techniques or will borrow from existing technologies. A historical perspective on the changing face of navigation may be found in Williams' article [28], which describes the history of air navigation and, in particular, the differences between air and marine navigation. For the historically minded reader, Kayton [9] also provides an interesting description of the history of land, sea, and air navigation and many excellent references.

[2] It is also impervious to jamming, and not detectable, both of which can be important military considerations.

require real-time reference navigation, for example, find the road perimeter and steer a course along it. The instruction, "for five miles" could be either dead reckoning or reference because it is formulated as a change in position rather than as an absolute position.

We begin this chapter with a discussion of two-dimensional position estimation based on the *simultaneous* measurement of the range or bearing to three known landmarks or more. The subject of triangulation and trilateration to determine position has been extensively studied. The first paper by Torrieri presents a thorough derivation of the principal algorithms for several common forms of triangulation/trilateration-based location systems, as well as an analysis of their performance. The paper also derives the relationship between position accuracy, measurement accuracy and geometric dilution of precision (GDOP). In the simplest case, the vehicle identifies at least three known reference points and then, when the distance (trilateration) or angle (triangulation) to each point is measured, simple trigonometry allows calculation of the vehicle's position. The accuracy of such systems is dependent on (1) the measurement error and (2) the geometric relationships between the vehicle and landmarks. See, for example, the early work of Burt et al. [2]. The latter geometric effect (*GDOP*) [8, 10, 11, 17] is defined as the ratio of the root-mean-square (rms) position error to the rms measurement error. For example, if the ranges to two beacons that are at measured 90° angles from the navigator are measured with independent errors of ± 10 m, the accuracy of the resulting position fix is the rms of errors, or ± 14 m. This is the best possible observation geometry, and leads to the smallest GDOP. As the angle becomes larger or smaller, position accuracy continually degrades. With poor geometry, position errors can easily be 10 times greater than range (measurement) errors [22].

Torrieri's paper addresses the problem of determining a vehicle's position from simultaneous measurements of sufficient observables to solve the navigation equations uniquely. It is not concerned with position estimation based upon a series of observations. Individual observations of a landmark can be incorporated to improve the position estimate, even if insufficient landmarks are visible to solve all the navigation equations. The combined position estimate is less noisy than any one of the sets of individual measurements *provided the error models are accurate*. As an example, consider a system that dead reckons between observation of the beacons. At each position fix, the discrepancy, called the *residual* or *innovation*, between the observation and the position as predicted by dead reckoning is computed. Then, based on estimates of the uncertainty in the dead reckoning (a function of such things as surface roughness or, wheel slippage) and estimates of the uncertainty in the position fix (a function of measurement noise and GDOP), the dead reckoning estimate is corrected by a weighted sum of the measurement residuals. Sensor integration can be routinely mechanized by use of the Kalman filter. Maybeck's paper is an overview of Kalman filter analyses; it appears in Maybeck's book *Stochastic Models, Estimation and Control* [14], which also provides many interesting case studies, including that of the Apollo spacecraft program.[3] The reader is also directed to Maybeck [13], Geier et al. [5], and McMillan [15], all of whom describe the performance analysis of Kalman filters designed to combine inertial navigation data (INS) with external reference data optimally. Gelb [6] also provides an excellent introduction to Kalman filtering, with particularly appropriate navigation examples provided in Chapter 4. Further references on Kalman filtering can be found in Bar-Shalom and Fortmann [1] and a special issue of the *IEEE Transactions on Automatic Control* [24].

Smith, Self, and Cheeseman's paper describes how to manipulate navigation frames optimally. It addresses such questions as whether a mobile robot is expected to

[3]A fascinating perspective on missile guidance technology can be found in the SPIE conference on image processing for missile guidance [20].

have a particular reference object within its field of view. The Kalman filter is used again, and, once more, it is shown how the uncertainty of a navigation frame can be reduced by additional sensing.

We conclude this chapter with two papers on autonomous navigation. Both of the papers describes vehicle navigation systems that use naturally occurring landmarks in the environment. The first, by Cox, describes a mobile robot navigation system in which an a priori map of the environment is provided. The vehicle senses its environment using an infrared range finder. The range data points are matched to the map using a novel matching algorithm that solves the correspondence problem (for small disparity) by associating data points with their closest (Euclidean) line segment and then applying a least-mean-square optimization procedure to determine the vehicle's position. The vehicle dead reckons between matches and the independent estimates are optimally (maximum likelihood) combined. The Kalman filter is not explicitly used, but equivalent results are obtained.

Finally, Ayache and Faugeras describe a scheme whereby a dead reckoning solution is updated by comparing TV images taken at each navigation interval. A range image is obtained via trinocular stereo, the vehicle moves a distance estimated to be D, and another range image is computed. Corresponding points in the two images are determined, taking into account the noise and measurement uncertainty, and a Kalman filter uses the corresponding change in range to refine the initial estimates of the vehicle's motion D. The process is iterated until a satisfactory convergence is obtained. The system not only refines the vehicle's position estimate but also constructs a model or map of its environment as it navigates.

The reader is also directed to the papers by Crowley [3] and Leonard and Durrant-Whyte [12], both of which describe navigation systems based on ultrasonic sensing and Kalman filtering. The latter paper is especially interesting because it attempts to apply an integrated framework to the problems of navigation, dynamic map construction, and their corresponding sensing strategies, based on a rigorous Kalman filter-based tracking methodology.

References

1. Bar-Shalom, Y., and Fortmann, T.E. *Tracking and Data Association.* Academic Press, Boston, Mass., 1988.

2. Burt, W.A., Kaplan, D.J., Keenly, R.R., Reeves, J.F., and Shaffer, F.B. Mathematical Considerations Pertaining to the Accuracy of Position Location and Navigation Systems - Part 1. AD 629 609, National Technical Information Service, 1966.

3. Crowley, J.L. World Modeling and Position Estimation for a Mobile Robot Using Ultrasonic Ranging. In *Proceedings of the IEEE International Conference on Robotics and Automation*, Vol. 2, IEEE, New York, 1989, pp. 674-680.

4. Durieu, C., Clergeot, H., and Monteil, F. Localization of a Mobile Robot with Beacons Taking Erroneous Data into Account. In *Proceedings of the IEEE International Conference on Robotics and Automation*, Vol. 2, IEEE, New York, 1989, pp. 1062-1068.

5. Geier, G.J., Cabak, A., and Sieh, L. Design of an Integrated Navigation System for Robotic Vehicle Applications. *Navigation 34*, 4 (1987-1988), 325-336.

6. Gelb, A. *Applied Optimal Estimation.* MIT Press, Cambridge, Mass., 1974.

7. Hyyppa, K. Lulea Turbo Turtle (LTT). In *Proceedings of the IEEE/RSJ International Workshop on Intelligent Robots and Systems*, IEEE/RSJ, New York, 1989, pp. 620-623.

8. Jorgensen, P.S. Navstar/Global Positioning System 18-Satellite Constellations. In *Collected Papers in GPS, Vol. 2*. Institute of Navigation, 1984.

9. Kayton, M. Navigation: Ships to Space. *IEEE Trans. Aerospace Electronic Systems 24*, 5 (1988), 474-519.

10. Lee, H.B. A Novel Procedure for Assessing the Accuracy of Hyperbolic Multilateration Systems. *IEEE Trans. Aerospace Electronic Systems AES-11*, 1 (1975), 2-15.

11. Lee, H.B. Accuracy Limitations of Hyperbolic Multilateration Systems. *IEEE Trans. Aerospace Electronic Systems AES-11*, 1 (1975), 16-29.

12. Leonard, J.J., and Durrant-Whyte, H.F. A Unified Approach to Mobile-Robot Navigation. (unpublished).

13. Maybeck, P.S. Performance Analysis of a Particularly Simple Kalman Filter. *J. Guidance Control 1* (1978), 391-396.

14. Maybeck, P.S. *Stochastic Models, Estimation, and Control*, vol. 1. Academic Press, Boston, 1979.

15. McMillan, J.C. An Integrated System for Land Navigation. *J. Instit. Navigation 34*, 1 (1987), 43-63.

16. Merriam-Webster. *Webster's Ninth New Collegiate Dictionary*, Merriam-Webster, Springfield, Mass., 1985.

17. Milliken, R.J., and Zoller, C.J. Principle of Operation of NAVSTAR and System Characteristics. In *Collected Papers in GPS, Vol. 1*. Institute of Navigation, 1980, pp. 3-14.

18. Payton, D. Circulation Maps: A Resource for Identifying Position Accuracy Requirements. In *Proceedings of the IEEE/RSJ International Workshop on Intelligent Robots and Systems*, IEEE/RSJ, New York, 1989, pp. 448-455.

19. *Proc. IEEE. Special Issue on Global Navigation Systems, 71*, 10 (1983).

20. *Proc. SPIE. Image Processing for Missile Guidance, 238* (1980).

21. Robins, M.P. Free-Ranging Automatic Guided-Vehicle System. *GEC Rev. 2*, 2 (1986).

22. Sampson, S.R. A Survey of Commercially Available Positioning Systems. *Navigation 32*, 2 (1985), 139-148.

23. Sinha, P., Barckley, K., and deDoes, D. Integrated Navigation System Design and Performance with Phase III GPS User Equipment. In *Proceedings of the Satellite Division (ION) International Tech. Meeting* Colorado Springs, Sept. 19-23) 1988, pp. 283-300.

24. H.W. Sorenson, (Ed.) *Special Issue on Applications of Kalman Filtering, Vol. AC-28*, (3.), IEEE Transactions on Automatic Control, 1983.

25. Sugihara, K. Some Location Problems for Robot Navigation Using a Single Camera. *Comput. Vision Graph. Image Process. 42*, 1 (1988), 112-129.

26. Tsumura, T. Survey of Automated Guided Vehicle in Japanese Factory. In *IEEE International Conference on Robotics and Automation*, IEEE, New York, 1986, pp. 1329-1334.

27. Van Brussel, H., et al. Frog - Free Ranging on Grids: New Perspectives on Automated Transport. In *Proceedings of Automated Guided Vehicle Systems*, IFS, 1988, pp. 223-232.

28. Williams, J.E.D. Air Navigation Systems. *J. Navigation 41*, 3 (1988), 375-406.

Statistical Theory of Passive Location Systems

DON J. TORRIERI
U.S. Army Countermeasures/Counter-Countermeasures Center

A derivation of the principal algorithms and an analysis of the performance of the two most important passive location systems for stationary transmitters, hyperbolic location systems and direction-finding location systems, are presented. The concentration ellipse, the circular error probability, and the geometric dilution of precision are defined and related to the location-system and received-signal characteristics. Doppler and other passive location systems are briefly discussed.

Manuscript received June 21, 1983.

Author's address: U.S. Department of the Army, Countermeasures/Counter-Countermeasures Center, 2800 Powder Mill Rd., Adelphi, MD 20783.

0018-9251/84/0300-0183 $1.00 © 1984 IEEE

I. INTRODUCTION

The position of a stationary transmitter or radiating emitter can be estimated from passive measurements of the arrival times, directions of arrival, or Doppler shifts of electromagnetic waves received at various sites. This paper presents a derivation of the principal algorithms and an analysis of the two most important passive location systems for stationary transmitters: hyperbolic location systems and direction-finding location systems [1, 2].

Hyperbolic location systems, often called time difference of arrival (TDOA) systems, locate a transmitter by processing signal arrival-time measurements at three or more stations. The measurements at the various stations are sent to a station that is designated the master station and does the processing. The arrival-time measurements at two stations are combined to produce a relative arrival time that, in the absence of noise and interference, restricts the possible transmitter location to a hyperboloid with the two stations as foci. Transmitter location is estimated from the intersections of three or more independently generated hyperboloids determined from at least four stations. If the transmitter and the stations lie in the same plane, location is estimated from the intersections of two or more hyperbolas determined from three or more stations. Fig. 1 illustrates two hyperbolas, each of which has two branches, derived from measurements at three stations. The two hyperbolas have two points of intersection. The resulting location ambiguity may be resolved by using a priori information about the location, bearing measurements at one or more of the stations, or a fourth station to generate an additional hyperbola.

Fig. 2 depicts an aircraft with a direction-finding location system that makes bearing measurements at three different points in its trajectory. The intersection of two bearing lines provides an estimate of the location of the transmitter, which may be on the surface of the Earth or airborne. In the presence of noise, more than two bearing lines will not intersect at a single point. However, the appropriate processing allows an improved estimate of the transmitter position.

The following three sections of this paper present the basic methods of estimation applicable to transmitter location and determine the accuracy of suitable estimators. Sections 5 and 6, respectively, consider passive location systems using arrival-time and bearing measurements. Section 7 summarizes the use of Doppler information. Since the next three sections provide the theoretical framework for the statistical analysis of any passive location system, the reader who is only interested in the applications may wish to omit this material, referring to it as necessary while reading Sections 5–7.

II. ESTIMATION METHODS

The components of an n-dimensional vector $\boldsymbol{x}$ that is to be estimated are the position coordinates in two or

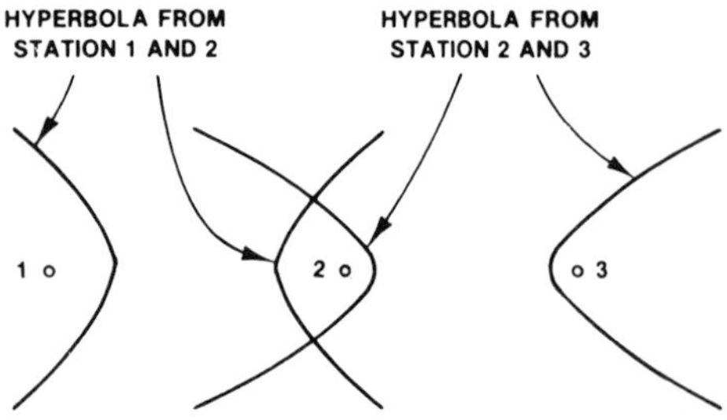

Fig. 1. Intersecting hyperbolas from three stations.

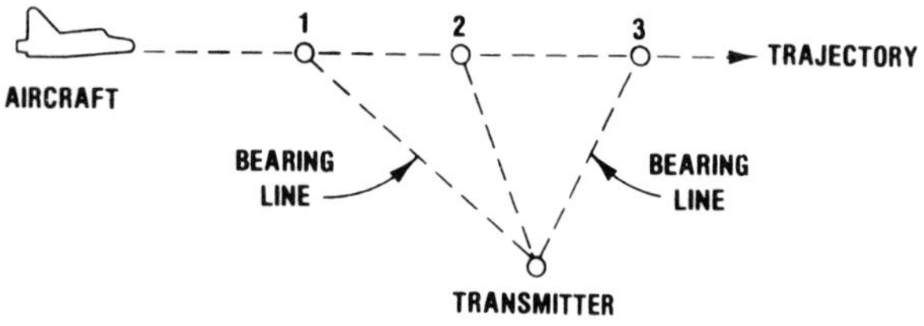

Fig. 2. Bearing lines from aircraft positions.

three dimensions and possibly other parameters such as the time of emission of the radiation. A set of N measurements r_i, $i = 1, 2, \ldots, N$, is collected at various positions. In the absence of random measurement errors, r_i is equal to a known function $f_i(x)$. In the presence of additive errors,

$$r_i = f_i(\boldsymbol{x}) + n_i, \qquad i = 1, 2, \ldots, N. \tag{1}$$

These N equations can be written as a single equation for N-dimensional column vectors:

$$\boldsymbol{r} = \boldsymbol{f}(\boldsymbol{x}) + \boldsymbol{n}. \tag{2}$$

The measurement error $\boldsymbol{n}$ is assumed to be a multivariate random vector with an $N \times N$ positive-definite covariance matrix

$$\boldsymbol{N} = E[(\boldsymbol{n} - E[\boldsymbol{n}])(\boldsymbol{n} - E[\boldsymbol{n}])^{\mathrm{T}}] \tag{3}$$

where $E[\ ]$ denotes the expected value and the superscript T denotes the transpose.

If $\boldsymbol{x}$ is regarded as an unknown but nonrandom vector and $\boldsymbol{n}$ is assumed to have a zero mean and a Gaussian distribution, then the conditional density function of $\boldsymbol{r}$ given $\boldsymbol{x}$ is

$$p(\boldsymbol{r}|\boldsymbol{x}) = \frac{1}{(2\pi)^{N/2}|\boldsymbol{N}|^{1/2}} \exp\{-(1/2)[\boldsymbol{r} - \boldsymbol{f}(\boldsymbol{x})]^{\mathrm{T}}\boldsymbol{N}^{-1}[\boldsymbol{r} - \boldsymbol{f}(\boldsymbol{x})]\} \tag{4}$$

where $|\boldsymbol{N}|$ denotes the determinant of $\boldsymbol{N}$ and the superscript -1 denotes the inverse. Because $\boldsymbol{N}$ is symmetric and positive definite, its inverse exists. The maximum likelihood estimator is that value of $\boldsymbol{x}$ which maximizes (4). Thus the maximum likelihood estimator minimizes the quadratic form

$$Q(\boldsymbol{x}) = [\boldsymbol{r} - \boldsymbol{f}(\boldsymbol{x})]^{\mathrm{T}}\boldsymbol{N}^{-1}[\boldsymbol{r} - \boldsymbol{f}(\boldsymbol{x})]. \tag{5}$$

Minimization of $Q(\boldsymbol{x})$ is a reasonable criterion for determination of an estimator even when the additive error cannot be assumed to be Gaussian. In this case, the resulting estimator is called the least squares estimator and $\boldsymbol{N}^{-1}$ is regarded as a matrix of weighting coefficients.

In general, $\boldsymbol{f}(\boldsymbol{x})$ is a nonlinear vector function. To determine a reasonably simple estimator, $\boldsymbol{f}(\boldsymbol{x})$ can be linearized by expanding it in a Taylor series about a reference point specified by the vector $\boldsymbol{x}_0$ and retaining the first two terms; that is, we use

$$\boldsymbol{f}(\boldsymbol{x}) \simeq \boldsymbol{f}(\boldsymbol{x}_0) + \boldsymbol{G}(\boldsymbol{x} - \boldsymbol{x}_0) \tag{6}$$

where $\boldsymbol{x}$ and $\boldsymbol{x}_0$ are $n \times 1$ column vectors and $\boldsymbol{G}$ is the $N \times n$ matrix of derivatives evaluated at $\boldsymbol{x}_0$:

$$\boldsymbol{G} = \begin{bmatrix} \left.\dfrac{\partial f_1}{\partial x_1}\right|_{\boldsymbol{x}=\boldsymbol{x}_0} & \cdots & \left.\dfrac{\partial f_1}{\partial x_n}\right|_{\boldsymbol{x}=\boldsymbol{x}_0} \\ \vdots & & \vdots \\ \left.\dfrac{\partial f_N}{\partial x_1}\right|_{\boldsymbol{x}=\boldsymbol{x}_0} & \cdots & \left.\dfrac{\partial f_N}{\partial x_n}\right|_{\boldsymbol{x}=\boldsymbol{x}_0} \end{bmatrix} \tag{7}$$

Each row of this matrix is the gradient vector of one of the components of $\boldsymbol{f}(\boldsymbol{x})$. The vector $\boldsymbol{x}_0$ could be an estimate of $\boldsymbol{x}$ determined from a previous iteration of the estimation procedure or based upon a priori information. It is assumed in the subsequent analysis that $\boldsymbol{x}_0$ is sufficiently close to $\boldsymbol{x}$ that (6) is an accurate approximation.

Combining (5) and (6) gives

$$Q(\boldsymbol{x}) = (\boldsymbol{r}_1 - \boldsymbol{G}\boldsymbol{x})^{\mathrm{T}}\boldsymbol{N}^{-1}(\boldsymbol{r}_1 - \boldsymbol{G}\boldsymbol{x}) \tag{8}$$

where

$$\boldsymbol{r}_1 = \boldsymbol{r} - \boldsymbol{f}(\boldsymbol{x}_0) + \boldsymbol{G}\boldsymbol{x}_0. \tag{9}$$

To determine the necessary condition for the estimator $\hat{\boldsymbol{x}}$ that minimizes $Q(\boldsymbol{x})$, we calculate the gradient of $Q(\boldsymbol{x})$, defined by

$$\nabla_x Q(\boldsymbol{x}) = \left[\frac{\partial Q}{\partial x_1}\ \frac{\partial Q}{\partial x_2} \cdots \frac{\partial Q}{\partial x_n}\right]^{\mathrm{T}} \tag{10}$$

and then solve for the $\boldsymbol{x}$ such that $\nabla_x Q(\boldsymbol{x}) = \boldsymbol{0}$. From its definition, $\boldsymbol{N}$ is a symmetric matrix; that is, $\boldsymbol{N}^{\mathrm{T}} = \boldsymbol{N}$. Since $(\boldsymbol{N}^{-1})^{\mathrm{T}} = (\boldsymbol{N}^{\mathrm{T}})^{-1}$, it follows that $(\boldsymbol{N}^{-1})^{\mathrm{T}} = \boldsymbol{N}^{-1}$, which implies that $\boldsymbol{N}^{-1}$ is a symmetric matrix. Therefore,

$$\nabla_x Q(\boldsymbol{x})|_{\boldsymbol{x}=\hat{\boldsymbol{x}}} = 2\boldsymbol{G}^{\mathrm{T}}\boldsymbol{N}^{-1}\boldsymbol{G}\hat{\boldsymbol{x}} - 2\boldsymbol{G}^{\mathrm{T}}\boldsymbol{N}^{-1}\boldsymbol{r}_1 = \boldsymbol{0}. \tag{11}$$

We assume that the matrix $\boldsymbol{G}^{\mathrm{T}}\boldsymbol{N}^{-1}\boldsymbol{G}$ is nonsingular. Thus the solution of (11) is

$$\hat{\boldsymbol{x}} = (\boldsymbol{G}^{\mathrm{T}}\boldsymbol{N}^{-1}\boldsymbol{G})^{-1}\boldsymbol{G}^{\mathrm{T}}\boldsymbol{N}^{-1}\boldsymbol{r}_1$$

$$= \boldsymbol{x}_0 + (\boldsymbol{G}^{\mathrm{T}}\boldsymbol{N}^{-1}\boldsymbol{G})^{-1}\boldsymbol{G}^{\mathrm{T}}\boldsymbol{N}^{-1}[\boldsymbol{r} - \boldsymbol{f}(\boldsymbol{x}_0)]. \tag{12}$$

Using (12), direct calculation shows that (8) can be written in the form

$$Q(x) = [x - \hat{x}]^T G^T N^{-1} G [x - \hat{x}] - r_1^T N^{-1} G (G^T N^{-1} G)^{-1} r_1 + r_1^T N^{-1} r_1 \quad (13)$$

where only the first term depends upon x. Since N is symmetric and positive definite, it has positive eigenvalues. If $Ne = \lambda e$, then $N^{-1} e = \lambda^{-1} e$. Thus if e is an eigenvector of N with eigenvalue λ, then e is also an eigenvector of N^{-1} with eigenvalue $1/\lambda$. Since it is symmetric and its eigenvalues are positive, N^{-1} is positive definite. Therefore, $x = \hat{x}$ minimizes $Q(x)$. The estimator of (12) is called the linearized least squares estimator.

Substituting (2) into (12) and rearranging terms, the expression for $\hat{x}$ can be written in the form

$$\hat{x} = x + (G^T N^{-1} G)^{-1} G^T N^{-1} [f(x) - f(x_0) - G(x - x_0) + n] \quad (14)$$

which shows how the estimator error is affected by the linearization error and the noise. The bias of the estimator $\hat{x}$ is defined as $b = E[\hat{x}] - x$. Using (14), we obtain

$$b = (G^T N^{-1} G)^{-1} G^T N^{-1} \{f(x) - f(x_0) - G(x - x_0) + E[n]\}. \quad (15)$$

If $f(x)$ is linear, as in (6), and $E[n] = \mathbf{0}$, then the least squares estimator is unbiased. If systematic errors occur in the measurements, then $E[n] \neq \mathbf{0}$. To minimize the estimator bias due to systematic errors, the magnitude of each $E[n_i]$ should be minimized through system calibrations. If some of the $E[n_i]$ are known functions of various parameters and N is sufficiently large, these parameters can be made components of the vector x and estimated along with the other components of x. The bias due to the nonlinearity of $f(x)$ can be estimated by expanding $f(x)$ in a Taylor series about x_0 and retaining second-order terms.

Let P denote the covariance matrix of $\hat{x}$. Equation (14) yields

$$P = E[(\hat{x} - E[\hat{x}])(\hat{x} - E[\hat{x}])^T] = (G^T N^{-1} G)^{-1}. \quad (16)$$

The diagonal elements of P give the variances of the errors in the estimated components of x. Since P is part of the estimator given by (12), one can compute both estimate and covariance simultaneously. If n is zero-mean Gaussian, the maximum likelihood or least squares estimator for the linearized model is the same as the minimum variance unbiased estimator [3].

The measurement error vector n is assumed to encompass all the contributions to error, including uncertainties in the system or physical parameters, such as the station coordinates or the speed of propagation. If q is a vector of the parameters, then the measurement vector r can often be expressed as

$$r = f_1(x,q) + n_1 \quad (17)$$

where $f_1(\ )$ is a vector function and n_1 is the random error due to causes unrelated to uncertainties in q. Let q_0 denote the assumed value of q. If q_0 is sufficiently close to q, then a Taylor series expansion yields

$$f_1(x,q) \simeq f_1(x,q_0) + G_1(q - q_0) \quad (18)$$

where G_1 is the matrix of derivatives with respect to q evaluated at q_0. Equation (2) results from making the identifications

$$f(x) = f_1(x,q_0), \qquad n = G_1(q - q_0) + n_1. \quad (19)$$

If q is nonrandom, then the parameter uncertainties ultimately contribute to the bias of the least squares estimator. If q is random, then the variance and possibly the bias are affected.

Any a priori information can be incorporated into the estimation procedure in several ways. It can be used to select an accurate reference point x_0 for the first iteration of the least squares estimator. If the transmitter is known to be located within a region, but the estimated position is outside this region, a logical procedure is to change the estimate to the point in the region that is closest to the original estimate. If an a priori distribution function for the transmitter position can be specified, a Bayesian estimator can be determined. However, the Bayesian estimator is usually too complex a mathematical function to yield a simple computational algorithm unless simplifying assumptions are made about the a priori distribution [4].

The location estimate can be continually refined if a sequence of measurements is taken. If successive measurements are uncorrelated, a new least squares estimate can be determined by combining new measurements with the old estimate [3]. Since measurements do not have to be stored after processing, a significant computational savings is sometimes possible.

III. ESTIMATOR ACCURACY

If r is a Gaussian random vector, then (12) indicates that $\hat{x}$ is a Gaussian random vector. Its probability density function is

$$f_{\hat{x}}(\xi) = [(2\pi)^{n/2} |P|^{1/2}]^{-1} \exp[-(1/2)(\xi - m)^T P^{-1} (\xi - m)] \quad (20)$$

where $m = E[\hat{x}]$ is the mean vector, and

$$P = E[(\hat{x} - m)(\hat{x} - m)^T] \quad (21)$$

is the covariance matrix given by (16). By definition, P is symmetric and positive semidefinite. Thus it has nonnegative eigenvalues. Equation (16) indicates that P^{-1} exists and in equal to $G^T N^{-1} G$. Therefore, P does not have zero as an eigenvalue. Thus P is positive definite.

The loci of constant density function values are described by equations of the form

$$(\xi - m)^T P^{-1} (\xi - m) = \kappa \quad (22)$$

where κ is a constant that determines the size of the n-dimensional region enclosed by the surface. In two dimensions, the surface is an ellipse; in three dimensions, it is an ellipsoid; in the general case of n dimensions, it may be considered a hyperellipsoid. Unless $\boldsymbol{P}$ is a diagonal matrix, the principal axes of the hyperellipsoids are not aligned with the coordinate axes.

The probability that $\hat{\boldsymbol{x}}$ lies inside the hyperellipsoid of (22) is

$$P_e(\kappa) = \int\int_R \cdots \int f_x(\boldsymbol{\xi})\, d\xi_1\, d\xi_2 \cdots d\xi_n \tag{23}$$

where the region of integration is

$$R = \{\boldsymbol{\xi} : (\boldsymbol{\xi} - \boldsymbol{m})^T \boldsymbol{P}^{-1}(\boldsymbol{\xi} - \boldsymbol{m}) \le \kappa\}. \tag{24}$$

To reduce (23) to a single integral, we perform a succession of coordinate transformations. First, we translate the coordinate system so that its origin coincides with $\boldsymbol{m}$ by making the change of variables $\boldsymbol{\gamma} = \boldsymbol{\xi} - \boldsymbol{m}$. Since the Jacobian is unity, we obtain

$$P_e(\kappa) = a \int\int_{R_1} \cdots \int \exp\left(-\frac{1}{2}\boldsymbol{\gamma}^T \boldsymbol{P}^{-1}\boldsymbol{\gamma}\right) d\gamma_1\, d\gamma_2 \cdots d\gamma_n \tag{25}$$

where

$$R_1 = \{\boldsymbol{\gamma} : \boldsymbol{\gamma}^T \boldsymbol{P}^{-1}\boldsymbol{\gamma} \le \kappa\} \tag{26}$$

$$a = [(2\pi)^{n/2}|\boldsymbol{P}|^{1/2}]^{-1}. \tag{27}$$

To simplify (25), we rotate the coordinate axes so that they are aligned with the principal axes of the hyperellipsoid. Because $\boldsymbol{P}$ is a symmetric positive-definite matrix, so is $\boldsymbol{P}^{-1}$. Therefore, an orthogonal matrix $\boldsymbol{A}$ (with eigenvectors as columns) exists that diagonalizes $\boldsymbol{P}^{-1}$. Thus $\boldsymbol{A}^T = \boldsymbol{A}$ and

$$\boldsymbol{A}^T\boldsymbol{P}^{-1}\boldsymbol{A} = \begin{bmatrix} \lambda_1^{-1} & & & 0 \\ & \lambda_2^{-1} & & \\ & & \ddots & \\ 0 & & & \lambda_n^{-1} \end{bmatrix} = [\lambda^{-1}] \tag{28}$$

where $\lambda_1, \lambda_2, \ldots, \lambda_n$ are the eigenvalues of $\boldsymbol{P}$. A rotation of axes results in new variables defined by

$$\boldsymbol{\zeta} = \boldsymbol{A}^T\boldsymbol{\gamma}. \tag{29}$$

Since $\boldsymbol{A}^T\boldsymbol{A} = \boldsymbol{I}$ and the determinant of the product of matrices is equal to the product of the determinants of the matrices, the determinant of $\boldsymbol{A}^T$, which is the Jacobian of the transformation, is unity. Substituting (28) and (29) into (25) and (26) yields

$$\begin{aligned} P_e(\kappa) &= a \int\int_{R_2} \cdots \int \exp\left(-\frac{1}{2}\boldsymbol{\zeta}^T[\lambda^{-1}]\boldsymbol{\zeta}\right) d\zeta_1\, d\zeta_2 \cdots d\zeta_n \\ &= a \int\int_{R_2} \cdots \int \exp\left(-\frac{1}{2}\sum_{i=1}^{n}\frac{\zeta_i^2}{\lambda_i}\right) d\zeta_1\, d\zeta_2 \cdots d\zeta_n \end{aligned} \tag{30}$$

where

$$R_2 = \left\{\boldsymbol{\xi} : \sum_{i=1}^{n} \frac{\zeta_i^2}{\lambda_i} \le \kappa\right\}, \tag{31}$$

and the ζ_i are the components of $\boldsymbol{\zeta}$. Regions R_2 is the interior of a hyperellipsoid with principal axes of lengths $2\sqrt{\kappa\lambda_i}$, $i = 1, 2, \ldots, n$. By introducing new variables

$$\eta_i = \zeta_i/\sqrt{\lambda_i}, \qquad i = 1, 2, \ldots, n \tag{32}$$

we can simplify (30) further. Since the determinant of $\boldsymbol{P}$ is equal to the product of the eigenvalues of $\boldsymbol{P}$, (27) and (30) to (32) give

$$P_e(\kappa) = (2\pi)^{-n/2} \int\int_{\sum_{i=1}^{n}\eta_i^2 \le \kappa} \cdots \int \exp\left(-\frac{1}{2}\sum_{i=1}^{n}\eta_i^2\right) d\eta_1\, d\eta_2 \cdots d\eta_n. \tag{33}$$

The region of integration, which is indicated below the integral signs, is the interior of a hypersphere.

It is shown below that the volume of an n-dimensional hypersphere of radius

$$\rho = \left(\sum_{i=1}^{n}\eta_i^2\right)^{1/2} \tag{34}$$

is

$$V_n(\rho) = \frac{\pi^{n/2}\rho^n}{\Gamma(n/2 + 1)} \tag{35}$$

where $\Gamma(\ )$ is the gamma function. Therefore, the differential volume between ρ and $\rho + d\rho$ is

$$dv = \frac{n\pi^{n/2}\rho^{n-1}}{\Gamma(n/2 + 1)}\, d\rho \tag{36}$$

and (33) can be reduced to

$$P_e(\kappa) = \frac{n}{2^{n/2}\Gamma(n/2 + 1)} \int_0^{\sqrt{\kappa}} \rho^{n-1} \exp\left(-\frac{\rho^2}{2}\right) d\rho. \tag{37}$$

For $n = 1$, 2, and 3, this integral can be expressed in simpler terms:

$$P_e(\kappa) = \text{erf}(\sqrt{\kappa/2}), \qquad n = 1 \tag{38}$$

$$P_e(\kappa) = 1 - \exp(-\kappa/2), \qquad n = 2 \tag{39}$$

$$P_e(\kappa) = \text{erf}(\sqrt{\kappa/2}) - \sqrt{2\kappa/\pi}\exp(-\kappa/2), \qquad n = 3 \tag{40}$$

where the error function is defined by

$$\text{erf}(x) = \frac{2}{\sqrt{\pi}}\int_0^x \exp(-t^2)\, dt. \tag{41}$$

Equation (40) is obtained by integration by parts.

To verify (35), we define the volume of a hypersphere of radius ρ by

$$V_n(\rho) = \underset{\sum_{i=1}^{n} x_i^2 \leq \rho^2}{\int\int \cdots \int} dx_1\, dx_2 \cdots dx_n. \tag{42}$$

A change of coordinates shows that

$$V_n(\rho) = \rho^n V_n(1) \tag{43}$$

where $V_n(1)$ is the volume of a unit hypersphere. Straightforward calculations give

$$V_1(1) = 2, \qquad V_2(1) = \pi \tag{44}$$

where the "volumes" are a length and an area, respectively. We define the sets

$$B = \{(x_1, x_2) : x_1^2 + x_2^2 \leq 1\} \tag{45}$$

$$C = \{(x_3, \ldots, x_n) : \sum_{i=3}^{n} x_i^2 \leq 1 - x_1^2 - x_2^2\}. \tag{46}$$

For $n \geq 3$, Fubini's theorem for interchanging the order of integrations [5] and a change of coordinates in (42) give

$$V_n(1) = \int\int_B dx_1\, dx_2 \int\int_C \cdots \int dx_3 \cdots dx_n$$
$$= \int\int_B V_{n-2}\left(\sqrt{1 - x_1^2 - x_2^2}\right) dx_1\, dx_2. \tag{47}$$

Equation (43) and further coordinate changes yield

$$V_n(1) = V_{n-2}(1) \int\int_B (1 - x_1^2 - x_2^2)^{(n-2)/2}\, dx_1\, dx_2$$
$$= V_{n-2}(1) \int_0^{2\pi} \int_0^1 (1 - r^2)^{(n-2)/2}\, r\, dr\, d\theta$$
$$= \pi V_{n-2}(1) \int_0^1 x^{(n-2)/2}\, dx$$
$$= 2\pi V_{n-2}(1)/n. \tag{48}$$

By induction, this recursion relation and (44) imply that

$$V_{2m}(1) = \frac{(2\pi)^m}{2 \cdot 4 \cdots (2m)};$$
$$V_{2m-1}(1) = \frac{2(2\pi)^{m-1}}{1 \cdot 3 \cdots (2m-1)}, \quad m = 1, 2, \ldots. \tag{49}$$

We can express $V_n(1)$ in terms of a compact formula by using the properties of the gamma function: $\Gamma(t+1) = t\Gamma(t)$; $\Gamma(1) = 1$; $\Gamma(1/2) = \sqrt{\pi}$. We obtain

$$V_n(1) = \frac{\pi^{n/2}}{\Gamma(n/2 + 1)}, \quad n = 1, 2, \ldots. \tag{50}$$

Combining (43) and (50) yields (35).

If P_e is specified, say $P_e = 1/2$; then (37), (38), (39), or (40) can be solved numerically to determine the corresponding value of κ, which in turn defines a hyperellipsoid by (22). The concentration ellipsoid corresponding to probability P_e is defined to be the particular hyperellipsoid for which P_e is the probability that $\hat{\boldsymbol{x}}$ lies inside it. Thus the concentration ellipsoid is a multidimensional measure of accuracy for an unbiased estimator.

A scalar measure of estimator accuracy is the root-mean-square error ϵ_r, which is defined by

$$\epsilon_r^2 = E\left[\sum_{i=1}^{n} (\hat{x}_i - x_i)^2\right]. \tag{51}$$

Expanding (51) and using (21), we obtain

$$\epsilon_r^2 = \mathrm{tr}(\boldsymbol{P}) + \sum_{i=1}^{n} b_i^2 \tag{52}$$

where $\mathrm{tr}(\boldsymbol{P})$ denotes the trace of $\boldsymbol{P}$ and $b_i = E[\hat{x}_i] - x_i$ denotes a component of the bias vector $\boldsymbol{b}$.

IV. TWO-DIMENSIONAL ESTIMATORS

For the estimator of a two-dimensional vector, such as position coordinates on the surface of the Earth, the bivariate covariance matrix can be expressed as

$$\boldsymbol{P} = \begin{bmatrix} \sigma_1^2 & \sigma_{12} \\ \sigma_{12} & \sigma_2^2 \end{bmatrix}. \tag{53}$$

A straightforward calculation yields the eigenvalues:

$$\lambda_1 = \frac{1}{2}\left[\sigma_1^2 + \sigma_2^2 + \sqrt{(\sigma_1^2 - \sigma_2^2)^2 + 4\sigma_{12}^2}\right] \tag{54}$$

$$\lambda_2 = \frac{1}{2}\left[\sigma_1^2 + \sigma_2^2 - \sqrt{(\sigma_1^2 - \sigma_2^2)^2 + 4\sigma_{12}^2}\right] \tag{55}$$

where the positive square root is used. By definition, $\lambda_1 \geq \lambda_2$.

Suppose that new coordinates are defined by rotating the axes of the old coordinate system counterclockwise through an angle θ, as shown in Fig. 3. A vector

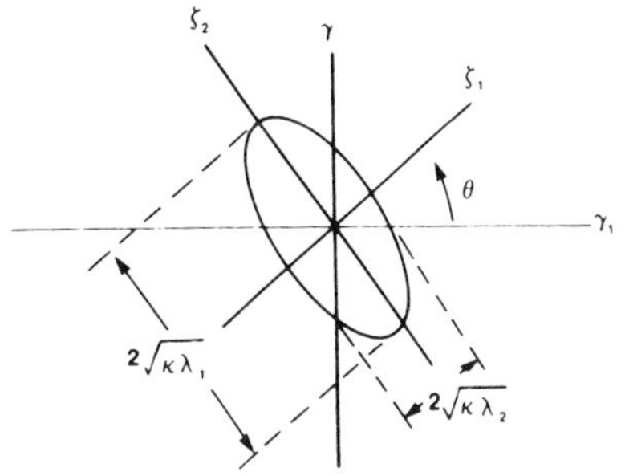

Fig. 3. Concentration ellipse and coordinate axes.

represented by $\boldsymbol{\gamma}$ in the old coordinates is represented in the new coordinates by $\boldsymbol{\zeta} = \boldsymbol{A}^T\boldsymbol{\gamma}$, where $\boldsymbol{A}$ is the orthogonal matrix

$$\boldsymbol{A} = \begin{bmatrix} \cos\theta & -\sin\theta \\ \sin\theta & \cos\theta \end{bmatrix}. \tag{56}$$

From (53) and (56), direct but lengthy calculation shows that $\boldsymbol{A}^T\boldsymbol{P}^{-1}\boldsymbol{A}$ is a diagonal matrix and the columns of $\boldsymbol{A}$ are eigenvectors if

$$\theta = \frac{1}{2}\tan^{-1}\left(\frac{2\sigma_{12}}{\sigma_1^2 - \sigma_2^2}\right), \qquad -\frac{\pi}{4} \leq \theta \leq \frac{\pi}{4}. \tag{57}$$

If $\sigma_1^2 = \sigma_2^2$ and $\sigma_{12} = 0$, we take $\theta = 0$. Since the determinant of a matrix is equal to the product of the eigenvalues, $\lambda_1\lambda_2 = \sigma_1^2\sigma_2^2 - \sigma_{12}^2$. Using this result, the diagonal matrix can be written in the form

$$[\lambda^{-1}] = \begin{bmatrix} \lambda_1^{-1} & 0 \\ 0 & \lambda_2^{-1} \end{bmatrix}, \qquad \sigma_1^2 \geq \sigma_2^2 \tag{58}$$

$$[\lambda^{-1}] = \begin{bmatrix} \lambda_2^{-1} & 0 \\ 0 & \lambda_1^{-1} \end{bmatrix}, \qquad \sigma_1^2 < \sigma_2^2. \tag{59}$$

Since $\boldsymbol{P}^{-1}$ exists according to (16), neither eigenvalue can equal zero and $[\lambda^{-1}]$ is well defined.

A concentration ellipse defined by $\boldsymbol{\gamma}^T\boldsymbol{P}^{-1}\boldsymbol{\gamma} = \kappa$ in the old coordinates is described by $(\zeta_1/\lambda_1)^2 + (\zeta_2/\lambda_2) = \kappa$ or by $(\zeta_1/\lambda_2)^2 + (\zeta_2/\lambda_1)^2 = \kappa$ in the new coordinates, a fact which indicates that the new axes coincide with the principal axes of the ellipse. Thus (57) represents the angular offset of one of the principal axes of the ellipse relative to the old coordinate axes. Fig. 3 depicts a concentration ellipse and the appropriate angle of axis rotation. Since $\lambda_1 \geq \lambda_2$, the major and minor axes have lengths $2\sqrt{\kappa\lambda_1}$ and $2\sqrt{\kappa\lambda_2}$, respectively. If the ellipse encloses a region that includes a Gaussian random vector with probability P_e, then (39) implies that

$$\kappa = -2\ln(1 - P_e). \tag{60}$$

Suppose that a two-dimensional Gaussian random vector describes the estimated location of a transmitter. A crude but simple measure of accuracy is the circular error probable (CEP). The CEP is defined as the radius of the circle that has its center at the mean and contains half the realizations of the random vector. The CEP is a measure of the uncertainty in the location estimator $\hat{\boldsymbol{x}}$ relative to its mean $E[\hat{\boldsymbol{x}}]$. If the location estimator is unbiased, the CEP is a measure of the estimator uncertainty relative to the true transmitter position. If the magnitude of the bias vector is bounded by B, then with a probability of one-half, a particular estimate is within a distance of B + CEP from the true position. The geometrical relations are depicted in Fig. 4.

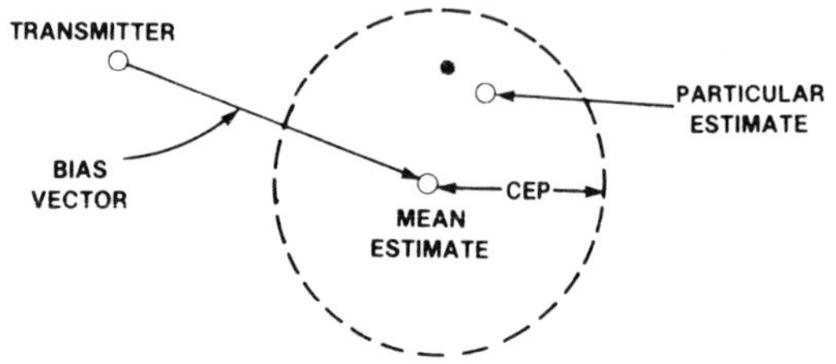

Fig. 4. Geometry of transmitter position, mean location estimate, CEP, estimator bias vector, and particular location estimate.

From the definition, it follows that we can determine the CEP by solving the equation

$$\frac{1}{2} = \iint_R f_{\hat{x}}(\boldsymbol{\xi})\, d\xi_1\, d\xi_2 \tag{61}$$

where

$$R = \{\boldsymbol{\xi} : |\boldsymbol{\xi} - \boldsymbol{m}| \leq \text{CEP}\}. \tag{62}$$

In a manner analogous to the derivation of (30), we successively translate and rotate coordinates to obtain

$$\frac{1}{2} = \frac{1}{2\pi\sqrt{\lambda_1\lambda_2}} \iint_{R_1} \exp\left(-\frac{1}{2}\sum_{i=1}^{2}\frac{\zeta_i^2}{\lambda_i}\right) d\zeta_1\, d\zeta_2 \tag{63}$$

where

$$R_1 = \{(\zeta_1, \zeta_2) : (\zeta_1^2 + \zeta_2^2)^{1/2} \leq \text{CEP}\} \tag{64}$$

and the λ_i are given by (54) and (55). Changing to polar coordinates by substituting $\zeta_1 = r\cos\theta$ and $\zeta_2 = r\sin\theta$, we get

$$\pi\sqrt{\lambda_1\lambda_2} = \int_0^{2\pi}\int_0^{\text{CEP}} r \exp\left[-\frac{r^2}{2}\left(\frac{\cos^2\theta}{\lambda_1} + \frac{\sin^2\theta}{\lambda_2}\right)\right] dr\, d\theta. \tag{65}$$

To simplify (65), we do some preliminary manipulations. The modified Bessel function of the first kind and zero order can be expressed as

$$I_0(x) = \frac{1}{2\pi}\int_0^{2\pi} \exp(x\cos\theta)\, d\theta. \tag{66}$$

Because of the periodicity of the integrand, we also have

$$I_0(x) = \frac{1}{2\pi}\int_{2\pi n}^{2\pi(n+1)} \exp(x\cos\theta)\, d\theta \tag{67}$$

for any integer n. Adding m equations of this form with successive values of n, we obtain

$$mI_0(x) = \frac{1}{2\pi}\int_0^{2\pi m} \exp(x\cos\theta)\, d\theta, \quad m = 1, 2, \ldots. \tag{68}$$

Changing coordinates with $\theta = m\phi$ gives

$$I_0(x) = \frac{1}{2\pi}\int_0^{2\pi} \exp(x\cos m\phi)\, d\phi, \quad m = 1, 2, \ldots. \tag{69}$$

Trigonometric identities yield

$$\frac{\cos^2\theta}{\lambda_1} + \frac{\sin^2\theta}{\lambda_2} = \frac{1}{2\lambda_1} + \frac{1}{2\lambda_2} + \left(\frac{1}{2\lambda_1} - \frac{1}{2\lambda_2}\right)\cos 2\theta. \tag{70}$$

Substituting (70) into (65) and using (69), we obtain

$$\frac{\sqrt{\lambda_1\lambda_2}}{2} = \int_0^{\text{CEP}} r \exp\left[-\left(\frac{1}{4\lambda_1}+\frac{1}{4\lambda_2}\right)r^2\right] I_0\left[\left(\frac{1}{4\lambda_2}-\frac{1}{4\lambda_1}\right)r^2\right] dr. \quad (71)$$

A final change of coordinates yields

$$\frac{1}{4\gamma^2}(1+\gamma^2) = \int_0^{[(\text{CEP})^2/4\lambda_2](1+\gamma^2)} \exp(-x) I_0\left(\frac{1-\gamma^2}{1+\gamma^2}x\right) dx, \quad \gamma^2 = \frac{\lambda_2}{\lambda_1}. \quad (72)$$

The form of this relation implies that the CEP has the form CEP $= \sqrt{\lambda_2} f(\gamma)$ for some function $f(\ )$. If $\sigma_{12} = 0$ and $\sigma_1 = \sigma_2 = \sigma$, then $\lambda_1 = \lambda_2 = \sigma^2$ and (72) can be solved to show that CEP $= 1.177\sigma$. In the general case where $\lambda_1 \neq \lambda_2$, numerical integration is necessary to solve for the CEP. A simple approximation that is consistent with the preceding observations is

$$\text{CEP} \approx 0.563\sqrt{\lambda_1} + 0.614\sqrt{\lambda_2} \quad (73)$$

which is accurate to within 1 percent for $\gamma \approx 0.3$ or larger, underestimates the CEP by less than 10 percent for $0.1 < \gamma < 0.3$, and underestimates by less than 20 percent elsewhere. Although approximations that are more accurate for small γ are easily produced, they are usually irrelevant because the eccentricity of the concentration ellipse for small γ may be too pronounced for the CEP to be an adequate performance measure. An approximation that is accurate to within approximately 10 percent for all values of γ is

$$\text{CEP} \approx 0.75\sqrt{\lambda_1+\lambda_2} = 0.75\sqrt{\sigma_1^2+\sigma_2^2} \quad (74)$$

where the last relation follows from the fact that the trace of a matrix is equal to the sum of its eigenvalues. Above $\gamma \approx 0.4$, this approximation underestimates the CEP; below $\gamma \approx 0.4$, it overestimates the CEP. For an unbiased estimator, (52) implies that CEP $\approx 0.75\,\epsilon_r$.

V. HYPERBOLIC LOCATION SYSTEMS

Suppose that the arrival times $t_1, t_2, \ldots, t_N$ of a signal transmitted at time t_0 are measured at N stations having positions specified by the column vectors $s_1, s_2, \ldots, s_N$. The geometrical configuration is illustrated in Fig. 5. If

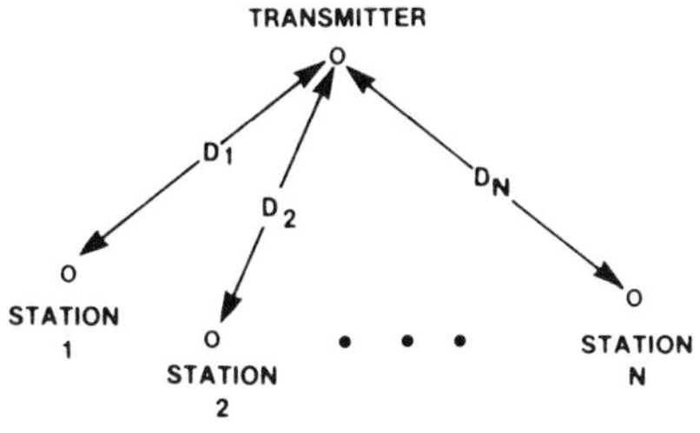

Fig. 5. Geometry of transmitter and N stations.

the signal velocity is c and if D_i is the propagation path length between the transmitter and station i, then

$$t_i = t_0 + D_i/c + \epsilon_i, \qquad i = 1, 2, \ldots, N. \quad (75)$$

The arrival-time measurement error ϵ_i accounts for propagation anomalies, receiver noise, and errors in the assumed station positions. In matrix form, (75) becomes

$$\boldsymbol{t} = t_0\mathbf{1} + \boldsymbol{D}/c + \boldsymbol{\epsilon} \quad (76)$$

where $\boldsymbol{t}$, $\boldsymbol{D}$, and $\boldsymbol{\epsilon}$ are N-dimensional column vectors with components t_i, D_i, and ϵ_i, $i = 1, 2, \ldots, N$, respectively, and $\mathbf{1}$ is a column vector of ones.

Suppose that we seek to estimate both t_0 and the column vector $\boldsymbol{R}$, with components x, y, and z, that specifies the transmitter position. Equation (76) has the form of (2) with $\boldsymbol{r} = \boldsymbol{t}$, $\boldsymbol{f}(\boldsymbol{x}) = t_0\mathbf{1} + \boldsymbol{D}/c$, $\boldsymbol{n} = \boldsymbol{\epsilon}$, and $\boldsymbol{x} = [t_0\ \ x\ \ y\ \ z]^T$. For line-of-sight propagation from the transmitter to the stations, $D_i = \|\boldsymbol{R} - \boldsymbol{s}_i\|$, where $\|\ \|$ represents the Euclidean norm. Let the column vector $\boldsymbol{R}_0$, with components x_0, y_0, and z_0, specify a reference point near the transmitter position. Let $D_{0i} = \|\boldsymbol{R}_0 - \boldsymbol{s}_i\|$ denote the distance from station i to the reference point. Using (7) with $\boldsymbol{x}_0 = [0\ \ x_0\ \ y_0\ \ z_0]^T$ after expressing each $\|\boldsymbol{R} - \boldsymbol{s}_i\|$ in terms of its components, we obtain

$$\boldsymbol{G} = [\mathbf{1}\ \ \boldsymbol{F}/c] \quad (77)$$

where

$$\boldsymbol{F} = \begin{bmatrix} (\boldsymbol{R}_0 - \boldsymbol{s}_1)^T/D_{01} \\ \vdots \\ (\boldsymbol{R}_0 - \boldsymbol{s}_N)^T/D_{0N} \end{bmatrix}. \quad (78)$$

Each row of $\boldsymbol{F}$ is the unit vector pointing from one of the stations to the reference point. Equation (12) with the above relations and substitutions gives the least squares or maximum likelihood estimator; (16) provides the covariance matrix of the estimator.

In hyperbolic systems, no attempt is made to estimate t_0. We eliminate it from consideration by measuring the relative arrival times:

$$t_i - t_{i+1} = (D_i - D_{i+1})/c + n_i, \qquad i = 1, 2, \ldots, N-1 \quad (79)$$

where n_i is the measurement error. Measuring time differences is not the only way to eliminate t_0, but it is the simplest. If the relative arrival times are determined by subtracting measured arrival times, then

$$n_i = \epsilon_i - \epsilon_{i+1}, \qquad i = 1, 2, \ldots, N-1. \quad (80)$$

The n_i have zero means if successive ϵ_i have equal means, even if the latter means are nonzero. A nonzero $E[n_i]$ may result from uncalibrated different time delays or unsynchronized clocks in two receivers. If the relative arrival times are determined by cross correlation, then (80) is not necessarily valid.

If the transmitter produces a sequence of pulses, the corresponding received pulses at stations i and $i+1$ must be correctly associated in measuring the time difference $t_i - t_{i+1}$. A potential ambiguity arises when the time difference exceeds the time between successive pulse transmissions. This ambiguity may be resolved by using bearing measurements or a priori information to eliminate associations that lead to impossible location estimates.

In matrix form, (79) becomes

$$\boldsymbol{Ht} = \boldsymbol{HD}/c + \boldsymbol{n} \tag{81}$$

where we use the $(N - 1) \times N$ matrix

$$\boldsymbol{H} = \begin{bmatrix} 1 & -1 & 0 & \cdots & 0 & 0 \\ 0 & 1 & -1 & \cdots & 0 & 0 \\ \vdots & \vdots & \vdots & & \vdots & \vdots \\ 0 & 0 & 0 & \cdots & 1 & -1 \end{bmatrix}. \tag{82}$$

If (80) is valid, then

$$\boldsymbol{n} = \boldsymbol{H\epsilon}. \tag{83}$$

Since we seek to estimate the position vector $\boldsymbol{R}$, (81) has the form of (2) with $\boldsymbol{r} = \boldsymbol{Ht}$, $\boldsymbol{f}(\boldsymbol{x}) = \boldsymbol{HD}/c$, and $\boldsymbol{x} = \boldsymbol{R}$. A direct calculation of $\boldsymbol{G}$ yields

$$\boldsymbol{G} = \boldsymbol{HF}/c \tag{84}$$

where $\boldsymbol{F}$ is defined by (78). Let $\boldsymbol{N}_\epsilon$ denote the covariance matrix of the arrival-time errors. If (83) holds, then the covariance matrix of the measurement errors, defined by (3), is related to $\boldsymbol{N}_\epsilon$ by

$$\boldsymbol{N} = \boldsymbol{HN}_\epsilon \boldsymbol{H}^{\mathrm{T}}. \tag{85}$$

Using (84), equation (12) implies that the least squares estimator is

$$\hat{\boldsymbol{R}} = \boldsymbol{R}_0 + c(\boldsymbol{F}^{\mathrm{T}}\boldsymbol{H}^{\mathrm{T}}\boldsymbol{N}^{-1}\boldsymbol{HF})^{-1}\boldsymbol{F}^{\mathrm{T}}\boldsymbol{H}^{\mathrm{T}}\boldsymbol{N}^{-1}(\boldsymbol{Ht} - \boldsymbol{HD}_0/c) \tag{86}$$

where $\boldsymbol{D}_0$ has components D_{0i}, $i = 1, 2, \ldots, N$. The estimator is unbiased if $\boldsymbol{n}$ is a zero-mean random variable and the linearization error is negligible. The covariance matrix of $\hat{\boldsymbol{R}}$, given by (16), is

$$\boldsymbol{P} = c^2(\boldsymbol{F}^{\mathrm{T}}\boldsymbol{H}^{\mathrm{T}}\boldsymbol{N}^{-1}\boldsymbol{HF})^{-1}. \tag{87}$$

Equation (86) is valid for line-of-sight propagation. If the signal propagation to the stations involves atmospheric reflections, the equations for the D_i change and thus the estimator changes.

In general, the least squares estimator requires knowledge of the statistics of the measurement errors. However, if (85) applies, if the covariances of the ϵ_i are zero, and if the variances of the ϵ_i have the common value σ_t^2, then cancellation in (86) leaves an estimator that is independent of σ_t^2. Equality of the variances is a reasonable assumption for stations with identical receivers that are much closer to each other than to the transmitter.

Let σ_{ti}^2 denote the variance of the measured arrival time t_i at station i. The mean-square ranging error is defined as $c^2\sigma_s^2$, where

$$\sigma_s^2 = \frac{1}{N}\sum_{i=1}^{N} \sigma_{ti}^2 \tag{88}$$

is the average variance of the arrival times. The geometric dilution of precision (GDOP) is defined as the ratio of the root-mean-square position error ϵ_r to the root-mean-square ranging error. It follows from (52) that the GDOP associated with an unbiased estimator and a hyperbolic system is

$$\mathrm{GDOP} = \sqrt{\mathrm{trace}[\boldsymbol{P}]}/c\sigma_s. \tag{89}$$

The GDOP indicates how much the fundamental ranging error, intuitively measured by $c\sigma_s$, is magnified by the geometric relation among the transmitter position and the stations. If the geometry is such that the arrival-time variances are nearly equal, then the GDOP is only weakly dependent on them. For the two-dimensional location problem, (74) and (89) yield

$$\mathrm{CEP} \approx (0.75c\sigma_s)\mathrm{GDOP}. \tag{90}$$

Since the arrival-time variance σ_t^2 is due primarily to the thermal and environmental noise, it is often reasonable to model ϵ_i as the sum of a constant bias plus zero-mean white Gaussian noise. The Cramer–Rao bound for an arrival-time estimate in the presence of white Gaussian noise gives [6]

$$\sigma_t^2 \geq [(2E/N_0)\,\beta_r^2]^{-1} \tag{91}$$

where E is the energy in the received signal, $N_0/2$ is the two-sided noise power spectral density, and β_r^2 is a function of the bandwidth of the signal. If $S(\omega)$ denotes the Fourier transform of the signal, then

$$\beta_r^2 = \frac{\displaystyle\int_{-\infty}^{\infty} \omega^2 |S(\omega)|^2\, d\omega}{\displaystyle\int_{-\infty}^{\infty} |S(\omega)|^2\, d\omega}. \tag{92}$$

If the received signal consists of pulses, then E is the sum of the energies of the individual pulses. An approximate model for many radar signals is a series of pulses, each of which results from passing a truncated sinusoid with an ideal rectangular envelope of duration T_p through an ideal rectangular bandpass filter of bandwidth B centered at the sinusoidal frequency. For each pulse and for the entire radar signal, (92) yields

$$\beta_r^2 \approx 2B/T_p, \qquad BT_p >> 1. \tag{93}$$

In contrast, for a signal with a uniform Fourier transform over a bandwidth B, (92) gives

$$\beta_r^2 = \pi^2 B^2/3. \tag{94}$$

This model might approximate a communications signal.

Let T denote the total signal duration, $R_s = E/T$ denote the average signal power at the receiver, and D denote the distance between the transmitter and the receiver. Over a large range of values of D, it is often possible to approximate R_s by [7]

$$R_s = K_E \exp(-\alpha D)/D^n \tag{95}$$

where α, n, and K_E are independent of D, but may be functions of other parameters such as the transmitter power, antenna gains, antenna heights, and the signal frequency. For optical and millimeter-wave frequencies, accurate modeling requires $\alpha > 0$, but we may usually set $\alpha = 0$ at other frequencies. Inequality (91) and (95) relate σ_t^2 to D.

As an important special case, we consider a transmitter and three stations in the same plane so that only two position coordinates are to be estimated. The planar model is reasonable if a transmitter and stations are near the surface of the Earth and close enough that the curvature of the Earth's surface can be neglected. One of the stations is designated the master station, and the other two are called slave stations. Arrival-time measurements at the slave stations are sent to the master station, where the time differences and then the position estimate are computed.

We assume that the ϵ_i are uncorrelated random variables so that

$$N_\epsilon = \begin{bmatrix} \sigma_{t1}^2 & 0 & 0 \\ 0 & \sigma_{t2}^2 & 0 \\ 0 & 0 & \sigma_{t3}^2 \end{bmatrix}. \tag{96}$$

The $\boldsymbol{H}$ matrix for $N = 3$ is

$$\boldsymbol{H} = \begin{bmatrix} 1 & -1 & 0 \\ 0 & 1 & -1 \end{bmatrix}. \tag{97}$$

Let ϕ_{0i} denote the bearing angle from station i at coordinates (x_i, y_i) to the reference point at coordinates (x_0, y_0), as illustrated in Fig. 6. Thus

$$\phi_{0i} = \tan^{-1}\left(\frac{y_0 - y_i}{x_0 - x_i}\right), \qquad i = 1, 2, 3. \tag{98}$$

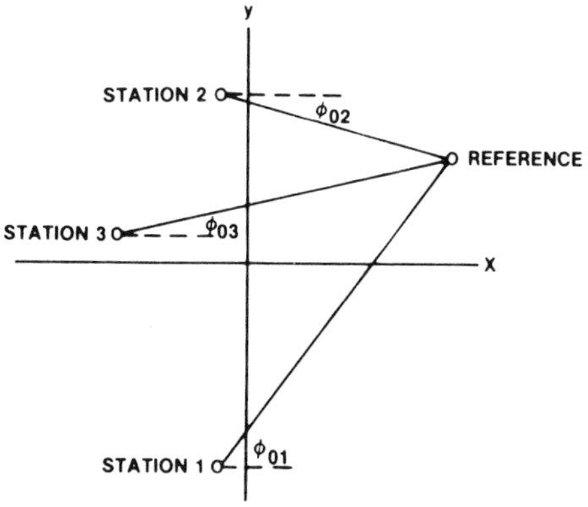

Fig. 6. Angle definitions for reference and three stations.

Equation (78) may be expressed as

$$\boldsymbol{F} = \begin{bmatrix} \cos\phi_{01} & \sin\phi_{01} \\ \cos\phi_{02} & \sin\phi_{02} \\ \cos\phi_{03} & \sin\phi_{03} \end{bmatrix}. \tag{99}$$

The covariance matrix $\boldsymbol{P}$ can be evaluated by substituting (85), (96), (97), and (99) into (87). The components of $\boldsymbol{P}$ defined by (53) are

$$\sigma_1^2 = \alpha[\sigma_{t1}^2(\sin\phi_{02} - \sin\phi_{03})^2 + \sigma_{t2}^2(\sin\phi_{01} - \sin\phi_{03})^2 + \sigma_{t3}^2(\sin\phi_{01} - \sin\phi_{02})^2] \tag{100}$$

$$\sigma_2^2 = \alpha[\sigma_{t1}^2(\cos\phi_{02} - \cos\phi_{03})^2 + \sigma_{t2}^2(\cos\phi_{01} - \cos\phi_{03})^2 + \sigma_{t3}^2(\cos\phi_{01} - \cos\phi_{02})^2] \tag{101}$$

$$\sigma_{12} = \alpha[\sigma_{t1}^2(\cos\phi_{03} - \cos\phi_{02})(\sin\phi_{02} - \sin\phi_{03}) + \sigma_{t2}^2(\cos\phi_{03} - \cos\phi_{01})(\sin\phi_{01} - \sin\phi_{03}) + \sigma_{t3}^2(\cos\phi_{02} - \cos\phi_{01})(\sin\phi_{01} - \sin\phi_{02})] \tag{102}$$

where

$$\alpha = c^2[(\cos\phi_{01} - \cos\phi_{02})(\sin\phi_{02} - \sin\phi_{03}) - (\cos\phi_{02} - \cos\phi_{03})(\sin\phi_{01} - \sin\phi_{02})]^{-2}. \tag{103}$$

If any two bearing angles are equal, then σ_1^2, σ_2^2, and $\sigma_{12} \to \infty$. These events correspond to reference points that lie along a line passing through two of the stations.

The least squares or maximum likelihood estimator, determined from (86), is

$$\hat{x} = x_0 + \sqrt{\alpha}\,[(t_1 - D_{01}/c)(\sin\phi_{02} - \sin\phi_{03}) + (t_2 - D_{02}/c)(\sin\phi_{03} - \sin\phi_{01}) + (t_3 - D_{03}/c)(\sin\phi_{01} - \sin\phi_{02})] \tag{104}$$

$$\hat{y} = y_0 + \sqrt{\alpha}\,[(t_1 - D_{01}/c)(\cos\phi_{03} - \cos\phi_{02}) + (t_2 - D_{02}/c)(\cos\phi_{01} - \cos\phi_{03}) + (t_3 - D_{03}/c)(\cos\phi_{02} - \cos\phi_{01})]. \tag{105}$$

To determine the transmitter range, which may be defined as the distance between the transmitter and the master station, it is convenient to align the x axis with the line between the master station and the reference point and to place the origin of the coordinate system at the master station. If the reference point is near the transmitter position, then $\hat{x}$ is a suitable range estimator and σ_1^2 approximates the variance of the range estimator; otherwise, the range can be estimated by $(\hat{x}^2 + \hat{y}^2)^{1/2}$. A suitable estimator for the bearing with respect to the x axis is

$$\hat{\phi} = \tan^{-1}(\hat{y}/\hat{x}). \tag{106}$$

The estimator bias can be determined from (15). Neglecting the linearization error and using (83), we obtain

$$b_1 = \sqrt{\alpha}\,\{E[\epsilon_1](\sin\phi_{02} - \sin\phi_{03}) + E[\epsilon_2](\sin\phi_{03} - \sin\phi_{01}) + E[\epsilon_3](\sin\phi_{01} - \sin\phi_{02})\} \tag{107}$$

$$b_2 = \sqrt{\alpha}\,\{E[\epsilon_1](\cos\phi_{03} - \cos\phi_{02}) + E[\epsilon_2](\cos\phi_{01} - \cos\phi_{03}) + E[\epsilon_3](\cos\phi_{02} - \cos\phi_{01})\}. \quad (108)$$

Nonzero values of the $E[\epsilon_i]$ are caused primarily by uncertainties in the station positions, synchronization errors, and the temperature dependence of the receiver delays and filter characteristics.

Assuming that $\boldsymbol{n} = \boldsymbol{H}\boldsymbol{\epsilon}$ has a Gaussian distribution, (54), (55), (73), and (100)–(103) give the CEP in terms of the bearing angles and the arrival-time variances. For a fixed deployment of stations, the locus of transmitter positions with a constant value of the CEP can be determined numerically. For this purpose, the equations may be expressed in terms of the Cartesian coordinates by using (98) and it is assumed that the reference point coincides with the transmitter position with negligible error so that $D_{0i} = DD_i$.

Let L denote the length of a linear array of three stations with coordinates $(0, -L/2)$, $(0,0)$, and $(0, L/2)$. Assuming that the lower bound of inequality (91) is nearly achieved and using (95), we obtain

$$\sigma_{ti}^2 \simeq \sigma_{tL}^2\,(D_{0i}/L)^n \exp[\alpha(D_{0i} - L)], \quad i = 1, 2, 3$$

$$\sigma_{tL}^2 = N_0 L^n \exp(\alpha L)/2\beta_r^2 T K_E \quad (109)$$

where σ_{tL} denotes the lower bound of σ_{ti} when $D_{0i} = L$. It is assumed that α, n, K_E, and hence σ_{tL} are identical for all three stations. We assume that the transmitter and the stations have omnidirectional antennas so that K_E does not depend upon the bearing angle to the transmitter. Figs. 7 and 8 depict loci of constant values of $\mathrm{CEP}/c\sigma_{tL}$

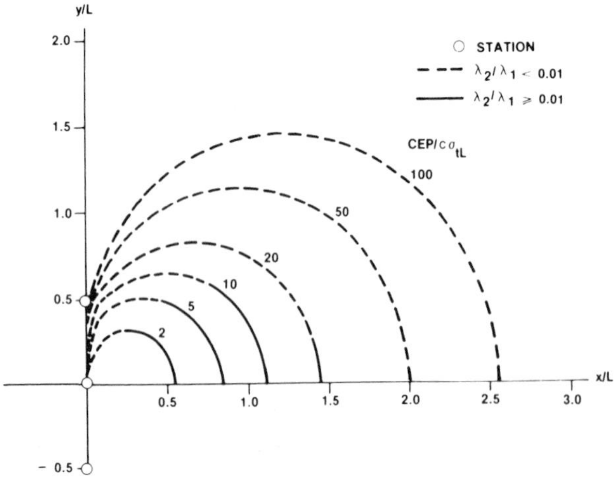

Fig. 7. Loci of constant $\mathrm{CEP}/c\sigma_{tL}$ for linear array of three stations with $n = 2$.

for $\alpha = 0$. Only the first quadrant is displayed because of the symmetry of the loci. Fig. 7 assumes $n = 2$, which corresponds to free-space propagation. Fig. 8 assumes $n = 4$, which might model VHF propagation near the Earth's surface.

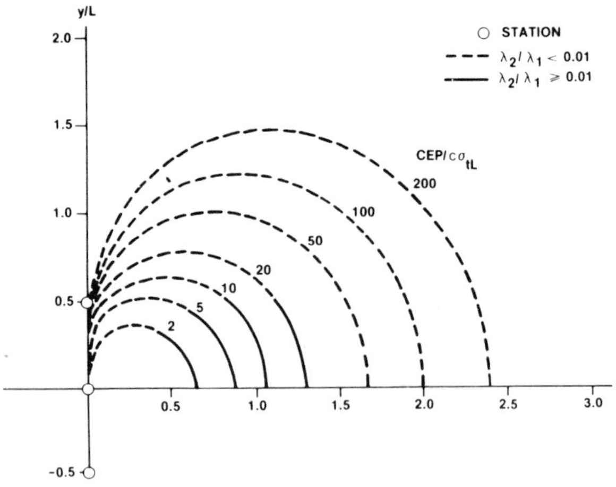

Fig. 8. Loci of constant $\mathrm{CEP}/c\sigma_{tL}$ for linear array of three stations with $n = 4$.

In Fig. 9, the stations form a nonlinear array with coordinates $(0, -L/2)$, $(-L/2, 0)$, and $0, L/2)$, respectively. The most significant features are the

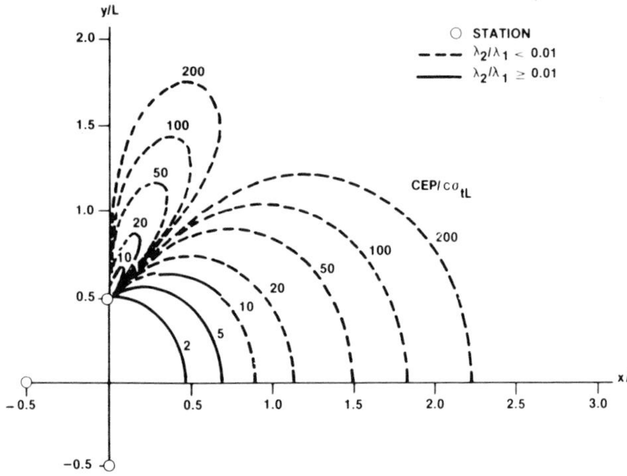

Fig. 9. Loci of constant $\mathrm{CEP}/c\sigma_{tL}$ for nonlinear array of three stations with $n = 4$.

singularities in the values of the CEP along the lines passing through two of the stations. Consequently, only a slight spatial nonlinearity is permissible if a broad field of view is required. However, other important factors in the choice of station positions are the needs to maintain line-of-sight paths from potential transmitter positions and to minimize the potential multipath interference.

In Figs. 7–9, the parts of the loci for which $\lambda_2 < 0.01\lambda_1$ are indicated by dotted lines. For these small values of λ_2/λ_1, the CEP is a questionable measure of performance of the passive location system. A more suitable measure may be the length of the major axis of the concentration ellipse,

$$L_e = 2\sqrt{\kappa\lambda_1}. \quad (111)$$

It follows from (73) that

$$\mathrm{CEP} \approx 0.563 L_e/2\sqrt{\kappa}, \qquad \lambda_2 < 0.01\lambda_1 \quad (112)$$

where κ is given by (60). Thus the dotted lines approximate the loci of constant values of $L_e/3.552\sqrt{\kappa c}\sigma_{tL}$.

VI. LOCATION USING BEARING MEASUREMENTS

The bearing measurements of passive direction-finding systems at two or more stations or points along an aircraft trajectory can be combined by a direction-finding location system to produce an estimate of transmitter position. The transmitted signal may be received at a station by line-of-sight propagation or after atmospheric reflection at a known altitude. A single bearing angle may be measured at each station of the location system. Alternatively, separate azimuth and elevation angle measurements, possibly made by orthogonal interferometers, can be used to determine transmitter position. In the absence of noise and interference, bearing lines from two or more stations will intersect to determine a unique location. In the presence of noise, more than two bearing lines will not intersect at a single point, as illustrated for a planar configuration in Fig. 10. Consequently, processing is

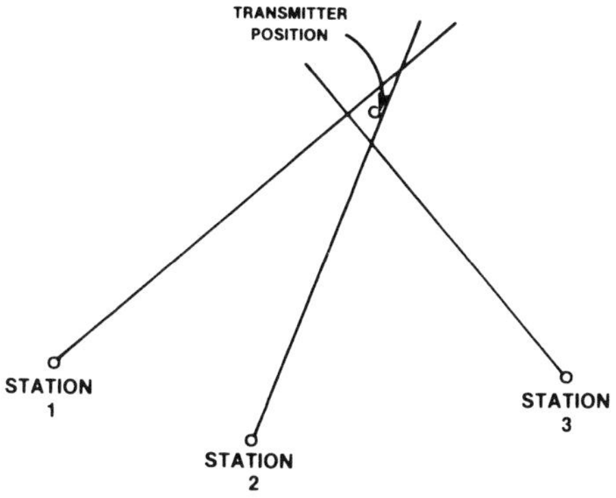

Fig. 10. Bearing lines from three direction-finding systems.

required to determine the optimal position estimate. Let θ_i denote the bearing angle measured at station i relative to a baseline in a three-dimensional coordinate system defined so that the x axis is parallel to the baseline, as shown in Fig. 11. If the coordinates of the station are

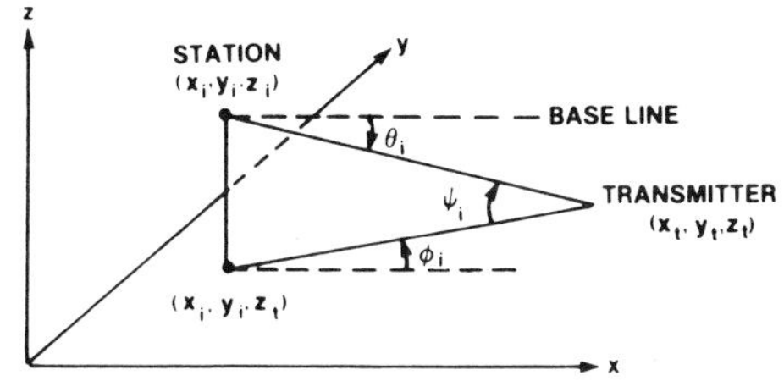

Fig. 11. Angle definitions for direction-finding systems.

(x_i, y_i, z_i) and the coordinates of the transmitter are (x_t, y_t, z_t), then in the absence of measurement errors, line-of-sight propagation implies that

$$\theta_i = \cos^{-1}\left[\frac{x_t - x_i}{\sqrt{(x_t - x_i)^2 + (y_t - y_i)^2 + (z_t - z_i)^2}}\right], \quad 0 \le \theta_i \le \pi. \quad (113)$$

In Fig. 11, the azimuth angle ϕ_i is defined in the plane passing through the transmitter and perpendicular to the z axis. It is positive in the counterclockwise direction relative to the positive x axis. If the elevation angle ψ_i of the station relative to the transmitter is known approximately or is estimated by a suitable means, such as a vertical interferometer, then ϕ_i may be calculated using the geometrical relation

$$\cos \theta_i = \cos \phi_i \cos \psi_i \quad (114)$$

which is easily derived from Fig. 11. If ψ_i is sufficiently small, the measured bearing is well approximated by the azimuth, which is defined by

$$\phi_i = \tan^{-1}\left(\frac{y_t - y_i}{x_t - x_i}\right). \quad (115)$$

In most applications, the transmitter is known to lie on the surface of the Earth or at a fixed altitude so that z_t is known and does not have to be estimated. Equation (115) is used in the estimation of the (x_t, y_t). The use of this equation is equivalent to the representation of the three-dimensional problem by a two-dimensional model. In the model, the transmitter and the stations are assumed to lie in the same plane so that the azimuths are identical to the bearings. If the transmitter and the stations actually lie on the Earth's surface, the model is an idealization that neglects the curvature of the surface. Two-dimensional position estimation using bearing information is often called triangulation.

We consider in detail the estimation of the two-dimensional column vector $\boldsymbol{R}$ having components x and y. Line-of-sight propagation is assumed. The measured bearing angle ϕ_i and the measurement error n_i satisfy

$$\phi_i = f_i(\boldsymbol{R}) + n_i, \qquad i = 1, 2, \ldots, N \quad (116)$$

where

$$f_i(\boldsymbol{R}) = \tan^{-1}\left(\frac{y - y_i}{x - x_i}\right), \qquad i = 1, 2, \ldots, N \quad (117)$$

and the station coordinates are x_i and y_i. In matrix form, we have

$$\boldsymbol{\phi} = \boldsymbol{f}(\boldsymbol{R}) + \boldsymbol{n}. \quad (118)$$

Let the column vector $\boldsymbol{R}_0$ with components x_0 and y_0 specify a reference point, which may be chosen to be in the middle of the polygon bounded by the measured bearing lines. Let ϕ_{0i} denote the bearing angle from station i to the reference point. Then

$$\sin \phi_{0i} = \frac{y_0 - y_i}{D_{0i}}; \qquad \cos \phi_{0i} = \frac{x_0 - x_i}{D_{0i}}, \qquad i = 1, 2, \ldots, N \quad (119)$$

where

$$D_{0i} = [(x_0 - x_i)^2 + (y_0 - y_i)^2]^{1/2}, \qquad i = 1, 2, \ldots, N. \quad (120)$$

From (7) with $\boldsymbol{x} = \boldsymbol{R}$ and $\boldsymbol{x}_0 = \boldsymbol{R}_0$, we obtain

$$\boldsymbol{G} = \begin{bmatrix} -(\sin\phi_{01})/D_{01} & (\cos\phi_{01})/D_{01} \\ \vdots & \vdots \\ -(\sin\phi_{0N})/D_{0N} & (\cos\phi_{0N})/D_{0N} \end{bmatrix}. \tag{121}$$

The least squares or maximum likelihood estimator is

$$\hat{\boldsymbol{R}} = \boldsymbol{R}_0 + (\boldsymbol{G}^T\boldsymbol{N}^{-1}\boldsymbol{G})^{-1}\,\boldsymbol{G}^T\boldsymbol{N}^{-1}\boldsymbol{\phi}_r \tag{122}$$

where $\boldsymbol{N}$ is the covariance matrix of the bearing measurement errors and

$$\boldsymbol{\phi}_r = \boldsymbol{\phi} - \boldsymbol{f}(\boldsymbol{R}_0). \tag{123}$$

The ith component of $\boldsymbol{\phi}_r$ is

$$\phi_{ri} = \phi_i - \phi_{0i} = \phi_i - \tan^{-1}\left(\frac{y_0 - y_i}{x_0 - x_i}\right), \quad i = 1, 2, \ldots, N \tag{124}$$

which is the bearing angle relative to the line between station i and the reference point, as depicted in Fig. 12.

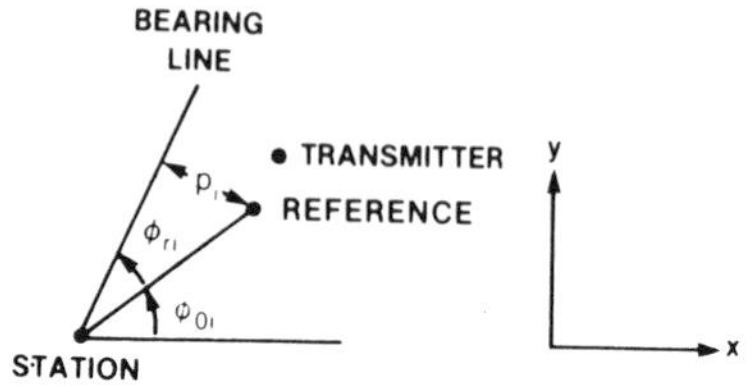

Fig. 12. Geometry of transmitter, reference point, and station.

If the bearing measurement errors are independent random variables with variances $\sigma_{\phi i}^2$, $i = 1, 2, \ldots, N$, then

$$\boldsymbol{N} = \begin{bmatrix} \sigma_{\phi 1}^2 & & 0 \\ & \ddots & \\ 0 & & \sigma_{\phi N}^2 \end{bmatrix}. \tag{125}$$

Direct calculation using (16), (121), and (125) establishes that the elements of the covariance matrix of $\hat{\boldsymbol{R}}$ are

$$\sigma_1^2 = E[(\hat{x} - x)^2] = \frac{\mu}{\mu\lambda - \nu^2} \tag{126}$$

$$\sigma_2^2 = E[(\hat{y} - y)^2] = \frac{\lambda}{\mu\lambda - \nu^2} \tag{127}$$

$$\sigma_{12} = E[(\hat{x} - x)(\hat{y} - y)] = \frac{\nu}{\mu\lambda - \nu^2} \tag{128}$$

where

$$\mu = \sum_{i=1}^{N} \frac{\cos^2\phi_{0i}}{D_{0i}^2\sigma_{\phi i}^2} \tag{129}$$

$$\lambda = \sum_{i=1}^{N} \frac{\sin^2\phi_{0i}}{D_{0i}^2\sigma_{\phi i}^2} \tag{130}$$

$$\nu = \sum_{i=1}^{N} \frac{\sin\phi_{0i}\cos\phi_{0i}}{D_{0i}^2\,\sigma_{\phi i}^2}. \tag{131}$$

It follows from (121), (122), and (125) that the components of the linearized least squares estimator are

$$\hat{x} = x_0 + \frac{1}{\mu\lambda - \nu^2}\sum_{i=1}^{N} \phi_{ri}\,\frac{(\nu\cos\phi_{0i} - \mu\sin\phi_{0i})}{D_{0i}\sigma_{\phi i}^2} \tag{132}$$

$$\hat{y} = y_0 + \frac{1}{\mu\lambda - \nu^2}\sum_{i=1}^{N} \phi_{ri}\,\frac{(\lambda\cos\phi_{0i} - \nu\sin\phi_{0i})}{D_{0i}\sigma_{\phi i}^2}. \tag{133}$$

Similarly, if the linearization error is negligible, the bias components are

$$b_1 = \frac{1}{\mu\lambda - \nu^2}\sum_{i=1}^{N} E[n_i]\,\frac{(\nu\cos\phi_{0i} - \mu\sin\phi_{0i})}{D_{0i}\sigma_{\phi i}^2} \tag{135}$$

$$b_2 = \frac{1}{\mu\lambda - \nu^2}\sum_{i=1}^{N} E[n_i]\,\frac{(\lambda\cos\phi_{0i} - \nu\sin\phi_{0i})}{D_{0i}\sigma_{\phi i}^2}. \tag{135}$$

The dependence of the estimator and bias on $\sigma_{\phi i}^2$, $i = 1, 2, \ldots, N$, is eliminated because of cancellation in (132) to (135) if these variances are all equal. This equality is a reasonable assumption if the receivers are identical and much closer to each other than to the transmitter.

Let p_i denote the shortest distance from the reference point to the measured bearing line at station i, as depicted in Fig. 12. Suppose that the reference point is close to the true transmitter position and that the measurement errors are small. Then

$$\phi_{ri} \approx p_i/D_{0i}, \qquad i = 1, 2, \ldots, N \tag{136}$$

$$\cos\phi_{0i} \approx \cos\phi_i; \qquad \sin\phi_{0i} \approx \sin\phi_i, \qquad i = 1, 2, \ldots, N. \tag{137}$$

Substituting (136) and (137) into (129) to (133), we obtain the components of an estimator that depends upon the measurements p_i and ϕ_i, $i = 1, 2, \ldots, N$. This estimator, called the Stansfield algorithm, was originally derived from heuristic arguments and the assumption of small bearing measurement errors [8]. If $\boldsymbol{R}_0$ is close to $\boldsymbol{R}$, then the linearized least squares estimator is preferable to the Stansfield algorithm, which produces a larger estimator bias unless the bearing errors are small. However, if the bearing errors are large, it may not be possible to choose $\boldsymbol{R}_0$ close to $\boldsymbol{R}$. In this case, it is not clear which estimator is preferable.

The mean-square ranging error associated with direction-finding systems is defined as the average variance of $D_{0i}\phi_{ri}$:

$$\sigma_d^2 = \frac{1}{N} \sum_{i=1}^{N} D_{0i}^2 \sigma_{\phi i}^2. \tag{138}$$

In analogy to (89), the GDOP associated with an unbiased estimator and a direction-finding location system is defined as

$$\text{GDOP} = \sqrt{\text{trace}[\boldsymbol{P}]}/\sigma_d. \tag{139}$$

If the geometry is such that the bearing variances are nearly equal, then the GDOP is only weakly dependent on them. From (74), it follows that

$$\text{CEP} \approx (0.75\sigma_d)\text{GDOP}. \tag{140}$$

The variance of a bearing estimator σ_ϕ^2 is due primarily to the thermal and environmental noise. Approximate expressions for σ_ϕ^2 are known for various direction-finding systems operating in white Gaussian noise [7]. In most cases, if E/N_0 is sufficiently large, σ_ϕ^2 can be expressed in the form

$$\sigma_\phi^2 \simeq \left(\frac{2E}{N_0} \beta_\phi^2\right)^{-1} \tag{141}$$

where β_ϕ^2 is a function of the system parameters other than E/N_0, and the variation of the signal energy with the distance to the transmitter can be determined from (95). For example, consider a planar configuration and a phase interferometer with its antennas pointing in the direction of the positive x axis. It can be shown that if the estimator bias is negligible, then [7]

$$\sigma_\phi^2 \geq \left(\frac{c}{2\pi f_0 d \cos\phi}\right)^2 \left(\frac{E}{N_0}\right)^{-1}, \qquad |\phi| < \frac{\pi}{2} \tag{142}$$

where f_0 is the carrier frequency of the received signal, d is the maximum separation between the interferometer antennas, and ϕ is the true bearing angle.

As a specific example, we consider identical stations that are symmetrically located with respect to the reference point so that

$$\phi_{0i} = -\phi_{0(N-i+1)}, \qquad i = 1, 2, \ldots, [N/2] \tag{143}$$

$$D_{0i}^2 \sigma_{\phi i}^2 = D_{0(N-i+1)}^2 \sigma_{\phi(N-i+1)}^2, \qquad i = 1, 2, \ldots, [N/2] \tag{144}$$

where $[x]$ denotes the largest integer in x. If N is odd, we further assume that

$$\phi_{0i} = 0, \qquad i = [N/2] + 1;\ N \text{ is odd}. \tag{145}$$

A possible configuration for $N = 5$ is illustrated in Fig. 13. This example is probably unrealistic for ground stations if $N \geq 4$, but might adequately represent location estimation by an aircraft that samples bearing data at evenly spaced points along its trajectory. Substitution of (143), (144), and (145) into (131) yields $\nu = 0$, which implies that $\sigma_{12} = 0$. We conclude that the symmetrical, but not necessarily linear, placement of the stations with respect to an accurately located reference leads to uncorrelated coordinate estimates. For an aircraft, we

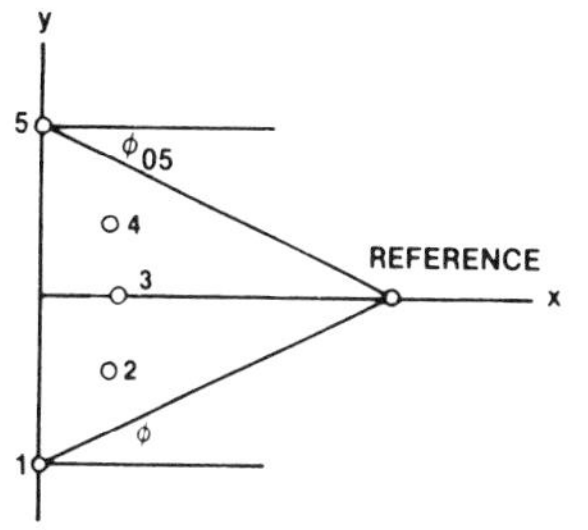

Fig. 13. Configuration of five symmetrically located stations.

may interpret σ_2^2 as the variance of the "cross-range" estimation error and σ_1^2 as the variance of the "down-range" estimation error.

If $N = 2$, then (126) to (133) give

$$\sigma_1^2 = \frac{D_{01}^2 \sigma_{\phi 1}^2}{2 \sin^2 \phi_{01}} \tag{146}$$

$$\sigma_2^2 = \frac{D_{01}^2 \sigma_{\phi 1}^2}{2 \cos^2 \phi_{01}} \tag{147}$$

$$\hat{x} = x_0 - \frac{D_{01}}{2 \sin \phi_{01}} (\phi_{r1} - \phi_{r2}) \tag{148}$$

$$\hat{y} = y_0 + \frac{D_{01}}{2 \cos \phi_{01}} (\phi_{r1} + \phi_{r2}). \tag{149}$$

If the reference point is located at the intersection of the two measured bearing lines, then $\phi_{r1} = \phi_{r2} = 0$. It follows that $(\hat{x}, \hat{y}) = (x_0, y_0)$, as expected. From (138), (139), (146), and (147), we obtain

$$\text{GDOP} = \frac{\sqrt{2}}{\sin 2\phi_{01}}. \tag{150}$$

The minimum value of the GDOP, equal to $\sqrt{2}$, is attained when $\phi_{01} = \pi/4$. Since $\sigma_{12} = 0$, (54), (55), and (73) give

$$\text{CEP} = 0.563 \max(\sigma_1, \sigma_2) + 0.614 \min(\sigma_1, \sigma_2). \tag{151}$$

If $N = 3$, the variance of $\hat{x}$ remains the same, but the variance of $\hat{y}$ becomes

$$\sigma_2^2 = \left(\frac{2\cos^2\phi_{01}}{D_{01}^2 \sigma_{\phi 1}^2} + \frac{1}{D_{02}^2 \sigma_{\phi 2}^2}\right)^{-1} \tag{152}$$

which shows that the extra station only improves the estimation of the y coordinate of the transmitter. As the transmitter range increases, ϕ_{01} decreases and thus σ_1^2/σ_2^2 increases.

If $\boldsymbol{n}$ in (118) has a Gaussian distribution, then (54), (55), (73), and (126)–(131) give the CEP in terms of the bearing angles and their variances. Assuming that the reference point coincides with the transmitter position so that $D_{0i} = D_i$ and ϕ_{0i} is equal to the bearing angle to the transmitter position, the locus of positions with a constant CEP can be determined numerically by using (119) and (120).

Consider a linear array of three stations with coordinates $(0, -L/2)$, $(0,0)$, and $(0,L/2)$. Each station has an interferometer with omnidirectional antennas pointing in the direction of the positive x axis. Let $\sigma_{\phi L}$ denote the value of $\sigma_{\phi i}$ when $D_{0i} = L$ and $\phi_{0i} = 0$. Assuming that $\sigma_{\phi L}$, n, and α are identical for all three stations and that the lower bound of inequality (142) is nearly achieved, (95) yields

$$\sigma_{\phi i}^2 \simeq \frac{\sigma_{\phi L}^2}{\cos^2 \phi_{0i}} \left(\frac{D_{0i}}{L}\right)^n \exp[\alpha(D_{0i} - L)],$$

$$|\phi_i| < \frac{\pi}{2};\ i = 1, 2, 3 \tag{153}$$

$$\sigma_{\phi L}^2 = \frac{c^2 N_0 L^n \exp(\alpha L)}{(2\pi f_0 d)^2 T K_E}. \tag{154}$$

Figs. 14 and 15 depict loci of constant values of CEP/$L\sigma_{\phi L}$ for $\alpha = 0$. The loci are similar in form to those for

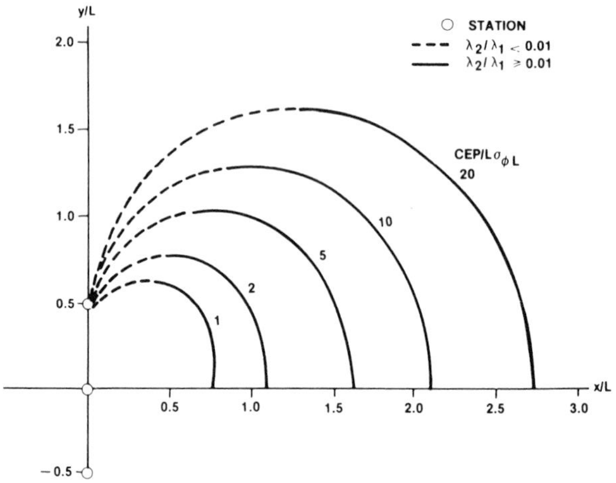

Fig. 14. Loci of constant CEP/$L\sigma_{\phi L}$ for linear array of three stations with $n = 2$.

hyperbolic location systems. From (112), it follows that the dotted lines approximate the loci of constant values of $L_e/3.552\sqrt{\kappa}\, L\sigma_{\phi L}$. At equal distances from the array, direction-finding location systems produce less eccentric

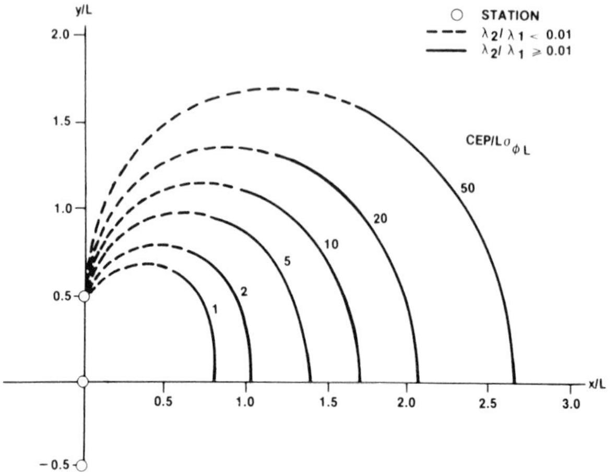

Fig. 15. Loci of constant CEP/$L\sigma_{\phi L}$ for linear array of three stations with $n = 4$.

concentration ellipses than similarly deployed hyperbolic location systems. This feature may be a significant factor in selecting the appropriate location systems for specific applications.

In Fig. 16, the stations form a nonlinear array with coordinates $(0,-L/2)$, $(-L/2,0)$, and $(0,L/2)$. A comparison with Fig. 9 indicates that the adverse effect of the nonlinear configuration is usually less for direction-finding systems than for hyperbolic systems.

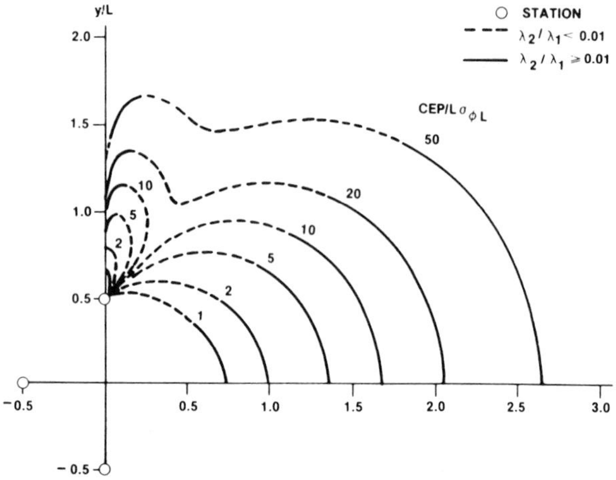

Fig. 16. Loci of constant CEP/$L\sigma_{\phi L}$ for nonlinear array of three stations with $n = 4$.

Fig. 17 plots the constant CEP/$L\sigma_{\phi L}$ loci for a linear array of five stations with coordinates $(0, -L/2)$, $(0, -L/4)$, $(0,0)$, $(0,l/4)$, and $(0,L/2)$. A comparison with Fig. 15 shows the CEP improvement from adding two stations while maintaining a constant baseline length equal to L. In general, the CEP is roughly inversely proportional to $\sqrt{N}$.

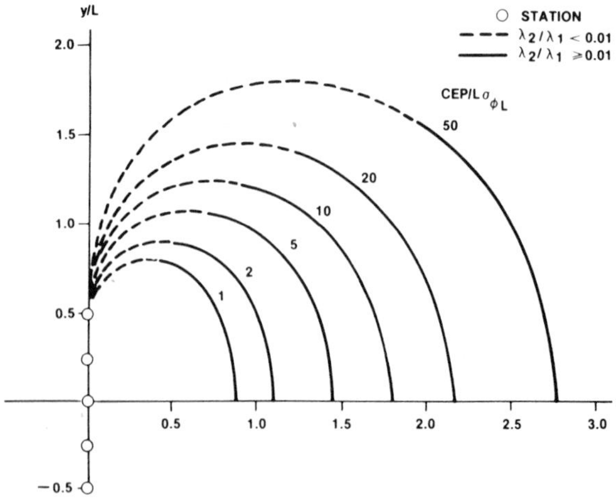

Fig. 17. Loci of constant CEP/$L\sigma_{\phi L}$ for linear array of five stations with $n = 4$.

For a two-dimensional transmitter location with three stations, a comparison of Figs. 14–16 with Figs. 7–9 indicates that for hyperbolic systems to offer a significant performance advantage over direction-finding location systems when $\alpha = 0$, it is necessary that

$$q c \sigma_{tL} < L \sigma_{\phi L} \tag{155}$$

where $q \approx 5$. Substituting (154) and (110) and assuming equal parameter values for the two systems, we obtain the criterion

$$\sqrt{2}\, \pi q f_0 d < L \beta_r . \tag{156}$$

Consequently, for the radar signal leading to (93), hyperbolic systems offer a potential advantage only if

$$T_p (\pi q f_0 d)^2 < B L^2 . \tag{157}$$

For the communications signal leading to (94), a significant advantage requires

$$\sqrt{6}\, q f_0 d < B L . \tag{158}$$

Inequalities (157) and (158) indicate that hyperbolic systems increase in desirability as the array length and signal bandwidth increase.

VII. OTHER LOCATION METHODS

When the receivers are moving, it may be possible to use the known receiver trajectories to enhance the accuracy of the transmitter location. For example, three bearing measurements and two turns by an aircraft can be used to greatly reduce the effect of strong unknown biases in the measurements [9].

Moving receivers can exploit the Doppler shift in several ways. In the absence of noise, the measured frequency at a receiver f_m is related to the transmitted frequency f_t by

$$f_m = f_t + f_t v_r / c = f_t + f_t v \cos \phi / c \tag{159}$$

where c is the signal velocity, v_r is the velocity of the receiver in the direction to the transmitter, v is the receiver velocity, and ϕ is the bearing angle to the transmitter relative to the velocity vector, as shown in Fig. 18.

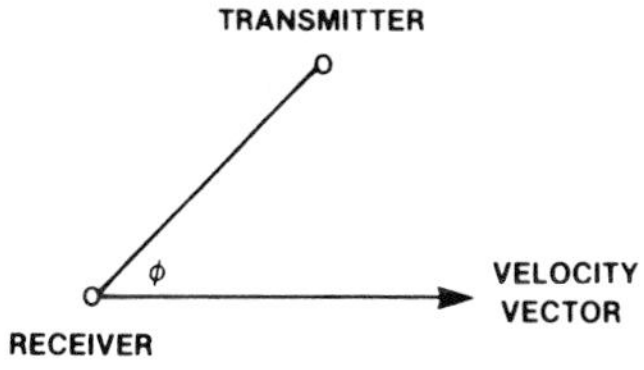

Fig. 18. Moving receiver.

Therefore, the bearing angle can be estimated if f_m is measured and f_t, v, and c are known. Bearing measurements from several receivers can be combined to obtain a transmitter location estimate, as is done in Section VI. Another approach, which may be less sensitive to inaccuracies in the assumed value of f_t, is to measure the Doppler difference, which is defined as

$$f_{m1} - f_{m2} = (f_t / c)(v_1 \cos \phi_1 - v_2 \cos \phi_2) \tag{160}$$

where the subscripts 1 and 2 refer to receivers 1 and 2. The differential Doppler is defined as the integral of $f_{m1} - f_{m2}$ over time. If f_t does not change too rapidly over the integration interval, the differential Doppler is

$$\int_{t_1}^{t_2} (f_{m1} - f_{m2})\, dt \simeq (f_{ta}/c) [D_1(t_2) - D_1(t_1) - D_2(t_2) + D_2(t_1)] \tag{161}$$

where f_{ta} is the average transmitted frequency and $D_i(t_j)$, $i, j = 1, 2$, is the distance of receiver i from the transmitter at time j. The right-hand sides of (160) and (161) can be expressed in terms of the transmitter coordinates. Thus, in the absence of noise, a Doppler difference or a differential Doppler measurement determines a surface on which the transmitter must lie. A location estimator can be derived in a manner analogous to the derivations of Sections V and VI. Because of the need for a precise estimate of f_t or f_{ta}, Doppler location systems appear to be most useful in the location of transmitters of narrowband signals.

Doppler, arrival-time, and bearing measurements at the same or different receivers can be combined in hybrid location systems. The combined measurements may allow a reduction in the number of receivers required for a given location accuracy and may facilitate the resolution of ambiguities.

To accommodate a moving transmitter, the observation interval can be decreased so that the transmitter is nearly stationary during the interval and points on the trajectory can be located. However, decreases in the observation interval eventually lead to unacceptably large estimation errors, and other methods must be adopted. If the trajectory can be described by a low-order polynomial in time and if a sufficient number of stations or measurements are available, it is possible to estimate the coefficients by expanding the dimension of the estimator $\hat{\mathbf{x}}$. Alternatively, if the differential equations of motion are known, Kalman filters can be used to track the transmitter movement [10, 11]. However, the implementation complexity of a passive location system with Kalman filters is usually considerably greater than that of a hyperbolic or direction-finding location system for stationary transmitters.

[1] Wegner, L.H. (1971)
On the accuracy analysis of airborne techniques for passively locating electromagnetic emitters.
Report R-722-PR, Rand Corp.; available from National Technical Information Service as ASTIA D.C. AD 729 767, 1971.

[2] Lee, H.B. (1975)
A novel procedure for assessing the accuracy of hyperbolic multilateration systems.
IEEE Trans. Aerosp. Electron. Syst., AES-11 (Jan. 1975), 2.

[3] Sage, A.P., and Melsa, J.L. (1971)
Estimation Theory with Applications to Communications and Control.
New York: McGraw-Hill, 1971.

[4] Butterly, P.J. (1972)
Position finding with empirical prior knowledge.
IEEE Trans. Aerosp. Electron. Syst., AES-8 (Mar. 1972), 142.

[5] Billingsley, P. (1979)
Probability and Measure.
New York: Wiley, 1979.

[6] Whalen, A.D. (1971)
Detection of Signals in Noise.
New York: Academic Press, 1971.

[7] Torrieri, D.J. (1981)
Principles of Military Communication Systems.
Dedham, Mass.: Artech House, 1981.

[8] Ancker, C.J. (1958)
Airborne direction finding—The theory of navigation errors.
IRE Trans. Aeronaut. Navig. Electron., ANE-5 (Dec. 1958), 199.

[9] Mangel, M. (1981)
Three bearing method for passive triangulation in systems with unknown deterministic biases.
IEEE Trans. Aerosp. Electron. Syst., AES-17 (Nov. 1981), 814.

[10] Gelb, A. (1974)
Applied Optimal Estimation.
Cambridge, Mass.: MIT Press 1974.

[11] Maybeck, P.S. (1979)
Stochastic Models, Estimation, and Control, vol. 1.
New York: Academic Press, 1979.

Estimating Uncertain Spatial Relationships in Robotics*

Randall Smith† **Matthew Self**‡ **Peter Cheeseman**§

SRI International
333 Ravenswood Avenue
Menlo Park, California 94025

In this paper, we describe a representation for spatial information, called the *stochastic map*, and associated procedures for building it, reading information from it, and revising it incrementally as new information is obtained. The map contains the estimates of relationships among objects in the map, and their uncertainties, given all the available information. The procedures provide a general solution to the problem of estimating uncertain relative spatial relationships. The estimates are probabilistic in nature, an advance over the previous, very conservative, worst-case approaches to the problem. Finally, the procedures are developed in the context of state-estimation and filtering theory, which provides a solid basis for numerous extensions.

1 Introduction

In many applications of robotics, such as industrial automation, and autonomous mobility, there is a need to represent and reason about spatial uncertainty. In the past, this need has been circumvented by special purpose methods such as precision engineering, very accurate sensors and the use of fixtures and calibration points. While these methods sometimes supply sufficient accuracy to avoid the need to represent uncertainty explicitly, they are usually costly. An alternative approach is to use multiple, overlapping, lower resolution sensors and to combine the spatial information (including the uncertainty) from all sources to obtain the best spatial estimate. This integrated information can often supply sufficient accuracy to avoid the need for the hard engineered approach.

In addition to lower hardware cost, the explicit estimation of uncertain spatial information makes it possible to decide *in advance* whether proposed operations are likely to

*The research reported in this paper was supported by the National Science Foundation under Grant ECS-8200615, the Air Force Office of Scientific Research under Contract F49620-84-K-0007, and by General Motors Research Laboratories.

†Currently at General Motors Research Laboratories
Warren, Michigan.

‡Currently at UC Berkeley.

§Currently at NASA Ames Research Center,
Moffett Field, California.

fail because of accumulated uncertainty, and whether proposed sensor information will be sufficient to reduce the uncertainty to tolerable limits. In situations utilizing inexpensive mobile robots, perhaps the *only* way to obtain sufficient accuracy is to combine the (uncertain) information from many sensors. However, a difficulty in combining uncertain spatial information arises because it often occurs in the form of uncertain *relative* information. This is particularly true where many different frames of reference are used, and the uncertain spatial information must be propagated between these frames. This paper presents a general solution to the problem of estimating uncertain spatial relationships, regardless of which frame the information is presented in, or in which frame the answer is required.

Previous methods for representing spatial uncertainty in typical robotic applications (e.g. [Taylor, 1976]) numerically computed min-max bounds on the errors. Brooks developed other methods for computing min-max bounds symbolically[Brooks, 1982]. These min-max approachs are very conservative compared to the probabilistic approach in this paper, because they combine many pieces of information, each with worst case bounds on the errors. More recently, a probabilistic representation of uncertainty was utilized by the HILARE mobile robot [Chatila, 1985] that is similar to the one presented here, except that it uses only a scalar representation of positional uncertainty instead of a multivariate representation of position and orientation. Smith and Cheeseman ([Smith, 1984], [Smith, 1985]), working on problems in off-line programming of industrial automation tasks, proposed operations that could reduce graphs of uncertain relationships (represented by multivariate probability distributions) to a single, best estimate of some relationship of interest. The current paper extends that work, but in the formal setting of estimation theory, and does not utilize graph transformations.

In summary, many important applications require a representation of spatial uncertainty. In addition, methods for combining uncertain spatial information and transforming such information from one frame to another are required. This paper presents a representation that makes explicit the uncertainty of each degree of freedom in the spatial relationships of interest. A method is given for combining uncertain information regardless of which frame it is presented in, and it allows the description of the spatial uncertainty of one frame relative to any other frame. The necessary procedures are presented in matrix form, suitable for efficient implementation. In particular, methods are given for incrementally building the best estimate "map" and its uncertainty as new pieces of uncertain spatial information are added.

2 The Stochastic Map

Our knowledge of the spatial relationships among objects is inherently uncertain. A man-made object does not match its geometric model *exactly* because of manufacturing tolerances. *Even if it did*, a sensor could not measure the geometric features, and thus locate the object *exactly*, because of measurement errors. And *even if it could*, a robot using the sensor cannot manipulate the object *exactly* as intended, because of hand positioning errors. These errors can be reduced to neglible limits for some tasks, by "pre-enginerring" the solution — structuring the working environment and using specially–suited high–precision equipment — but at great cost of time and expense. However, rather than treat spatial uncertainty as a side issue in geometrical reasoning, we believe it must be treated as an intrinsic part of spatial representations. In this paper, uncertain spatial relationships will be tied together in a representation called the *stochastic map.* It contains estimates of the spatial relationships, their uncertainties, and their inter-dependencies.

First, the map structure will be described, followed by methods for extracting information from it. Finally, a procedure will be given for building the map *incrementally*, as new spatial information is obtained. To illustrate the theory, we will present an example of a mobile robot acquiring knowledge about its location and the organization of its environment by making sensor observations at different times and in different places.

2.1 Representation

In order to formalize the above ideas, we will define the following terms. A *spatial relationship* will be represented by the vector of its *spatial variables*, $\mathbf{x}$. For example, the position and orientation of a mobile robot can be described by its coordinates, x and y, in a two dimensional cartesian reference frame and by its orientation, ϕ, given as a rotation about the z axis:

$$\mathbf{x} = \begin{bmatrix} x \\ y \\ \phi \end{bmatrix}.$$

An *uncertain* spatial relationship, moreover, can be represented by a *probability distribution* over its spatial variables — i.e., by a probability density function that assigns a probability to each particular combination of the spatial variables, $\mathbf{x}$:

$$P(\mathbf{x}) = f(\mathbf{x})d\mathbf{x}.$$

Such detailed knowledge of the probability distribution is usually unneccesary for making decisions, such as whether the robot will be able to complete a given task (e.g. passing through a doorway). Furthermore, most measuring devices provide only a nominal value of the measured relationship, and we can estimate the average error from the sensor specifications. For these reasons, we choose to model an uncertain spatial relationship by estimating the first two moments of its probability distribution—the *mean*, $\hat{\mathbf{x}}$ and the *covariance*, $\mathbf{C}(\mathbf{x})$, defined as:

$$\begin{aligned} \hat{\mathbf{x}} &\triangleq E(\mathbf{x}), \\ \tilde{\mathbf{x}} &\triangleq \mathbf{x} - \hat{\mathbf{x}}, \\ \mathbf{C}(\mathbf{x}) &\triangleq E(\tilde{\mathbf{x}}\tilde{\mathbf{x}}^T). \end{aligned} \tag{1}$$

where E is the expectation operator, and $\tilde{\mathbf{x}}$ is the deviation from the mean. For our mobile robot example, these are:

$$\hat{\mathbf{x}} = \begin{bmatrix} \hat{x} \\ \hat{y} \\ \hat{\phi} \end{bmatrix}, \qquad \mathbf{C}(\mathbf{x}) = \begin{bmatrix} \sigma_x^2 & \sigma_{xy} & \sigma_{x\phi} \\ \sigma_{xy} & \sigma_y^2 & \sigma_{y\phi} \\ \sigma_{x\phi} & \sigma_{y\phi} & \sigma_\phi^2 \end{bmatrix}.$$

The diagonal elements of the covariance matrix are just the variances of the spatial variables, while the off-diagonal elements are the covariances between the spatial variables. It is useful to think of the covariances in terms of their correlation coefficients, ρ_{ij}:

$$\rho_{ij} \triangleq \frac{\sigma_{ij}}{\sigma_i \sigma_j} = \frac{E(\tilde{x}_i \tilde{x}_j)}{\sqrt{E(\tilde{x}_i^2) E(\tilde{x}_j^2)}}, \qquad -1 \leq \rho_{ij} \leq 1.$$

Similarly, to model a system of n uncertain spatial relationships, we construct the vector of *all* the spatial variables, which we call the *system state vector*. As before, we will estimate the mean of the state vector, $\hat{\mathbf{x}}$, and the *system covariance matrix*, $\mathbf{C}(\mathbf{x})$:

$$\mathbf{x} = \begin{bmatrix} \mathbf{x}_1 \\ \mathbf{x}_2 \\ \vdots \\ \mathbf{x}_n \end{bmatrix}, \quad \hat{\mathbf{x}} = \begin{bmatrix} \hat{\mathbf{x}}_1 \\ \hat{\mathbf{x}}_2 \\ \vdots \\ \hat{\mathbf{x}}_n \end{bmatrix},$$

$$\mathbf{C}(\mathbf{x}) = \begin{bmatrix} \mathbf{C}(\mathbf{x}_1) & \mathbf{C}(\mathbf{x}_1, \mathbf{x}_2) & \cdots & \mathbf{C}(\mathbf{x}_1, \mathbf{x}_n) \\ \mathbf{C}(\mathbf{x}_2, \mathbf{x}_1) & \mathbf{C}(\mathbf{x}_2) & \cdots & \mathbf{C}(\mathbf{x}_2, \mathbf{x}_n) \\ \vdots & \vdots & \ddots & \vdots \\ \mathbf{C}(\mathbf{x}_n, \mathbf{x}_1) & \mathbf{C}(\mathbf{x}_n, \mathbf{x}_2) & \cdots & \mathbf{C}(\mathbf{x}_n) \end{bmatrix} \tag{2}$$

where:

$$\begin{aligned} \mathbf{C}(\mathbf{x}_i, \mathbf{x}_j) &\triangleq E(\tilde{\mathbf{x}}_i \tilde{\mathbf{x}}_j^T), \\ \mathbf{C}(\mathbf{x}_j, \mathbf{x}_i) &= \mathbf{C}(\mathbf{x}_i, \mathbf{x}_j)^T. \end{aligned} \tag{3}$$

Here, the $\mathbf{x}_i$'s are the vectors of the spatial variables of the individual uncertain spatial relationships, and the $\mathbf{C}(\mathbf{x}_i)$'s are the associated covariance matrices, as discussed earlier. The $\mathbf{C}(\mathbf{x}_i, \mathbf{x}_j)$'s are the cross-covariance matrices between the uncertain spatial relationships. These off–diagonal sub–matrices encode the dependencies between the estimates of the different spatial relationships, and provide the mechanism for updating all the relational estimates that depend on those that are changed.

In our example, each uncertain spatial relationship is of the same form, so $\mathbf{x}$ has $m = 3n$ elements, and we may write:

$$\mathbf{x}_i = \begin{bmatrix} x_i \\ y_i \\ \phi_i \end{bmatrix}, \quad \hat{\mathbf{x}}_i = \begin{bmatrix} \hat{x}_i \\ \hat{y}_i \\ \hat{\phi}_i \end{bmatrix}, \quad \mathbf{C}(\mathbf{x}_i, \mathbf{x}_j) = \begin{bmatrix} \sigma_{x_i x_j} & \sigma_{x_i y_j} & \sigma_{x_i \phi_j} \\ \sigma_{x_i y_j} & \sigma_{y_i y_j} & \sigma_{y_i \phi_j} \\ \sigma_{x_i \phi_j} & \sigma_{y_i \phi_j} & \sigma_{\phi_i \phi_j} \end{bmatrix}.$$

Thus our "map" consists of the current estimate of the mean of the system state vector, which gives the nominal locations of objects in the map with respect to the world reference frame, and the associated system covariance matrix, which gives the uncertainty of each point in the map and the inter-dependencies of these uncertainties.

2.2 Interpretation

For some decisions based on uncertain spatial relationships, we must assume a particular distribution that fits the estimated moments. For example, a robot might need to be able to calculate the probability that a certain object will be in its field of view, or the probability that it will succeed in passing through a doorway.

Given only the mean, $\mathbf{x}$, and covariance matrix, $\mathbf{C}(\mathbf{x})$, of a multivariate probability distribution, the principle of maximum entropy indicates that the distribution which assumes the least information is the normal distribution. Furthermore if the spatial relationship is calculated by combining many different pieces of information the central limit theorem indicates that the resulting distribution will tend to a normal distribution:

$$P(\mathbf{x}) = \frac{\exp\left[-\frac{1}{2}(\mathbf{x}-\hat{\mathbf{x}})^T\mathbf{C}^{-1}(\mathbf{x})(\mathbf{x}-\hat{\mathbf{x}})\right]}{\sqrt{(2\pi)^m|\mathbf{C}(\mathbf{x})|}}d\mathbf{x}. \tag{4}$$

We will graph uncertain spatial relationships by plotting contours of constant probability from a normal distribution with the given mean and covariance information. These contours are concentric ellipsoids (ellipses for two dimensions) whose parameters can be calculated from the covariance matrix, $\mathbf{C}(\mathbf{x}_i)$ [Nahi, 1976]. It is important to emphasize that we do not assume that the uncertain spatial relationships are described by normal distributions. We estimate the mean and variance of their distributions, and use the normal distribution only when we need to calculate specific probability contours.

In the figures in this paper, the plotted points show the *actual* locations of objects, which are known only by the simulator and displayed for our benefit. The robot's information is shown by the ellipses which are drawn centered on the estimated mean of the relationship and such that they enclose a 99.9% confidence region (about four standard deviations) for the relationships.

2.3 Example

Throughout this paper we will refer to a two dimensional example involving the navigation of a mobile robot with three degrees of freedom. In this example the robot performs the following sequence of actions:

- The robot senses object #1
- The robot moves.
- The robot senses an object (object #2) which it determines cannot be object #1.
- Trying again, the robot succeeds in sensing object #1, thus helping to localize itself, object #1, and object #2.

Figure 1 shows two examples of uncertain spatial relationships — the sensed location of object #1, and the end-point of a planned motion for the robot. The robot is initially sitting at a landmark which will be used as the world reference location. There is enough information in our stochastic map at this point for the robot to be able to decide how likely a collision with the object is, if the motion is made. In this case the probability is vanishingly small. The same figure shows how this spatial knowledge can be presented from the robot's new reference frame after its motion. As expected, the uncertainty in the location of object #1 becomes larger when it is compounded with the uncertainty in the robot's motion.

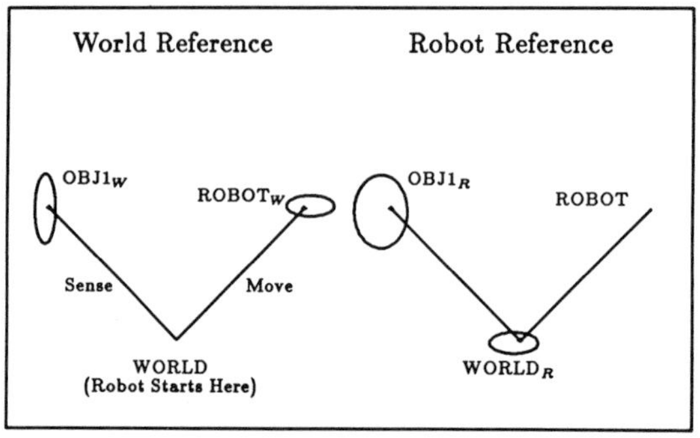

Figure 1: The Robot Senses Object 1 and Moves

From this new location, the robot senses object #2 (Figure 2). The robot is able to determine with the information in its stochastic map that this must be a new object and is not object #1 which it observed earlier.

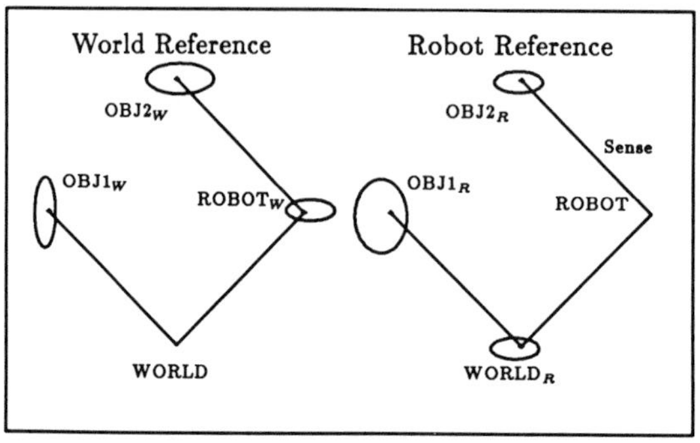

Figure 2: The Robot Senses Object 2

In Figure 3, the robot senses object #1 again. This new sensor measurement acts as a constraint and is incorporated into the map, reducing the uncertainty in the locations of the robot, object #1 *and* Object #2 (Figure 4).

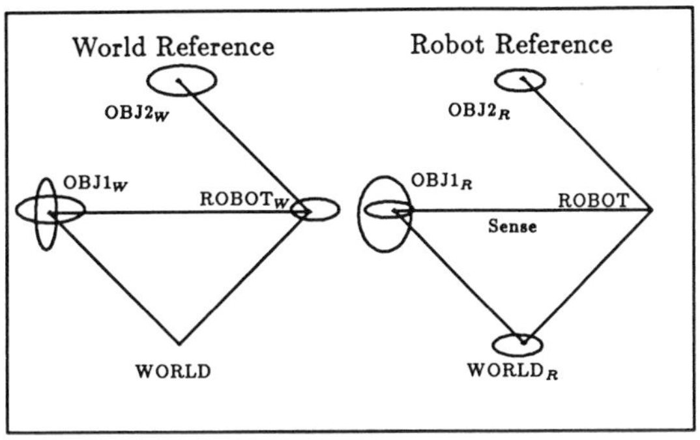

Figure 3: The Robot Senses Object 1 Again

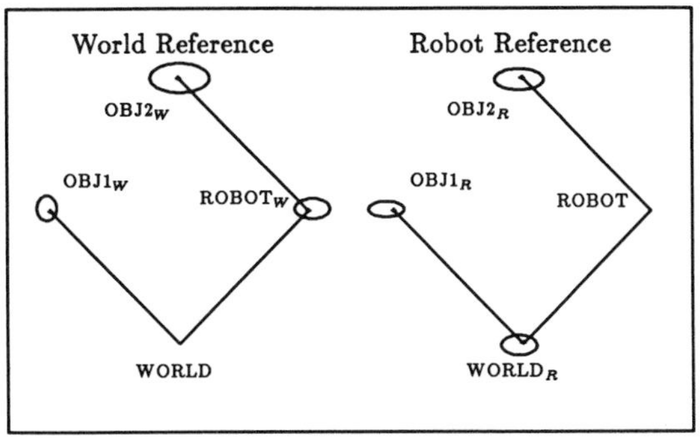

Figure 4: The Updated Estimates After Constraint

3 Reading the Map

3.1 Uncertain Relationships

Having seen how we can represent uncertain spatial relationships by estimates of the mean and covariance of the system state vector, we now discuss methods for estimating the first two moments of unknown multivariate probability distributions. See [Papoulis, 1965] for detailed justifications of the following topics.

3.1.1 Linear Relationships

The simplest case concerns relationships which are linear in the random varables, e.g.:

$$\mathbf{y} = \mathbf{M}\mathbf{x} + \mathbf{b},$$

where, $\mathbf{x}$ $(n \times 1)$ is a random vector, $\mathbf{M}$ $(r \times n)$ is the *non*-random coefficient matrix, $\mathbf{b}$ $(r \times 1)$ is a constant vector, and $\mathbf{y}$ $(r \times 1)$ is the resultant random vector. Using the definitions from (1), and the linearity of the expectation operator, E, one can easily verify that the mean of the relationship, $\hat{\mathbf{y}}$, is given by:

$$\hat{\mathbf{y}} = \mathbf{M}\hat{\mathbf{x}} + \mathbf{b}, \tag{5}$$

and the covariance matrix, $\mathbf{C}(\mathbf{y})$, is:

$$\mathbf{C}(\mathbf{y}) = \mathbf{M}\mathbf{C}(\mathbf{x})\mathbf{M}^T. \tag{6}$$

We will also need to be able to compute the covariance between $\mathbf{y}$ and some other relationship, $\mathbf{z}$, given the covariance between $\mathbf{x}$ and $\mathbf{z}$:

$$\begin{aligned} \mathbf{C}(\mathbf{y},\mathbf{z}) &= \mathbf{M}\mathbf{C}(\mathbf{x},\mathbf{z}), \\ \mathbf{C}(\mathbf{z},\mathbf{y}) &= \mathbf{C}(\mathbf{z},\mathbf{x})\mathbf{M}^T. \end{aligned} \tag{7}$$

The first two moments of the multivariate distribution of $\mathbf{y}$ are computed exactly, given correct moments for $\mathbf{x}$. Further, if $\mathbf{x}$ follows a normal distribution, then so does $\mathbf{y}$.

3.1.2 Non-Linear Relationships

The first two moments computed by the formulae below for non-linear relationships on random variables will be first-order estimates of the true values. To compute the actual values requires knowledge of the *complete* probability density function of the spatial variables, which will not generally be available in our applications. The usual approach is to approximate the non-linear function

$$\mathbf{y} = \mathbf{f}(\mathbf{x})$$

by a Taylor series expansion about the estimated mean, $\hat{\mathbf{x}}$, yielding:

$$\mathbf{y} = \mathbf{f}(\hat{\mathbf{x}}) + \mathbf{F}_{\mathbf{x}}\tilde{\mathbf{x}} + \cdots,$$

where $\mathbf{F}_{\mathbf{x}}$ is the matrix of partials, or Jacobian, of $\mathbf{f}$ evaluated at $\hat{\mathbf{x}}$:

$$\mathbf{F}_{\mathbf{x}} \triangleq \frac{\partial \mathbf{f}(\mathbf{x})}{\partial \mathbf{x}}(\hat{\mathbf{x}}) \triangleq \begin{bmatrix} \frac{\partial f_1}{\partial x_1} & \frac{\partial f_1}{\partial x_2} & \cdots & \frac{\partial f_1}{\partial x_n} \\ \frac{\partial f_2}{\partial x_1} & \frac{\partial f_2}{\partial x_2} & \cdots & \frac{\partial f_2}{\partial x_n} \\ \vdots & \vdots & \ddots & \vdots \\ \frac{\partial f_r}{\partial x_1} & \frac{\partial f_r}{\partial x_2} & \cdots & \frac{\partial f_r}{\partial x_n} \end{bmatrix}_{\mathbf{x}=\hat{\mathbf{x}}}.$$

This terminology is the extension of the f_x terminology from scalar calculus to vectors. The Jacobians are always understood to be evaluated at the estimated mean of the given variables.

Truncating the expansion for $\mathbf{y}$ after the linear term, and taking the expectation produces the linear estimate of the mean of $\mathbf{y}$:

$$\hat{\mathbf{y}} \approx \mathbf{f}(\hat{\mathbf{x}}). \tag{8}$$

Similarly, the first-order estimate of the covariances are:

$$\begin{aligned} \mathbf{C}(\mathbf{y}) &\approx \mathbf{F_x C(x) F_x^T}, \\ \mathbf{C}(\mathbf{y},\mathbf{z}) &\approx \mathbf{F_x C(x,z)}, \\ \mathbf{C}(\mathbf{z},\mathbf{y}) &\approx \mathbf{C(z,x) F_x^T}. \end{aligned} \tag{9}$$

Though not utilized in our application, the second order term may be included in the Taylor series expansion to improve the mean estimate:

$$\mathbf{y} = \mathbf{f}(\hat{\mathbf{x}}) + \mathbf{F_x}\tilde{\mathbf{x}} + \frac{1}{2}\mathbf{F_{xx}}(\tilde{\mathbf{x}}\tilde{\mathbf{x}}^T) + \cdots,$$

We denote the (3 dimensional) matrix of second partials of $\mathbf{f}$ by $\mathbf{F_{xx}}$. To avoid uneccesary complexity, we simply state that the ith element of the vector produced when $\mathbf{F_{xx}}$ is multiplied on the right by a matrix $\mathbf{A}$ is defined by:

$$(\mathbf{F_{xx}A})_i = trace\left[\left(\left.\frac{\partial^2 f_i}{\partial x_j \partial x_k}\right|_{\mathbf{x}=\hat{\mathbf{x}}}\right)\mathbf{A}\right].$$

The second-order estimate of the mean of $\mathbf{y}$ is then:

$$\hat{\mathbf{y}} \approx \mathbf{f}(\hat{\mathbf{x}}) + \frac{1}{2}\mathbf{F_{xx}C(x)},$$

and the second-order estimate of the covariance is:

$$\mathbf{C}(\mathbf{y}) \approx \mathbf{F_x C(x) F_x^T} - \frac{1}{4}\mathbf{F_{xx} C(x) C(x)^T F_{xx}^T}.$$

In the remainder of this paper we consider only first order estimates, and the symbol "$\approx$" should read as "linear estimate of."

3.2 Spatial Relationships

We now consider the actual spatial relationships which are most often encountered in robotics applications. We will develop our presentation about the three degree of freedom formulae, since they suit our examples concerning a mobile robot. Formulae for the three dimensional case with six degrees of freedom are given in Appendix A.

We would like to take a chain of relationships, starting at an initial coordinate frame, passing through several intermediate frames to a final frame, and estimate the resultant relationship between initial and final frames. Since frame relationships are directed, we will need the ability to invert the sense of some given relationships during the calculation. The formulae needed for calculating these estimates are given in the following sections.

3.2.1 Compounding

Given two spatial relationships, $\mathbf{x}_{ij}$ and $\mathbf{x}_{jk}$, as in Figure 1 (under Robot Reference), we wish to compute the resultant relationship $\mathbf{x}_{ik}$. The formula for computing $\mathbf{x}_{ik}$ from $\mathbf{x}_{ij}$ and $\mathbf{x}_{jk}$ is:

$$\mathbf{x}_{ik} \stackrel{\triangle}{=} \mathbf{x}_{ij} \oplus \mathbf{x}_{jk} = \begin{bmatrix} x_{jk}\cos\phi_{ij} - y_{jk}\sin\phi_{ij} + x_{ij} \\ x_{jk}\sin\phi_{ij} + y_{jk}\cos\phi_{ij} + y_{ij} \\ \phi_{ij} + \phi_{jk} \end{bmatrix}.$$

We call this operation compounding, and it is used to calculate the resultant relationship from two given relationships which are arranged head-to-tail. It would be used, for instance, to determine the location of a mobile robot after a sequence of relative motions. Remember that these transformations involve rotations, so compounding is not merely vector addition.

Utilizing (8), the first-order estimate of the mean of the compounding operation is:

$$\hat{\mathbf{x}}_{ik} \approx \hat{\mathbf{x}}_{ij} \oplus \hat{\mathbf{x}}_{jk}.$$

Also, from (9), the first-order estimate of the covariance is:

$$\mathbf{C}(\mathbf{x}_{ik}) \approx \mathbf{J}_{\oplus} \begin{bmatrix} \mathbf{C}(\mathbf{x}_{ij}) & \mathbf{C}(\mathbf{x}_{ij},\mathbf{x}_{jk}) \\ \mathbf{C}(\mathbf{x}_{jk},\mathbf{x}_{ij}) & \mathbf{C}(\mathbf{x}_{jk}) \end{bmatrix} \mathbf{J}_{\oplus}^{T}.$$

where the Jacobian of the compounding operation, $\mathbf{J}_{\oplus}$ is given by:

$$\mathbf{J}_{\oplus} \stackrel{\triangle}{=} \frac{\partial(\mathbf{x}_{ij} \oplus \mathbf{x}_{jk})}{\partial(\mathbf{x}_{ij},\mathbf{x}_{jk})} = \frac{\partial \mathbf{x}_{ik}}{\partial(\mathbf{x}_{ij},\mathbf{x}_{jk})} = \begin{bmatrix} 1 & 0 & -(y_{ik}-y_{ij}) & \cos\phi_{ij} & -\sin\phi_{ij} & 0 \\ 0 & 1 & (x_{ik}-x_{ij}) & \sin\phi_{ij} & \cos\phi_{ij} & 0 \\ 0 & 0 & 1 & 0 & 0 & 1 \end{bmatrix}.$$

Note how we have utilized the resultant relationship $\mathbf{x}_{ik}$ in expressing the Jacobian. This results in greater computational efficiency than expressing the Jacobian only in terms of the compounded relationships $\mathbf{x}_{ij}$ and $\mathbf{x}_{jk}$. We can always estimate the mean of an uncertain relationship and then use this result when evaluating the Jacobian to estimate the covariance of the relationship.

If the two relationships being compounded are independent ($\mathbf{C}(\mathbf{x}_{ij},\mathbf{x}_{jk}) = \mathbf{0}$), we can rewrite the first-order estimate of the covariance as:

$$\mathbf{C}(\mathbf{x}_{ik}) \approx \mathbf{J}_{1\oplus}\mathbf{C}(\mathbf{x}_{ij})\mathbf{J}_{1\oplus}^{T} + \mathbf{J}_{2\oplus}\mathbf{C}(\mathbf{x}_{jk})\mathbf{J}_{2\oplus}^{T}$$

where $\mathbf{J}_{1\oplus}$ and $\mathbf{J}_{2\oplus}$ are the left and right halves (3×3) of the compounding Jacobian (3×6):

$$\mathbf{J}_{\oplus} = \begin{bmatrix} \mathbf{J}_{1\oplus} & \mathbf{J}_{2\oplus} \end{bmatrix}.$$

3.2.2 The Inverse Relationship

Given a relationship $\mathbf{x}_{ij}$, the formula for the coordinates of the inverse relationship $\mathbf{x}_{ji}$, as a function of $\mathbf{x}_{ij}$ is:

$$\mathbf{x}_{ji} \stackrel{\triangle}{=} \ominus\mathbf{x}_{ij} \stackrel{\triangle}{=} \begin{bmatrix} -x_{ij}\cos\phi_{ij} - y_{ij}\sin\phi_{ij} \\ x_{ij}\sin\phi_{ij} - y_{ij}\cos\phi_{ij} \\ -\phi_{ij} \end{bmatrix}.$$

We call this the reverse relationship. Using (8) we get the first-order mean estimate:

$$\hat{\mathbf{x}}_{ji} \approx \ominus\hat{\mathbf{x}}_{ij}.$$

From (9) the first-order covariance estimate is:

$$\mathbf{C}(\mathbf{x}_{ji}) \approx \mathbf{J}_{\ominus}\mathbf{C}(\mathbf{x}_{ij})\mathbf{J}_{\ominus}^{T},$$

where the Jacobian for the reversal operation, $\mathbf{J}_{\ominus}$ is:

$$\mathbf{J}_{\ominus} \stackrel{\triangle}{=} \frac{\partial\mathbf{x}_{ji}}{\partial\mathbf{x}_{ij}} = \begin{bmatrix} -\cos\phi_{ij} & -\sin\phi_{ij} & y_{ji} \\ \sin\phi_{ij} & -\cos\phi_{ij} & -x_{ji} \\ 0 & 0 & -1 \end{bmatrix}.$$

Note that the uncertainty is not inverted, but rather expressed from the opposite (reverse) point of view.

3.2.3 Composite Relationships

We have shown how to compute the resultant of two relationships which are arranged head-to-tail, and also how to reverse a relationship. With these two operations we can calculate the resultant of any sequence of relationships. For example, the resultant of a chain of relationships arranged head-to-tail can be computed recursively by:

$$\begin{aligned} \mathbf{x}_{il} &= \mathbf{x}_{ij} \oplus \mathbf{x}_{jl} = \mathbf{x}_{ij} \oplus (\mathbf{x}_{jk} \oplus \mathbf{x}_{kl}) \\ &= \mathbf{x}_{ik} \oplus \mathbf{x}_{kl} = (\mathbf{x}_{ij} \oplus \mathbf{x}_{jk}) \oplus \mathbf{x}_{kl} \end{aligned}$$

Note, the compounding operation is associative, but not commutative. We have denoted the reversal operation by $\ominus$ so that by analogy to conventional $+$ and $-$ we may write:

$$\mathbf{x}_{ij} \ominus \mathbf{x}_{kj} \stackrel{\triangle}{=} \mathbf{x}_{ij} \oplus (\ominus\mathbf{x}_{kj}).$$

This is the head-to-head combination of two relationships. The tail-to-tail combination arises quite often (as in Figure 1, under World Reference), and is given by:

$$\mathbf{x}_{jk} = \ominus\mathbf{x}_{ij} \oplus \mathbf{x}_{ik}$$

To estimate the mean of a complex relationship, such as the tail-to-tail combination, we merely solve the estimate equations recursively:

$$\hat{\mathbf{x}}_{jk} = \hat{\mathbf{x}}_{ji} \oplus \hat{\mathbf{x}}_{ik} = \ominus\hat{\mathbf{x}}_{ij} \oplus \hat{\mathbf{x}}_{ik}.$$

The covariance can be estimated in a similar way:

$$\begin{aligned}\mathbf{C}(\mathbf{x}_{jk}) &\approx \mathbf{J}_{\oplus}\begin{bmatrix}\mathbf{C}(\mathbf{x}_{ji}) & \mathbf{C}(\mathbf{x}_{ji},\mathbf{x}_{ik})\\ \mathbf{C}(\mathbf{x}_{ik},\mathbf{x}_{ji}) & \mathbf{C}(\mathbf{x}_{ik})\end{bmatrix}\mathbf{J}_{\oplus}^{T}\\ &\approx \mathbf{J}_{\oplus}\begin{bmatrix}\mathbf{J}_{\ominus}\mathbf{C}(\mathbf{x}_{ij})\mathbf{J}_{\ominus}^{T} & \mathbf{J}_{\ominus}\mathbf{C}(\mathbf{x}_{ij},\mathbf{x}_{ik})\\ \mathbf{C}(\mathbf{x}_{ik},\mathbf{x}_{ij})\mathbf{J}_{\ominus}^{T} & \mathbf{C}(\mathbf{x}_{ik})\end{bmatrix}\mathbf{J}_{\oplus}^{T}.\end{aligned}$$

This method is easy to implement as a recursive algorithm. An equivalent method is to precompute the Jacobians of useful combinations of relationships such as the tail-to-tail combination by using the chain rule. Thus, the Jacobian of the tail-to-tail relationship, ${}_{\ominus}\mathbf{J}_{\oplus}$, is given by:

$${}_{\ominus}\mathbf{J}_{\oplus} \triangleq \frac{\partial \mathbf{x}_{jk}}{\partial(\mathbf{x}_{ij},\mathbf{x}_{ik})} = \frac{\partial \mathbf{x}_{jk}}{\partial(\mathbf{x}_{ji},\mathbf{x}_{ik})}\frac{\partial(\mathbf{x}_{ji},\mathbf{x}_{ik})}{\partial(\mathbf{x}_{ij},\mathbf{x}_{ik})} = \mathbf{J}_{\oplus}\begin{bmatrix}\mathbf{J}_{\ominus} & \mathbf{0}\\ \mathbf{0} & \mathbf{I}\end{bmatrix} = \begin{bmatrix}\mathbf{J}_{1\oplus}\mathbf{J}_{\ominus} & \mathbf{J}_{2\oplus}\end{bmatrix}.$$

Comparison will show that these two methods are symbolically equivalent, but the recursive method is easier to program, while pre-computing the composite Jacobians is more computationally efficient. Even greater computational efficiency can be achieved by making a change of variables such that the already computed mean estimate is used to evaluate the Jacobian, much as described earlier and in Appendix A.

It may appear that we are calculating first-order estimates of first-order estimates of ..., but actually this recursive procedure produces *precisely* the same result as calculating the first-order estimate of the composite relationship. This is in contrast to min-max methods which make conservative estimates at each step and thus produce *very* conservative estimates of a composite relationship.

If we now assume that the cross-covariance terms in the estimate of the covariance of the tail-to-tail relationship are zero, we get:

$$\mathbf{C}(\mathbf{x}_{jk}) \approx \mathbf{J}_{1\oplus}\mathbf{J}_{\ominus}\mathbf{C}(\mathbf{x}_{ij})\mathbf{J}_{\ominus}^{T}\mathbf{J}_{1\oplus}^{T} + \mathbf{J}_{2\oplus}\mathbf{C}(\mathbf{x}_{ik})\mathbf{J}_{2\oplus}^{T}$$

The Jacobians for six degree-of-freedom compounding and reversal relationships are given in Appendix A.

3.2.4 Extracting Relationships

We have now developed enough machinery to describe the procedure for estimating the relationships between objects which are in our map. The map contains, by definition, estimates of the locations of objects with respect to the world frame; these relations can be extracted directly. Other relationships are implicit, and must be extracted, using methods developed in the previous sections. For any general spatial relationship among world locations we can write:

$$\mathbf{y} = \mathbf{g}(\mathbf{x}).$$

The estimated mean and covariance of the relationship are given by:

$$\begin{aligned}\hat{\mathbf{y}} &\approx \mathbf{g}(\hat{\mathbf{x}}),\\ \mathbf{C}(\mathbf{y}) &\approx \mathbf{G}_{\mathbf{x}}\mathbf{C}(\mathbf{x})\mathbf{G}_{\mathbf{x}}^{T}.\end{aligned}$$

In our mobile robot example we will need to be able to estimate the relative location of one object with respect to the coordinate frame of another object in our map. In this case, we would simply substitute the tail-to-tail operation previously discussed for $\mathbf{g}()$,

$$\mathbf{y} = \mathbf{x}_{ij} = \ominus\mathbf{x}_i \oplus \mathbf{x}_j.$$

4 Building the Map

Our map represents uncertain spatial relationships among objects referenced to a common world frame. Entries in the map may change for two reasons:

- An object moves.
- New spatial information is obtained.

To change the map, we must change the two components that define it — the (mean) estimate of the system state vector, $\hat{\mathbf{x}}$, and the estimate of the system variance matrix, $\mathbf{C}(\mathbf{x})$. Figure 5 shows the changes in the system due to moving objects, or the addition of new spatial information (from sensing).

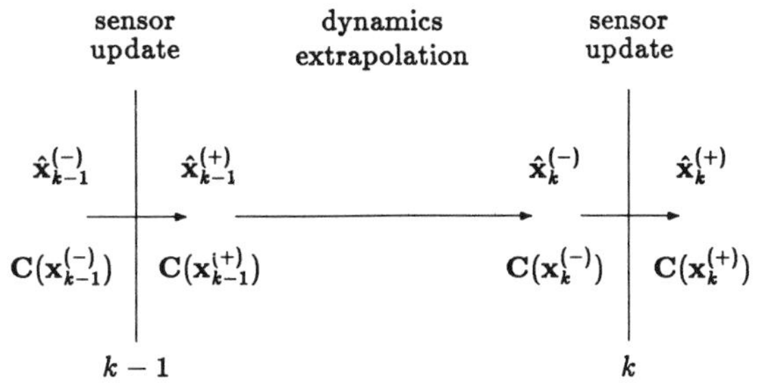

Figure 5: The Changing Map

We will assume that new spatial information is obtained at discrete moments, marked by states k. The update of the estimates at state k, based on new information, is considered to be instantaneous. The estimates, at state k, *prior* to the integration of the new information are denoted by $\hat{\mathbf{x}}_k^{(-)}$ and $\mathbf{C}(\mathbf{x}_k^{(-)})$, and *after* the integration by $\hat{\mathbf{x}}_k^{(+)}$ and $\mathbf{C}(\mathbf{x}_k^{(+)})$.

In the interval between states the system may be changing dynamically — for instance, the robot may be moving. When an object moves, we must define a process to extrapolate the estimate of the state vector and uncertainty at state $k-1$, to state k to reflect the changing relationships.

4.1 Moving Objects

Before describing how the map changes as the mobile robot moves, we will present the general case, which treats any processes that change the state of the system.

The *system dynamics model*, or process model, describes how components of the system state vector change (as a function of time in a continuous system, or by discrete transitions). Between state $k-1$ and k, no measurements of external objects are made. The new state is determined only by the process model, $\mathbf{f}$, as a function of the old state, and any control variables applied in the process (such as relative motion commands sent to our mobile robot). The process model is thus:

$$\mathbf{x}_k^{(-)} = \mathbf{f}\left(\mathbf{x}_{k-1}^{(+)}, \mathbf{y}_{k-1}\right), \tag{10}$$

where $\mathbf{y}$ is a vector comprised of control variables, $\mathbf{u}$, corrupted by mean-zero process noise, $\mathbf{w}$, with covariance $\mathbf{C}(\mathbf{w})$. That is, $\mathbf{y}$ is a noisy control input to the process, given by:

$$\mathbf{y} = \mathbf{u} + \mathbf{w}. \tag{11}$$

$$\hat{\mathbf{y}} = \mathbf{u}, \qquad \mathbf{C}(\mathbf{y}) = \mathbf{C}(\mathbf{w}).$$

Given the estimates of the state vector and variance matrix at state $k-1$, the estimates are extrapolated to state k by:

$$\hat{\mathbf{x}}_k^{(-)} \approx \mathbf{f}\left(\hat{\mathbf{x}}_{k-1}^{(+)}, \hat{\mathbf{y}}_{k-1}\right), \tag{12}$$

$$\mathbf{C}(\mathbf{x}_k^{(-)}) \approx \mathbf{F}_{(\mathbf{x},\mathbf{y})} \begin{bmatrix} \mathbf{C}(\mathbf{x}_{k-1}^{(+)}) & \mathbf{C}(\mathbf{x}_{k-1}^{(+)}, \mathbf{y}_{k-1}) \\ \mathbf{C}(\mathbf{y}_{k-1}, \mathbf{x}_{k-1}^{(+)}) & \mathbf{C}(\mathbf{y}_{k-1}) \end{bmatrix} \mathbf{F}_{(\mathbf{x},\mathbf{y})}^T.$$

where,

$$\mathbf{F}_{(\mathbf{x},\mathbf{y})} = \begin{bmatrix} \mathbf{F}_\mathbf{x} & \mathbf{F}_\mathbf{y} \end{bmatrix} \triangleq \frac{\partial \mathbf{f}(\mathbf{x},\mathbf{y})}{\partial(\mathbf{x},\mathbf{y})} \left(\hat{\mathbf{x}}_{k-1}^{(+)}, \hat{\mathbf{y}}_{k-1}\right)$$

If the process noise is uncorrelated with the state, then the off-diagonal sub-matrices in the matrix above are $\mathbf{0}$ and the covariance estimate simplifies to:

$$\mathbf{C}(\mathbf{x}_k^{(-)}) \approx \mathbf{F}_\mathbf{x}\mathbf{C}(\mathbf{x}_{k-1}^{(+)})\mathbf{F}_\mathbf{x}^T + \mathbf{F}_\mathbf{y}\mathbf{C}(\mathbf{y}_{k-1})\mathbf{F}_\mathbf{y}^T.$$

The new state estimates become the current estimates to be extrapolated to the next state, and so on.

In our example, only the robot moves, so the process model need only describe its motion. A continuous dynamics model can be developed given *a particular robot*, and the above equations can be reformulated as functions of time (see [Gelb, 1984]). However, if the robot only makes sensor observations at discrete times, then the discrete motion approximation is quite adequate. When the robot moves, it changes its relationship, $\mathbf{x}_R$, with the world. The robot makes an uncertain relative motion, $\mathbf{y}_R = \mathbf{u}_R + \mathbf{w}_R$, to reach a final world location $\mathbf{x}'_R$. Thus,

$$\mathbf{x}'_R = \mathbf{x}_R \oplus \mathbf{y}_R.$$

Only a small portion of the map needs to be changed due to the change in the robot's location from state to state — specifically, the Rth element of the estimated mean of the state vector, and the Rth row and column of the estimated variance matrix. Thus, $\hat{\mathbf{x}}_{k-1}^{(+)}$ becomes $\hat{\mathbf{x}}_k^{(-)}$ and $\mathbf{C}(\mathbf{x}_{k-1}^{(+)})$ becomes $\mathbf{C}(\mathbf{x}_k^{(-)})$, as shown below:

$$\hat{\mathbf{x}}_{k-1}^{(+)} = \begin{bmatrix} \\ \hline \hat{\mathbf{x}}_R \\ \hline \\ \\ \end{bmatrix}, \quad \hat{\mathbf{x}}_k^{(-)} = \begin{bmatrix} \\ \hline \hat{\mathbf{x}}'_R \\ \hline \\ \\ \end{bmatrix}, \quad \mathbf{C}(\mathbf{x}_k^{(-)}) = \left[\begin{array}{c|c|c} & \mathbf{B}'^T & \\ \hline \mathbf{B}' & \mathbf{A}' & \\ \hline & & \\ & & \end{array}\right]$$

where

$$\hat{\mathbf{x}}'_R \approx \hat{\mathbf{x}}_R \oplus \hat{\mathbf{y}}_R,$$

$$\mathbf{A}' = \mathbf{C}(\mathbf{x}'_R) \approx \mathbf{J}_{1\oplus}\mathbf{C}(\mathbf{x}_R)\mathbf{J}_{1\oplus}^T + \mathbf{J}_{2\oplus}\mathbf{C}(\mathbf{y}_R)\mathbf{J}_{2\oplus}^T,$$

$$\mathbf{B}'_i = \mathbf{C}(\mathbf{x}'_R, \mathbf{x}_i) \approx \mathbf{J}_{1\oplus}\mathbf{C}(\mathbf{x}_R, \mathbf{x}_i).$$

$\mathbf{A}'$ is the covariance matrix representing the uncertainty in the new location of the robot. $\mathbf{B}'$ is a row in the system variance matrix. The ith element is a sub-matrix — the cross-covariance of the robot's estimated location and the estimated location of the ith object, as given above. If the estimates of the two locations were not dependent, then that sub-matrix was, and remains $\mathbf{0}$. The newly estimated cross-covariance matrices are transposed, and written into the Rth column of the system variance matrix, marked by $\mathbf{B}'^T$.

4.2 New Spatial Information

The second process which changes the map is the update that occurs when new information about the system state is incorporated. New spatial information might be given, determined by sensor measurements, or even deduced as the consequence of applying a geometrical constraint. For example, placing a box on a table reduces the degrees of freedom of the box and eliminates the uncertainties in the lost degrees of freedom (with respect to the table coordinate frame). In our example, state information is obtained as prior knowledge, or through measurement.

There are two cases which arise when adding new spatial information about objects to our map:

- I: A new object is added to the map,
- II: A (stochastic) constraint is added between objects already in the map.

We will consider each of these cases in turn.

4.2.1 Case I: Adding New Objects

When a new object is added to the map, a new entry must be made in the system state vector to describe the object's *world* location. A new row and column are also added to the system variance matrix to describe the uncertainty in the object's estimated location, and the inter-dependencies of this estimate with estimated locations of other objects. The expanded system is:

$$\hat{\mathbf{x}}^{(+)} = \left[\begin{array}{c} \hat{\mathbf{x}}^{(-)} \\ \hline \hat{\mathbf{x}}_{n+1} \end{array}\right], \quad \mathbf{C}(\mathbf{x}^{(+)}) = \left[\begin{array}{c|c} \mathbf{C}(\mathbf{x}^{(-)}) & \mathbf{B}^T \\ \hline \mathbf{B} & \mathbf{A} \end{array}\right],$$

where $\hat{\mathbf{x}}_{n+1}$, $\mathbf{A}$, and $\mathbf{B}$ will be defined below.

We divide Case I into two sub-cases: I-a, the estimate of the new object's location is *independent* of the estimates of other object locations described in the map; or I-b, it is *dependent* on them.

Case I-a occurs when the estimated location of the object is given directly in world coordinates — i.e., $\hat{\mathbf{x}}_{new}$ and $\mathbf{C}(\mathbf{x}_{new})$ — perhaps as prior information. Since the estimate is independent of other location estimates:

$$\begin{aligned} \mathbf{x}_{n+1} &= \mathbf{x}_{new}, \\ \hat{\mathbf{x}}_{n+1} &= \hat{\mathbf{x}}_{new}, \\ \mathbf{A} = \mathbf{C}(\mathbf{x}_{n+1}) &= \mathbf{C}(\mathbf{x}_{new}), \\ \mathbf{B}_i = \mathbf{C}(\mathbf{x}_{n+1}, \mathbf{x}_i) &= \mathbf{C}(\mathbf{x}_{new}, \mathbf{x}_i) = \mathbf{0}. \end{aligned} \tag{13}$$

where $\mathbf{A}$ is a covariance matrix, and $\mathbf{B}$ is a row of cross-covariance matrices, as before. $\mathbf{B}$ is identically $\mathbf{0}$, since the new estimate is independent of the previous estimates, by definition.

Case I-b occurs when the *world* location of the new object is determined as a function, $\mathbf{g}$, of its spatial relation, $\mathbf{z}$, to other object locations estimated in the map. The relation might be measured or given as prior information. For example, the robot measures the location of a new object relative to itself. Clearly, the uncertainty in the object's *world* location is correlated with the uncertainty in the robot's (world) location. For Case I-b:

$$\begin{aligned} \mathbf{x}_{n+1} &= \mathbf{g}(\mathbf{x}, \mathbf{z}), \\ \hat{\mathbf{x}}_{n+1} &= \mathbf{g}(\hat{\mathbf{x}}, \hat{\mathbf{z}}), \\ \mathbf{A} = \mathbf{C}(\mathbf{x}_{n+1}) &= \mathbf{G_x}\mathbf{C}(\mathbf{x})\mathbf{G}_\mathbf{x}^T + \mathbf{G_y}\mathbf{C}(\mathbf{z})\mathbf{G_y}, \\ \mathbf{B}_i &= \mathbf{C}(\mathbf{x}_{n+1}, \mathbf{x}_i), \\ \mathbf{B} &= \mathbf{G_x}\mathbf{C}(\mathbf{x}). \end{aligned} \tag{14}$$

We see that Case I-a is the special case of Case I-b, where estimates of the world locations of new objects are independent of the old state estimates and are given exactly by the measured information. That is, when:

$$\mathbf{g}(\mathbf{x}, \mathbf{z}) = \mathbf{z}.$$

4.2.2 Case II: Adding Constraints

When new information is obtained relating objects *already in the map*, the system state vector and variance matrix do not increase in size; i.e., no new elements are introduced. However, the old elements are *constrained* by the new relation, and their values will be changed. Constraints can arise in a number of ways:

- A robot measures the relationship of a *known* landmark to itself (i.e., estimates of the world locations of robot and landmark already exist).
- A geometric relationship, such as colinearity, coplanarity, etc., is given for some set of the object location variables.

In the first example the constraint is noisy (because of an imperfect measurement). In the second example, the constraint could be absolute, but could also be given with a tolerance. The two cases are mathematically similar, in that they have to do with uncertain relationships on a number of variables — either measured, or hypothesized. A "rectangularity" constraint is discussed later in the example.

When a constraint is introduced, there are two estimates of the geometric relationship in question — our current best estimate of the relation, which can be extracted from the map, and the new information. The two estimates can be compared (in the same reference frame), and together should allow some improved estimate to be formed (as by averaging, for instance).

For each sensor, we have a *sensor model* that describes how the sensor maps the spatial variables in the state vector into sensor variables. Generally, the measurement, $\mathbf{z}$, is described as a function, $\mathbf{h}$, of the state vector, corrupted by mean-zero, additive noise $\mathbf{v}$. The covariance of the noise, $\mathbf{C}(\mathbf{v})$, is given as part of the model.

$$\mathbf{z} = \mathbf{h}(\mathbf{x}) + \mathbf{v}. \tag{15}$$

The *conditional* sensor value, given the state, and the *conditional covariance* are easily estimated from (15) as:

$$\hat{\mathbf{z}} \approx \mathbf{h}(\hat{\mathbf{x}}).$$

$$\mathbf{C}(\mathbf{z}) \approx \mathbf{H_x}\mathbf{C}(\mathbf{x})\mathbf{H_x}^T + \mathbf{C}(\mathbf{v}),$$

where:

$$\mathbf{H_x} \triangleq \frac{\partial \mathbf{h}_k(\mathbf{x})}{\partial \mathbf{x}} \left(\hat{\mathbf{x}}_k^{(-)}\right)$$

The formulae describe what values we *expect* from the sensor under the circumstances, and the likely variation; it is our current best estimate of the relationship to be measured.

The actual sensor values returned are usually assumed to be conditionally independent of the state, meaning that the noise is assumed to be independent in each measurement, even when measuring the same relation with the same sensor. The actual sensor values, corrupted by the noise, are the second estimate of the relationship.

For simplicity, in our example we assume that the sensor measures the relative location of the observed object in Cartesian coordinates. Thus the sensor function becomes the tail-to-tail relation of the location of the sensor and the sensed object, described in Section 3.2.3. (Formally, the sensor function is a function of all the variables in the state vector, but the unused variables are not shown below):

$$\mathbf{z} = \mathbf{x}_{ij} = \ominus \mathbf{x}_i \oplus \mathbf{x}_j.$$

$$\hat{\mathbf{z}} = \hat{\mathbf{x}}_{ij} = \ominus \hat{\mathbf{x}}_i \oplus \hat{\mathbf{x}}_j.$$

$$\mathbf{C}(\mathbf{z}) = {}_{\ominus}\mathbf{J}_{\oplus} \begin{bmatrix} \mathbf{C}(\mathbf{x}_i) & \mathbf{C}(\mathbf{x}_i, \mathbf{x}_j) \\ \mathbf{C}(\mathbf{x}_j, \mathbf{x}_i) & \mathbf{C}(\mathbf{x}_j) \end{bmatrix} {}_{\ominus}\mathbf{J}_{\oplus}^T + \mathbf{C}(\mathbf{v}).$$

Given the sensor model, the conditional estimates of the sensor values and their uncertainties, and an actual sensor measurement, we can update the state estimate using the Kalman Filter equations [Gelb, 1984] given below, and described in the next section:

$$\hat{\mathbf{x}}_k^{(+)} = \hat{\mathbf{x}}_k^{(-)} + \mathbf{K}_k \left[\mathbf{z}_k - \mathbf{h}_k(\hat{\mathbf{x}}_k^{(-)}) \right],$$

$$\mathbf{C}(\mathbf{x}_k^{(+)}) = \mathbf{C}(\mathbf{x}_k^{(-)}) - \mathbf{K}_k \mathbf{H}_{\mathbf{x}} \mathbf{C}(\mathbf{x}_k^{(-)}), \tag{16}$$

$$\mathbf{K}_k = \mathbf{C}(\mathbf{x}_k^{(-)}) \mathbf{H}_{\mathbf{x}}^T \left[\mathbf{H}_{\mathbf{x}} \mathbf{C}(\mathbf{x}_k^{(-)}) \mathbf{H}_{\mathbf{x}}^T + \mathbf{C}(\mathbf{v})_k \right]^{-1}.$$

4.2.3 Kalman Filter

The updated estimate is a weighted average of the two estimates, where the weighting factor (computed in the weight matrix **K**) is proportional to the prior covariance in the state estimate, and inversely proportional to the conditional covariance of the measurement. Thus, if the measurement covariance is large, compared to the state covariance, then $\mathbf{K} \to \mathbf{0}$, and the measurement has little impact in revising the state estimate. Conversely, when the prior state covariance is large compared to the noise covariance, then $\mathbf{K} \to \mathbf{I}$, and nearly the entire difference between the measurement and its expected value is used in updating the state.

The Kalman Filter generally contains a system dynamics model defined less generally than presented in (10); in the standard filter equations the process noise is additive:

$$\mathbf{x}_k^{(-)} = \mathbf{f}\left(\mathbf{x}_{k-1}^{(+)}, \mathbf{u}_{k-1}\right) + \mathbf{w}_{k-1} \tag{17}$$

in that case $\mathbf{F_y}$ of (10) is the identity matrix, and the estimated mean and covariance take the form:

$$\hat{\mathbf{x}}_k^{(-)} \approx \mathbf{f}\left(\hat{\mathbf{x}}_{k-1}^{(+)}, \mathbf{u}_{k-1}\right), \tag{18}$$

$$\mathbf{C}(\mathbf{x}_k^{(-)}) \approx \mathbf{F_x} \mathbf{C}(\mathbf{x}_{k-1}^{(+)}) \mathbf{F_x}^T + \mathbf{C}(\mathbf{w}_{k-1}).$$

If the functions $\mathbf{f}$ in (17) and $\mathbf{h}$ in (15) are *linear* in the state vector variables, then the partial derivative matrices $\mathbf{F}$ and $\mathbf{H}$ are simply constants, and the update formulae (16) with (17), (15), and (18), represent the Kalman Filter [Gelb, 1984].

If, in addition, the noise variables are drawn from normal distributions, then the Kalman Filter produces the *optimal minimum-variance Bayesian estimate*, which is equal to the mean of the *a posteriori conditional density function* of $\mathbf{x}$, given the prior statistics of $\mathbf{x}$, and the statistics of the measurement $\mathbf{z}$. No non-linear estimator can produce estimates with smaller mean-square errors. If the noise does not have a normal distribution, then the Kalman Filter is not optimal, but produces the optimal *linear* estimate.

If the functions $\mathbf{f}$ and $\mathbf{h}$ are *non-linear* in the state variables, then $\mathbf{F}$ and $\mathbf{H}$ will have to be evaluated (they are not constant matrices). The given formulae then represent the Extended Kalman Filter, a sub-optimal non-linear estimator. It is one of the most widely used non-linear estimators because of its similarity to the optimal linear filter, its simplicity of implementation, and its ability to provide accurate estimates in practice. The error in the estimation due to the non-linearities in $\mathbf{h}$ can be greatly reduced by iteration, using the Iterated Extended Kalman Filter equations [Gelb, 1984]:

$$\hat{\mathbf{x}}_{k,i+1}^{(+)} = \hat{\mathbf{x}}_k^{(-)} + \mathbf{K}_{k,i}\left[\mathbf{z}_k - \left(\mathbf{h}_k(\hat{\mathbf{x}}_{k,i}^{(+)}) + \mathbf{H}_{\mathbf{x}}(\hat{\mathbf{x}}_k^{(-)} - \hat{\mathbf{x}}_{k,i}^{(+)})\right)\right],$$

$$\mathbf{C}(\mathbf{x}_{k,i+1}^{(+)}) = \mathbf{C}(\mathbf{x}_k^{(-)}) - \mathbf{K}_{k,i}\mathbf{H}_{\mathbf{x}}\mathbf{C}(\mathbf{x}_k^{(-)}),$$

$$\mathbf{K}_{k,i} = \mathbf{C}(\mathbf{x}_k^{(-)})\mathbf{H}_{\mathbf{x}}^T\left[\mathbf{H}_{\mathbf{x}}\mathbf{C}(\mathbf{x}_k^{(-)})\mathbf{H}_{\mathbf{x}}^T + \mathbf{C}(\mathbf{v}_k)\right]^{-1},$$

where:

$$\mathbf{H}_{\mathbf{x}} \triangleq \frac{\partial \mathbf{h}_k(\mathbf{x})}{\partial \mathbf{x}}\left(\hat{\mathbf{x}}_{k,i}^{(-)}\right)$$

$$\hat{\mathbf{x}}_{k,0}^{(+)} \triangleq \hat{\mathbf{x}}_k^{(-)}.$$

Note that the original measurement value, $\mathbf{z}$, and the prior estimates of the mean and covariance of the state, are used in each step of the iteration. The ith estimate of the state is used to evaluate the weight matrix, $\mathbf{K}$, and is the argument to the non-linear sensor function, $\mathbf{h}$. Iteration can be carried out until there is little further improvement in the estimate. The final estimate of the covariance need only be computed at the end of iteration, rather than at each step, since the intermediate system covariance estimates are not used.

5 Developed Example

The methods developed in this paper will now be applied to the mobile robot example in detail. We choose the world reference frame to be the initial location of the robot, without loss of generality. The robot's initial location with respect to the world frame is then the identity relationship (of the compounding operation), with no uncertainty.

$$\hat{\mathbf{x}} = [\hat{\mathbf{x}}_R] = [\mathbf{0}],$$

$$\mathbf{C}(\mathbf{x}) = [\mathbf{C}(\mathbf{x}_R)] = [\mathbf{0}].$$

Note, that the normal distribution corresponding to this covariance matrix (from (4)) is singular, but the limiting case as the covariance goes to zero is a dirac delta function centered on the mean estimate. This agrees with the intuitive interpretation of zero covariance implying no uncertainty.

Step 1: When the robot senses object #1, the new information must be added into the map. Normally, adding new information relative to the robot's position would fall under case I-b, but since the robot's frame is the same as the world frame, it falls under case I-a. The sensor returns the mean location and variance of object #1 ($\hat{\mathbf{z}}_1$ and $\mathbf{C}(\mathbf{z}_1)$). The new system state vector and variance matrix are:

$$\hat{\mathbf{x}} = \begin{bmatrix} \hat{\mathbf{x}}_R \\ \hat{\mathbf{x}}_1 \end{bmatrix} = \begin{bmatrix} \mathbf{0} \\ \hat{\mathbf{z}}_1 \end{bmatrix},$$

$$\mathbf{C}(\mathbf{x}) = \begin{bmatrix} \mathbf{C}(\mathbf{x}_R) & \mathbf{C}(\mathbf{x}_R, \mathbf{x}_1) \\ \mathbf{C}(\mathbf{x}_1, \mathbf{x}_R) & \mathbf{C}(\mathbf{x}_1) \end{bmatrix} = \begin{bmatrix} \mathbf{0} & \mathbf{0} \\ \mathbf{0} & \mathbf{C}(\mathbf{z}_1) \end{bmatrix}.$$

where $\mathbf{x}_1$ is the location of object #1 with respect to the world frame.

Step 2: The robot moves from its current location to a new location, where the relative motion is given by $\mathbf{y}_R$. Since this motion is also from the world frame, it is a special case of the dynamics extrapolation.

$$\hat{\mathbf{x}} = \begin{bmatrix} \hat{\mathbf{x}}_R \\ \hat{\mathbf{x}}_1 \end{bmatrix} = \begin{bmatrix} \hat{\mathbf{y}}_R \\ \hat{\mathbf{z}}_1 \end{bmatrix},$$

$$\mathbf{C}(\mathbf{x}) = \begin{bmatrix} \mathbf{C}(\mathbf{x}_R) & \mathbf{C}(\mathbf{x}_R, \mathbf{x}_1) \\ \mathbf{C}(\mathbf{x}_1, \mathbf{x}_R) & \mathbf{C}(\mathbf{x}_1) \end{bmatrix} = \begin{bmatrix} \mathbf{C}(\mathbf{y}_R) & \mathbf{0} \\ \mathbf{0} & \mathbf{C}(\mathbf{z}_1) \end{bmatrix}.$$

We can now transform the information in our map from the world frame to the robot's new frame to see how the world looks from the robot's point of view:

$$\begin{aligned} \hat{\mathbf{x}}_{RW} &= \ominus \hat{\mathbf{x}}_R, \\ \mathbf{C}(\mathbf{x}_{RW}) &\approx \mathbf{J}_\ominus \mathbf{C}(\mathbf{x}_R) \mathbf{J}_\ominus^T, \\ \hat{\mathbf{x}}_{R1} &= \ominus \hat{\mathbf{x}}_R \oplus \hat{\mathbf{x}}_1, \\ \mathbf{C}(\mathbf{x}_{R1}) &\approx \mathbf{J}_{1\oplus} \mathbf{J}_\ominus \mathbf{C}(\mathbf{x}_R) \mathbf{J}_\ominus^T \mathbf{J}_{1\oplus}^T + \mathbf{J}_{2\oplus} \mathbf{C}(\mathbf{x}_1) \mathbf{J}_{2\oplus}^T. \end{aligned}$$

Step 3: The robot now senses an object from its new location. The new measurement, $\mathbf{z}_2$, is of course, relative to the robot's location, $\mathbf{x}_R$.

$$\begin{aligned} \hat{\mathbf{x}} &= \begin{bmatrix} \hat{\mathbf{x}}_R \\ \hat{\mathbf{x}}_1 \\ \hat{\mathbf{x}}_2 \end{bmatrix} = \begin{bmatrix} \hat{\mathbf{y}}_R \\ \hat{\mathbf{z}}_1 \\ \hat{\mathbf{y}}_R \oplus \hat{\mathbf{z}}_2 \end{bmatrix}, \\ \mathbf{C}(\mathbf{x}) &= \begin{bmatrix} \mathbf{C}(\mathbf{x}_R) & \mathbf{C}(\mathbf{x}_R, \mathbf{x}_1) & \mathbf{C}(\mathbf{x}_R, \mathbf{x}_2) \\ \mathbf{C}(\mathbf{x}_1, \mathbf{x}_R) & \mathbf{C}(\mathbf{x}_1) & \mathbf{C}(\mathbf{x}_1, \mathbf{x}_2) \\ \mathbf{C}(\mathbf{x}_2, \mathbf{x}_R) & \mathbf{C}(\mathbf{x}_2, \mathbf{x}_1) & \mathbf{C}(\mathbf{x}_2) \end{bmatrix} \\ &= \begin{bmatrix} \mathbf{C}(\mathbf{y}_R) & \mathbf{0} & \mathbf{C}(\mathbf{y}_R) \mathbf{J}_{1\oplus}^T \\ \mathbf{0} & \mathbf{C}(\mathbf{z}_1) & \mathbf{0} \\ \mathbf{J}_{1\oplus} \mathbf{C}(\mathbf{y}_R) & \mathbf{0} & \mathbf{C}(\mathbf{x}_2) \end{bmatrix}. \end{aligned}$$

where:

$$\mathbf{C}(\mathbf{x}_2) = \mathbf{J}_{1\oplus}\mathbf{C}(\mathbf{y}_R)\mathbf{J}_{1\oplus}^T + \mathbf{J}_{2\oplus}\mathbf{C}(\mathbf{z}_2)\mathbf{J}_{2\oplus}^T.$$

Step 4: Now, the robot senses object #1 again. In practice one would probably calculate the world location of a new object, and only after comparing the new object to the old ones could the robot decide that they are likely to be the same object. For this example, however, we will assume that the sensor is able to identify the object as being object #1 and we don't need to map this new measurement into the world frame before performing the update. The symbolic expressions for the estimates of the mean and covariance of the state vector become too complex to reproduce as we have done for the previous steps. Also, if the iterated methods are being used, there is no symbolic expression for the results.

Notice that the formulae presented in this section are correct for *any* network of relationships which has the same topology as this example. This procedure can be completely automated, and is very suitable for use in off-line robot planning.

As a further example of some of the possibilities of this stochastic map method, we will present an example of a geometric constraint — four points known to be arranged in a rectangle. Figure 6 shows the estimated locations of the four points with respect to the world frame, before and after introduction of the information that they are the vertices of a rectangle. The improved estimates are overlayed on the original estimates in the "after" diagram. One way to specify the "rectangularity" of four points — $\mathbf{x}_i, \mathbf{x}_j, \mathbf{x}_k, \mathbf{x}_l$ is as follows:

$$\mathbf{h} = \begin{bmatrix} x_i - x_j + x_k - x_l \\ y_i - y_j + y_k - y_l \\ (x_i - x_j)(x_k - x_j) + (y_i - y_j)(y_k - y_j) \end{bmatrix}.$$

The first two elements of $\mathbf{h}$ are zero when opposite sides of the closed planar figure represented by the four vertices are parallel; the last element of $\mathbf{h}$ is zero when the two sides forming the upper–right corner are perpendicular.

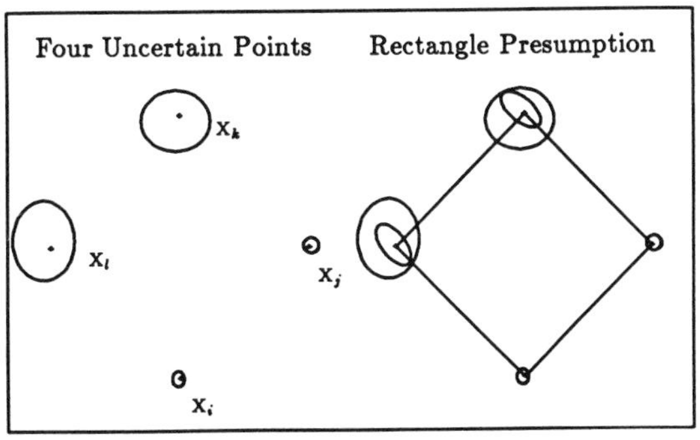

Figure 6: Application of a Rectangular Shape Constraint

We model the rectangle constraint similarly to a sensor, except that we hypothesize rather than measure the relationship. Just as the sensor model included measurement noise, this shape constraint could be "noisy", but here the "noise" describes random tolerances in the shape parameters, possibly given in the geometric model of the object:

$$\mathbf{z} = \mathbf{h}(\mathbf{x}) + v.$$

Given four estimated points, their nominal rectangularity ($\hat{\mathbf{z}}$) and the estimated covariance can be computed. The new information — the presumed shape — is chosen with shape parameters from a distribution with mean $\mathbf{0}$ and covariance $\mathbf{C}(\mathbf{v})$. We might as well choose the most likely a priori value, $\mathbf{0}$.

If we are going to impose the constraint that the four points are precisely in a rectangle — i.e., there is no shape uncertainty, and $\mathbf{C}(\mathbf{v}) = \mathbf{0}$ — then we can choose $\mathbf{h}$ to be *any* function which is zero only when the four points are in a rectangle. If, however, we wish to impose a *loose* rectangle constraint, we must formulate the function $\mathbf{h}$ such that $\mathbf{z}$ is a useful measure of *how* the four points fail to be rectangular.

6 Discussion and Conclusions

This paper presents a general theory for estimating uncertain relative spatial relationships between reference frames in a network of uncertain spatial relationships. Such networks arise, for example, in industrial robotics and navigation for mobile robots, because the system is given spatial information in the form of sensed relationships, prior constraints, relative motions, and so on. The theory presented in this paper allows the efficient estimation of these uncertain spatial relations. This theory can be used, for example, to compute *in advance* whether a proposed sequence of actions (each with known uncertainty) is likely to fail due to too much accumulated uncertainty; whether a proposed sensor observation will reduce the uncertainty to a tolerable level; whether a sensor result is so unlikely given its expected value and its prior probability of failure that it should be ignored, and so on. This paper applies state estimation theory to the problem of estimating parameters of an entire spatial configuration of objects, with the ability to transform estimates into any frame of interest.

The estimation procedure makes a number of assumptions that are normally met in practice. These assumptions are detailed in the text, but the main assumptions can be summarized as follows:

- The angular errors are "small". This requirement arises because we linearize inherently nonlinear relationships. In Monte Carlo simulations[Smith, 1985], angular errors with a standard deviation as large as $5°$ gave estimates of the means and variances to within 1% of the correct values.
- Estimating only two moments of the probability density functions of the uncertain spatial relationships is adequate for decision making. We believe that this is the case since we will most often model a sensor observation by a mean and variance, and the relationships which result from combining many pieces of information become rapidly Gaussian, and thus are accurately modelled by only two moments.

Although the examples presented in this paper have been solely concerned with *spatial* information, there is nothing in the theory that imposes this restriction. Provided that functions are given which describe the relationships among the components to be estimated, those components could be forces, velocities, time intervals, or other quantities in robotic and non-robotic applications.

Appendix A

In this paper we presented formulae for computing the resultant of two spatial relationships in two dimensions (three degrees of freedom). The Jacobians for the three-dimensional transformations are described below. In three dimensions, there are six degrees of freedom: translations in x, y, z and three orientation angles: ϕ, θ, ψ. For computational reasons, orientation is often expressed as a rotation matrix composed of orthogonal column vectors (one per Cartesian axis):

$$\mathbf{R} = \begin{bmatrix} \mathbf{n} & \mathbf{o} & \mathbf{a} \end{bmatrix} = \begin{bmatrix} n_x & o_x & a_x \\ n_y & o_y & a_y \\ n_z & o_z & a_z \end{bmatrix}$$

A primitive rotation is a rotation about one of the axes, and can be represented by a primitive rotation matrix with the above form (see [Paul, 1981] for definitions). For example, $Rot(z, a)$ describes the rotation by a radians about the z axis. Primitive rotation matrices can be multiplied together to produce a rotation matrix describing the final orientation. Orientation will be represented by rotation matrices in the following. There are two common interpretations of the orientation angles—Euler angles and roll, pitch, and yaw.

Relationships Using Euler Angles

Euler angles are defined by:

$$Euler(\phi, \theta, \psi) = Rot(z, \phi) Rot(y', \theta) Rot(z'', \psi) =$$

$$\begin{bmatrix} \cos\phi\cos\theta\cos\psi - \sin\phi\sin\psi & -\cos\phi\cos\theta\sin\psi - \sin\phi\cos\psi & \cos\phi\sin\theta \\ \sin\phi\cos\theta\cos\psi + \cos\phi\sin\psi & -\sin\phi\cos\theta\sin\psi + \cos\phi\cos\psi & \sin\phi\sin\theta \\ -\sin\theta\cos\psi & \sin\theta\sin\psi & \cos\theta \end{bmatrix}.$$

The head to tail relationship, $\mathbf{x}_3 = \mathbf{x}_1 \oplus \mathbf{x}_2$, is then given by:

$$\mathbf{x}_3 = \begin{bmatrix} x_3 \\ y_3 \\ z_3 \\ \phi_3 \\ \theta_3 \\ \psi_3 \end{bmatrix} = \begin{bmatrix} \mathbf{T}_E \\ \mathbf{A}_E \end{bmatrix}$$

where $\mathbf{T}_E$ and $\mathbf{A}_E$ are

$$\mathbf{T}_E = \mathbf{R}_1 \begin{bmatrix} x_2 \\ y_2 \\ z_2 \end{bmatrix} + \begin{bmatrix} x_1 \\ y_1 \\ z_1 \end{bmatrix},$$

$$\mathbf{A}_E = \begin{bmatrix} atan2(a_{y_3}, a_{x_3}) \\ atan2(a_{x_3}\cos\phi_3 + a_{y_3}\sin\phi_3, a_{z_3}) \\ atan2(-n_{x_3}\sin\phi_3 + n_{y_3}\cos\phi_3, -o_{x_3}\sin\phi_3 + o_{y_3}\cos\phi_3) \end{bmatrix}.$$

The matrix $\mathbf{R}_1$, representing the orientation angles of $\mathbf{x}_1$, has the same definition as the Euler rotation matrix defined above (with angles subscripted by 1). The terms a_{x_3} etc. are the elements of the compound rotation matrix $\mathbf{R}_3$, whose values are defined by $\mathbf{R}_3 = \mathbf{R}_1\mathbf{R}_2$. Note that the inverse trignometric function *atan2* is a function of two arguments, the ordinate y and the abscissa x. This function returns the correct result when either x or y are zero, and gives the correct answer over the entire range of possible inputs [Paul, 1981]. Also note that the solution for ϕ_3 is obtained first, and then used in solving for the other two angles.

The Jacobian of this relationship, $\mathbf{J}_\oplus$, is:

$$\mathbf{J}_\oplus = \frac{\partial \mathbf{x}_3}{\partial(\mathbf{x}_1, \mathbf{x}_2)} = \begin{bmatrix} \mathbf{I}_{3\times3} & \mathbf{M} & \mathbf{R}_1 & \mathbf{0}_{3\times3} \\ \mathbf{0}_{3\times3} & \mathbf{K}_1 & \mathbf{0}_{3\times3} & \mathbf{K}_2 \end{bmatrix}$$

where

$$\mathbf{M} = \begin{bmatrix} -(y_3 - y_1) & (z_3 - z_1)\cos\phi_1 & o_{x_1}x_2 - n_{x_1}y_2 \\ x_3 - x_1 & (z_3 - z_1)\sin\phi_1 & o_{y_1}x_2 - n_{y_1}y_2 \\ 0 & -x_2\cos\theta_1\cos\psi_1 + y_2\cos\theta_1\sin\psi_1 - z_2\sin\theta_1 & o_{z_1}x_2 - n_{z_1}y_2 \end{bmatrix},$$

$$\mathbf{K}_1 = \begin{bmatrix} 1 & [\cos\theta_3\sin(\phi_3 - \phi_1)]/\sin\theta_3 & [\sin\theta_2\cos(\psi_3 - \psi_2)]/\sin\theta_3 \\ 0 & \cos(\phi_3 - \phi_1) & \sin\theta_2\sin(\psi_3 - \psi_2) \\ 0 & \sin(\phi_3 - \phi_1)/\sin\theta_3 & [\sin\theta_1\cos(\phi_3 - \phi_1)]/\sin\theta_3 \end{bmatrix},$$

$$\mathbf{K}_2 = \begin{bmatrix} [\sin\theta_2\cos(\psi_3 - \psi_2)]/\sin\theta_3 & [\sin(\psi_3 - \psi_2)]/\sin\theta_3 & 0 \\ \sin\theta_2\sin(\psi_3 - \psi_2) & \cos(\psi_3 - \psi_2) & 0 \\ [\sin\theta_1\cos(\phi_3 - \phi_1)]/\sin\theta_3 & [\cos\theta_3\sin(\psi_3 - \psi_2)]/\sin\theta_3 & 1 \end{bmatrix}.$$

Note that this Jacobian (and similarly, the one for RPY angles) has been simplified by the use of final terms (e.g. x_3, ψ_3). Since the final terms are computed routinely in determining the mean relationship, they are available to evaluate the Jacobian. Examination of the elements indicates the possibility of a singularity; as the mean values of the angles approach a singular combination, the accuracy of the covariance estimates using this Jacobian will decrease. Methods for avoiding the singularity during calculations are being explored.

The inverse relation, $\mathbf{x}'$, in terms of the elements of the relationship $\mathbf{x}$, using the Euler angle definition, is:

$$\mathbf{x}' = \begin{bmatrix} x' \\ y' \\ z' \\ \phi' \\ \theta' \\ \psi' \end{bmatrix} = \begin{bmatrix} -(n_x x + n_y y + n_z z) \\ -(o_x x + o_y y + o_z z) \\ -(a_x x + a_y y + a_z z) \\ -\psi \\ -\theta \\ -\phi \end{bmatrix}$$

where n_x etc. are the elements of the rotation matrix $\mathbf{R}$ associated with the angles in the given transformation to be inverted, $\mathbf{x}$. The Jacobian of the this relationship, $\mathbf{J}_\ominus$, is:

$$\mathbf{J}_\ominus = \frac{\partial \mathbf{x}'}{\partial \mathbf{x}} = \begin{bmatrix} -\mathbf{R}^T & \mathbf{N} \\ \mathbf{0}_{3\times3} & \mathbf{Q} \end{bmatrix}, \qquad \mathbf{Q} = \begin{bmatrix} 0 & 0 & -1 \\ 0 & -1 & 0 \\ -1 & 0 & 0 \end{bmatrix},$$

$$\mathbf{N} = \begin{bmatrix} n_y x - n_x y & -n_z x\cos\phi - n_z y\sin\phi + z\cos\theta\cos\psi & y' \\ o_y x - o_x y & -o_z x\cos\phi - o_z y\sin\phi - z\cos\theta\sin\psi & -x' \\ a_y x - a_x y & -a_z x\cos\phi - a_z y\sin\phi + z\sin\theta & 0 \end{bmatrix}.$$

Relationships Using Roll, Pitch and Yaw Angles

Roll, pitch, and yaw angles are defined by:

$$RPY(\phi,\theta,\psi) = Rot(z,\phi)Rot(y',\theta)Rot(x'',\psi) =$$

$$\begin{bmatrix} \cos\phi\cos\theta & \cos\phi\sin\theta\sin\psi - \sin\phi\cos\psi & \cos\phi\sin\theta\cos\psi + \sin\phi\sin\psi \\ \sin\phi\cos\theta & \sin\phi\sin\theta\sin\psi + \cos\phi\cos\psi & \sin\phi\sin\theta\cos\psi - \cos\phi\sin\psi \\ -\sin\theta & \cos\theta\sin\psi & \cos\theta\cos\psi \end{bmatrix}$$

The head to tail relationship, $\mathbf{x}_3 = \mathbf{x}_1 \oplus \mathbf{x}_2$, is then given by:

$$\mathbf{x}_3 = \begin{bmatrix} x_3 \\ y_3 \\ z_3 \\ \phi_3 \\ \theta_3 \\ \psi_3 \end{bmatrix} = \begin{bmatrix} \mathbf{T}_{RPY} \\ \mathbf{A}_{RPY} \end{bmatrix}$$

where $\mathbf{T}_{RPY}$ and $\mathbf{A}_{RPY}$ are defined by:

$$\mathbf{T}_{RPY} = \mathbf{R}_1 \begin{bmatrix} x_2 \\ y_2 \\ z_2 \end{bmatrix} + \begin{bmatrix} x_1 \\ y_1 \\ z_1 \end{bmatrix},$$

$$\mathbf{A}_{RPY} = \begin{bmatrix} atan2(n_{y_3}, n_{x_3}) \\ atan2(-n_{z_3}, n_{x_3}\cos\phi_3 + n_{y_3}\sin\phi_3) \\ atan2(a_{x_3}\sin\phi_3 - a_{y_3}\cos\phi_3, -o_{x_3}\sin\phi_3 + o_{y_3}\cos\phi_3) \end{bmatrix}.$$

The matrix $\mathbf{R}_1$ is the rotation matrix for the RPY angles in $\mathbf{x}_1$. The Jacobian of the head-to-tail relationship is given by:

$$\mathbf{J}_\oplus = \frac{\partial \mathbf{x}_3}{\partial(\mathbf{x}_1, \mathbf{x}_2)} = \begin{bmatrix} \mathbf{I}_{3\times3} & \mathbf{M} & \mathbf{R}_1 & \mathbf{0}_{3\times3} \\ \mathbf{0}_{3\times3} & \mathbf{K}_1 & \mathbf{0}_{3\times3} & \mathbf{K}_2 \end{bmatrix}$$

where

$$\mathbf{M} = \begin{bmatrix} -(y_3 - y_1) & (z_3 - z_1)\cos(\phi_1) & a_{x_1}y_2 - o_{x_1}z_2 \\ x_3 - x_1 & (z_3 - z_1)\sin(\phi_1) & a_{y_1}y_2 - o_{y_1}z_2 \\ 0 & -x_2\cos\theta_1 - y_2\sin\theta_1\sin\psi_1 - z_2\sin\theta_1\cos\psi_1 & a_{z_1}y_2 - o_{z_1}z_2 \end{bmatrix},$$

$$\mathbf{K}_1 = \begin{bmatrix} 1 & [\sin\theta_3 \sin(\phi_3 - \phi_1)]/\cos\theta_3 & [o_{x_2}\sin\psi_3 + a_{x_2}\cos\psi_3]/\cos\theta_3 \\ 0 & \cos(\phi_3 - \phi_1) & -\cos\theta_1 \sin(\phi_3 - \phi_1) \\ 0 & [\sin(\phi_3 - \phi_1)]/\cos\theta_3 & [\cos\theta_1 \cos(\phi_3 - \phi_1)]/\cos\theta_3 \end{bmatrix},$$

$$\mathbf{K}_2 = \begin{bmatrix} [\cos\theta_2 \cos(\psi_3 - \psi_2)]/\cos\theta_3 & [\sin(\psi_3 - \psi_2)]/\cos\theta_3 & 0 \\ -\cos\theta_2 \sin(\psi_3 - \psi_2) & \cos(\psi_3 - \psi_2) & 0 \\ [a_{x_1}\cos\phi_3 + a_{y_1}\sin\phi_3]/\cos\theta_3 & [\sin\theta_3 \sin(\psi_3 - \psi_2)]/\cos\theta_3 & 1 \end{bmatrix}.$$

The inverse relation, $\mathbf{x}'$, in terms of the elements of $\mathbf{x}$, using the RPY angle definition, is:

$$\mathbf{x}' = \begin{bmatrix} x' \\ y' \\ z' \\ \phi' \\ \theta' \\ \psi' \end{bmatrix} = \begin{bmatrix} -(n_x x + n_y y + n_z z) \\ -(o_x x + o_y y + o_z z) \\ -(a_x x + a_y y + a_z z) \\ atan2(o_x, n_x) \\ atan2(-a_x, n_x \cos\phi' + o_x \sin\phi') \\ atan2(n_z \sin\phi' - o_z \cos\phi', -n_y \sin\phi' + o_y \cos\phi') \end{bmatrix}$$

where n_x etc. are the elements of the rotation matrix, $\mathbf{R}$, for the RPY angles in $\mathbf{x}$. The Jacobian of the inverse relationship is:

$$\mathbf{J}_\ominus = \frac{\partial \mathbf{x}'}{\partial \mathbf{x}} = \begin{bmatrix} -\mathbf{R}^T & \mathbf{N} \\ \mathbf{0}_{3\times3} & \mathbf{Q} \end{bmatrix}$$

where

$$\mathbf{N} = \begin{bmatrix} n_y x - n_x y & -n_z x \cos\phi - n_z y \sin\phi + z\cos\theta & 0 \\ o_y x - o_x y & -o_z x \cos\phi - o_z y \sin\phi + z \sin\theta \sin\psi & z' \\ a_y x - a_x y & -a_z x \cos\phi - a_z y \sin\phi + z \sin\theta \cos\psi & -y' \end{bmatrix},$$

$$\mathbf{Q} = \begin{bmatrix} -a_z/(1 - a_{x^2}) & -a_y \cos\phi/(1 - a_{x^2}) & n_z a_x/(1 - a_{x^2}) \\ a_y/(1 - a_{x^2})^{1/2} & -a_z \cos\phi/(1 - a_{x^2})^{1/2} & o_x/(1 - a_{x^2})^{1/2} \\ a_z a_x/(1 - a_{x^2}) & -o_x \cos\psi/(1 - a_{x^2}) & -n_x/(1 - a_{x^2}) \end{bmatrix}.$$

References

Brooks, R. A. 1982. Symbolic Error Analysis and Robot Planning. *Int. J. Robotics Res.* 1(4):29-68.

Chatila, R. and Laumond, J-P. 1985. Position Referencing and Consistent World Modeling for Mobile Robots. *Proc. IEEE Int. Conf. Robotics and Automation.* St. Louis: IEEE, pp. 138-145.

Gelb, A. 1984. *Applied Optimal Estimation.* M.I.T. Press

Nahi, N. E. 1976. *Estimation Theory and Applications.* New York: R.E. Krieger.

Papoulis, A. 1965. *Probability, Random Variables, and Stochastic Processes.* McGraw-Hill.

Paul, R. P. 1981. *Robot Manipulators: Mathematics, Programming and Control.* Cambridge: MIT Press.

Smith, R. C., and Cheeseman, P. 1985. On the Representation and Estimation of Spatial Uncertainty. SRI Robotics Lab. Tech. Paper, and to appear *Int. J. Robotics Res.* 5(4): Winter 1987.

Smith, R. C., *et al.* 1984. Test-Bed for Programmable Automation Research. Final Report-Phase 1, SRI International, April 1984.

Taylor, R. H. 1976. A Synthesis of Manipulator Control Programs from Task-Level Specifications. AIM-282. Stanford, Calif.: Stanford University Artificial Intelligence Laboratory.

The Kalman Filter: An Introduction to Concepts

PETER S. MAYBECK

Before we delve into the details of the text, it would be useful to see where we are going on a conceptual basis. Therefore, the rest of this chapter will provide an overview of the optimal linear estimator, the Kalman filter. This will be conducted at a very elementary level but will provide insights into the underlying concepts. As we progress through this overview, contemplate the ideas being presented: try to conceive of graphic *images* to portray the concepts involved (such as time propagation of density functions), and to generate a *logical structure* for the component pieces that are brought together to solve the estimation problem. If this basic conceptual framework makes sense to you, then you will better understand the need for the details to be developed later in the text. Should the idea of where we are going ever become blurred by the development of detail, refer back to this overview to regain sight of the overall objectives.

First one must ask, what is a Kalman filter? A Kalman filter is simply an *optimal recursive data processing algorithm.* There are many ways of defining *optimal,* dependent upon the criteria chosen to evaluate performance. It will be shown that, under the assumptions to be made in the next section, the Kalman filter is optimal with respect to virtually any criterion that makes sense. One aspect of this optimality is that the Kalman filter incorporates all information that can be provided to it. It processes all available measurements, regardless of their precision, to estimate the current value of the variables of interest, with use of (1) knowledge of the system and measurement device dynamics, (2) the statistical description of the system noises, measurement errors, and uncertainty in the dynamics models, and (3) any available information about initial conditions of the variables of interest. For example, to determine the velocity of an aircraft, one could use a Doppler radar, or the velocity indications of an inertial navigation system, or the pitot and static pressure and relative wind information in the air data system. Rather than ignore any of these outputs, a Kalman filter could be built to combine all of this data and knowledge of the various systems' dynamics to generate an overall best estimate of velocity.

The word *recursive* in the previous description means that, unlike certain data processing concepts, the Kalman filter does not require all previous data to be kept in storage and reprocessed every time a new measurement is taken. This will be of vital importance to the practicality of filter implementation.

The "filter" is actually a *data processing algorithm*. Despite the typical connotation of a filter as a "black box" containing electrical networks, the fact is that in most practical applications, the "filter" is just a computer program in a central processor. As such, it inherently incorporates discrete-time measurement samples rather than continuous time inputs.

Figure 1.1 depicts a typical situation in which a Kalman filter could be used advantageously. A system of some sort is driven by some known controls, and measuring devices provide the value of certain pertinent quantities. Knowledge of these system inputs and outputs is all that is explicitly available from the physical system for estimation purposes.

The *need* for a filter now becomes apparent. Often the variables of interest, some finite number of quantities to describe the "state" of the system, cannot be measured directly, and some means of inferring these values from the available data must be generated. For instance, an air data system directly provides static and pitot pressures, from which velocity must be inferred. This inference is complicated by the facts that the system is typically driven by inputs other than our own known controls and that the relationships among the various "state" variables and measured outputs are known only with some degree of uncertainty. Furthermore, any measurement will be corrupted to some degree by noise, biases, and device inaccuracies, and so a means of extracting valuable information from a noisy signal must be provided as well. There may also be a number of different measuring devices, each with its own particular dynamics and error characteristics, that provide some information about a particular variable, and it would be desirable to combine their outputs in a systematic and optimal manner. A Kalman filter combines all available measurement data, plus prior knowledge about the system and measuring devices, to produce an estimate of the desired variables in such a manner that the error is minimized

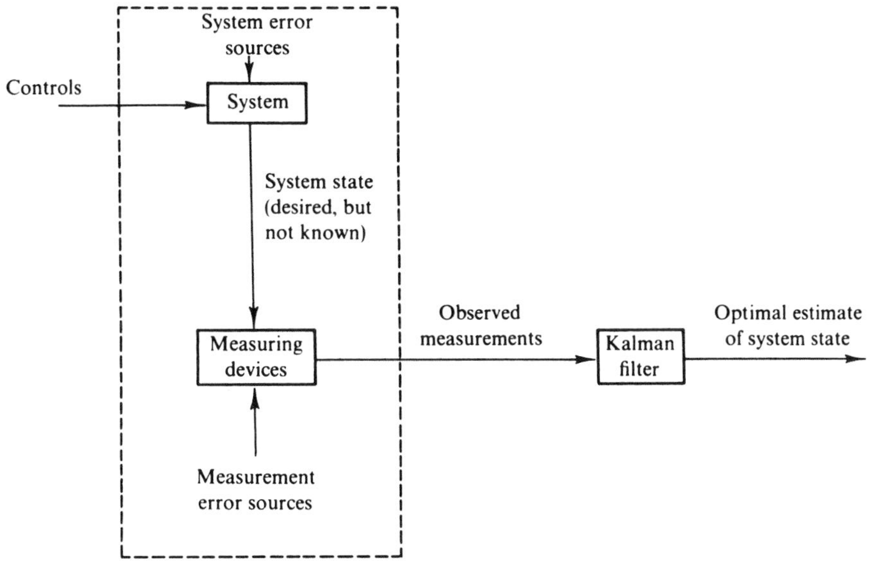

FIG. 1.1 Typical Kalman filter application.

statistically. In other words, if we were to run a number of candidate filters many times for the same application, then the average results of the Kalman filter would be better than the average results of any other.

Conceptually, what any type of filter tries to do is obtain an "optimal" estimate of desired quantities from data provided by a noisy environment, "optimal" meaning that it minimizes errors in some respect. There are many means of accomplishing this objective. If we adopt a Bayesian viewpoint, then we want the filter to propagate the *conditional probability density* of the desired quantities, conditioned on knowledge of the actual data coming from the measuring devices. To understand this concept, consider Fig. 1.2, a portrayal of a conditional probability density of the value of a scalar quantity x at time instant i ($x(i)$), conditioned on knowledge that the vector measurement $\mathbf{z}(1)$ at time instant 1 took on the value $\mathbf{z}_1$ ($\mathbf{z}(1) = \mathbf{z}_1$) and similarly for instants 2 through i, plotted as a function of possible $x(i)$ values. This is denoted as $f_{x(i)|\mathbf{z}(1), \mathbf{z}(2), \ldots, \mathbf{z}(i)}(x|\mathbf{z}_1, \mathbf{z}_2, \ldots, \mathbf{z}_i)$. For example, let $x(i)$ be the one-dimensional position of a vehicle at time instant i, and let $\mathbf{z}(j)$ be a two-dimensional vector describing the measurements of position at time j by two separate radars. Such a conditional probability density contains all the available information about $x(i)$: it indicates, for the given value of all measurements taken up through time instant i, what the probability would be of $x(i)$ assuming any particular value or range of values.

It is termed a "conditional" probability density because its shape and location on the x axis is dependent upon the values of the measurements taken. Its shape conveys the amount of certainty you have in the knowledge of the value of x. If the density plot is a narrow peak, then most of the probability "weight" is concentrated in a narrow band of x values. On the other hand, if the plot has a gradual shape, the probability "weight" is spread over a wider range of x, indicating that you are less sure of its value.

Once such a conditional probability density function is propagated, the "optimal" estimate can be defined. Possible choices would include

(1) the *mean*—the "center of probability mass" estimate;

(2) the *mode*—the value of x that has the highest probability, locating the peak of the density; and

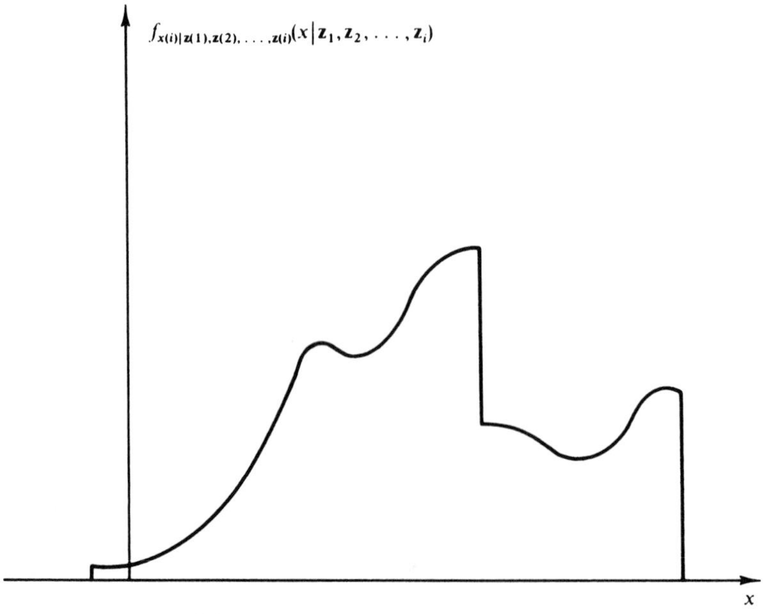

FIG. 1.2 Conditional probability density.

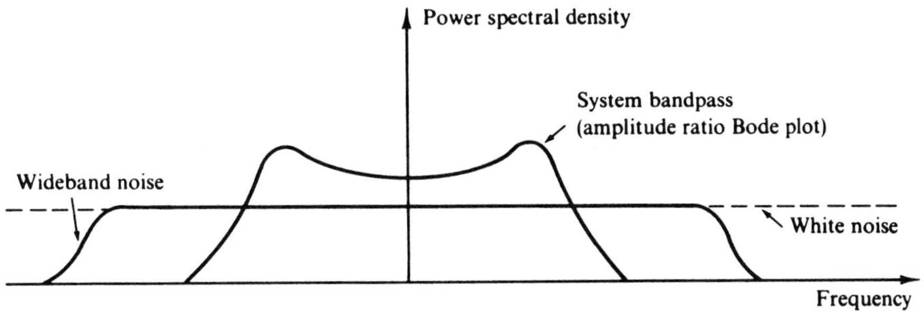

FIG. 1.3 Power spectral density bandwidths.

(3) the *median*—the value of x such that half of the probability weight lies to the left and half to the right of it.

A Kalman filter performs this conditional probability density propagation for problems in which the system can be described through a *linear* model and in which system and measurement noises are *white* and *Gaussian* (to be explained shortly). Under these conditions, the mean, mode, median, and virtually any reasonable choice for an "optimal" estimate all coincide, so there is in fact a unique "best" estimate of the value of x. Under these three restrictions, the Kalman filter can be shown to be the best filter of any conceivable form. Some of the restrictions can be relaxed, yielding a qualified optimal filter. For instance, if the Gaussian assumption is removed, the Kalman filter can be shown to be the best (minimum error variance) filter out of the class of linear unbiased filters. However, these three assumptions can be justified for many potential applications, as seen in the following section.

1.4 BASIC ASSUMPTIONS

At this point it is useful to look at the three basic assumptions in the Kalman filter formulation. On first inspection, they may appear to be overly restrictive and unrealistic. To allay any misgivings of this sort, this section will briefly discuss the physical implications of these assumptions.

A linear system model is justifiable for a number of reasons. Often such a model is adequate for the purpose at hand, and when nonlinearities do exist, the typical engineering approach is to linearize about some nominal point or trajectory, achieving a perturbation model or error model. Linear systems are desirable in that they are more easily manipulated with engineering tools, and linear system (or differential equation) theory is much more complete and practical than nonlinear. The fact is that there are means of extending the Kalman filter concept to some nonlinear applications or developing nonlinear filters directly, but these are considered only if linear models prove inadequate.

"Whiteness" implies that the noise value is not correlated in time. Stated more simply, if you know what the value of the noise is now, this knowledge does you no good in predicting what its value will be at any other time. Whiteness also implies that the noise has equal power at all frequencies. Since this results in a noise with infinite power, a white noise obviously cannot really exist. One might then ask, why even consider such a concept if it does not exist in real life? The answer is twofold. First, any physical system of interest has a certain frequency "bandpass"—a frequency range of inputs to which it can respond. Above this range, the input either has no effect, or the system so severely attentuates the effect that it essentially does not exist. In Fig. 1.3, a typical system bandpass curve is drawn on a plot of "power spectral density" (interpreted as the amount of power content at a certain frequency) versus frequency. Typically a system will be driven by wideband noise—one having power at frequencies above the system bandpass, and essentially constant power at all frequencies within the system bandpass—as shown in the figure. On this same plot, a white noise would merely extend this constant power level out across all frequencies. Now, within the bandpass of the system of interest, the fictitious white noise looks identical to the real wideband noise. So what has been gained? That is the second part of the answer to why a white noise model is used. It turns out that the mathematics involved in the filter are vastly simplified (in fact, made tractable) by replacing the real wideband noise with a white noise which, from the system's "point of view," is identical. Therefore, the white noise model is used.

One might argue that there are cases in which the noise power level is not constant over all frequencies within the system bandpass, or in which the noise is in fact time correlated. For such instances, a white noise put through a small linear system can duplicate virtually any form of time-correlated noise. This small system, called a "shaping filter," is then added to the original system, to achieve an overall linear system driven by white noise once again.

Whereas whiteness pertains to time or frequency relationships of a noise, Gaussianness has to do with its amplitude. Thus, at any single point in time, the probability density of a Gaussian noise amplitude takes on the shape of a normal bell-shaped curve. This assumption can be justified physically by the fact that a system or measurement noise is typically caused by a number of small sources. It can be shown mathematically that when a number of independent random variables are added together, the summed effect can be described very closely by a Gaussian probability density, regardless of the shape of the individual densities.

There is also a practical justification for using Gaussian densities. Similar to whiteness, it makes the mathematics tractable. But more than that, typically an engineer will know, at best, the first and second order statistics (mean and variance or standard deviation) of a noise process. In the absence of any higher order statistics, there is no better form to assume than the Gaussian density. The first and second order statistics completely determine a Gaussian density, unlike most densities which require an endless number of orders of statistics to specify their shape entirely. Thus, the Kalman filter, which propagates the first and second order statistics, includes *all* information contained in the conditional probability density, rather than only some of it, as would be the case with a different form of density.

The particular assumptions that are made are dictated by the objectives of, and the underlying motivation for, the model being developed. If our objective were merely to build good descriptive models, we would not confine our attention to linear system models driven by white Gaussian noise. Rather, we would seek the model, of whatever form, that best fits the data generated by the "real world." It is our desire to build estimators and controllers based upon our system models that drives us to these assumptions: other assumptions generally do not yield tractable estimation or control problem formulations. Fortunately, the class of models that yields tractable mathematics also provides adequate representations for many applications of interest. Later, the model structure will be extended somewhat to enlarge the range of applicability, but the requirement of model usefulness in subsequent estimator or controller design will again be a dominant influence on the manner in which the extensions are made.

1.5 A SIMPLE EXAMPLE

To see how a Kalman filter works, a simple example will now be developed. Any example of a single measuring device providing data on a single variable would suffice, but the determination of a position is chosen because the probability of one's exact location is a familiar concept that easily allows dynamics to be incorporated into the problem.

Suppose that you are lost at sea during the night and have no idea at all of your location. So you take a star sighting to establish your position (for the sake of simplicity, consider a one-dimensional location). At some time t_1 you determine your location to be z_1. However, because of inherent measuring device inaccuracies, human error, and the like, the result of your measurement is somewhat uncertain. Say you decide that the precision is such that the standard deviation (one-sigma value) involved is σ_{z_1} (or equivalently, the variance, or second order statistic, is $\sigma_{z_1}^2$). Thus, you can establish the conditional probability of $x(t_1)$, your position at time t_1, conditioned on the observed value of the measurement being z_1, as depicted in Fig. 1.4. This is a plot of $f_{x(t_1)|z(t_1)}(x|z_1)$ as a function of the location x: it tells you the probability of being in any one location, based upon the measurement you took. Note that σ_{z_1} is a direct measure of the uncertainty: the larger σ_{z_1} is, the broader the probability peak is, spreading the probability "weight" over a larger range of x values. For a Gaussian density, 68.3% of the probability "weight" is contained within the band σ units to each side of the mean, the shaded portion in Fig. 1.4.

Based on this conditional probability density, the best estimate of your position is

$$\hat{x}(t_1) = z_1 \tag{1-1}$$

and the variance of the error in the estimate is

$$\sigma_x{}^2(t_1) = \sigma_{z_1}^2 \tag{1-2}$$

Note that $\hat{x}$ is both the mode (peak) and the median (value with $\frac{1}{2}$ of the probability weight to each side), as well as the mean (center of mass).

Now say a trained navigator friend takes an independent fix right after you do, at time $t_2 \cong t_1$ (so that the true position has not changed at all), and obtains a measurement z_2 with a variance $\sigma_{z_2}^2$. Because he has a higher skill, assume the variance in his measurement to be somewhat smaller than in yours. Figure 1.5 presents the conditional density of your position at time t_2, based only on the measured value z_2. Note the narrower peak due to smaller variance, indicating that you are rather certain of your position based on his measurement.

At this point, you have two measurements available for estimating your position. The question is, how do you combine these data? It will be shown

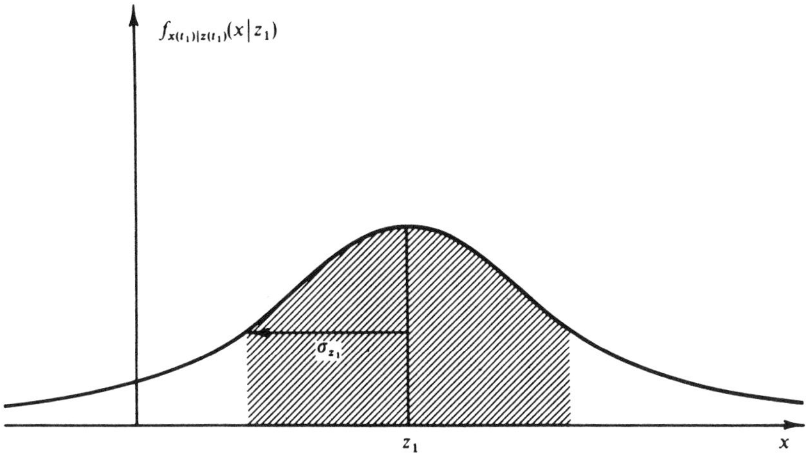

FIG. 1.4 Conditional density of position based on measured value z_1.

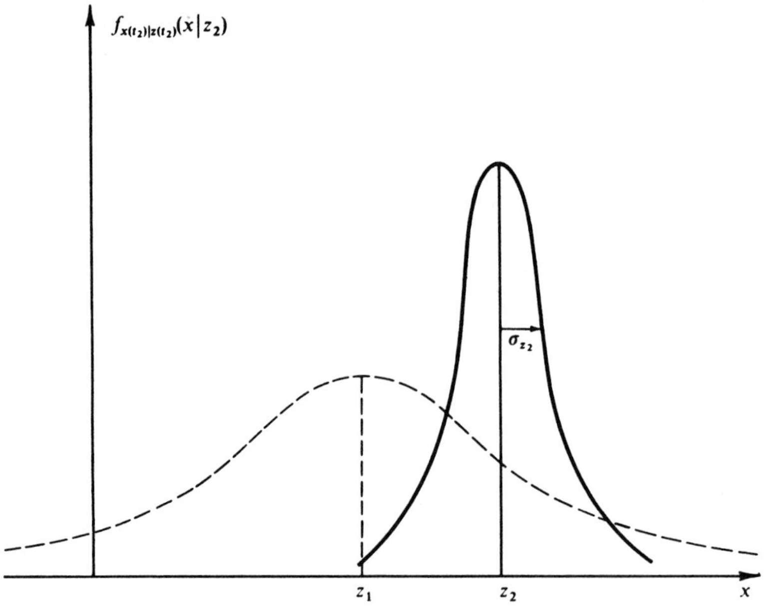

FIG. 1.5 Conditional density of position based on measurement z_2 alone.

subsequently that, based on the assumptions made, the conditional density of your position at time $t_2 \cong t_1$, $x(t_2)$, given *both* z_1 *and* z_2, is a Gaussian density with mean μ and variance σ^2 as indicated in Fig. 1.6, with

$$\mu = [\sigma_{z_2}^2/(\sigma_{z_1}^2 + \sigma_{z_2}^2)]z_1 + [\sigma_{z_1}^2/(\sigma_{z_1}^2 + \sigma_{z_2}^2)]z_2 \tag{1-3}$$

$$1/\sigma^2 = (1/\sigma_{z_1}^2) + (1/\sigma_{z_2}^2) \tag{1-4}$$

Note that, from (1-4), σ is less than either σ_{z_1} or σ_{z_2}, which is to say that the uncertainty in your estimate of position has been decreased by combining the two pieces of information.

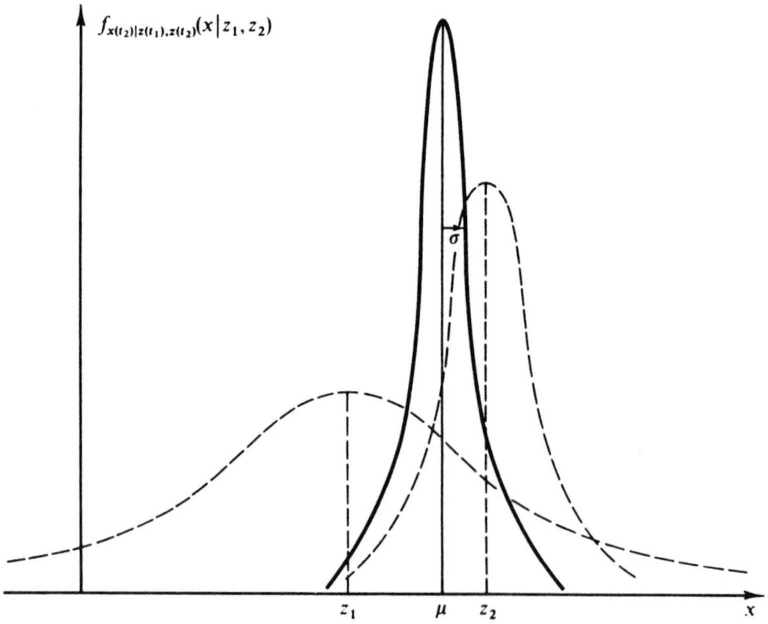

FIG. 1.6 Conditional density of position based on data z_1 and z_2.

Given this density, the best estimate is

$$\hat{x}(t_2) = \mu \tag{1-5}$$

with an associated error variance σ^2. It is the mode and the mean (or, since it is the mean of a conditional density, it is also termed the conditional mean). Furthermore, it is also the maximum likelihood estimate, the weighted least squares estimate, and the linear estimate whose variance is less than that of any other linear unbiased estimate. In other words, it is the "best" you can do according to just about any reasonable criterion.

After some study, the form of μ given in Eq. (1-3) makes good sense. If σ_{z_1} were equal to σ_{z_2}, which is to say you think the measurements are of equal precision, the equation says the optimal estimate of position is simply the average of the two measurements, as would be expected. On the other hand, if σ_{z_1} were larger than σ_{z_2}, which is to say that the uncertainty involved in the measurement z_1 is greater than that of z_2, then the equation dictates "weighting" z_2 more heavily than z_1. Finally, the variance of the estimate is less than σ_{z_1} even if σ_{z_2} is very large: even poor quality data provide some information, and should thus increase the precision of the filter output.

The equation for $\hat{x}(t_2)$ can be rewritten as

$$\begin{aligned}\hat{x}(t_2) &= [\sigma_{z_2}^2/(\sigma_{z_1}^2 + \sigma_{z_2}^2)]z_1 + [\sigma_{z_1}^2/(\sigma_{z_1}^2 + \sigma_{z_2}^2)]z_2 \\ &= z_1 + [\sigma_{z_1}^2/(\sigma_{z_1}^2 + \sigma_{z_2}^2)][z_2 - z_1]\end{aligned} \tag{1-6}$$

or, in final form that is actually used in Kalman filter implementations [noting that $\hat{x}(t_1) = z_1$],

$$\hat{x}(t_2) = \hat{x}(t_1) + K(t_2)[z_2 - \hat{x}(t_1)] \tag{1-7}$$

where

$$K(t_2) = \sigma_{z_1}^2/(\sigma_{z_1}^2 + \sigma_{z_2}^2) \tag{1-8}$$

These equations say that the optimal estimate at time t_2, $\hat{x}(t_2)$, is equal to the best prediction of its value before z_2 is taken, $\hat{x}(t_1)$, plus a correction term of an optimal weighting value times the difference between z_2 and the best prediction of its value before it is actually taken, $\hat{x}(t_1)$. It is worthwhile to understand this "predictor–corrector" structure of the filter. Based on all previous information, a prediction of the value that the desired variables and measurement will have at the next measurement time is made. Then, when the next measurement is taken, the difference between it and its predicted value is used to "correct" the prediction of the desired variables.

Using the $K(t_2)$ in Eq. (1-8), the variance equation given by Eq. (1-4) can be rewritten as

$$\sigma_x^2(t_2) = \sigma_x^2(t_1) - K(t_2)\sigma_x^2(t_1) \tag{1-9}$$

Note that the values of $\hat{x}(t_2)$ and $\sigma_x^2(t_2)$ embody all of the information in $f_{x(t_2)|z(t_1), z(t_2)}(x|z_1, z_2)$. Stated differently, by propagating these two variables, the conditional density of your position at time t_2, given z_1 and z_2, is completely specified.

Thus we have solved the static estimation problem. Now consider incorporating dynamics into the problem.

Suppose that you travel for some time before taking another measurement. Further assume that the best model you have of your motion is of the simple form

$$dx/dt = u + w \tag{1-10}$$

where u is a nominal velocity and w is a noise term used to represent the uncertainty in your knowledge of the actual velocity due to disturbances, off-nominal conditions, effects not accounted for in the simple first order equation, and the like. The "noise" w will be modeled as a white Gaussian noise with a mean of zero and variance of $\sigma_w{}^2$.

Figure 1.7 shows graphically what happens to the conditional density of position, given z_1 and z_2. At time t_2 it is as previously derived. As time progresses, the density travels along the x axis at the nominal speed u, while simultaneously spreading out about its mean. Thus, the probability density starts at the best estimate, moves according to the nominal model of dynamics, and spreads out in time because you become less sure of your exact position due to the constant addition of uncertainty over time. At the time $t_3{}^-$, just before the measurement is taken at time t_3, the density $f_{x(t_3)|z(t_1),z(t_2)}(x|z_1, z_2)$ is as shown in Fig. 1.7, and can be expressed mathematically as a Gaussian density with mean and variance given by

$$\hat{x}(t_3{}^-) = \hat{x}(t_2) + u[t_3 - t_2] \tag{1-11}$$

$$\sigma_x{}^2(t_3{}^-) = \sigma_x{}^2(t_2) + \sigma_w{}^2[t_3 - t_2] \tag{1-12}$$

Thus, $\hat{x}(t_3{}^-)$ is the optimal prediction of what the x value is at $t_3{}^-$, before the measurement is taken at t_3, and $\sigma_x{}^2(t_3{}^-)$ is the expected variance in that prediction.

Now a measurement is taken, and its value turns out to be z_3, and its variance is assumed to be $\sigma_{z_3}^2$. As before, there are now two Gaussian densities available that contain information about position, one encompassing all the information available before the measurement, and the other being the information provided by the measurement itself. By the same process as before, the density with mean $\hat{x}(t_3{}^-)$ and variance $\sigma_x{}^2(t_3{}^-)$ is combined with the density with mean z_3 and variance $\sigma_{z_3}^2$, to yield a Gaussian density with mean

$$\hat{x}(t_3) = \hat{x}(t_3{}^-) + K(t_3)[z_3 - \hat{x}(t_3{}^-)] \tag{1-13}$$

and variance

$$\sigma_x{}^2(t_3) = \sigma_x{}^2(t_3{}^-) - K(t_3)\sigma_x{}^2(t_3{}^-) \tag{1-14}$$

where the gain $K(t_3)$ is given by

$$K(t_3) = \sigma_x{}^2(t_3{}^-)/[\sigma_x{}^2(t_3{}^-) + \sigma_{z_3}^2] \tag{1-15}$$

The optimal estimate, $\hat{x}(t_3)$, satisfies the same form of equation as seen previously in (1-7). The best prediction of its value before z_3 is taken is corrected by an optimal weighting value times the difference between z_3 and the predic-

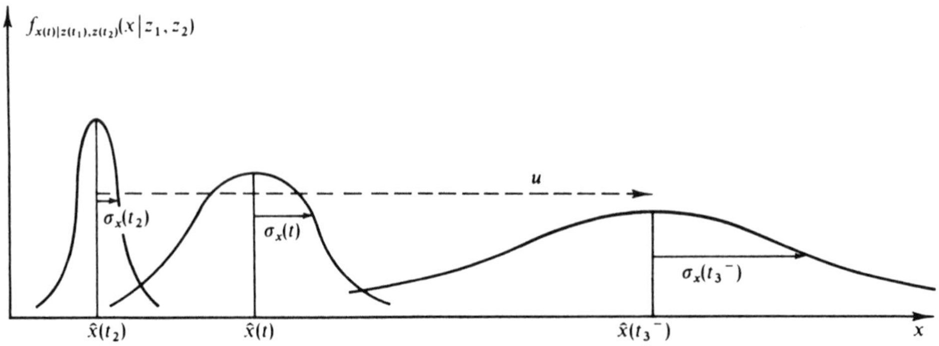

FIG. 1.7 Propagation of conditional probability density.

tion of its value. Similarly, the variance and gain equations are of the same form as (1-8) and (1-9).

Observe the form of the equation for $K(t_3)$. If $\sigma_{z_3}^2$, the measurement noise variance, is large, then $K(t_3)$ is small; this simply says that you would tend to put little confidence in a very noisy measurement and so would weight it lightly. In the limit as $\sigma_{z_3}^2 \to \infty$, $K(t_3)$ becomes zero, and $\hat{x}(t_3)$ equals $\hat{x}(t_3^-)$: an infinitely noisy measurement is totally ignored. If the dynamic system noise variance σ_w^2 is large, then $\sigma_x^2(t_3^-)$ will be large [see Eq. (1-12)] and so will $K(t_3)$; in this case, you are not very certain of the output of the system model within the filter structure and therefore would weight the measurement heavily. Note that in the limit as $\sigma_w^2 \to \infty$, $\sigma_x^2(t_3^-) \to \infty$ and $K(t_3) \to 1$, so Eq. (1-13) yields

$$\hat{x}(t_3) = \hat{x}(t_3^-) + 1 \cdot [z_3 - \hat{x}(t_3^-)] = z_3 \tag{1-16}$$

Thus in the limit of absolutely no confidence in the system model output, the optimal policy is to ignore the output and use the new measurement as the optimal estimate. Finally, if $\sigma_x^2(t_3^-)$ should ever become zero, then so does $K(t_3)$; this is sensible since if $\sigma_x^2(t_3^-) = 0$, you are absolutely sure of your estimate before z_3 becomes available and therefore can disregard the measurement.

Although we have not as yet derived these results mathematically, we have been able to demonstrate the reasonableness of the filter structure.

1.6 A PREVIEW

Extending Eqs. (1-11) and (1-12) to the vector case and allowing time varying parameters in the system and noise descriptions yields the general Kalman filter algorithm for propagating the conditional density and optimal estimate from one measurement sample time to the next. Similarly, the Kalman filter update at a measurement time is just the extension of Eqs. (1-13)–(1-15). Further logical extensions would include estimation with data beyond the time when variables are to be estimated, estimation with nonlinear system models rather than linear, control of systems described through stochastic models, and both estimation and control when the noise and system parameters are not known with absolute certainty. The sequel provides a thorough investigation of those topics, developing both the theoretical mathematical aspects and practical engineering insights necessary to resolve the problem formulations and solutions fully.

GENERAL REFERENCES

The following references have influenced the development of both this introductory chapter and the entirety of this text.

1. Aoki, M., *Optimization of Stochastic Systems—Topics in Discrete-Time Systems*. Academic Press, New York, 1967.
2. Åström, K. J., *Introduction to Stochastic Control Theory*. Academic Press, New York, 1970.
3. Bryson, A. E. Jr., and Ho, Y., *Applied Optimal Control*. Blaisdell, Waltham, Massachusetts, 1969.
4. Bucy, R. S., and Joseph, P. D., *Filtering for Stochastic Processes with Applications to Guidance*. Wiley, New York, 1968.
5. Deutsch, R., *Estimation Theory*. Prentice-Hall, Englewood Cliffs, New Jersey, 1965.
6. Deyst, J. J., "Estimation and Control of Stochastic Processes," unpublished course notes. M.I.T. Dept. of Aeronautics and Astronautics, Cambridge, Massachusetts, 1970.
7. Gelb, A. (ed.), *Applied Optimal Estimation*. M.I.T. Press, Cambridge, Massachusetts, 1974.
8. Jazwinski, A. H., *Stochastic Processes and Filtering Theory*. Academic Press, New York, 1970.
9. Kwakernaak, H., and Sivan, R., *Linear Optimal Control Systems*. Wiley, New York, 1972.
10. Lee, R. C. K., *Optimal Estimation, Identification and Control*. M.I.T. Press, Cambridge, Massachusetts, 1964.

11. Liebelt, P. B., *An Introduction to Optimal Estimation*. Addison-Wesley, Reading, Massachusetts, 1967.
12. Maybeck, P. S., "The Kalman Filter—An Introduction for Potential Users," TM-72-3. Air Force Flight Dynamics Laboratory, Wright-Patterson AFB, Ohio, June 1972.
13. Maybeck, P. S., "Applied Optimal Estimation—Kalman Filter Design and Implementation," notes for a continuing education course offered by the Air Force Institute of Technology, Wright-Patterson AFB, Ohio, semiannually since December 1974.
14. Meditch, J. S., *Stochastic Optimal Linear Estimation and Control*. McGraw-Hill, New York, 1969.
15. McGarty, T. P., *Stochastic Systems and State Estimation*. Wiley, New York, 1974.
16. Sage, A. P., and Melsa, J. L., *Estimation Theory with Application to Communications and Control*. McGraw-Hill, New York, 1971.
17. Schweppe, F. C., *Uncertain Dynamic Systems*. Prentice-Hall, Englewood Cliffs, New Jersey, 1973.
18. Van Trees, H. L., *Detection, Estimation and Modulation Theory*, Vol. 1. Wiley, New York, 1968.

Maintaining Representations of the Environment of a Mobile Robot

NICHOLAS AYACHE AND OLIVIER D. FAUGERAS

Abstract—**In this paper we describe our current ideas related to the problem of building and updating 3-D representation of the environment of a mobile robot that uses passive Vision as its main sensory modality. Our basic tenet is that we want to represent both geometry and uncertainty. We first motivate our approach by defining the problems we are trying to solve and give some simple didactic examples. We then present the tool that we think is extremely well-adapted to solving most of these problems: the extended Kalman filter (EKF). We discuss the notions of minimal geometric representations for 3-D lines, planes, and rigid motions. We show how the EKF and the representations can be combined to provide solutions for some of the problems listed at the beginning of the paper, and give a number of experimental results on real data.**

I. Introduction

In the last few years, Computer Vision has gone extensively into the area of three-dimensional (3-D) analysis from a variety of sensing modalities such as stereo, motion, range finders, and sonars. A book that brings together some of this recent work is [24].

Most of these sensing modalities start from pixels which are then converted into 3-D structures. A characteristic of this work as compared to previous work (like in image restoration, for example) where images were the starting and the ending point is that noise in the measurements is, of course, still present but, contrary to what has happened in the past, it has to be taken into account all the way from pixels to 3-D geometry.

Another aspect of the work on 3-D follows from the observation that if noise is present, it has to be evaluated, i.e., we need models of sensor noise (sensor being taken here in the broad sense of sensory modality), and reduced. This reduction can be obtained in many ways. The most important ones are as follows:

- First, the case of one sensor in a fixed position: it can repeat its measurements and thus maybe obtain better estimations.
- Second, the case of a sensor that can be moved around: given its measurements in a given position, what is the best way to move in order to reduce the uncertainty and increase the knowledge of the environment in a way that is compatible with the task at hand.
- Third, is the case of several different sensors that have to combine their measurements in a meaningful fashion.

Interesting work related to those issues has already emerged which is not reported in [24]. In the area of robust estimation procedures and models of sensors noise, Hager and Mintz [22] and McKendall and Mintz [27] have started to pave the ground. Bolle and Cooper [12] have developed maximum likelihood techniques to combine range data to estimate object positions. Darmon [16] applies the Kalman filter formalism to the detection of moving objects in sequences of images. Durrant-Whyte [18], in his Ph.D. dissertation has conducted a thorough investigation of the problems posed by multi-sensory systems. Applications to the navigation of a mobile robot have been discussed by Crowley [15], Smith and Cheeseman [32], and Matthies and Shafer [28]. The problem of combining stereo views has been attacked by Ayache and Faugeras [3], [4], [19], Porril *et al.* [30], and Kriegman [25]. It also appears that the linearization paradigm extensively used in this paper has been already used in the photogrammetry field [26].

Several problems related to these preliminary studies need more attention. Modeling sensor noise in general and more specifically visual sensor noise appears to us an area where considerable progress can be achieved; relating sensor noise to geometric uncertainty and the corresponding problem of representing geometric information with an eye toward describing not only the geometry but also the uncertainty on this geometry are key problems to be investigated further as is the problem of combining uncertain geometric information produced by different sensors.

II. What are the Problems that we are Trying to Solve

We have been focusing on a number of problems arising in connection with a robot moving in an indoor environment and using passive vision and proprioceptive sensory modalities such as odometry. Our mid-term goals are to incrementally build on the robot an increasing set of sensing and reasoning capabilities such as:

- build local 3-D descriptions of the environment,
- use the descriptions to update or compute motion descriptions where the motion is either the robot's motion or others,
- fuse the local descriptions of neighboring places into more global, coherent, and accurate ones,
- "discover" interesting geometric relations in these descriptions,
- "discover" semantic entities and exhibit "intelligent" behavior.

We describe how we understand each of these capabilities and what are the underlying difficulties.

Manuscript received April 5, 1988; revised May 2, 1989. This work was partially supported by the Esprit Project P940.

The authors are with INRIA—Rocquencourt, Domaine de Voluceau, Rocquencourt, B. P. 105—78153, Le Chesney Cedex, France.

IEEE Log Number 8930460

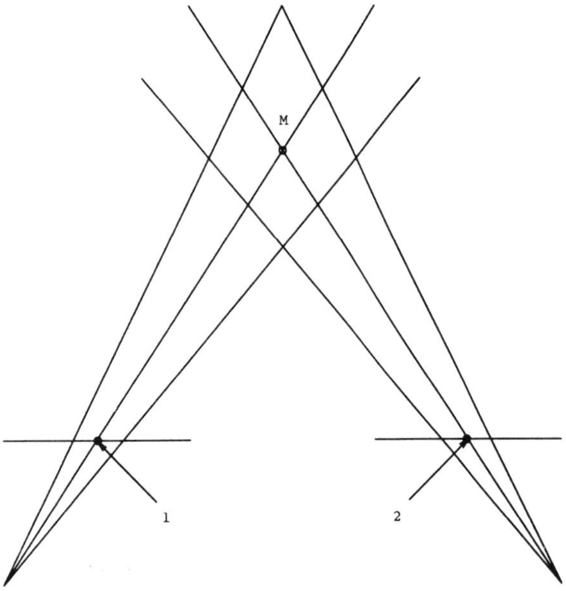

Fig. 1. Effect of pixel noise on 3-D reconstruction.

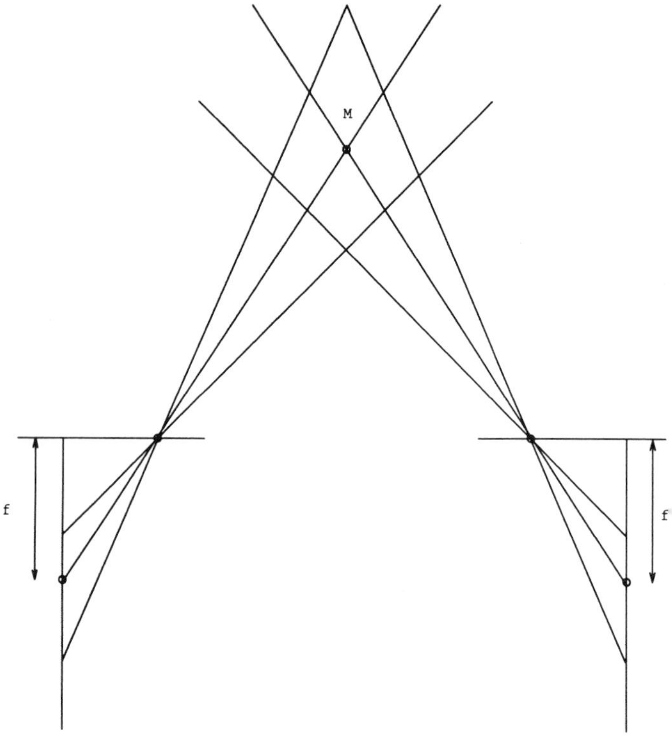

Fig. 2. Effect of calibration errors on 3-D reconstruction.

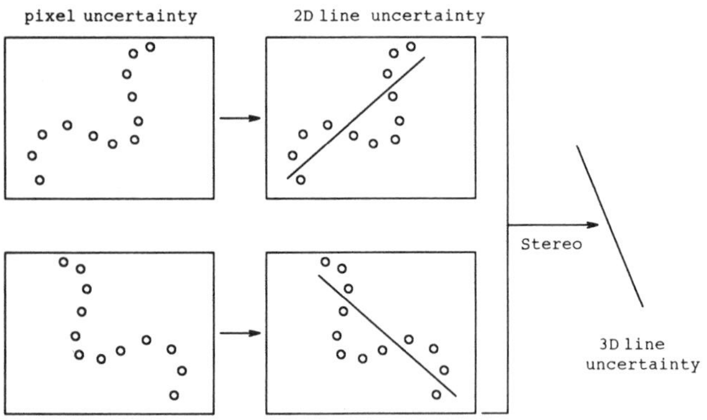

Fig. 3. From pixel uncertainty to 3-D line uncertainty.

A. Build Local 3-D Descriptions of the Environment

Up until now, our main source of 3-D information has been Stereo [5], [9] even though we have made considerable progress toward the use of structure from motion as well [21]. In any case, the problems are very similar for both sensing modalities and we concentrate on Stereo. As announced in the Introduction, our main concern is to track uncertainty all the way from pixel noise to geometric descriptions. Fig. 1 shows, for example, that in a Stereo system, if pixels positions are imperfectly known, then the corresponding 3-D point varies in an area with a quite anisotropic diamond shape. This is a clear example of a relation between pixel uncertainty and geometric (the position of point M) uncertainty. Another source of uncertainty in Stereo is the calibration uncertainty. In a stereo rig, intrinsic parameters of the cameras such as focal length, and extrinsic parameters such as relative position and orientation of the cameras have to be calculated. Fig. 2 shows the effect on the reconstruction of a point M of an uncertainty on the focal lengths of the two cameras. Again, M varies in a diamond-like shape. Of course, this source of uncertainty adds to the previous pixel uncertainty.

Another example of the propagation of uncertainty is given in Fig. 3 where pixels in left and right images are grouped into line segments: pixel uncertainty is converted into 2-D line uncertainty. Line segments are then matched and used to reconstruct 3-D line segments: 2-D line uncertainty and calibration uncertainty are converted into 3-D uncertainty.

Yet another set of examples of this kind of propagation is shown in Fig. 4 where coplanar and cocylindrical line segments are grouped together; again, the question is, what is the uncertainty on the plane or on the cylinder? (the uncertainty on the position of the lines, plane, and cylinder is represented on the picture in a symbolic manner by ellipses).

From these examples, we see that the main problem that needs to be solved in order to build local 3-D descriptions of the environment is how geometric uncertainty propagates when we build more complex primitives from simpler ones. This, in turn, generates two questions:

1) How do we represent geometric primitives?
2) How do we represent uncertainty on these primitives?

B. Update Position and Motion Information

Fig. 5 shows a measurement of a physical point made in two positions 1 and 2 of a mobile vehicle. In position 1, it "sees" M with some uncertainty represented by the ellipse around it. In position 2, it "sees" P with another uncertainty. Assuming that the displacement between 1 and 2 is exactly known, it is possible to express P and M in the same coordinate system. If the displacement estimate is wrong, as it is in Fig. 5, the two zones of uncertainty do not intersect and it is very unlikely that the observer will realize that the points M and P are instances of the same physical point. If we now take into account the uncertainty on the displacement (assuming that we can estimate it) we have Fig. 6 where the combination of

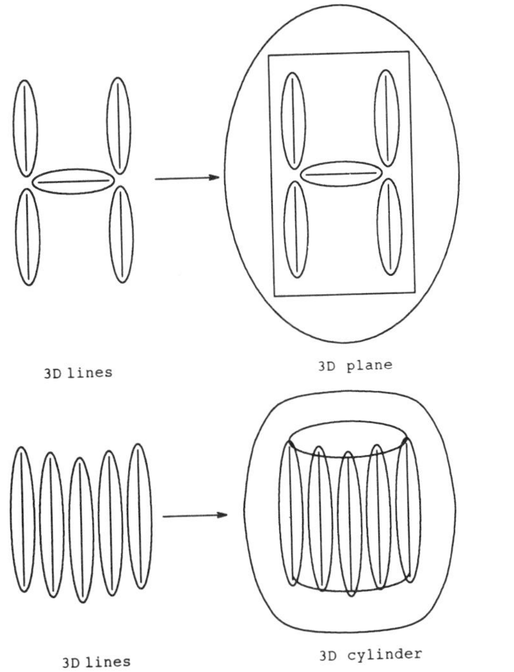

Fig. 4. From 3-D line uncertainty to 3-D surface uncertainty.

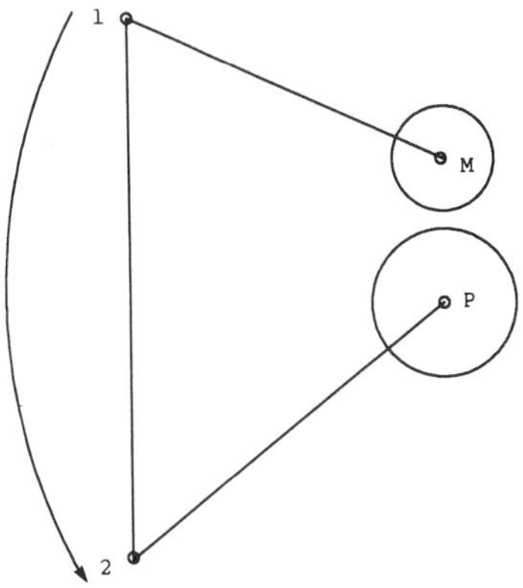

Fig. 5. Measuring a point in two positions (wrong displacement estimation).

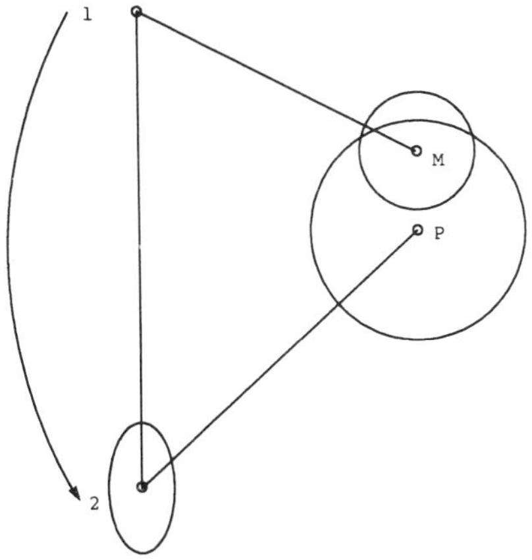

Fig. 6. Measuring a point in two positions (displacement and uncertainty estimation).

displacement uncertainty and measurement uncertainty produces a larger ellipse around P which intersects the one around M: the observer can now infer that the probability of M and P being the same physical point is quite high and use the two measurements to obtain a better estimate of the displacement

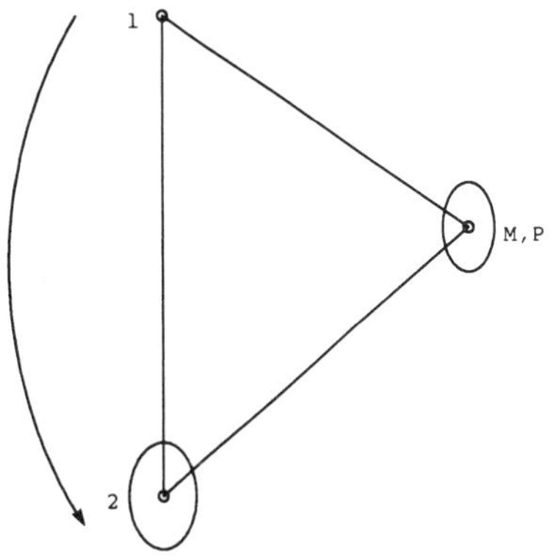

Fig. 7. Improving the estimation of the points position.

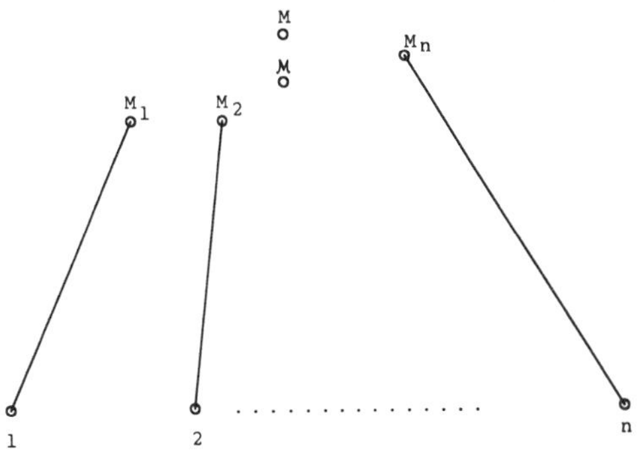

Fig. 8. Fusing n points measured from different positions.

and reduce its uncertainty. We explain how to do this in Section V. The measurements can also be used to produce better estimates of the positions (Fig. 7). This is related to what we call geometric fusion.

C. Fusing Geometric Entities

Fig. 8 shows a slightly more general case than what is depicted in Fig. 7. The mobile vehicle has measured the physical point M in n positions numbered from 1 and n. Each measurement yields a point M_i, $i = 1, \cdots, n$ and some uncertainty in the coordinate system attached to the robot. Displacement uncertainty is also available. Using the ideas described in Section V, we can improve the estimates of the displacements and reduce their uncertainty by discovering that points $M_1, \cdots, M_n$ are all instanciations of the same point. We can also use this observation to reduce the uncertainty on, let us say M_1, by combining the n measurements and produce a point $\mathfrak{M}$, fusion of $M_1, \cdots, M_n$, as well as its related uncertainty. The points $M_1, \cdots, M_n$ can then be erased from the representation of the environment, they can be forgotten. What remains is the point $\mathfrak{M}$ expressed in the coordinate system attached to position 1, for example, and the displacement from 1 to 2, 2 to 3, etc...., which allows us to express $\mathfrak{M}$ in the other coordinate systems.

Fusing geometric entities is therefore the key to ''intelligent'' forgetting which, in turn, prevents the representation of the environment from growing too large.

D. Discovering ''Interesting'' Geometric Relations

Using this approach also allows us to characterize the likelihood that a given geometric relation exists between a

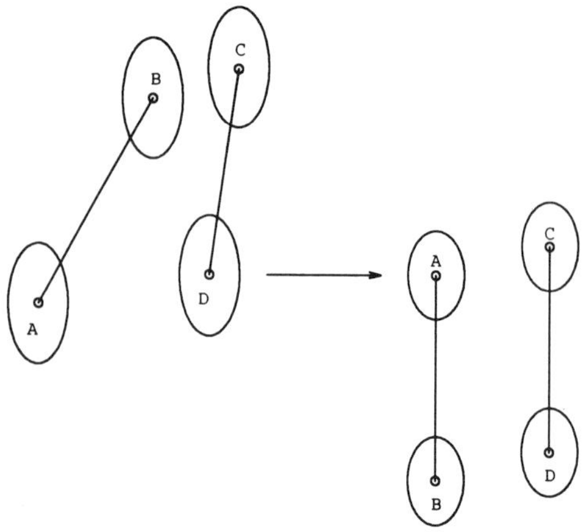

Fig. 9. Discovering that AB and CD are parallel.

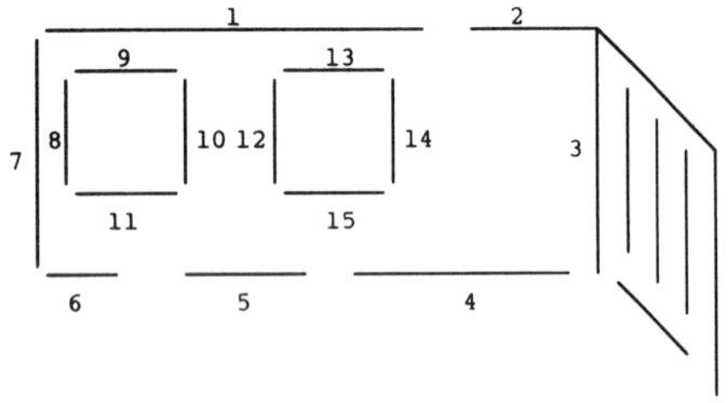

Fig. 10. Hypothesizing walls, windows, and doors.

number of geometric entities and to use this information to obtain better estimates of these entities and reduce their uncertainty. For example, as shown in Fig. 9, segments AB and CD which have uncertainty attached to their endpoints have a high likelihood to be parallel. Assuming that they are, we can update their position (they become more parallel) and reduce the uncertainty of their endpoints. The same reasoning can be used, for the relation "to be perpendicular."

E. Discovering Semantic Entities

Fig. 10 shows the kind of "semantic" grouping that is of interest to us in the context of a mobile robot moving indoors, to combine geometry and some *a priori* description of the environment. The line segments numbered from 1 to 15 are found, using the ideas described in Section II-D, to be coplanar with a high probability; the corresponding plane is found to be vertical with a very high probability which can be deduced from the geometric uncertainty of the line segments. This observation can then be used to infer that the plane has a high probability to be a wall. If we also observe that segments 8 to 11 and 12 to 15 form approximately two rectangles this can be used to infer that they have a high probability to be parts of a window or a door.

III. What is the Tool that we are Using

In this section, we introduce the Extended Kalman Filter (EKF) formalism which is applied in Sections IV and V to solve the problems we have just listed in Section II.

A. Unifying the Problems

In all of these previously listed problems, we are confronted with the estimation of an unknown parameter $\boldsymbol{a} \in R^n$ given a set of k possibly nonlinear equations of the form

$$f_i(\boldsymbol{x}_i, \boldsymbol{a}) = \boldsymbol{0} \tag{1}$$

where $\boldsymbol{x}_i \in R^m$ and f_i is a function from $R^m \times R^n$ into R^p. The vector $\boldsymbol{x}_i$ represents some random parameters of the function f_i in the sense that we only measure an estimate $\hat{\boldsymbol{x}}_i$ of them, such that

$$\hat{\boldsymbol{x}}_i = \boldsymbol{x}_i + \boldsymbol{v}_i \tag{2}$$

where $\boldsymbol{v}_i$ is a random error. The only assumption we make on $\boldsymbol{v}_i$ is that its mean is zero, its covariance is known, and that it is a white noise

$$E[\boldsymbol{v}_i] = 0$$

$$E[\boldsymbol{v}_i \boldsymbol{v}_i^t] = \boldsymbol{\Lambda}_i \geq \boldsymbol{0}$$

$$E[\boldsymbol{v}_i \boldsymbol{v}_j^t] = 0 \qquad \forall i \neq j.$$

These assumptions are reasonable. If the estimator is biased, it is often possible to subtract its mean to get an unbiased one. If we do not know the covariance of the error (or at least an upper bound of it), the estimator is meaningless. If two measurements $\hat{\boldsymbol{x}}_i$ and $\hat{\boldsymbol{x}}_j$ are correlated, we take the concatenation of them $\hat{\boldsymbol{x}}_k = (\hat{\boldsymbol{x}}_i, \hat{\boldsymbol{x}}_j)$ and the concatenated vector function $f_k = [f_i^t, f_j^t]^t$. The problem is to find the optimal estimate $\hat{\boldsymbol{a}}$ of $\boldsymbol{a}$ given the function f_i and the measurements $\hat{\boldsymbol{x}}_i$.

B. Linearizing the Equations

The most powerful tools developed in parameter estimation are for linear systems. We decided to apply these tools to a linearized version of our equations. This is the EKF approach that we now develop.

For each nonlinear equation $f_i(\boldsymbol{x}_i, \boldsymbol{a}) = \boldsymbol{0}$ we need to know an estimate $\hat{\boldsymbol{a}}_{i-1}$ of the sougth parameter $\boldsymbol{a}$, and again a measure S_i of the confidence we have in this estimate.[1] Actually, we model probalistically the current estimate $\hat{\boldsymbol{a}}_{i-1}$ of $\boldsymbol{a}$ by assuming that

$$\hat{\boldsymbol{a}}_{i-1} = \boldsymbol{a} + \boldsymbol{w}_i \tag{3}$$

where $\boldsymbol{w}_i$ is a random error. The only assumptions we make on $\boldsymbol{w}_i$ are the same as for $\boldsymbol{v}_i$, i.e.,

$$E[\boldsymbol{w}_i] = 0$$

$$E[\boldsymbol{w}_i \boldsymbol{w}_i^t] = S_i \geq \boldsymbol{0}$$

where S_i is a given non-negative matrix. Here again, no assumption of gaussianness is required.

Having an estimate $\hat{\boldsymbol{a}}_{i-1}$ of the solution, the equations are linearized by a first-order Taylor expansion around $(\hat{\boldsymbol{x}}_i, \hat{\boldsymbol{a}}_{i-1})$

$$f_i(\boldsymbol{x}_i, \boldsymbol{a}) = \boldsymbol{0} \approx f_i(\hat{\boldsymbol{x}}_i, \hat{\boldsymbol{a}}_{i-1}) + \frac{\widehat{\partial f_i}}{\partial \boldsymbol{x}}(\boldsymbol{x}_i - \hat{\boldsymbol{x}}_i) + \frac{\widehat{\partial f_i}}{\partial \boldsymbol{a}}(\boldsymbol{a} - \hat{\boldsymbol{a}}_{i-1}) \tag{4}$$

where the derivatives $\widehat{\partial f_i}/\partial \boldsymbol{x}$ and $\widehat{\partial f_i}/\partial \boldsymbol{a}$ are estimated at $(\hat{\boldsymbol{x}}_i, \hat{\boldsymbol{a}}_{i-1})$.

[1] In practice, we shall see that only an initial estimate $(\hat{\boldsymbol{a}}_0, S_0)$ of $\boldsymbol{a}$ is required prior to the first measurement $\hat{\boldsymbol{x}}_1$, while the next ones $(\hat{\boldsymbol{a}}_i, S_i)$ are provided automatically by the Kalman filter itself.

Equation (4) can be rewritten as

$$y_i = M_i a + u_i \tag{5}$$

where

$$y_i = -f_i(\hat{x}_i, \hat{a}_{i-1}) + \widehat{\frac{\partial f_i}{\partial a}} \hat{a}_{i-1}$$

$$M_i = \widehat{\frac{\partial f_i}{\partial a}}$$

$$u_i = \widehat{\frac{\partial f_i}{\partial x}} (x_i - \hat{x}_i).$$

Equation (5) is now a linear measurement equation, where y_i is the new measurement, M_i is the linear transformation, u_i is the random measurement error. Both y_i and M_i are readily computed from the actual measurement $\hat{x}_i$, the estimate $\hat{a}_{i-1}$ of a, the function f_i, and its first derivative. The second-order statistics of u_i are derived easily from those of v_i

$$E[u_i] = 0$$

$$W_i \triangleq E[u_i u_i^t] = \widehat{\frac{\partial f_i}{\partial x}} \Lambda_i \widehat{\frac{\partial f_i^t}{\partial x}}.$$

C. Recursive Kalman Filter

When no gaussianness is assumed on the previous random errors u_i, v_i, and w_i, the Kalman filter equations provide the best (minimum variance) linear unbiased estimate of a. This means that among the estimators which seek a_k as a linear combination of the measurements $\{y_i\}$, it is the one which minimizes the expected error norm squared

$$E[(\hat{a}_k - a)^t(\hat{a}_k - a)]$$

while verifying

$$E[\hat{a}_k] = a.$$

The recursive equations of the Kalman filter which provide a new estimate $(\hat{a}_i, S_i)$ of a from $(\hat{a}_{i-1}, S_{i-1})$ are as follows [23]:

$$\hat{a}_i = \hat{a}_{i-1} + K_i(y_i - M_i\hat{a}_{i-1}) \tag{6}$$

$$K_i = S_{i-1}M_i^t(W_i + M_iS_{i-1}M_i^t)^{-1} \tag{7}$$

$$S_i = (I - K_iM_i)S_{i-1} \tag{8}$$

or equivalently

$$S_i^{-1} = S_{i-1}^{-1} + M_i^t W_i^{-1} M_i. \tag{9}$$

One can see that the previously estimated parameter $\hat{a}_{i-1}$ is corrected by an amount proportional to the current error $y_i - M_i\hat{a}_{i-1}$ called the innovation. The proportionality factor K_i is called the Kalman gain. At the end of the process, $\hat{a}_k$ is the final estimate and S_k represents the covariance of the estimation error

$$S_k = E[(\hat{a}_k - a)(\hat{a}_k - a)^t].$$

The recursive process is initialized by $\hat{a}_0$, an initial estimate of a, and S_0, its error covariance matrix. Actually, the criterion minimized by the final estimate $\hat{a}_k$ is

$$C = (a - \hat{a}_0)^t S_0^{-1}(a - \hat{a}_0) + \sum_{i=1}^{k} (y_i - M_i a)^t W_i^{-1}(y_i - M_i a). \tag{10}$$

It is interesting to note that the first term of (10) measures the squared distance of a from an initial estimate, weighted by its covariance matrix, while the second term is nothing else but the classical least square criterion, i.e., the sum of the squared measurement errors weighted by the covariance matrices. Indeed, initializing the process with an arbitrary $\hat{a}_0$ and $S_0^{-1} = 0$, criterion (10) provides the classical least square estimate $\hat{a}_k$ obtained from the measurements only, while the initial estimate does not play any role.

The enormous advantage of such a recursive solution is that if we decide, after a set of k measurements $\{\hat{x}_i\}$, to stop the measures, we only have to keep $\hat{a}_k$ and $\hat{S}_k$ as the whole memory of the measurement process. If we decide later to take into account additional measurements, we simple have to initialize $\hat{a}_0 \sim \hat{a}_k$ and $S_0 \sim S_k$ and to process the new measurements to obtain exactly the same solution as if we had processed all the measurements together.

D. Gaussian Assumption

Up to now, we did not introduce any Gaussian assumptions on the random measurement errors $v_i = x_i - \hat{x}_i$ of (2) and on the prior estimate error $w_0 = a - \hat{a}_0$ of (3). However, in practice, these errors usually come from a sum of independent random processes, which tend toward a Gaussian process (Central Limit theorem). If we actually identify v_i and w_0 with Gaussian processes, i.e.,

$$v_i \sim N(0, \Lambda_i)$$

$$w_0 \sim N(0, S_0)$$

then, it follows that the noise u_i in (5) is also Gaussian, i.e.,

$$u_i \sim N(0, W_i)$$

and that all the successive estimates provided by the recursive Kalman filter are also Gaussian, with mean a and covariance S_k

$$\hat{a}_k \sim N(a, S_k).$$

Moreover, in this case, the Kalman filter provides the best (minimum variance) unbiased estimate $\hat{a}_k$ among all, even nonlinear, filters. This estimate $\hat{a}_k$ is also the maximum likelihood estimator of a. This comes from the fact that in the Gaussian case, the solution is the conditional mean $\hat{a}_k = E[a/y_1, \cdots, y_k]$ which both minimizes the variance and maximizes the likelihood while being expressed as a linear combination of the measurements y_i. Therefore, in this case, the minimum variance and minimum variance linear estimates are the same; namely, the estimate $\hat{a}_k$ provided by the Kalman filter [23].

In conclusion, in the Gaussian case, the Kalman filter provides the best estimate with the advantage of preserving gaussianness of all the implied random variables, which means that no information on the probability density functions of the parameters is lost while keeping only their mean and covariance matrix.

E. Rejecting Outlier Measurements

At iteration i, we have an estimate $\hat{a}_{i-1}$ and an attached covariance matrix S_{i-1} for parameter a. We also have a noisy measurement $(\hat{x}_i, \Lambda_i)$ of x_i and we want to test the plausibility of this measurement with respect to the equation $f_i(x_i, a) = 0$.

If we consider again a first-order expansion of $f_i(x_i, a)$ around $(\hat{x}_i, \hat{a}_{i-1})$ (4), considering that $(\hat{x}_i - x_i)$ and $(\hat{a}_{i-1} - a)$ are independent centered Gaussian processes, we see that $f_i(\hat{x}_i, \hat{a}_{i-1})$ is also (up to a linear approximation) a centered Gaussian process whose mean and covariance are given by

$$E[f_i(\hat{x}_i, \hat{a}_{i-1})] = 0$$

$$Q_i = E[f_i(\hat{x}_i, \hat{a}_{i-1}) f_i(\hat{x}_i, \hat{a}_{i-1})^t] = \widehat{\frac{\partial f_i}{\partial x}} \Lambda_i \widehat{\frac{\partial f_i^t}{\partial x}} + \widehat{\frac{\partial f_i}{\partial a}} S_{i-1} \widehat{\frac{\partial f_i^t}{\partial a}}.$$

Therefore, if the rank of Q_i is q, the generalized Mahalanobis distance

$$d(\hat{x}_i, \hat{a}_{i-1}) = [f_i(\hat{x}_i, \hat{a}_{i-1})]^t Q_i^{-1} [f_i(\hat{x}_i, \hat{a}_{i-1})] \tag{11}$$

has a χ^2 distribution with q degrees of freedom.[2]

Looking at a χ^2 distribution table, it is therefore possible to reject an outlier measurement $\hat{x}_i$ at a 95-percent confidence rate by setting an appropriate threshold ϵ on the Mahalanobis distance, and by keeping only those measurements $\hat{x}_i$ which verify

$$d(\hat{x}_i, \hat{a}_{i-1}) < \epsilon. \tag{12}$$

We shall see in the experimental section at the end of this paper how this formalism can be used in practice, and how well it fits with reality.

IV. Geometric Representations

In this section, we give the details of the geometric representations that we have found useful at various stages of our work. It is first important to note that we have been dealing so far only with points, lines, and planes, i.e., with affine geometric entities. This may appear to be quite a restriction on the type of environments that we can cope with. This is indeed the case but there are a number of reasons why we think that our approach is quite reasonable.

1) The obvious one is that for the kind of environment that our mobile robot moves into, these primitives are very likely to cover most of the geometric features of importance.

2) A second reason is that more complicated curved features can be first approximated with affine primitives which are then grouped into more complicated nonaffine primitives.

3) A third reason is that we believe that the techniques we have developed for representing and combining uncertainty of affine primitives are generic and directly applicable to nonaffine primitives.

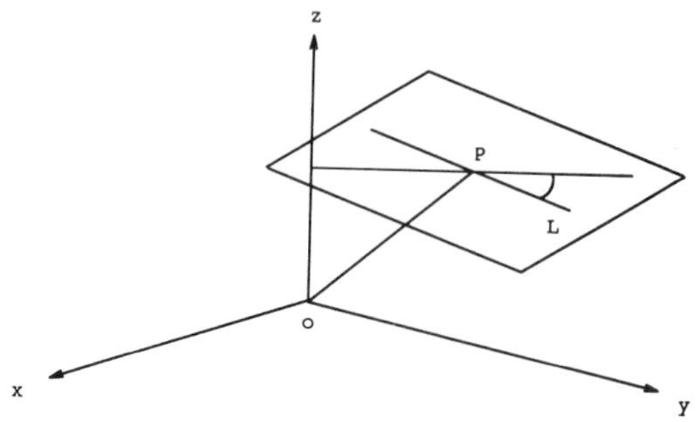

Fig. 11. A possible 3-D line representation.

Let us now discuss specifically lines, planes, and rigid displacements.

A. Line Segments

The 3-D segments that we deal with are usually constructed from stereo [4], [9]. Their endpoints may be quite unreliable, even though they can be of some use from time to time, and we largely depend on the infinite lines supporting those line segments.

We concentrate here on how to represent 3-D lines. The obvious representation we mention here only for pedagogical reasons, is by two points; this representation is six-dimensional and, as we will see next, not minimal. Another way to represent a line is to choose a point on it (three parameters), and a unit vector defining its direction (two parameters). The corresponding representation is five-dimensional and, again, not minimal. In fact, the set of affine 3-D lines is a manifold of dimension 4 for which we will exhibit later an atlas of class C^∞.[3] This implies that a minimal representation of a straight line has four parameters.

One such representation can be obtained by considering the normal to the line from the origin (if the line goes through the origin it is the same as a vector line and can be defined by two parameters only). The point of intersection between the normal and the line is represented by three parameters. If we now consider (see Fig. 11) the plane normal at P to OP, the line is in that plane and can be defined by one more parameter, its angle with an arbitrary direction, for example, the line defined by P, and one of the axis of coordinates (in Fig. 11, the z axis). Of course, when the line is parallel to the xy plane this direction is not defined and we must use either the x or the y axis. This brings up an interesting point, namely, that a global minimal representation for affine lines, i.e., one which can be used for all such lines, does not exist. We must choose the representation as a function of the line orientation. Mathematically, this means that the manifold of the affine straight lines cannot be defined with only one map. This is quite common and is also true for affine planes and rotations of R^3, as will be shown next.

The previous representation for a line is not in fact the one

[2] If $q < p$ = the size of the measurement vector f_i, Q_i^{-1} is the pseudo-inverse of Q_i.

[3] Grossly speaking, a manifold of dimension d is a set that can be defined locally by d parameters. When the functions that transform one set of parameters into another are p times differentiable, the manifold is said to be of class C^p. For more details, see [14].

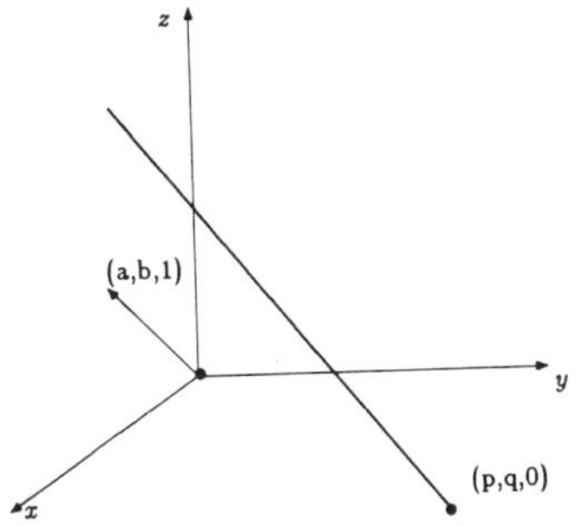

Fig. 12. A better 3-D line representation.

we have been using. In effect, the parameters involved in the previous representation are usually combined in a highly nonlinear manner in the measurement equations expressing geometric relationships between geometric entities (cf. next section), which is not good for the extended Kalman filtering approach. Also, the angular parameter must be assigned some fixed bounds (for instance $]0, \pi[$), which might cause some problems during a recursive evaluation with the Kalman filter. This latter constraint also appears in the representation recently proposed by Roberts [31].

Therefore, we prefer the following representation in which the retained parameters are usually combined linearly in the measurement equations, and are not constrained to any bounded interval. This representation considers a line (not perpendicular to the z axis) as the intersection of a plane parallel to the y axis, and a plane parallel to the x axis

$$\begin{cases} x = az + p \\ y = bz + q. \end{cases} \tag{13}$$

The intersection is represented by the four-dimensional vector $\boldsymbol{L} = [a, b, p, q]^T$ which has the following geometric interpretation (see Fig. 12): The direction of the line is that of the vector $[a, b, 1]^T$, and the point of intersection of the line with the xy plane has coordinates p and q. Since the last coordinate of the direction vector is equal to 1, the line cannot be perpendicular to the z axis or parallel to the xy plane. If we want, and we do in practice, represent such lines, we must choose another representation, for example

$$\begin{cases} y = ax + p \\ z = bx + q \end{cases} \tag{14}$$

which cannot represent lines parallel to the yz plane, or perpendicular to the x axis, or

$$\begin{cases} z = ay + p \\ x = by + q \end{cases} \tag{15}$$

which excludes lines parallel to the zx plane.

Each representation defines a one-to-one mapping between R^4 and a subset (in fact an open subset) of the set of affine 3-D lines and it can be shown that these three mappings define on this set a structure of C^∞ manifold for which they form an atlas. In practice, this means the representation is not exactly four-dimensional, but is made of the four numbers a, b, p, and q and an integer i taking the values 1, 2, and 3 to indicate which map 13, 14, or 15 we are currently using.

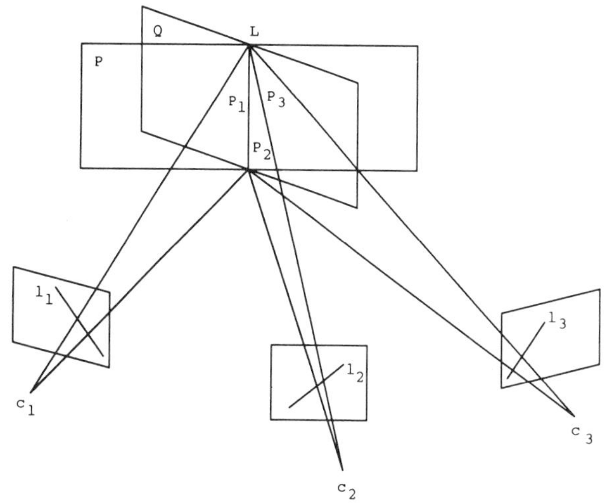

Fig. 13. Reconstruction of 3-D lines.

The fact that the set of affine 3-D lines has been given a structure of C^∞ manifold implies that the a', b', p', q' of a given representation are C^∞ functions of the a, b, p, q of another representation for all lines for which the two representations are well defined (for example, all lines not parallel to the xy and yz planes). The representation of a line also includes a 4×4 covariance matrix $\boldsymbol{\Delta}_L$ on the vector $\boldsymbol{L}$.

It is interesting at this stage to trace the computation of this covariance matrix all the way from pixel to 3-D. In order to do this, we must briefly explain how 3-D lines are computed in our current Stereo system [9]. We use three cameras as indicated in Fig. 13. In theory, the three planes defined by the 2-D lines l_1, l_2, and l_3 and the optical centers C_1, C_2, and C_3 belong to the same pencil and intersect along the 3-D line L. In practice they do not because of noise, and we have to find the "best" line satisfying the measurements, i.e., l_1, l_2, and l_3. This can be done by using the idea of pencil of planes, described more fully in [21]. We assume that in the coordinate system attached to camera 1, for example, the equation of the ith plane P_i, $i = 1, 2, 3$, is given by

$$u_i x + v_i y + w_i z + r_i = 0$$

where the three four-vectors $\boldsymbol{P}_i = [u_i, v_i, w_i, r_i]^T$ are known, as well as their covariance matrix $\boldsymbol{\Lambda}_{P_i}$ (we show later how to compute them). If we use representation (13) for the 3-D line, it is represented as the intersection of the two planes P of the equation $x = az + p$ and Q of the equation $y = bz + q$. Writing that the five planes P, Q, and P_i, $i = 1, 2, 3$, form a pencil allows us to write six equations

$$\begin{cases} w_i + au_i + bv_i = 0, \\ r_i + pu_i + qu_i = 0, \end{cases} \qquad i = 1, 2, 3$$

in the four unknowns a, b, p, and q.

We can apply directly the Kalman formalism to these measurement equations and choose $\boldsymbol{a} = [a, b, p, q]^T$, and x_i as the four-vector $\boldsymbol{P}_i$. We can therefore simply compute an estimate $\hat{\boldsymbol{a}}$ of $\boldsymbol{a}$ and its covariance matrix $\boldsymbol{\Lambda}_{\hat{a}}$ from the $\boldsymbol{P}_i$'s and $\boldsymbol{\Lambda}_{P_i}$'s.

Let us now show how we can compute the $\boldsymbol{P}_i$'s and $\boldsymbol{\Lambda}_{P_i}$'s. Each line l_i, $i = 1, 2, 3$, is obtained by fitting a straight line to

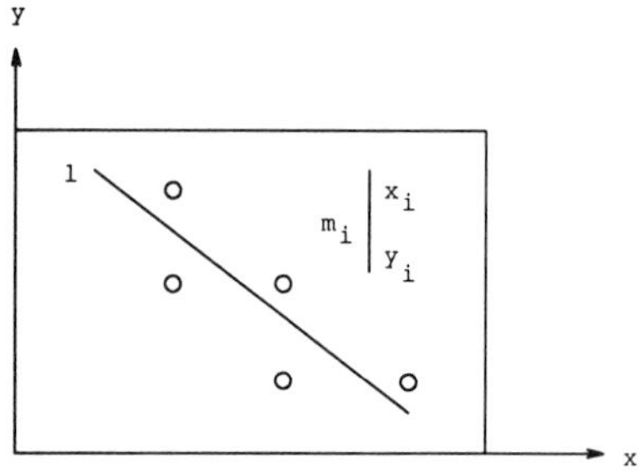

Fig. 14. 2-D line approximation.

a set of edge pixels which have been detected using a modified version of the Canny edge detector [13], [17]. Looking at Fig. 14, let $x \cos \theta + y \sin \theta - p = 0$ be the equation of the line l which is fit to the edge pixels m_i of coordinates x_i, y_i, $(0 \leq \theta < 2\pi, \rho \geq 0)$. We assume that the measured edge pixels are independent and corrupted by a gaussian isotropic noise and take the parameter $\boldsymbol{a}$ equal to $[\theta, \rho]^T$ and the measurement $\boldsymbol{x}$ as the vector $[x, y]^T$. The measurement equation is therefore

$$f(\boldsymbol{x}, \boldsymbol{a}) = x \cos \theta + y \sin \theta - \rho.$$

Applying the EKF formalism to the n-edge pixels forming the line provides the best estimate $\hat{\boldsymbol{a}}$ of the line parameters and its covariance matrix. Having done this for all three cameras, it is easy to deduce the equations of the three planes P_i and the covariance matrices on their coefficients.

B. Planes

Planes can receive pretty much the same treatment as lines. A plane is defined by three parameters, and this is minimal. A possible representation is the representation by the normal $\hat{\boldsymbol{n}}$ (a unit norm vector), and the distance d to the origin. The problem with this representation is that it is not unique since $(-\boldsymbol{n}, -d)$ represents the same plane. It is possible to fix that problem by assuming that one component of $\boldsymbol{n}$, say n_z, is positive, i.e., we consider planes not parallel to the z axis. For these planes we must choose another convention, for example, that n_z is positive. Again, this works well for planes not parallel to the x axis. The third possible representation is to assume n_y positive which excludes planes parallel to the y axis.

So, we have three one-to-one mappings of open subsets of the product $S_2 \times R$, where S_2 is the usual Gaussian sphere into open subsets of the set of planes

$$(\boldsymbol{n}, d),\ n_z > 0 \rightarrow \text{planes not parallel to } Oz$$

$$(\boldsymbol{n}, d),\ n_x > 0 \rightarrow \text{planes not parallel to } Ox$$

$$(\boldsymbol{n}, d),\ n_y > 0 \rightarrow \text{planes not parallel to } Oy.$$

It is easy to show that these three mappings define on the set of 3-D planes a structure of C^∞ manifold of dimension 3.

One practical disadvantage of the previous representations is that the normal $\boldsymbol{n}$ is constrained to lie on the unit sphere S_2, i.e., it must satisfy the constraint $\|\boldsymbol{n}\| = 1$. A possibly simpler representation is obtained by considering the mapping from R^3 to the set of 3-D planes defined by

$$p_1 : (a, b, c) \rightarrow ax + by + z + c = 0. \tag{16}$$

This can represent all planes except those parallel of Oz and it is a one-to-one continuous mapping from R^3 to the open subset of the set of 3-D planes constituted of the planes not parallel to the z axis. In order to obtain all possible planes, we must also consider the mappings

$$p_2 : (a, b, c) \rightarrow x + ay + bz + c = 0 \tag{17}$$

$$p_3 : (a, b, c) \rightarrow bx + y + az + c = 0. \tag{18}$$

p_2 (respectively, p_3) excludes planes to the x axis (respectively, the y axis). It is easy to show that p_1, p_2, p_3 also define on the set of 3-D planes a structure of C^∞ manifold of dimension 3.

C. Rigid Displacements

In a previous paper [4], [6] we have proposed the use of the exponential representation of rotations. This is the same as saying that a rotation is defined by its axis $\boldsymbol{u}$ (a unit vector) and its angle θ. The vector $\boldsymbol{r} = \theta \boldsymbol{u}$ can be used to represent the rotation and we have

$$\boldsymbol{R} = e^{\boldsymbol{H}}$$

where $\boldsymbol{H}$ is an antisymmetric matrix representing the cross product with the vector $\boldsymbol{r}$ (i.e., $\boldsymbol{Hx} = \boldsymbol{r} \times \boldsymbol{x}$, for all $\boldsymbol{x}$). In this case, the rotation is represented by the three coordinates of $\boldsymbol{r}$, i.e., by three independent numbers. There are several other possible representations for rotations, the most widely known being the one with orthogonal matrices or quaternions. Their main disadvantage is that an orthogonal matrix is defined by nine numbers subject to six quadratic constraints, whereas a quaternion is defined by four numbers subject to one quadratic constraint. These constraints are not easy to deal with in the EKF formalism and, moreover, these two representations are more costly than the exponential one.

Let us see how we can define a structure of manifold on the set of rotations using this representation. If we allow θ to vary over the semi-open interval $[0, 2\pi[$, the vector $\boldsymbol{r}$ can vary in the open ball $B(0, 2\pi)$ of R^3 of radius 2π. But the mapping f: $B(0, 2\pi)$ into the set of rotations is not one to one because $(\boldsymbol{u}, \pi)$ and $(-\boldsymbol{u}, \pi)$ represent the same rotation. To enforce uniqueness we can assume that one of the coordinates, for example u_z, of the rotation axis $\boldsymbol{u}$ is positive. We can then represent uniquely the open subset of the set of rotations for which the axis is not perpendicular to the z axis, and has a positive component along the axis, and the mapping is continuous. If we consider the open set of rotations defined by $(\boldsymbol{u}, \theta)$, $u_z < 0$, we have another one-to-one continuous mapping. With these two mappings, we cannot represent rotations with an axis perpendicular to the z axis. In order to obtain all possible rotations, we have to introduce the other four mappings defined by $(\boldsymbol{u}, \theta)$ and $u_x > 0$ (respectively, $u_x < 0$, $u_y > 0$, $u_y < 0$) which represent rotations with an axis not perpendicular to the x axis (respectively, the y axis). We are sill missing the null vector, i.e., we have no representation for the null rotation, the identity matrix. In order to include it, we have to add a seventh map by considering for example the rotations defined by the "small" open ball $B(0, \epsilon)$ where ϵ

must be smaller than π. These seven mappings define on the set of rotations a structure of C^∞ manifold of dimension 3.[4]

It is interesting that in all three cases (3-D lines, planes, and rotations), unique global representation does not exist and that we must deal with at least three local mappings.

It is now instructive to study how the group of rigid displacements operates on the representations for lines and planes.

1) Applying Rigid Displacement to Lines: The easiest way to derive how representation (13) changes under rotation and translation is by considering that the line is defined by two points M_1 and M_2 of coordinates (x_1, y_1, z_1) and (x_2, y_2, z_2). It is then easy to verify that

$$a=\frac{x_2-x_1}{z_2-z_1} \quad b=\frac{y_2-y_1}{z_2-z_1}$$

$$p=\frac{z_2x_1-z_1x_2}{z_2-z_1} \quad q=\frac{z_2y_1-z_1y_2}{z_2-z_1}.$$

Introducing the vector $M_1M_2 = [A, B, C]^T$, we have $a = A/C$, and $b = B/C$. a and b are therefore only sensitive to rotation

$$M_1M_2 \to \boldsymbol{R}\, M_1M_2$$

$$\begin{bmatrix} A \\ B \\ C \end{bmatrix} \to \begin{bmatrix} A' \\ B' \\ C' \end{bmatrix} = \boldsymbol{R} \begin{bmatrix} A \\ B \\ C \end{bmatrix}$$

$$\begin{bmatrix} a \\ b \end{bmatrix} \to \begin{bmatrix} a' \\ b' \end{bmatrix} = \begin{bmatrix} A'/C' \\ B'/C' \end{bmatrix}.$$

This yields

$$a' = \frac{\boldsymbol{r}_1 \cdot \boldsymbol{m}}{\boldsymbol{r}_3 \cdot \boldsymbol{m}} \qquad b' = \frac{\boldsymbol{r}_2 \cdot \boldsymbol{m}}{\boldsymbol{r}_3 \cdot \boldsymbol{m}}$$

where $\boldsymbol{m} = [a, b, 1]^T$, and the $\boldsymbol{r}_i$'s are the row vectors of matrix $\boldsymbol{R}$. This is true only if $\boldsymbol{r}_3 \cdot \boldsymbol{m} \neq 0$; if $\boldsymbol{r}_3 \cdot \boldsymbol{m} = 0$, the transformed line is perpendicular to the z axis and representation (14) or (15) must be used.

To treat the case of p and q, let us introduce $P = p(z_2 - z_1) = pC$ and $Q = q(z_2 - z_1) = qC$. It is easy to show that

$$\begin{bmatrix} P \\ Q \end{bmatrix} = \begin{bmatrix} 0 & -1 & 0 \\ 1 & 0 & 0 \end{bmatrix} OM_1 \times OM_2 = H(OM_1 \times OM_2).$$

This allows us to study how P and Q change under rotation and translation

$$OM_1 \to \boldsymbol{R}\, OM_1 + \boldsymbol{t} \qquad OM_2 \to \boldsymbol{R}\, OM_2 + \boldsymbol{t}.$$

Therefore

$$\begin{bmatrix} P \\ Q \end{bmatrix} \to \begin{bmatrix} P' \\ Q' \end{bmatrix} = H(\boldsymbol{R}(OM_1 \times OM_2) + \boldsymbol{t} \times \boldsymbol{R}\, M_1M_2).$$

Using the previous notations, $M_1M_2 = [A, B, C]^t$, and $OM_1 \times OM_2 = [Q, -P, X]^t$ where X is unknown. But noticing that $M_1M_2 \cdot (OM_1 \times OM_2) = 0$ we have

$$AQ - BP + CX = 0$$

and therefore

$$X = \frac{BP - AQ}{C} = bP - aQ.$$

C is not equal to 0 since by definition, the line is not perpendicular to the z axis. Putting everything together

$$\begin{bmatrix} P' \\ Q' \end{bmatrix} = CH\left(\boldsymbol{R}\begin{pmatrix} q \\ -p \\ bp-aq \end{pmatrix} + \boldsymbol{t} \times \boldsymbol{Rm}\right).$$

Finally

$$\begin{bmatrix} p' \\ q' \end{bmatrix} = \begin{bmatrix} P'/C' \\ Q'/C' \end{bmatrix} = \frac{C}{C'} \boldsymbol{H}\left(\boldsymbol{R}\begin{pmatrix} q \\ -p \\ bp-aq \end{pmatrix} + \boldsymbol{t} \times \boldsymbol{Rm}\right)$$

and we know from the previous derivation that $C/C' = 1/\boldsymbol{r}_3 \cdot \boldsymbol{M}$, therefore

$$\begin{bmatrix} p' \\ q' \end{bmatrix} = \frac{1}{\boldsymbol{r}_3 \cdot \boldsymbol{m}} H(\boldsymbol{Rp} + \boldsymbol{t} \times \boldsymbol{Rm})$$

where we have taken $\boldsymbol{p} = [q, -p, bp - aq]^t$.

2) Applying Rigid Displacements to Planes: Given a plane represented by its normal n and its distance to the origin d, if we apply to it a rotation along an axis going through the origin represented by a matrix $\boldsymbol{R}$ followed by a translation represented by a vector $\boldsymbol{t}$, the new plane is represented by $\boldsymbol{Rn}$ and $d - \boldsymbol{t} \cdot \boldsymbol{Rn}$ [20].

This allows us to compute how the representation (16), for example, is transformed by the rigid displacement. From the previous observation:

$$\begin{pmatrix} a \\ b \\ 1 \end{pmatrix} \to \boldsymbol{R}\begin{pmatrix} a \\ b \\ 1 \end{pmatrix} \quad \text{and} \quad c \to c - \boldsymbol{t} \cdot \boldsymbol{R}\begin{pmatrix} a \\ b \\ 1 \end{pmatrix}.$$

Introducing the three row vectors $\boldsymbol{r}_1, \boldsymbol{r}_2, \boldsymbol{r}_3$ of matrix $\boldsymbol{R}$, we have, assuming that $\boldsymbol{r}_3 \cdot \boldsymbol{m} \neq 0$.

$$a' = \frac{\boldsymbol{r}_1 \cdot \boldsymbol{m}}{\boldsymbol{r}_3 \cdot \boldsymbol{m}} \qquad b' = \frac{\boldsymbol{r}_2 \cdot \boldsymbol{m}}{\boldsymbol{r}_3 \cdot \boldsymbol{m}} \qquad c' = \frac{c - \boldsymbol{t} \cdot \boldsymbol{Rm}}{\boldsymbol{r}_3 \cdot \boldsymbol{m}}$$

if $\boldsymbol{r}_3 \cdot \boldsymbol{m} = 0$, this means that we cannot use the same representation for the transformed plane since it is parallel to the z axis, therefore, we must choose the representation (17) or (18).

V. Registration, Motion, and Fusion of Visual Maps

In this section we show how to solve the problems listed in Section II within the formalism and the representations detailed in Sections III and IV.

A. Initial Assumptions

We are given two visual maps $\mathcal{V}$ and $\mathcal{V}'$, each of them attached to a coordinate reference frame $\mathfrak{F}$ and $\mathfrak{F}'$ (see Fig. 15).

[4] In [10] and [11] one can find an atlas of rotations with only four maps.

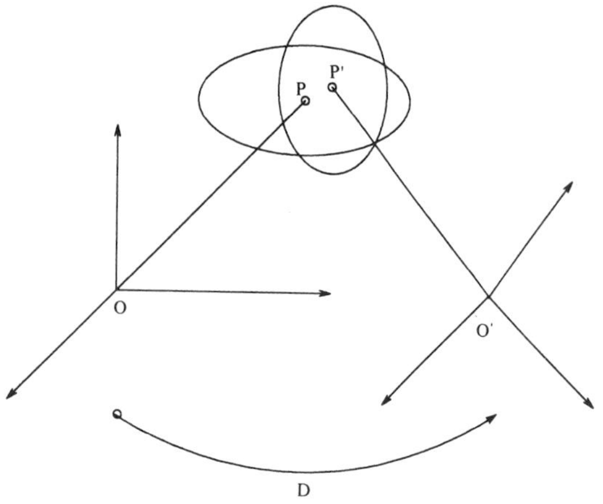

Fig. 15. The general registration motion fusion problem.

TABLE I
RELATIONS BETWEEN THE PRIMITIVES

Relations	Points	Lines	Planes
Points	$\equiv$	$\subset$	$\subset$
Lines		$\equiv \parallel \perp$	$\subset \parallel \perp$
Planes			$\equiv \parallel \perp$

Each visual map $\mathcal{V}$ is composed of primitives $\mathcal{P}$, described by a parameter vector $\boldsymbol{P}$. We have an estimate $\hat{\boldsymbol{P}}_0$ of $\boldsymbol{P}$ and an error covariance matrix $\boldsymbol{W}_{P_0}$.

The coordinate frames $\mathcal{F}$ and $\mathcal{F}'$ are related by a rigid displacement $\mathcal{D}$ such that each point M' of $\mathcal{F}'$ is related to a point M of $\mathcal{F}$ by the relation

$$O'M' = \boldsymbol{R}\ OM + \boldsymbol{t}$$

where $\boldsymbol{R}$ is the rotation matrix and $\boldsymbol{t}$ the translation vector of the displacement $\mathcal{D}$. We also have an estimate $\hat{\boldsymbol{D}}_0$ of $\boldsymbol{D}$, with an error covariance matrix $\boldsymbol{W}_{D_0}$.

B. Defining Geometric Relations

We define a set of geometric relations between the primitive $\mathcal{P}$ and $\mathcal{P}'$ of two visual maps $\mathcal{V}$ and $\mathcal{V}'$. These relations are given in Table I.

The list of relations/primitives is not exhaustive but only demonstrative. The relation "identical" expresses the fact that the primitives $\mathcal{P}$ and $\mathcal{P}'$ represented in $\mathcal{V}$ and $\mathcal{V}'$ actually describe the same physical primitive. The relation "included" expresses that $\mathcal{P}$ describes a physical primitive which is part of the physical primitive described by $\mathcal{P}'$. The relations "parallel" and "orthogonal" are interpreted in a similar fashion.

Each geometric relation can be expressed by a vector equation of the form

$$f_i(\boldsymbol{P}, \boldsymbol{P}', \boldsymbol{D}) = \boldsymbol{0}. \qquad (19)$$

C. Expressing Geometric Relations

We rewrite (19) for the geometric relations of Table I. We denote by $\bar{\boldsymbol{P}}$ the parameters of the primitive $\bar{\mathcal{P}} = \boldsymbol{D}(\mathcal{P})$, the image of $\mathcal{P}$ by the rigid displacement $\boldsymbol{D}$. The computation of $\bar{\boldsymbol{P}}$ from $\boldsymbol{P}$ is, in the case of points, $\overline{OM} = \boldsymbol{R}\ OM + \boldsymbol{t}$. The case of lines and planes was detailed in the previous section. The measurement equations are as follows:

Point-Point:

$$\text{relation} \equiv : O'M' - \overline{OM} = \boldsymbol{0}.$$

Point-Line: assuming the line is not orthogonal to the z axis:

$$\text{relation} \subset : \begin{cases} \bar{x} - a'\bar{z} - p' = 0 \\ \bar{y} - b'\bar{z} - q' = 0. \end{cases}$$

Point-Plane: assuming the plane is not parallel to the z axis:

$$\text{relation} \subset : a'\bar{x} + b'\bar{y} + \bar{z} + c' = 0.$$

Line-Line: assuming the two lines are not orthogonal to the z axis:

$$\text{relation} \equiv : (a', b', c', d')^t - (\bar{a}, \bar{b}, \bar{c}, \bar{d})^t = \boldsymbol{0}$$

$$\text{relation} \parallel : (a', b')^t - (\bar{a}, \bar{b})^t = \boldsymbol{0}$$

$$\text{relation} \perp : a'\bar{a} + b'\bar{b} + 1 = 0.$$

Line-Plane: assuming the line is not orthogonal and the plane not parallel to the z axis:

$$\text{relation} \subset : \begin{cases} a'\bar{a} + b'\bar{b} + 1 = 0 \\ a'\bar{p} + b'\bar{q} + c' = 0 \end{cases}$$

$$\text{relation} \parallel : a'\bar{a} + b'\bar{b} + 1 = 0$$

$$\text{relation} \perp : (a', b')^t - (\bar{a}, \bar{b})^t = \boldsymbol{0}.$$

Plane-Plane: assuming the plane is not parallel to the z axis:

$$\text{relation} \equiv : (a', b', c')^t - (\bar{a}, \bar{b}, \bar{c})^t = \boldsymbol{0}$$

$$\text{relation} \parallel : (a', b')^t - (\bar{a}, \bar{b})^t = \boldsymbol{0}$$

$$\text{relation} \perp : a'\bar{a} + b'\bar{b} + 1 = 0.$$

This approach should be compared to that of [29].

D. Registration

1) Principle: The registration (or matching) of two primitives $\mathcal{P}$ and $\mathcal{P}'$ consists in detecting that their parameters $\boldsymbol{P}$ and $\boldsymbol{P}'$ verify (19) for one of the above listed geometric relations, with respect to the current noisy estimates $(\hat{\boldsymbol{P}}_0, \boldsymbol{W}_{P_0})$, $(\hat{\boldsymbol{P}}'_0, \boldsymbol{W}_{P'_0})$, and $(\hat{\boldsymbol{D}}_0, \boldsymbol{W}_{D_0})$ of $\boldsymbol{P}$, $\boldsymbol{P}'$, and $\boldsymbol{D}$.

This "detection" is done by computing between each pair of primitive the generalized Mahalanohis distance given by (11), and by matching a pair of primitives each time the χ^2 acceptance test given by inequality (12) is verified, i.e., when

$$d(\hat{\boldsymbol{P}}_0, \boldsymbol{W}_{P_0}, \hat{\boldsymbol{P}}'_0, \boldsymbol{W}_{P'_0}, \hat{\boldsymbol{D}}_0, \boldsymbol{W}_{D_0}) < \epsilon. \qquad (20)$$

2) Reliability: The above-described registration procedure detects what would be called *plausible* matches between geometric primitives. When the uncertainty attached to the primitives parameters is large, it may happen that a plausible match is false. In order to improve the reliability of the procedure, one can use a strategy (inspired by [16]) which starts by registrating primitives whose parameters have a small covariance matrix, or primitives which can be matched unambiguously. Such a strategy is exemplified in the experimental results section of this paper and also in another paper [8].

3) Efficiency: In order to avoid a $O(n^2)$ complexity algorithm, it is of course possible to use additional control structures to select a subset of candidate primitives for each test. For instance, to test the relation "$\equiv$" between points or lines, bucketing techniques can be used with efficiency (see for instance [1], [7]).

E. Motion

Having registered two primitives $\mathcal{P}$ and $\mathcal{P}'$, the motion problem consists in reducing the uncertainty on the *motion parameters* $\boldsymbol{D}$ while taking into account the uncertainty on the parameters $\boldsymbol{P}$, $\boldsymbol{P}'$, and $\boldsymbol{D}$.

This is done by setting $\boldsymbol{a} = \boldsymbol{D}$ and $\boldsymbol{x} = (\boldsymbol{P}, \boldsymbol{P}')^t$, and by using the relation equation (19) as a measurement equation (1)

$$f_i(\boldsymbol{x}, \boldsymbol{a}) \equiv f_i((\boldsymbol{P}, \boldsymbol{P}'), \boldsymbol{D}) = \boldsymbol{0}.$$

Starting from the initial estimate $\hat{\boldsymbol{a}}_0 = \hat{\boldsymbol{D}}_0$, $\boldsymbol{S}_0 = \boldsymbol{W}_{D_0}$, and using the measurement $\hat{\boldsymbol{x}}_1 = (\hat{\boldsymbol{P}}_0, \boldsymbol{P}_0')^t)$ with

$$W_1 = \begin{pmatrix} W_{P_0} & \boldsymbol{0} \\ \boldsymbol{0} & W_{P_0'} \end{pmatrix}$$

one applies the EKF formalism to obtain a new estimate $\hat{\boldsymbol{a}}_1$ of the motion with a reduced covariance matrix $\boldsymbol{S}_1 < \boldsymbol{S}_0$. (In the sense $\boldsymbol{S}_0 - \boldsymbol{S}_1$ is nonnegative).

This process is recursively repeated: at iteration i, if a new pair of primitives can be registered with the new motion estimate $(\hat{\boldsymbol{a}}_{i-1}, \boldsymbol{S}_{i-1})$, the additional measurement equations they bring lead to a new better estimate $\hat{\boldsymbol{a}}_i$ of the motion with a still reduced covariance matrix $\boldsymbol{S}_i$. This process ends after the matching of k primitives with a final estimate $(\hat{\boldsymbol{a}}_k, \boldsymbol{S}_k)$ of the motion parameter $\boldsymbol{D}$.

F. Fusion

1) General Fusion: The fusion problem is exactly the dual of the motion problem, as it consists, after the registration of two primitives, in reducing the uncertainty on the *primitive parameters* $\boldsymbol{P}$ and $\boldsymbol{P}'$ while taking into account the uncertainty on the parameters $\boldsymbol{P}$, $\boldsymbol{P}'$, and $\boldsymbol{D}$.

This is done by "switching the attention," i.e., by choosing $\boldsymbol{a} = (\boldsymbol{P}, \boldsymbol{P}')^t$ and $\boldsymbol{x} = \boldsymbol{D}$ while using again the relation equation (19) as a measurement equation (1)

$$f_i(\boldsymbol{x}, \boldsymbol{a}) \equiv f_i(\boldsymbol{D}, (\boldsymbol{P}, \boldsymbol{P}')) = 0.$$

The initial estimate is taken as $\hat{\boldsymbol{a}}_0 = (\hat{\boldsymbol{P}}_0, \hat{\boldsymbol{P}}_0')^t$ and

$$S_0 = \begin{pmatrix} W_{P_0} & \boldsymbol{0} \\ \boldsymbol{0} & W_{P_0'} \end{pmatrix}$$

and one uses the measurement $\hat{\boldsymbol{x}}_1 = \hat{\boldsymbol{D}}_0$ with $\boldsymbol{W}_1 = \boldsymbol{W}_{D_0}$ to apply the EKF formalism and obtain a new estimate $\hat{\boldsymbol{a}}_1$ of the primitive parameters with a reduced covariance matrix $\boldsymbol{S}_1 < \boldsymbol{S}_0$.

If additional relations hold between these primitives and other ones, the same treatment allows for a further reduction in their parameters uncertainty, and therefore a more accurate estimation of the primitive parameters.

2) Forgetting Primitives: After the treatment of a constraint, the parameters $\boldsymbol{P}_1$ and $\boldsymbol{P}_1'$ of the primitives are usually correlated, which means that the covariance matrix

$$\text{cov}(\hat{\boldsymbol{P}}_1, \hat{\boldsymbol{P}}_1') = \begin{pmatrix} W_{P_1} & W_{P_1 P_1'} \\ W_{P_1' P_1} & W_{P_1} \end{pmatrix}$$

contains $W_{P_1' P_1} = W^t_{P_1 P_1'} \neq 0$.

Therefore, it is no longer possible to treat independently $\mathcal{P}$ and $\mathcal{P}'$ in successive measurement equations. One has to consider them as a new primitive, either by keeping only one of them, or the union of them.

For instance if one updates the parameters of $\mathcal{P}'$ with those of an "identical" primitive $\mathcal{P}$ observed in a previous visual map, one keeps only the updated parameters of $\mathcal{P}'$ in the new map, with their covariance matrix $W_{p'}$, forgetting the previous parameters $\mathcal{P}$ after having used them.

On the other hand, if one updates the parameters of two lines by detecting that they are orthogonal, one keeps the new primitive formed by the union of the updated two lines, with the corresponding covariance matrix. One must use this kind of relation carefully, in order to control the size of the state parameter $\boldsymbol{a}$.

3) Autofusion: In the special case where $\mathcal{V} \equiv \mathcal{V}'$, all primitives come from the same visual map, and the motion parameters vanish as they correspond to the identity transform and are perfectly known.

Nevertheless, one can still detect the previous geometric relations between pairs of primitives $\mathcal{P}$ and $\mathcal{P}'$, and use them to reduce the uncertainty on the primitives parameters.

VI. Experimental Results

The basic principles presented in this paper were tested on a variety of synthetic and real data. The interested reader can find registration and motion results with real points and lines in [3], registration and fusion results with synthetic and real points and lines in [2], and results on the building of global 3-D maps from passive stereovision in [9]. In this paper we only present results of the motion estimation from two 3-D maps from passive stereovision in [9]. In this paper we only present results of the motion estimation from two 3-D maps, the fusion of several inaccurate 3-D maps, and the detection of colinearity within a single 3-D map (what we called "autofusion"). In each of these examples, the 3-D map is made of 3-D lines.

A. Registration and Motion

Fig. 16 shows the edges of a triplet of images taken by the mobile robot in a first position. From these edges, the trinocular stereovision system computes a set of 3-D segments. Each 3-D segment is represented by the parameters (a, b, p, q) of the 3-D line supporting it and by the error covariance computed—as explained in Section IV-B—from the uncertainty on the edge points in the three images (we took an isotropic Gaussian density function of covariance 1 pixel around each edge point). Each 3-D line is bounded by two endpoints obtained from the endpoints measured in the three images which are projected on the reconstructed 3-D line.

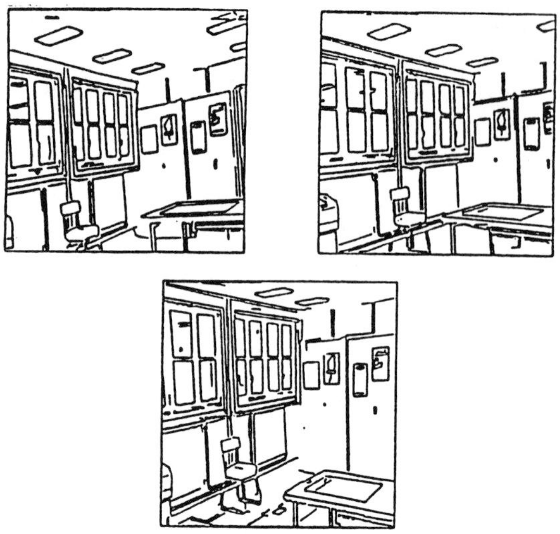

Fig. 16. Triplet of images taken in position 1.

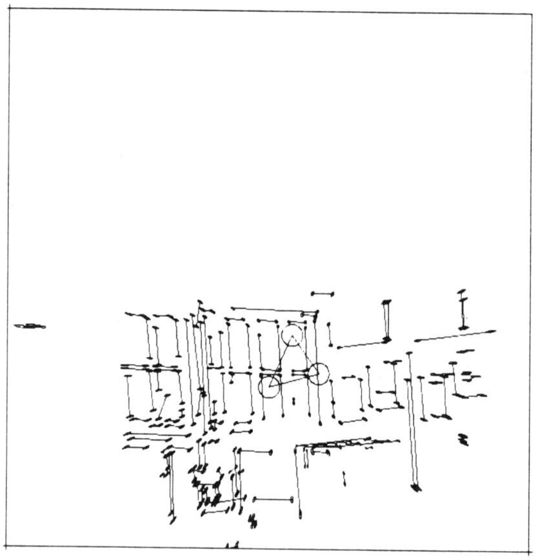

Fig. 17. Front view of reconstructed 3-D lines.

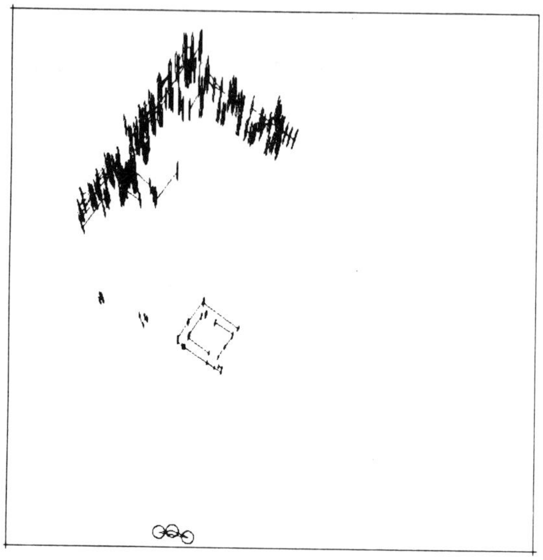

Fig. 18. Top view of reconstructed 3-D lines.

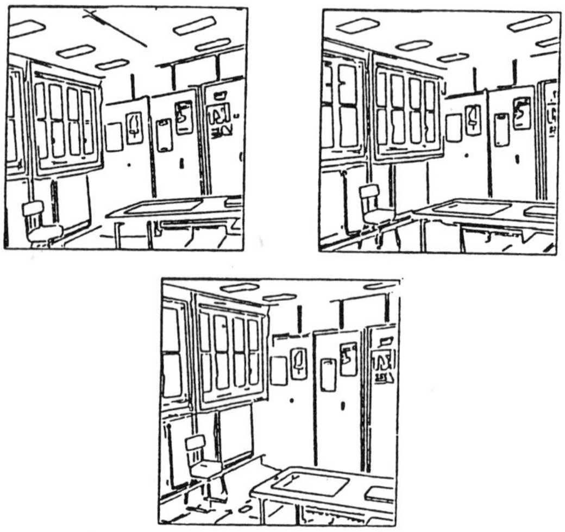

Fig. 19. Triplet of images taken in position 2.

We show in Figs. 17 and 18, respectively, the horizontal and vertical projections of the reconstructed 3-D segments. We also show the uncertainty attached to the reconstructed 3-D lines by showing the uncertainty it produces on the coordinates of their endpoints. The 95-percent confidence regions of the endpoints positions are ellipsoids whose projections are the ellipses shown in Figs. 17 and 18. One can see the anisotropic distribution of the uncertainty on the three coordinates of the points and its variation as a function of their position relative to the cameras (the projections of the three optical centers of the cameras correspond to the vertices of the triangle located grossly in the middle of the front view and at the bottom of the top view. Also, the circles around these vertices have been given an arbitrary radius of 20 cm to allow the reader to estimate the uncertainty attached to the other primitives).

The robot now moves a little, a new triplet of images is taken (Fig. 19) and another set of 3-D lines is computed. Initially, the robot is given a very crude estimate of its motion between the two views. Applying this crude estimate to the 3-D lines obtained in position 1, and projecting them in one of the images obtained in position 2 (the image of camera 3), one obtains the crude superimposition observed in Fig. 20. Solid lines are the transformed 3-D segments computed in position 1, while the dotted lines are the 2-D segments observed in position 2.

We now ask the system to discover the relation "≡" between the 3-D lines (see Section V-C) reconstructed in position 1 and 2, given the initial crude motion estimate and its uncertainty. The program takes each 3-D line in position 1, applies the noisy current motion estimate to place it in the 3-D map obtained in position 2 with a new covariance matrix (combining the initial uncertainty with the motion uncertainty), and computes its Mahalanobis distance (11) to all the other lines of position 2 (see Section V-D).

The program detects a match each time a pair of lines passes the χ^2 test of (12). If a line in position 1 can be matched to several lines in position 2, this is an *ambiguous* match, and nothing is done. On the other hand, each time an *unambiguous* match is found, the parameters of the motion are updated

Fig. 20. Superimposition of 3-D segments of position 1 with 2-D edges of position 2 (crude initial motion estimate).

Fig. 21. Superimposition of 3-D segments of position 1 with 2-D edges of position 2 (final motion estimate).

as it is explained in Section V-E. As the uncertainty on motion decreases after each new match, some previously ambiguous matches can now become unambiguous. Therefore, the entire matching process is repeated until no more lines can be matched (three iterations in this example). The final estimate of the motion is very accurate as can be seen in Fig. 21 where the obtained superimposition is now almost perfect.

Applying exactly the same technique to a set of six triplets of views taken during the motion of the robot (Figs. 22–27), the system was able to build a global 3-D map of the room shown in Fig. 28 where rotating segments at the bottom right are the computed successive robot positions. Fig. 29 gives a hand sketched semantic interpretation of this global map.

B. Registration, Motion, and Fusion

In this experiment, the robot is looking from four different positions at a regular pattern (Figs. 30 and 31) formed by vertical lines floating in front of horizontal lines, and builds in each position a local 3-D map. Exactly the same technique as in the previous example was used to register each successive

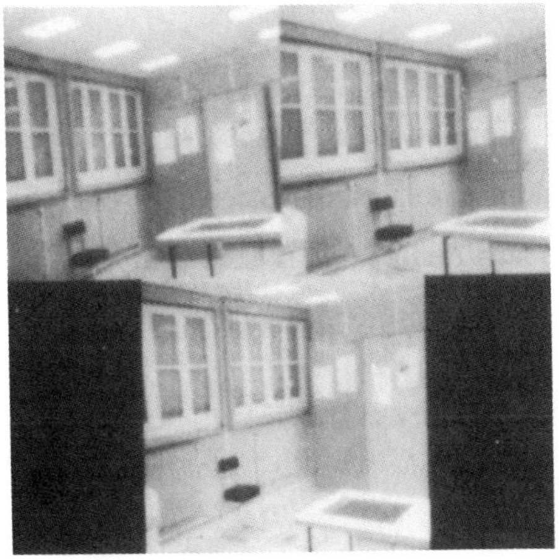

Fig. 22. First triplet of laboratory images.

Fig. 23. Second triplet of laboratory images.

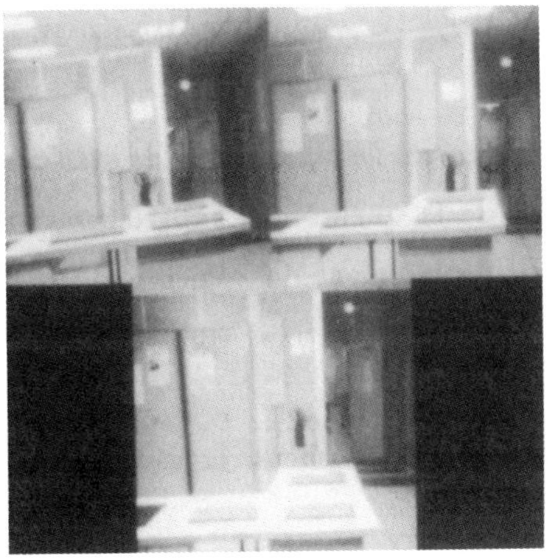

Fig. 24. Third triplet of laboratory images.

local 3-D map, and put all of them in a single absolute reference frame. Fig. 32 shows the resulting 3-D map before fusion. Fusion is achieved by discovering the relation "≡" computed between lines in the global 3-D map, and taking into account the uncertainty on the 3-D lines due to their reconstruction and to the successive motion estimations. Fusion yields a reduction from 1808 to 650 segments and improves accuracy, as can be seen by looking at the front and

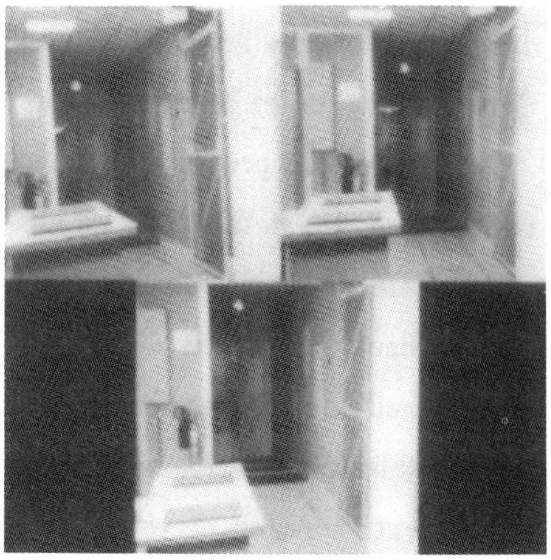

Fig. 25. Fourth triplet of laboratory images.

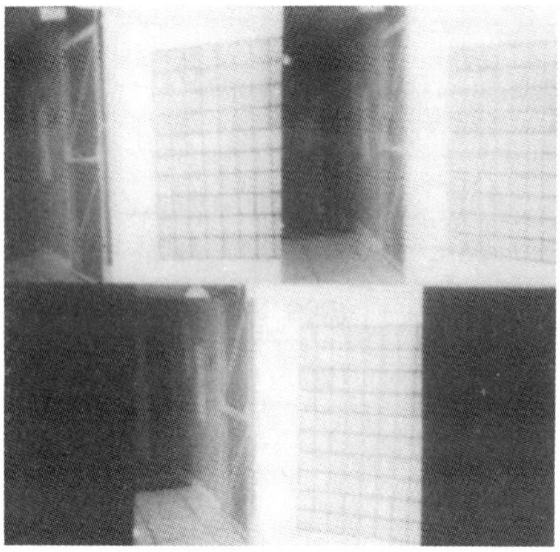

Fig. 26. Fifth triplet of laboratory images.

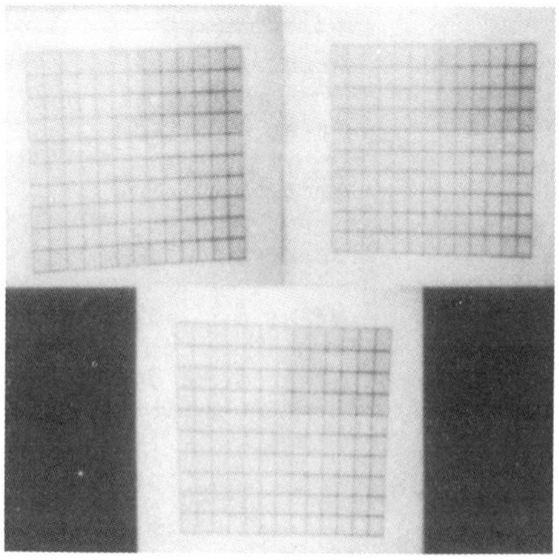

Fig. 27. Sixth triplet of laboratory images.

top view of the reconstructed 3-D pattern after fusion (Fig. 33).

C. Detecting Colinearity in Space

In this experiment, the robot is looking at the regular pattern only once. We show in Fig. 34 the vertical and horizontal projections of the initially reconstructed 3-D segments. We also show in Fig. 35 the uncertainty attached to reconstructed 3-D lines by showing the uncertainty it produces on the

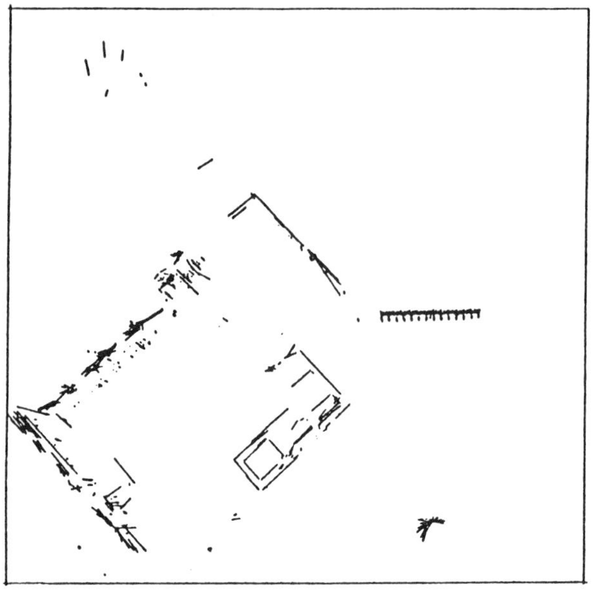

Fig. 28. Top view of a global 3-D map of the room computed from six local 3-D maps.

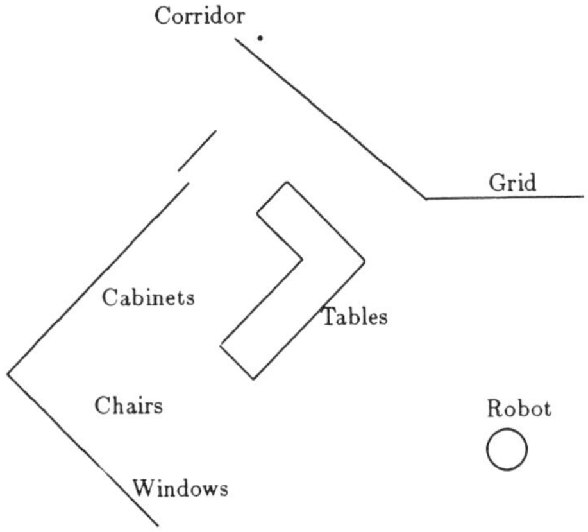

Fig. 29. Semantic interpretation of the previous global 3-D map.

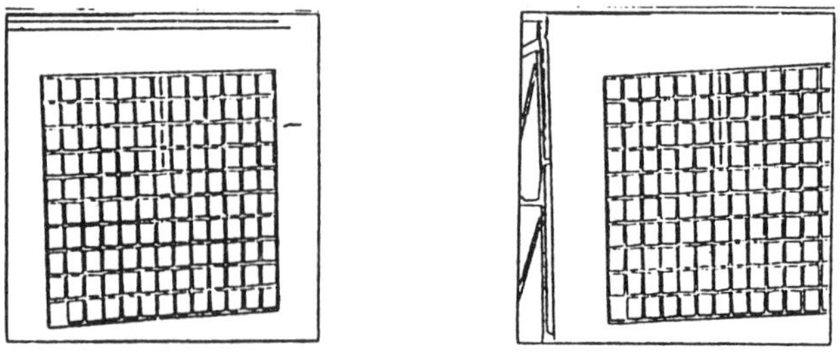

Fig. 30. A regular grid observed from position 1 and 2.

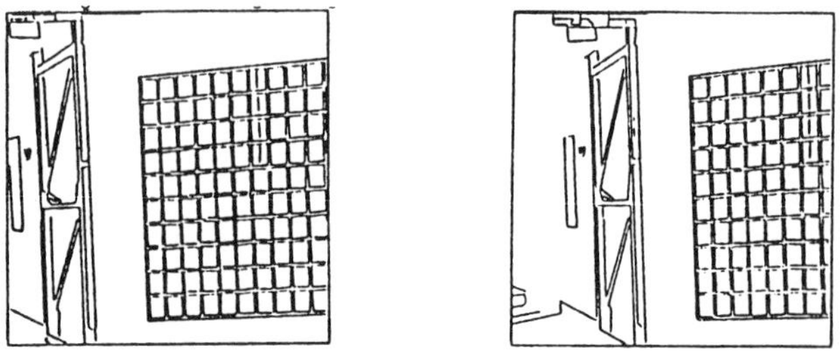

Fig. 31. A regular grid observed from positions 3 and 4.

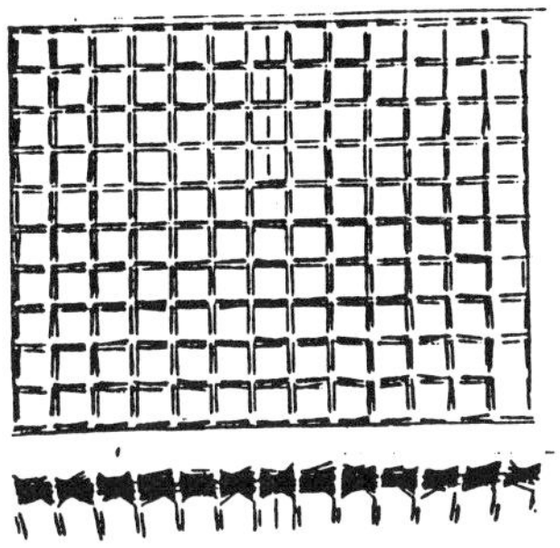

Fig. 32. Front an top view of reconstructed 3-D lines before identical lines are detected.

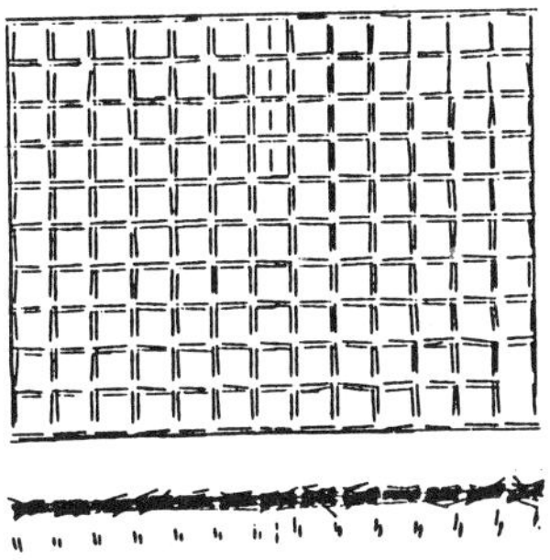

Fig. 33. Front and top view of 3-D lines after the fusion of identical lines.

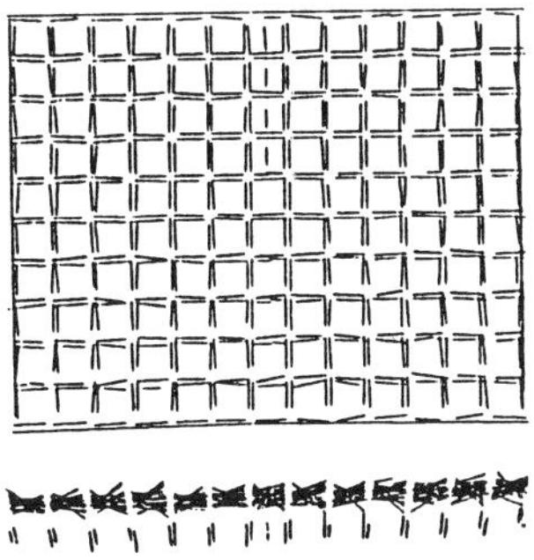

Fig. 34. Front and top view of reconstructed 3-D lines, before colinearity is detected.

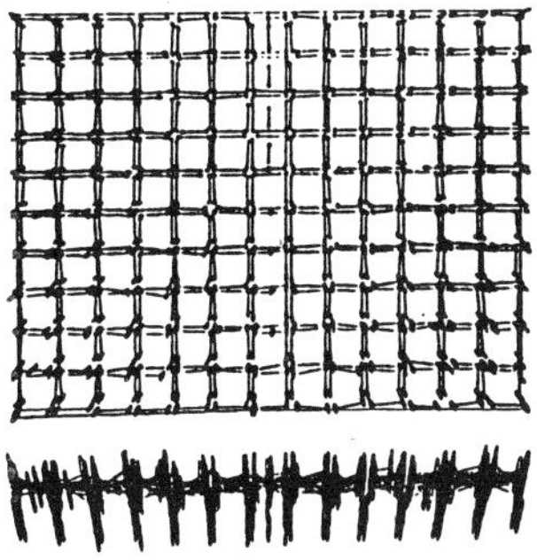

Fig. 35. Initial uncertainty attached to 3-D lines endpoints.

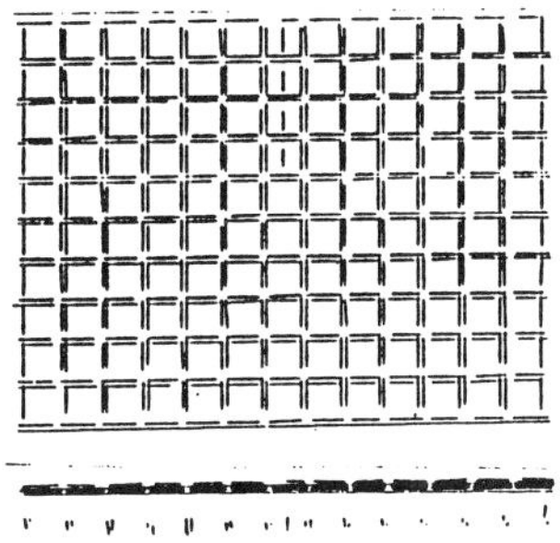

Fig. 36. Front and top view of 3-D lines when colinearity is discovered and enforced.

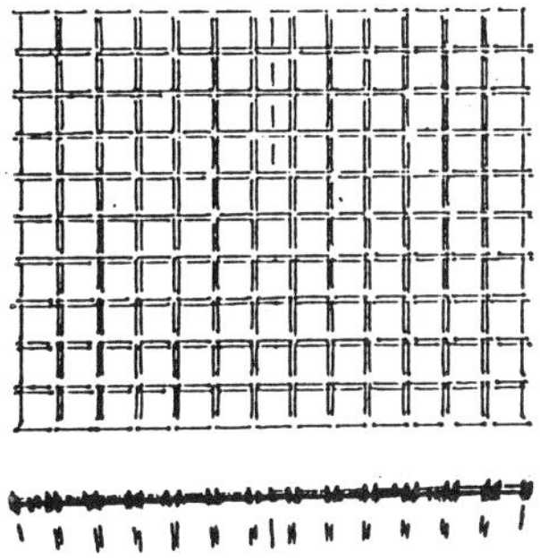

Fig. 37. Uncertainty attached to 3-D lines endpoints after the fusion of colinear segments.

coordinates of their endpoints (in the same way as in the first experiment).

We now ask the system to discover the relation "≡" between the 3-D lines (see Section V-C). The program takes a first 3-D line, computes its Mahalanobis distance (11) to all the other lines of the scene, and accepts the first line which passes the χ^2 test of (12) (see Section V-D). The two lines are fused using the technique of Section V-F and one keeps only the parameters of the optimal line representing both of them with an updated covariance matrix. The remaining lines are now compared to this new virtual line still with the Mahalanobis distance of (11) but with the new updated covariance matrix, while the χ^2 test of (12) remains unchanged. This process is repeated until no more lines can be matched with the first one, and then repeated with all the remaining unmatched lines.

The result is a reduced set of virtual lines on which the endpoints of the original segments have been projected, as shown in Fig. 36. The uncertainty on the line parameters has been greatly reduced: Fig. 37 shows the resulting uncertainty on the lines endpoints, which agrees very well with the reality.

VII. Conclusion

In this paper we have proposed a methodology for building and maintaining a geometric representation of the environment of a mobile robot. This methodology has the following salient features:

Representation: 1) We use geometric primitives to describe the environment and rigid displacements to describe the motion. These entities are described with a minimal number of parameters. 2) Uncertainty is modeled by a probability density function of these parameters. 3) Relationships between geometric entities are represented by algebraic equations on their parameters.

Algorithms: detecting geometric relationships, computing or updating the parameters of geometric entities (both for primitives and displacements) is done by recursive prediction-and-verification algorithms including the Extended Kalman Filter. These algorithms, better detailed in [8], [10], and [11], take into account prior knowledge to compute and propagate uncertainties.

Finally, the experimental results showed that the major approximations we made (linearization of the algebraic equations, second-order approximation of the probability density functions) were valid in a number of practical cases. Of course, a lot of theoritical and experimental work is still necessary to extend the approach to a wider class of problems for which such approximations cannot be made. This might be a good direction for future research.

ACKNOWLEDGMENT

The authors want to thank N. Gaudechoux for her precious help in the preparation of this paper and the reviewers for their helpful comments.

REFERENCES

[1] N. Ayache and O. D. Faugeras, "Hyper: A new approach for the recognition and positioning of two-dimensional objects," *IEEE Trans. Pattern Anal. Machine Intell.*, vol. PAM1-8, no. 1, pp. 44–54, Jan. 1986.

[2] ——, "Building a consistent 3d representation of a mobile robot environment by combining multiple stereo views," in *Proc. Int. Joint Conf. on Artificial Intelligence* (Milano, Italy, Aug. 1987).

[3] ——, "Building, registrating and fusing noisy visual maps," in *Proc. Int. Conf. on Computer Vision* (London, UK, June 1987), pp. 73–82. Also an INRIA Int. Rep. 596, 1986.

[4] ——, "Maintaining representations of the environment of a mobile robot," in *Int. Symp. on Robotics Research* (Santa Cruz, CA, Aug. 1987), pp. 337–350.

[5] N. Ayache and B. Faverjon, "Efficient registration of stereo images by matching graph descriptions of edge segments," *Int. J. Computer Vision,* vol. 1, no. 2, Apr. 1987.

[6] N. Ayache and O. D. Faugeras, "Building, registrating and fusing noisy visual maps," *Int. J. Robotics Res.*, vol. 7, no. 6, pp. 45–65, Dec. 1988 (Special Issue on Sensor Data Fusion).

[7] N. Ayache, O. D. Faugeras, and B. Faverjon, "Matching depth maps obtained by passive stereovision," in *Proc. 3rd Workshop on Computer Vision: Representation and Control,* pp. 197–204, Oct. 1985.

[8] N. Ayache, O. D. Faugeras, F. Lustman, and Z. Zhang, "Visual navigation of a mobile robot," in *IEEE Int. Workshop on Intelligent Robots and Systems (IROS'88)* (Tokyo, Japan, Oct. 1988).

[9] N. Ayache and F. Lustman, "Fast and reliable passive trinocular stereovision," in *Proc. Int. Conf. on Computer Vision* (London, UK, June 1987), pp. 422–427.

[10] N. Ayache, "Construction et fusion de représentations visuelles tridimensionnelles; applications à la robotique mobile," Thèse d'Etat, Université de Paris-Sud, Orsay, May 1988. INRIA Int. Rep.

[11] ——, *Vision Stéréoscopique et Perception Multisensorielle—Application à la Robotique Mobile.* Inter-Editions, 1989. English translation will be available from MIT Press in 1990.

[12] R. M. Bolle and D. B. Cooper, "On optimally combining pieces of information, with application to estimating 3D complex-object position from range data," *IEEE Trans. Pattern Anal. Machine Intell.*, vol. PAMI-8, pp. 619–638, 1986.

[13] J. Canny, "A computational approach to edge detection," *IEEE Trans. Pattern Anal. Machine Intell.*, vol. PAMI-8, no. 6, pp. 679–698, 1986.

[14] M. P. do Carmo, *Differential Geometry of Curves and Surfaces.* Englewood Cliffs, NJ: Prentice Hall, 1976.

[15] J. L. Crowley, "Representation and maintenance of a composite surface model," in *Proc. Int. Conf. on Robotics and Automation* (San Francisco, CA, Apr. 1986), pp. 1455–1462.

[16] C. Darmon, "A new recursive method to detect moving objects in a sequence of images," in *Proc. IEEE Conf. on Pattern Recognition and Image Processing,* pp. 259–261, June 1982.

[17] R. Deriche, "Using Canny's criteria to derive an optimal edge detector recursively implemented," *Int. J. Computer Vision,* vol. 2, Apr. 1987.

[18] H. F. Durrant-Whyte, "Consistent integration and propagation of disparate sensor observations," in *Proc. Int. Conf. on Robotics and Automation* (San Francisco, CA, Apr. 1986), pp. 1464–1469.

[19] O. D. Faugeras, N. Ayache, and B. Faverjon, "Building visual maps by combining noisy stereo measurements," in *Proc. Int. Conf. on Robotics and Automation* (San Francisco, CA, Apr. 1986), pp. 1433–1438.

[20] O. D. Faugeras and M. Hébert, "The representation, recognition, and locating of 3d objects," *Int. J. Robotics Res.*, vol. 5, no. 3, pp. 27–52, 1986.

[21] O. D. Faugeras, F. Lustman, and G. Toscani, "Motion and structure from motion from point and line matches," in *Proc. Int. Conf. on Computer Vision* (London, UK, June 1987), pp. 25–34.

[22] Hager and M. Mintz, "Estimation procedures for robust sensor control, in the integration of sensing with actuation to form a robust intelligent control system," GRASPLAB Rep. 97, Dep. Comput. Informat. Sci., Moore School of Elec. Eng. Univ. of Pennsylvania, Mar. 1987.

[23] A. M. Jazwinsky, *Stochastic Processes and Filtering Theory.* New York, NY: Academic Press, 1970.

[24] T. Kanade, *Three-Dimensional Machine Vision.* New York, NY: 1987.

[25] D. J. Kriegman, E. Triendl, and T. O. Binford, "A mobile robot: Sensing, planning, and locomotion," in *Proc. Int. Conf. on Robotics and Automation* (Raleigh, NC 1987), pp. 402–408.

[26] E. M. Mikhail, *Observations and Least Squares.* University Press of America, 1976.

[27] McKendall and M. Mintz, "Models of sensor noise and optimal algorithms for estimation and quantization in vision systems," GRASPLAB Rep. 97, Dep. Comput. Informat. Sci., Moore School of Elec. Eng., Univ. of Pennsylvania, Mar. 1987.

[28] L. Matthies and S. A. Shafer, "Error modelling in stereo navigation," *IEEE J. Robotics Automat.*, vol. RA-3, no. 3, pp. 239–248, 1987.

[29] J. L. Mundy, "Reasoning about 3-d space with algebraic deduction," in O. D. Faugeras and G. Giralt, Eds., *Robotics Research, The Third International Symposium.* Cambridge, MA: MIT Press, 1986, pp. 117–124.

[30] J. Porrill *et al.*, "Optimal combination and constraints for geometrical sensor data," Mar. 1987, to be published.

[31] K. Roberts, "A new representation for a line," in *Proc. Int. Conf. on Computer Vision and Pattern Recognition,* pp. 635–640, 1988.

[32] R. C. Smith and P. Cheeseman, "On the representation and estimation of spatial uncertainty," *Int. J. Robotics Res.*, vol. 5, no. 4, pp. 56–68, 1987.

Blanche: Position Estimation for an Autonomous Robot Vehicle

Ingemar J. Cox

AT&T Bell Laboratories
Murray Hill, New Jersey 07974

ABSTRACT

This paper describes the position estimation system for an autonomous robot vehicle called Blanche, which is designed for use in structured office or factory environments. Blanche is intended to be low cost, depending on only two sensors, an optical rangefinder and odometry. Briefly, the position estimation system consists of odometry supplemented with a fast, robust matching algorithm which determines the congruence between the range data and a 2D map of its environment. This is used to correct any errors existing in the odometry estimate. The integration of odometry with fast, robust matching allows for accurate estimates of the robot's position and accurate estimates of the robot's position allow for fast, robust matching. That is, the system is self sustaining.

The vehicle and associated algorithms have all been implemented and tested within a structured office environment. There is no recourse to passive or active beacons placed in the environment. The entire autonomous vehicle is self contained, all processing being performed on board. We believe this vehicle is significant not just because of the sensing and algorithms to be described, but also because its implementation represents a high level of performance at very low cost.

1. INTRODUCTION

A key to autonomy is navigation, i.e. an accurate knowledge of position. By position, we mean the vehicle's (x,y,θ) configuration with respect to either a global or local coordinate frame, *not* topological position, e.g. to the left of the wall. Dead reckoning using inertial guidance and odometry sensors drift with time so that the estimated position of the vehicle becomes increasingly poor. This complicates the process of position estimation. In order to correct for these cumulative errors, the vehicle must sense its environment and at least recognize key landmarks. Using sensory information to locate the robot in its environment is the most fundamental problem to providing a mobile robot with autonomous capabilities.

This paper describes the position estimation system employed by the vehicle. Section (2) begins with a brief overview of the vehicle and its guidance system. Section (3) then discusses in detail the navigation (position estimation) subsystem employed by the vehicle. Included in this is a description of the robot's map representation, Section (3.1), the algorithm used to match the range data to this map, Section (3.3), as well as some implementation details. Section (3.4) discusses how the odometry position estimate is combined with the correction estimated by the matcher. The system has been implemented on the vehicle and experimental results are included in Section (4).

2. OVERVIEW OF BLANCHE

Blanche [3] is an experimental vehicle intended to operate autonomously within a structured office or factory environment. It is designed to be low cost, depending on only two sensors, an optical rangefinder and odometry. Blanche, shown in Figure (1), has a tricycle configuration consisting of a single steerable drive wheel at the front and two passive rear wheels. The vehicle is powered by two sealed 12V 55Ah batteries which, in the current configuration, provide a useful lifetime of approximately seven hours. Control of the cart is based on a Multibus system consisting of a MC68020 microprocessor with MC68881 math coprocessor, 2 Mbyte of memory, an ethernet controller, a custom two-axis motor controller and an analogue-to-digital convertor. The Motorola 68020 runs a real-time UNIX® derived executive called NRTX [10].

The cart is equipped with two primary sensors: odometry on each of

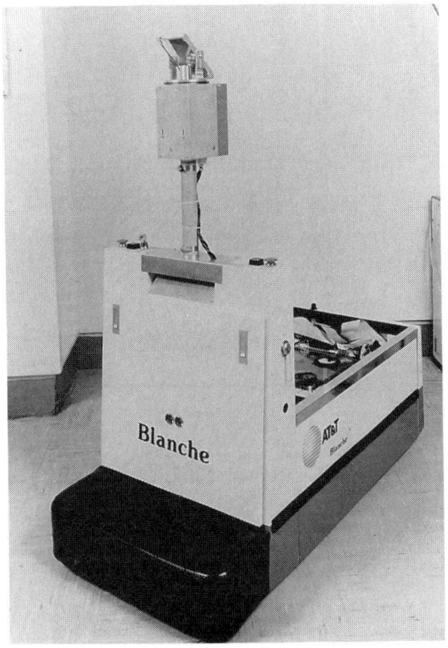

Figure 1: Blanche, an autonomous robot vehicle.

the two rear wheels and an optical rangefinder. Both sensors are extremely low cost (under $1000 each for components), and together provide all the navigational sensing information available to the cart. The advantage of odometry is, of course, that it is both simple and inexpensive. However it is prone to several sources of errors. First, surface roughness and undulations may cause the distance to be over estimated. Second, wheel slippage can cause distance to be under estimated. Finally, variations in load can distort the odometer wheels and introduce additional errors. If the load can be measured, then the distortion can be modelled and corrected for [17]. Where appropriate, a simple and more accurate alternative is to provide a pair of knife edge, non-load bearing wheels solely for odometry. Blanche uses this approach. In addition, even very small errors in the vehicle's initial position can lead to gross errors over a long enough path. Consequently, it is imperative that the vehicle's environment be sensed.

A simple low cost time-of-flight optical rangefinder has been developed specifically for cart navigation [12]. The rangefinder uses an approximately 1″ diameter beam and a rotating mirror to provide 360° polar coordinate coverage of both distance and reflectance out to about 15 feet. Both radial and range resolution correspond to about 1 inch at a ranging distance of 5 feet, with an overall bandwidth of approximately 1 kHz. Figure (2) shows a typical range map of a room obtained from a single scan of the rangefinder. A scan typically takes about one second. Each point is represented by its corresponding region of uncertainty, denoted by a circle of radius twice the standard deviation in the measurement. It should be pointed out that the error due to assuming that the range output is linear with distance, may sometimes exceed the error due to noise in the rangefinder. This systematic error can be removed by using a table look up technique to accurately map range output into distance.

The control of Blanche can be classified into three main components; path planning, guidance (trajectory generation and low level control) and navigation (position estimation). These components are illustrated in Figure (3).

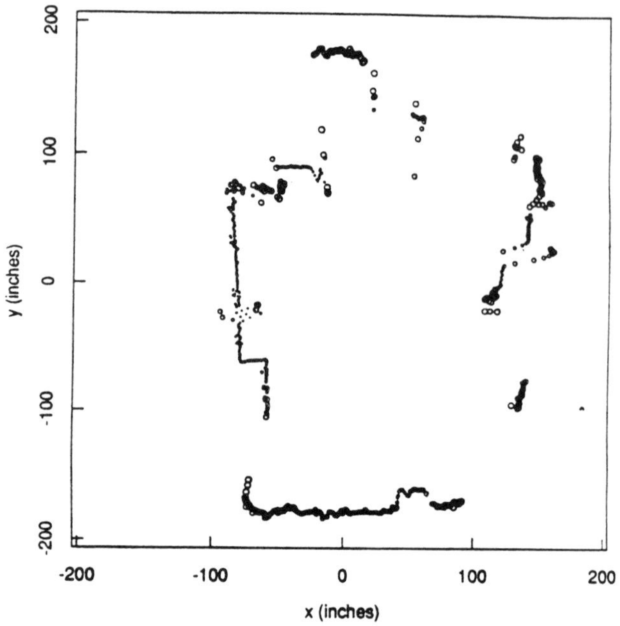

Figure 2: A range data scan obtained in a typical room. (Each point is denoted by twice its standard deviation).

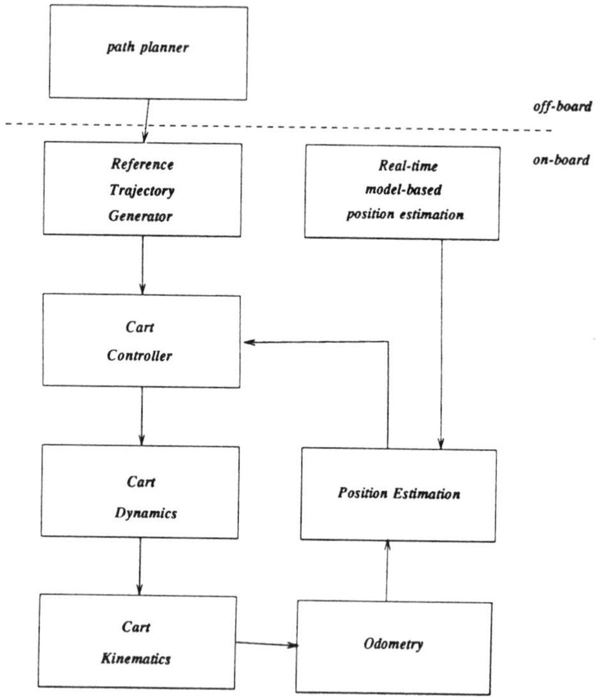

Figure 3: Block diagram of the overall control system.

The path planner [19] is an off-line program which generates a series of collision free maneuvers, consisting of line and arc segments, to move the vehicle from a current to a desired position. Since the cart has a minimum turning radius (approximately 2 feet), it is not possible to simply turn on the spot and vector to the desired position as is the case for differentially driven vehicles.

This path is downloaded to the vehicle, which then navigates along the commanded route. The line and arc segments specifications are sent to control software consisting of low level trajectory generation and closed-loop motion control [13]. Briefly, the reference state generator takes each segment specification from which it generates a reference vector at each control update cycle (every 0.1 secs). The cart controller controls the front steering angle and drive velocity using conventional feedback compensation to maintain small errors between the reference and measured states.

3. NAVIGATION (POSITION ESTIMATION)

Navigation can be broadly separated into two distinct phases, reference and dead reckoning, as discussed in [6]. Reference guidance refers to navigation with respect to a coordinate frame based on visible external landmarks. Dead reckoning refers to navigation with respect to a coordinate frame that is an integral part of the guidance equipment. Dead reckoning has the advantage that it is totally self contained. Consequently, it is always capable of providing the vehicle with an estimate of its position. Its disadvantage is that the position error grows without bound unless an independent reference is used to periodically reduce the error. Reference guidance has the advantage that the position errors are bounded, but detection of external references or landmarks and real-time position fixing may not always be possible. Clearly inertial and external reference navigation are complementary and combinations of the two approaches can provide very accurate positioning systems.

Position estimation based on the *simultaneous* measurement of the range or bearing to three or more known landmarks is well understood [15]. However, recognizing naturally occurring reference points within a robot's environment is not always easy due to noise and/or difficulties in interpreting the sensory information. Placing easy to recognize beacons in the robots workspace is one way to alleviate this problem. Many different types of beacons have been investigated including (i) corner cubes and laser scanning system, (ii) bar-code, spot mark or infra-red diodes [16] and associated vision recognition systems [9] and (iii) sonic or laser beacon systems. We chose not to rely on beacons, believing that the ability to operate in an unmodified environment was preferable from a user standpoint.

There have been many efforts to use high level vision to navigate by, particularly stereo vision [1], [8]. However, conventional vision systems were ruled out because of the large computational and associated hardware costs: We want the vehicle to be economic.

Figure (4) is an overview of Blanche's position estimation system. It consists of:

1. An *a priori* map of its environment.
2. A combination of odometry and optical range sensing to sense its environment.
3. An algorithm for matching the sensory data to the map [4].
4. An algorithm to estimate the precision of the corresponding match/correction [5] which allows the correction to be optimally (in a maximum likelihood sense) combined with the current odometric position to provide an improved estimate of the vehicle's position.

Provided the error models are accurate, the combined position estimate is less noisy than any one of the sets of individual measurements. The sensor integration process can, of course, be routinely mechanized by use of the Kalman filter [11]. The Kalman filter is not explicitly used in this system, but equivalent results are obtained.

3.1 Map Representation

Many spatial representations have been proposed. However, it is worth reflecting on the purpose of a map representation. Our purpose is to compare sensed range data to a map in order to refine our position estimate. The map is *not* intended to be used for path planning, it is not even necessarily intended to be updated by sensory data. It's sole purpose is for position estimation in an absolute coordinate frame. While many spatial representations have been proposed few appear to have been tested on a real vehicle. One major exception is occupancy grids [7]. Occupancy grids represent space as a 2– or 3D array of cells, each cell hold an estimate of the confidence that it is occupied. A major reason given for not using a more geometric representation is that sensor data is very noisy making geometric interpretation difficult.

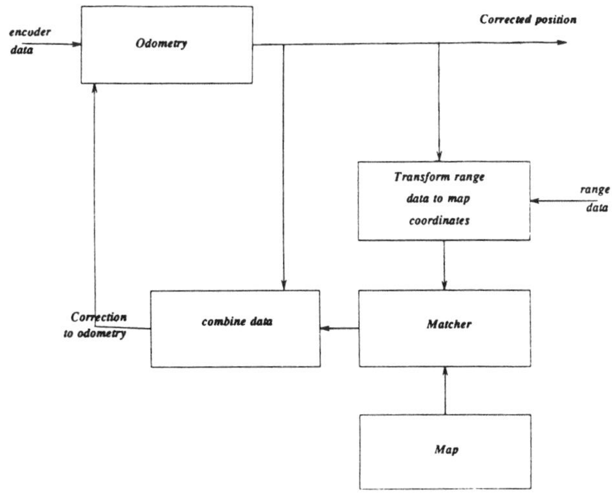

Figure 4: Block diagram of the navigation system.

This is especially true of sonar data (which Elfes and others used) and is one reason why sonar was not used on Blanche.

The infrared range data is much less noisy. This, combined with the fact that factory or office buildings are easily described by collections of line segments also influenced our choice of representation. We represent the environment as a collection of discrete line segments in the plane. A 2D representation was chosen because (i) much of the robot's environment is uniform in the vertical direction; there is not much to be gained from a 3D representation, (ii) the range sensor currently provides only 2D information (r, θ), (iii) matching sensor data to 2D maps is significantly simpler than matching to 3D maps. But above all else, a line segment description was chosen because a matching algorithm had been developed that used line segments. Moreover, the matching algorithm use (see Section 3.3) does not require the explicit extraction of any features from the image.

3.2 Sensor data

Sensing of the environment is quite straightforward, relying on odometry and an infrared rangefinder, as previously mentioned. The choice of a rangefinder was based primarily on the belief that ultrasonic rangefinders were poor, with severe problems due to specular reflection. A characteristic of the position estimation system described here is that the vehicle does not move under odometric control for a period, then stop to locate beacons, update its position and continue. Rather, its position is constantly being updated as the vehicle moves. Since range data is collected while the vehicle is moving, there is a need to remove any motion distortion that may arise. This is done by reading the odometric position at each each range sample. The current position and range value are then used to convert the data point into world coordinates for matching to the map.

If (r, α) is the range, angle pair from the rangefinder's local coordinate frame, it's position in world coordinates is given by

$$\begin{bmatrix} x \\ y \\ 1 \end{bmatrix} = \mathbf{C}\ \mathbf{R} \begin{bmatrix} r\cos(\alpha) \\ r\sin(\alpha) \\ 1 \end{bmatrix}$$

where **R** is the homogeneous transformation describing the relative position of the rangefinder with respect to the cart and **C** describes the vehicle's position with respect to the base coordinate system.

3.3 Matcher

We now describe how matching of range data to the map is achieved. First, an example. The solid line in Figure (5a) is the line segment description of a laboratory room in which Blanche is moved. The model, consisting of 24 line segments, was constructed very simply from measurements based on the one foot square floor tile grid. Item (a) is a tall desk, item (b) is a ventilation duct, item (c) is some cardboard posters, item (d) is a small refrigerator and item (e) a large cupboard. It should be pointed out that the model is very simple; some items in the room are not modeled, and others are only roughly approximated, such as item (c), which in fact is made up of several cardboard sections all at different distances from the wall. The dotted points show the image based on range data acquired from a single scan of the sensor. In this example there is a small degree of rotation and a large translation, of the order of nine feet in x and eight feet in y. Figure (5b) shows the results of applying the algorithm. It is evident that the correct congruence has been found.

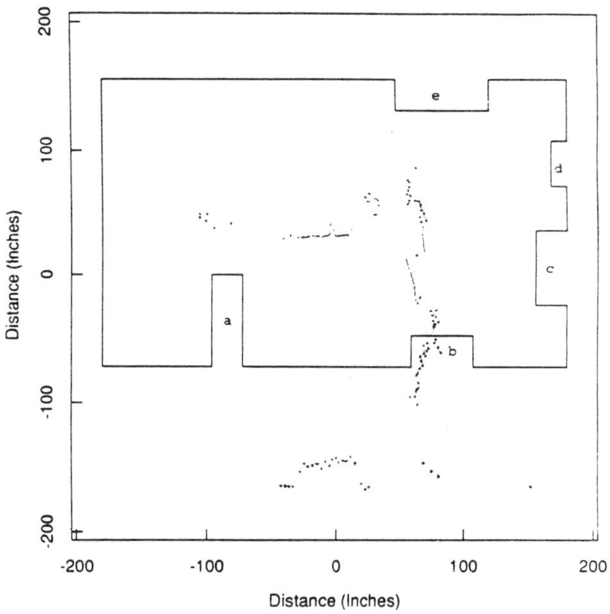

Figure 5a: Range image and associated map of room.

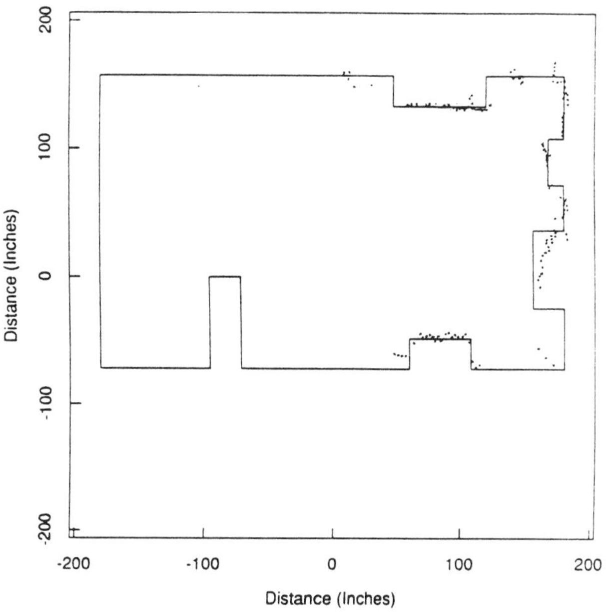

Figure 5b: Registered range image and associated map of room.

The general matching problem has been extensively studied by the computer vision community. Most work has centered upon the general problem of matching an image of arbitrary position and orientation relative to a model. Matching is achieved by first extracting features followed by determination of the correct correspondence between image and model features, usually by some form of constrained search.

Once the correspondence is known, determination of the congruence is straightforward.

The following observations motivated the derivation of the matching algorithm described below.

1. The displacement of the image relative to the model is small, i.e. we roughly know where we are at all times. This assumption is almost always true for practical autonomous vehicle applications, particularly if position updating occurs frequently.

2. Feature extraction can be difficult and noisy. Ideally, we would like a matcher which does not require a feature extraction preprocessing stage, but works directly on the range data points.

This section is taken from (Cox and Kruskal 1988 & Cox, Kruskal and Wallach 1988) [4], [5]. If the displacement between image and model is small, then for each point in the image its corresponding line segment in the model is very likely to be its closest line segment in the model. Determination of the (partial) correspondence between image points and model lines reduces to a simple search to find the closest line to each point. The central feature of the original conception was devising a method to use the approximately correct correspondence between image *points* and model *lines* to find a congruence that greatly reduces the displacement. Iterating this method leads to the following algorithm (for actual use, the algorithm incorporates certain computational simplifications as described below):

1. For each point in the image, find the line segment in the model which is nearest to the point. Call this the *target*.

2. Find the congruence that minimizes the total squared distance between the image points and their target lines.

3. Move the points by the congruence found in (2).

4. Repeat steps (1-3) until the procedure converges. The composite of all the step (3) congruences is the desired total congruence.

The rationale for our method of performing step (2) is explained with the aid of Figure (6). The model shown there consists of two line segments. The image consists of several points from both segments which have been displaced by translation up and to the right and by a slight rotation counterclockwise. For every point in the image, its target segment is the correct segment, as one would hope. It is natural to seek the congruence that minimizes the total squared distance from the model points to their target segments, i.e., the minimizing procedure tries to move the image points from each segment so that they lie on that segment, but it doesn't care where on the segment they go. In this case it is possible to reduce the total squared distance to 0, and there is a unique congruence that does so, namely, the inverse of the displacement used in forming the image, and one application of steps (1-3) perfectly recovers the desired congruence. If the original displacement was larger and/or the image contained points from near the ends of the line segments, the correspondence of image points to line segments might be imperfect, and several iterations might be required. If the image points contain some error, a potentially infinite iteration process might be required, though in practice only a few iterations usually achieve convergence to sufficient accuracy.

Step (2) is computationally the most complex step. For computational efficiency, two modifications are introduced and the new version is called step (2′). First, each target is changed from a line segment to the infinite line containing the segment. Note, however, that step (1) continues to use the finite segments. Second, the dependence of the moved points on θ is non-linear, so this dependence is approximated by the first order terms in θ. Such an approximate congruence is called a pseudo-congruence. Note, however, that step (3) continues to use a real congruence, namely, the congruence with the same (t_x, t_y, θ) as the pseudo-congruence found in step (2′). Since the algorithm is iterative and the final iterations involve vanishingly small displacements, the approximations involved in these two modifications do not cause error in the final result.

Any congruence can be described as a rotation by some angle θ followed by a translation by **t**, and can thus be denoted by $(\mathbf{t}, \theta)$.

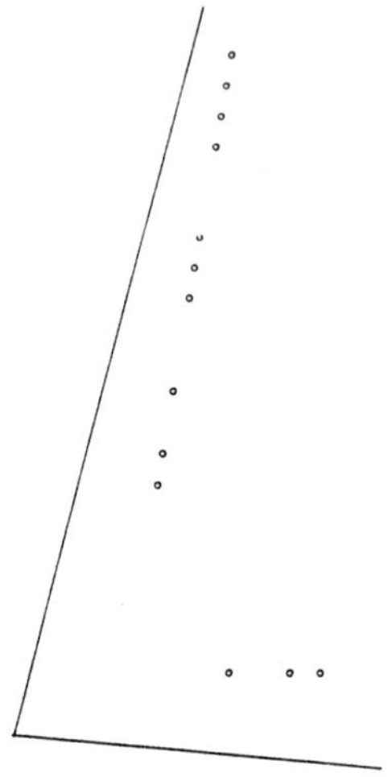

Figure 6: Simple model and displaced image.

However, instead of taking rotations around an arbitrary origin, we will take them around the center of gravity **c** of the image, for reasons that will be noted shortly. Thus the congruence $(\mathbf{t}, \theta)$ maps any point

$$\mathbf{x} \to R(\theta)(\mathbf{x}-\mathbf{c})+(\mathbf{c}+\mathbf{t}),$$

where $R(\theta)$ is a clockwise rotation of angle θ, and can be denoted by the matrix

$$R(\theta) = \begin{bmatrix} \cos(\theta) & -\sin(\theta) \\ \sin(\theta) & \cos(\theta) \end{bmatrix}.$$

In practice, **c** will be the mean point (center of gravity) of the image. Using a center that moves with the image (such as its mean point or its *k*-th point) has the advantage that $(\mathbf{t}_1, \theta_1)$ followed by $(\mathbf{t}_2, \theta_2)$ is the same motion of the image as $(\mathbf{t}_1+\mathbf{t}_2, \theta_1+\theta_2)$. This does *not* hold for rotations around a point, such as the origin, that does not move with the image. Furthermore, an approximation that will be introduced shortly has an error that is proportional to the distance from the center to the point being rotated. Use of the mean point of the image is desirable to minimize this error.

We wish to find the value of $(\mathbf{t}, \theta)$ that minimizes

$$S = \sum_i ([R(\theta)(\mathbf{v}_i-\mathbf{c})+(\mathbf{c}+\mathbf{t})]' \, \mathbf{u}_i - r_i)^2.$$

where $\mathbf{v}_i$ are the image points, $\mathbf{u}_i$ their corresponding target lines and ′ denotes transpose. However, because θ enters the image of $\mathbf{v}_i$ in a non-linear way, it is not evident how to do this in a rapid simple manner. Hence we introduce a linear approximation, namely,

$$R(\theta) = \begin{bmatrix} 1 & -\theta \\ \theta & 1 \end{bmatrix}.$$

Geometrically, this means that a point **x** is not moved in a circle around **c**, but is instead moved along the tangent to the circle. If **x** is at distance *d* from **c**, then instead of moving along an arc of angle θ (in radians), it moves along the tangent a distance of $d\,\theta$.

Now we proceed as we did above. The vector derivative of S with respect to **t** and the derivative of S with respect to θ are set equal to 0. After simplification, the following equations are obtained,

$$\begin{bmatrix} M_2 & \mathbf{m}_1 \\ \mathbf{m}_1' & m_0 \end{bmatrix} \begin{bmatrix} \mathbf{t} \\ \theta \end{bmatrix} = \begin{bmatrix} \mathbf{d}_1 \\ d_0 \end{bmatrix},$$

where

$$M_2 = \sum \mathbf{u}_i \, \mathbf{u}_i' \text{ (matrix)},$$

$$\mathbf{m}_1 = \sum \mathbf{u}_i \, (\mathbf{u}_i' \, \mathbf{v}_i) \text{ (vector)},$$

$$m_0 = \sum (\mathbf{u}_i{}' \, \mathbf{v}_i)^2 \text{ (scalar)},$$

$$\mathbf{d}_1 = \sum \mathbf{u}_i (r_i - \mathbf{u}_i{}' \, \mathbf{v}_i) \text{ (vector)},$$

$$d_0 = \sum (r_i - \mathbf{u}_i{}' \, \mathbf{v}_i)(\mathbf{u}_i{}' \, \mathbf{v}_i) \text{ (scalar)}.$$

3.3.1 Implementation details

The algorithm as described so far is intrinsically robust against incompleteness of the image, i.e. missing data points. To make it robust against spurious data, e.g. people walking by, and incompleteness of the model, it is modified further by deleting from consideration in step (2′) points whose distance to their target segments exceed some limit. Note that for the general matching problem, in which the assumption of small displacement is not necessarily valid, such an approach would not be possible. Our experience is that robustness is very desirable, yet very hard to achieve. The robustness of this algorithm is one of its major features.

Step (1) of the algorithm requires that for each point, its corresponding (closest) line segment be found. Since the map is currently unordered, this entails exhaustively calculating the distance of *every* line from each point in order to find the closest line segment. If there are n point and m line segments, the complexity of this operation is $O(nm)$. However, if the map is first preprocessed to compute the corresponding Voronoi diagram (a step which takes $m\log(m)$ time) determination of the closest line segment can then be achieved in $O(n\log(m))$ [14] Presently, we have not implemented the Voronoi solution.

Since determination of the partial correspondence can quickly dominate the processing time, it pays to keep N and M small. We therefore restrict the number of image points to approximately 180 out of a possible 1000. Originally the map was small enough that we did not have to worry about the size of M, the number of line segments. However, this is no longer the case. Fortunately, our knowledge of the vehicle's position allows us to extract only those lines within say a 20′ radius. It is this subset of the entire map, which often only numbers four or five, that is then used by the matching algorithm.

Currently, position updates occur approximately every 8 seconds for an image size of 180 points and a map of 24 line segments. This is sufficient for our requirements. However, it should be noted that the matching is the *lowest* level priority process running on the uniprocessor system. A considerable improvement in speed would be gained by simply dedicating a processor board to this task.

Finally, we note that the matcher returns an (x, y, θ) correction based on an origin at the centroid of the image/range scan. A coordinate transformation is therefore required to convert the matcher's centroid centered correction to a correction based on the vehicle's current position.

3.4 Integrating Odometric and Matched Positions

It is assumed that the x, y and θ position and orientation are independent of one another. We therefore describe how the x position is updated, y and θ being identical. We first need to estimate the standard deviation in the measurement of x for both the matcher and odometry.

3.4.1 Estimating matcher accuracy

This section is taken from (Cox, Kruskal and Wallach 1988) [5]. Let $\mathbf{b}$ be the vector of parameters $\mathbf{b} \equiv (t_x, t_y, \theta)'$ that describes a congruence or a pseudo-congruence. An approach to estimating the variances of the three parameter estimators arises from the method described in Section 3 for forming the estimators. This method can also estimate the covariances as well.

The covariance matrix of the estimator $\hat{\mathbf{b}}$ of $\mathbf{b}$ is a 3×3 symmetric matrix whose diagonal elements are the variances of the three parameter estimators and whose off-diagonal elements are their covariances.

Now consider the method described in Section 3. Each time step (2′) is performed, an estimate $\hat{\mathbf{b}}$ must be found. This estimate is calculated by solving a matrix equation

$$X'\mathbf{y} = (X'X)\hat{\mathbf{b}} \,,$$

where X and $\mathbf{y}$ are calculated from the data in an elementary way.

It is explained in Section (3.3) that after convergence is complete, the desired congruence is the composite of the $\hat{\mathbf{b}}$ vectors found in all repetitions of step (2′). The covariance matrix of this composite $\hat{\mathbf{b}}$ vector is the same as the covariance matrix of the final $\hat{\mathbf{b}}$ vector. This covariance matrix is estimated by $s^2(X'X)^{-1}$, where s^2 is the usual estimate of σ^2,

$$s^2 = (\mathbf{y}-X\hat{\mathbf{b}})\cdot(\mathbf{y}-X\hat{\mathbf{b}})/(n-4) \,.$$

The covariance matrix is calculated as part of the overall calculation for x [5].

3.4.2 Estimating odometry accuracy

Estimation of the standard deviation for odometry is less rigorous. Presently, it is assumed to be directly proportional to the distance moved. Empirically, we have set the constant of proportionality to 0.01, i.e. after 100′ the standard deviation is 1′. For a more rigorous treatment of odometric error see (C. Ming Wang 1988) [18].

3.4.3 Integration

Given (x_o, σ_o) and (x_m, σ_m) the position and standard deviation from odometry and matching, respectively, we can optimally combine these using the formulas [2], [11]

$$x_c = x_o + \frac{\sigma_o}{\sigma_o + \sigma_m}(x_m - x_o)$$

$$\frac{1}{\sigma_c^2} = \frac{1}{\sigma_o^2} + \frac{1}{\sigma_m^2}$$

where (x_c, σ_c) are the updated values for the x position and its corresponding standard deviation. It is clear that if the standard deviation in the odometry is small compared with the matcher, the error term $(x_m - x_o)$ has little effect. Alternatively, if the standard deviation in the odometry is large compared with the matcher, almost all the error term is added to the corrected value. This is intuitively correct.

This updated value is fed back to the odometry where it is used as the new value from which the current position is estimated. This is referred to as an *indirect feedback* configuration [11] in comparison to an *indirect feedforward* configuration, since the error correction is fed back into the odometry subsystem. In this way, the odometry errors are not allowed to grow without bound. Further, since the odometry estimate is corrected after each match or registration, failure of the rangefinder or matching subsystem does not lead to immediate failure of the robot system. Instead, the robot is free to continuing navigating for some period using only odometry until such time as the odometric estimates of the standard deviation in position exceed some unacceptable limits.

4. EXPERIMENTAL RESULTS

Figure (7) is a sequence of range images taken by the robot vehicle as it moved along a predetermined path. At initialization the vehicle was given its initial position. In practice, the vehicle had been displaced approximately 3″ in its (x, y) position. Its orientation is approximately correct.

Table (1) tabulates the partial set of corrections made by the vehicle as it traverses its path.

TABLE 1. A sequence of corrections estimated by the matcher

correction	x (inches)	y (inches)	θ (degrees)
1	0.0	2.8	0.7
.	.	.	.
.	.	.	.
.	.	.	.
8	0.0	-0.1	0.3
9	-1.2	0.1	-0.6
10	6.7	-0.1	-0.5
11	-3.0	0.1	0.3

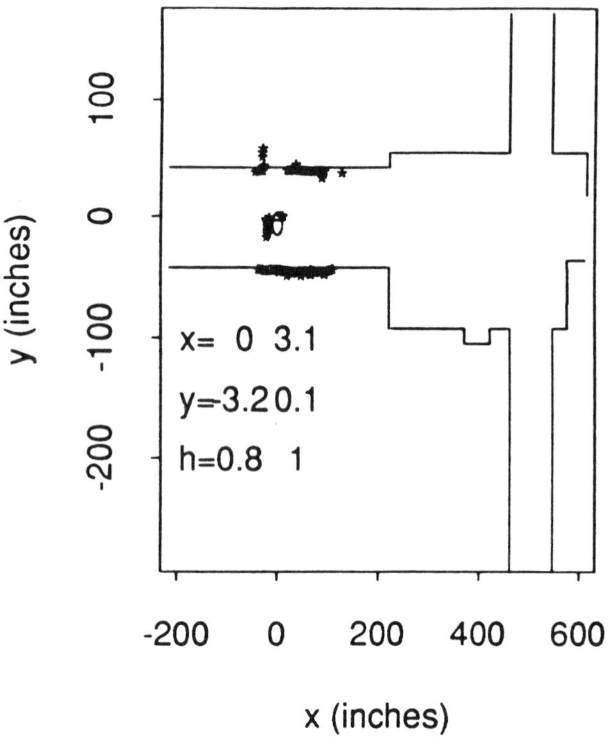

Figure 7a: Sequence (1) of range images taken as vehicle navigated along a corridor.

The first correction correctly determines the error in the y direction. A correction of 0.7° is also applied to the heading, but, examination of the heading corrections suggests that this is noise. However, there is no correction in the x direction. This is because the vehicle cannot tell where along the corridor it is[1]. The matcher has an infinite standard deviation along the x direction. Only when the vehicle comes out into the elevator bay — see scan (9) of Figure (7) — is it able to begin to refine its x position.

Large collections of range points not associated with the model usually indicate people walking by. In addition, maintenance and cleaning personnel may occasionally leave unmodeled objects in the environment. However, almost all these data points are rejected as spurious because their vicinity to a modelled line segment is not within preset thresholds.

The matcher's estimates of the standard deviations in (x, y, θ) assume a gaussian noise form with zero mean. However, the prototype rangefinder appears to have systematic errors: range values have been seen to depend on the strength of the received signal. For example, very bright objects appear further away. We believe that the large corrections in the x direction at scans 9 through 11 are due in part to this problem. An improved rangefinder is currently under test which we hope will solve this problem. Other potential sources of systematic errors include errors in the map and errors in the alignment of the rangefinder relative to the vehicle. A misalignment of only 1° would cause an almost 2″ error in the y direction as the vehicle travelled 9′ along the x direction. This may account for why the sum of the corrections in the y direction totals to 7.8″ rather than to the approximately 3″ error expected.

1. Remember that doorways are not modeled. Locally, the environment appears as two parallel lines.

5. CONCLUSIONS

This paper described the position estimation system of an autonomous robot vehicle for use in structured office or factory environments. The navigation system consists of (i) an *a priori* map of its environment consisting of a collection of discrete line segments in the plane, (ii) a combination of odometry and optical range sensing to sense its environment, (iii) an algorithm for matching the sensory data to the map. The precision of the corresponding match/correction is determined which allows the correction to be optimally combined with the current odometric position to improve the estimate of the vehicle's position. There is no recourse to passive or active beacons placed in the environment.

A 2D representation based on collections of line segments in the plane was chosen because much of the robot's environment is uniform in the vertical direction, the range sensor currently provides only 2–D information and matching sensor data to 2–D maps is significantly simpler than matching to 3–D maps.

The matching algorithm assumes that the displacement between image and model is small. This is a very reasonable assumption since odometry is providing the vehicle with a good estimate of position and matching occurs very frequently. The algorithm is intrinsically robust against incompleteness of the image, i.e. missing data points. Spurious data, e.g. people walking by, and incompleteness of the model, is dealt with by deleting from consideration any points whose distance to their target segments exceed some limit. Note that for the general matching problem, in which the assumption of small displacement is not necessarily valid, such an approach would not be possible.

In contrast to many papers presenting results of simulations, the vehicle and associated algorithms have all been implemented and tested within a structured office environment. The entire autonomous vehicle is self contained, all processing being performed on board. We believe this vehicle is significant not just because of the sensing and algorithms described, but also because its implementation represents a high level of performance at low cost. There also appears to be a self sustaining property to this configuration: Accurate knowledge of position allows for fast robust matching which leads to accurate knowledge of position.

There are several areas of in which the vehicle might be improved. First, any map has so far been assumed to be perfectly accurate, all errors being considered due to sensor noise or errors in the matcher. In practice, the map also has an associated accuracy which should also be modeled. Second, the need to provide a map of the environment can become tedious and it would therefore be desirable to automate this step, i.e. allow the vehicle to construct its own map. Third, the covariance matrix estimated by the matcher assumes there is no correlation between scans of the rangefinder. This is not necessarily true in practice because of (i) errors in the map and (ii) sensor noise may be correlated with particular wall surfaces. Future work is directed to addressing these problems.

Acknowledgements: The development of "Blanche" would not have been possible without the help of many individuals. It is a pleasure to thank J. B. Kruskal whose collaboration with the matching algorithm was critical to the success of the project. Also special thanks to G. L. Miller, and E. R. Wagner for the optical ranger. Finally, R. A. Boie, W. J. Kropfl, D. A. Kapilow, J. E. Shopiro, W. L. Nelson, F. W. Sinden, and G. T. Wilfong all provided help and assistance of one kind or another. Thank you.

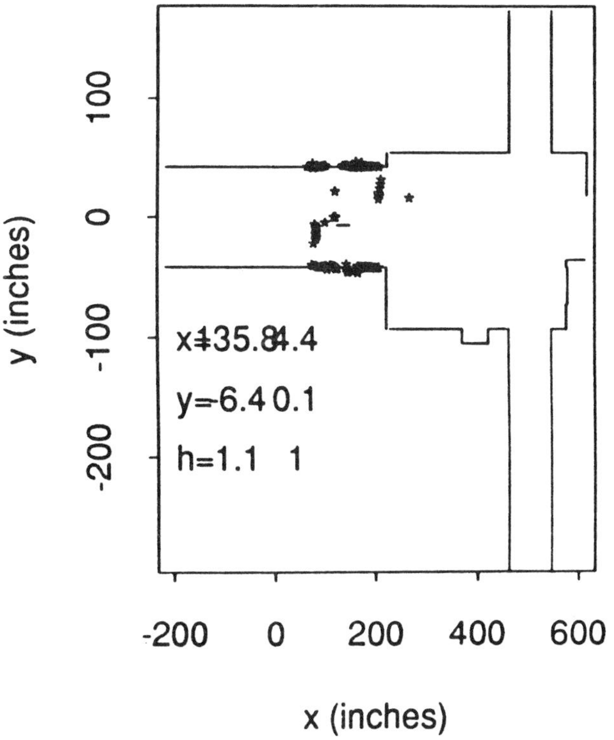

Figure 7b: Sequence (8) of range images taken as vehicle navigated along a corridor.

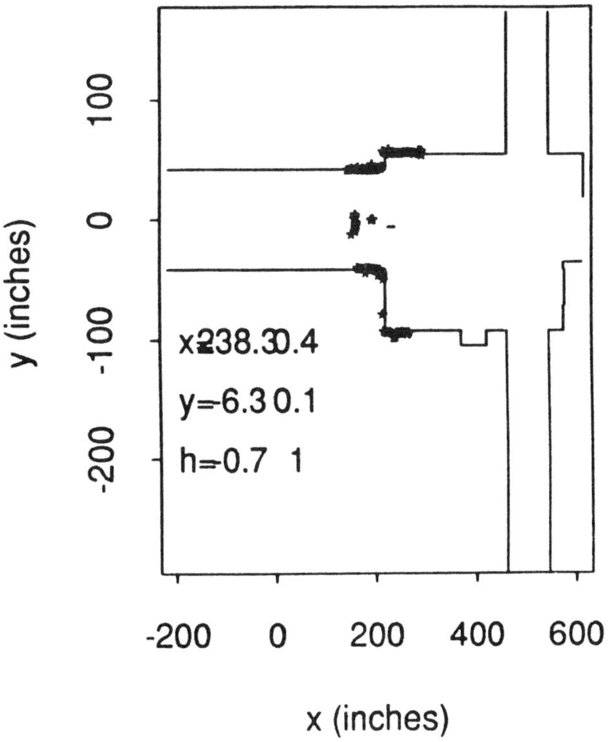

Figure 7d: Sequence (10) of range images taken as vehicle navigated along a corridor.

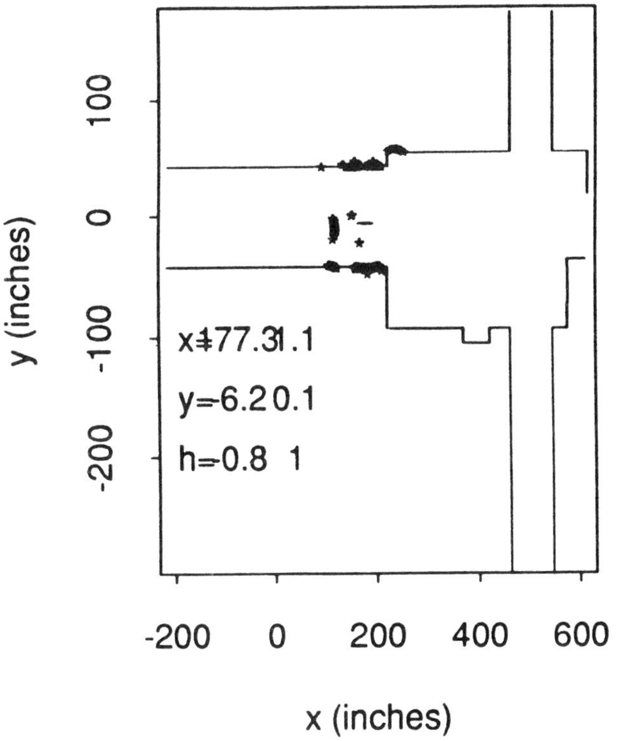

Figure 7c: Sequence (9) of range images taken as vehicle navigated along a corridor.

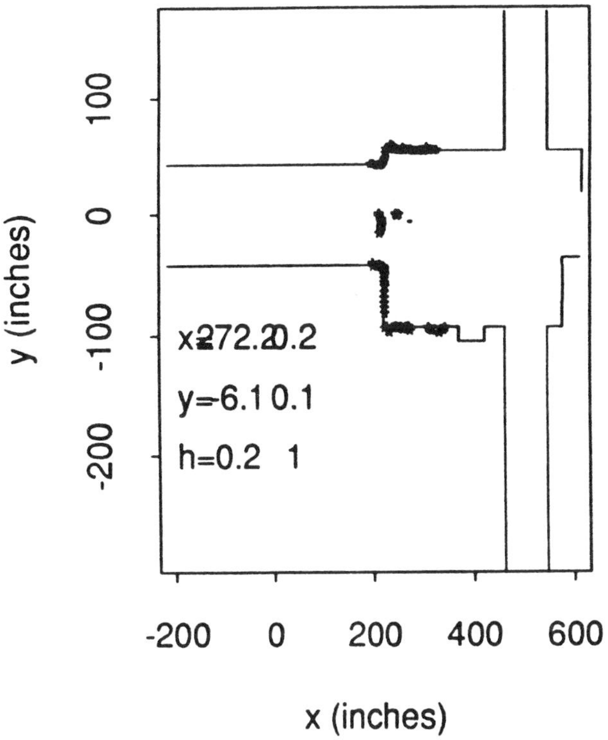

Figure 7e: Sequence (11) of range images taken as vehicle navigated along a corridor.

REFERENCES

1. Ayache, N. and Faugeras, O. Building, Registrating, and Fusing Noisy Visual Maps. In *Int. Conf. Computer Vision*, IEEE, London, UK, 1987, pp. 73-79.

2. Bar-Shalom, Y. and Fortmann, T.E. *Tracking and Data Association*. Academic Press, 1988.

3. Cox, I.J. Blanche: An Autonomous Robot Vehicle for Structured Environments. In *IEEE Int. Conf. on Robotics and Automation*, IEEE, 1988, pp. 978-982.

4. Cox, I.J. and Kruskal, J.B. On the Congruence of Noisy Images to Line Segment Models. In *Int Conf. Computer Vision*, IEEE, 1988.

5. Cox, I.J., Kruskal, J.B., and Wallach, D.A. Predicting and Estimating the Performance of a Subpixel Registration Algorithm. 1988.

6. Cox, I.J. and Wilfong, G.T., Eds. *Autonomous Robot Vehicles*. Springer-Verlag, New York, to be published.

7. Elfes, A. Sonar-Based Real-World Mapping and Navigation. *IEEE J. of Robotics and Automation RA-3*, 3 (1987), 249-265.

8. Elfes, A. and Matthies, L. Sensor Integration for Robot Navigation: Combining Sonar and Stereo Range Data in a Grid-Based Representation. In *IEEE Conference on Decision and Control*, IEEE, 1987.

9. Giralt, G., Chatila, R., and Vaisset, M. An Integrated Navigation and Motion Control System for Autonomous Multisensory Mobile Robots. In *1st Int. Symp. on Robotics Research*, Bretton Woods, NH, USA, 1983, pp. 191-214.

10. Kapilow, D.A. Real-Time Programming in a UNIX Environment. *1985 Symposium on Factory Automation and Robotics* (1985), 28-29.

11. Maybeck, P.S. *Stochastic Models, Estimation, and Control*. Vol. 1. Academic Press, 1979.

12. Miller, G.L. and Wagner, E.R. An Optical Rangefinder for Autonomous Robot Cart Navigation. In *SPIE Mobile Robots II*, Vol. 852, SPIE, 1987, pp. 132-144.

13. Nelson, W.L. and Cox, I.J. Local path Control of an Autonomous Vehicle. In *IEEE Int. Conf. Robotics and Automation*, IEEE, 1988, pp. 1504-1510.

14. Preparata, F.P. and Shamos, M.I. *Computational Geometry: An Introduction*. Springer-Verlag, New York, 1985.

15. Torrieri, D.J. Statistical Theory of Passive Location Systems. *IEEE Trans. on Aerospace and Electronic Systems AES-20*, 2 (1984), 183-198.

16. Tsumura, T. Survey of Automated Guided Vehicle in Japanese Factory. In *IEEE Int. Conf. on Robotics and Automation*, IEEE, 1986, pp. 1329-1334.

17. Tsumura, T., Fujiwara, N., and Hashimoto, M. An Experimental System for Self-Contained Position and Heading Measurement of Ground Vehicle. In *Int. Conf. Advanced Robotics*, 1983, pp. 269-276.

18. Wang, C.M. Location Estimation and Uncertainty Analysis for Mobile Robots. In *IEEE Int. Conf. Robotics and Automation*, Vol. 2, IEEE, 1988, pp. 1230-1235.

19. Wilfong, G.T. Motion Planning For An Autonomous Vehicle. In *IEEE Conf. Robotics and Automation*, IEEE, 1988, pp. 529-533.

4. Map Representation

> *"What's the good of Mercator's North Poles and Equators, Tropics, Zones and Meridian Lines?" So the Bellman would cry: and the crew would reply, "They are merely conventional signs!"*
>
> Lewis Carroll, *The Hunting of the Snark*, Fit II, st. 3 (1876)

The use of external landmarks for navigation, explicitly or implicitly, involves the use of a map. A *map* is a collection of spatial relationships. It consists of a set of information, such as population densities and political boundaries, together with a format for representation, for example, contour and perspective. Many of the issues involved in conventional cartography are relevant to the design of a map for use by autonomous vehicles [16].

The information represented in a map can vary considerably with the intended application. Demographic and topographic maps are common examples and, analogously, autonomous robot vehicles may have several different maps containing quite different information for different applications. For instance, a map used for navigation might contain information on the position and configuration of a set of active transponders for triangulation/trilateration. In contrast, a map for path planning purposes is likely to contain information on the positions and shapes of obstacles. Furthermore, the detail in a map is constrained not just by the available measurement accuracy but also by considerations of cost. Here cost refers to the storage requirements and the time needed to manipulate the data. The level of detail in a map is also influenced by the intended application. For example, a map showing only the major highways is useful for intercity traveling, whereas a detailed street map becomes important once the destination city has been reached. This example suggests that a hierarchy of maps of increasing level of detail may prove useful. The value of having a large number of different maps, each containing different types of information, is illustrated by the thickness of most popular atlases.

Applications affect not only the information content but also the manner in which this information is represented. That is, the same information may be represented quite differently depending on the intended application. For example, the information content of a map representation called the *Voronoi diagram*, as described in the paper by Canny and Donald, is essentially the shape and position of the obstacles. The Voronoi diagram represents this information as a collection of points that are equidistant from the two nearest obstacles. This representation is extremely useful for path planning, since it leads to paths that are optimally distant from obstacles. However, a more explicit description of the obstacles' positions and features would be more useful for purposes of navigation.

This chapter presents papers in which various maps and methods for their construction, representation, and manipulation are discussed. The first paper by Elfes illustrates the usefulness of having a hierarchy of maps available, depending on the needs of the computation involved. In deciding on the level of detail to include in a map, there is often a trade-off between offering sufficient detail for the vehicle to perform its duties and too much detail, which makes computations infeasible. In fact, Elfes chooses to maintain multiple hierarchies based on (1) abstraction, (2) geography, and (3) resolution. The basic representation is a two-dimensional grid of cells. Each cell contains the occupancy status (unknown, empty, or occupied) along with an uncertainty factor. The information needed to construct this grid is obtained by using overlapping sonar readings as the vehicle traverses the region. Ikegami et al. [9] have proposed a similar representation in which the grid cells contain the distance to the closest border.[1] Another method for the representation of occupancy grid information is a quadtree (or octree in three dimensions). A good survey of the literature on octrees can be found in [17].

The second paper in this chapter describes the Etak Navigation System. Although the system was developed for use in automobiles, the techniques can easily be used for autonomous vehicles. The key to the system is the digital map used for navigation and path planning. The map is in the form of an electronic database of topographical information that has been constructed from several sources including U.S. Census Bureau Dual Incidence Matrix Encoded (DIME) files, U.S. Geological Survey quadrangles, and aerial photographs. The underlying structure of the map is a cellular planar partition, see [6], that is, the map is partitioned into two-dimensional regions by a collection of one-dimensional edges that meet at zero-dimensional vertices. An edge is usually a section of a political boundary or a section of a road. The records in the database contain information on the edges, such as the endpoints of the edge and the regions on either side of the edge. A method for maintaining a representation of a planar partition so that modifications such as adding or removing edges or vertices can be done efficiently is found in [14]. For an overview of automobile navigational aides, see [7].

The paper by Lozano-Pérez describes a map called *configuration space*. For a system of vehicles with n degrees of freedom in their positioning, *configuration space* is the space of all n-tuples (called *configurations*), where each coordinate of a tuple represents a value for a particular degree of freedom. The subspace of configuration space containing the configurations in which the vehicles are in collision-free positions is called *free space*. That is, if the vehicles are in the positions as defined by a configuration in free space, then the vehicles are not colliding with one another or with any obstacle in the environment. Since a path in free space corresponds to a collision-free motion of the vehicles, this map has proved to be extremely useful for path planning; see also Chapter 6. Much research has involved exploring methods of computing configuration space efficiently and proving bounds on the complexity of configuration space, for example, see [12] and [8].

Although maps may traditionally be thought of as two-dimensional representations, free space has as many dimensions as there are degrees of freedom in positioning the object: for a robot vehicle it is three dimensional. A Voronoi diagram is a lower dimensional representation of the connectivity of free space. Since connectivity is the essential property needed for path planning, the Voronoi diagram is a useful representation for this purpose. The paper by Canny and Donald describes a variant on the standard definition of a Voronoi diagram (see Chapter 6) that has the feature that the algebraic degree of the diagram is lower than that of the standard one.

A useful method for representing free space as a collection of (overlapping) generalized cones is described in the paper by Brooks. A *generalized cone* can be thought of as the free space between a pair of obstacle walls, and the spine is the bisector of the cone. The collection of spines are a lower dimensional representation of free space and are similar to the Voronoi diagram in that points on a spine tend to be positions that are locally maximally distant to obstacles. In a subsequent paper [5] Brooks discusses extending this representation to include not only *highways*, that is, generalized cones but also *meadows*, convex areas of free space.

---

[1] Such a representation is very similar to that used in chamfer matching in computer vision [1, 3, 4].

For additional reading, there is a survey article [19] on map representations that contains a good selection of references. The papers by Ayache and Faugeras [2] and Smith et al. [18], reprinted in the previous section, and the paper by Payton [13] describe alternative map representations. The reader is also directed to a special issue of *AI Magazine* [10]. A good source on the fundamentals of map making can be found in [16]. For a historical perspective on map making and other features of navigation, see [11]. There is also a great deal of literature in the field of computational geometry dealing with the complexity of reasoning about objects that can be interpreted as representations of environments [15]. The path planning literature is a rich source of specialized map representations, since the essential feature of most path planning algorithms (see Chapter 6) is a clever representation of free space as well as efficient methods for constructing and searching the representation.

References

1. Anderson, A., and Borgefors, G. Hierarchical Surface Matching in 3D Images. NTIS PB87-193132, National Defense Research Institute, Linkoping, Sweden, 1987.
2. Ayache, N., and Faugeras, O.D. Maintaining Representations of the Environment of a Mobile Robot. *IEEE Trans. Robotics Automation RA-5*, 6 (1989), 804-819.
3. Barrow, H.G., Tenenbaum, J.M., Bolles, R.C., and Wolf, H.C. Parametric Correspondence and Chamfer Matching: Two New Techniques for Image Processing. In *Proceedings of the International Joint Conference on Artificial Intelligence*, 1977, pp. 659-663.
4. Borgefors, G. Distance Transformations in Digital Images. *Comput. Vision, Graph., Image Process. 34* (1986), 344-371.
5. Brooks, R.A. Visual Map making for a Mobile Robot. In *IEEE International Conference on Robotics and Automation*, IEEE, New York, 1985, pp. 824-829.
6. Edelsbrunner, H. *Algorithms in Computational Geometry.* Springer-Verlag, Heidelberg, Germany, 1987.
7. French, R. Automobile Navigation: Where is it Going? *IEEE Aerospace Electronic Systems Magazine 2*, 5 (1987), 6-12.
8. Guibas, L., Sharir, M., and Sifrony, S. On the General Motion Planning Problem with Two Degrees of Freedom. In *Proceedings of the 4th Annual Symposium on Computational Geometry*, ACM, New York, 1988, pp. 289-298.
9. Ikegami, T., Kato, J., and Ozono, S. Sensor Data Integration Based on the Border Distance Model. In *Proceedings of the IEEE/RSJ International Workshop on Intelligent Robots and Systems*, IEEE/RSJ, New York, 1989, pp. 32-39.
10. A. Kak, (Ed.) *AI Magazine*, *9*, 2, (1988).
11. Kayton, M. Navigation: Ships to Space. *IEEE Trans. Aerospace Electronic Systems 24*, 5 (1988), 474-520.
12. Kim, M.S. Motion Planning with Geometric Models. Ph.D. thesis, Pohang Institute of Science and Technology, (unpublished), 1989.

13. Payton, D. Circulation Maps: A Resource for Identifying Position Accuracy Requirements. In *Proceedings of the IEEE/RSJ International Workshop on Intelligent Robots and Systems*, IEEE/RSJ, New York, 1989, pp. 448-455.

14. Preparata, F., and Tamassia, R. Fully Dynamic Techniques for Point Location and Transitive Closure in Planar Structures. In *Symposium on the Foundations of Computer Science*, IEEE, New York, 1988, pp. 558-567.

15. *Proceedings of the ACM Annual Conference on Computational Geometry.*

16. Robinson, A., and Sale, R. *Elements of Cartography.* Wiley, New York, 1953.

17. Samet, H. Octree Survey. *ACM Comput. Surv. 16*, 2 (1984), 187-260.

18. Smith, R., Self, M., and Cheeseman, P. Estimating Uncertain Spatial Relationships in Robotics. *Uncertainty Artif. Intell. 2* (1988), 435-461.

19. Soldo, M. Survey of Map Representations. *ACM Comput. Surv.* (to be published).

Sonar-Based Real-World Mapping and Navigation

ALBERTO ELFES, MEMBER, IEEE

Abstract—**A sonar-based mapping and navigation system developed for an autonomous mobile robot operating in unknown and unstructured environments is described. The system uses sonar range data to build a multileveled description of the robot's surroundings. Sonar readings are interpreted using probability profiles to determine empty and occupied areas. Range measurements from multiple points of view are integrated into a sensor-level sonar map, using a robust method that combines the sensor information in such a way as to cope with uncertainties and errors in the data. The resulting two-dimensional maps are used for path planning and navigation. From these sonar maps, multiple representations are developed for various kinds of problem-solving activities. Several dimensions of representation are defined: the abstraction axis, the geographical axis, and the resolution axis. The sonar mapping procedures have been implemented as part of an autonomous mobile robot navigation system called Dolphin. The major modules of this system are described and related to the various mapping representations used. Results from actual runs are presented, and further research is mentioned. The system is also situated within the wider context of developing an advanced software architecture for autonomous mobile robots.**

I. Introduction

To WIDEN the range of application of robotic devices, both in industrial and research applications, it is necessary to develop systems with high levels of autonomy and able to operate in unstructured environments with little *a priori* information. To achieve this degree of independence, the robot system must have an understanding of its surroundings, by acquiring and manipulating a rich model of its environment of operation. For that, it needs a variety of sensors to be able to interact with the real world and mechanisms to extract meaningful information from the data being provided. Systems will little or no sensing capability are usually limited to fixed sequence operations in highly structured working areas and cannot provide any substantial degree of autonomy or adaptability.

A central need, both for manipulators and for mobile robots, is the ability to acquire and handle information about the existence and localization of objects and empty spaces in the environment of operation of the device. This is crucial for fundamental operations that envolve spatial and geometric reasoning. Typically, due to limitations intrinsic to any kind of sensor, it is important to compose information coming from multiple readings, and build a coherent world-model that reflects the information acquired and the hypotheses proposed so far. This world model can then serve as a basis for essential operations such as path planning, obstacle avoidance, landmark identification, position and motion estimation, etc.

Building up such a description involves the complex task of extracting range information from the real world. Several range measurement systems have been proposed in the literature [21]. Of particular interest for mobile robot research [22], [24] are stereo vision systems [29], [31], [25], [39] and active rangefinding devices [14], [15], [19], since these do not require artificial environments or contrived lighting.

In this paper, we explore the use of one specific active rangefinding device, an ultrasonic range transducer, to build a dense two-dimensional map of the robot's surroundings. Each sonar distance reading provides information concerning *empty* and *occupied* volumes in a cone in front of the sensor. The reading is interpreted using probability profiles that are projected onto a rasterized map, where *unknown, occupied,* and *empty* areas are explicitly represented.

Range measurements from multiple points of view (taken from multiple sensors on the robot and from different positions occupied by the robot) are integrated into the sonar map, using a robust method that combines the sensor information in such a way as to cope with uncertainties and errors in the data. Overlapping empty areas reinforce each other, the same happening with occupied areas. Additionally, the empty spaces serve to narrow down the profiles of occupied spaces. The result is that the sonar map grows in coverage and its definition improves as more readings are added. The resulting sonar map shows regions probably occupied, probably empty, and unknown areas. The method works effectively in cluttered environments, and the dense maps resulting can be used for motion planning, position estimation, landmark recognition, and navigation.

These probabilistic sensor-level sonar maps serve as the basis for a multilevel description of the robot's operating environment. These multiple descriptions are developed for various kinds of problem-solving activities. Several dimensions of representation are defined: the *abstraction* axis, the *geographical* axis, and the *resolution* axis.

The sonar mapping method has been implemented as part of an autonomous mobile robot navigation system called Dolphin. This system is intended to provide sonar-based mapping and navigation for an autonomous mobile robot operating in unknown and unstructured environments. The system is completely autonomous in the sense that it has no *a priori*

Manuscript received June 29, 1986; revised December 3, 1986. This research has been supported in part by the Office of Naval Research under Contract N00014-81-K-0503, in part by Denning Mobile Robotics, Inc., in part by the Western Pennsylvania Advanced Technology Center, and in part by USAETL, DARPA, DoD, under contract DACA 76-85-C-0003. The author is supported in part by the Conselho Nacional de Desenvolvimento Científico e Tecnológico-CNPq, Brazil, under Grant 200.986.80, in part by the Instituto Tecnológico de Aeronáutica-ITA, Brazil, in part by The Robotics Institute, Carnegie-Mellon University, and in part by the Design Research Center-CMU.

The author is with the Mobile Robot Laboratory, The Robotics Institute, Carnegie-Mellon University, Pittsburgh, PA 15213, USA.

IEEE Log Number 8715000.

VIII. Supervisor

- Global Supervision of System Behaviour
- User Interface

VII. Global Planning

- Task-Level Planning to provide sequences of sensory, actuator and processing (software) actions
- Simulation
- Error-Recovery and Replanning in case of failure or unexpected events

VI. Control

- Scheduling of Activities
- Integration of Plan-Driven with Data-Driven Activities

V. Navigation

- Navigation Modules provide services such as Path-Planning and Obstacle Avoidance

IV. Real-World Modelling

- Integration of local pieces of correlated information into a Global Real-World Model that describes the robot's environment of operation
- Matching acquired information against stored maps
- Object Identification
- Landmark Recognition

III. Sensor Integration

- Information provided by different Sensor Modules is correlated and abstracted
- Common representations and compatible frames of reference are used

II. Sensor Interpretation

- Acquisition of Sensor Data (Vision, Sonar, Laser Rangefinder, etc.)
- Interpretation of Sensor Data

I. Robot Control

- Set of Primitives for Robot Operation
- Actuator Control (e.g., Locomotion)
- Sensor Control
- Internal Sensor Monitoring

Fig. 1. Conceptual processing levels in mobile robot software architecture.

model or knowledge of its surroundings and also carries no user-provided map. It incrementally builds sonar maps that are used to plan safe paths and navigate the vehicle towards a given goal and may be coupled to other systems, such as vision, that would locate landmarks that would serve as long-range destinations. The system has been tested both indoors and outdoors using the Neptune and Terregator mobile robots at CMU.

In the sequence of this paper, we will briefly identify some of the conceptual processing levels needed for mobile robot software and relate the present system to this framework, describe the sonar mapping method, discuss the multiple representations developed for mapping information, present the overall system architecture, and show some results from actual runs. We finish with an outline of further research.

II. Conceptual Processing Levels for an Autonomous Mobile Robot

The sonar mapping and navigation system described in this paper is part of a wider investigation into issues related to the development of a software architecture for an autonomous mobile robot. In this section, we briefly outline a conceptual framework within which the sonar system is situated by characterizing the *conceptual processing levels* into which the various problem-solving activities of a mobile robot software architecture can be classified. The levels include the robot control, sensor interpretation, sensor integration, real-world modeling, navigation, control, global planning, and supervisor levels (Fig. 1), and are briefly described below.

Robot Control: This level takes care of the physical control of the different sensors and actuators available to the robot. It provides a set of primitives for locomotion, actuator and sensor control, data acquisition, etc., that serve as the robot interface, freeing the higher levels of the system from low-level details. It includes activities such as vehicle-based motion estimation and monitoring of internal sensors. *Internal sensors* provide information on the status of the different physical subsystems of the robot, while *external sensors* are used to acquire data about the outside world.

Sensor Interpretation: On this level the acquisition of sensor data and its interpretation by sensor modules is done. Each sensor module is specialized in one type of sensor or even in extracting a specific kind of information from the sensor data. The modules provide information to the higher levels using a common representation and compatible frames of reference.

Sensor Integration: Due to the intrinsic limitations of any sensory device, it is essential to integrate information coming from qualitatively different sensors, such as stereo vision systems, sonar devices, laser range sensors, etc. Specific assertions provided by the sensor modules are correlated to each other on this level. For example, the geometric boundaries extracted from an obstacle detected by sonar can be used to provide connectivity information concerning a set of scattered three-dimensional (3D) points generated by the stereo vision subsystem. On this level, information is aggregated and assertions about specific portions of the environment can be made.

Real-World Modeling: To achieve any substantial degree of autonomy, a robot system must have an understanding of its surroundings, by acquiring and manipulating a rich model of its environment of operation. This model is based on assertions composed from the various sensors, and reflects the data obtained and the hypotheses proposed so far. On this level, local pieces of information are used in the incremental construction of a coherent global real-world model; this model can then be used for several other activities, such as landmark recognition, matching of newly acquired information against previously stored maps, and generation of expectations and goals.

Navigation: For autonomous locomotion, a variety of problem-solving activities are necessary, such as short-term and long-term path planning, obstacle avoidance, detection of emergencies, etc. These different activities are performed by modules that provide specific services.

Control: This level is responsible for the scheduling of the different activities and for combining plan-driven and data-driven activities in an integrated manner to achieve coherent behavior. In other words, this level tries to execute the task-level plan that was handed to it, while adapting to changing real-world conditions as detected by the sensors.

Global Planning: To achieve a global goal proposed to the robot, this level provides task-level planning for autonomous generation of sequences of actuator, sensor, and processing actions. Other activities needed include simulation, error detection, diagnosis and recovery, and replanning in the case of unexpected situations or failures.

Supervisor: Finally, on this level a supervisory module controls the various activities and provides an interface to a human overseer.

By identifying these areas of activity, we are not implying that communication among processing modules is only possible between adjacent levels. On the contrary, experience with real systems shows that usually there are very complex interconnections and interdependencies between the various subsystems, with multiple flows of control and data. Additionally, a specific module (such as stereo vision or sonar mapping) may be a very complex system in itself, with sophisticated control, planning and problem-solving activities.

Clearly, none of the presently existing mobile robot systems cover all of the levels described. This conceptual structure provides, however, a context within which some of our research is situated [10], [31] and has influenced in particular the design of the Dolphin sonar-based mapping and navigation system, as mentioned in Section V.

III. Sonar-Based Mapping

This section describes a sonar-based mapping method developed for mobile robot navigation [32], [12]. We discuss the relative merits of sonar sensors, describe the interpretation of sonar data and the map-building process, and present some experimental results.

A. Range Sensors

Several methods for obtaining range data have been reported in the literature. The survey provided by Jarvis [21] discusses, among others, contrived lighting techniques (including striped lighting and grid coding), depth from occlusion, texture gradient and focusing, some "shape from" methods, range from stereo or motion, and triangulation-based and time-of-flight rangefinders. Of these, contrived lighting techniques are not generally useful to systems operating in unstructured environments, while triangulation-based rangefinders suffer from "gaps" in the data due to occlusions.

Of particular interest for natural unstructured environments are stereo vision systems and active rangefinding devices. In the time-of-flight category, the main two representatives are ultrasonic and laser rangefinders. Sonar systems have lower resolution; on the other hand, they are orders of magnitude less expensive than laser-based sensors. Phase-shift-based laser rangefinders are subject to a 2π uncertainty. In the case of the ERIM sensor [19], this limits the useful range of the sensor to 64 ft, as compared to 35 ft for the Polaroid sonar sensor. Both sensors suffer from absorption and specular reflection problems, while measurement precision is obviously much higher with laser rangefinders.

B. Stereo Vision and Sonar

One of the traditional approaches in mobile robot research has been the use of stereo vision systems to extract range information from pairs of images [29]. One of the difficulties in applying these techniques in real-world navigation is the fact that the intrinsic computational expense of extracting three-dimensional (3D) information from stereo pairs of images limits the number of points that can be tracked [40].

Real-time constraints preclude the obtention of dense 3D descriptions such as those presented in [16], [35], though recent work in the area shows very promising results [34]. Additionally, traditional stereo vision systems relied upon specific features such as high-contrast edges or points that could be easily tracked along several images [30].

As a result, practical real-world stereo vision navigation systems such as the ones described in [29], [40], [25] only build sparse depth maps of their surroundings, selecting points to be matched and tracked using an interest operator. Handling on the order of 30–50 points, the system described in [41] takes 30–60 s to generate a 3D map (on a VAX-11/780 processor).

These limitations led us to explore the use of an alternative kind of sensor that could deliver range information directly. Direct sonar range measurements promised to provide basic navigation and denser maps with considerably less computation. Additionally, we also became interested in exploring the composition of information from qualitatively different sensors, such as a sonar array and a stereo pair of cameras into a more complex and rich description of the robot's environment.

C. Sonar Applications in Robotics

Sonar range sensing is a mature technology but few applications until recently involved detailed map building. Traditionally, active and passive sonar systems have been used for military intelligence applications in marine environments. Active sonar systems are used for communications, navigation, detection, and tracking, while passive sonar systems are used in surveillance [3]. Some of this research is obviously classified. More recently, with the rapid increase of commercial and civilian activity in the oceans, nonmilitary marine sonar systems have spread. Typical applications include navigation, charting, and fishing. Generally speaking, these systems are characterized by detailed knowledge of the physical properties of the marine environment, by sophisticated sonar signal generation and processing devices [3], and by addressing the problem of localization and tracking of objects [17], as opposed to determining their shape.

An important area of application is medical imaging [8]. Ultrasound systems used in medicine are active and build maps for human perusal, but again depend on accurate physical models of the tissues that the sound traverses [20], and work with very small beam widths, about 1–3°. Typical frequencies used span the range from 1 to 20 MHz, and distances on the order of 10–50 cm are measured.

Other applications of sonar sensors include rangefinding devices for industrial control applications, as well as camera autofocus systems [38].

In the robotics area, ultrasonic range transducers have recently attracted increasing attention [1], [2]. This is due in part to their simplicity, low cost, and the fact that distance measurements are provided directly. Some research has focused specifically on the development of more elaborate beam-forming and detection devices (see, for example, [26]), or the utilization of phased array techniques [2], such as are used in an advanced side-looking sonar system for submersibles. Other efforts investigate the application of highly sophisticated signal processing techniques [3] to complex sonar signals.

Specific applications of sonar sensors in robotics include simple distance measurements [15], localizing object surfaces [4], determining the position of a robot in a given environment, and some *ad hoc* navigation schemes [5].

Miller [27], [28] uses sonar sensors to determine the position of a robot. It assumes that an accurate map of the environment is known and performs a search to determine where the robot would have to be to explain a given set of distance readings. The method does not take into account the errors that occur in actual sonar data. A similar approach is used by Drumheller [9] who also pressuposes an accurate map of the environment, but is able to cope with noisy data.

An independent CMU sonar mapping and navigation effort [6], [7] used a narrower beam, formed by a parabolic reflector, to build a line-based description of the robot's surroundings. The sonar readings are interpreted by fitting line segments to the points detected and matching these to an *a priori* map of the environment of the robot. One difficulty with this approach was that the geometric interpretation is done very early and is difficulted by noise and uncertainty in the data.

D. Sonar-Based Mapping for Mobile Robots

By contrast, our own work has centered on the development of a system for sonar-based mapping and navigation for an autonomous mobile robot operating in unknown and unstructured environments. The system has no *a priori* map of its surroundings. Instead, it acquires data from the real world through a set of sonar sensors and uses the interpreted data to build a sonar map of the robot's operating environment.

In applying sonar range sensors to mobile robot mapping and navigation, we expected to obtain dense maps of the robot's environment with regions classified as *empty*, *occupied*, and *unknown*, with sufficient precision and detail so as to be useful for autonomous navigation. This includes path planning, obstacle avoidance, motion solving, and landmark recognition. Additionally, we planned to develop a hierarchy of representations, from data intensive maps where position details are stored, to symbolic representations suitable for high-level planning.

1) Approach: Our method starts with range measurements obtained from sonar units whose position with respect to the robot is known. Each sonar range measurement is interpreted as providing information about *probably empty* and *somewhere occupied* volumes in the space subtended by the sonar beam (in our case, a 30° cone in front of the sensor). This occupancy information is modelled by probability profiles that are projected onto a rasterized two-dimensional horizontal map, where empty, occupied, and unknown areas are represented. Sets of range measurements taken from multiple sensors on the robot and from different positions occupied by the robot as it travels provide multiple views that are systematically integrated into the sonar map. In this way, the accuracy and extent of the sonar map are incrementally improved and the uncertainty in the positions of objects is reduced. Overlapping empty areas reinforce each other, the

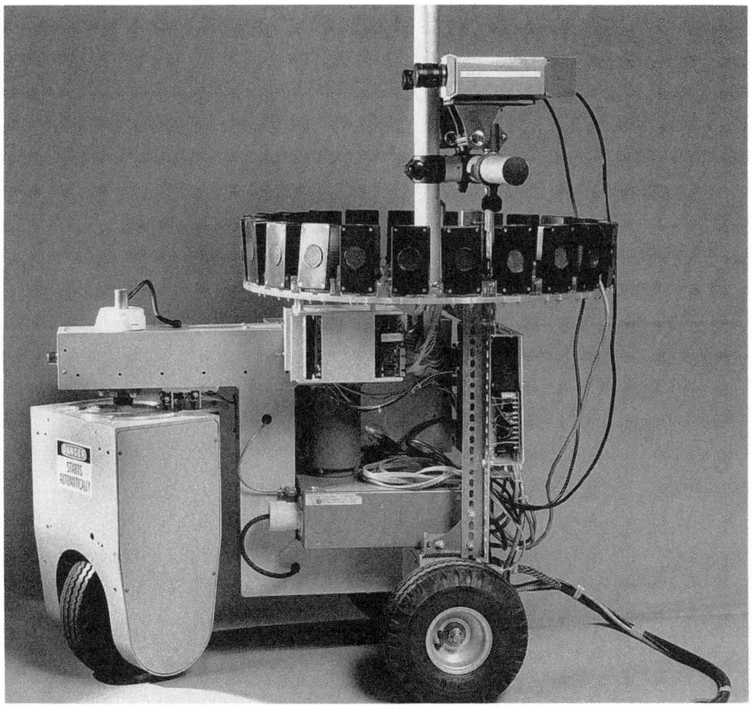

Fig. 2. Neptune mobile robot, with pair of cameras and sonar ring.

same happening with occupied areas. Additionally, empty regions serve to sharpen the boundaries of the occupied regions, so that the map definition improves as more readings are added. The final map shows probably occupied, probably empty, and unknown regions. This method deals effectively with noisy data and with the limitations intrinsic to the sensor.

For positional update as well as recognition of previously mapped areas, we developed a way of matching two sonar maps by convolving them. The method gives the displacement and rotation that best brings one map into registration with the other, along with a measure of the goodness of the match.

E. The Sonar System

1) The Sonar Sensor: The sonar devices being used are Polaroid laboratory grade ultrasonic range transducers [38]. These sensors have a useful measuring range of 0.9–35.0 ft. The main lobe of the sensitivity function is contained within a solid angle Ω of 30° and falls off to -38 dB. The beamwidth ω at -3 dB is approximately 15°. Experimental results showed that the range accuracy of the sensors is on the order of ± 0.1 ft. We are using the control circuitry provided with the unit, which is designed to return the distance to the nearest sound reflector in its field of view. The sensor interface does not provide information on multiple echos nor on the phase shift or the intensity of the detected echo.

2) The Sonar Sensor Array: The sonar sensor array, built at Denning Mobile Robotics, Inc., consists of a ring of 24 Polaroid ultrasonic transducers, spaced 15° apart. A Z80 controlling microprocessor selects and fires the sensors, times the returns and provides the corresponding range value. Over a serial link, this information is sent to a VAX mainframe, where currently the interpretation of the sonar data and the higher level mapping and navigation functions are performed.

3) The Robots: We have conducted several experiments by mounting the sonar sensor array on two currently available robots. The Neptune mobile robot [36] was developed at the Mobile Robot Laboratory of the Robotics Institute, Carnegie–Mellon University (CMU) (see Fig. 2). It has been used successfully in several areas of research, including stereo vision navigation [25], [41] and path planning [42]. Mounted on this vehicle, the sonar sensors are at an height of 31 in above the ground.

The Terregator robot (Fig. 3) is a larger vehicle developed by the Civil Engineering Robotics Construction Laboratory, CMU [23]. It has been used in several outdoor experiments, including road-following [43] and outdoor sonar navigation [12].

A new mobile robot, Uranus, incorporating an innovative omnidirectional design [37], is currently nearing completion and will be used for several current and new projects in the Mobile Robot Lab [31]. We plan to continue the sonar work using this vehicle.

F. Sonar Mapping

The sonar mapping process incorporates various stages. Initially, the sensor data are preprocessed, screened, and annotated with the corresponding sensor position. In sequence, the readings are interpreted using the probability density functions. A set of readings taken from one position of the robot is used to build a *view,* which stores the *empty, occupied,* and *unknown* areas as seen from that position. This view is then combined with the sonar map.

1) Problems with the Sonar Data: A number of problems are inherent to the data obtained from the sonar device and intrinsic to the sensor itself.

- The timing circuitry limits the range precision.

Fig. 3. Terregator outdoors robot.

- The detection sensitivity of the sensor varies with the angle of the reflecting object to the main beam axis.
- Sonar beams can suffer multiple reflections or specular reflections away from the sensor, giving false distance readings.
- Because of the relatively wide angle of the sonar beam, an isolated sonar reading imposes only a loose constraint on the position of the detected object.

These problems preclude a direct interpretation of the sonar readings and led us to consider a probabilistic approach to the interpretation of range data.

2) Preprocessing the Sonar Data: We begin by preprocessing the incoming data to remove easily detectable incorrect readings. This includes rejecting data below the lower range threshold $R_{\min}$ (usually due to faulty sensors) or above the useful range R_u of the device, as well as averaging multiple readings taken from the same sensor at the same robot position. The useful range is defined as $R_u = \min(R_g, R_{\max})$, with R_g being the range distance from the sensor to the ground and $R_{\max}$ the maximum sensor range.

3) Occupancy Probabilities: Due to the wide beamwidth, the data obtained from the sonar sensor provide only indirect information about the location of the detected objects. We combine the constraints from individual readings to reduce this uncertainty. Our inferences are represented as probabilities in a discrete grid.

A range reading is interpreted as making an assertion about two volumes in 3D space: one that is *probably empty* and one that is *somewhere occupied.* We model the sonar beam by two probability density functions, f_E and f_O, defined over these volumes. Informally, these functions measure our confidence concerning a point inside the cone of the beam being empty, and our uncertainty about the location of the surface patch that caused the echo, somewhere on the range surface of the cone. The probability density functions are defined based on the geometry of the beam and the spatial sensitivity pattern of the sonar sensor. They are parameterized by the range reading and the beamwidth.

Consider a point $\boldsymbol{P} = (x, y, z)$ belonging to the volume swept by the sonar beam. Define the following:

R range measurement returned by the sonar sensor,
ϵ maximum sonar measurement error,
ω sensor beamwidth,
Ω solid angle subtending the main lobe of the sensitivity function,
$\boldsymbol{S}$ $= (x_s, y_s, z_s)$, position of the sonar sensor,
δ distance from $\boldsymbol{P}$ to $\boldsymbol{S}$,
θ angle between the main axis of the beam and $\boldsymbol{P}$ as seen from $\boldsymbol{S}$.

We now define two volumes of space in the sonar beam.

Probably empty region: This includes points inside the sonar beam ($\delta < R - \epsilon$ and $\theta \leq \Omega/2$) that have a probability $p_E = f_E(\delta, \theta)$ of being empty.

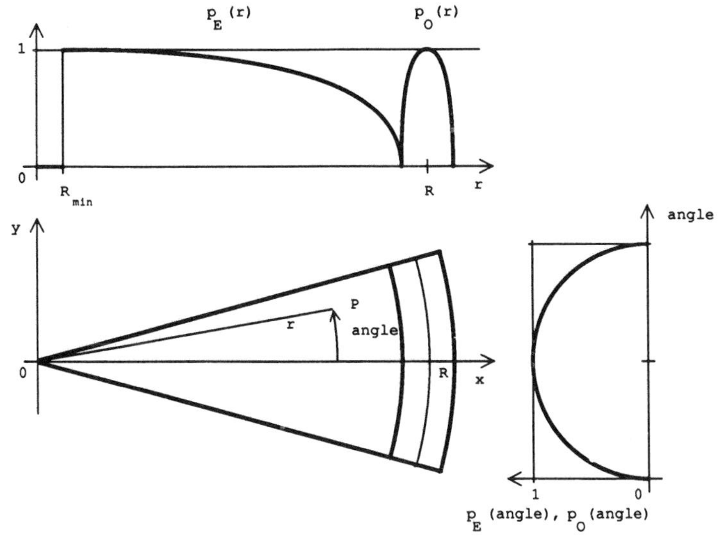

Fig. 4. Probability profiles corresponding to Probably Empty and Somewhere Occupied regions in sonar beam. Profiles correspond to horizontal cross section of beam.

Somewhere occupied region: This includes points on the sonar beam front ($\delta \in [R - \epsilon, R + \epsilon]$ and $\theta \le \Omega/2$), that have a probability $p_O = f_O(\delta, \theta)$ of being occupied.

The *empty* probability density function for a point $\boldsymbol{P}$ inside the sonar beam is given by

$$p_E(x, y, z) = p \text{ [point } (x, y, z) \text{ is empty]}$$
$$= E_r(\delta) \cdot E_a(\theta) \tag{3.1}$$

where

$$E_r(\delta) = 1 - ((\delta - R_{\min})/(R - \epsilon - R_{\min}))^2, \quad \text{for } \delta \in [R_{\min}, R - \epsilon], \tag{3.2}$$
$$E_r(\delta) = 0, \quad \text{otherwise}$$

and

$$E_a(\theta) = 1 - (2\theta/\Omega)^2 \quad \text{for } \theta \in [-\Omega/2, \Omega/2]. \tag{3.3}$$

The *occupied* probability density function for a point $\boldsymbol{P}$ on the beam front is given by

$$p_O(x, y, z) = p \text{ [position } (x, y, z) \text{ is occupied]}$$
$$= O_r(\delta) \cdot O_a(\theta) \tag{3.4}$$

where

$$O_r(\delta) = 1 - ((\delta - R)/\epsilon)^2, \quad \text{for } \delta \in [R - \epsilon, R + \epsilon], \tag{3.5}$$
$$O_r(\delta) = 0, \quad \text{otherwise},$$

and

$$O_a(\theta) = 1 - (2\theta/\Omega)^2, \quad \text{for } \theta \in [-\Omega/2, \Omega/2]. \tag{3.6}$$

Note that, strictly speaking, p_E and p_O are not true probability density functions.

Fig. 4 shows the empty and occupied probability distributions for a sonar beam that returned a range reading R. The profiles shown correspond to a horizontal cross section of the sonar beam ($z = z_s$). For map building, these probability density functions are evaluated for each reading and projected on a horizontal two-dimensional grid.

4) Representing occupancy maps: Sonar maps are two-dimensional arrays of cells corresponding to a horizontal grid imposed on the area to be mapped. The grid has $M \times N$ cells, each of size $\Delta \times \Delta$. Each cell in the final sonar map contains its occupancy status (unknown, empty, or occupied) with an associated certainty factor, using the following convention:

unknown	0
empty	$[-1, 0)$
occupied	$(0, 1]$.

A cell is considered unknown if no information concerning it is available. Cells can be empty with a certainty factor emp (x_i, y_j) (ranging from 0 to -1) and occupied with a certainty factor occ (x_i, y_j) (ranging from 0 to 1). The final map is computed from two separate arrays derived from the empty and occupied probability distributions introduced above. In the arrays themselves the empty and occupied probabilities are maintained as values ranging from zero to one.

5) Composing information from several readings: The sonar map is built by computing the empty and occupied sonar beam probability distributions for each range readings, projecting these probabilities onto the discrete cells of a view, and combining the view with the sonar map, which already stores the information derived from other readings. The position and orientation of the sonar sensors is used to register the view with the map.

Each sonar reading provides partial evidence about a map cell being occupied or empty. Different readings asserting that a cell is empty will confirm each other, as will readings implying that the cell is occupied. On the other hand, evidence that the cell is empty will weaken the certainty of it being occupied and vice versa.

The operations performed on the empty and occupied probabilities are not symmetrical. The probability distribution for empty areas represents a space whose totality is probably empty, while the occupied probability distribution for a single reading represents a lack of knowledge concerning the location of the reflecting object, somewhere on the front of the beam. Empty regions are simply combined using a probabilistic addition formula. The occupied probabilities for a single reading, on the other hand, are initially weakened by conflicting data and then normalized to make their sum unity. Only after this narrowing process are the occupied probabilities from each reading combined, using again a probabilistic addition formula.

One range measurement contains only a small amount of information. By combining the evidence from many readings as the robot moves in its environment, the area known to be empty is expanded. Possibly occupied regions also increase, while the fuzziness of these regions decreases. The overall effect as more readings are added is a gradually increasing coverage along with an increasing precision in object locations. Correct information is incrementally enhanced and wrong data are progressively canceled out. Typically, after a few hundred readings (and less than a second of computer time) the method is able to "condense out" a comprehensive map covering a thousand square feet with up to 0.1-ft position accuracy in the position of detected objects. Note that such a result does not violate information theoretic or degree of freedom constraints, since the detected boundaries of objects are linear, not quadratic in the dimensions of the map. A thousand square feet map may contain only a hundred linear feet of boundary.

Formally, the evidence combination process proceeds along the following steps.

1) Initialization: The sonar map is set to unknown.

2) Superposition of empty areas: The probabilities corresponding to empty areas are combined using the probability addition formula p_E (cell) $:=$ p_E (cell) $+$ p_E (reading) $-$ p_E (cell) $\times$ p_E (reading).

3) Superposition of occupied areas: The probabilities corresponding to occupied areas are initially weakened by the evidence of the empty certainty factors, using p_O (reading) $:=$ p_O (reading)$\cdot(1 - p_E$ (cell)$)$. We then normalize the occupied probabilities over the beam front. Finally, the occupied probabilities are combined using p_O (cell) $:=$ p_O (cell) $+$ p_O (reading) $-$ p_O (cell) $\times$ p_O (reading).

4) Thresholding: The final occupation value attributed to a cell in the sonar map is given by comparing the relative strengths of the empty and occupied values.

G. Results

Fig. 5 shows a typical sonar map obtained using the method outlined, with the conflicting information still superposed. The corresponding occupied and empty certainty factor distributions are shown in Figs. 6 and 7. These are the maps obtained before the thresholding step. The final maps obtained after thresholding are shown in Figs. 8–10. These maps correspond to the Mobile Robot Lab at CMU. In Figs. 5 and 8 the outline of the room and the major objects is shown, as well as the positions of the robot from where the readings were taken. The 3D plots correspond to a view from the lower left corner of the room.

The resulting sonar maps are very useful for navigation and landmark recognition. They are much denser than the ones generated by our stereo vision programs and computationally about an order of magnitude faster to produce. We have demonstrated an autonomous navigation system [12], discussed in Section V, that uses an A*-based path planner to obtain routes in these maps. The system was tested in cluttered indoor environments using Neptune, and outdoors in open spaces, operating among trees, using the Terregator.

H. Matching

One useful capability in robot navigation is the ability to match sets of observations against each other. Possible applications include landmark recognition and updating the robot's estimate of its position and orientation.

Towards this end, Moravec developed a method that can match two maps and report the displacement and rotation that best takes one into the other [32]. The sonar maps described in the previous section are used.

A measure of the goodness of the match between two maps at a trial displacement and rotation is found by computing the sum of products of corresponding cells in the two maps. An occupied cell falling on an occupied cell contributes a positive increment to the sum, as does an empty cell falling on an empty cell (the product of two negatives). An empty cell falling on an occupied one reduces the sum, and any comparison involving an unknown value causes neither an increase nor a decrease. This naive approach is very slow. Applied to maps with a linear dimension of n, each trial position requires $O(n^2)$ multiplications. Each search dimension (two axes of displacement and one of rotation) requires $O(n)$ trial positions. The total cost of the approach thus grows as $O(n^5)$. With a typical n of 50, this approach can burn up a good fraction of an hour of VAX time.

Considerable savings come from the observation that most of the information in the maps is in the occupied cells alone. Typically, only $O(n)$ cells in the map, corresponding to wall and object boundaries, are labeled occupied. A revised matching procedure compares maps A and B through trial transformation T (represented by a 2×2 rotation matrix and a two-element displacement vector) by enumerating the occupied cells of A and transforming the coordinates of each such cell in A through T to find a corresponding cell in B. The $[A, B]$ pairs obtained this way are multiplied and summed, as in the original procedure. The occupied cells in B are enumerated and multiplied with corresponding cells in A, found by transforming the B coordinates through T^{-1} (the inverse function of T), and these products are also added to the sum. The result is normalized by dividing by the total number of terms. This procedure is implemented efficiently by preprocessing each sonar map to give both a raster representation and a linear list of the coordinates of occupied cells. The cost grows as $O(n^4)$, and the typical VAX running time is down to a few minutes.

A further speedup is achieved by generating a hierarchy of

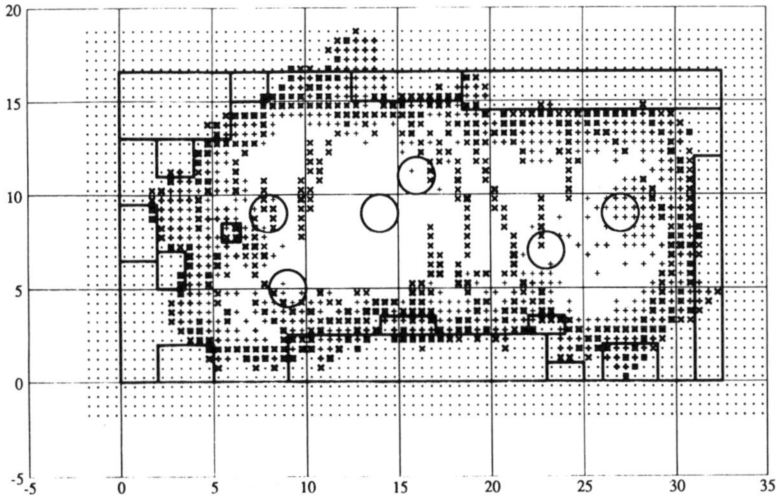

Fig. 5. Two-dimensional sonar map. Each symbol represents square cell 0.5 ft on side. Empty areas with high certainty factors are represented by white space; lower certainty factors by + symbols of increasing thickness. Occupied areas are represented by × symbols, and unknown areas by ·. This map still shows conflicting information superposed. Robot positions where scans were taken are shown by circles and outline of room and of major objects by solid lines. Experiment was done in Mobile Robot Lab.

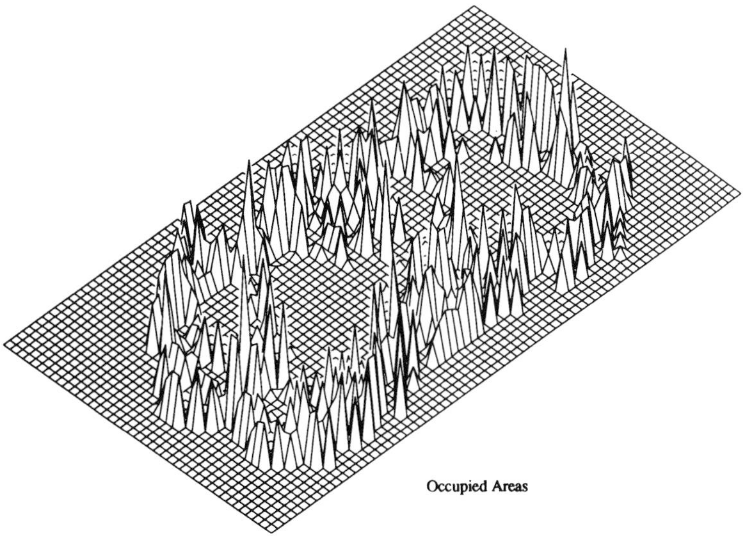

Fig. 6. Occupied areas in sonar map. This 3D view shows certainty factors occ (X, Y).

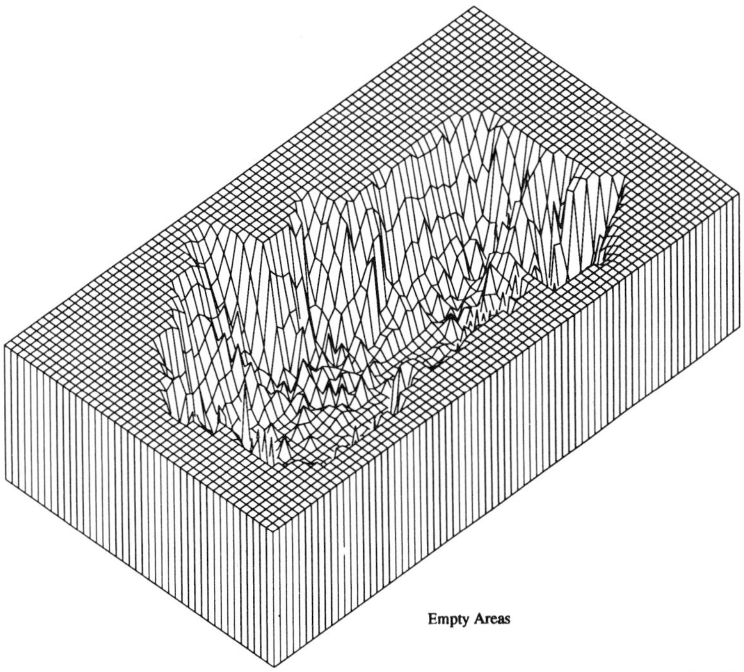

Fig. 7. Empty areas in sonar map. This 3D view shows certainty factors emp (X, Y).

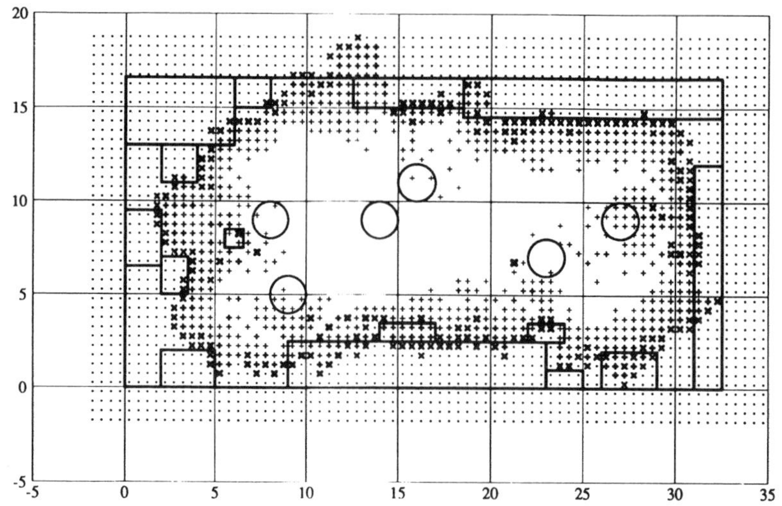

Fig. 8. Two-dimensional sonar map after thresholding.

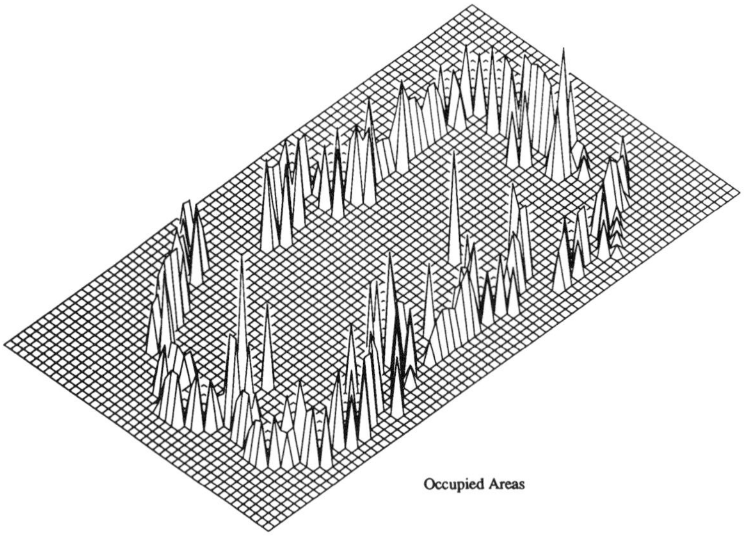

Fig. 9. Occupied areas in sonar map after thresholding.

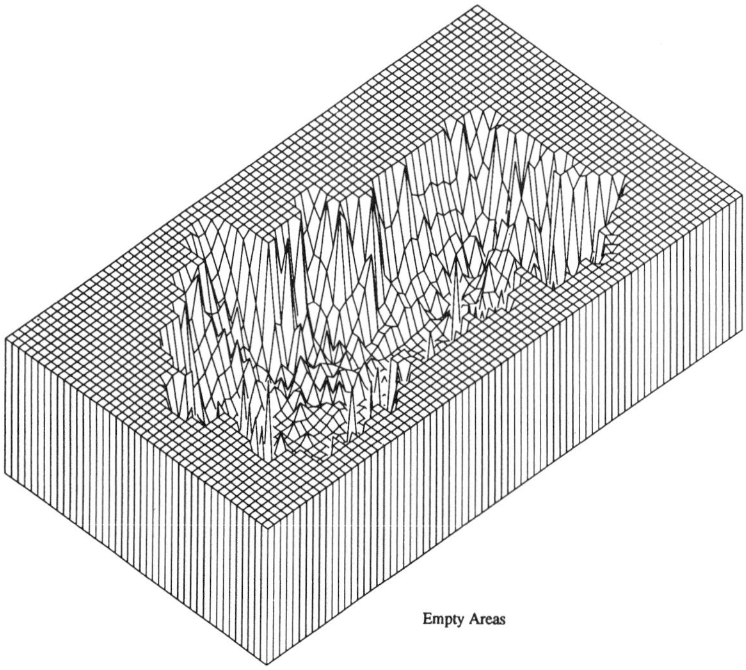

Fig. 10. Empty areas in sonar map after thresholding.

reduced resolution versions of each map. A coarser map is produced from a finer one by converting two by two subarrays of cells in the original into single cells of the reduction. Our existing programs assign the maximum value found in the subarray as the value of the result cell, thus preserving occupied cells. If the original array has dimension n, the first reduction is of size $n/2$, the second of $n/4$ and so on. A list of occupied cell locations is produced for each reduction level so that the matching method of the previous paragraph can be applied. The maximum number of reduction levels is $\log_2 n$. A match found at one level can be refined at the next finer level by trying only about three values of each of the two translational and one rotational parameters, in the vicinity of the values found at the coarser level, for a total of 27 trials. With a moderate *a priori* constraint on the transformation this amount of search is adequate even at the first (coarsest) level. Since the cost of a trial evaluation is proportional to the dimension of the map, the coarse matches are inexpensive in any case. Applied to its fullest, this method brings the matching cost down to slightly larger than $O(n)$, and practical VAX matching times to under a second.

We found one further preprocessing step is required to make the matching process work in practice. Raw maps at typical resolutions (6-in cells) produced from moderate numbers on sonar measurements (about 100) have narrow bands of cells labeled "occupied." In separately generated maps of the same area the relative positions of these narrow bands shift by as much as several pixels, making good registration of the occupied areas of the two maps impossible. This can be explained by saying that the high spatial frequency component of the position of the bands is noise and only the lower frequencies carry information. The problem can be fixed by filtering (blurring) the occupied cells to remove the high-frequency noise. Experiments suggest that a map made from 100 readings should be blurred with a spread of about 2 ft, while for a map made from 200 readings a 1-ft smear is adequate. Blurring increases the number of cells labeled "occupied." So as not to increase the computational cost from this effect, only the final raster version of the map is blurred. The occupied cell list used in the matching process is still made from the unfiltered raster. With the full process outlined here, maps with about 3000 6-in cells made from 200 well-spaced readings of a cluttered 20 by 30-ft room can be matched with an accuracy of about 6 in displacement and 3° rotation in one second of VAX time.

IV. Multiple Axis of Representation of Sonar Mapping Information

In the previous section we have shown how the range data acquired from the real world through a sonar sensor array are interpreted and used to build a sonar map. These probabilistic local maps, described earlier, are the starting point for building a multileveled and multifaceted description of the robot's operating environment. In this section, we briefly describe these multiple axes of representation of mapping information and mention how they are used in different kinds of navigational activities [11]. We define the following axes of representation (Fig. 11).

1) The Abstraction Axis: Along this axis we move from a sensor-based low-level data-intensive representation to increasingly higher levels of interpretation and abstraction. Three levels are defined as the sensor level, the geometric level, and the symbolic level.

2) The Geographical Axis: Along this axis we define views, local maps, and global maps, depending on the extent and characteristics of the area covered.

3) The Resolution Axis: Sonar maps are generated at different values of grid resolution for different applications. Some computations can be performed satisfactorily at low levels of detail, while others have to be done at high or even multiple degrees of resolution.

A. The Abstraction Axis

The first kind of sonar map built from the sonar range data uses the *probabilistic* representation described earlier. A two-dimensional grid covering a limited area of interest is used. This map is derived directly from the interpretation of the sensor readings and is, in a sense, the description closest to the real world. It serves as the basis from which other kinds of representations are derived. Along the abstraction axis, this data-intensive representation is also called the *sensor level map*.

The second level is called the *geometric level*. It is built by scanning the sensor level map and identifying groups of cells with high occupied confidence factors. These are merged into uniquely labeled objects with explicitly represented polygonal boundaries (see Fig. 14). If needed, the same can be done with empty areas.

The third is the *symbolic level,* where maps of larger areas (typically global maps) are described using a framelike representation. This description bears only a topological equivalence to the real world. Nodes may represent "interesting" areas, where more detailed mapping information is necessary or available, or may correspond to simpler or "uninteresting" areas (navigationally speaking), such as corridors.

Different kinds of problem-solving activities are better performed on different levels of abstraction. For example, global path planning (such as how to get from one building wing to another) would be done on the symbolic level, while navigation through a specific office or lab uses the sensor-level map, where all the detailed information about objects and free space, as well as the associated certainty factors, is stored.

B. The Geographical Axis

To be able to focus on specific geographical areas and to handle portions of as well as complete maps, we define a hierarchy of maps with increasing degrees of coverage. Progressing along the geographical axis, we start with a *view,* which is a map generated from scans taken from the current position and which describes the area visible to the robot from that position. As the vehicle moves, several views are acquired and combined into a *local map.* The latter corresponds to physically delimited spaces such as labs or offices, which define a connected region of visibility. Global Maps are sets of several local maps and cover wider spaces such as a whole wing of a building, with labs, offices, open areas, corridors,

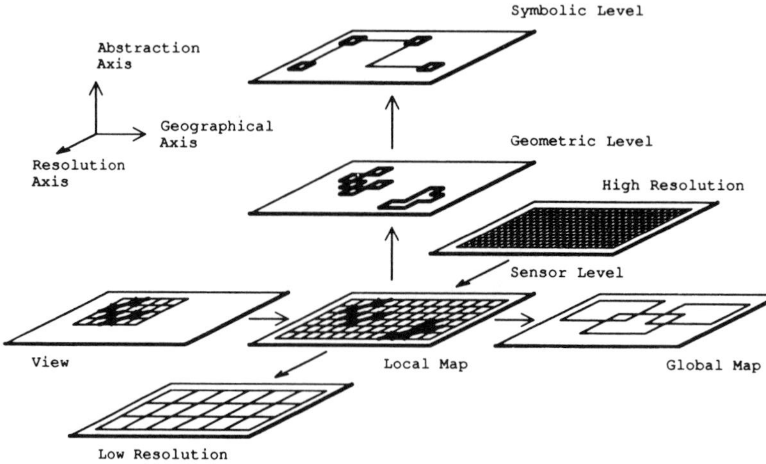

Fig. 11. Multiple axis of representation of sonar maps.

etc. The global map stores information about the whole known environment of operation of the robot.

C. The Resolution Axis

Finally, along the resolution axis, we again start with the sensor-level local map and generate a progression of maps with increasingly less detail. This allows certain kinds of processing to be performed at lower levels of resolution with correspondingly less computational expense. Alternatively, it enables operations at coarser levels to guide the problem-solving activities at finer levels of resolution.

The most detailed sonar maps that can be obtained from the method outlined in Section III-F (considering the intrinsic limitations of the sensors) have a cell size of 0.1×0.1 ft. For navigation purposes, we have typically been using a 0.5-grid for indoors and a 1.0-ft grid for outdoors. Nevertheless, several operations on the maps are expensive and are done more quickly at even lower levels of resolution. For these cases we reduce higher resolution maps by an averaging process that produces a coarser description, as discussed in Section III-H.

V. System Architecture

To provide a context for the multiple descriptions introduced above, we present in this Section the overall architecture of the Dolphin sonar-based mapping and navigation system. The functions of the major modules and their interaction with the various sonar map representations are discussed, and the results of an actual run are shown.

The Dolphin system is intended to provide sonar-based mapping and navigation for an autonomous mobile robot operating in unknown and unstructured environments. Conceptually, two modes of operation are possible: in the cruising mode, the system acquires data, builds maps, plans paths, and navigates towards a given goal. In the exploration mode, it can wander around and collect enough information so as to be able to build a good description of its environment. The system is intended for indoor as well as outdoor use; it may be coupled to other systems, such as vision [23], to locate landmarks that would serve as long-range destinations.

A. Sonar-Based Mapping and Navigation System Architecture

The overall architecture of the sonar mapping and navigation system is shown in Fig. 12. The function of the major modules and their interaction with the different sonar map representations are described below [11]:

- sonar control — interfaces to and runs the sonar sensor array, providing range readings;
- scanner — preprocesses and filters the sonar data and annotates it with the position and orientation of the corresponding sensor, based on the robot's motion estimate;
- mapper — using the information provided by the scanner, generates a view obtained from the current position of the robot (this view is then integrated into a local map);
- cartographer — aggregates sets of local maps into global maps and provides map handling and bookkeeping functions;
- matcher — matches a newly acquired sonar map against already stored local maps for operations such as landmark identification or update of the robot's position estimate;
- object extraction — provides geometric information about obstacles (objects are extracted by merging regions of occupied cells and determining the corresponding polygonal boundaries); a region-coloring approach is used for unique labeling;
- graph building — generates a frame-based symbolic description of the environment;
- path planning — can occur on three different levels: *symbolic* path planning is done over wider areas (global maps) and at a higher level of abstraction (symbolic maps); *geometric* path planning can be used as an intermediary stage, when the uncertainty in local maps is low, and has the advantage of being faster than finding routes in the sensor map; finally, *sensor map* path planning generates detailed safe paths (the latter performs an A* search [18] over the map cells, with the cost function

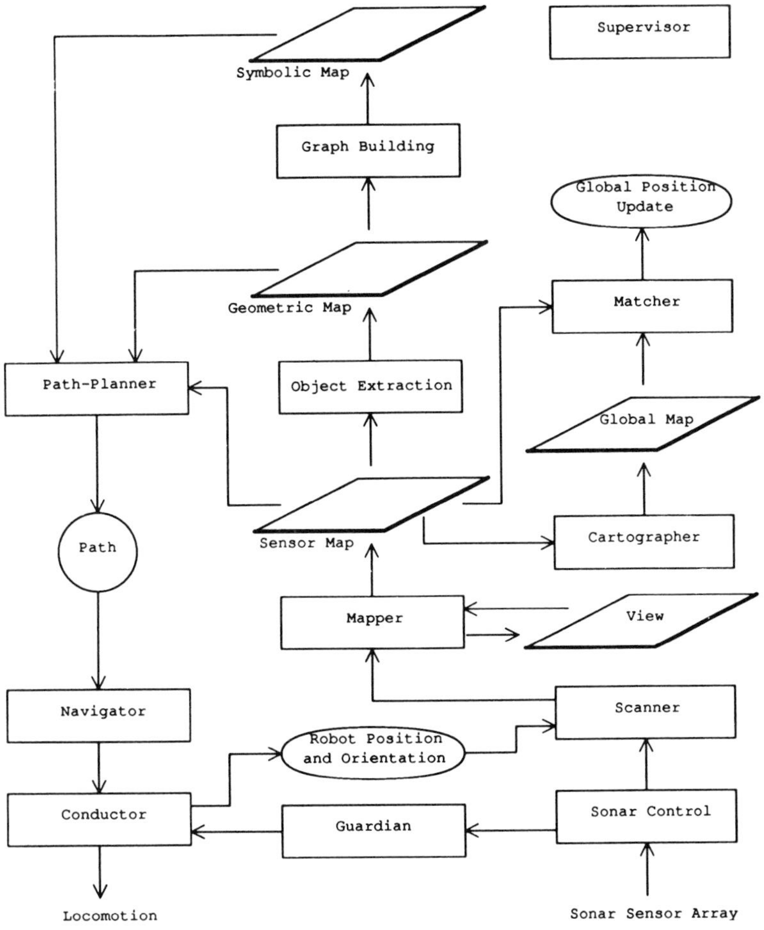

Fig. 12. Architecture of Dolphin sonar mapping and navigation system.

taking into account the occupied and empty certainty factors, as well as the unknown areas and the distance to the goal; the path found by this module is provided to the navigator);

navigator takes care of the overall navigation issues for the vehicle (includes examining already planned paths to determine whether they are still usable, invoking the path-planner to provide new paths, setting intermediary goals, overseeing the actual locomotion, etc.);

conductor controls the physical locomotion of the robot along the planned path (the latter is smoothed and approximated by sequences of line segments, using a line-fitting approach; this module also returns an estimate of the new position and orientation of the robot);

guardian during actual locomotion, this module continuously checks the incoming sonar readings and signals a stop if the robot is coming too close to a (possibly moving) obstacle not detected previously. It serves as a "sonar bumper";

supervisor oversees the operation of the various modules and takes care of the overall control of the system. It also provides a user interface.

We are currently working on a computational definition of the symbolic level and the corresponding mechanisms to extract this description from the sonar maps. All other subsystems mentioned have been implemented.

Comparing this architecture with the conceptual framework outlined in Section II, we can identify an immediate correspondence between the subsystems of the Dolphin system and some of the processing levels described previously: the sonar control and conductor modules belong to level I; scanning and mapping provide functions on level II; the object extraction, graph building, cartographer, and matcher operate on level IV; path planning, navigation, and the guardian are situated in level V; and the supervisor is on level VIII.

B. Tests of the System

The Dolphin system described in this section was tested in several indoor runs in cluttered environments using the Neptune mobile robot. Additionally, it was also tested in outdoor environments, operating among trees, using the Terregator robot, as part of the CMU ALV project. The system operated successfully in both kinds of environments, navigating the robot towards a given destination.

In Fig. 13, an example run is given. The sequence of maps

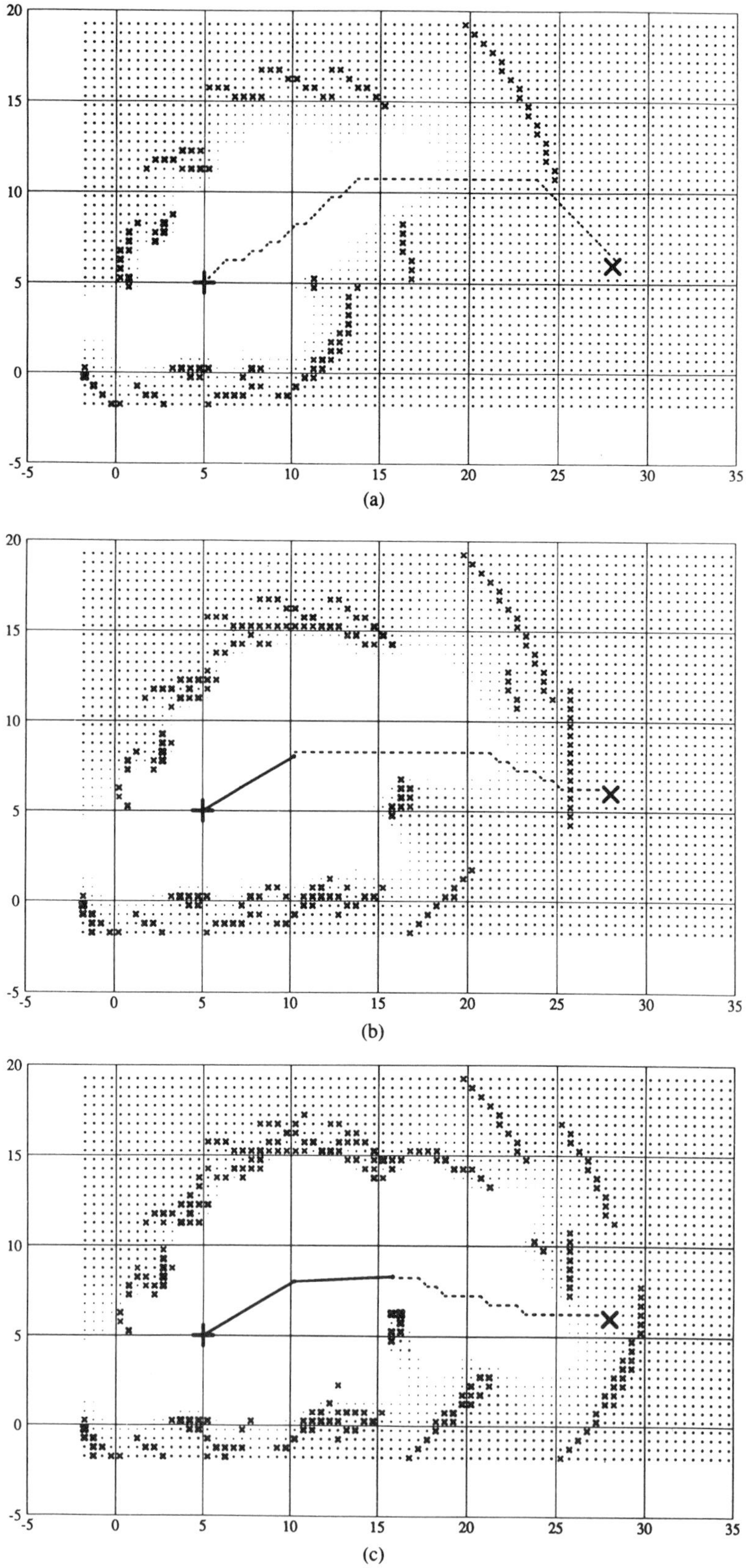

Fig. 13. Example run. This run was performed indoors, in Mobile Robot Lab. Distances are in ft. Grid size is 0.5 ft. Planned path is shown as dotted line, and route actually followed by robot as solid line segments. Starting point is solid + and goal, solid ×.

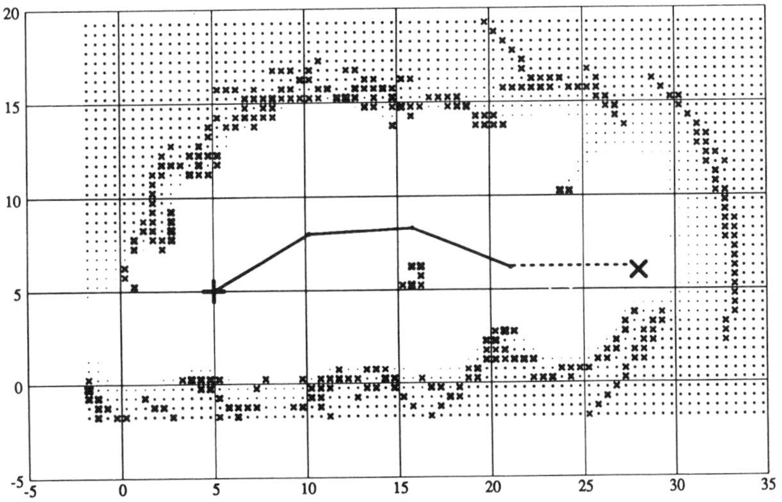

(d)

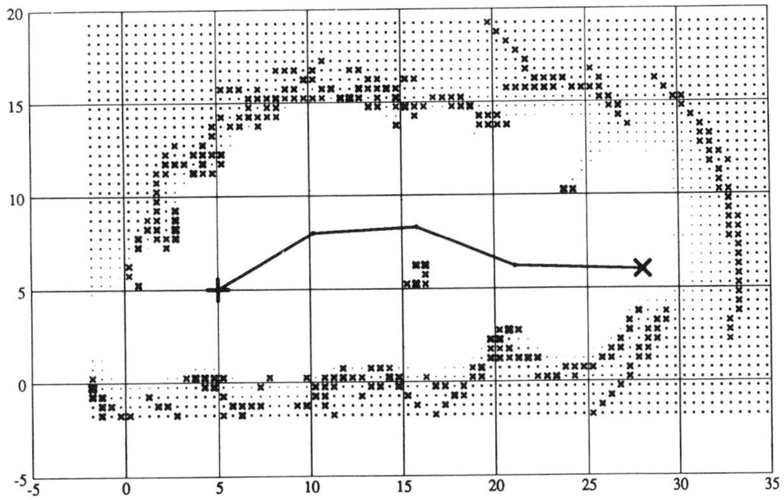

(e)

Fig. 13. (*Continued*).

shows how the sonar map becomes gradually more detailed and how the path is improved as more information is gathered. This example corresponds to an indoor run carried out in the Mobile Robot Lab. A distance of approximately 25 ft was covered; the grid size is 0.5 ft. Objects present in the lab included chairs, tables, boxes, workstations, filing cabinets, etc.

In Fig. 14, an outdoor run, together with an example of the object extraction module, is shown. The geometric map is build by scanning the sonar map and using a region-coloring technique to extract the obstacles. The objects are uniquely labeled and the polygonal boundaries are determined. The map shown corresponds to an outdoor run in Schenley Park, among trees. A distance of approximately 50 ft was traversed. The grid size was 1.0 ft, which proved adequate for navigation but did not allow a more precise description of the actual boundaries of the detected objects.

VI. Further Research

We are currently extending the research described in this paper by pursuing several topics. These include

- introducing the robot position uncertainty into the view registration and map making process;
- performing motion solving by matching a set of readings taken from a new robot position against the sonar map constructed so far;
- investigating issues in sensor integration in the specific context of combining sonar maps with 2D stereo data [39];
- exploring better sonar beam models by using Gaussian distributions in the empty and occupied probability density functions, and by taking into account the dependency of the beamwidth on the range measured;
- implementation of a distributed version of the Dolphin system as an actual test of the distributed control system described in [13].

We also mention that plans currently exist to extend the sonar-mapping method to do three-dimensional modeling. For that, a sonar sensor array that covers part of the surface of a sphere will be constructed [33].

VII. Conclusion

Research in mobile autonomous vehicles provides a very rich environment for the development and tests of advanced

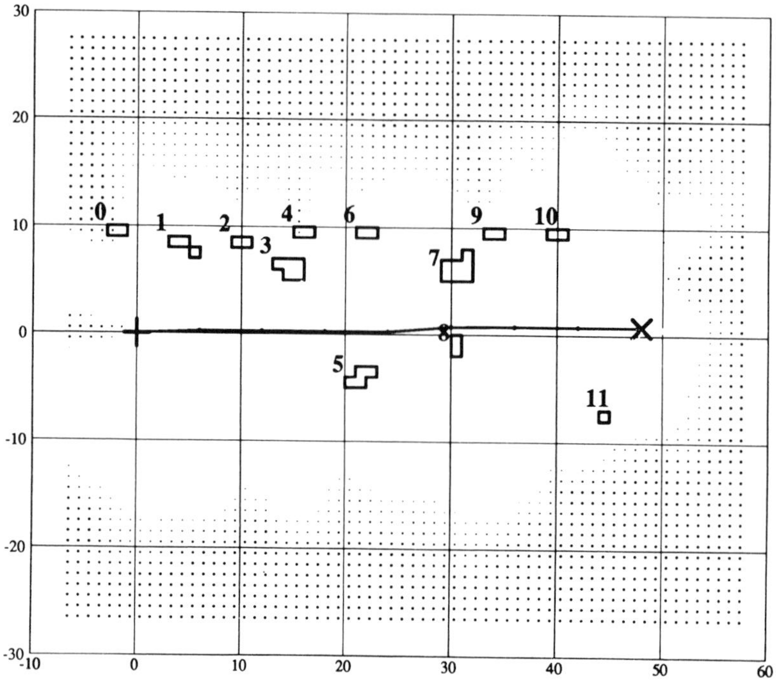

Fig. 14. Objects extracted from sonar map. Objects are uniquely labeled and their polygonal boundaries are shown. This map shows outdoor run, and objects are trees. Distances are in ft. Grid size is 1.0 ft.

concepts in a variety of areas such as robotics, artificial intelligence, sensor understanding and integration, real-world modeling, planning, and high-level control. Much work in mobile robots has either concentrated on very specific subareas, such a path planning, or has been conceptual work that never made it to a real application. This is partially due to the complexity of the overall task, the unavailability of adequate testbeds and the difficulty of doing actual experiments in something close to real time. This paper has presented a sonar-based mapping and navigation system that is able to operate in unknown and unstructured environments. It provides a sufficiently rich description of the robot's environment to call for more complex tasks; additionally, this is done in a sufficiently real-time situation so as to allow actual experiments to be done within a reasonable time frame.

Frequently, the real-world information available through sensors is quite inadequate. Stereo systems used in mobile robot research build a very sparse description of the robot's surroundings. Some of the range-based systems build denser descriptions but do not cope well with measurement errors. The representation chosen for sonar maps provides a way of explicitly describing unknown, empty, and occupied areas and of reasoning about them. This, coupled with a robust method for combining multiple sensor readings, allows us to cope with uncertainties and errors in the data and provides dense sonar maps, useful for navigation.

Acknowledgment

Hans P. Moravec contributed several key insights into the work described in this paper. I would like to thank him for his interest and support. I would also like to thank Gregg W. Podnar for providing assistance with the Neptune robot, and Richard Redpath for assistance in outdoor runs.

References

[1] U. Ahrens, "Möglichkeiten und Probleme der Anwendung von Luft-Ultraschallsensoren in der Montage-und Handhabungstechnik," *Robotersysteme,* vol. 1, 1, 1985.

[2] ——, "Möglichkeiten und Grenzen des Einsatzes von Luft-Ultraschallsensoren in der Montage-und Handhabungstechnik," *Robotersysteme,* vol. 1, 4, 1985.

[3] A. B. Baggeroer, *Sonar Signal Processing,* in *Applications of Digital Signal Processing,* Signal Processing Series. Englewood Cliff, NJ: Prentice-Hall, 1978.

[4] M. K. Brown, "Locating object surfaces with an ultrasonic range sensor," in *Proc. 1985 IEEE Int. Conf. Robotics and Automation,* St. Louis, MO, Mar. 1985.

[5] A. Chattergy, "Some heuristics for the navigation of a robot," Robotics Res. Lab., Dep. Elec. Eng., Univ. of Hawaii, Honolulu, 1984.

[6] J. L. Crowley, "Position estimation for an intelligent mobile robot," in *1983 Annu. Research Rev.,* Robotics Inst., Carnegie-Mellon Univ., Pittsburgh, PA, 1984.

[7] ——, "Dynamic world modelling for an intelligent mobile robot using a rotating ultra-sonic ranging device," in *Proc. 1985 IEEE Int. Conf. Robotics and Automation,* St. Louis, MO, Mar. 1985.

[8] G. B. Devey and P. N. T. Wells, "Ultrasound in medical diagnosis," *Sci. Amer.,* vol. 238, May 1978.

[9] M. Drumheller, "Mobile robot localization using sonar," Artificial Intelligence Lab., Mass. Inst. Technol., AI-M-826, Jan. 1985.

[10] A. Elfes and S. N. Talukdar, "A distributed control system for the CMU rover," in *Proc. 8th Int. Joint Conf. Artificial Intelligence,* Karlsruhe, Germany, Aug. 1983.

[11] A. Elfes, "Multiple levels of representation and problem-solving using maps from sonar data," in *Proc. DOE/CESAR Workshop on Planning and Sensing for Autonomous Navigation,* Oak Ridge Nat. Lab., Univ. California, Los Angeles, Aug. 18–19, 1985.

[12] ——, "A sonar-based mapping and navigation system," in *1986 IEEE Int. Conf. Robotics and Automation,* San Francisco, CA, Apr. 7–10, 1986.

[13] ——, "A distributed control architecture for an autonomous mobile robot," *Int. J. Artificial Intelligence in Eng.,* vol. 1, Oct. 1986.

[14] O. D. Faugeras, "Object representation, identification, and positioning from range data," presented at the 1st Int. Symp. Robotics Research, Cambridge, MA, 1984.

[15] G. Giralt, R. Chatila, and M. Vaisset, "An integrated navigation and motion control system for autonomous multisensory mobile robots,"

presented at the 1st Int. Symp. Robotics Research, Cambridge, MA, 1984.

[16] W. E. L. Grimson, *From Images to Surfaces: A Computational Study of the Human Early Visual Systems.* Cambridge, MA: MIT Press, 1981.

[17] J. Hallam, "Resolving observer motion by object tracking," in *Proc. 8th Int. Joint Conf. Artificial Intelligence,* Karlsruhe, Germany, Aug. 1983, pp. 792–798.

[18] P. E. Hart, N. J. Nilsson, and B. Raphael, "A formal basis for the heuristic determination of minimum cost paths," *IEEE Trans. Syst., Sci., Cybern.,* vol. SSC-4, 1968.

[19] M. Herbert and T. Kanade, "Outdoor scene analysis using range data," in *Proc. 1986 IEEE Int. Conf. Robotics and Automation,* San Francisco, CA, Apr. 7–10, 1986.

[20] M. Hussey, *Diagnostic Ultrasound: An Introduction to the Interactions between Ultrasound and Biological Tissues.* London: Blackie, 1975.

[21] R. A. Jarvis, "A perspective on range finding techniques for computer vision," *IEEE Trans. Pattern Anal. Machine Intell.,* vol. PAMI-5, Mar. 1983.

[22] M. Julliere and L. Marce, "Contribution a l'autonomie des robots mobiles," Lab. d'Applications des Techniques Electroniques Avancées, Inst. Nat. des Sciences Appliqués, Rennes, France, 1982.

[23] T. Kanade and C. E. Thorpe, "CMU strategic computing vision project report: 1984 to 1985," Robotics Inst., Carnegie-Mellon Univ., Pittsburgh, PA, CMU-RI-TR-86-2, Nov. 1985.

[24] M. Lionel, "Contribution a l'autonomie des robots mobiles," Ph.D. dissertation, L'Institut National des Sciences Appliquées de Rennes et L'Université de Rennes I, Rennes, France, July 1984.

[25] L. H. Matthies and C. E. Thorpe, "Experience with visual robot navigation," in *Proc. IEEE Oceans 84,* Washington, DC, Aug. 1984.

[26] G. L. Miller, R. A. Boie, and M. J. Sibilia, "Active damping of ultrasonic transducers for robotic applications," in *Proc. Int. Conf. Robotics,* Atlanta, GA, Mar. 1984.

[27] D. Miller, "Two dimensional mobile robot positioning using onboard sonar," in *Pecora IX Remote Sensing Symp. Proc.,* Sioux Falls, SD, Oct. 1984.

[28] ——, "A spatial representation system for mobile robots," in *Proc. 1985 IEEE Int. Conf. Robotics and Automation,* St. Louis, MO, Mar. 1985.

[29] H. P. Moravec, "Obstacle avoidance and navigation in the real world by a seeing robot rover," Ph.D. dissertation, Stanford Univ., Sept. 1980 (also available as Stanford AIM-340, Cs-80-813 and CMU-RI-TR-01-82, 1982; and published as *Robot Rover Visual Navigation.* Ann Arbor, MI: UMI Research Press, 1981.

[30] ——, "The Stanford cart and the CMU rover," *Proc. IEEE,* vol. 71, July 1983.

[31] H. P. Moravec *et al.,* "Towards autonomous vehicles," in *1985 Robotics Research Review,* Robotics Inst., Carnegie-Mellon Univ., Pittsburgh, PA, 1985.

[32] H. P. Moravec and A. Elfes, "High resolution maps from wide angle sonar," presented at the Int. Conf. Robotics and Automation, IEEE, Mar. 1985.

[33] H. P. Moravec, "Three-dimensional imaging with cheap sonar," in *Autonomous Mobile Robots: Annual Report 1985,* Mobile Robot Lab., Pittsburgh, PA, Tech. Rep. CMU-RI-TR-86-4, Feb. 1986.

[34] H. K. Nishihara and T. Poggio, "Stereo vision for robotics," presented at the 1st Int. Symp. Robotics Research, Cambridge, MA, 1984.

[35] Y. Ohta and T. Kanade, "Stereo by intra- and inter-scanline search using dynamic programming," *IEEE Trans. Pattern Anal. Machine Intell.,* vol. PAMI-7, Mar. 1985.

[36] G. W. Podnar, M. K. Blackwell, and K. Dowling, "A functional vehicle for autonomous mobile robot research," CMU Robotics Inst., Apr. 1984.

[37] G. Podnar, "The Uranus mobile robot," Autonomous Mobile Robots: Ann. Rep. 1985, Mobile Robot Lab., Pittsburgh, PA, Tech. Rep. CMU-RI-TR-86-4, Feb. 1986.

[38] *Ultrasonic Range Finders.* Polaroid Corporation, 1982.

[39] B. Serey and L. Mattheis, "Obstacle avoidance using 1-D stereo vision," to be published.

[40] C. E. Thorpe, "The CMU rover and the FIDO vision and navigation system," presented at the Symp. Autonomous Underwater Robots, Univ. New Hampshire, Marine Systems Engineering Lab., May 1983.

[41] C. E. Thorpe, "FIDO: Vision and navigation for a robot rover," Ph.D. dissertation, Dep. of Comput., Sci., Carnegie-Mellon Univ., Pittsburgh, PA, Dec. 1984.

[42] C. E. Thorpe, "Path relaxation: Path planning for a mobile robot," CMU Robotics Inst., CMU-RI-TR-84-5, Apr. 1984, also in *Proc. IEEE Oceans 84,* Washington, DC, Aug. 1984, and *Proc. AAAI-84,* Austin, TX, Aug. 1984.

[43] R. S. Wallace, K. Matsuzaki, Y. Goto, J. Webb, J. Crisman, and T. Kanade, "Progress in robot road following," in *Proc. 1986 IEEE Int. Conf. Robotics and Automation,* San Francisco, CA, Apr. 1986.

Cartographic Databases

by Stanley K. Honey
and Marvin S. White

CD ROM is an ideal storage medium for digital maps. Data volumes are very large, and random access is required for many applications. Etak has developed a digital map database which is used in conjunction with a vehicle navigation system. This database can be licensed for use in other applications. With CD ROM storage, the variety and depth of such applications is expected to expand dramatically.

The Etak digital street map database currently includes the San Francisco Bay Area, southern California, Detroit, Chicago, and the Eastern Seaboard from Boston through New York and Baltimore through Washington D.C. It will soon include Houston, Dallas-Ft. Worth, Atlanta, Seattle, and Miami, and within a year every metropolitan area in the United States.

Accuracy and Contents

Every map feature in Etak's digital map database is located by latitude-longitude coordinates to within 40 feet of its true location as determined by survey, but more importantly, within 20 feet of its location relative to other nearby features. Errors and omissions occur, and maps are always partially out of date because building is continuous. But Etak's digital map requirements specify that 98 percent of the coordinates are within the stated limits and 95 percent of the street network connections are correct.

Having detailed map features located accurately is insufficient for a digital map to be useful; other data is necessary as well. Streets must be prioritized according to travel characteristics: limited access, highway, arterial, collector, local street, ramp, and so on. Numeric addresses must be stored at every intersection to allow destinations to be found by matching street addresses entered by the user. Other types of information must be included, such as city names, zip codes, SMSAs, census tracts, elevation, and even information about the local magnetic field.

Topology

For a digital map covering a large area to be useful, more is required than merely coordinates, street names, classifications, and addresses. The final requirement is really the first requirement; it is at the core of any digital map database, and failure to provide it is the cause of many failures of automated mapping applications. This requirement is topology. A digital map must be able to provide answers to the questions a person viewing a paper map could easily answer, such as, "What cities are adjacent to Menlo Park?" To do this, it must explicitly contain a complete topological description of the map.

The central idea of topology, which is a branch of mathematics, is the *neighborhood.* In the neighborhood of a downtown intersection are the incident streets and the adjacent city blocks; in the neighborhood of a street segment are the endpoints (intersections or cul de sacs) and blocks on both sides; and in the neighborhood of a city block are the bounding streets or waterways, intersections, and adjacent city blocks. This is all information immediately grasped by geometric intuition without explicit statement (when viewing a paper map), but this information is required in exact and complete detail for automated use.

DIME Data Interchange Format

Etak's map database contains all this data in a format that includes topological relationships. This is the familiar (to digital map aficionados) DIME (Dual Incidence Matrix Encoded) format. Every record in a DIME file represents a line segment on the map and contains data about the bounding endpoints and the cobounding areas left and right. The topological name for a point is 0-cell, line is 1-cell, and area is 2-cell.

Nongeometric data is associated with the appropriate cells. For example, a city consists of many 2-cells and those 2-cells carry the city code. Likewise, a street is a chain of 1-cells and those 1-cells carry the street name. And 0-cells carry address information. For a detailed description of how the data is represented in Etak's DIME files, consult the File Specifications section at the end of this article.

Vehicle Navigation Considerations

Any database application has performance requirements in addition to data content requirements, but vehicle navigation is more demanding than most. The onboard computer must be able to find destinations quickly, maintain a map cache in DRAM that covers the vehicle's current location and its neighborhood, update that map for changes in position as the vehicle moves at highway speeds, and display that map in great detail in the immediate vicinity and decreasing detail as one views more distant regions. Topology is essential in efficiently performing these tasks.

Etak has developed a database structure that, like the DIME structure, maintains topological information but is specially designed for efficiency in the navigation application. While this structure is essential for the real-time requirements of vehicle navigation, its specialization and complexity put it beyond the scope of this discussion.

CD ROM and Vehicle Navigation

Having detailed map features located accurately is insufficient for a digital map to be useful; other data is necessary as well.

The present Etak Navigator™ uses a high-speed tape drive that has been specially designed to perform in the environmental rigors of the automobile. While tape storage allows the Navigator to be produced now, and offers useful map coverage per tape, we are eagerly looking forward to the improvements that CD ROM promises to provide. These improvements include faster data transfer and access to 100 times the amount of data. The CD ROM-based navigator will operate more quickly, particularly in destination finding, but the truly dramatic difference will be due to the two orders of magnitude increase in data capacity.

CD ROMs will permit coverage of entire states or regions of the country on a single Etakmap™. Indeed, we could squeeze the whole U.S. street map onto a single CD ROM. But there are many other more promising uses of the space, such as online geographically linked Yellow Pages, tourist information, roadside services, and motel and restaurant guides.

The geographic Yellow Pages is a fascinating example. A driver so equipped could find all the nearby hardware stores and select one based on proximity, operating hours, or other information retrieved from the CD. The automobile could become a fabulous information center, and considering its linkage to geography and mobility, a fabulously useful one.

Digital Maps Outside the Vehicle

Since maps are necessary to navigation, Etak has focused its attention on in-vehicle uses of maps, but useful applications of the same digital map database in the office and home abound. With a standard CD ROM reader attached as a peripheral to your personal computer, it won't be necessary to trudge out to your car to scan the Yellow Pages or review a map.

Etak's digital map database is unique in positional accuracy and broad coverage. By the end of 1986 every major U.S. city will be digitized with the relative accuracy of a car length. The database highlights major streets and highways and shows every local street and ramp. This, combined with relatively complete address data, presents an extremely useful general-purpose map database. We expect a significant part of our business will be licensing the digital map database for myriad uses beyond navigation.

Consider, for example, the services a travel agent could provide, armed with a digital map of all the major metropolitan areas combined with hotel and motel locations. The agent could call up a plot on a graphics monitor showing the location of your appointment and its environs. The display could also show hotels or rental car offices in the area and permit immediate selection. The agent could even print a map highlighting the important locations on a simple dot-matrix printer.

Any database application has performance requirements in addition to data content requirements, but vehicle navigation is more demanding than most.

Out-of-plant facilities management, such as maintaining utility company equipment, is an exploding applications area for digital maps. The kernel of an information system for utilities is the so called land-base, a digital map. The building of such a land base is the major expense, major cause of delay, and the cause of many failures in the implementation of facilities management systems. The availability of Etak's digital map database eliminates these difficulties. The continuing availability in future years of Etak's map database with updated information removes a burden that utility companies were forced to bear just to do their jobs.

Automatic vehicle monitoring of emergency service fleets, public transportation, and private service and delivery fleets is becoming cost-effective. Whether navigation is performed by augmented dead reckoning (à la Etak), Loran-C, or GPS (Global Positioning System), a digital map with automated address matching is essential to efficient use.

Geographically based information systems in many areas can benefit from the map database. For example, a large daily newspaper is using Etak's map database to build a computer system for home and business delivery, routing of trucks, market penetration analysis, locally targeted advertising, and billing.

Etak's Vehicle Navigation System

As mentioned earlier, the Etak digital map database was developed to support vehicle navigation. As an example of one novel use of digital maps, the Etak Navigator will be described. Central to the Navigator is a graphic display which continuously shows a vehicle's position on a map of the surrounding area. An arrowhead symbol in the center of the display represents the position of the vehicle and points toward the top of the map, indicating the direction the vehicle is heading. As the vehicle is driven, the map rotates and shifts about the arrowhead, in order to keep the area that is geographically "in front of" the car at the top of the map. Different-size streets are represented by lines of varying brightness, with streets and key landmarks labeled.

The operator can change the map's scale. As the display scale is increased or decreased, features and labels appear and disappear in an orderly fashion according to their priority. At the most detailed scale, even the smallest streets are displayed and labeled.

The driver selects destinations by entering the street address and scrolling through an index to choose the street name. Alternately, a destination can be entered by selecting two intersecting streets. The Navigator then displays a map on which both the car and the destination are shown. The direction and distance to the destination are continuously indicated, which helps orient the driver if the scale of the map display is changed so that the destination is no longer shown on the screen.

Hardware

The Etak navigation system is packaged in three major parts: a processor, a cassette tape drive, and a display. The processor can be mounted in the trunk or other accessible location. The tape drive is small enough to mount under the dashboard or in the glove compartment. Access to the tape drive is required only to change cassettes. The display, which uses vector graphics for high resolution, is mounted on a flexible stalk near the dashboard.

In addition, a solid-state compass and two wheel sensors are required. The compass is approximately half the size of an audio cassette and mounts either inside the vehicle's roof or on the rear window. Sensors are installed on two nondriven wheels.

Specially manufactured cassettes, similar to audio cassettes, are used for map data and program storage. The primary advantages of this approach are 3.5-megabyte storage capacity, low media cost, low drive cost, tolerance to the automotive environment, tolerable access time, and present availability. Each map cassette covers an area comparable to that covered by two typical paper street maps. For example, one cassette allows navigation on any street throughout one third of the San Francisco Bay Area. Highways over an extended area are also included on each cassette.

Navigation

Dead reckoning. Dead reckoning (DR) is the ancient technique of advancing a known position from measurements of courses and distances traveled. The Navigator uses dead reckoning, with wheel sensors to measure distance, and differential wheel sensors and compass to measure heading. The nondriven wheels are sensed rather than the driven wheels, which are more subject to errors during high-speed driving or driving in conditions of poor traction. An adaptive filter combines relative turn information from the differential wheel sensors with absolute heading information from the compass. This signal processing allows effects of magnetic anomalies and wheel skids to be ignored. As with any DR system, errors in position accumulate proportionally to the distance traveled and proportionally to the inaccuracy of the sensors.

Augmented dead reckoning. The Navigator takes advantage of the fact that automobile drivers tend to drive on roads if there are roads nearby. By matching the vehicle's track to the digital map, the Navigator eliminates the accumulated error that results from dead reckoning.

The central idea of topology, which is a branch of mathematics, is the neighborhood.

The Navigator uses other parameters to make decisions to update the road network stored on the map. These parameters include the connectivity of the road network and analysis of ambiguous update options. The key to system performance is proper updating. The algorithm uses information available in the map to make certain of update corrections.

Self-calibration. The Navigator additionally uses the comparisons between the map and the DR track to continually improve the calibration of both the wheel and compass sensors. For example, if the DR track generally is "long" compared to the map, the wheel calibration will be corrected.

Display

The display is adaptively configured by an algorithm designed to select only the information that the driver is likely to need and to present it in a form which is readable at a glance.

Limited complexity. The driver is able to change the scale of the map display, thereby adjusting the area shown. As the scale of the map is changed, road priority information encoded in the map database is used to select the roads to be displayed, and still keep the display complexity limited.

Selective labels. Only the streets most likely to be of interest to the driver are labeled. These include cross streets ahead, high-priority streets ahead, the street being driven, and streets near a selected destination. Labels are always written near-upright and at a constant size, regardless of map scale.

Heading up presentation. The *heading up* display orientation causes the display to correspond to the drivers orientation and therefore align with what is seen out the window. This allows the driver to quickly grasp information sought from the map display.

Conclusion

In order to produce vehicle navigation products, Etak has developed a nationwide digital map. While this digital map has already been licensed in other applications, CD ROMs may allow a far greater range of digital map applications to be developed for both office and home. In particular, CD ROMs allow large databases such as the described digital map to become available to personal computers.

The storage capacity and high speed (relative to tape) of CD ROMs makes them attractive for use in vehicle navigation. New applications are easily offered to owners of Etak Navigators because programs are stored on the map media along with the digital map and other data. This technique will become even more powerful with the greater storage capacity of CD ROMs. We are convinced that many important applications will be invented, and to encourage innovation, Etak will gladly discuss ways to help developers of digital map-related products for Navigators, or for other applications in the car, office, or home.

File Specifications

Record Contents

Each record includes the following data:

Data Item	Description
class	classification (local street, major road, etc.)
name	feature name (street name, highway name, etc.)
from–x	longitude + – ddd.dddd
from–y	latitude + – ddd.dddd
to–x	longitude + – ddd.dddd
to–y	latitude + – ddd.dddd
left–fradd	left from address
left–toadd	left to address
left–geography	geographic codes (city, etc)
right–fradd	right from address
right—toadd	right to address
right–geography	geographic codes (city, etc)
shape–n	count of following shape points

A *shape record* is the same size as a DIME record but contains shape point coordinates and reference to the corresponding DIME record. As many shape records as necessary follow a DIME record.

Data Item	Description
shape x 1	x coordinate for 1st shape pt
shape y 1	y coordinate for 1st shape pt
shape x 2	x coordinate for 2nd shape pt
shape y 2	y coordinate for 2nd shape pt
. . .	
shape x n	x coordinate for nth shape pt
shape y n	y coordinate for nth shape pt

Record Classification

In the table below, the first column of class codes contains the most common codes, which apply to fully verified data. The second column contains corresponding codes for data that has not yet been verified.

Class		Description
1	A	Interstate highway or equivalent
2	B	Major highway (limited access or partly limited access)
3	C	Access road (providing access to limited access highways) or minor highway
4	D	Arterial or collector
5	E	Local street
6	F	Unpaved
8		Railroad
9		Ramps (provides access from class 3 to class 2 and 1)
S		Shoreline
P		Political boundary
Q		High-speed ramps
N		Incomplete feature; coordinates are not correct but street name and address are presumed correct.

Feature classes indicate the type of feature or class of road. The classifications are meant to correspond to differences in driving and may vary in exact meaning from locality to locality. Classes 1 and 2 provide the pathways to travel across town. Class 3 connects the local net to the highway net, usually via ramps.

Geography Fields

The geography fields contain the following data:

city	place (city) code
zip	zip (postal) code
smsa	Standard Metropolitan Statistical Area
tract	census tract
block	census block

Spatial Planning: A Configuration Space Approach

TOMÁS LOZANO-PÉREZ

Abstract—This paper presents algorithms for computing constraints on the position of an object due to the presence of other objects. This problem arises in applications that require choosing how to arrange or how to move objects without collisions. The approach presented here is based on characterizing the position and orientation of an object as a single point in a configuration space, in which each coordinate represents a degree of freedom in the position or orientation of the object. The configurations forbidden to this object, due to the presence of other objects, can then be characterized as regions in the configuration space, called *configuration space obstacles.* The paper presents algorithms for computing these configuration space obstacles when the objects are polygons or polyhedra.

Index Terms—Computational geometry, obstacle avoidance, robotics.

I. Introduction

INCREASINGLY, computer applications deal with models of two- and three-dimensional objects. Partly because of this, there has been rapid growth of interest in efficient algorithms for geometric problems. For example, research has focused on algorithms for 1) computing convex hulls [16], [32], 2) intersecting convex polygons and polyhedra [6], [27], [36], [38], 3) intersecting half-spaces [11], [33] 4) decomposing polygons [35], and 5) closest-point problems [37].[1] Another class of geometric problems involves placing an object among other objects or moving it without colliding with nearby objects. We call this class of problems: *spatial planning* problems. The following are representative applications where spatial planning plays an important role:

1) the layout of templates on a piece of stock [1]–[3], [13] so as to minimize the area of stock required:

2) machining a part using a numerically controlled machine tool [50], which requires plotting the path of one or more cutting surfaces so as to produce the desired part;

3) the layout of an IC chip [48] to minimize area, subject to geometric design constraints;

4) automatic assembly using an industrial robot [22], [23], [43], which requires grasping objects, moving them without collisions, and ultimately bringing them into contact.

One common spatial planning problem is to determine where an object A can be placed, inside some specified region R, so that it does not collide with any of the objects B_j already

Manuscript received August 15, 1980; revised June 29, 1981 and June 15, 1982. This work was supported in part by the Office of Naval Research under Contract N00014-81-K-0334 and in part by the Defense Advanced Research Projects Agency under Office of Naval Research Contracts N00014-80-C-0505, N00014-82-K-0494.

The author is with the Artificial Intelligence Laboratory, Massachusetts Institute of Technology, Cambridge, MA 02139.

[1] The references cited here are representative of the current literature; they are by no means a complete survey.

placed there. We call this the **Findspace** problem. Finding where to place another suitcase in the trunk of a car is an example of Findspace, where the new suitcase is A, the previous suitcases are the B_j, and the inside of the trunk is R. A related problem is to determine how to move A from one location to another without causing collisions with the B_j. We call this the **Findpath** problem. For example, moving the suitcase mentioned above from its initial position outside the trunk to the desired position in the trunk, requires computing a path for the suitcase (and the mover's arms) that avoids the rest of the car. These two geometric problems, Findspace and Findpath, are the subject of this paper. Previous work on Findspace and Findpath is surveyed in Section VIII.

Findspace and Findpath can be defined more formally as follows.

Definition: Let R be an object that completely contains k_B other, possibly intersecting, objects B_j.

1) *Findspace*—Find a position for A, inside R, such that for all B_j, $A \cap B_j = \emptyset$. This is called a *safe position.*

2) *Findpath*—Find a path for A from position s to position g such that A is always in R and all positions of A on the path are safe. This is called a *safe path.*

Throughout this paper, the objects R and B_j are fixed convex polyhedra (or polygons). We take A to be the set union of k_A (possibly intersecting) convex polyhedra (or polygons) A_i. For example, A may be a convex decomposition of a nonconvex polyhedron [35]. Fig. 1 illustrates the definitions of Findspace and Findpath for convex polygons.

The algorithm presented here for the Findspace and Findpath problems has two main steps: 1) building a data structure that captures the geometric constraints and 2) searching the data structure to find the solution. In this paper we focus on algorithms for constructing the appropriate data structure. In this sense, the approach is similar to many geometric search algorithms, for example, the Voronoi polygon approach to closest-point problems [37]. In the Findspace and Findpath algorithms described here, we build geometric objects, called *configuration space obstacles*, that represent all the positions of the object A that cause collisions with the B_j. Given these objects, Findspace and Findpath correspond to the simpler problems of finding a single point (a position of A) or a path (a sequence of positions of A), outside of the configuration space obstacles. The advantage of this formulation is that the intersection of a point relative to a set of objects is easier to deal with than the intersection of objects among themselves.

Representing the positions of rigid objects requires specifying all their degrees of freedom, both translations and rotations. We will use the notion of configuration to unify our treatment of degrees of freedom. The *configuration* of a polyhedron is a set of independent parameters that characterize the position of every point[2] in the object. The configuration of a polyhedron is defined relative to an initial configuration. In this initial configuration, by convention, a fixed vertex of the polyhedron coincides with the origin of the global coordinate frame. For a polyhedron A, this vertex is called the *reference vertex of* A, or rv_A.

[2] In what follows, all geometric entities—points, lines, edges, planes, faces, and objects—will be treated as (infinite) sets of points. All of these entities will be in some $\mathcal{R}^n$, an n-dimensional real Euclidean space. a, b, x, and y shall denote points of $\mathcal{R}^n$, as well as the corresponding vectors. A, B, and C shall denote sets of points in $\mathcal{R}^n$, while I and K shall denote sets of integers. γ, θ, and β, shall denote reals, while i, j, k, l, m, n shall be used for integers. The coordinate representation of a point $c \in \mathcal{R}^n$, shall be $c = (\gamma_i) = (\gamma_1, \cdots, \gamma_n)$.

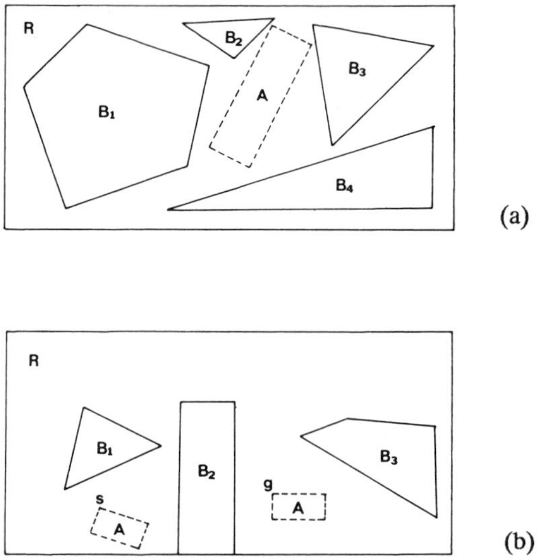

Fig. 1. R, B_j, and A for Findspace and Findpath problems in two dimensions. (a) The Findspace problem is to find a configuration for A where A does not intersect any of the B_j. (b) The Findpath problem is to find a path for A from s to g that avoids collisions with the B_j.

The number of parameters required to specify the configuration of a k-dimensional polyhedron, A, relative to its initial configuration, is d, where $d = k + \binom{k}{2}$ [7, p. 10]; k parameters are required to specify the position of rv_A in $\mathcal{R}^k$ and $\binom{k}{2}$ are required to specify the orientation[3] of A. Thus, the configuration of A can be regarded as a point $x \in \mathcal{R}^d$; this d-dimensional space of configurations of A is denoted $Cspace_A$. A in configuration x is $(A)_x$; A in its initial configuration is $(A)_0$. When an object's configuration is fixed, e.g., the B_j mentioned earlier, we leave it unspecified.

If A is a polygon in $\mathcal{R}^2$, the configuration of A is specified by (x, y, θ), where (x, y) is the position of rv_A and θ is the rotation of A, about rv_A, relative to $(A)_0$. That is, for polygons in $\mathcal{R}^2$, $k = 2$, configurations are elements of $\mathcal{R}^3$, $d = 2 + 1$. If the orientation of A is fixed, (x, y) alone is sufficient to specify the polygons configuration; therefore, $Cspace_A$ is simply the (x, y) plane. If A is a polyhedron in $\mathcal{R}^3$, $k = 3$, the configurations of A are elements of $\mathcal{R}^6$, $d = 3 + 3$. That is, three translations and three rotations are needed to specify the position and orientation of a rigid three-dimensional object [7].

Not all possible configurations in $Cspace_A$ represent legal configurations of A; in particular, configurations of A where $A \cap B_j \neq \emptyset$ are illegal because they would cause collisions.

[3] The relative rotation of one coordinate system relative to another can be specified in terms of $\binom{k}{2}$ angles usually referred to as Euler angles [7]. These angles indicate the magnitude of three successive rotations about specified axes. Many conventions for the choice of axes exist, any of which is suitable for our purposes.

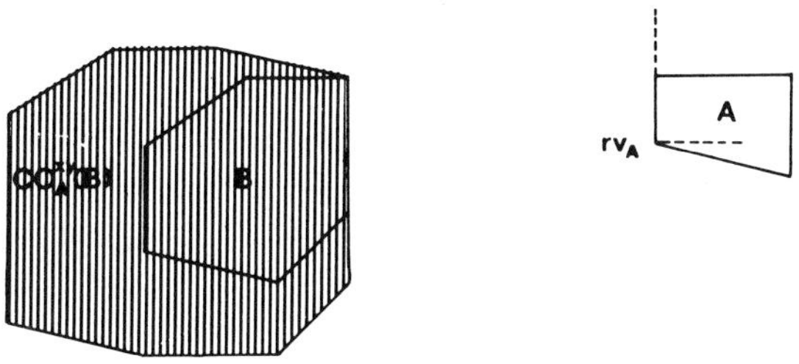

Fig. 2. The $Cspace_A$ obstacle due to B, for fixed orientation of A.

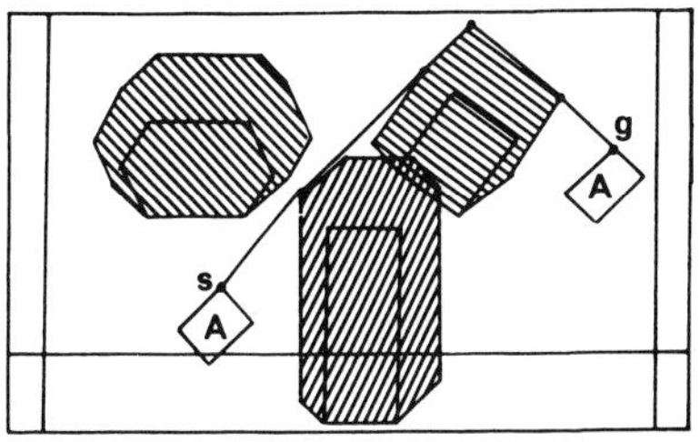

Fig. 3. The Findpath problem and its formulation using the $CO_A^{xy}(B_j)$. The shortest collision-free paths connect the origin and the destination via the vertices of the $CO_A^{xy}(B_j)$ polygons.

These illegal configurations are the result of a mapping of the B_j into $Cspace_A$. This mapping exploits two fundamental properties of objects: 1) their *rigidity*, which allows their configurations to be characterized by a few parameters and 2) their *solidity*, which requires that a point not be inside more than one object.

Definition: The $Cspace_A$ obstacle due to B, denoted $CO_A(B)$, is defined as follows:

$$CO_A(B) \equiv \{x \in Cspace_A \mid (A)_x \cap B \neq \emptyset\}.$$

Thus, if $x \in CO_A(B)$ then $(A)_x$ intersects B, therefore x is not safe. Conversely, any configuration $x \notin CO_A(B_j)$ (for all objects B_j) is safe. If A is a convex polygon with fixed orientation, the presence of another convex polygon B constrains the configuration of A, in this case simply the position of rv_A, to be outside of $CO_A(B)$, a larger convex polygon, shown as the shaded region in Fig. 2. The choice of a different vertex as rv_A would result in translating $CO_A(B)$ relative to B in the figure.

Just as $CO_A(B)$ defines those configurations for which A intersects B, $CI_A(B)$ defines those configurations for which A is completely inside B.

Definition: The $Cspace_A$ *interior* of B, denoted $CI_A(B)$, is defined as follows:

$$CI_A(B) \equiv \{x \in Cspace_A \mid (A)_x \subseteq B\}.$$

Clearly, $CI_A(B) \subseteq CO_A(B)$. Moreover, it is easy to see that for A to be inside B, it must be outside of B's complement. Therefore, letting $-X$ represent the complement of the set X, $CI_A(B) = -CO_A(-B)$.

A superscript to $CO_A(B)$ and $CI_A(B)$ will be used to indicate the coordinates of the configurations in the sets, e.g., $CO_A^{xy}(B)$ and $CO_A^{xy\theta}(B)$ denote sets of (x, y) and (x, y, θ) values, respectively. When no superscript is used, as in $CO_A(B)$, we mean sets of configurations in the complete $Cspace_A$ for a polyhedron of A's dimension, e.g., $\mathcal{R}^6$ for a three-dimensional polyhedron.

Using the definitions of *Cspace* obstacle and *Cspace* interior, Findspace and Findpath can be expressed as equivalent problems that involve placing one point, the configuration of A, relative to the $Cspace_A$ objects $CO_A(B_j)$ and $CI_A(R)$. In general, these problems are equivalent to finding either a single configuration of A or a connected sequence of configurations of A (a path), outside all of the $CO_A(B_j)$, but inside $CI_A(R)$

If A and all of the B_j are polygons and if the orientation of A is fixed, then the $CO_A^{xy}(B_j)$ are also polygons. In that case, the shortest[4] safe paths for A are piecewise linear paths connecting the start and the goal configurations via the vertices of the $CO_A^{xy}(B_j)$ polygons; see Fig. 3. Therefore, Findpath can be formulated as a graph search problem. The graph is formed by connecting all pairs of $CO_A^{xy}(B_j)$ vertices (and the start and goal) that can "see" each other, i.e., can be connected by a straight line that does not intersect any of the obstacles. The shortest path from the start to the goal in this *visibility graph* (**Vgraph**) is the shortest safe path for A among the B_j [24]. This algorithm solves two-dimensional Findpath problems when the orientation of A is fixed, but the paths it finds are very susceptible to inaccuracies in the object model. These paths touch the $Cspace_A$ obstacles; therefore, if the model were exact, an object moving along this type of path would just touch the obstacles. Unfortunately, an inaccurate model or a slight error in the motion may result in a collision. Furthermore, the Vgraph algorithm does not find optimal paths among three-dimensional obstacles [24]. Alternative techniques for pathfinding are treated in [23].

Here is a brief summary of the rest of the paper. Section II presents algorithms for computing $CO_A^{xy}(B)$. Section III characterizes $CO_A^{xy\theta}(B)$, the $Cspace_A$ obstacle for polygons that are allowed to rotate. Section IV describes an algorithm for computing $CO_A^{xyz}(B)$, the $Cspace_A$ obstacle for polyhedra with fixed orientation. Section V characterizes $CO_A(B)$, the $Cspace_A$ obstacle for polyhedra that are allowed to rotate. Section VI deals with slice projection, an approximation technique for higher dimensional $Cspace_A$ obstacles, for example, those obtained when a polyhedron is allowed to rotate. Section VII discusses the extensions to the Findspace and Findpath algorithms needed to plan the motions of industrial robots. Section VIII discusses related work in spatial planning.

II. Computing $CO_A^{xy}(B)$

The crucial step in the *Cspace* approach to Findspace and Findpath is computing the $Cspace_A$ obstacles for the B_j. Thus far, we have only provided an implicit definition of $CO_A(B)$; we now provide, in Theorem 1, a characterization of $CO_A^{xy}(B)$ and $CI_A^{xy}(B)$ in terms of set sums that will lead us to an efficient algorithm for computing $Cspace_A$ obstacles.

Set sum, *set difference*, and *set negation* are defined on sets of points, eqivalently vectors, in $\mathcal{R}^n$ as follows:

[4] This assumes Euclidean distance as a metric. For the optimality conditions using a rectilinear (Manhattan) metric, see [19].

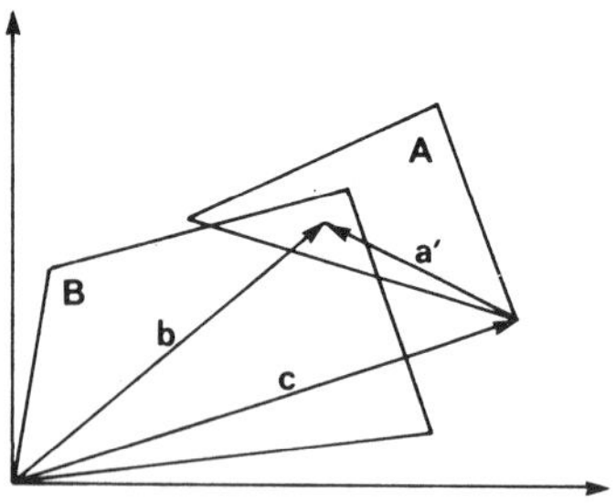

Fig. 4. Illustration of Theorem 1. Any location of rv_A, denoted c, for which A and B have a point in common (expressible as b and a'), can be expressed as $c = b - a'$. Therefore, $CO_A^{xy}(B) = B \ominus (A)_0$.

$$A \oplus B = \{a + b \mid a \in A, b \in B\}$$
$$A \ominus B = \{a - b \mid a \in A, b \in B\}$$
$$\ominus A = \{-a \mid a \in A\}.$$

If a set A consists of a single point a, then $a \oplus B = \{a\} \oplus B = A \oplus B$. Also, $A \ominus B = A \oplus (\ominus B)$. Note that, typically $A \oplus A \neq \{2a \mid a \in A\}$ and $A \ominus A \neq \emptyset$, although $A \oplus B = B \oplus A$.

We can characterize the $Cspace_A$ obstacle for objects with fixed orientation as a set difference of the objects' point sets:

Theorem 1: For A and B, sets in $\mathcal{R}^2$, $CO_A^{xy}(B) = B \ominus (A)_0$.

Proof: If c is an (x, y) configuration of A, then $(A)_c = c \oplus (A)_0$. Therefore, if $a \in (A)_c$, then there exists $a' \in (A)_0$ such that $a = c + a'$; see Fig. 4. Thus, if $b \in B \cap (A)_c$, and therefore $b \in (A)_c$, then, for some $a' \in (A)_0$, $b = c + a'$ and therefore $c = b - a'$. Clearly, the converse is also true. ■

This theorem extends naturally to higher dimensions, e.g., $CO_A^{xyz}(B)$, as long as the orientation of A is fixed.

If A and B are convex, then $A \oplus B$ and $A \ominus B$ are also convex [16, p. 9]; therefore $CO_A^{xy}(B)$ is convex. In fact, if A and B are convex polygons, then $CO_A^{xy}(B)$ is also a convex polygon. We can now use well-known properties of convex polygons and the set sum operation [5] to state an $O(n)$ algorithm for computing $CO_A^{xy}(B)$ when A and B are convex n-gons.

Let $\pi(A, u)$ denote the *supporting line* [5] of A with outward normal u. All of A is in one of the closed half-spaces bounded by $\pi(A, u)$ and u points away from the interior of A. $\pi(A, u) \cap A$ is the set of boundary points of A on the supporting line.

Lemma 1 [5]*:* If A and B are convex sets and u is an arbitrary unit normal, then

$$\pi(A \oplus B, u) \cap (A \oplus B) = (\pi(A, u) \cap A) \oplus (\pi(B, u) \cap B). \quad (1)$$

Fig. 5 illustrates this lemma.

Lemma 2[5]*:*

a) Let $s(a_1, a_2)$ be a line segment and b a point, then $s(a_1, a_2) \oplus b = s(a_1 + b, a_2 + b)$ is a line segment parallel to $s(a_1, a_2)$ and of equal length. See Fig. 6(a).

b) Let $s(a_1, a_2)$ and $s(b_1, b_2)$ be parallel line segments such that $(a_2 - a_1) = k(b_2 - b_1)$ for $k > 0$. Then $s(a_1, a_2) \oplus s(b_1, b_2) = s(a_1 + b_1, a_2 + b_2)$ and the length of the sum is the sum of the lengths of the summands. See Fig. 6(b).

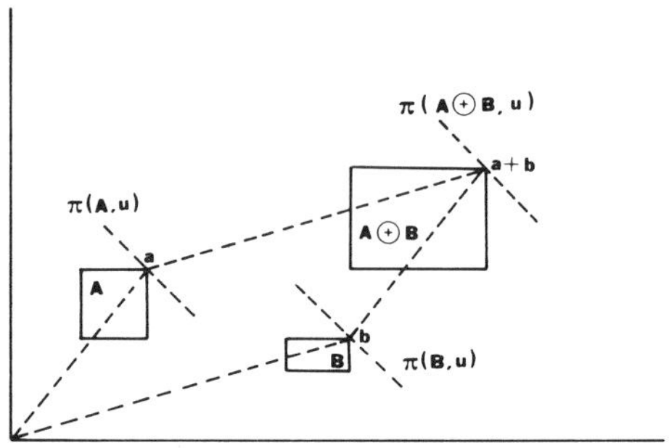

Fig. 5. Illustration for Lemma 1.

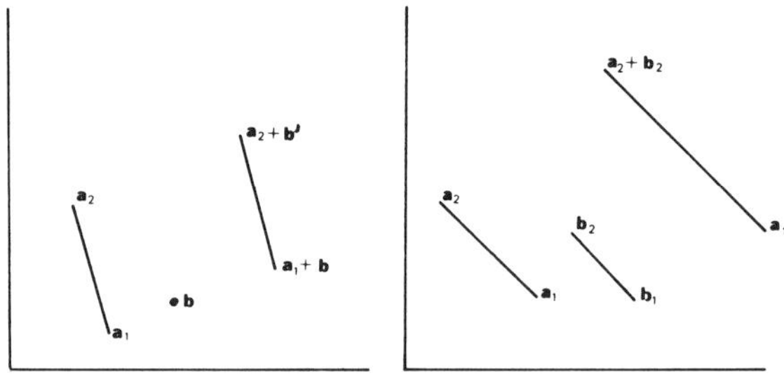

Fig. 6. Illustration for Lemma 2.

Theorem 2: For A a convex n-gon and B a convex m-gon, $CO_A^{xy}(B)$ can be computed in time $O(n + m)$.

Proof: For a polygon P, assume the jth edge, $s(v_j, v_{j+1})$, makes the angle θ_j with the x-axis. If $\theta(u)$ is the angle u makes with the x axis, then

$$\pi(P, \mathrm{u}) \cap P = \begin{cases} v_j, & \text{if } \theta_{j-1} < \theta(u) < \theta_j \\ s(v_j, v_{j+1}), & \text{if } \theta(u) = \theta_j \\ v_{j+1}, & \text{if } \theta_j < \theta(u) < \theta_{j+1}. \end{cases}$$

We now apply Lemmas 1 and 2. Depending on the angle $\theta(u)$, each term on the right-hand side of (1) is either a line segment (edge) or a single point (vertex). It follows from Lemma 2 that the term on the left of (1) is one of:

a) a new vertex, when two vertices are combined;

b) a displaced edge, when an edge and a vertex are combined (Lemma 2a);

c) an edge, corresponding to a pair of displaced end-to-end edges, when two edges are combined (Lemma 2b).

As u rotates counterclockwise, the boundary of $A \oplus B$ is formed by joining a succession of these elements. Note that, because of the convexity of A and B, each edge is encountered exactly once [25, p. 13].

Polygons are stored as lists of vertices in the same order as they are encountered by the counterclockwise sweep of u. This is equivalent to a total order on the edges, based on the angle that the edge makes with the x axis. These lists for A and B can be merged into a single total order on the angle in linear time, as they are traversed. At each step, we construct a new vertex (edges need not be represented explicitly) by the method indicated in the lemmas. The time for constructing the new vertices is bounded by a constant, since it involves at most two

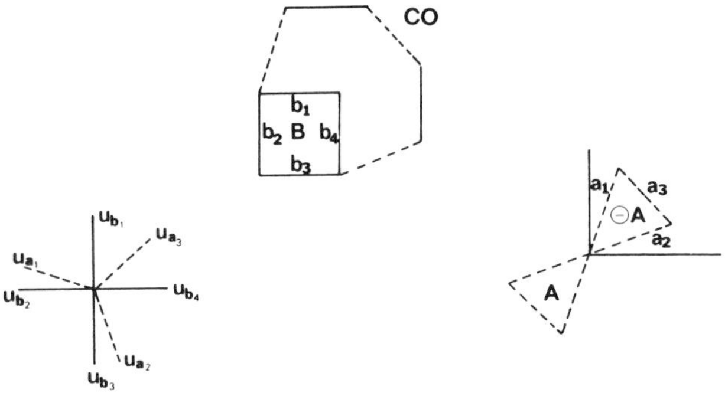

Fig. 7. The edges of $B \ominus (A)_0$, when A and B are convex polygons, are found by merging the edge lists of B and $\ominus(A)_0$, ordered on the angle their normals make with the positive x axis.

vector additions. Thus $A \oplus B$ can be computed in linear time during a scan of the vertices of A and B; see Fig. 7. An implementation of this algorithm is included in Appendix I. Thus, $B \ominus (A)_0$ can be computed in linear time by first converting each vertex a_i to $rv_A - a_i$; see Fig. 7. ■

This algorithm is similar to Shamos' diameter algorithm using antipodal pairs [36], but instead of dealing with two supporting lines on one polygon, it deals with two polygons and one supporting line on each. An algorithm, essentially identical to the one in Theorem 2, has been recently described in [39]. The proof Theorem 2, however, will be used in subsequent sections to derive characterizations and algorithms for other *Cspace* entities.

When A or B are nonconvex polygons, $CO_A^{xy}(B)$ can be computed by an extension of the algorithm above. The extension relies on decomposing the boundaries of the polygons into a sequence of polygonal arcs whose internal angles, i.e., the angle facing the inside of the polygon, are each less than π. The algorithm of Theorem 2 can then be applied to pairs of arcs; the result is a polygon whose boundary, in general, intersects itself. The algorithm requires, in the worst case, $O(n \times m)$ steps.

An alternative method of computing $CO_A^{xy}(B)$ for nonconvex A and B can be used when convex decompositions of A and B are available, e.g., the objects may have been designed by set operations on convex primitives. If A is represented as the union of k_A objects A_i, and B is the union of k_B objects B_j, then Theorem 3 follows directly from the definition of $CO_A(B)$.

Theorem 3: If $A = \bigcup_{i=1}^{k_A} A_i$ and $B = \bigcup_{j=1}^{k_B} B_j$:

$$CO_A(B) = \bigcup_{i=1}^{k_A} \bigcup_{j=1}^{k_B} CO_{A_i}(B_j).$$

This theorem simply says that the set of configurations that cause a collision between A and B are those for which any part of A intersects any part of B.

III. Characterizing $CO_A^{xy\theta}(B)$

We have so far restricted our attention to cases where A remains in a fixed orientation. In these cases, all the geometric constraints for spatial planning are embodied in $CO_A^{xy}(B)$. However, $CO_A^{xy}(B)$ is only the cross section, for fixed θ, of the three-dimensional full configuration space obstacle for polygons, $CO_A^{xy\theta}(B)$. In this section, we consider $CO_A^{xy\theta}(B)$, when

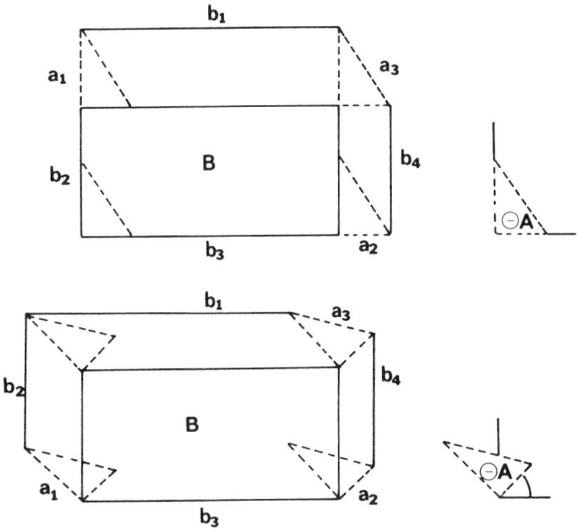

Fig. 8. When A (and $\ominus(A)_0$) rotates by θ, the e_i^a rotate around b_j and the e_j^b are displaced. When an e_i^a is aligned with an e_j^b for some θ, any additional rotation of A will interchange the order in which they are encountered during the counterclockwise scan of Theorem 2. For example, in the top figure a_1 appears before b_2 on the boundary of $CO_A^{xy}(B)$ and is (nearly) aligned with b_2; in the bottom figure, after additional rotation of A, a_1 appears after b_2 on the boundary of $CO_A^{xy}(B)$. Therefore, the top figure illustrates A in the orientation $\theta_{i,j}^*$.

A and B are convex, by examining changes in its cross section as θ changes.

For fixed θ, we know from Theorem 2 that the edges of $CO_A^{xy}(B)$ are either displaced edges of A or displaced edges of B. Therefore, for a small rotation of A, we expect that those edges of $CO_A^{xy}(B)$ corresponding to the edges of A will rotate, but the edges corresponding to edges in B will not. As A rotates, however, the rate of displacement of these edges changes discontinuously when edges of A and B become parallel, as illustrated below.

Let $vert(B)$ denote the set of vertices of a polygon B, b_j be the position vector of the jth member of $vert(B)$, and $a_i(\theta)$ be the position of the ith member of $vert(\ominus(A)_0)$, which depends on θ. Assume that A and B have no parallel edges. For fixed θ, the proof of Theorem 2 shows that each edge of $CO_A^{xy}(B)$ can be expressed as one of

$$e_i^a = b_j \oplus s(a_i(\theta), a_{i+1}(\theta)) \qquad (2a)$$

$$e_j^b = a_i(\theta) \oplus s(b_j, b_{j+1}). \qquad (2b)$$

The order in which the a_i and b_j are encountered in the counterclockwise scan described in Theorem 2 determines the (i, j) pairings of vertices and edges. For example, in (2b), $a_i(\theta)$ is the vertex of $\ominus(A)_0$ that is on the supporting line of $\ominus(A)_0$ which is parallel to $s(b_j, b_{j+1})$, i.e., if u_j is the normal to $s(b_j, b_{j+1})$, then $a_i = \pi(\ominus(A)_0, u_j) \cap \ominus(A)_0$.

Equation (2) shows that, for a given pairing of edges and vertices, the e_i^a rotate around b_j, while the e_j^b are simply displaced by the vector $a_i(\theta)$; see Fig. 8. The discontinuous changes occur at values of θ, denoted $\theta_{i,j}^*$, where the ith edge of A becomes parallel to the jth edge of B. For values of θ just greater than these $\theta_{i,j}^*$, some pair of edges has a different order in the scan of Theorem 2 from what they had when θ was just less than $\theta_{i,j}^*$; see Fig. 8. Therefore, at each $\theta_{i,j}^*$, the pairings between edges and vertices change. For A a convex n-gon and

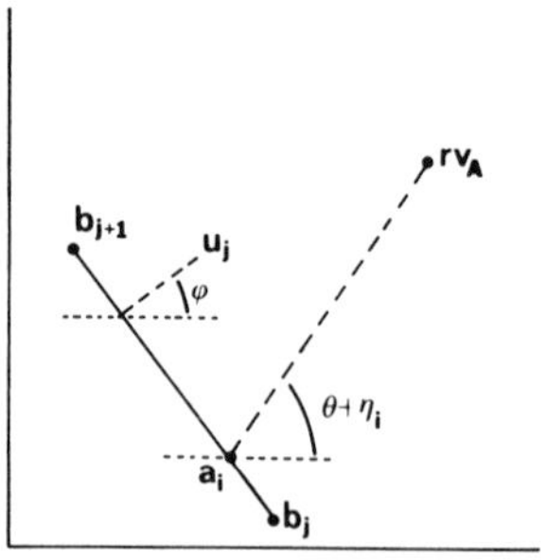

Fig. 9. Illustration of terms used in (3).

B a convex m-gon, there are $O(n \times m)$ such $\theta^*_{i,j}$ in $CO_A^{xy\theta}(B)$.

Between discontinuities, the lines defined by e_j^b edges have a simple dependence on θ. The edge $s(b_j, b_{j+1})$ is on a line whose vector equation[5] is: $\langle u_j, x\rangle = \langle u_j, b_j\rangle$ where u_j is the constant unit normal to $s(b_j, b_{j+1})$. Let $a_i(\theta)$ make the angle $\theta + \eta_i$ with the x axis, with η_i constant, and u_j make the angle ϕ_j with the x axis. Then, if $\|a_i\|$ represents the vector magnitude of a_i, the equation for the line including e_j^b is

$$\begin{aligned}\langle u_j, x\rangle &= \langle u_j, a_i(\theta) + b_j\rangle \\ &= \|a_i\| \cos(\theta + \eta_i - \phi_j) + \langle b_j, u_j\rangle. \quad (3)\end{aligned}$$

The terms are illustrated in Fig. 9. This equation holds between discontinuities.

The equation for the e_i^a edges is not as simple, because the orientation of the edge changes with θ, i.e., the edge is a cross section of a curved surface in (x, y, θ) space. Let $v_i(\theta)$ be the normal vector to $s(a_i(\theta), a_{i+1}(\theta))$; then the vector equation of this curved surface is

$$\langle v_i(\theta), x\rangle = \langle v_i(\theta), a_i(\theta) + b_j\rangle. \quad (4)$$

This equation also holds only between discontinuities, i.e., for each pairing of vertices and edges.

Equations (3) and (4) define the shape of $CO_A^{xy\theta}(B)$. Since the resulting object is not polyhedral, however, it cannot be used for the Vgraph algorithm. Section VI discusses a technique for constructing lower dimension polyhedral approximations of $Cspace_A$ obstacles and an extended Vgraph algorithm to use them (see also [23]). These equations for $CO_A^{xy\theta}(B)$ are the basis for the Findpath algorithm described in [10].

IV. Algorithm for $CO_A^{xyz}(B)$

Theorem 1 applies also to $CO_A^{xyz}(B)$, but the algorithm of Theorem 2 cannot be extended to polyhedra, since there is no similar total ordering of the faces of a polyhedron. However, Theorem 4 below follows easily from Theorem 1 and provides a way to compute $CO_A^{xyz}(B)$ for convex polyhedra A and B. The method of Theorem 4 also applies to polygons, but is much less efficient than the linear algorithm of Theorem 2. Theorem 4 provides the basis for approximating $CO_A^{xyz}(B)$ when A and B are nonconvex, simply by replacing A or B by their convex hulls.

Let $conv(A)$ denote the *convex hull* of a polygon A, i.e., the smallest convex polygon enclosing A. We know that $conv(A)$, for a nonempty set $A \subseteq \mathcal{R}^d$, is $\{\sum_{i=1}^n \gamma_i x_i \mid x_i \in A, \gamma_i \geq 0, \sum_{i=1}^n \gamma_i = 1\}$, for some n [16, p. 15]. This definition says that every point in the convex hull of A can be written as a convex linear combination of points in A.

[5] The *scalar (dot) product* of vectors a and b will be denoted $\langle a, b\rangle$.

Theorem 4: For polyhedra A and B,

$$\begin{aligned}conv(A \oplus B) &= conv(A) \oplus conv(B) \\ &= conv(vert(A) \oplus vert(B)).\end{aligned}$$

Proof: First show that $conv(A \oplus B) = conv(A) \oplus conv(B)$.

($\supseteq$): By the definition of $\oplus$, if $x \in conv(A) \oplus conv(B)$, then there exist $a \in conv(A)$ and $b \in conv(B)$ such that $x = a + b$. The definition of convex hull states that any $a \in conv(A)$ can be expressed as a convex linear combination of points in A; likewise for any $b \in conv(B)$. Therefore, there exist $a_i \in A$, $b_i \in B$, $\sum_i \gamma_i = 1$, $\gamma_i \geq 0$, $\sum_j \beta_j = 1$, and $\beta_j \geq 0$ such that $a = \sum_i \gamma_i a_i$ and $b = \sum_j \beta_j b_j$ and thus

$$\begin{aligned}x = a + b &= \left(\sum_i \gamma_i a_i\right) + b = \sum_i \gamma_i(a_i + b) \\ &= \sum_i \gamma_i\left(a_i + \sum_j \beta_j b_j\right) \\ &= \sum_i \gamma_i\left(\sum_j \beta_j a_i + \sum_j \beta_j, b_j\right) \\ &= \sum_i \gamma_i \sum_j \beta_j(a_i + b_j) = \sum_i \sum_j \gamma_i\beta_j(a_i + b_j).\end{aligned}$$

But, since $\sum_i \sum_j \gamma_i\beta_j = 1$ and $\gamma_i\beta_i \geq 0$, x is a convex linear combination of points in $A \oplus B$ and therefore belongs to its convex hull. Therefore $conv(A) \oplus conv(B) \subseteq conv(A \oplus B)$.

($\subseteq$): If $x \in conv(A \oplus B)$, then there exist $\gamma_i \geq 0$ with $\sum_i \gamma_i = 1$, $a_i \in A$, and $b_i \in B$, such that

$$x = \sum_i \gamma_i(a_i + b_i) = \sum_i \gamma_i a_i + \sum_i \gamma_i b_i.$$

Therefore, $x \in conv(A) \oplus conv(B)$.

This establishes that $conv(A \oplus B) = conv(A) \oplus conv(B)$. Replacing A by $vert(A)$ and B by $vert(B)$, and using the fact that $conv(A) = conv(vert(A))$ [16], shows that $conv(vert(A) \oplus vert(B)) = conv(A) \oplus conv(B)$. ■

Corollary: For convex polyhedra A and B, $CO_A^{xyz}(B) = conv(vert(B) \ominus vert((A)_0))$.

Proof: $A \oplus B$ is convex, when A and B are both convex; thus, $A \oplus B = conv(A \oplus B)$. By Theorem 4, $A \oplus B = conv(vert(A) \oplus vert(B))$. Using Theorem 1 establishes the corollary. ■

Several algorithms exist for finding the convex hull of a finite set of points on the plane, e.g., [15] and [32]. The latter [32] also describes an efficient algorithm for points in $\mathcal{R}^3$. These algorithms are known to run in worst case time $O(v \log v)$, where v is the size of the input set. Therefore, Theorem 4 leads immediately to an algorithm for computing $CO_A^{xyz}(B)$ and an upper bound on the computational complexity of the problem or convex polyhedra.

Theorem 5: For convex polyhedra $A, B \subseteq \mathcal{R}^3$, each with

$O(n)$ vertices, $CO_A^{xyz}(B)$ can be computed in time $O(n^2 \log n)$.

Proof: The set $vert(B) \ominus vert((A)_0)$ is of size $O(n^2)$. Applying an $O(v \log v)$ convex hull algorithm to this set gives an $O(n^2 \log n)$ algorithm for computing $CO_A^{xyz}(B)$. ■

The Vgraph algorithm discussed in Section I can be extended directly to deal with three-dimensional $Cspace_A$ obstacles, $CO_A^{xyz}(B_j)$. However, the paths found are not, in general, optimal paths [24]. Furthermore, with three-dimensional obstacles, the Vgraph algorithm is not even guaranteed to find a solution when one exists. This happens when the vertices of the $CO_A^{xyz}(B_j)$ are inaccessible, because they are outside of $CI_A^{xyz}(R)$. In that case, there may exist collision-free paths (via edges of the $CO_A^{xyz}(B_j)$), but the Vgraph algorithm will not find them. An alternative suboptimal, but complete, path searching strategy is described in [23]. A path searching algorithm based on mathematical optimization of a path along a fixed set of edges is described in [12].

V. Characterizing $CO_A(B)$

The surfaces of $CO_A(B)$, when A is three-dimensional and allowed to rotate, can be characterized in the same manner as the surfaces of $CO_A^{xy\theta}(B)$ were characterized in Section III. There are three types of surfaces that need to be considered, rather than two types as in two-dimensional objects. Let $f_i(\Theta)$ be the ith face of the convex polyhedron A, with Θ being the vector of three Euler angles indicating the orientation of A relative to its initial orientation. Similarly, let g_j be the jth face of the convex polyhedron B. As before, we let $a_i(\Theta)$ and b_j denote vertices of A and B, respectively. Each face of $CO_A(B)$ can be expressed as one of

$$f_i^a = b_j \oplus f_i(\Theta) \tag{5a}$$

$$f_j^b = a_i(\Theta) \oplus g_j \tag{5b}$$

$$f_{i,j}^{a \times b} = s(a_i(\Theta), a_{i+1}(\Theta)) \oplus s(b_j, b_{j+1}). \tag{5c}$$

The faces defined by (5a) and (5b) are parallel to the faces of A and B, respectively. Each face defined by (5c) is a parallelogram, with edges parallel to the edges of A and B that give rise to the face. The vector equation for each type of surface follows the pattern of (3) and (4) above:

$$\langle N, x \rangle = \langle N, a_i(\Theta) + b_j \rangle \tag{6}$$

where N is 1) the normal to $f_i(\Theta)$ for (5a) faces, 2) the normal to g_j for (5b) faces, or 3) the cross product of the vectors along $s(a_i(\Theta), a_{i+1}(\Theta))$ and $s(b_j, b_{j+1})$ for (5c) faces. As above, this characterization only holds between discontinuities.

VI. Approximating High-Dimension *Cspace* Obstacles

We have seen that when A is a three-dimensional solid which is allowed to rotate, $CO_A(B)$ is a complicated curved object in a six-dimensional $Cspace_A$. An alternative to computing these objects directly is to use a sequence of low-dimensional projections of the high-dimensional $Cspace_A$ obstacles. For example, a three-dimensional (x, y, θ) $Cspace_A$

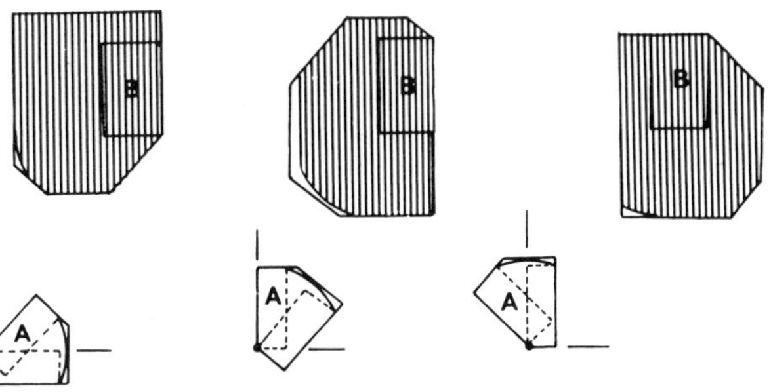

Fig. 10. Slice projections of $Cspace_A$ obstacles computed using the (x, y)-area swept out by A over a range of θ values. Each of the shaded obstacles is the (x, y)-projection of a θ-slice of $CO_A(B)$. The figure also shows a polygonal approximation to the slice projection and the polygonal approximation to the swept volume from which it derives.

obstacle can be simply approximated by its projection on the (x, y) plane, and any path of A that avoided the projection would be safe for all orientations of A. On the other hand, there may be no paths that completely avoid the projection. A better approach is to divide the complete range of θ values into k smaller ranges and, for each of these ranges, find the section of the (x, y, θ) obstacle in that range of θ. These are called θ-slices of the obstacle. The projection of these slices serves as an approximation of the obstacle. Paths that avoid individual slices are safe for orientations of A in the θ-range defining the slice.

The shaded areas in Fig. 10 are the (x, y) projection of θ-slices of $CO_A(B)$ when A and B are rectangles. These slices represent configurations where A overlaps B for some orientation of A in the specified range of θ. We will show that these slice projections are the $Cspace_A$ obstacles of the area swept out by A over the range of orientations of the slice. The swept area under rotation of a polygon is not pologonal. To use the $CO_A^{xy}(B)$ algorithm developed earlier, we approximate the swept area as the union of polygons [23]. This polygonal approximations leads to a polygonal approximation for the projected slices, as shown in Fig. 10. Similar considerations apply to polyhedra.

The crucial properties of slice projection are: 1) a solution to a Findspace or Findpath problem in any of the slices is a solution to the original problem, although not all actual solutions can be found in the slices; and 2) the degree of approximation can be controlled by choosing the range of parameters of the slice, in particular the approximation need not be uniform across the range of parameters.

The Vgraph algorithm for Findpath, has been extended [24], by means of slice projection, to find paths when A and all the B_j are three-dimensional polyhedra that are allowed to rotate. Fig. 11 illustrates the basic idea of this algorithm. An alternative path-searching technique, also using slice projection is described in [23]. Because the slice projections are approximations to the $Cspace_A$ obstacles, neither of these algorithms is guaranteed to find solutions to Findpath problems. Paths found by Findpath algorithms that use slice projections are composed of sequences of translations interspersed with rotations, but where the rotations happen in quantized increments corresponding to the ranges of orientations that define the slices. Not all paths can be expressed in this fashion. For

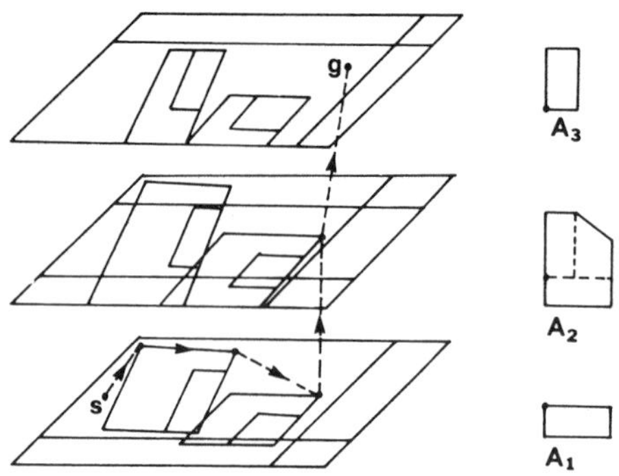

Fig. 11. An illustration of the Findpath algorithm using slice projection described by Lozano-Pérez and Wesley in [24]. A number of slice projections of the *Cspace* obstacles are constructed for different ranges of orientations of A. The problem of planning safe paths in the high-dimensional $C\,space_A$ is decomposed into 1) planning safe paths via $CO_A^{xyz}(B_j)$ vertices within each slice projection and 2) moving between slices, at configurations that are safe in both slices. A_1 represents A in its initial configuration, A_3 represents A in its final configuration and A_2 is a simple polyhedral approximation to the swept volume of A between its initial and final orientation.

example, the classic problem of moving a rectangular sofa through a rectangular bend in a hallway that just fits the sofa requires continuous rotation during translation. However, a large class of useful problems can be solved using slice projection.

In the rest of the section we show how slice projections of $CO_A(B)$ may be computed using the $CO_A^{xy}(B)$ and $CO_A^{xyz}(B)$ algorithms of Section IV. The idea is simply that if a collision would occur for A in some orientation, it would also occur for a swept volume of A that includes A in that orientation.

More formally, a j-slice of an object $C \in \mathcal{R}^n$ is defined to be $\{(\beta_1, \cdots, \beta_n) \in C \mid \gamma_j \le \beta_j \le \gamma_j'\}$, where γ_j and γ_j' are the lower and upper bounds of the slice, respectively. Then, if $I = \{1, \cdots, n\}$ and $K \subseteq I$, then a K-slice is the intersection of the j-slices for $j \in K$. Note that a K-slice of C is an object of the same dimension as C. Slices can then be *projected* onto those coordinates in I not in K, i.e., $I - K$, to obtain objects of lower dimension. A j-cross section is a j-slice whose lower and upper bounds are equal, e.g., $CO_A^{xy}(B)$, for some orientation of A, is the projection on the (x, y) plane of a θ-cross-section of $CO_A^{xy\theta}(B)$.

Slice projections are related to cross-section projections by the *swept volume* of an object. Intuitively, the swept volume of A is all the space that A covers when moving within a range of configurations. In particular, given two configurations for A, called c and c', then the union of $(A)_a$ for all $c \le a \le c'$ is the swept volume of A over the configuration range $[c, c']$. Generally, c and c' differ only on some subset, K, of the configuration coordinates. For example, if c and c' are of the form $(\beta_1, \beta_2, \beta_3)$ and $K = \{3\}$, then the swept volume of A over the range $[c, c']_K$ refers to the union of A over a set of configurations differing only on β_3. The swept volume of A over a configuration range is denoted $A[c, c']_K$.

If $A[c, c']_K$ overlaps some object B then, for some configuration a in that range, $(A)_a$ overlaps B. The converse is also true. $CO_{A[c,c']_K}(B)$ is the set of $I - K$ projections of those configurations of A within $[c, c']_K$ for which A overlaps B.

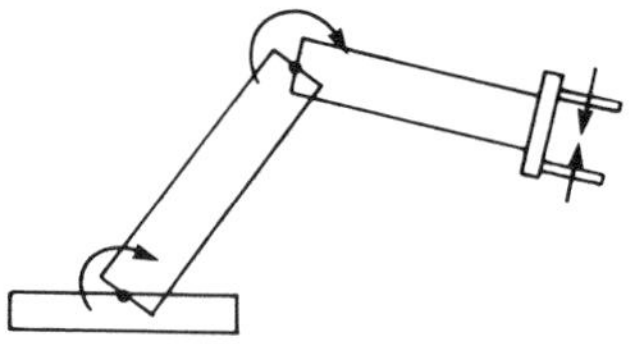

Fig. 12. Linked polyhedra can be used to model the gross geometry of industrial robot manipulators.

Equivalently, $CO_{A[c,c']_K}(B)$ is the $I - K$ projection of the $[c, c']_K$ slice of $CO_A(B)$. If the configurations of the swept volume are one of (x, y), (x, y, z), or (x, y, θ) then the algorithms of the previous sections can be used to compute $CO_{A[c,c']_K}(B)$ and thereby compute the required slice projections.[6]

A formal statement and proof of this result is included in Appendix II as Theorem 6. This theorem is of practical importance since it provides the mechanism underlying the Findspace and Findpath implementations described in [23] and [24].

VII. Automatic Planning of Industrial Robot Motions

One application of the algorithms for Findspace and Findpath developed above is in the automatic planning of industrial robot motions [23], [43]. However, some extensions of the results for polyhedra are needed. In this section, we briefly discuss these extensions.

Industrial robots are open kinematic chains in which adjacent links are connected by prismatic or rotary joints, each with one degree of freedom [29]. We model them by *linked polyhedra*, kinematic chains with polyhedral links, each of which has either a translational or rotational degree of freedom relative to the previous joint in the chain; see Fig. 12. The relative position and orientation of adjacent links, A_i and A_{i+1}, is determined by the ith *joint parameter* [29], [7]. The set of joint parameters of a linked polyhedron completely specifies the position and orientation of all the links. This type of model is clearly an approximation to the actual geometry; in particular, the shape of the joints is not represented and some values of the joint parameters may cause overlap of adjacent links.

The natural $Cspace_A$ for a linked polyhedron is that defined by the set of joint parameters. A point in this space determines the shape of the linked polyhedron and the configuration of each of its links. Unfortunately, the presence of rotary joints prevents the use of the $CO_A^{xyz}(B)$ algorithm of Section IV to plan the motions of linked polyhedra. However, there is an increasingly popular class of industrial robots, known as *Cartesian* robots, where the translational degrees of freedom of the robot are separate from the rotational. With this class of robots, we can use the $CO_A^{xyz}(B)$ algorithm and slice projection approach to plan collision-free paths and to plan how to grasp objects [23]. Actually doing this requires constructing the swept volume, over the rotational parameters, of the linked

[6] Of course, this requires computing a convex polyhedral approximation to the swept volume of A. Simple approximations are not difficult to compute [23], but this is an area where better algorithms are required. Nevertheless, the swept volume computation is a three-dimensional operation which can be defined and executed without recourse to six-dimensional constructs.

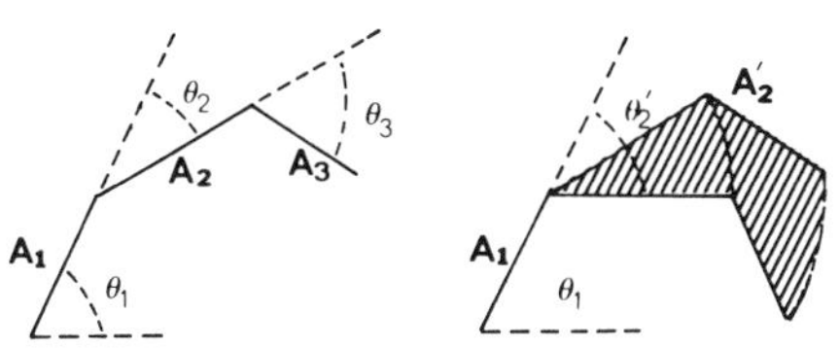

Fig. 13. Changes in the second joint angle from θ_2 to θ_2' causes changes in the configurations of both link A_2 and link A_3.

polyhedron modeling the robot. The resulting swept volume can be viewed as a polyhedron with only translational degrees of freedom, for which the $CO_A^{xyz}(B)$ algorithm is applicable.

Using swept volumes of linked polyhedra for slice projection requires taking into consideration the interdependence of the joint parameters. Note that for a linked polyhedron, the position of link j typically depends on the positions of links $k < j$, which are closer to the base than link j. Let $K = \{j\}$, $c = (\theta_i)$, $c' = (\theta_i')$, and $[c, c']_K$ define a range of configurations differing on the jth $Cspace_A$ parameter. Since joint j varies over a range of values, links $l \geq j$ will move over a range of positions which depend on the values of c and c', as shown in Fig. 13. The union of each of the link volumes over its specified range of positions is the swept volume of the linked polyhedron. The swept volume of links j through n can be taken as defining a new jth link. The first $j - 1$ links and this new jth link define a new manipulator whose configuration can be described by the first $j - 1$ joint parameters. On the other hand, the shape of the new link j depends not only on the K-parameters of c and c', i.e., θ_j and θ_j', but also on θ_l for $l > j$. This implicit dependence on parameters of c and c' that are not in K is undesirable, since it means that the *shape* of the new jth link will vary as the swept volume is displaced, i.e., the $(I - K)$-parameters are changed. If we let $K = \{j, \cdots, n\}$, then the shape of the swept volume depends only on the K-parameters of c and c', while its configuration is determined by the $(I - K)$-parameters. A swept volume that satisfies this property is called *displaceable.* This property plays a crucial role in proving the fact, mentioned in Section VI, that slice projections of a $Cspace_A$ obstacle can be computed as the $Cspace_A$ obstacles of the swept volume of A (see Theorem 6 in Appendix II).

In summary, with the extensions discussed in this section, the spatial planning algorithms developed for the case of rigid polyhedra can serve as the basis for planning the motions of industrial robots.

VIII. Related Work in Spatial Planning

The definition of the Findspace problem used here is based on that in [49]. One previous approach to this problem is described in [30]; it is an application of the Warnock algorithm [47] for hidden line elimination. The idea is to recursively subdivide the workspace until an area "large enough" for the object is found. This approach has several drawbacks: 1) any nonoverlapping subdivision strategy will break up potentially useful areas and 2) the implementation of the predicate "large enough" is not specified (in general, the $CI_A(B)$ computation is required to implement this predicate). However, once the $Cspace_A$ obstacles have been computed, the Warnock search provides a good way of solving Findspace; since we only need space for a point, any free area is "large enough" [10].

The work by Udupa, reported in [45], [46], was the first to approach Findpath by explicitly using transformed obstacles and a space where the moving object is a point. Udupa used only rough approximations to the actual *Cspace* obstacles and had no direct method for representing constraints on more than three degrees of freedom. A survey of previous heuristic approaches to the Findpath problem for manipulators, for example, [20], [31], has been given in [46]. An early paper on Shakey [28] describes a technique for Findpath using a simple object transformation that defines safe points for a circular approximation to the mobile robot and uses a graph search formulation of the problem. More recent papers on navigation of mobile robots are also relevant to two-dimensional Findpath [14], [26] [44]. An early paper [18] reports on a program for planning the path of a two-dimensional sofa through a corridor. This program does a brute-force graph search through a quantized *Cspace.*

The *Cspace* approach to Findspace and Findpath described here is an extension of that reported in [24]. In that paper, an approximate algorithm for $CO_A^{xyz}(B)$ is described and the Vgraph algorithm for high-dimensional Findpath is first represented. An application of the Findpath and Findspace approach described in the current paper to automatic planning of manipulator motions is described in [23]. Alternative approaches to path searching in the presence of obstacles are described in [19], [12], [23]. The visibility computation needed in Vgraph is treated, in the context of hidden-line elimination, in [4], [51].

The basic idea of representing position constraints as geometric figures, e.g., $CO_A^{xy}(B)$, has been used (independently) in [1]–[3], where an algorithm to compute $CO_A^{xy}(B)$ for nonconvex polygons is used in a technique for two-dimensional layout. The template packing approach described in [13] uses a related computation based on a chain-code description of figure boundaries. Algorithms for packing of parallelopipeds, in the presence of forbidden volumes, using a construct equivalent to the $CO_A^{xyz}(B)$, but defined as "the hodograph of the close positioning function" are reported in [42]. The only use of this construct in the paper is for computing $CO_A^{xyz}(B)$ for aligned rectangular prisms.

An extension of the approach in [24] to the general Findpath problem is proposed in [34]. The proposal is based on the use of an exact representation of the high-dimensional *Cspace* obstacles. The basic approach is to define the general configuration constraints as a set of multinomials in the position parameters of A. However, the proposal still requires elaboration. It defines the configuration space constraints in terms of the relationships of vertices of one object to the faces of the other. This is adequate for polygons, but the equations in the paper only express the constraints necessary for vertices of A to be outside of B, i.e., they are of the form of (3). They do not account for the positions of A where vertices of B are in contact with A [see (4)]. The new equations will have terms of the form $x \cos\theta$ and $y \cos\theta$. Furthermore, the approach of defining the configuration constraints by examining the interaction of vertices and faces does not generalize to three-dimensional polyhedra. It is not enough to consider the interaction of vertices and faces; the interaction of edges and faces must also be taken into account (see Section V and [8]).

Two recent papers describe solutions for the Findpath problem with rotations in two [40] and three dimensions [41]. In [41], the *Cspace* surfaces are represented as algebraic manifolds in a 12-dimensional space; in this way the surfaces can be described as polynomials, allowing the use of some powerful mathematical machinery. The resulting algorithm has (large) polynomial time complexity, for fixed dimensionality of the *Cspace*.

A *Cspace* algorithm is described in [10] for solving Findpath, allowing rotations of the moving objects. The algorithm is based on recursively subdividing *Cspace* until a path of cells completely outside of the obstacles is found.

An alternative approach to two-dimensional Findpath with rotations is described in [9]. The algorithm is based on representing the empty space outside the objects B_j explicitly as *generalized cones*. Motions of A are restricted to be along the spines of the cones. The algorithm bounds the moving object by a convex polygon and characterizes the legal rotations of the bounding polygon along each spine.

APPENDIX I

ALGORITHM FOR $CO_A^{xy}(B)$

This Appendix shows an algorithm for computing $A \oplus B$, called SET-SUM(A, B, C), when A and B are convex polygons. Section II used this operation to compute $CO_A^{xy}(B)$.

Each polygon is described in terms of its vertices and the angles that the edges make with the positive x axis. The edges and vertices are ordered in counterclockwise order, i.e., by increasing angle. The implementation assumes that a POLYGON record is available with the following components:

1) size—number of edges in the polygon.

2) vert [1:size + 1, 1:2]—an array of vectors representing the coordinates of a vertex. The ith edge, $i = 1, \cdots$, size, has the endpoints vert[i, k] and vert[i + 1, k], for $k \in \{1, 2\}$. Note that vert[size + 1, k] = vert[1, k].

3) angle [0:size]—the edge normal's angle (in the range $[0, 2\pi]$) with the x axis, monotonically increasing. For convenience angle [0] = 2π − angle [size].

References to the components of a polygon, a, are written as one of a.size, a.vert, and a.angle.

The algorithm implements the angle scan in the proof of Theorem 2; in particular, the edges of the input polygons, a and b, are examined in order of angle. The algorithm determines the position of the vertices of c. It is clear that vertices can occur only at angles where there is either a vertex of a, or a vertex of b, or both. From Lemmas 1 and 2, it is easy to see that the position of the vertex of c is the sum of the positions of the corresponding vertices of a and b. The algorithm starts the scan at the angle determined by the first edge of b, the first loop in the program below serves to find the edges of a that straddle that angle. From there, the algorithm increments the edge index into a or b depending on which makes the smaller angle increment. In general, the algorithm requires incrementing the angle beyond 2π so as to consider all the edges of a before the edge found by the first loop of the program. Since the edges are stored with angles between 0 and 2π, an offset variable is used to add 2π to the angle when the wraparound on polygon a is detected.

```
PROCEDURE setsum (a, b, c);
    POLYGON a, b, c;
    BEGIN INTEGER ea, eb, vc, i;
        REAL ang, offset;
        COMMENT Initialize an index into a, one into b, and one into c.
            The value of offset will be either 0 or 2*pi, and it is used
            to handle angle wraparound as described above;
        ea := 1;
        eb := 1;
        vc := 1;
        offset := 0;
        COMMENT Find adjacent edges in a whose angles straddle the angle
            of the first edge of b;
        WHILE (a.angle[ea] <= b.angle[1] OR a.angle[ea − 1] >= b.angle[1])
            DO ea := ea + 1;
        FOR i := 1 STEP 1 UNTIL 2 DO
            c. vert[1, i] := a.vert[ea, i] + b.vert[1, i];
        COMMENT This loop implements the scan of Theorem 2 in the body of the
            paper. The result of the loop is to fill the vertex array of c;
        WHILE (eb <= b.size) DO
            BEGIN
                vc := vc + 1;
                ang := offset + a.angle[ea];
                IF (ang <= b.angle[eb])
```

```
                THEN IF (ea >= a.size)
                        THEN BEGIN offset := 2*pi; ea := 1 END
                        ELSE ea := ea + 1;
                IF (ang >= b.angle[eb])
                THEN eb := eb + 1;
                FOR i := 1 STEP 1 UNTIL 2 DO
                     c.vert[vc, i] := a.vert[ea, i] + b.vert[eb, i]
            END;
        c.size := vc;
        FOR i := 1 STEP 1 UNTIL 2 DO
            c.vert[vc + 1, i] := c.vert[1, i];
END
```

Appendix II

Proof of Theorem 6

Assume that $Cspace_A \subseteq \mathcal{R}^d$, let $I = \{1, 2, \cdots, d\}$ and $K \subseteq$ '. Let I, K and $I - K$ denote sets of indexes for the coordinates of $a \in Cspace_A$. Define the following vectors, all in $Cspace_A$: $b = (\beta_i)$, $c = (\gamma_i)$ and $c' = (\gamma'_i)$ for $i \in I$. Then,

$$\Phi_K(c, c') \equiv \left\{ b \in \mathcal{R}^d \middle| \bigwedge_{k \in K} \gamma_k \leq \beta_k \leq \gamma'_k \right\}$$

$$\Phi_K(c) \equiv \Phi_K(c, c)$$

$$\Theta_K(c, c') \equiv \Phi_K(c, c') \cap \Phi_{I-K}(c, c)$$

These definitions are illustrated in Fig. 14.

The *projection operator*, denoted $P_K[\cdot]$: $\mathcal{R}^d \rightarrow \mathcal{R}^{|K|}$ is defined, for vectors and sets of vectors, by

$$P_K[b] = (\beta_k) \qquad k \in K$$

$$P_K[B] = \{P_K[b] | b \in B\}$$

Superscripts on vectors indicate projection, e.g., $b^K = P_K[b]$. In addition, the vector in $\mathcal{R}^{|I|}$ composed from one vector in $\mathcal{R}^{|K|}$ and one in $\mathcal{R}^{|I-K|}$ is denoted $(a^{I-K}: b^K)$, where $P_{I-K}[(a^{I-K}: b^K)] = a^{I-K}$ and $P_K[(a^{I-K}: b^K)] = b^K$.

In this notation, precise definitions for the notions of *cross section projection* and *slice projection* can be provided. The cross section projection of a $Cspace_A$ obstacle is written as follows:

$$P_{I-K}[CO_A(B) \cap \Phi_K(c)].$$

The slice projection, is similar to the cross section projection,

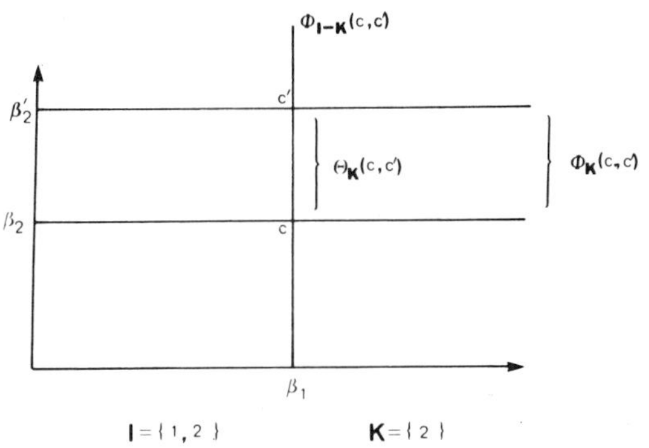

Fig. 14. Illustration of the definition of $\Phi_K(c,c')$ and $\Theta_K(c,c')$.

but carried out for all configurations between two cross sections:

$$P_{I-K}[CO_A(B) \cap \Phi_K(c, c')].$$

The K-parameters of the two configurations, c and c', define the bounds of the slice. Similarly, the swept volume can be defined in this notation.

Definition: The *swept volume* of A over the configuration range $[c, c']_K$ is

$$(A[c, c']_K)_c \equiv \bigcup_{a \in \Theta_K(c,c')} (A)_a.$$

The requirement discussed in Section VII that the swept volume of A be *displaceable* is embodied in the following condition:

$$\forall a: \bigcup_{x \in \Theta_K(c,c')} (A)_{(a^{I-K}:x^K)} = (A[c, c']_K)_{(a^{I-K}:c^K)} \qquad (7)$$

Note that the $I - K$ parameters may be changed, as in (7), but not those parameters in K. Therefore, $(A[c, c']_K)_a$ is defined only if $a \in \Phi_K(c)$.

Theorem 6: If (7) holds, then

$$P_{I-K}[CO_A(B) \cap \Phi_K(c, c')] = P_{I-K}[CO_{A[c,c']_K}(B) \cap \Phi_K(c)]$$

Proof of Theorem: Assume that the configuration a is in the slice projection of $CO_A(B)$, that is,

$$a \in P_{I-K}[CO_A(B) \cap \Phi_K(c, c')].$$

This assumption and the definition of the projection operator allows us to deduce that some configuration in $Cspace_A$, whose $I - K$-projection is a, is in $CO_A(B)$:

$$\Leftrightarrow \exists x_1 \in \Phi_K(c, c'):((a^{I-K}:x_1^K) \in CO_A(B)).$$

In fact, since we are only interested in the K-parameters of x_1 and $\Theta_K(c, c') \subseteq \Phi_K(c, c')$, we can assume without loss of generality that x_1 is in the smaller set, i.e.,

$$\Leftrightarrow \exists x_1 \in \Theta_K(c, c'):((a^{I-K}:x_1^K) \in CO_A(B)).$$

Simply using the definition of $CO_A(B)$, it follows that

$$\Leftrightarrow (A)_{(a^{I-K}:x_1^K)} \cap B \neq \emptyset,$$

but if A in this configuration intersects B, then any set including A in that configuration will also intersect B. In particular,

$$\Leftrightarrow \bigcup_{x \in \Theta_K(c,c')} (A)_{(a^{I-K}:x^K)} \cap B \neq \emptyset.$$

We are assuming that swept volumes are displaceable, i.e., that (7) holds. Therefore, using (7), we get

$$\Leftrightarrow (A[c, c']_K)_{(a^{I-K}:c^K)} \cap B \neq \emptyset$$

Hence, by the definitions of $CO_A(B)$ and $\Phi_K(c)$,

$$\Leftrightarrow (a^{I-K}:c^K) \in CO_{A[c,c']}(B) \text{ and } (a^{I-K}:c^K) \in \Phi_K(c)$$

Applying the definition of the projection operator completes the proof:

$$\Leftrightarrow a \in P_{I-K}[CO_{A[c,c']_K}(B) \cap \Phi_K(c)]. \quad \blacksquare$$

Acknowledgment

The author would like to thank M. Brady, R. Brooks, J. Hollerbach, B. Horn, T. Johnson, M. Mason, P. Winston, B. Woodham, and the referees for their suggestions on the content and presentation of this paper. J. Hollerbach suggested the proof of Theorem 6 in the current version of the paper, which is much simpler than the original proof.

References

[1] M. Adamowicz, "The optimum two-dimensional allocation of irregular, multiply-connected shapes with linear, logical and geometric constraints," Ph.D. dissertation, Dep. Elec. Eng., New York Univ., 1970.

[2] M. Adamowicz and A. Albano, "Nesting two-dimensional shapes in rectangular modules," *Computer Aided Design*, vol. 8, pp. 27–32, Jan. 1976.

[3] A. Albano and G. Sapuppo, "Optimal allocation of two-dimensional irregular shapes using heuristic search methods," *IEEE Trans. Syst., Man., Cybern.*, vol. SMC-10, pp. 242–248, May 1980.

[4] D. Avis and G. T. Toussaint, "An optimal algorithm for determining the visibility of a polygon from an edge," Sch. Comput. Sci., McGill Univ., Rep. SOCS-80.2, Feb. 1980.

[5] R. V. Benson, *Euclidean Geometry and Convexity.* New York: McGraw-Hill, 1966.

[6] J. I. Bentley and T. Ottman, "Algorithms for reporting and counting geometric intersections," *IEEE Trans. Comput.*, vol. C-28, Sept. 1979.

[7] O. Bottema and B. Roth, *Theoretical Kinematics.* Amsterdam: North-Holland, 1979.

[8] J. W. Boyse, "Interference detection among solids and surfaces," *Commun. Ass. Comput. Mach.*, vol. ACM 22, pp. 3–9, Jan. 1979.

[9] R. A. Brooks, "Solving the Find-Path problem by representing free space as generalized cones," M.I.T. Artificial Intell. Lab., Rep. AIM-674, May 1982.

[10] R. A. Brooks and T. Lozano-Pérez, "A subdivision algorithm in configuration space for Findpath with rotation," M.I.T. Artificial Intell. Lab., Rep. AIM-684, Dec. 1982.

[11] K. Q. Brown, "Fast intersection of half spaces," Dep. Comput. Sci., Carnegie-Mellon Univ., Rep. CS-78-129, June 1978.

[12] C. E. Campbell and J. Y. S. Luh, "A preliminary study on path planning of collision avoidance for mechanical manipulators," Sch. Elec. Eng., Purdue Univ., Rep. TR-EE 80-48, Dec. 1980.

[13] H. Freeman, "On the packing of arbitrary-shaped templates," in *Proc. 2nd USA-Japan Comput. Conf.*, 1975, pp. 102–107.

[14] G. Giralt, R. Sobek, and R. Chatila, "A multilevel planning and navigation system for a mobile robot," in *Proc. 6th Int. Joint Conf. Artificial Intell.*, Tokyo, Japan, Aug. 1979, pp. 335–338.

[15] R. L. Graham, "An efficent algorithm for determining the convex hull of a finite planar set," *Inform. Processing Lett.*, vol. 1, pp. 132–133, 1972.

[16] B. Grunbaum, *Convex Polytopes.* New York: Wiley-Interscience, 1967.

[17] L. J. Guibas and F. F. Yao, "On translating a set of rectangles," in *Proc 12th Annu. ACM Symp. Theory of Computing*, Los Angeles, CA, Apr 1980, pp. 154–160.

[18] W. E. Howden, "The sofa problem," *Comput. J.*, vol. 11, pp. 299–301 Nov. 1968.

[19] R. C. Larson and V. O. K. Li, "Finding minimum rectilinear distance paths in the presence of obstacles," *Networks*, vol. 11, pp. 285–304, 1981.

[20] R. A. Lewis, "Autonomous manipulation on a robot: Summary of manipulator software functions," Jet Propulsion Lab., California Inst. Technol., TM 33-679, Mar. 1974.

[21] L. Lieberman and M. A. Wesley, "AUTOPASS: An automatic programming system for computer controlled assembly," *IBM J. Res. Develop.*, vol. 21, July 1977.

[22] T. Lozano-Pérez, "The design of a mechanical assembly system," M.I.T. Artificial Intell. Lab., Rep. TR-397, Dec. 1976.

[23] ——, "Automatic planning of manipulator transfer movements," *IEEE Trans. Syst., Man., Cybern.*, vol. SMC-11, pp. 681–698, Oct. 1981.

[24] T. Lozano-Pérez and M. A. Wesley, "An algorithm for planning collision-free paths among polyhedral obstacles," *Commun. Ass. Comput. Mach.*, vol. ACM 22, pp. 560–570, Oct. 1979.

[25] L. Lyusternik, *Convex Figures and Polyhedra.* Dover, 1963, original copyright, Moscow, 1956.

[26] H. P. Moravec, "Visual mapping by a robot rover," in *Proc. 6th Int. Joint Conf. Artificial Intell.*, Tokyo, Japan, Aug. 1979.

[27] D. Mueller and F. Preparata, "Finding the intersection of two convex polyhedra," Coordinated Sci. Lab., Univ. Illinois, Urbana, Rep. R-793, Oct. 1977.

[28] N. Nilsson, "A mobile automaton: An application of artificial intelligence techniques," in *Proc. 2nd Int. Joint Conf. Artificial Intell.*, 1969, pp. 509–520.

[29] R. P. Paul, "Manipulator Cartesian path control," *IEEE Trans. Syst., Man., Cybern.*, vol. SMC-9, pp. 702–711, Nov. 1979.

[30] G. Pfister, "On solving the FINDSPACE problem, or how to find where things aren't," M.I.T. Artificial Intell. Lab., Working Paper 113, Mar. 1973.

[31] D. L. Pieper, "The kinematics of manipulators under computer control," Stanford Artificial Intell. Lab., AIM-72, Oct. 1968.

[32] F. Preparata and S. Hong, "Convex hulls of finite sets of point in two and three dimensions," *Commun. Ass. Comput. Mach.*, vol. 20, pp. 87–93, Feb. 1977.

[33] F. Preparata and D. Mueller, "Finding the intersection of a set of half spaces in time $O(n \log n)$," Coordinated Sci. Lab., Univ. Illinois, Urbana, Rep. R-803, Dec. 1977.

[34] J. Reif, "On the movers problem," in *Proc. 20th Annu. IEEE Symp. Foundation of Comput. Sci.*, 1979, pp. 421–427.

[35] B. Schachter, "Decomposition of polygons into convex sets," *IEEE Trans. Comput.*,vol. C-27, pp. 1078–1082, Nov. 1978.

[36] M. I. Shamos, "Geometric complexity," in *Proc. 7th Annu. ACM Symp. Theory of Computing*, 1975, pp. 224–233.

[37] M. I. Shamos and D. Hoey, "Closest-point problems," in *Proc. 16th Annu. IEEE Symp. Foundation of Comput. Sci.*, 1975, pp. 151–161.

[38] ——, "Geometric intersection problems," in *Proc. 17th Annu. IEEE Symp. Foundation of Comput. Sci.*, 1976, pp. 208–215.

[39] J. T. Schwartz, "Finding the minimum distance between two convex polygons," *Inform. Processing Lett.*, vol. 13, pp. 168–170, 1981.

[40] J. T. Schwartz and M. Sharir, "On the piano movers' problem I. The case of a two-dimensional rigid polygonal body moving amidst polygonal barriers," Dep. Comput. Sci., Courant Inst. Math. Sci., New York Univ., Rep. 39, 1981.

[41] ——, "On the piano movers' problem II. General techniques for computing topological properties of real manifolds," Dep. Comput. Sci., Courant Inst. Math. Sci., New York Univ., Rep. 41, 1982.

[42] Y. G. Stoyan and L. D. Ponomarenko, "A rational arrangement of geometric bodies in automated design problems," *Eng. Cybern.*, vol. 16, Jan. 1978.

[43] R. Taylor, "A synthesis of manipulator control programs from task-level specifications," Stanford Artificial Intell. Lab., Rep. AIM-282, July 1976.

[44] A. M. Thompson, "The navigation system of the JPL robot," in *Proc. 5th Int. Joint Conf. Artificial Intell.*, Massachusetts Inst. Technol., 1977.

[45] S. Udupa, "Collision detection and avoidance in computer controlled manipulators," in *Proc. 5th Int. Joint Conf. Artificial Intell.*, Massachusetts Inst. Technol., 1977.

[46] ——, "Collision detection and avoidance in computer controlled manipulators," Ph.D. dissertation, Dep. Elec. Eng., California Inst. Technol., 1977.

[47] J. E. Warnock, "A hidden line algorithm for halftone picture representation," Dep. Comput. Sci., Univ. Utah, Rep. TR4-5, May 1968.

[48] J. Williams, "STICKS—A new approach to LSI design," S.M. thesis, Dep. Elec. Eng., Massachusetts Inst. Technol., 1977.

[49] T. Winograd, *Understanding Natural Language.* New York: Aademic, 1972.

[50] T. C. Woo, "Progress in shape modelling," *IEEE Computer*, vol. 10, Dec. 1977.

[51] F. F. Yao, "On the priority approach to hidden-surface algorithms," in *Proc. 21st Symp. Foundation of Comput. Sci.*, 1980, pp. 301–307.

Simplified Voronoi Diagrams*

John Canny[1] and Bruce Donald[2]

Artificial Intelligence Laboratory, Massachusetts Institute of Technology, Cambridge, MA 02139, USA

Abstract. We are interested in Voronoi diagrams as a tool in robot path planning, where the search for a path in an r-dimensional space may be simplified to a search on an $(r-1)$-dimensional Voronoi diagram. We define a Voronoi diagram V based on a measure of distance which is not a true metric. This formulation has lower algebraic complexity than the usual definition, which is a considerable advantage in motion-planning problems with many degrees of freedom. In its simplest form, the measure of distance between a point and a polytope is the maximum of the distances of the point from the half-spaces which pass through faces of the polytope. More generally, the measure is defined in configuration spaces which represent rotation. The Voronoi diagram defined using this distance measure is no longer a strong deformation retract of free space, but it has the following useful property: any path through free space which starts and ends on the diagram can be continuously deformed so that it lies entirely on the diagram. Thus it is still complete for motion planning, but it has lower algebraic complexity than a diagram based on the Euclidean metric.

1. Introduction

The Voronoi diagram has proved to be a useful tool in a variety of contexts in computational geometry. Our interest here is in using the diagram to simplify

* This report describes research done at the Artificial Intelligence Laboratory of the Massachusetts Institute of Technology. Support for the Laboratory's Artificial Intelligence research is provided in part by the Office of Naval Research under Office of Naval Research Contract N00014-81-K-0494 and in part by the Advanced Research Projects Agency under Office of Naval Research Contracts N00014-85-K-0124 and N00014-82-K-0334. John Canny was supported by an IBM fellowship. Bruce Donald was funded in part by a NASA fellowship administered by the Jet Propulsion Laboratory.

[1] Current address: Computer Science Division, University of California, Berkeley, CA 94720, USA.

[2] Current address: Computer Science Department, Upson Hall, Cornell University, Ithaca, NY 14853, USA.

the planning of collision-free paths for a robot among obstacles, the so-called generalized movers' problem. The Voronoi diagram, as usually defined, is a *strong deformation retract* of free space so that free space can be continuously deformed onto the diagram. This means that the diagram is complete for path planning, i.e., searching the original space for paths can be reduced to a search on the diagram. Reducing the dimension of the set to be searched usually reduces the time complexity of the search. Secondly, the diagram leads to robust paths, i.e., paths that are maximally clear of obstacles.

The Voronoi diagram generated by a set of points in a Euclidean space partitions the space into convex regions which have a single nearest point under some (usually L_2) metric. A generalized Voronoi diagram can be defined for points and line segments in the plane (Lee and Drysdale, 1981) which partitions the plane into (generally nonconvex) regions. In both cases the diagram is defined to be the set of points equidistant from two or more generators under the appropriate metric. This construction has proved to be useful for motion planning among a set of obstacles in configuration space (see Ó'Dúnlaing and Yap (1985), Ó'Dúnlaing *et al.* (1984), Yap (1984), and the textbook of Schwartz and Yap (1986) for an introduction and review of the use of Voronoi diagrams in motion planning). Its virtue for motion planning is that the diagram is a strong deformation retract of free space. i.e., the space outside the obstacles can be continuously deformed onto the diagram. To find a path between two points in free space, it suffices to find a path for each point onto the diagram, and to join these points with a path that lies wholly on the diagram.

The simplified diagram has lower algebraic complexity than the L_2 diagram. For example, in $\mathfrak{R}^3$ the L_2 diagram about polyhedral obstacles consists of quadric sheets; the simplified diagram is piecewise linear. In $\mathfrak{R}^2$ the simplified diagram for polygonal obstacles is a graph of straight lines, see Fig. 4. In general, the simplified diagram has the same degree as the algebraic obstacle constraints. However, it may not have linear size in the worst case.

One useful aspect of the simplified Voronoi diagram is that it is naturally defined for the six-dimensional configuration space of an arbitrary three-dimensional polyhedron moving amidst three-dimensional polyhedral obstacles. Our definition elaborates a suggestion of Donald (1984) and Canny (1985), who describe certain Voronoi-like properties of the algebraic set $\bigcup_{i \neq i'} \ker(C_i - C_{i'})$ for a set of algebraic constraints $\{C_i\}$. In this paper we consider the configuration space of a polyhedral object with translational and rotational degrees of freedom. The simplified Voronoi diagram has the same algebraic complexity as the resulting configuration space obstacle boundaries. The completeness property holds for the simplified diagram when the defining algebraic obstacle constraints in the configuration space have unit gradients. The diagram has the same degree as these normalized constraints. (The *degree* of the diagram is the degree of the defining equations). Thus in $\mathfrak{R}^2$ and $\mathfrak{R}^3$ the simplified diagram has degree 1 whereas the Euclidean diagram has degree 2. However, note that the Euclidean diagram in $\mathfrak{R}^3$ has curves of degree 4 and vertices of degree 8, whereas the simplified diagram is piecewise linear. In the configuration space $\mathfrak{R}^2 \times S^1$ of a planar polygon, the simplified diagram has degree 3, whereas the Euclidean diagram has degree 6. In the six-dimensional configuration space $\Re^3 \times SO(3)$ of a three-dimensional polyhedron, both the simplified and Euclidean diagram have degree 10. A one-dimensional skeleton of the simplified diagram (in six dimensions) can be computed in time $O(n^7 \log n)$; this is the cost of computing an obstacle-avoiding path along the diagram.

The completeness proof in this paper holds for the normalized six-dimensional case, and specializes to the lower-dimensional cases. The simplified diagram can also be defined for unnormalized configuration space constraints. In the lower-dimensional configuration spaces, the constraints may be normalized with no increase in algebraic complexity. In the six-dimensional case, a quaternion representation of rotation gives constraints of degree 3 in the configuration parameters. In fact the constraints are simultaneously quadric in the quaternion components and linear in the position components. Some of these constraints do not *a priori* satisfy the normalization condition, but they can be normalized by dividing by a polynomial factor. Since this increases the degree of the equations defining the diagram (to an effective degree of 10, the same as the Euclidean diagram), we suggest instead that they be left unnormalized, and we offer the completeness proof in this paper as heuristic evidence that the property holds for a reasonable class of unnormalized systems of constraints.

2. Object–Obstacle Constraints

We briefly derive conditions for overlap of two polyhedral objects A and B. A more complete derivation of an equivalent condition is given by Canny (1986a). The form we derive here is different from (Canny, 1986a) in that it uses a local test for nonoverlap, rather than overlap. We assume first that A and B are convex, and then generalize to the nonconvex case by taking the conjunction of pairwise nonoverlap predicates between convex pieces.

The overlap predicates in Canny (1986a) generate a shallow (depth 2) AND-OR predicate tree, whose root is a disjunction. It will be advantageous to make the predicate tree as deep as possible and it is also desirable for the root to be a conjunction. So instead we use the following test based on conditions for nonoverlap.

Definition. For any face f of a convex polyhedron A, the *affine hull* $\bar{f}$ of f is the plane which contains f. The affine hull of a face f defines two closed half-spaces, one of which contains A. We call this half-space the *interior half-space*, and denote it $\bar{f}^-$. Finally, we define the *wedge* of an edge e of A as the intersection of the two interior half-spaces of the faces which cobound e. The wedge of e is denoted $\hat{e}$, and it contains A.

Lemma 2.1. *Two convex polyhedra A and B are nonoverlapping iff either all edges of A are outside some wedge of an edge of B, or all edges of B are outside of some wedge of an edge of A.*

Proof. This condition is clearly sufficient for nonoverlap, by convexity of A and B. Conversely, if A and B are disjoint, then there is a (not necessarily unique) nonzero shortest vector between them. Let p_A and p_B be the endpoints of this vector in A and B, respectively. If one of these points lies in the interior of a face f, then the test succeeds for any wedge of an edge in the boundary of f. If one of the points lies in the interior of an edge e, then the test succeeds for $\hat{e}$.

The only case remaining is where p_A and p_B are both vertices. Let f be a face adjacent to p_A, and such that p_B lies outside the interior half-space $\bar{f}^-$ (there must be at least one such f, or p_B would be contained in A). Then if B is also outside this half-space, the test succeeds for the wedge of any edge that cobounds f.

Otherwise, let W_{p_B} be the intersection of the wedges of all edges that cobound p_B. Let S be the plane passing through p_B and normal to the vector $(p_B - p_A)$. S defines two closed half-spaces, one (S_A) containing A and the other (S_B) containing both B and W_{p_B}, i.e., S separates A and B. Then the intersection $(W_{p_B} \cap \bar{f})$ lies in $(S_B \cap \bar{f})$. Let p_0 be the closest point in $(W_{pB} \cap \bar{f})$ to $(S \cap \bar{f})$. Then p_0 is in the boundary of some wedge $\hat{e}$ of an edge e that cobounds p_B. Now $(\hat{e} \cap \bar{f}) \subset (S_B \cap \bar{f})$ and so by projection from p_B, $(\hat{e} \cap \bar{f}^-) \subset (S_B \cap \bar{f}^-)$. But $A \subset (S_A \cap \bar{f}^-)$, so $(\hat{e} \cap A) = \varnothing$, and the test succeeds for $\hat{e}$. □

Thus we can define the following predicate for nonoverlap $F_{A,B}$ of A and B from the above test:

$$F_{A,B} = \left(\bigvee_{\substack{e_j \in \\ \text{edges}(B)}} \bigwedge_{\substack{e_i \in \\ \text{edges}(A)}} F_{e_i, \hat{e}_j} \right) \vee \left(\bigvee_{\substack{e_i \in \\ \text{edges}(A)}} \bigwedge_{\substack{e_j \in \\ \text{edges}(B)}} F_{\hat{e}_i, e_j} \right), \tag{1}$$

where $F_{\hat{e}_i, e_j} = ((\hat{e}_i \cap e_j) = \varnothing)$. The corresponding condition for overlap is in the form of a conjunction of disjunctions, as desired, and this is the form we will use in our development. If the object consists of several convex pieces A_i, as do the obstacles B_j, then the nonoverlap predicate is the conjunction of pairwise predicates

$$F = \bigwedge_i \bigwedge_j F_{A_i, B_j}. \tag{2}$$

We must now decompose the nonoverlap predicate for a wedge $\hat{e}_A$ of A and an edge e_B of B into simple geometric predicates that can be computed directly. These geometric predicates are $A_{f,p}$, $C^+_{e_A,e_B}$, and $C^-_{e_A,e_B}$. $A_{f,p}$ indicates nonoverlap of a vertex p of B and the interior half-space of a face f of A and is given by

$$A_{f,p} = (\mathbf{n}_f \cdot p - c_f > 0), \tag{3}$$

where $\mathbf{n}_f$ is the outward normal of f, and c_f is its distance from the origin. For $C^+_{e_A,e_B}$ and $C^-_{e_A,e_B}$ we need the following definitions: let d_A and d_B be the vector directions of e_A and e_B, respectively, and let p_A be any point on e_A. And let H and T be the head and tail vertex, respectively, of e_B, then we have

$$\begin{aligned} C^+_{e_A,e_B} &= (H - p_A) \cdot (d_A \times d_B) > 0, \\ C^-_{e_A,e_B} &= (H - p_A) \cdot (d_A \times d_B) < 0. \end{aligned} \tag{4}$$

Now we can define $F_{\hat{e}_A,e_B}$ in terms of the above predicates, and L and R which are the left and right faces, respectively, which cobound e_B (left and right are determined here by viewing e_B from outside $\hat{e}_B$ with d_B upward).

Lemma 2.2. *The following predicate indiates nonoverlap of the wedge $\hat{e}_A$ and the edge e_B:*

$$F_{\hat{e}_A,e_B} = ((A^+_{L,H} \wedge (A^+_{L,T} \vee (A^+_{R,T} \wedge C^+_{e_A,e_B}))) \\ \vee (A^+_{R,H} \wedge (A^+_{R,T} \vee (A^+_{L,T} \wedge C^-_{e_A,e_B})))). \tag{5}$$

Proof. The proof is by analysis of all possible predicate values. Firstly, the predicate $A^+_{L,H}$ requires the head of e_B to be above the plane of the left face of e_A. For the rest of this paragraph, we assume that $A^+_{L,H}$ is true, so that H is above the plane of L. The subcases are itemized below:

- If $A^+_{L,T}$ is true, then the tail T of e_B is also above the plane of L, i.e., the entire edge e_B is outside of $\hat{e}_A$ and the predicate correctly returns true.
- If $A^+_{L,T}$ is false, and so is $A^+_{R,T}$, then the vertex T lies inside $\hat{e}_A$, and the predicate $F_{\hat{e}_A,e_B}$ correctly returns false.
- The only case remaining is where $A^+_{L,T}$ is false and $A^+_{R,T}$ is true, so that both vertices are outside of $\hat{e}_A$, and H is above the plane of L and T is below that plane. This case is illustrated in Fig. 1. In this case $d_A \times d_B$ is a vector which points out of $\hat{e}_A$, and is normal to d_A and d_B. Then e_B is outside of $\hat{e}_A$ if and only if the inner product of $(H - p_A)$ with this vector is positive, a condition which is indicated by $C^+_{e_A,e_B}$.

The subcases for $A^+_{R,H}$ true are similar, with left and right faces interchanged. The only case remaining is where both $A^+_{L,H}$ and $A^+_{R,H}$ are false, but here the point H is inside the object and the predicate $F_{\hat{e}_A,e_B}$ correctly returns a false value. □

The predicate $F_{\hat{e}_A,e_B}$ can be written in the equivalent form:

$$F_{\hat{e}_A,e_B} = (A_{L,H} \vee A_{R,H}) \wedge (A_{L,T} \vee A_{R,T}) \\ \wedge (A_{L,H} \vee A_{R,T} \vee C^+_{e_A,e_B}) \wedge (A_{R,H} \vee A_{L,T} \vee C^-_{e_A,e_B}). \tag{6}$$

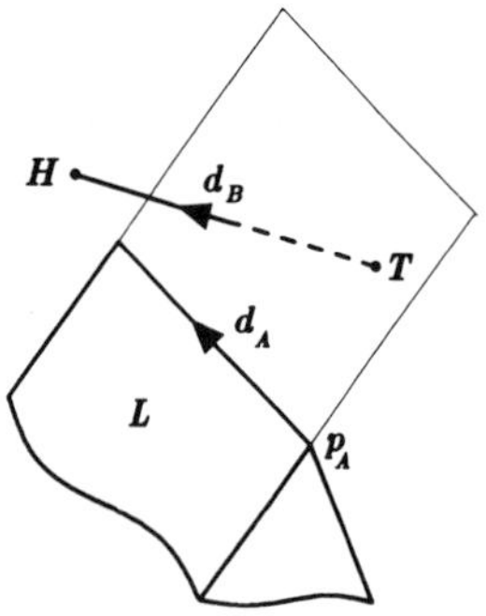

Fig. 1. Position of edge e_B outside the wedge $\hat{e}_A$, where inside/outside is determined by $C^+_{e_A,e_B}$.

There is a similar form for the predicate $F_{e_A,\hat{e}_B}$. From (1), (2), and (6) we see that the overall form of the nonoverlap predicate can be written as

$$F = \bigwedge_i \bigvee_j \bigwedge_k \bigvee_l (C_{ijkl} > 0). \tag{7}$$

3. The Voronoi Diagram

For motion planning, the configuration of object A is variable. For a polyhedron in three-dimensional space, the configuration contains a position $\mathbf{x} \in \Re^3$ and an orientation $\mathbf{q} \in SO(3)$ component, where $SO(3)$ is the group of three-dimensional rotations. Thus the face normals, vertex locations, and edge directions of A are all functions of $\mathbf{x}$ and $\mathbf{q}$. The predicate (7) is now also a function of the configuration $(\mathbf{x}, \mathbf{q})$, i.e.,

$$F(\mathbf{x}, \mathbf{q}) = \bigwedge_i \bigvee_j \bigwedge_k \bigvee_l (C_{ijkl}(\mathbf{x}, \mathbf{q}) > 0). \tag{8}$$

The forms of the functions $C_{ijkl}(\mathbf{x}, \mathbf{q})$ are given explicitly in Canny (1986a), and they are algebraic if, say, a quaternion representation of rotation is used. The set of overlap configurations is called the *configuration obstacle* and is denoted $CO = \{(\mathbf{x}, \mathbf{q}) \mid \neg F(\mathbf{x}, \mathbf{q})\}$. It may be thought of as a physical obstacle in configuration space to be avoided by a path planner. We now observe that by letting positive real values represent logical one, and nonpositive values represent logical zero, that the min function implements logical AND, and the max function implements logical OR. Thus an equivalent form to (8) is

$$F(\mathbf{x}, \mathbf{q}) = \left(\left(\min_i(\max_j(\min_k(\max_l C_{ijkl}(\mathbf{x}, \mathbf{q}))))\right) > 0\right) \tag{9}$$

which suggests that the quantity

$$\rho(\mathbf{x}, \mathbf{q}) = \min_i\left(\max_j(\min_k(\max_l C_{ijkl}(\mathbf{x}, \mathbf{q})))\right) \tag{10}$$

can be used as a measure of distance from the configuration obstacle, because it varies continuously through configuration space, is positive at configurations outside CO, and nonpositive at configurations inside CO. Thus the configuration obstacle can be rewritten as $CO = \{\mathbf{p} \mid \rho(\mathbf{p}) \leq 0\}$, and its complement, the set of points in free space can be written $F = \{\mathbf{p} \mid \rho(\mathbf{p}) > 0\}$.

In order to define the Voronoi diagram under the distance measure ρ, we need a notion of closest feature. The closest features to a configuration $(\mathbf{x}, \mathbf{q})$ are those C_{ijkl} which are *critical* in determining the value of $\rho(\mathbf{x}, \mathbf{q})$, that is, small changes in the value of C_{ijkl} cause indentical changes in the value of ρ.

Definition. A constraint $C_{i_0j_0k_0l_0} \in \{C_{ijkl}\}$ is *critical* at a configuration $(\mathbf{x}, \mathbf{q})$ if the value of $C_{i_0j_0k_0l_0}(\mathbf{x}, \mathbf{q})$ equals the maximum (or minimum) value of every max (resp. min) ancestor of $C_{i_0j_0k_0l_0}$ in the min-max tree in (10), i.e.,

$$C_{i_0j_0k_0l_0}(\mathbf{x}, \mathbf{q}) = \max_l C_{i_0j_0k_0l}(\mathbf{x}, \mathbf{q}) = \min_k\left(\max_l C_{i_0j_0kl}(\mathbf{x}, \mathbf{q})\right) = \cdots \tag{11}$$

Now we have

Definition. The *simplified Voronoi diagram* V is the set of configurations in free space F at which at least two distinct constraints are critical.

It should be clear from the definition of criticality that V is semialgebraic if the constraints C_{ijkl} are algebraic. V is closed as a subset of free space, although it is not closed in configuration space. Notice that V has no interior, since it is contained inside a finite set of bisectors, each of which has no interior. A bisector is the zero set of $(C_{i_0j_0k_0l_0} - C_{i_1j_1k_1l_1})$ for some pair of distinct constraints. It will prove useful to subdivide the Voronoi diagram into two parts:

Definition. The *concave part* of the Voronoi diagram V denoted conc(V) is the set of configurations in F where two distinct constraints are critical, and the lowest common ancestor of these constraints in the min-max tree of (10) is a min node. The *convex part* of the Voronoi diagram V denoted conv(V) is the set of configurations in F where two distinct constraints are critical, and the lowest common ancestor of these constraints is a max node.

Notice that these two definitions are not mutually exclusive, because there may be points where more than two constraints are critical, and which satisfy both definitions. Thus conv(V) and conc(V) may overlap.

4. Completeness for Motion Planning

Our key result is that any path in F with endpoints in V can be deformed (in F) to a path with the same endpoints lying entirely in V. We start with a path $p: I \rightarrow \Re^3 \times SO(3)$ lying in free space, $p(I) \subset F$, where $I = [0, 1]$ is the unit interval.

First we assume without loss of generality that p intersects V at a finite number of points. We can do this because, as defined in the previous section, V is a semialgebraic set, and by Whitney's (1957) result, it can be split into a finite number of manifolds, or *strata*. Since V has no interior, all these manifolds have codimension at least 1. For *any* path p there is a path p' arbitrarily close to p, which is an embedding of I, and so $p'(I)$ is a 1-manifold. Almost every perturbation of p' intersects all of the strata transversally, and therefore at a finite number of points. We can choose such a perturbation to be arbitrarily small, in particular, smaller than the minimum distance from $p'(I)$ to CO. Such a perturbation gives a new path p'' which is path homotopic to p, and which has finite intersection with V.

So we assume that p has m intersections with V, and that these occur at points $p(x_i)$ with $x_1, \ldots, x_m \in I$, and $x_1 = 0$ and $x_m = 1$. We then break each interval $[x_i, x_{i+1}]$ in half, giving us two intervals sharing an endpoint. Thus we now have $2m-2$ intervals each of which intersects V at only one endpoint. Below we give homotopies for each of these intervals which continuously deform the image of the interval onto V. Since all these homotopies agree at their endpoints, they can be pasted together to give us a global homotopy which deforms p onto V. For simplicity, we assume the path segment is parametrized in the range $I=[0,1]$ and that $p(0) \in V$.

The motion constraints C_{ijkl} are either A or C predicates, (3) and (4), and all can be written in the general form below, called the parametrized plane equations by Lozano-Pérez (1983):

$$C_{ijkl}(\mathbf{x}, \mathbf{q}) = \mathbf{N}_{ijkl}(\mathbf{q}) \cdot \mathbf{x} + c_{ijkl}(\mathbf{q}), \tag{12}$$

where $\mathbf{N}_{ijkl}(\mathbf{q}) \in \Re^3$ and $c_{ijkl}(\mathbf{q}) \in \Re$ are both continuous functions of $\mathbf{q}$. We assume that the C_{ijkl} are normalized so that $|\mathbf{N}_{ijkl}(\mathbf{q})| = 1$ for all $\mathbf{q}$. Our objective is to deform the path p continuously onto the diagram, and we use the $\mathbf{N}_{ijkl}$ as "normals" to push a point on the path $p(I)$ away from the critical C_{ijkl}. We assume that the set of positions is bounded by some set of constraint "walls" of the same form as (1), so that a point can be displaced only a finite distance in free space. We also assume that the workspace has unit diameter.

General Position Assumptions. The construction requires the following general position assumptions. The second assumption requires an arbitrarily small perturbation of the constraints that are used to define the diagram, and this should be done as a preprocessing step before computation of the diagram. First, suppose C_{ijkl} is type C predicate (4). Then $\mathbf{N}_{ijkl}(\mathbf{q}) = d_A(\mathbf{q}) \times d_B$. To normalize $\mathbf{N}_{ijkl}$ we must divide by its magnitude, which must remain nonzero. Hence the set of configurations

$$\{\mathbf{q} \mid d_A(\mathbf{q}) \times d_B = 0\} \subset SO(3) \tag{13}$$

must not intersect the image of p. However, (13) is clearly of codimension 2 in $SO(3)$, and hence by Sard's lemma there is always an arbitrarily small perturbation of p which avoids (13).

Similarly, the set

$$\{(\mathbf{x}, \mathbf{q}) \mid (\mathbf{N}_{ijkl}(\mathbf{q}) = \pm \mathbf{N}_{i'j'k'l'}(\mathbf{q}))\} \tag{14}$$

is also of codimension 2 in $\Re^3 \times SO(3)$, and the proof is given below for the "+" case. For the − case, the argument below can be applied by simply negating one of the normals.

The singular set where two (unit) normals are equal has codimension 2 for the following reason. We consider a map from $\Re^3 \times SO(3)$ to the product $S^2 \times S^2$ which represents the values of the two normals. The diagonal of this space is the set where the two normals agree, and has codimension 2. If the maps are generic (specifically, if they are transversal to the diagonal set), then the preimage of the diagonal set (which is the "bad" set) will also have codimension 2.

We must define two different homotopies depending on whether $p(0) \in \text{conc}(V)$, which is the simplest case, or $p(0) \in \text{conv}(V)$. Notice that since $p(0)$ is the only point on the path which is in V, there is exactly one constraint which is critical at all configurations in $p(0, 1]$ (since constraints change value continuously along the path and for another constraint to become critical, it must first equal the first constraint, which can only occur at points on the diagram). Let the critical constraint be C_{ijkl}. We define a homotopy $J_0 : I \times I \to \Re^3 \times SO(3)$ as

$$J_0(t, u) = p(t) + u\mathbf{N}_{ijkl}(\pi_q(p(t))), \tag{15}$$

where $\pi_q(p(t))$ is the orientation component of $p(t)$ and the addition symbol means we add the vector quantity $u\mathbf{N}_{ijkl}(\pi_q(p(t)))$ to the position component of $p(t)$. The deformation above pushes points beyond the diagram, so we define a second homotopy $J_1 : I \times I \to F$:

$$J_1(t, u) = J_0(t, \min(u, u_c(t))) \qquad \text{where} \quad u_c(t) = \inf\{s \mid J_0(t, s) \in \text{conc}(V)\}. \tag{16}$$

Note that u_c is bounded above by 1 since the workspace has unit diameter. Recall from the definition of homotopy, that J_1 is a homotopy of p and a path p' if J_1 is continuous and $J_1(t, 0) = p(t)$ and $J_1(t, 1) = p'(t)$. The homotopy J_1 suffices to map paths with one endpoint in $\text{conc}(V)$ onto V:

Lemma 4.1. *Let $p : I \to \Re^3 \times SO(3)$ be a path having $p(0) \in V$ and no other points in V. Then J_1 is a homotopy of p and a path p' such that $p'(I) \subset \text{conc}(V)$. Furthermore, $p'(0) = p(0)$ if $p(0) \in \text{conc}(V)$.*

Proof. From the definition of J_1 we have $J_1(t, 0) = p(t)$ and $J_1(t, 1) \in \text{conc}(V)$. Also if $p(0) \in \text{conc}(V)$ then $u_c(0) = 0$ *so* $p'(0) = p(0)$. It remains to show that J_1 is continuous. First we notice that J_0 is continuous. Continuity of J_1 follows if we can show that $u_c(t)$ is continuous. Now $J_0(t, u_c(t))$ is contained in the zero sets of all bisectors $\{(C_{i'j'k'l'} - C_{ijkl})\}$. Let $u'_c(t)$ be a deformation onto a *particular* bisector:

$$u'_c(t) = \inf\{u \mid J_0(t, u) \in \ker(C_{i'j'k'l'} - C_{ijkl})\}, \tag{17}$$

then u'_c is continuous because by definition

$$(C_{ijkl} - C_{i'j'k'l'})(J_0(t, u'_c(t))) = 0 \tag{18}$$

and, differentiating with respect to t, we obtain

$$\frac{\partial}{\partial t}(C_{ijkl} - C_{i'j'k'l'})(J_0(t, u'_c)) + \left(\frac{\partial u'_c}{\partial t}\right)\frac{\partial}{\partial u'_c}(C_{ijkl} - C_{i'j'k'l'})(J_0(t, u'_c)) = 0 \tag{19}$$

which can be rearranged to yield

$$\frac{\partial u_c'}{\partial t} = -\frac{(\partial/\partial t)(C_{ijkl} - C_{i'j'k'l'})(J_0(t, u_c'))}{(\partial/\partial u_c')(C_{ijkl} - C_{i'j'k'l'})(J_0(t, u_c'))} \tag{20}$$

and therefore $(\partial/\partial t)u_c'(t)$ is finite, because the denominator above is nonzero by our general position assumption. Now we observe that $u_c(t)$ can be constructed by pasting together segments of $u_c'(t)$ for various bisectors in conc(V), and we must show that they agree at their endpoints. The proof is by contradiction. Suppose we had

$$u_c(t) = \begin{cases} u_c'(t) & \text{for } t \in (t_0, t_1]; \\ u_c''(t) & \text{for } t \in (t_1, t_2); \end{cases} \quad \text{with} \quad u_c'(t_1) \neq u_c''(t_1) \tag{21}$$

for u' and u'' derived from distinct bisectors. Then since u_c is the minimum of all such u, we must have $u_c'(t_1) < u_c''(t_1)$.

But this means that as u increases, $J_0(t_1, u)$ crosses two bisectors in conc(V) between C_{ijkl} and other constraints. This is impossible because all constraints C have $|\mathbf{N}(\mathbf{q})| = 1$, and it follows that

$$\begin{aligned} &\frac{\partial}{\partial u} C_{i'j'k'l'}(J_0(t_1, u)) \leq 1, \\ &\frac{\partial}{\partial u} C_{ijkl}(J_0(t_1, u)) = 1. \end{aligned} \tag{22}$$

That is, all constraints increase no faster than C_{ijkl} with the deformation parameter u. By our general position assumption if $C_{ijkl} \neq C_{i'j'k'l'}$, then the inequality in (22) is strict, and so if the clause $C_{i'j'k'l'}$ becomes critical at u_c', it shares a min node lowest ancestor with C_{ijkl}. Since all constraints in the tree increase more slowly than C_{ijkl}, the value of this min node will be less than the value of C_{ijkl} for $u > u_c'(t_1)$. Therefore C_{ijkl} cannot be critical for any $u > u_c'(t_1)$, contradicting the assumption that $u_c''(t_1)$ is distinct from $u_c'(t_1)$. So all bisectors in conc(V) between C_{ijkl} and other constraints agree at their endpoints, and $u_c(t)$ is continuous by pasting. This shows that J_1 is continuous. □

The homotopy described above is illustrated in Fig. 2. The direction of the gradient of the distance function is shown by the arrows. Each point on the path moves in the gradient direction until it hits a concave bisector.

If $p(0) \in$ conv(V) the situation is more complicated, because the deformation J_1 pushes $p(0, 1]$ away from $p(0)$. To correct this, we first compress the first half of p to a point:

Lemma 4.2. *Let p: $I \to F$ be a path in free space. Then p is homotopic to a path p' such that*:

(i) $p'([0, \frac{1}{2}]) = p(0)$.
(ii) $p'(1) = p(1)$.
(iii) $p'(I) = p(I)$.

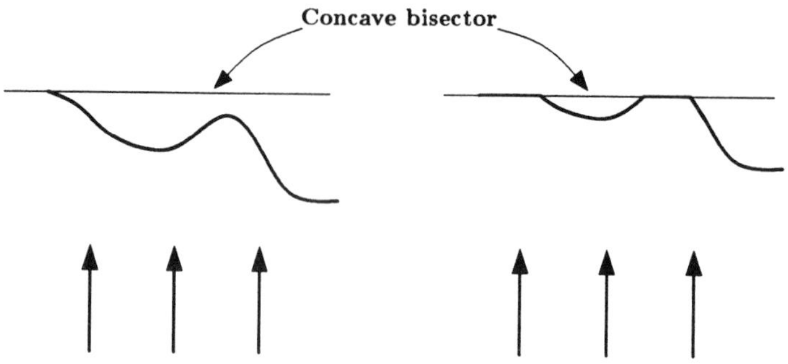

Fig. 2. Illustration of the homotopy of Lemma 4.2.

Proof. The required homotopy is

$$J(t,u)=\begin{cases} p(0) & \text{if } t\leq \frac{1}{2}u; \\ p\left(\dfrac{2t-u}{2-u}\right) & \text{otherwise;} \end{cases} \tag{23}$$

so $J(t,0)=p(t)$, and $J(t,1)=p'(t)$. □

We apply the homotopy of Lemma 4.2 to p to give us a path p'. Applying the homotopy J_1 to the path segment $p'|_{[1/2,1]}$ continuously deforms this path segment onto conc(V). Then we define a new homotopy which slowly "unravels" $p'|_{[0,1/2]}$ from $p(0)$. All the points in this homotopy have the same orientation, and for each value of the deformation parameter u, the path consists of a finite number of straight line segments.

We now construct a third homotopy. The construction is inductive and we start with a homotopy that gives us two straight line segments. If the orientation at the configuration $p(0)$ is $\mathbf{q}_0$, every point on the joining path will also have orientation $\mathbf{q}_0$. We define two vectors in position space $\mathbf{N}_0$ and $\mathbf{M}_0$ which will be used to define the joining path segment.

Let C_{ijkl} and $C_{i'j'k'l'}$ be the two constraints that are critical at $p(0)$, then $\mathbf{N}_0$ lies in the plane of the bisector of C_{ijkl} and $C_{i'j'k'l'}$. $\mathbf{N}_0$ is normalized so that $\mathbf{N}_0\cdot\mathbf{N}_{ijkl}(\mathbf{q}_0)=1$ and it follows that $\mathbf{N}_0\cdot\mathbf{N}_{i'j'k'l'}(\mathbf{q}_0)=1$. A second vector $\mathbf{M}_0$ is chosen so that $\mathbf{M}_0+\mathbf{N}_{ijkl}(\mathbf{q}_0)=\mathbf{N}_0$, and so

$$\mathbf{N}_0=\frac{\mathbf{N}_{ijkl}+\mathbf{N}_{i'j'k'l'}}{1+\mathbf{N}_{ijkl}\cdot\mathbf{N}_{i'j'k'l'}}, \tag{24}$$

$$\mathbf{M}_0=\frac{\mathbf{N}_{i'j'k'l'}-(\mathbf{N}_{ijkl}\cdot\mathbf{N}_{i'j'k'l'})\mathbf{N}_{ijkl}}{1+\mathbf{N}_{ijkl}\cdot\mathbf{N}_{i'j'k'l'}}, \tag{25}$$

where $\mathbf{N}_{ijkl}$ (shorthand for $\mathbf{N}_{ijkl}(\mathbf{q}_0)$) is the normal vector to the critical constraint

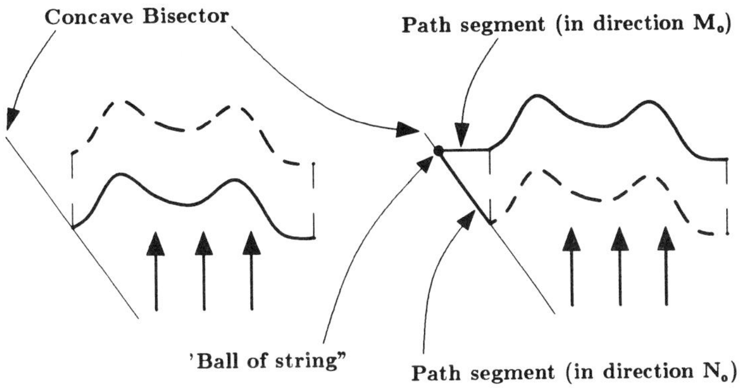

Fig. 3. Homotopy to link a deforming path continuously to a point on a convex bisector.

C_{ijkl} at orientation $\mathbf{q}_0$, and $\mathbf{N}_{i'j'k'l'} = \mathbf{N}_{i'j'k'l'}(\mathbf{q}_0)$ is the vector normal to the critical constraint $C_{i'j'k'l'}$. (Note that $\mathbf{N}_{ijkl}$ cannot equal $-\mathbf{N}_{i'j'k'l'}$ by our general position assumptions.)

We can now define the joining homotopy $J_2: I \times I \to \mathfrak{R}^3 \times SO(3)$ which deforms the path segment $p'|_{[0,1/2]}$ (with p' reparametrized so that its domain is I):

$$J_2(t, u) = \begin{cases} p(0) + 2t\mathbf{N}_0, & \text{if } t \in [0, \tfrac{1}{2}u]; \\ p(0) + u\mathbf{N}_0, & \text{if } t \in [\tfrac{1}{2}u, 1 - \tfrac{1}{2}u]; \\ p(0) + (2-2t)\mathbf{M}_0 + u\mathbf{N}_{ijkl}(\mathbf{q}_0), & \text{if } t \in [1 - \tfrac{1}{2}u, 1]. \end{cases} \tag{26}$$

The action of this homotopy on a path segment is illustrated in Fig. 3. All points not on a bisector move in the direction of the gradient of the distance function. Points which are on the bisector do not move, except the single corner point. This point may be thought of as a "ball of string," which continuously unravels and allows points on the two straight line segments to move as described above.

Now $\mathbf{N}_0$ lies in the plane of the bisector of the two constraints that are critical at $p(0)$, but it is possible that as u increases, $J_2(\frac{1}{2}, u)$ leaves the convex part of the diagram before reaching the concave part. That is, there may be bends in the convex part of the diagram which must be tracked. We must therefore stop the deformation at this point by defining

$$u_v = \sup\{u \mid C_{ijkl} \text{ is critical throughout } J_2(\tfrac{1}{2}, [0, u])\} \tag{27}$$

and once again we define a homotopy which stops points when they reach the diagram:

$$J_3(t, u) = J_2(t, \min(u, u_v, u_c(t))) \quad \text{where} \quad u_c(t) = \inf\{u \mid J_2(t, u) \in \operatorname{conc}(V)\}. \tag{28}$$

Then we have:

Lemma 4.3. *Let* $p: I \to \Re^3 \times SO(3)$ *be a path having* $p(I) = p(0) \in \operatorname{conv}(V)$. *Then* J_3 *is a homotopy of* p *and a path* p' *such that*:

(i) $p'(0) = p(0)$.
(ii) $p'([0, 1 - \frac{1}{2}u_v]) \subset \operatorname{conv}(V)$.
(iii) $p'((1 - \frac{1}{2}u_v, 1]) \cap V \subset \operatorname{conc}(V)$.

Proof. Parts (i) and (ii) follow immediately from the definition of J_2 and J_3. Part (iii) states that points in the interval $(1 - \frac{1}{2}u_v, 1]$ that are mapped into the diagram by p' are mapped into the concave part of the diagram. For this we notice that

$$\frac{\partial J_2(t, u)}{\partial u} = \mathbf{N}_{ijkl}(\mathbf{q}_0) \tag{29}$$

for $t > 1 - \frac{1}{2}u$, while $J_2(t, u) \in \operatorname{conv}(V)$ if $t \leq u$. Therefore for $t > 1 - \frac{1}{2}u$ the following is true:

$$\begin{aligned} \frac{\partial C_{i'j'k'l'}(J_2(t_1, u))}{\partial u} &\leq 1, \\ \frac{\partial C_{ijkl}(J_2(t_1, u))}{\partial u} &= 1. \end{aligned} \tag{30}$$

So as u increases, all constraints increase no faster than C_{ijkl}. So another constraint can only become critical if its lowest common ancestor with C_{ijkl} is a min node, and such a configuration must be in $\operatorname{conc}(V)$.

For continuity of J_3, first we notice that J_2 is continuous, and J_3 will be continuous if $u_c(t)$ defined in (28) is continuous. For this we notice that the use of u_v in the min function in (28) guarantees that C_{ijkl} is critical at configuration $J_3(t, u)$ for all t and u. The rest of the proof of continuity is identical to the proof of continuity of $u_c(t)$ in Lemma 4.1, using the rate of change condition in (30). □

Lemma 4.4. *If* $p: I \to F$ *is a path having* $p([0, \frac{1}{2}]) = \{p(0)\} \subset \operatorname{conv}(V)$ *then* p *is homotopic to a path* $p': I \to V$ *such that* $p'(0) = p(0)$.

Proof. The proof is inductive. We define a sequence of partial homotopies, that is, maps $J^n: I \times [u^{n-1}, u^n] \to F$ such that $J^n(t, u^n) = J^{n+1}(t, u^n)$. Then we show that the number m of partial homotopies required is finite. (We employ a superscript notation which will prove convenient in our inductive argument.)

Inductive Hypothesis. The input to our construction is a path p^n, a value of $u^n \in I$, and points t_0^n and t_1^n in I such that:

(i) $p^n(t) \in \operatorname{conv}(V)$ for $t \leq t_0^n$.
(ii) $p^n([t_0^n, t_1^n]) = \{p^n(t_0^n)\} \subset \operatorname{conv}(V)$.
(iii) C_{ijkl} is critical on $p^n(I)$.

From C_{ijkl} we use Lemma 4.1 to define a homotopy J_1 for the path segment $p^n|_{[t_1^n, 1]}$ (reparametrized to I). Similarly, let $C_{i'j'k'l'}$ be a constraint which is critical at $p^n(t_0^n)$ and whose lowest common ancestor with C_{ijkl} is a max node. For these two constraints, we use Lemma 4.3 to define an unravelling homotopy J_3 of the path segment $p^n|_{[t_0^n, t_1^n]}$ (reparametrized to I). This gives us a value of u_v as in (27), and we define $u^{n+1} = u^n + u_v$. Now J^{n+1} can be defined on the range $u \in [u^n, u^{n+1}]$ as

$$J^{n+1} = \begin{cases} p^n(t) & \text{for } t \in [0, t_0^n]; \\ J_3\left(\dfrac{t - t_0^n}{t_1^n - t_0^n}, u - u^n\right) & \text{for } t \in [t_0^n, t_1^n]; \\ J_1\left(\dfrac{t - t_1^n}{1 - t_1^n}, u - u^n\right) & \text{for } t \in [t_1^n, 1]; \end{cases} \tag{31}$$

then J^{n+1} is a homotopy because J_3 and J_1 agree on their intersection, and $J_3(0, u) = p^n(t_0^n)$. We can define a new path $p^{n+1}(t) = J^{n+1}(t, u^{n+1})$ and points

$$t_0^{n+1} = t_0^n + \frac{u_v}{2}(t_1^n - t_0^n) \quad \text{and} \quad t_1^{n+1} = t_1^n - \frac{u_v}{2}(t_1^n - t_0^n) \tag{32}$$

and it is readily verified that these satisfy the inductive hypothesis.

For the base case we set $p^0 = p$, $u^0 = 0$, and $t_0^0 = 0$, $t_1^0 = \frac{1}{2}$, which clearly satisfies the inductive hypothesis.

Finiteness. For termination we must show that after a finite number of steps m, $p^m(t_0^m) \in \text{conc}(V)$. Suppose $p^n(t_0^n) \notin \text{conc}(V)$, and let $C_{i'j'k'l'}$ be the constraint (along with C_{ijkl}) which is critical at $p^n([t_0^{n-1}, t_0^n])$, i.e., these are the constraints used to define the homotopy J_3 for J^n. A third constraint $C_{i''j''k''l''}$ is also critical at $p^n(t_0^n)$, by (27). This constraint has a max node lowest common ancestor with $C_{i'j'k'l'}$, which implies that

$$\frac{\partial}{\partial u} C_{i'j'k'l'}(J_3(\tfrac{1}{2}, u)) < \frac{\partial}{\partial u} C_{i''j''k''l''}(J_3(\tfrac{1}{2}, u)) \tag{33}$$

and from (24) it follows that

$$(N_{ijkl} + N_{i'j'k'l'}) \cdot N_{i'j'k'l'} < (N_{ijkl} + N_{i'j'k'l'}) \cdot N_{i''j''k''l''} \tag{34}$$

which implies

$$N_{ijkl} \cdot N_{i'j'k'l'} < N_{ijkl} \cdot N_{i''j''k''l''} \tag{35}$$

but condition (35) defines a total ordering on the constraints distinct from C_{ijkl}. That is, as u increases and $p^n(t_0^n)$ is deformed according to some J_2, a new constraint $C_{i''j''k''l''}$ can only become critical if it satisfies condition (35). Once $C_{i''j''k''l''}$ has become critical, $C_{i'j'k'l'}$ can never again become critical. Thus we need to define homotopies J^n at most once for each constraint, and so their number m is bounded by the number of constraints.

We then construct J_1 for the path segment $p^m|_{[t_1^m,1]}$ reparametrized to I. The final homotopy is defined for the range $u \in [u^m, 1]$ as

$$J^{m+1}(t, u) = \begin{cases} p^m(t) & \text{for } t \in [0, t_1^m]; \\ J_1\left(\dfrac{t - t_1^m}{1 - t_1^m}, u - u^m\right) & \text{for } t \in [t_1^m, 1]; \end{cases} \tag{36}$$

then the homotopy

$$J_4(t, u) = J^n(t, u) \quad \text{with} \quad u \in [u^{n-1}, u^n] \quad \text{for} \quad n = 1, \ldots, m+1 \tag{37}$$

is continuous, and defines a homotopy between $p(t) = J_4(t, 0)$ and a path $p'(t) = J_4(t, 1)$ such that $p'(I) \subset V$. □

Theorem 4.5. *Let $p: I \to F$ be a path with endpoints in V. Then p is path homotopic in F to a path p' with the same endpoints which lies entirely in V.*

Proof. We first apply the homotopy of Lemma 4.2 to all path segments $p|_{[t_n, t_{n+1}]}$ with an endpoint in conv(V). This does not displace endpoints in V. Then we construct a global homotopy J by pasting together homotopies J_1 for path segments with an endpoint in conc(V), and homotopies J_4 for the remaining path segments. The resulting homotopy is continuous if these homotopies agree at their endpoints. Firstly, both J_1 and J_4 do not displace endpoints in V. Therefore they agree at endpoints in V. At free endpoints both satisfy the *free endpoint condition*: assume that after reparametrization, $p(1)$ is a free endpoint. Then

$$J_n(1, u) = p(1) + \min(u, u_c)\mathbf{N}_{ijkl} \qquad \text{for} \quad n = 1, 4 \tag{38}$$

with

$$u_c = \inf\{u \mid p(1) + u\mathbf{N}_{ijkl} \in \text{conc}(V)\}$$

thus J is continuous, and we define $p'(t) = J(t, 1)$. Since $J_1(I, 1) \subset V$ and $J_4(I, 1) \subset V$, we have $p'(I) \subset V$. □

Finally, suppose that in a motion-planning problem we are given a start configuration $(\mathbf{x}, \mathbf{q})$ which is not on V. Then exactly one constraint C_{ijkl} is critical there. We apply the homotopy J_1 to the constant path at $(\mathbf{x}, \mathbf{q})$ to attain the diagram; that is, we plan a straight-line path in direction $\mathbf{N}_{ijkl}(\mathbf{q})$ to reach V from the start.

The completeness condition for motion planning has the following simple algebraic formulation. Let $i: V \hookrightarrow F$ be the inclusion map. Then if V is a Euclidean Voronoi diagram, then it is a strong deformation retract of F, and hence i induces an isomorphism of fundamental groups. In our case we have the weaker completeness condition that i induces an epimorphism:

Corollary 4.6 (Algebraic Formulation of the Completeness Condition for Motion Planning). *Let $\imath : V \hookrightarrow F$ be the inclusion map of the simplified Voronoi diagram in free space, with $y_0 \in V$, and let $\pi_1(X, x)$ denote the fundamental group of X with base point x. Then the induced homomorphism $i_* : \pi_1(V, y_0) \to \pi_1(F, y_0)$ is surjective.*

Hence the fundamental group of F is isomorphic to the quotient group $\pi_1(V, y_0)/\ker i_*$. This quotient measures the structural difference between F and V.

5. Complexity Bounds

We have given a definition of the simplified Voronoi diagram V in the configuration space of a polyhedron in 3-space. This definition does not constitute an algorithm, so our bounds depend on the algorithm used with the diagram. We assume that the diagram will be used as input to a version of the roadmap algorithm of Canny (1986b). This algorithm computes one-dimensional skeletons of semialgebraic sets in time $(d^{O(r^2)} n^r \log n)$ for a semialgebraic set defined by n polynomials of degree d in r variables. In our case the number of variables and the degree of the equations are constants.

A naive bound on the complexity of computing a skeleton of V would be $O(n^{12} \log n)$ if we are given n constraints, because the diagram is a subset of the zero sets of all $O(n^2)$ bisectors of constraints. This bound can be reduced to $O(n^7 \log n)$ by noticing that the diagram has a simple stratification (decomposi-

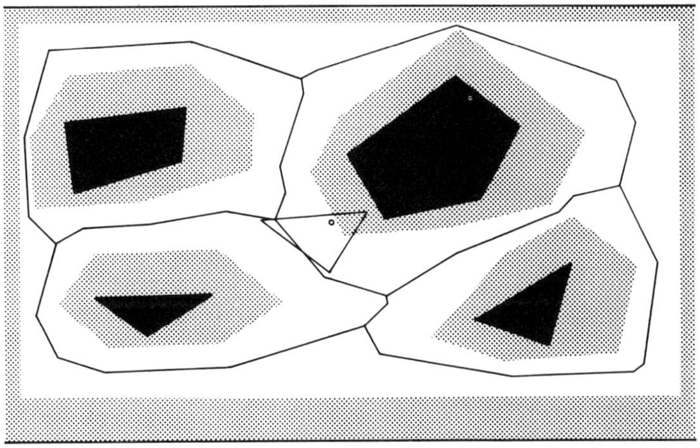

Fig. 4. The simplified Voronoi diagram in the plane. The black polygons are the real-space obstacles. The triangle is the moving "robot." The shaded polygons are the configuration-space obstacles. The simplified Voronoi diagram is a network of straight line segments.

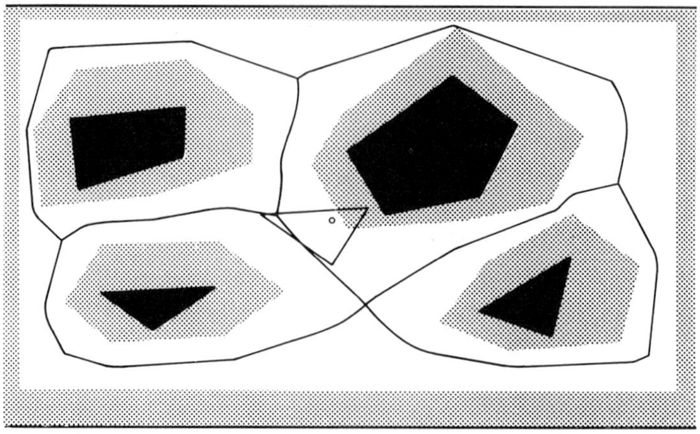

Fig. 5. True Voronoi diagram for the same obstacles.

tion into a union of disjoint manifolds). The diagram is a subset of the set of all m-sectors, where an m-sector is the set of points where m constraints have the same value. If the constraints are in general position, each m-sector is a manifold of codimension $m-1$. There are $O(n^m)$ m-sectors and, by the codimension condition, m must be less than or equal to 7. The complexity of computing the skeleton of this stratification is $O(n^7 \log n)$.

While its worst-case bounds are poor, the actual performance of the algorithm is expected to be much better, because V approximates the Euclidean Voronoi diagram, as shown in Figs. 4 and 5. The evidence for this is that the complexity of the Euclidean Voronoi diagram for a set of n points in r dimensions is $O(n^{\lfloor (r+1)/2 \rfloor}$, and the Euclidean Voronoi diagram for disjoint line segments in the plane has linear size.

This conjecture is supported by some experimental evidence. We have implemented an algorithm for constructing the simplified Voronoi diagram for the following configuration spaces: $\Re^2$, the case of an arbitrary polygon translating in the plane amidst polygonal obstacles, and $\Re^2 \times S^1$, which allows the moving polygon to rotate as well as translate. In many cases the size of V has been observed to remain roughly linear, as in Fig. 4, which our implementation produced.

References

Canny, J. F., A Voronoi method for the piano-movers problem, *Proc. IEEE Int. Conf. Robotics and Automation*, St. Louis, MO, March 1985.

Canny, J. F., Collision detection for moving polyhedra, *IEEE Trans. PAMI* (1986a) **8**.

Canny, J. F., Constructing roadmaps of semi-algebraic sets, *Proc. Int. Workshop on Geometric Reasoning*, Oxford University, June 1986b.

Donald, B. R., Motion Planning with Six Degrees of Freedom, MIT AI-TR-791, MIT Artificial Intelligence Lab., 1984.

Lee, D. T. and Drysdale, R. L., Generalization of Voronoi diagrams in the plane, *SIAM J. Comput.* **10** (1981), 73-87.

Lozano-Pérez, T., Spatial planning: a configuration space approach, *IEEE Trans. Comput.* **32** (1983), 108-120.

Lozano-Pérez, T. and Wesley, M., An algorithm for planning collision-free paths among polyhedral obstacles, *Comm. ACM* **22** (1979), 560-570.

Ó'Dúnlaing, C., Sharir, M., and Yap, C., Generalized Voronoi Diagrams for Moving a Ladder: I Topological Analysis, Robotics Lab. Tech. Report No. 32, NYU-Courant Institute, (1984).

Ó'Dúnlaing, C., Sharir, M., and Yap, C., Generalized Voronoi Diagrams for Moving a Ladder: II Efficient Construction of the Diagram, Robotics Lab. Tech. Report No. 33, NYU-Courant Institute (1984).

Ó'Dúnlaing, C. and Yap, C., A retraction method for planning the motion of a disc, *J. Algorithms*, **6** (1985), 104-111.

Schwartz, J. and Sharir, M., On the "Piano Movers" Problem, II. General Techniques for Computing Topological Properties of Real Algebraic Manifolds, Report No. 41, Comp. Sci. Dept., New York University 1982.

Schwartz, J. and Yap, C. K., *Advances in Robotics*, Lawrence Erlbaum Associates, Hillside, New Jersey, 1986.

Whitney, H., Elementary structure of real algebraic varieties, *Ann. of Math.* **66** (1957), 3.

Yap, C., Coordinating the Motion of Several Discs, Robotics Lab. Tech. Report No. 16, NYU-Courant Institute (1984).

Received December 22, 1986, and in revised form October 26, 1987.

Solving the Find-Path Problem by Good Representation of Free Space

RODNEY A. BROOKS

Abstract—Free space is represented as a union of (possibly overlapping) generalized cones. An algorithm is presented which efficiently finds good collision-free paths for convex polygonal bodies through space littered with obstacle polygons. The paths are good in the sense that the distance of closest approach to an obstacle over the path is usually far from minimal over the class of topologically equivalent collision-free paths. The algorithm is based on characterizing the volume swept by a body as it is translated and rotated as a generalized cone, and determining under what conditions one generalized cone is a subset of another.

I. Introduction

THE FIND-PATH problem is well-known in robotics. Given an *object* with an initial location and orientation, a goal location and orientation, and a set of *obstacles* located in space, the problem is to find a continuous path for the object from the initial position to the goal position which avoids collisions with obstacles along the way. This paper presents a new representation for free space as natural "freeways" between obstacles, and a new algorithm to solve find-path using that representation. The algorithm's advantages over those previously presented are that it is quite fast and it finds good paths which generously avoid obstacles rather than barely avoiding them. It does not find possible paths in extremely cluttered situations, but it can be used to provide direction to more computationally expensive algorithms in those cases.

A. Approaches to the Find-Path Problem

A common approach to the find-path problem, used with varying degrees of sophistication, is the configuration space approach. Lozano-Pérez [5] gives a thorough mathematical treatment of configuration space. The idea is to determine those parts of free space which a reference point of the moving object can occupy without colliding with any obstacles. A path is then found for the reference point through this truly free space. Dealing with rotations turns out to be a major difficulty with the approach, requiring complex geometric algorithms which are computationally expensive.

Manuscript received June 11, 1982; revised September 28, 1982. This paper was originally presented at and appeared in the Proceedings of the Second National Conference of the American Association for Artificial Intelligence, Pittsburgh, PA, August 1981. This work was supported in part by the Office of Naval Research under Contract N00014-81-K-0494, and in part by the Advanced Research Projects Agency under Office of Naval Research Contract N00014-80-C-0505.

The author is with The Artificial Intelligence Laboratory, Massachusetts Institute of Technology, 545 Technology Square, Cambridge, MA 02139.

Conceptually one can view configuration space as shrinking the object to a point while at the same time expanding the obstacles to the shape of the moving object. This approach works well when the moving object is not allowed to rotate. If it can rotate, then the grown obstacles must be embedded in a higher dimensional space—one extra dimension for each degree of rotational freedom. Furthermore the grown obstacles have nonplanar surfaces, even when the original problem was completely polygonal or polyhedral. Typically implementors have approximated the grown obstacles in order to be able to deal with rotations.

Moravec [6] had to solve the find-path problem in two dimensions. He bounded all obstacles and the moving object by circles. Then the grown obstacles were all perpendicular cylinders and the problem could be projected back into two dimensions where rotations could be ignored. This method missed all paths that required rotational maneuvering. Lozano-Pérez [5] split the rotation range into a fixed number of slices, and within each slice bounded the ground obstacles by polyhedra. Earlier Udupa [8] had used much cruder bounding polyhedra in a similar way.

Recently Brooks and Lozano-Pérez [3] developed a method for computing more directly with the curved surfaces of the grown obstacles. The algorithm can be quite expensive although given enough time and space it can find a path if one exists.

B. Good Representations Simplify Find-Path

The algorithm presented in this paper is based on a different idea, i.e., that of using representations of space which capture the essential effects of translating and rotating a body through space. Free space is represented as overlapping generalized cones since they are descriptions of swept volumes. The cones provide a high-level plan for moving the object through space. The volume swept by the particular object as it is translated and rotated is characterized as a function of its orientation. Find-path then reduces to comparing the swept volume of the object with the sweepable volumes of free space. This is done by inverting the characterization of the volume swept by an object, to determine its valid orientation as it moves through a generalized cone. Throughout this paper we restrict our attention to the two-dimensional problem where the object to be moved is a convex polygon and the obstacles are represented as unions of convex polygons.

II. Describing Free Space as Generalized Cones

Generalized cones are a commonly used representation of volume for modeling objects in artificial intelligence-based computer vision systems. They were first introduced by Binford [2]. A generalized cone is formed by sweeping a two-dimensional *cross section* along a curve in space, called a *spine*, and deforming it according to a *sweeping rule*.

We will consider a two-dimensional specialization of generalized cones (although we will still refer to them as volumes). The spines will be straight and the cross sections will be line segments held perpendicular to the spine with left and right radii. The sweeping rule will independently control the magnitudes of the left and right radii as piecewise linear functions.

Free space is represented as overlapping generalized cones. Natural "freeways," elongated regions through which the object might be moved, are represented as generalized cones. The generalized cones overlap at intersections of these natural freeways. Fig. 1 illustrates a few of the generalized cones describing the free space around three obstacles in a confined workspace.

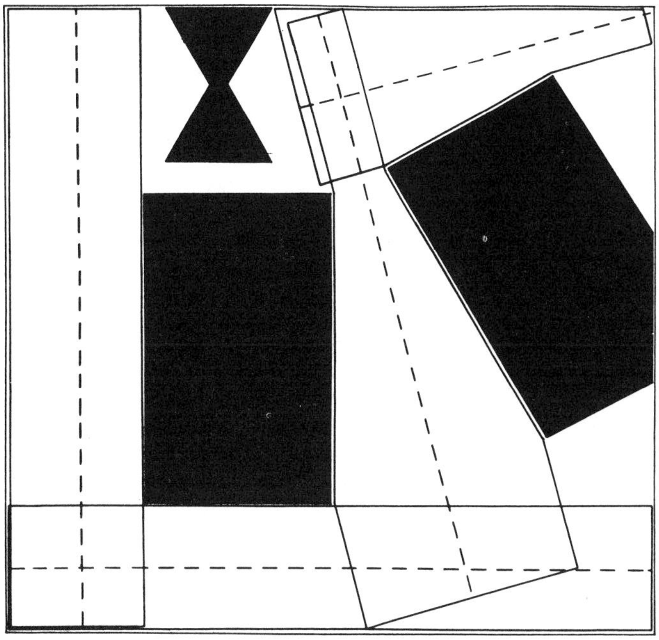

Fig. 1. Generalized cones generated by two obstacles and workspace boundary.

A. Candidate Generalized Cones

A representation of free space is constructed by examining all pairs of edges of obstacle polygons. If the edges define a natural "freeway" through space they are used to construct a generalized cone. Each edge has a "free" side and a "full" side, where the "free" side is the outside of the particular polygon with which it is associated. Each edge has an outward pointing normal, pointing into the "free" side. Two edges are accepted as defining a *candidate generalized cone* if they meet the following requirements. 1) At least one vertex of each edge should be on the "free" side of the other, and 2) the dot product of the outward pointing normals should be negative. The second condition ensures that the "free" sides of the two edges essentially face each other. Thus in this initial stage there is a complexity factor of $O(n^2)$, where n is the number of edges of obstacle polygons.

Given a candidate pair of edges, a spine is constructed for the generalized cone. It is the bisector of the space which is on the "free" side of both edges. (Thus if the edges are parallel the spine is parallel and equidistant to them both, or if the edges are not parallel, then the spine bisects the angle they form.) The generalized cone occupies the volume between the two defining edges. At each vertex of the two edges (if the vertex is on the "free" side of the other edge) the cone is extended parallel to the spine.

The generalized cone so defined may not lie entirely in free space. There may be other obstacles which intersect it. Each obstacle is compared to the generalized cone. A polygon can be intersected with the cone in time $O(n)$, where n is the number of edges of the polygon. If the intersection is empty, then nothing further need be done. If not, then the intersection is projected normally onto the spine of the generalized cone. This is illustrated in Fig. 2. Again this is an $O(n)$ operation in the number of vertices (and hence edges). The result of comparing all obstacles to the generalized cone is a set of regions of the spine where there is no obstacle which intersects the cone in a slice normal to the spine. Each disjoint slice which includes parts of the original two edges is then accepted as a generalized cone describing part of free space. Clearly the complete operation of describing free space is at most $O(n^3)$ in the number of edges in the obstacle polygons.

B. Representation Used for Generalized Cones

Fig. 3 shows the complete representation used for cones describing parts of free space. There is a straight spine, parameterized over the range $t \in [0, l]$ where l is the length of the cone. If the sides of the cone are not parallel to the spin, then $t = 0$ corresponds to the wider end.

On both the left and right the maximal radii achieved over the length of the cone occurs at $t = 0$, and are denoted b_l and b_r, respectively, describing the "big" end of the cone. The minimal radii achieved occur at $t = l$ and are denoted s_l and s_r, describing the "small" end of the cone. If $b_l = s_l$ and $b_r = s_r$, then the two sides of the cone are parallel to the spine. If not, then there is a symmetric thinning of the cone (which may start, and end, at different values of t on the left and right), where the left and right radii of the thinning parts of the cone are both given by the expression $mt + c$ where m and c are constants. Note that it is always the case that $m \leqslant 0$ and $c \geqslant 0$.

In summary the seven constants l, b_l, b_r, s_l, s_r, m, and c completely specify the shape and size of the generalized cone. In addition its location and orientation must be determined concurrently with computing these parameters.

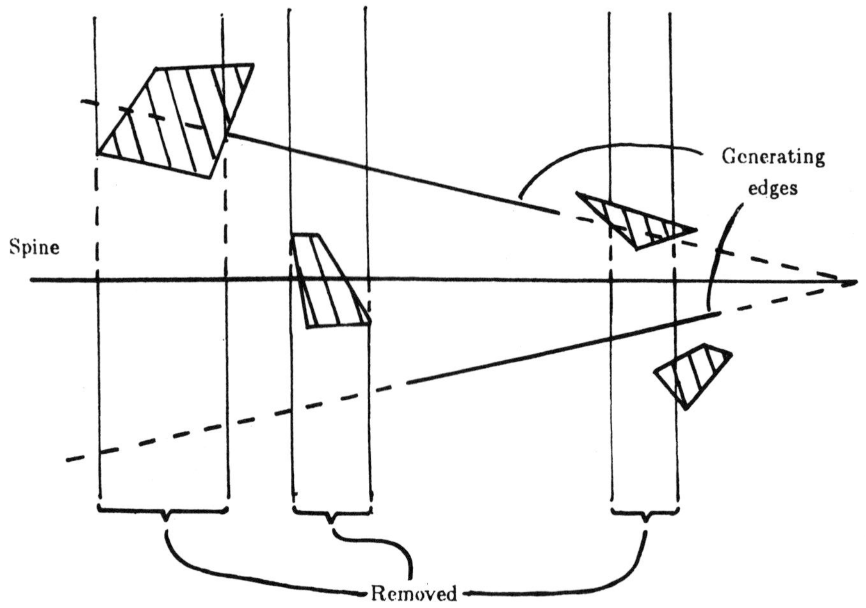

Fig. 2. Slices of generalized cone removed due to obstacles.

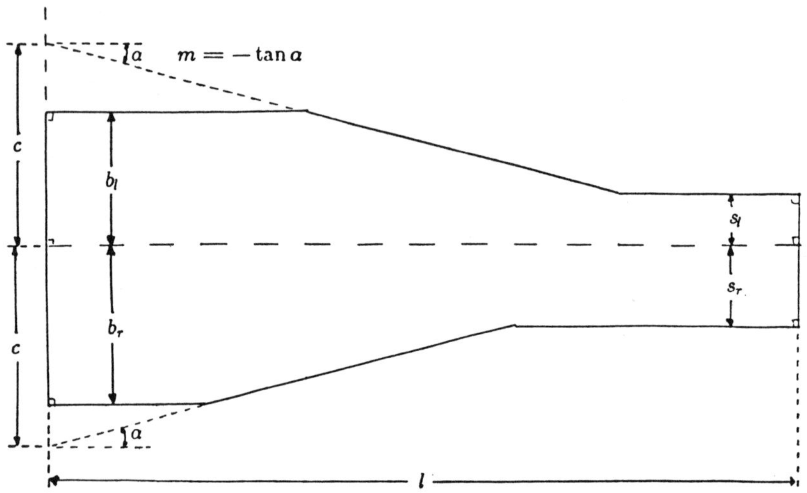

Fig. 3. Generalized cone defined by l, b_l, b_r, s_l, s_r, m, and c.

III. Determining Legal Orientations

Let the moving object be called A. It is a convex polygon and has vertices $a_1, a_2, \cdots, a_n$. Choose an origin and x and y axes. Let d_i be the distance of vertex a_i from the origin. For optimal performance of the find-path algorithm, the origin should be chosen so that $\max_{1 \leqslant i \leqslant n} d_i$ is minimized over possible origins. The direction of the x axis can be chosen arbitrarily.

Orientations relative to A are defined in terms of an angle relative to A's x axis. Thus a point's angle relative to A is the angle made by a ray from the origin to the point, relative to the x axis. A direction relative to A is the angle made by a vector in that direction with A's x axis. Let $\eta_1, \eta_2, \cdots, \eta_n$ be the angles of the vertices $a_1, a_2, \cdots, a_n$.

A. Radius Function of a Polygon

Consider first the problem of determining the volume swept by a polygon as it is moved in a straight line with fixed orientation. Let the object be moving in direction θ. The swept volume depends on the object's cross section perpendicular to that angle. The cross section can be broken into two components, that on the left of the direction of movement and that on the right. (See Fig. 4.)

Define a *radius function* $R(\xi)$ as the infimum of the distance from the origin that a line normal to a ray at angle ξ can be without intersecting the interior of A. (See Fig. 5). (The radius function is closely related to the support of a convex polygon, e.g., see [1].) The magnitudes of the left and right cross sections of Fig. 4 can then be denoted $R(\theta + \pi/2)$ and $R(\theta - \pi/2)$, respectively.

Fig. 6 shows the geometric construction of $R(\xi)$ for a given object A and angle ξ. The darkened outline in Fig. 7 shows the radius function R in polar coordinates for the same object. Thus R can be defined by

$$R(\xi) = \max_{1 \leqslant i \leqslant n} d_i \cos(\xi - \eta_i).$$

The major interest in functions of this form will be in their

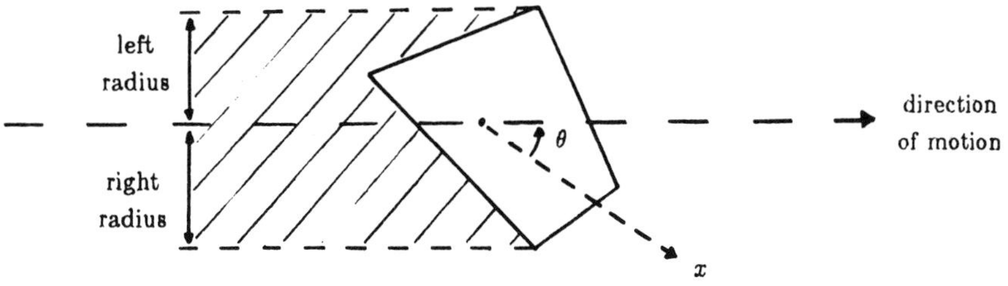

Fig. 4. Volume swept by object during pure translation.

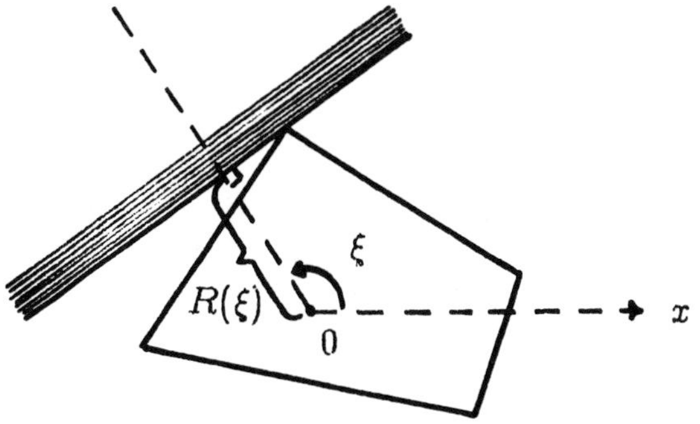

Fig. 5. Definition of $R(\xi)$.

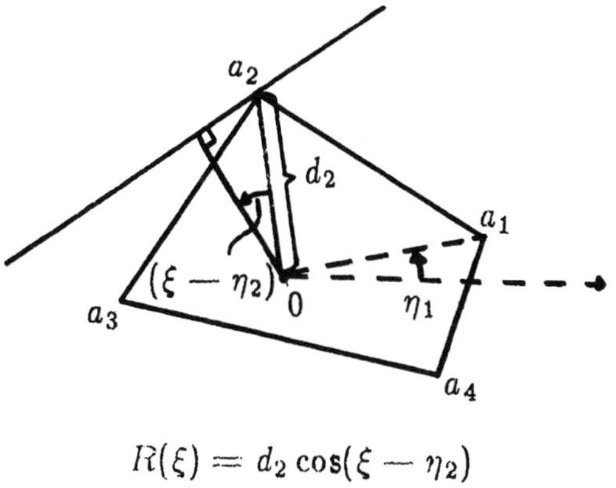

Fig. 6. Geometric construction of $R(\xi)$.

Fig. 7. $R(\xi)$ in polar coordinates.

inverse images of intervals of the form $(-\infty, r]$. Let

$$R^{\perp}(r) = \{\xi | R(\xi) \leqslant r\}.$$

Thus $R^{\perp}(r)$ is the range of angles at which the cross section of object A is no more than r. The inverse image can be easily computed by using the two values possible for $\arccos(r/d_i)$ for each i to form an interval containing η_i, and subtracting it from the interval $[0, 2\pi]$. We restrict our attention to moving objects which are convex polygons precisely because their radius functions can be so easily inverted.

If we write $R_{\alpha}(\xi) = R(\xi + \alpha)$, then the legal orientations of A relative to the direction of its movement down the center of a strip of diameter d can be written as

$$L = R^{\perp}_{\pi/2}(d/2) \cap R^{\perp}_{-\pi/2}(d/2).$$

Note that while L may consist of more than one interval, the set of orientations taken by A in some trajectory along the length of the cone must form a connected set—i.e., it must be a single interval, and so must be contained in a single interval of L.

Finally note that the sum of two radius functions has the same form as a radius function, and so the " $\perp$ " operation can be computed just as easily on the sums of such functions as on single functions, i.e., with R as above and

$$S(\xi) = \max_{1 \leqslant i \leqslant m} e_i \cos(\xi - \mu_i),$$

then

$$\begin{aligned} &[R + S](\xi) \\ &\quad = R(\xi) + S(\xi) \\ &\quad = \max_{1 \leqslant i \leqslant n, 1 \leqslant j \leqslant m} \left(d_i \cos(\xi - \eta_i) + e_j \cos(\xi - \mu_j)\right) \end{aligned}$$

and each term inside the "max" can be written in the form $f_{ij} \cos(\xi - \nu_{ij})$.

B. Bounding Polygons with an Appropriate Rectangle

In general we wish to find the legal orientations for object A when its origin is at a point with spine parameter t on the spine of a generalized cone C. A conservative set of orientations can be found by enclosing A in a rectangle which has two of its sides parallel to the spine of C. When A is oriented so that the spine has angle θ relative to it, the bounding rectangle has the following properties. Its extent forward (in the direction of increasing spine parameter) from the origin is $R(\theta)$. Its extents left and right are $R(\theta + \pi/2)$ and $R(\theta - \pi/2)$, respectively. Its extent to the rear is (rather conservatively) $d = \max_{1 \leqslant i \leqslant n} d_i$. (See Fig. 8.) The problem now is to invert this bounding operation, i.e., to find for what range of θ the bounding rectangle is within the generalized cone. That is, for each $t \in [0, l]$ we wish to find a range of valid θ, denoted $V(t)$ for orientations of the object A.

The rear of the bounding rectangle simply implies $V(t) = \phi$ outside of the interval $[d, l]$. The forward bound

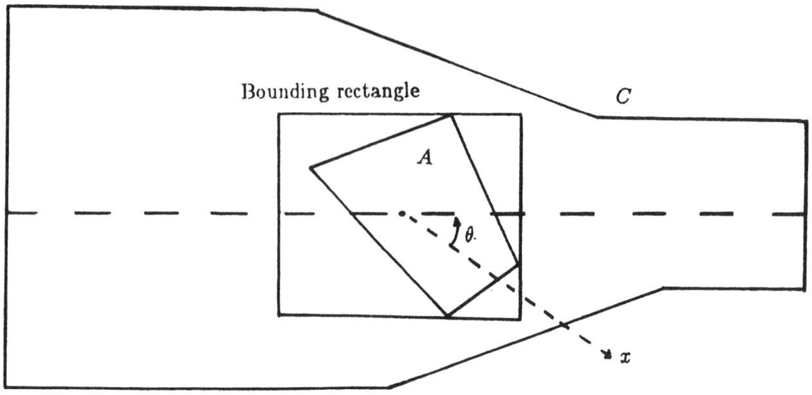

Fig. 8. Object A bounded by rectangle aligned with spine of cone C.

implies

$$V(t) \subseteq R^{\perp}(l-t).$$

The parallel sides of the "big" part of the generalized cone imply upper bounds on the width of the bounding rectangle. Thus

$$V(t) \subseteq \left(R^{\perp}(b_l) - \pi/2\right) \cap \left(R^{\perp}(b_r) + \pi/2\right).$$

Note that the terms on the right involve adding a constant to a subset of $[0, 2\pi]$ in the obvious way. Let V' be defined as

$$V'(t) = R^{\perp}(l-t) \cap \left(R^{\perp}(b_l) - \pi/2\right) \cap \left(R^{\perp}(b_r) + \pi/2\right)$$

over the interval $[d, l]$.

Subject to the above three constraints the rectangle can be as big as one which will fit in the "small" part of the generalized cone. Thus

$$\left(\left(R^{\perp}(s_l) - \pi/2\right) \cap \left(R^{\perp}(s_r) + \pi/2\right)\right) \cap V'(t) \subseteq V(t).$$

Furthermore, subject to $V'(t)$, any orientation of A is valid which results in a bounding rectangle that is within the bounds given by the decreasing boundaries of the generalized cones. Recall that at each point of the spine, with parameter t, these boundaries have both left and right radius of $mt + c$ and that $m \leqslant 0$. Thus sufficient conditions are that

$$R(\theta + \pi/2) \leqslant m \times (t + R(\theta)) + c$$
$$R(\theta - \pi/2) \leqslant m \times (t + R(\theta)) + c$$

are both true. These can be expressed as

$$\left[R_{\pi/2} - mR\right](\theta) \leqslant (mt + c)$$
$$\left[R_{-\pi/2} - mR\right](\theta) \leqslant (mt + c)$$

whence

$$\left(\left[R_{\pi/2} - mR\right]^{\perp}(mt + c) \cap \left[R_{-\pi/2} \quad mR\right]^{\perp}(mt + c)\right) \cap V'(t) \subseteq V(t).$$

In summary then, over the range $t \in [d, l]$, a valid subset of orientations for A can be expressed as

$$V(t) = R^{\perp}(l-t) \cap \left(R^{\perp}(b_l) - \pi/2\right) \cap \left(R^{\perp}(b_r) + \pi/2\right) \cap \Big(\left(\left(R^{\perp}(s_l) - \pi/2\right) \cap \left(R^{\perp}(s_r) + \pi/2\right)\right) \cup \Big(\left[R_{\pi/2} - mR\right]^{\perp}(mt + c) \cap \left[R_{-\pi/2} - mR\right]^{\perp}(mt + c)\Big)\Big)$$

and elsewhere $V(t) = \phi$. The key property of $V(t)$ is the following.

Lemma: For $t_1, t_2 \in [d, l]$, $t_1 \leqslant t_2$ implies $V(t_2) \subseteq V(t_1)$.

Note that $V(t)$ does not necessarily include all possible orientations for object A to be contained in cone C at point with parameter t along the spine. However it is the set of possibilities which the find-path algorithm described here is willing to consider. Its key property is that as stated by the lemma the set of valid orientations (amongst those to be considered) does not increase as the object is moved from the "big" end to the "small" end of a generalized cone.

IV. Searching for a Path

The algorithm to find a collision-free path proceeds as follows. First the generalized cones which describe free space are computed (this can be done in time $O(n^3)$). Next the cones are pairwise examined to determine if and where their spines intersect. Up to this stage the algorithm is independent of the object A to be moved through the environment. Each spine intersection point for each cone is now annotated with the subset of $[0, 2\pi]$, which describes the orientations for the object A in which it is guaranteed to be completely contained in the generalized cone. The final stage of the algorithm is to find a path from the initial position to the goal position following the spines of the generalized cones, and changing from cone to cone at spine intersection points. If the object is kept at valid orientations at each point on each spine, then the path will be collision free.

A. A Searchable Graph

A graph is built, then searched with the A^* algorithm (see Nilsson [7]). The cost function used is the distance traveled through space. The nodes of the graph consist of points on generalized cone spines which correspond to points of intersection, along with a single orientation subinterval of $[0, 2\pi]$ (except that intervals may wrap around from 2π to 0). Thus for an intersection point with parameter t on the spine of cone C, a node for each interval in $V(t)$ is built.

The arcs in the graph arise in two ways—those that correspond to transfer from one intersection point on a spine to another, and those that correspond to transfer from one spine to another at their common intersection point. All such candidate pairs of nodes are checked for connectivity. The lemma of Section III guarantees that it suffices to check if the orientation intervals of the two nodes, for intracone arcs, have a nonempty intersection. If so, then it is certainly possible to move the object A from one point to the other using any orientation within that intersection. For intercone pairs of nodes it also suffices to check for nonempty intersection of their orientation intervals, as the points are coincident in space. The graph can now be searched to find a path consisting of an ordered set of nodes from initial position to goal position.

B. Intermixing Rotations and Translations

If the graph search is successful it indicates a whole class of collision-free paths. It remains to chose one particular trajectory. The class of paths found share a common trajectory for the origin of the moving object. The orientation interval associated with each node of the path through the graph constrains the valid orientations at that point. It is also necessary to ensure that no invalid orientation is assumed during translation and rotation along the spine of a generalized cone.

It may well be the case that rotations are expensive, and it is worth searching the class of paths resulting from the graph search to find the optimal path in terms of least rotation necessary. This could be carried out using the A^* algorithm. Every node in the path found in the first search would be turned into a collection of nodes. Each pair of adjacent nodes in the path found above would have their orientation ranges intersected. Combine each end value of each intersection, and the initial and final orientations into a set of possible orientations. Each orientation which is valid for a particular path node, along with that node, generates a node in the new graph. Adjacency would be inherited from adjacency of generating nodes in the original path. The A^* algorithm is used to search this new graph.

Fig. 9 and 10 illustrate two paths found by a much simpler method. Rotations are restricted to points corresponding to nodes of the path, i.e., the object is held with fixed orientation during each translation. The orientation at each point is the midpoint of the orientations given by

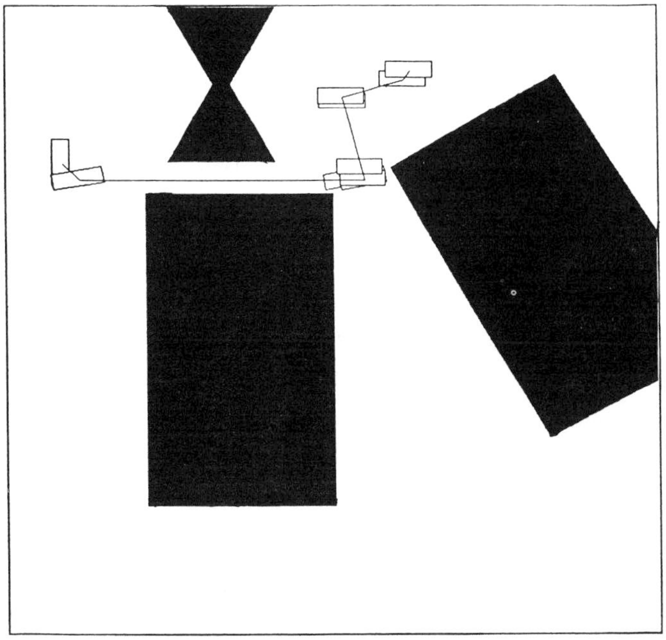

Fig. 9. Path found by algorithm.

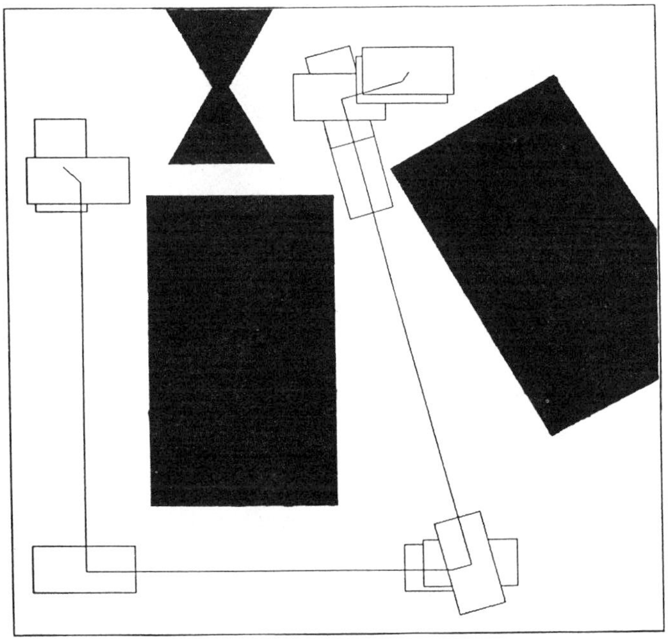

Fig. 10. Path found by algorithm.

intersecting the allowed orientation intervals of the corresponding path node, and the subsequent node. Thus traversal of intracone arcs of the graph built above are rotation free, and traversal of intercone arcs are pure rotations. The lemma of Section III guarantees that this strategy leads to valid collision-free paths.

V. Usefulness and Applicability

The major drawback of the find-path algorithm presented in this paper is that paths are restricted to follow the spines of the generalized cones chosen to represent free space. It typically does not work well in tightly constrained spaces as there are insufficient generalized cones to provide a rich choice of paths. Furthermore in a tight space the interior points of spines are always near to the spine ends. Therefore the need to extend the bounding rectangles rearward to the maximum possible distance for the moving object (so that the lemma of Section III will hold) means that typically there are very few points of intersections of generalized cones where the object's bounding rectangle in each orientation is contained within the appropriate generalized cone. This situation could be improved significantly by developing a better algorithm for decomposing free space into generalized cones. The current pruning technique is often too drastic; if an obstacle intersects a generalized cone it may be better to simply reduce the cone radius, rather than the current practice of slicing out all of the spine onto which the obstacle projects.

A. Advantages

In relatively uncluttered environments the algorithm is extremely fast. Furthermore it has the following properties which are absent from some or all of the algorithms mentioned in Section I.

1) The paths found, and the amount of computation needed, are completely independent of the chosen world coordinate system.
2) Obstacles affect the representation of free space only in their own locality. Thus additional obstacles spatially separated from a free path found when they are not present cannot affect the ability to find that free path when they are present. (Resource limited, and approximating algorithms which cut free space into rectangloid cells often suffer from this problem.)
3) Free space is represented explicitly rather than as the complement of forbidden space.
4) The paths tend to be equally far from all objects, out in truly free space, rather than scraping as close as possible to obstacles, making the instantiation of the paths by mechanical devices subject to failure due to mechanical imperfections.

We plan to use this algorithm in conjunction with an implementation of an algorithm [3] which finds paths with rotations based on a configuration-space approach. The generalized cone method will be used to find the easy parts of the path (the other method is computationally expensive both for easy and hard paths) leaving the more expensive method for the hard parts. The algorithm presented here can also provide some direction for the hard parts as it can quickly compute all topologically possible paths.

B. Three Dimensions

Clearly the eventual goal of this work is to develop algorithms which solve the find-path problem in three dimensions, with three rotational degrees of freedom for each link of general articulated devices. In this section we discuss approaches to extending the presented algorithm to the case of a single convex polyhedron moving through a space of polyhedral obstacles.

Three-dimensional generalized cones can be constructed by considering all triples of faces of obstacle polyhedra. The construction algorithm (including intersection with obstacle polyhedra) has complexity $O(n^4)$ in the number of obstacle faces. The generalized cones so constructed will have triangular cross sections. It is not clear how to pairwise intersect the generalized cones, as in general their spines will not intersect. The "path-class" idea presented below may solve this problem.

Rather than a bounding rectangle, a right triangular prism should be used where the triangle is similar to the cross section of the generalized cone being traversed. If more than one rotational degree of freedom is assumed there may be problems inverting this bounding volume. Rather than $V(t)$ having a set of intervals as a value, it will have a set of three-dimensional polyhedra.

For both the two-dimensional and three-dimensional problems it is worth investigating using paths other than the spine through a generalized cone. Then V is parameterized in terms of t and the two end points of a whole class of paths through individual cones. This increases the complexity of the search but will lead to more paths than simply using the spine.

C. Complexity Issues

In discussions with Tomás Lozano-Pérez it became evident that there exists an $O(n^2 c^{\sqrt{\log n}})$ algorithm which can find the same generalized cones as does the $O(n^3)$ algorithm presented here. The spines used for generalized cones correspond to the Voronoi boundaries between obstacles (see Drysdale [4] for a complete solution to the problem), and they can be found in $O(nc^{\sqrt{\log n}})$ time. There remains a factor of $O(n)$ to do the intersection of the hypothesized generalized cones with other obstacles. It is unlikely that such an algorithm would perform better than the $O(n^3)$ algorithm presented for practical sized problems.

The complexity of pairwise intersecting the generalized cones was not addressed in the body of the paper. An upper bound of $O(n^4)$ can easily be obtained as there are at most $O(n^2)$ generalized cones to be considered. However it seems hard to find a constant where cn^2 generalized cones can be achieved constructively as the number of obstacle polygons is increased. For any c it seems that space eventually tends to become too cluttered for all the cones to be constructed. Thus it seems likely that a better bound than $O(n^4)$ exists. (In fact the Voronoi complexity above suggests a bound of $O(n^2 c^{2\sqrt{\log n}})$.)

Acknowledgment

The presentation of this paper has benefitted greatly from careful readings of drafts by J. Michael Brady and Tomás Lozano-Peréz.

References

[1] Russell V. Benson, *Euclidean Geometry and Convexity*. New York: McGraw-Hill, 1966.

[2] Thomas O. Binford, "Visual perception by computer," presented at IEEE Systems, Science, and Cybernetics Conf., Miami, FL, Dec. 1971.

[3] Rodney A. Brooks and Tomás Lozano-Pérez, "A subdivision algorithm in configuration space for findpath with rotation," M.I.T. Tech Rep. AIM-684, in preparation, 1982.

[4] R. L. (Scot) Drysdale, "Generalized Voronoi diagrams and geometric searching," Stanford, CS Rep. STAN-CS-79-705, Stanford, CA, Jan. 1979.

[5] Tomás Lozano-Pérez, "Automatic planning of manipulator transfer movements, *IEEE Trans. Syst., Man, Cybern.*, vol. SMC-11, pp. 681–698, 1981.

[6] Hans P. Moravec, "Obstacle avoidance and navigation in the real world by a seeing robot rover," Stanford, Univ. Tech Rep., AIM-340, Sep. 1980.

[7] Nils J. Nilsson, *Problem-Solving Methods in Artificial Intelligence*. New York: McGraw-Hill, 1971.

[8] Shriram M. Udupa, "Collision detection and avoidance in computer controlled manipulators," Proc. IJCAI-5, MIT, Cambridge, MA, Aug. 1977, pp. 737–748.

5. Sensing Strategies

If a little knowledge is dangerous, where is the man who has so much as to be out of danger?

T. H. Huxley, *On Elemental Instruction in Physiology* (1877)

A map may not always be complete or remain current. An important problem then becomes how best to collect data to update the map database. Suppose the map is to be used for optimal path planning, such as finding shortest paths, and, initially, the environment is unknown. Then the map should contain the locations of as many of the features, for example, obstacles or walls, as possible so that potential routes are not missed. Often, the only method available for gathering data is to obtain local information from the vehicle's sensors, as opposed to obtaining a global view of the area from a remote sensor such as a satellite. This raises the question of how best to route the vehicle to ensure that the entire area is covered. Also, some computation is usually necessary to extract the map information from the sensor readings. It may therefore be necessary to minimize the number of sensor readings so as not to overload the computational resources of the vehicle. Perhaps more importantly, it may be necessary to construct a map with the minimum of time and energy. This may be particularly true for interplanetary exploration in which energy considerations are a constant worry. Under these conditions, a strategy for deciding where to acquire additional sensor data so as to maximize the amount of new information becomes important. The topic of sensing strategies for data acquisition is still very much in its infancy. However, some results have been obtained, and we suspect that this topic may become increasingly important as robot vehicles become more autonomous.

The class of problems referred to as *art gallery problems* is a collection of problems dealing with placing sensors in a workspace so as to ensure that the combined views of the sensors covers the entire workspace, see [16]. The motivation for these problems is the question of where to place guards in an art gallery so that each location in the gallery can be viewed by at least one guard. Usually the guard is modeled as a fixed point that can see in every direction and the region that is visible to a guard is constrained by the walls of the gallery but not by distance. The gallery is modeled as a simple polygon, that is, a polygon without holes. Chvatal [8] proves that the problem can always be solved using at most $\lfloor n / 3 \rfloor$ guards in a polygonal gallery with n walls. O'Rourke [17] has shown that $\lfloor n / 4 \rfloor$ guards suffice if each guard is allowed to move along a wall rather than being fixed at one point. Although it is assumed in these problems that the workspace is known in advance, the results obtained can be instructive in the case in which the workspace is unknown. For example, the results for fixed position guards show that the guards' positions can be constrained to be at corners of the polygon. Suppose that the locations of corners can easily be extracted from the vehicle's sensory data. Then one possible sensor acquisition strategy would be to move the vehicle from corner to corner, using the sensor to obtain a radial map from each corner. Clearly, the entire workspace could then be mapped using the n radial maps. This naive method of gathering data uses a number of partial maps that is only a factor of three off the theoretical minimum number that suffices.

Fisk [10] was able to demonstrate a simpler proof of Chvatal's bound on the number of guards that suffice to cover an area. Fisk's proof was used by Avis and Toussaint [2] to develop an efficient algorithm to actually compute placements of guards capable of covering the gallery. Unfortunately, although the algorithm is guaranteed never to produce more than $\lfloor n / 3 \rfloor$ guard locations, it cannot guarantee that the number of guard locations is optimal. The question of computing the minimum number of guards necessary to cover a gallery is studied in the paper by Lee and Lin. They show that several variations on the problem are all NP-complete.[1] Therefore (if P ≠ NP), there is no polynomial time algorithm to compute an optimal strategy for placing the guards, even if the gallery is a completely known simple polygon. In the case of a vehicle moving in an unknown environment, it is unlikely that a computationally feasible method can be devised to compute an optimal strategy for taking sensor readings. However, this does not exclude the possibility of heuristic methods that work well in practice.

Consider the problem of finding a "good" route for the vehicle to follow that guarantees that at the end of the route the entire workspace has been mapped. Suppose that for any fixed position of the vehicle, the onboard sensor is capable of obtaining a radial map of the workspace visible from that position, where a point in the workspace is considered visible if the ray from the sensor to the point is not blocked by any obstacle. Then the goal is to find a route within the workspace such that if the vehicle traveled along that route and stopped periodically along the route to obtain radial maps, the combination of all the acquired maps would be a map of the entire workspace. In [7], the problem of finding the shortest such route is shown to be NP-hard, even if the workspace is completely known and is a polygon containing polygonal obstacles. Thus again, there is little hope of finding an optimal route in the case in which the environment is completely unknown initially. Lumelsky et al. [14] studied this problem in the case in which the sensor had a limited range and the measure of the route length was stated in terms of the perimeters of the objects in the two-dimensional workspace. They proposed a scheme for mapping the environment such that the length of the route followed by the vehicle is bounded by a quadratic in the perimeters of the obstacles.

Sensing strategies also have applications in terms of location estimation. A great deal of study, especially in the vision research community, has focused on the problem of object recognition; see [5]. The problem is to develop a strategy for obtaining a minimal sample of (noisy) sensor data that is sufficient to recognize or determine the pose, that is, the position and orientation, of an object. The basic assumption made in this work is that the object to be recognized or located is one of a collection of known objects. Therefore the techniques involved are probably not applicable to the problem of mapping a completely arbitrary environment. However, the ideas should be applicable to determining which of several candidate landmarks is currently being sensed. This is potentially helpful in aiding navigation, see Chapter 3. Determining the pose of an object, given the sensor data and the known pose of the sensor, is the dual problem of position estimation, in which the goal is to determine the pose of the sensor (equivalently, the pose of the vehicle), using the known pose of some landmarks and a sample of sensor data. Therefore, the techniques developed for the problem of minimizing the number of sensor readings needed to determine the pose of an object may be applicable to the position location problem. The paper by Grimson describes a robust sensing strategy for the recognition (or location) problem for objects that are modeled as polyhedrons. The reader is also directed to Cameron and Durrant-Whyte [6] who describe a Bayesian approach to optimal sensor placement.

Another case of object recognition or location occurs when the object being

[1]See [11] for the meaning of NP-complete.

sensed is completely unknown. That is, the object is not restricted to be one of a finite number of objects. This problem is related to the mapmaking problem in that the issue is to obtain a description of an object (or the environment) using as few sensor readings as possible. Cole and Yap [9] have shown that for a tactile finger probe sensor, at least $3n - 1$ probes are necessary to obtain a description of an arbitrary, n-sided convex polygon. In other words, if the object is an n-sided convex polygon and fewer than $3n - 1$ probes are used, then there is insufficient information to describe the object reliably. Note that this lower bound fails to hold if the number of sides of the polygon is known in advance. Cole and Yap also give an algorithm to choose probe directions adaptively and show that the algorithm uses $3n$ different directions and that these suffice to determine the object. A number of other papers, for example, [1, 4, 12, 15], have studied similar problems. The paper by Dobkin et al. in this chapter considers the problem of determining a convex polytope in d-dimensional space using a number of different probe models. Although these strategies produce good upper bounds on the number of probes sufficient to determine the body, the amount of computation needed to decide the next probe direction adaptively may be high. However, this may well be offset by the time and energy saved in reducing the amount of probing required.

The Kalman filter can also be used to provide a strategy for sensing based on the prediction phase of the filter cycle [3, 13]. Finally, Payton's work on circulation maps [18] offers information regarding position accuracy requirements upon which a sensing strategy could be based.

References

1. Alevizos, P., Boissonnat, J., and Yvinec, M. An Optimal $O(n \log n)$ Algorithm for Contour Reconstruction from Rays. In *Proceedings of the 3rd Annual Symposium on Computational Geometry*, ACM, New York, 1987, pp. 162-170.

2. Avis, D., and Toussaint, G. An Efficient Algorithm for Decomposing a Polygon into Star-Shaped Polygons. *Pattern Recog. 13* (1981), 395-398.

3. Ayache, N., and Faugeras, O.D. Maintaining Representations of the Environment of a Mobile Robot. *IEEE Trans. Robotics Automation RA-5*, 6 (1989), 804-819.

4. Bernstein, H. Determining the Shape of a Convex n-sided Polygon by Using $2n + k$ Tactile Probes. Robotics Research Technical Report 29, NYU-Courant Institute, New York, June 1984.

5. Besl, P., and Jain, R. 3-Dimensional Object Recognition. *ACM Comput. Surv. 17*, 1 (1985), 75-154.

6. Cameron, A., and Durrant-Whyte, H. A Bayesian Approach to Optimal Sensor Placement. OUEL 1759/89, Oxford University, 1989.

7. Chin, W., and Ntafos, S. Optimum Watchman Routes. In *Proceedings of the 2nd Annual Symposium on Computational Geometry*, ACM, New York, 1986, pp. 24-33.

8. Chvatal, V. A Combinatorial Theorem in Plane Geometry. *J. Comb. Theory, Ser. B. 18* (1975), 39-41.

9. Cole, R., and Yap, C. Shape from Probing. Robotics Research Technical Report 15, NYU-Courant Institute, New York, Dec. 1983.

10. Fisk, S. A Short Proof of Chvatal's Watchman Theorem. *J. Comb. Theory, Ser. B. 24* (1978), 374.

11. Garey, M., and Johnson, D. *Computers and Intractability: A Guide to the Theory of NP-Completeness.* W.H. Freeman and Co., San Francisco, Calif., 1979.

12. Gaston, P., and Lozano-Pérez, T. Tactile Recognition and Localization Using Object Models: The Case of Polyhedra on a Plane. *IEEE Trans. Pattern Anal. Machine Intell. PAMI-6*, 3 (1984), 257-265.

13. Leonard, J.J., and Durrant-Whyte, H.F. A Unified Approach to Mobile-Robot Navigation. (to be published).

14. Lumelsky, V., Mukhopadhyay, S., and Sun, K. Sensor-Based Terrain Acquisition: A "Seed Spreader" Strategy. In *Procedings of the IEEE/RSJ International Workshop on Intelligent Robots and Systems '89*, IEEE, New York, 1989, pp. 62-67.

15. Natarajan, B. On Detecting the Orientation of Polygons and Polyhedra. In *Proceedings of the 3rd Annual Symposium on Computational Geometry*, ACM, New York, 1987, pp. 147-152.

16. O'Rourke, J. *Art Gallery Theorems and Algorithms.* Oxford University Press, New York, NY, 1987.

17. O'Rourke, J. Galleries Need Fewer Mobile Guards: A Variation on Chvatal's Theorem. *Geometicae Dedicata 14* (1983), 273-283.

18. Payton, D. Circulation Maps: A Resource for Identifying Position Accuracy Requirements. In *Proceedings of the IEEE/RSJ International Workshop on Intelligent Robots and Systems*, IEEE/RSJ, New York, 1989, pp. 448-455.

Computational Complexity of Art Gallery Problems

D. T. LEE, SENIOR MEMBER, IEEE, AND ARTHUR K. LIN, MEMBER, IEEE

Abstract—We study the computational complexity of the art gallery problem originally posed by Klee, and its variations. Specifically, the problem of determining the minimum number of vertex guards that can see an n-wall simply connected art gallery is shown to be NP-hard. The proof can be modified to show that the problems of determining the minimum number of edge guards and the minimum number of point guards in a simply connected polygonal region are also NP-hard. As a byproduct, the problem of decomposing a simple polygon into a minimum number of star-shaped polygons such that their union is the original polygon is also shown to be NP-hard.

I. Introduction

THE ART GALLERY problem of determining how many guards are sufficient to see every point in the interior of an n-wall art gallery room was posed by Klee [11]. Conceptually, the room is a simple polygon P with n vertices, and the guards are stationary points in P that can see any point of P connected to them by a straight line segment that lies entirely within P. It has been proven by Chvatal [6] that $\lfloor n/3 \rfloor$ guards are always sufficient, and this number is the best possible in some cases. Fisk [9] later found a simpler proof which lends itself to an $O(n \log n)$ algorithm developed by Avis and Toussaint [3] for locating these $\lfloor n/3 \rfloor$ stationary guards. If the polygon is rectilinear, that is, the edges of the polygon are either horizontal or vertical, Kahn *et al.* [12] have shown that $\lfloor n/4 \rfloor$ guards are sufficient and sometimes necessary. O'Rourke [14] later gave a completely different and somewhat simpler proof of this result. Sack [17] and Edelsbrunner, *et al.* [7] have, based on the results of [12] and [14], respectively, devised an $O(n \log n)$ algorithm for locating these $\lfloor n/4 \rfloor$ guards.

Presumably, the guards can be arbitrarily placed in the interior of the art gallery. However, the aforementioned results indicate that the guards can be restricted to the vertices of the polygon without affecting the outcome. Hence we shall refer to this problem as the *vertex guard problem* and the former as the *point guard problem*. The vertex guard problem can be treated as a polygon decomposition problem in which a polygon is to be decomposed according to vertex visibility. For polygon decompositions into various types of "primitives" and their complexities see, for example, [2], [5], [8], and [13]. In this paper we address the computational complexity of the *minimum vertex guard problem*, that is, determining the minimum number of vertex guards for an n-edge simple polygon, and we show that this problem and the *minimum point guard problem* are NP-hard.

In [4] Avis and Toussaint studied and provided an $O(n)$ algorithm for the following edge visibility problem, that is, given an n-edge simple polygon P and a guard, called an *edge guard*, patrolling along an edge of P, decide if every point of P can be seen by the guard. In the context of the art gallery problem, O'Rourke [15] showed that $\lfloor n/4 \rfloor$ such mobile guards, where the guards can move along fixed line segments (the edges of the polygon, for example) are always sufficient and sometimes necessary. We shall show that the *minimum edge guard problem*, that is, determining the minimum number of edge guards needed to see any point of P for an n-edge simple polygon, is also NP-hard. For definitions of NP-hardness and NP-completeness, refer to [10].

The terminology and notation will be given in Section II, and in Sections III and IV the main results of NP-hardness of the aforementioned problems are presented.

II. Definitions and Notation

A simple polygon $P = (v_0, v_1, \cdots, v_n)$ is a closed plane figure whose boundary bd(P) is composed of straight line edges (v_i, v_{i+1}), $i = 0, 1, \cdots, n-1$, $v_0 = v_n$, and where no two nonconsecutive edges intersect. The edges of P are oriented in the counterclockwise sense, that is, when they are traversed, the interior of P always lies to the left. If P contains holes that are themselves represented as simple polygons (with no holes), then P is said to be *multiply connected*; if P contains no holes, then it is said to be *simply connected*. Unless otherwise specified, the term polygon refers to a simply connected simple polygon, and P refers to the region enclosed in bd(P).

Two points p and q of polygon P are said to be visible from each other if the line segment $\overline{pq}$ connecting p and q lies completely in P. A point p of P is said to be visible from an edge (u, v) of P if there exists a point q on (u, v) such that p and q are visible. For any point v of P the

Manuscript received May 24, 1984; revised June 17, 1985. This work was supported in part by the National Science Foundation under Grant MCS8342682.

D. T. Lee is with the Department of Electrical Engineering and Computer Science, Northwestern University, Evanston, IL 60201.

A. K. Lin was with the Department of Electrical Engineering and Computer Science, Northwestern University, Evanston, IL. He is now with Gerber System Technology, 40 Gerber Road East, South Winsor, CT 06074.

IEEE Log Number 8406631.

v-cover is the locus of points of P that are visible from v, and the (u, v)*-cover*, where (u, v) is an edge of P, is defined similarly. The v-cover is also known as the *visibility polygon* for v in P and the (u, v)-cover is known as the *edge-visibility polygon* for (u, v) in P (see, for example, [4]). A polygon P is *star-shaped* if a point v in P exists such that the v-cover is the polygon P. Thus any v-cover for any point v in P is a star-shaped polygon. The kernel of a polygon P is the region in P such that for any point q in P, q is visible from any point in this region.

III. Vertex Guard and Edge Guard Problems

Let V denote a certain subset of vertices of P. We say that a *V-cover* of P exists or P is *coverable* by V if a subset $T \subseteq V$ of vertices exists such that $\bigcup_{v \in T} v\text{-cover} = P$. T is said to be a *minimum cover* if $|T|$ is minimum among all V-covers of P.

We are concerned with the following problems. Given P and V, we must determine if P is coverable by V and, if so, find a minimum cover. The former problem can be solved in polynomial time, since for any v in V the v-cover and the union of any two covers (polygonal regions) can be computed in polynomial time. The latter, however, is NP-hard as we shall show.

An application of this problem is the following. We have a region P to be monitored by a number of guards that are placed at positions specified in V, and we want to find if the entire region of P is indeed covered by the guards, and, if so, we want to minimize the number of guards needed.

Vertex Guard Problem of a Polygonal Region

Instance: A polygonal region $P = (v_0, v_1, \cdots, v_{n-1}, v_n)$ and a positive integer $K < n$ exist.

Question: Does a subset $T \subseteq V$ exist with $|T| \le K$, such that

$$\bigcup_{v_i \in T} v_i\text{-cover} = P$$

where $V = \{v_0, v_1, \cdots, v_{n-1}\}$?

Proposition 1: The vertex guard problem for simply connected polygonal regions (VGSCP) is NP-complete.

Proof: It is easy to see that VGSCP is in NP since a nondeterministic algorithm needs only to guess a subset $V' \subseteq V$ of K vertices and check in polynomial time if

$$\bigcup_{a \in V'} a\text{-cover} = P.$$

To show NP-completeness we reduce the following NP-complete problem to the VGSCP problem.

Boolean Three Satisfiability (3SAT)

Instance: A set $U = \{u_1, u_2, \cdots, u_n\}$ of Boolean variables and a collection $C = \{c_1, c_2, \cdots, c_m\}$ of clauses over U exist such that $c_i \in C$ is a disjunction of precisely three literals.

Question: Does a satisfying truth assignment for C, that is, a truth assignment to the n variables in U, exist such that the conjunctive normal form $c_1 c_2 \cdots c_m$ is true?

We will show that 3SAT is polynomially transformable to VGSCP. The goal is to accept an instance of 3SAT as input and to construct in polynomial time a simply connected polygonal region P such that P is coverable by K or fewer vertices of P if and only if the instance of 3SAT is satisfiable. We first introduce some basic constructs on which the simply connected polygonal region is built and identify a number of distinguished points in this polygonal region such that no two different distinguished points can be covered by a single vertex. In the following construction the dotted lines shown in the figures indicate where these basic constructs are to be "attached" to the main polygonal region. Let the bound K used in the VGSCP be $3m + n + 1$.

Literal Patterns: The literal pattern is shown in Fig. 1. The dot shown is a distinguished point associated with the pattern. The important characteristic of the pattern is that only vertex a or vertex b can cover the entire region defined by the pattern. Three such patterns per clause will exist in the final construction, each of which corresponds to one literal.

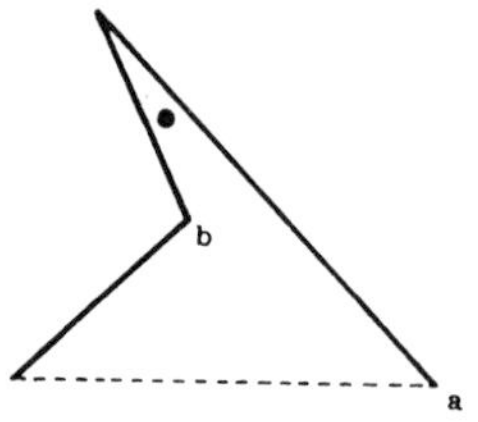

Fig. 1. Literal pattern.

Clause Junctions: Without loss of generality, consider the clause $C = A + B + D$, where $A \in \{u_i, \overline{u_i}\}$, $B \in \{u_j, \overline{u_j}\}$, and $D \in \{u_k, \overline{u_k}\}$ are literals, and u_i, u_j, and u_k are variables in U. The basic pattern for the clause junction C_h is shown in Fig. 2.

Let Δabc denote the triangle area determined by points a, b, and c, and let $(a_1, \cdots, a_h)$ denote that points a_1, $a_2, \cdots$, and a_h are collinear. Thus in the pattern of Fig. 2 we have $(g_{h2}, g_{h8}, a_{h4}, a_{h1}, b_{h4}, b_{h1}, d_{h4}, d_{h1}, g_{h9}, g_{h1})$,

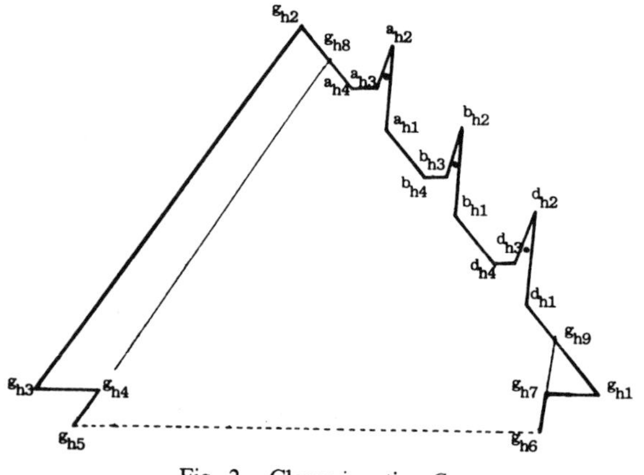

Fig. 2. Clause junction C_h.

$(g_{h3}, g_{h4}, g_{h7}, g_{h1})$, (g_{h8}, g_{h4}, g_{h5}), and (g_{h9}, g_{h7}, g_{h6}). Moreover, $|(g_{h2}, g_{h8})| = |(g_{h8}, a_{h4})|$, and $|(d_{h1}, g_{h9})| = |(g_{h9}, g_{h1})|$, where $|(u, v)|$ denotes the length of the edge (u,v).

Since none of g_{hi}, $i = 1,2,\cdots,7$ can cover $\Delta a_{h1}a_{h2}a_{h3}$, $\Delta b_{h1}b_{h2}b_{h3}$, and $\Delta d_{h1}d_{h2}d_{h3}$ in the pattern for C_h, they cannot be used to cover these three triangles as shown in Fig. 2. Furthermore, any two vertices of each literal pattern in the clause are not sufficient by themselves to cover the region defined by the pattern C_h. We therefore have the following.

Lemma 1: At least three vertices of C_h are required to cover the region defined by the pattern C_h shown in Fig. 2.

Lemma 2: Only seven three-vertex covers exist that can cover the region defined by the pattern C_h.

Proof: Since none of a_{h4}, b_{h4}, d_{h4}, and g_{hi}, $i = 1,2,\cdots,7$, can cover any of the three literal patterns, they cannot be chosen as possible candidates in the cover for Fig. 2. Therefore, only $H_i = \{a_{hi}, b_{hi}, d_{hi}\}$, $i = 1,2,3$, need be considered. If a_{h2} is chosen, $\Delta a_{h1}a_{h3}a_{h4}$ cannot be covered no matter which vertices are selected for the literal patterns B and D. This implies that vertices in H_2 cannot be candidates either. Vertices a_{h1} and a_{h3} should not be selected at the same time. If they were selected, we would need two more vertices to cover the literal patterns for B and D. This is also true for b_{h1} and b_{h3}, and d_{h1} and d_{h3}. This means that eight possible three-vertex covers exist from the sets H_1 and H_3 of vertices. Since the three vertices in H_3 cannot cover $\Delta g_{h1}g_{h2}g_{h3}$, we are left with seven three-vertex covers. It is easy to check that each of these seven three-vertex covers, that is, $\{a_{h1}, b_{h1}, d_{h1}\}$, $\{a_{h1}, b_{h1}, d_{h3}\}$, $\{a_{h1}, b_{h3}, d_{h1}\}$, $\{a_{h1}, b_{h3}, d_{h3}\}$, $\{a_{h3}, b_{h1}, d_{h1}\}$, $\{a_{h3}, b_{h1}, d_{h3}\}$, and $\{a_{h3}, b_{h3}, d_{h1}\}$, can indeed cover the region defined by the pattern C_h.

Labeling Vertices: We shall label vertices a_{h1} and a_{h3} of literal pattern A as follows. If $A = u_i$, a_{h1} and a_{h3} are labeled as T and F (true and false), respectively; if $A = \overline{u_i}$, a_{h1} and a_{h3} are labeled as F and T, respectively. In other words, vertex a_{h1} represents truth value *true* for literal A, and a_{h3} represents *false*. Vertices b_{h1} and b_{h3} of literal pattern B, and d_{h1} and d_{h3} of literal pattern D are labeled in the same manner.

We say that the truth value assignment to the variables u_i, u_j, and u_k in clause C_h is a true assignment if the resulting truth value of C_h is true; otherwise, it is a false assignment. The key idea in the labeling of vertices in a clause junction C_h is that the vertices that are selected according to the truth value assignment of $\{u_i, u_j, u_k\}$ can cover the region defined by $g_{h1}g_{h2} \cdots g_{h7}$ "free," that is, without increasing the number of vertices needed to cover the region defined by the pattern C_h, if and only if the assignment is a true assignment. Specifically, we have the following.

Lemma 3: The three vertices selected from the clause pattern $C_h = A + B + D$ cover the region defined by C_h if and only if the truth values represented by the labels of these vertices give a true assignment for C_h.

Proof: C_h has a truth value false if and only if the truth values of literals A, B, and D are false. According to the vertex labels, this implies that the vertices selected must be a_{h3}, b_{h3}, and d_{h3}. From Lemma 2 exactly seven three-vertex covers exist that can cover the region defined by C_h. We claim that the labels of each vertex in the three-vertex cover correspond to a true assignment to the variables u_i, u_j, and u_k for C_h. Since the verification for each of these seven three-vertex covers of the claim is straightforward, we leave out the details.

Variable Patterns: The variable pattern for variable u_i is as shown in Fig. 3. Note that a distinguished point exists in $\Delta t_{i1}t_{i2}t_{i3}$ and that $\Delta t_{i1}t_{i2}t_{i3}$ can only be covered by t_{i1}, t_{i2}, t_{i3}, t_{i5}, t_{i6}, or t_{i8}. One such pattern will exist per variable in the final construction. We shall henceforth refer to the two legs of the variable pattern as *rectangles* or *rectangular regions*, although they are not really rectangles. Now we consider how to put variable patterns and clause junctions together.

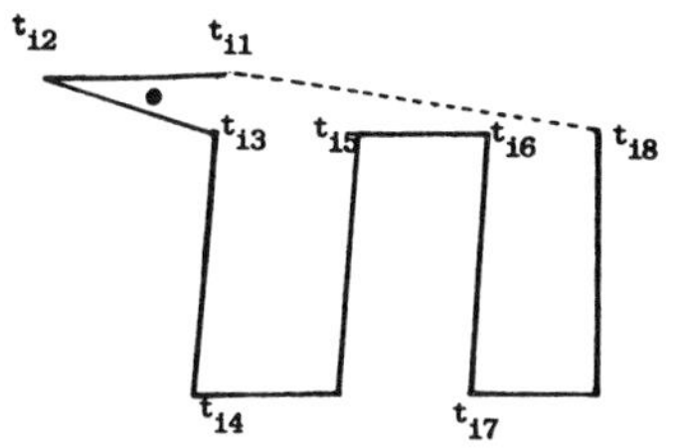

Fig. 3. Variable pattern; $t_{i1}t_{i2}$ and $(t_{i3}, t_{i5}, t_{i6}, t_{i8})$ are parallel.

Complete Construction of a Simply Connected Polygonal Region: Three steps are needed.

a) Putting variable patterns and clause junctions together: We put literal patterns and clause junctions together as shown in Fig. 4. In Fig. 4 we have that 1) vertex W can cover the n literal patterns, except $\Delta t_{i1}t_{i2}t_{i3}$, $i = 1,2,\cdots,n$, and 2) $(W, g_{15}, g_{16}, g_{25}, g_{26},\cdots, g_{m5}, g_{m6})$, $(t_{11}, g_{h5}, g_{h4}, g_{h8})$, and $(t_{n8}, g_{h6}, g_{h7}, g_{h9})$, $h = 1,2,\cdots,m$.

b) Augmenting variable patterns with "spikes": Suppose variable u_i appears in clause C_h. If u_i itself is in C_h, then the two spikes, $\overline{pq}$ and $\overline{p_1q_1}$ as shown in Fig. 5, where (p, q, t_{i5}, a_{h1}) and $(p_1, q_1, t_{i8}, a_{h3})$, are added to Fig. 4. If its complement $(\overline{u_i})$ is in C_h, then the two spikes, $\overline{pq}$ and $\overline{p_1q_1}$ as shown in Fig. 6, where (p, q, t_{i5}, a_{h3}) and $(p_1, q_1, t_{i8}, a_{h1})$, are added to Fig. 4. This is equivalent to labeling vertex t_{i5} F and vertex t_{i8} T. That is, vertices t_{i5} and t_{i8} represent, respectively, truth values false and true for variable u_i.

From the previous construction the picture is not a simply connected polygonal region because the new added spikes are line segments. For each spike $\overline{pq}$, p and q are collinear with b and t, where b is a vertex (a_{h1}, for example) of a literal pattern and t is a vertex (t_{i8}, for example) of a variable pattern. Therefore, $\overline{pq}$ can be covered by p, q, b, and t. The next step is to construct a polygonal region out of $\overline{pq}$ such that this polygonal region can only be covered by b, t, and its vertices.

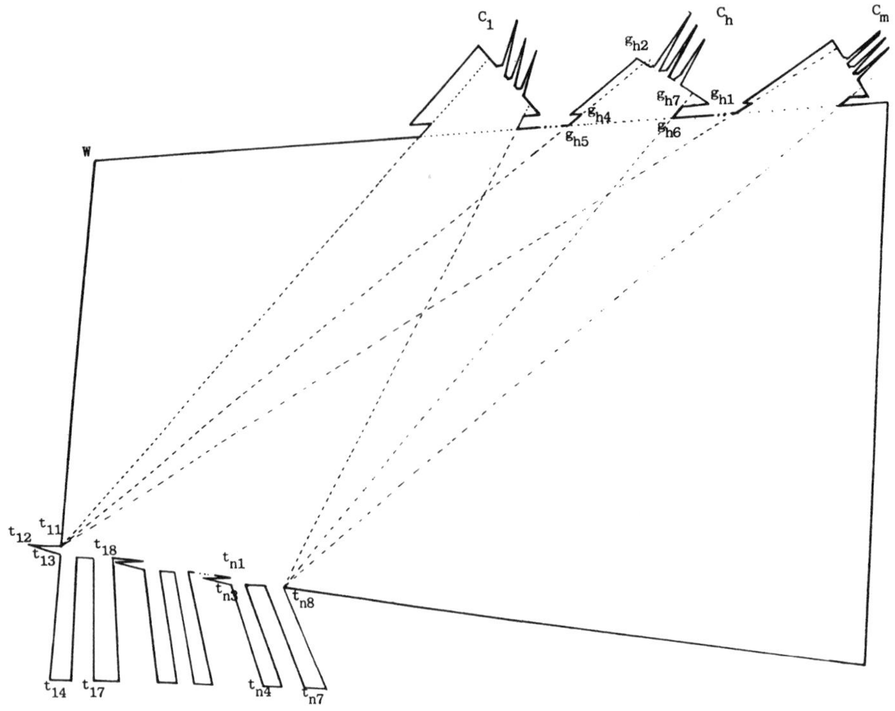

Fig. 4. Putting variable patterns and clause junctions together.

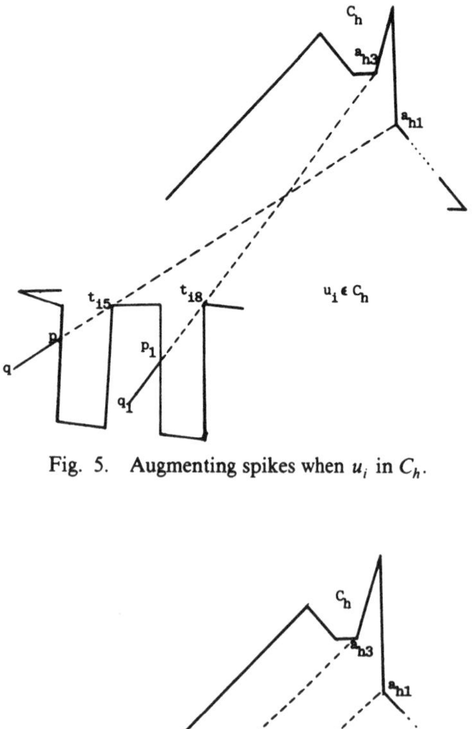

Fig. 5. Augmenting spikes when u_i in C_h.

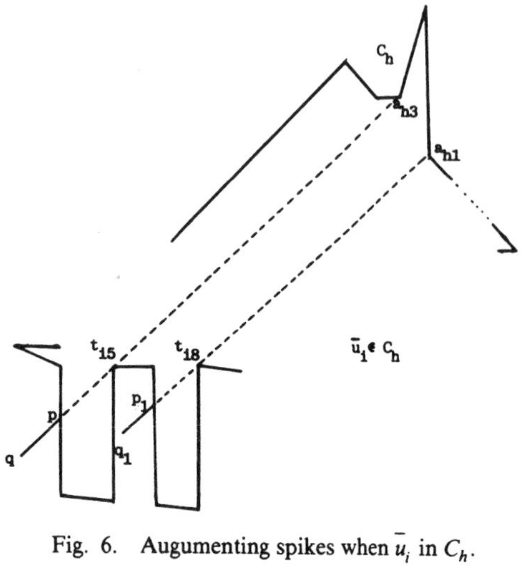

Fig. 6. Augumenting spikes when $\bar{u}_i$ in C_h.

c) Replacing spikes by polygonal regions: Suppose a is in literal pattern A, and A is in clause junction C_h. If a corresponds to a_{h1}, the shaded area in Fig. 7 is the polygonal region to replace $\overline{pq}$. More precisely, we have in Fig. 7 (a_{h1}, t_{i5}, p, q) and (a_{h1}, e, f). That is, we have a triangle $\Delta a_{h1}fq$. We replace $\overline{pq}$ similarly if a corresponds to a_{h3}. We call the polygonal regions created as described the *consistency-check patterns*.

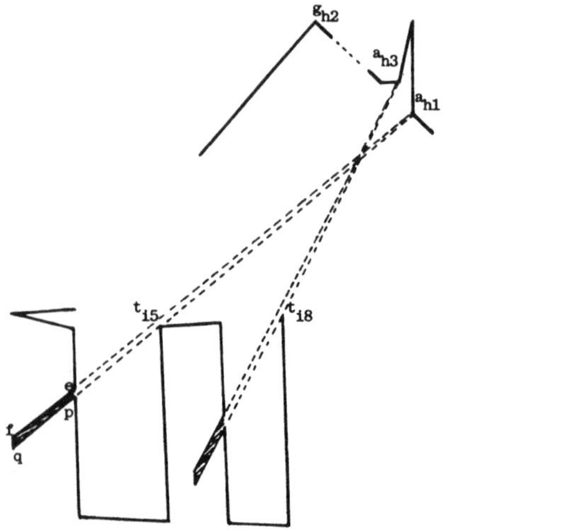

Fig. 7. Each spike is replaced with small region.

Fig. 8 is an example for converting the Boolean formula $F = (u_1 + u_2 + u_3) \wedge (u_1 + \bar{u}_2 + u_3) \wedge (u_1 + \bar{u}_2 + \bar{u}_3)$, and the dots are vertices in the minimum cover. From the minimum cover obtained we conclude that the truth values of u_1, u_2, and u_3 are true, true, and false, respectively.

Lemma 4: At least $K = 3m + n + 1$ vertices are needed for covering the simply connected polygonal region.

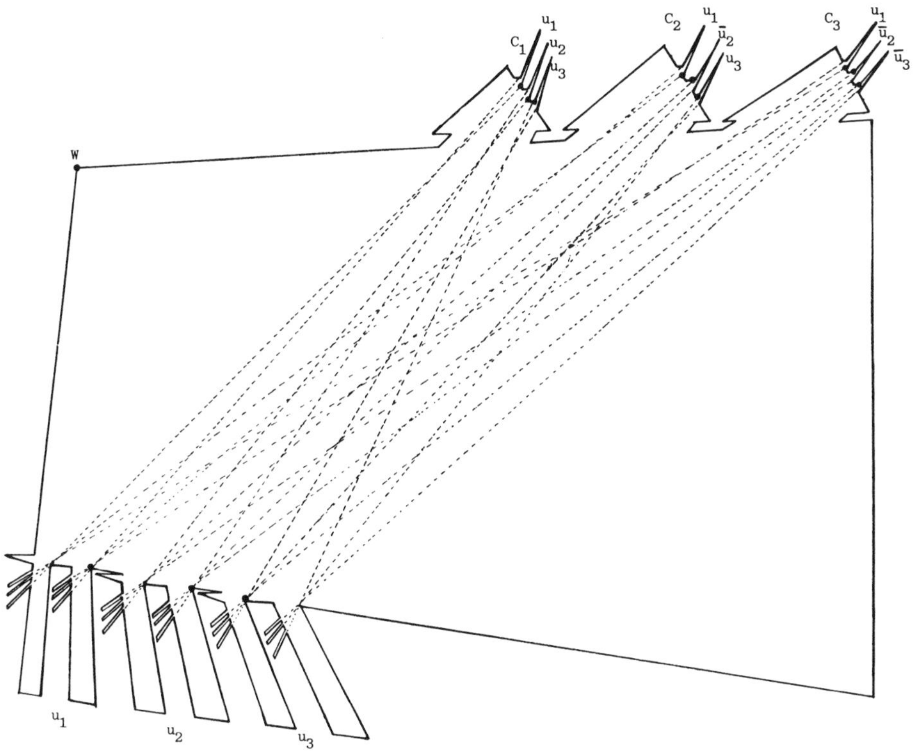

Fig. 8. Example and minimum cover.

Proof: At least $3m + n$ vertices are needed for covering $3m$ literal patterns, and $\Delta t_{i1}t_{i2}t_{i3}$, $i = 1, 2, \cdots, n$. At least one more vertex is needed to cover all the variable patterns' rectangles. Therefore, the lemma follows.

Lemma 5: The minimum number of vertices needed to cover the simply connected polygonal region is $K = 3m + n + 1$ if and only if C is satisfiable.

Proof: ($\leftarrow$) If C is satisfiable, then a truth value assignment to the variables exists such that each of the clauses in C has a truth value true. If u_i is true, then t_{i8} of the variable pattern u_i is included in the cover T, and either a_{h1} or a_{h3} of the literal pattern is included in T depending, respectively, on whether u_i or $\bar{u}_i$ appears in the clause C_h. If u_i is false, then t_{i5} is included in T and either a_{h1} or a_{h3} of the literal pattern is included depending, respectively, on whether $\bar{u}_i$ or u_i appears in the clause C_h. From the construction the regions defined by the consistency-check patterns and literal patterns are covered with $3m + n$ vertices. The remaining rectangular regions defined by the variable patterns can be covered by vertex W. Therefore, by including W in T we have a minimum cover with $K = 3m + n + 1$ vertices.

($\rightarrow$) Suppose a cover T with $K = 3m + n + 1$ vertices exists to cover the polygonal region. If W is not in T, K vertices are in no way sufficient to cover the regions defined by the $3m$ literal patterns and the $2n$ rectangles contained in the n variable patterns. Thus $K - 1 = 3m + n$ vertices are left to be considered.

In the polygonal region $3m$ literal patterns exist and n triangles $\Delta t_{i1}t_{i2}t_{i3}$, $i = 1, 2, \cdots, n$, each of which contains a distinguished point. We know that any vertex that covers a distinguished point cannot cover other distinguished points. Therefore, at least $3m + n$ vertices are needed to cover the above $3m + n$ subregions. To get a K-vertex cover each of the above $3m + n$ polygonal regions is allowed to be covered by one vertex. This means that for any literal pattern A, supposing A in C_h, either a_{h1} or a_{h3} must be included in T.

However, we cannot arbitrarily include in T any $3m$ vertices in these $3m$ literal patterns for they may make the n variable patterns inconsistent. The definition of consistency follows.

We say the any variable pattern L is *consistent* if all the consistency-check patterns connected to one of its two rectangles are covered by the $3m$ vertices in literal patterns and those connected to the other rectangle are not covered at all by the same $3m$ vertices; it is *inconsistent* otherwise.

In any variable pattern L, the number of consistency-check patterns connected to the first rectangle and second rectangle is the same; let it be q. The total number of consistency-check pattens in the variable pattern L is $2q$. From the construction the $3m$ vertices included in T can cover only q consistency-check patterns in L. If variable pattern L is not consistent, then some of the consistency-check patterns connected to the first rectangle and some connected to the second rectangle will not be covered. Then at least two extra vertices of L are needed to cover L. On the other hand, if it is consistent, one extra vertex is enough. Therefore, to get a K-vertex cover consistency of all variable patterns is required.

Since these $3m$ vertices of the literal patterns must satisfy the consistency requirement, the remaining n vertices chosen from the variable patterns can be determined. We know that these n vertices and vertex W cannot cover any $\Delta g_{h1}g_{h2}g_{h3}$, $h = 1, 2, \cdots, m$. Therefore, $\Delta g_{h1}g_{h2}g_{h3}$, $h = 1, 2, \cdots, m$, must be coverable by the $3m$ vertices. From Lemma 2, if they are covered free, then each c_i is satisfiable. This implies C is satisfiable. Once it is known that the instance of 3SAT is satisfiable, the truth value assignment to the variables can easily be determined from the consistency property possessed by the minimum cover.

Thus we have the following main results.

Theorem 1: The minimum vertex guard problem for simply connected polygonal regions is NP-hard.

Proof: It follows from Lemma 5 and from the fact that the construction of the simply connected polygonal region, as described earlier, from a given instance of 3SAT takes polynomial time.

Edge Guard of a Polygonal Region

In this section we consider the minimum edge guard problem for a simply connected polygonal region where the guard can move along an edge. Since the NP-hardness proof parallels that for the minimum vertex guard problem, we summarize it as follows.

Theorem 2: The minimum edge guard problem for simply connected polygonal regions is also NP-hard.

Proof: Use similar arguments except that slight modifications to the literal patterns, variable patterns, and W are needed. They are shown in Figs. 9–11. In Fig. 9 only edges $\overline{a_{h1}a_{h8}}$ and $\overline{a_{h3}a_{h12}}$ can cover the entire literal pattern a_h. In Fig. 10 edges $\overline{t_{i5}t_{i6}}$ and $\overline{t_{i8}t_{i9}}$ are the only edges that can cover $\Delta t_{i1}t_{i2}t_{i3}$ and one of the rectangles in the variable pattern t_i. Change W into $\overline{W_1W_2}$. W_1 is just like W in the proof of vertex guard problem. W_2 covers $\Delta W_2W_3W_4$.

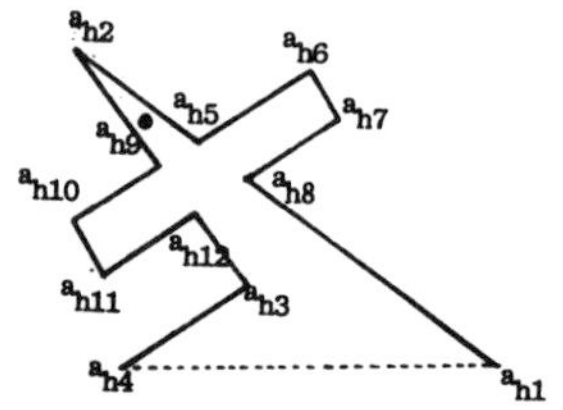

Fig. 9. Literal pattern for edge guards.

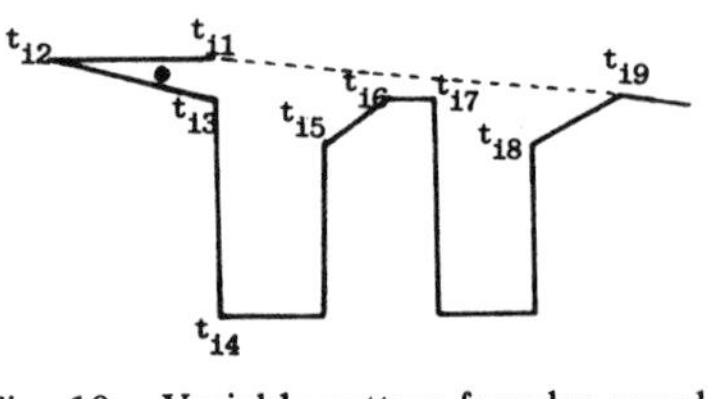

Fig. 10. Variable pattern for edge guards.

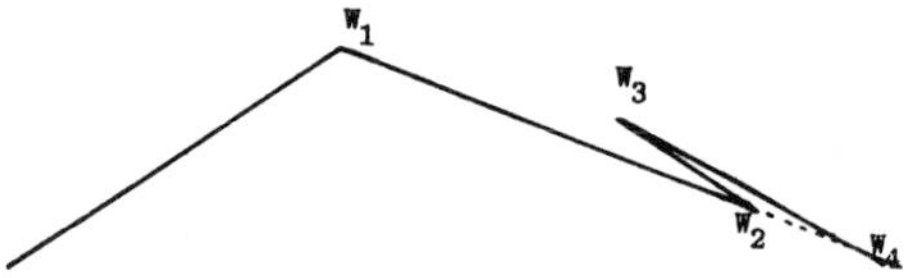

Fig. 11. Vertex W for edge guards.

IV. Point Guard Problem

In this section we consider the minimum point guard problem for simply connected polygons, which is similar to the minimum vertex guard problem except that the guards need not necessarily be fixed on the vertices. This problem can also be shown to be NP-hard using similar arguments as given earlier with some modifications [1]. The roles played in the proof by vertices a_{h1}, a_{h3}, t_{i5}, and t_{i8} for variable u_i should be replaced by their corresponding kernels. The shaded regions around t_{i5} and t_{i8} in Fig. 12, referred to as the kernels, are regions such that any point in these regions can cover $\Delta t_{i1}t_{i2}t_{i3}$ and the consistency-check patterns visible from t_{i5} and t_{i8}, respectively. We note that the kernels of the literal pattern a_h and the consistency-check patterns visible from a_{h1} and a_{h3}, respectively, are the vertices a_{h1} and a_{h3} themselves. Thus the point guards in the minimum cover are restricted to vertex W and these kernels in the simply connected polygonal region.

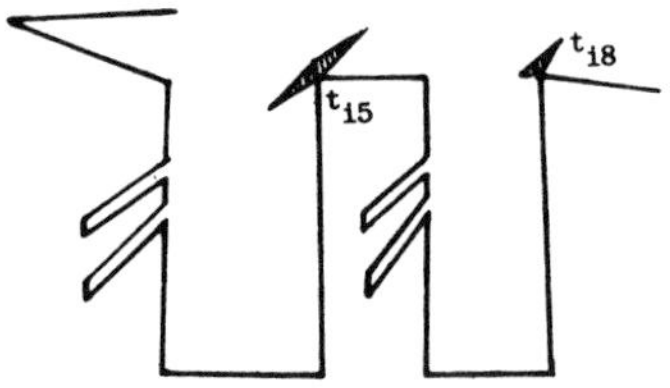

Fig. 12. Variable pattern for point guards.

Theorem 3: The minimum point guard problem for simply connected polygonal regions is NP-hard.

Since each point guard covers a region that is star-shaped, we have the following corollary.

Corollary: The problem of decomposing a simple polygon into the minimum number of star-shaped polygons whose union is the given polygon is NP-hard.

V. Conclusion

We have shown that the minimum vertex guard problem for a simply connected polygon is NP-hard. With slight modifications to the construction of the proof of the NP-hard result we have also shown that the minimum edge guard problem and minimum point guard problem for simple polygons are NP-hard. As a result of the NP-hardness of the minimum point guard problem, the problem of decomposing a simple polygon P into a minimum number of star-shaped polygons such that their union is P is also NP-hard.

Acknowledgment

The authors wish to thank Alok Aggarwal at Johns Hopkins University for useful comments on an earlier draft of this paper and for providing insight into the complexity result of the minimum point guard problem in Section IV, and the referees for constructive suggestions that greatly helped the presentation of the paper.

References

[1] A. Aggarwal, personal communication.
[2] N. Ahuja, R. T. Chien, R. Yen, and N. Birdwell, "Interface detection and collision avoidance among three dimensional objects," in *Proc. 1st Nat. Conf. Artificial Intelligence,* 1980, pp. 44–48.
[3] D. Avis and G. T. Toussaint, "An efficient algorithm for decomposing a polygon into star-shaped polygons," *Pattern Recognition*, vol. 13, pp. 395–398, 1981.
[4] ——, "An optimal algorithm for determining the visibility of a polygon from an edge," *IEEE Trans. Comput.*, vol. C-30, pp. 910–914, Dec. 1981.
[5] B. M. Chazelle and D. Dobkin, "Decomposing a polygon into its convex parts," in *Proc. 11th ACM Symp. Theory Comput.*, 1979, pp. 38–48.
[6] V. Chvatal, "A combinatorial theorem in plane geometry," *J. Comb. Theory, Ser. B*, vol. 18, pp. 39–41, 1975.
[7] H. Edelsbrunner, J. O'Rourke, and E. Welzl, "Stationing guards in rectilinear art galleries," *Comput. Vision, Graphics, Image Processing*, vol. 27, pp. 167–176, 1984.
[8] H.-Y. Feng and T. Pavlidis, "Decomposition of polygons into simpler components: Feature generation for syntactic pattern recognition," *IEEE Trans. Comput.*, vol. C-24, pp. 636–650, June 1975.
[9] S. Fisk, "A short proof of Chvatal's watchman theorem," *J. Comb. Theory Ser. B*, vol. 24, p. 374, 1978.
[10] M. R. Garey and D. S. Johnson, *Computers and Intractability: A Guide to the Theory of NP-Completeness*. San Francisco, CA: Freeman, 1979.
[11] R. Honsberger, *Mathematical Gems II*. Mathematical Assoc. of America, 1976.
[12] J. Kahn, M. Klawe, and D. Kleitman, "Traditional galleries require fewer watchmen," *SIAM J. Alg. Disc. Meth.*, vol. 4, pp. 194–206, June 1983.
[13] J. O'Rourke, "The complexity of computing minimum convex covers for polygons," in *Proc. 20th Allerton Conf.*, 1982, pp.75–84.
[14] ——, "An alternate proof of the rectilinear art gallery theorem," *J. Geometry*, vol. 211, pp. 118–130, 1983; also in Dep. Elec. Eng. and Comput. Sci., Johns Hopkins Univ., Baltimore, MD, Tech. Rep. 82/15, Dec. 1982.
[15] ——, "Galleries need fewer mobile guards: A variation on Chvatal's theorem," *Geometricae Dedicata*, vol. 14, pp. 273–283, 1983.
[16] J. O'Rourke and K. Supowit, "Some NP-hard polygon decomposition problems," *IEEE Trans. Inform. Theory*, vol. IT-29, pp. 181–190, Mar. 1983.
[17] J. Sack, "An $O(n \log n)$ algorithm for decomposing simple rectilinear polygons into convex quadrilaterals," in *Proc. 20th Allerton Conf.*, 1982, pp. 64–74.

Sensing Strategies for Disambiguating Among Multiple Objects in Known Poses

W. ERIC L. GRIMSON, MEMBER, IEEE

Abstract—**The need for intelligent interaction of a robot with its environment frequently requires sensing of the environment. Further, the need for rapid execution requires that the interaction between sensing and action take place using as little sensory data as possible, while still being reliable. Previous work has developed a technique for rapidly determining the feasible poses of an object from sparse, noisy, occluded sensory data. Techniques for acquiring position and surface orientation data about points on the surfaces of objects are examined with the intent of selecting sensory points that will force a unique interpretation of the pose of the object with as few data points as possible. Under some simple assumptions about the sensing geometry, we derive a technique for predicting optimal sensing positions. The technique has been implemented and tested. To fully specify the algorithm, estimates of the error in estimating the position and orientation of the object are needed. Analytic expressions for such errors in the case of one particular approach to object recognition are derived.**

Problem Definition

A ROBOT often must recognize and locate objects in its workspace, or more informally, must use sensory information to determine what objects are where, in order to manipulate them. Since speed of operation is also an important consideration in robotics applications, the interaction of sensing and action should take place using a minimal amount of sensory data. This requires methods that optimally (or near optimally) select positions at which sensory data are obtained. Clearly, the notion of optimal selection of new data points will in part be tied to the specific recognition engine used to interpret those data points. In previous papers [10], [14]–[16], we have presented a constraint based recognition and localization technique that uses sparse, noisy, occluded measurements of the position and orientation of small patches of an object's surface as input, which are obtained from any of several sensing modalities. Applying this recognition system to such sensory input data results in a small set of object poses; that is, a set of transformations taking a known object model from an intrinsic coordinate system into a coordinate system defined relative to the sensor. In this paper, we consider the problem of disambiguating among this fixed set of object poses. Note that the set of poses could include poses corresponding to different objects. To disambiguate among a set of interpretations, we need to acquire sensory data that will clearly distinguish one pose of an object from another, using as few additional sensory points as possible. Thus our problem is to optimally select places at which to obtain the needed sensory data.

While we use the recognition system developed in [14]–[16] as the basis for investigating sensing strategies for disambiguation, we expect that some of the results of this investigation should have application in more general situations of recognition and localization. To illustrate this, we begin with a set of examples of the use of sensing strategies.

Manuscript received November 26, 1985; revised June 20, 1986. This report describes research done at the Artificial Intelligence Laboratory of the Massachusetts Institute of Technology. Support for the Laboratory's Artificial Intelligence research is provided in part by a grant from the System Development Foundation, and in part by the Advanced Research Projects Agency under Office of Naval Research contracts N00014-80-C-0505 and N00014-82-K-0334.

The author is with the Artificial Intelligence Laboratory, Massachusetts Institute of Technology, 545 Technology Square, Cambridge, MA 02139, USA.

IEEE Log Number 8610325.

Example I: Disambiguating Multiple Interpretations

Suppose we are given a sparse set of sensory data points, each recording the position and orientation of a small patch of some surface in the workspace of a robot. Our goal is to determine what objects, from a set of known objects, are consistent with this data, together with the pose (position and orientation) of the object that leads to such a consistent interpretation. In the case of sensory data known to all lie on one object, we take consistent to mean that a rigid transformation of the object will cause all of the data points to lie on the object, with the correct surface orientation (to within some known error bounds). In the case of sensory data that may come from more than one object, we take consistent to mean that a maximum subset of the data satisfies the above condition. In this case, of course, other interpretations of consistent are possible.

In [14]–[16] we described an efficient constrained search technique for matching the sensory data to faces of an object model, in order to find the interpretations of the data. The sensory data consist of measurements of the position and surface orientation of small patches of object surfaces. The objects are modeled by sets of planar face equations. The technique uses efficient constraints between data elements and model elements to determine the set of interpretations of the data consistent with the model; that is, the set of poses of the object that agree with the input data. Empirical testing, as well as theoretical analysis [12], indicates that in general, there will be only one consistent interpretation of the data. It is possible, however, that more than one pose of the object will be consistent with the data, even for non-symmetric objects and even if the object is known (see Fig. 1). To determine the correct pose, we will need additional sensing.

The simplest method for obtaining the supplementary

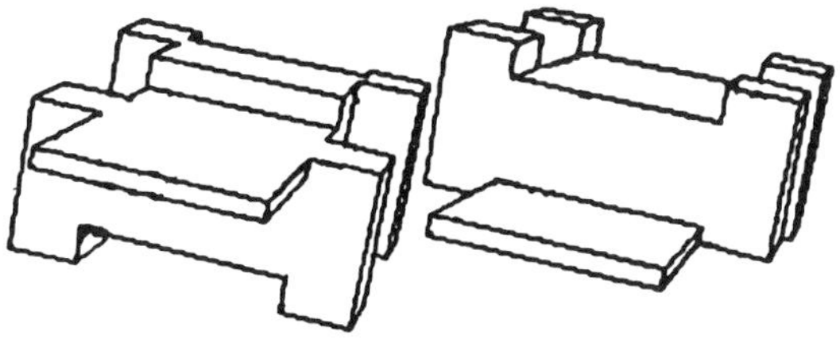

Fig. 1. Example of multiple interpretations. Given only a sparse set of isolated data points, multiple interpretations such as those indicated may be possible.

TABLE I
HISTOGRAMS OF POINTS FOR DISAMBIGUATION[1]

No. of Points	9	10	11	12	13	14	15	16	17	18	19	20	>20
No. of Trials	59	15	6	5	4	1	1	2	1	1	2		3

[1] Each column indicates the number of sensory points needed to force a unique interpretation, and the number indicated in the column is the number of trials, out of 100, for which that number of data points was required.

sensory data is to sample the object at random. If the sensing process is fast enough, and if, on average, only a few additional points are required in order to remove the ambiguity, then such a random sensing strategy could suffice. It is easy, however, to find situations in which a random sensing strategy would be ineffective in disambiguating between possible interpretations, and in general one expects a random sensing strategy to have a very slow convergence. Moreover, some sensing modalities, for example, tactile sensing, are inherently sparse, and require considerable expense to obtain additional sensing points. In this case, it is particularly desirable to perform recognition with minimal sensory interaction.

For example, Table I lists histograms of the number of additional, randomly chosen, sensing points needed to uniquely disambiguate several consistent interpretations. We generated an initial set of nine points of data, all lying on a single object (shown in Fig. 1), and we determined the set of consistent interpretations of that data, using the system described in [15], [16]. We then generated additional sense points until only one interpretation remained consistent with the data. This process was repeated for 100 trials, and the number of sense points needed to disambiguate the interpretations was histogrammed. The results are recorded in Table I, where each entry is the number of trials terminating with the indicated number of sense points. The sensory data was generated by randomly choosing approach directions towards the object, and sensing for contact along them, much as might occur in tactile sensing. It can be seen that choosing sensing directions at random may have a slow convergence towards a unique interpretation, especially since in this case we are only dealing with the simple case of data from a single known object.

In general, one would expect a tradeoff between random sensing strategies and feature driven sensing strategies. Given two possible interpretations of the data, consider constructing the volume difference, consisting of all points contained in one but not both of the interpretations. If the size of this volume relative to the volume of the object is large, then in general, one would expect randomly generated additional sensing points to quickly disambiguate the situation. On the other hand, if the relative volume is small, one would expect that a large number of additional sense points would be needed before one of them struck this volume difference. In this case, a more directed sensing strategy is likely to be more effective.

Example II: Localization with Minimal Sensing

In the previous example, we discussed the problem of generating additional sensory data, given some initial set of data and the interpretations consistent with it. A related problem is to consider the optimal acquisition of all of the sensory data, rather than just that needed to disambiguate interpretations. For example, consider a situation in which a known object, with a fixed set of known stable positions is being sensed. This might be the case, for example, when considering objects in pallets, or feeders. We would like to determine the pose of the object with as few sensory points as possible. Here, the initial set of interpretations is the set of stable configurations of the object. Given this set of stable configurations, we want to determine the optimal sensing directions for distinguishing that set of configurations.

Example III: Simple Inspection

The problem of determining sensing positions can also arise in simple inspection tasks. Suppose we are given an object pose, and a set of distinctive points defined on the object model. In this case, we may be able to use the techniques developed below to choose the sensing rays needed to test that the designated distinctive model points are in fact present in the sensed object.

ASSUMPTIONS

The problem to be addressed in this paper is finding effective and rigorous sensing strategies for deciding between a set of possible poses of an object, or multiple objects. We will assume that the following are given.

- *Set of Interpretations:* Some initial set of possible interpretations is assumed. This could be either from the application of some recognition process to a set of initial sensed points or from assumptions about the object to be sensed, in particular that it is lying in one of a known number of stable positions. In each case, the interpretation includes a computed transformation giving the pose of the model in sensor coordinates.

- *Set of Sensing Directions:* It is assumed that the initial sensory data were generated by sampling along a set of known directions. For example, in the case of visual sensing these could be given by the orientation of the cameras relative to the workspace. In general, determining optimal sensing rays is a four-degree-of-freedom problem. In this paper, we assume that the two rotational degrees-of-freedom are restricted to a small set of possibilities by the sensing geometry, such as the given camera orientations. We then optimize over the remaining two degrees of freedom.
- *Polyhedral Object Models:* We assume that the objects to be sensed have been modeled as polyhedra, although the objects themselves need not be polyhedral. Any deviations between curved objects and their polyhedral models will simply contribute to a small amount of error in the sensory data, to which a recognition system should be insensitive.

The goal is to disambiguate between the set of interpretations by determining positions at which to obtain subsequent sensory information. These positions should be such that by sensing along one of the possible directions, the recorded information will disambiguate between the set of possible interpretations (or some subset of the interpretations) in the presence of possible error in the computed transformations associated with each of the interpretations.

In the examples given above, we assumed that we had available techniques for acquiring the sensory data, and techniques for solving the recognition and localization problem. There are, of course, many techniques for obtaining information about the three-dimensional positions of points on an object, as well as the local surface normals at those points. Typical examples of such measurement processes include tactile sensing [17], [18], [32], [35]–[36]; binocular stereo [1], [3], [11], [13], [26], [27], [31]; photometric stereo [23], [42]–[44]; laser range-finding [24], [30]; and structured-light systems [34], [38]. These methods can provide information about the three-dimensional positions of points on the object, as well as the local surface normals at those points, usually with some error in the measurements.

A number of different techniques have been developed for model-based recognition and localization. If one views recognition as a search for a consistent match between data elements and model elements, then much of the variation between existing recognition schemes can be accounted for by the choice of what descriptive tokens to match. Examples of techniques relying on sparse distinctive features include the use of a few extended features [2], [33]; the use of one feature as a focus, with the search restricted to a few nearby features [7], [8], [19], [40], [41]; matching of high-level descriptions [4], [5], [25], [28], [29]; and the use of geometric relationships between simple descriptors [6], [9], [10], [14], [20], [21], [22], [39]. The basis for the present work is the approach presented in [10], [14]–[16].

For the purposes of this paper, we will assume that such techniques are available. Our concentration is on the problem of choosing optimal sensing strategies for interacting with such techniques.

An Algorithm for Computing Sensing Directions

To demonstrate the approach of computing sensing directions, we first look at an example in two dimensions (see Fig. 2), where the object has three positional degrees-of-freedom (one rotational and two translational).

After our recognition and localization process has been applied to a sparse set of data points, we are left with some set of poses of the object consistent with that data. We are given a set of sensing directions, that is, a set of unit vectors $\hat{s}_i$ indexed over $i \in I$, such that sensing can occur along directions parallel to any of these unit vectors, for some set of initial positions. For example, in Fig. 3, if o is an offset vector, where $o \cdot \hat{s}_i = 0$, then we can sense along the ray $o + \alpha\hat{s}_i$ as α varies. Equivalently, we can think of this as having some finite portion of a plane perpendicular to $\hat{s}_i$, such that for any point on the plane, we can sense along a ray through that point in the direction of $\hat{s}_i$. We are also given some bounds on the sensitivity of the sensing device in measuring surface normals and surface positions. In particular, we define ϵ_n and ϵ_d in the following manner, illustrated in Fig. 4. If $\hat{n}_{\text{true}}$ is the surface normal at some point on an object, measured in sensor coordinates, and $\hat{n}_{\text{sense}}$ is the normal measured by the sensing device, then

$$\hat{n}_{\text{sense}} \cdot \hat{n}_{\text{true}} > \epsilon_n.$$

If p_{true} is the actual position of a point on an object, measured in sensor coordinates, and p_{sense} is the position measured by the sensing device, then

$$|p_{\text{sense}} - p_{\text{true}}| < \epsilon_d.$$

Thus ϵ_n and ϵ_d describe the range of uncertainty in the measurements of normals and distances, respectively.

The basic idea is that over the set of all given sensing directions $\{\hat{s}_i | i \in I\}$, we want to find a particular direction $\hat{s}_{i0}$, and an offset position o, such that sensing along the ray $o + \alpha\hat{s}_{i0}$ will distinguish the poses. By distinguish, we mean that for all pairs of possible poses, either the difference in the expected normals of the faces that intersect the ray, or the difference in the expected positions of the points of intersection of the ray with the corresponding faces of the poses, is greater than the sensitivity of the sensing device.

We note that a sensing ray that does not intersect exactly one of the possible poses is acceptable. Indeed, in the case of two possible poses, sensing rays that would contact only one of the poses are likely to be among the best candidates for disambiguating the two poses. Secondly, we note that if there are many possible poses, it may not be possible to find one sensing ray that will distinguish between all of them. Instead, we may have to use a series of measurements to determine the correct pose. The number of such measurements will be bounded above by the number of poses, however.

The main problem to be faced in finding good sensing rays

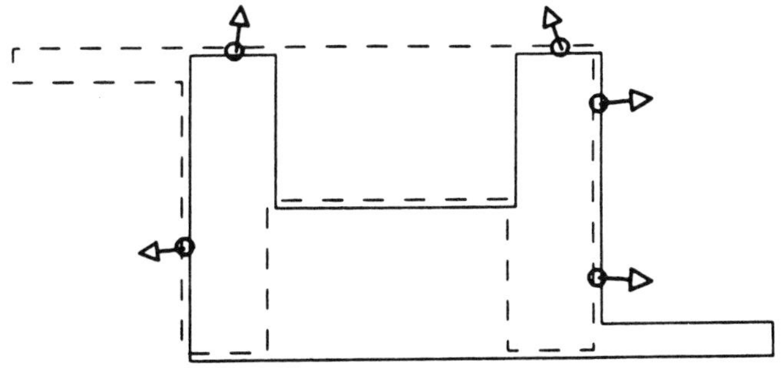

Fig. 2. Two-dimensional example of multiple poses. Both poses are consistent with the sensory data indicated by the small surface normals and the points of contact.

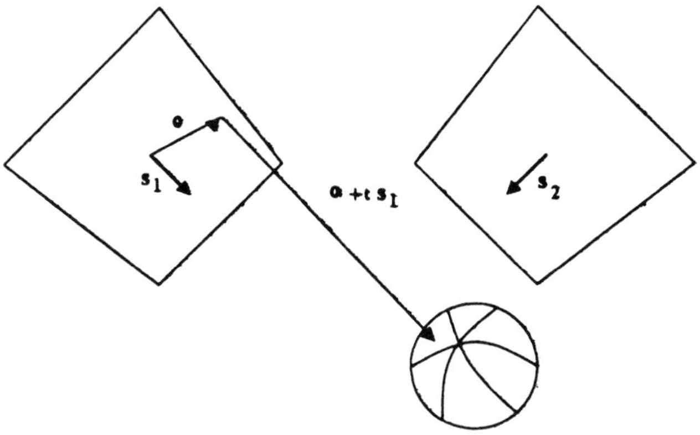

Fig. 3. Examples of the sensing geometry. Each vector s defines a sensing direction. The actual sensing ray is defined by specifying an offset vector o relative to the origin of the sensing plane through which the ray must pass, parallel to s.

Fig. 4. Error bounds. The true surface normal is known to lie within a specified cone of the measured normal, while the true position is known to lie within a specified ball about the measured position.

is the existence of error in the computed transformations associated with each pose. Thus, for the sensing strategy to be effective, the ray must both distinguish the poses and be insensitive to errors in the position and orientation of the poses.

The proposed method is quite simple and is illustrated in Fig. 5, in which two poses of the object are shown, one in solid lines, the other in hashed lines. The steps of the method are as follows.

1) Pick a particular sensing direction $\hat{s}$ (we will assume the convention that $\hat{s}$ points from the sensor towards the object). In the two-dimensional case, we can define a line perpendicular to the sensing direction, which we will call the *sensing line,* with origin at the point on the line closest to the origin of the sensor space. In three dimensions, this would be a *sensing plane.* This is shown in Fig. 5(a).

2) We fix the position of this line at some arbitrary reference point, for example by specifying the minimum distance of the line from the origin of the space to be d. This is shown in Fig. 5(b).

3) Now consider one of the poses, for example, the one shown in hashed lines in the figure. For each face f_i in the model,. with corresponding model unit normal $\hat{\boldsymbol{n}}_{m,i}$, we let ${}^s\hat{\boldsymbol{n}}_{m,i}$ denote the unit normal rotated into sensor coordinates, i.e., corresponding to the orientation of the face relative to the pose of the object. If the face points towards the sensor ($\hat{\boldsymbol{s}} \cdot {}^s\hat{\boldsymbol{n}}_{m,i} < 0$), we project the boundaries of the face onto the sensing line, as shown for example in Fig. 5(c). In other words, each end point $\boldsymbol{e}$ of the edge is projected to a point on the sensing line, $\boldsymbol{e} + (d - \boldsymbol{e} \cdot \hat{\boldsymbol{s}})\hat{\boldsymbol{s}}$. In three dimensions, this would entail the projection of the edges of a face onto the sensing plane.

4) We can label the resulting segment of the $\hat{s}$-line with the surface normal ${}^s\boldsymbol{n}_{m,i}$ and with the range of distances from the object face to the $\hat{s}$-line. That is, if $\boldsymbol{v}$ is a point on the edge, in sensor coordinates, then $\boldsymbol{v} \cdot \hat{\boldsymbol{s}} - d$ is the distance from the point to the $\hat{s}$-line. We let $\alpha_{\min}$ and $\alpha_{\max}$ denote the extreme values taken on by $\boldsymbol{v} \cdot \hat{\boldsymbol{s}} - d$ as $\boldsymbol{v}$ ranges over the edge, and the segment is labeled by

$$\{{}^s\hat{\boldsymbol{n}}_{m,i}, \{\alpha_{\min,\, i}, \alpha_{\max,\, i}\}\}.$$

5) When all the visible faces of a pose have been projected onto the $\hat{s}$-line, we can perform hidden surface removal to reduce the set of possibly overlapping segments of the $\hat{s}$-line to a set of disjoint segments. Each segment will be labeled by the surface orientation of the corresponding face, in sensor coordinates, and the range of distances to points on the face.

6) We can perform this operation for each pose, obtaining a different disjoint partition of the $\hat{s}$-line, labeled by the appropriate surface normals and distance ranges, as shown in Fig. 5(d) (slightly offset for graphical clarity).

7) Next, we intersect the set of all such partitions. That is, we define a new partition of the $\hat{s}$-line with two properties. First, each segment of this new partition lies within exactly one segment of each of the partitions of the $\hat{s}$-line corresponding to a pose. Second, this new partition is the smallest (in terms of number of segments) such partition. The label associated with each segment of the new partition is the union of the labels of the corresponding segments of the individual partitions.

8) This partition can now be analyzed for distinguishability. More precisely, given a segment of the partition, the set of normals

$$\{\hat{\boldsymbol{n}}_j \mid j \in J\}$$

associated with that segment is *distinguishable* if

$$\max_{i,j \in I, i \neq j} (\hat{\boldsymbol{n}}_i \cdot \hat{\boldsymbol{n}}_j) < 2\epsilon_n.$$

In other words, given a measurement of the actual object in this region, we can uniquely determine to which pose it corresponds. Similarly, the set of distance measurements

$$\{(\alpha_{\min,\, j}, \alpha_{\max,\, j}) \mid j \in J\}$$

is *distinguishable* if

$$\max_{i \neq j} \{ |\alpha_{\min,\, i} - \alpha_{\max,\, j}| \} > 2\epsilon_d.$$

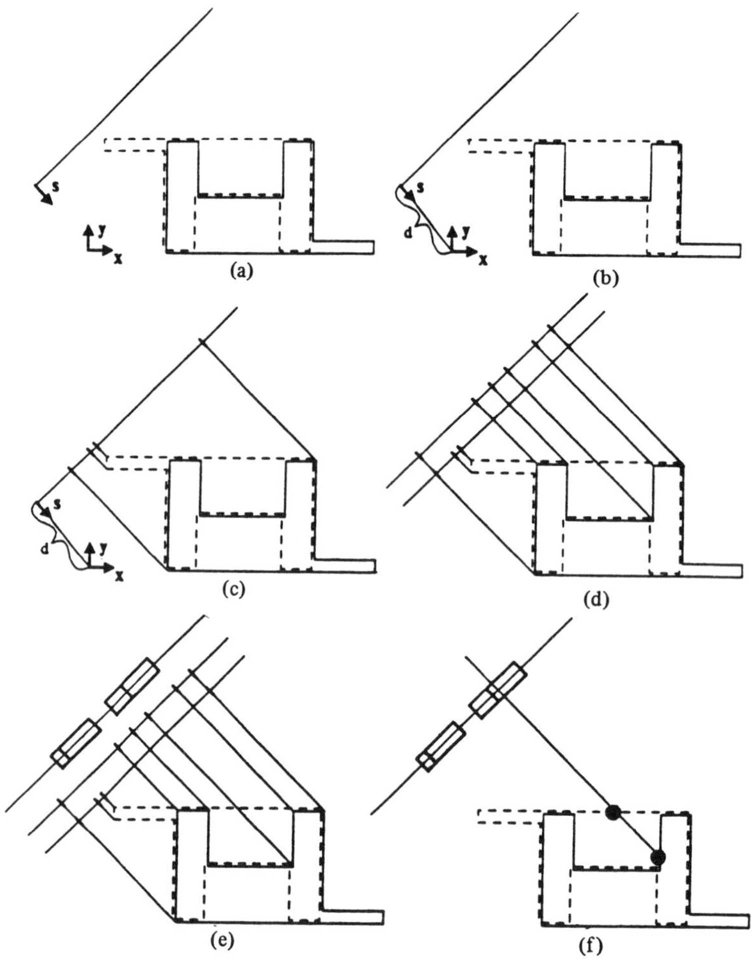

Fig. 5. Projection of poses onto the sensing line. (a) Sensing line is indicated, orthogonal to the sensing direction. (b) This line is fixed at a distance d from the origin. (c) Visible faces of one of the poses are projected onto the sensing line defined by the sensing ray. This projection defines a partition of the sensing line. Here s is the sensing ray and d is the distance to the origin. (d) Visible faces of the second pose are also projected onto the sensing line. (e) Respective partitions are tested for distinguishability, based on differences in expected surface orientation and differences in expected position, and the distinguishable regions of the sensing line are marked. (f) Using a sensing ray through the midpoint of either of the two marked regions would enable one to disambiguate the two poses in which the expected sensory points are indicated.

We can collect all such distinguishable segments of the partition, thereby determining the set of possible sensing points along the particular choice of $\hat{s}$. This is illustrated in Fig. 5(e).

If there were no error in the transformations associated with the poses, we would be done, since any point in this set would disambiguate the poses, (see Fig. 5(f) for an example). To account for possible error in the transformations associated with the poses, however, we need to be somewhat judicious in our choice of sensing point. The basic idea is to choose a point such that the face with which contact is made remains the same over small perturbations in the transformation. In two dimensions this is most easily done by choosing the midpoint of the longest segment. In three dimensions, the easiest way to choose such a point, from among the set of distinguishable polygons on the sensing plane, is by applying the notion of a Chebychev point, defined as follows. Suppose we are given a polygon on a plane, each of whose edges is defined by a pair $(\hat{n}_j, d_j)$, where $\hat{n}_j$ is a unit normal lying in the plane, and d_j is a constant such that points along the edge are defined by

$$\{\boldsymbol{v} \mid \boldsymbol{v} \cdot \hat{n}_j - d_j = 0\}.$$

Then the distance from any point $\boldsymbol{v}$ to an edge is given by

$$\boldsymbol{v} \cdot \hat{n}_j - d_j.$$

The Chebychev point of a polygon is the point that maximizes the minimum distance from the point to any edge of the polygon; that is, the point $\boldsymbol{v}$ that satisfies

$$\min_j (\boldsymbol{v} \cdot \hat{n}_j - d_j) \geq \min_j (\boldsymbol{u} \cdot \hat{n}_j - d_j), \quad \forall \boldsymbol{u}$$

where the value taken by this expression at the Chebychev point is called the Chebychev value of the polygon. Clearly, the polygon with the maximum Chebychev value will be the least sensitive to perturbations in the computed transformations, and thus the Chebychev point with the maximum Chebychev value, as measured over the set of all distinguishable polygons, defines the best sensing position. Note that we can improve the reliability of the sensing strategy even further by choosing the maximum Chebychev point as measured over connected sequences of distinguishable segments.

9) We repeat this process over all sensing directions $\hat{s}_i$, choosing the direction that best distinguishes the feasible poses.

While this analysis has been done in two dimensions, it clearly extends to the general three dimensional case. Here, the visible faces are projected into polygons on a sensing plane, and the intersection of the projections over all poses gives a partition of this plane, which can be tested for distinguishability.

Implementation of the Technique

In testing the proposed algorithm, we have chosen a slightly modified implementation of the technique, which avoids some of the difficulties of performing hidden surface removal and of intersecting polygonal partitions of a plane. One means of circumventing these difficulties is to use a regular grid tesselation of the plane.

In particular, suppose that we partition the $\hat{s}$-plane with a rectangular grid whose elements have sides of length h. Rather than trying to compute polygonal regions on the $\hat{s}$-plane that are distinguishable, we shall examine each grid segment within the bounds of the projected object, seeking those segments that are themselves distinguishable, and then we will piece these grid elements back together.

The steps of the new algorithm, many of which are identical to those of the previous solution, are sketched below.

- Initially, mark all grid segments as active.
- Given a pose, and a sense direction $\hat{s}$, test each face for visibility. If the normal of the face, in sensor coordinates, is given by ${}^s\hat{n}_{m,i}$, then a face is visible if $\hat{s} \cdot {}^s\hat{n}_{m,i} < 0$.
- For each visible face, project its vertices onto the $\hat{s}$-plane, resulting in a set of new vertices that define a polygon on the plane.
- Given this polygon on the sensing plane, compute the smallest bounding rectangle composed of an integral number of grid elements which encloses the enscribed polygon. This rectangle has no intrinsic merit, but is simply a convenient means of restricting the search process.
- For each grid element lying in this enclosing rectangle, apply the following test. If the grid segment lies entirely outside of the polygon, nothing is done. If some edge of the polygon passes through the segment, this segment is marked as inactive. If the grid segment is still active and lies entirely within the polygon, a label is attached to the grid segment. This label is composed of two elements. The first is the normal of the face whose projection resulted in the current polygon on the sensing plane, measured in sensor coordinates. The second is the range of possible positions that could be achieved by intersecting a sensing ray passing through a point in this grid segment with the face of the underlying interpretation. If the vector $\boldsymbol{o}$, lying in the $\hat{s}$-plane, defines the midpoint of the grid segment, this range is given by

$$\alpha_0 \pm \frac{\sqrt{2}h}{2} \tan \theta$$

where α_0 is the value of α for which the ray $o + \alpha\hat{s}$ intersects the face of the pose, h is the size of the grid segment, θ is given by $\cos \theta = {}^s\hat{n} \cdot \hat{s}$, and ${}^s\hat{n}$ is the normal of the face in sensor coordinates.
- Repeat this process for all visible faces. This results in a set of active grid segments, each of which is labeled by possibly several labels of the type described above. This set of labeled active grid segments represents the equivalent of the partition of the sensing plane described in the ideal solution. Note that we have avoided the hidden surface problem by incorporating multiple labels for a grid segment, from a single pose. This may reduce the number of distinguishable segments, by applying additional constraints on the criteria of distinguishability, but it also greatly reduces the computational expense of the process.
- Once a partition of the grid is obtained for each pose, test the grid segments for distinguishability. First, only grid segments that are active in all poses are considered. Such a segment is considered distinguishable if for all pairs of sets of labels, either all the face normals of one label are distinguishable from all the face normals of the other (in the sense defined in the previous section), or all the distance ranges of one label are distinguishable from all the distance ranges of the other (also in the sense defined in the previous section).
- Finally, collect the set of distinguishable grid segments into convex connected components.
- Compute the best sensing position as the center of the largest square (with sides an integral number of grid segments) that can be placed entirely within the set of distinguishable grid segments. Note that if the square has sides of size s then the Chebychev value for the segment is at least $s/2$. This process can be repeated over all sensing directions, and the midpoint of the largest such connected convex collection of distinguishable grid segments can be used to define the best sensing position. To save on computation, it is also possible to define a minimum size for an acceptable convex connected component, and to only apply this process until the first such acceptable component is obtained.

In Fig. 6, we illustrate the above technique on the multiple poses of Fig. 1. Note that each of the small circles denotes a point on the grid of the sensing plane that is distinguishable. We can then determine the best sensing position by finding the largest square area filled by such distinguishable points. The figure illustrates the computation for each of three different sensing rays.

Bounds on Transform Errors

In order to use such an algorithm, we need to determine values for two parameters. First, errors in the computed transformation associated with a pose will affect the threshold needed to determine distinguishability. For example, if there is no error in the computed transformation, then two surface normals are distinguishable if the angle between them exceeds the range of error in measuring such normals. When an error

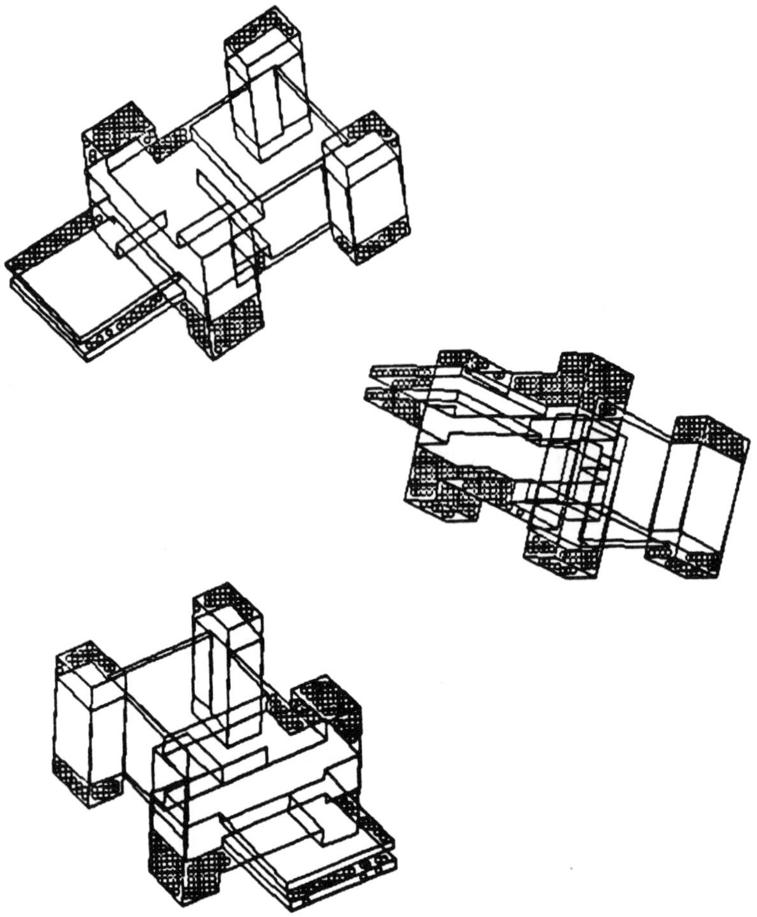

Fig. 6. Example of distinguishable points. For the multiple poses of Fig. 1, we show three different sensing plane projections. The small circles mark positions, on the defined grid, that are distinguishable in these poses.

is present in the transformation, its effect on the expected surface normals must be added to this threshold, thereby reducing the set of distinguishable normals. Second, we need a bound on the minimum Chebychev value (or its approximation) such that errors in the computed transformation will not affect the expected values of the sensor along the sensing ray. In order to deal with these parameters, in this section we will derive theoretical bounds on the possible errors in the computed transformations. In doing so, we will also derive criteria that can be imposed on the computation of the transformation from model coordinates to sensor coordinates in order to reduce the range of possible error. Depending on the sensor data available, it may not always be possible to satisfy these criteria, in which case higher possible errors will have to be tolerated.

Computing the Transform

There are many different methods for determining the transformation from model coordinates to sensor coordinates, and the errors associated with that computation will clearly be dependent on the specific method. To illustrate the disambiguation technique developed here, we choose one particular scheme, and derive specific error bounds on the model transformation for that scheme. This will then allow us to actually test our disambiguation algorithm. We begin by reviewing the process used in [14] for computing the transformation from model coordinates to sensor coordinates.

We are given a set of possible poses of the sensed data, each one consisting of a set of triples $(\boldsymbol{p}_i, \hat{\boldsymbol{n}}_i, \boldsymbol{f}_i)$, where $\boldsymbol{p}_i$ is the vector representing the sensed position, $\hat{\boldsymbol{n}}_i$ is the vector representing the sensed normal, and $\boldsymbol{f}_i$ is the face assigned to this sensed data for that particular pose. We want to determine the actual transformation from model coordinates to sensor coordinates, corresponding to the pose [14].

We assume that a vector in the model coordinate system is transformed into a vector in the sensor coordinate system by the following transformation:

$$\boldsymbol{v}_s = R\boldsymbol{v}_m + \boldsymbol{v}_0$$

where R is a rotation matrix, and $\boldsymbol{v}_0$ is some translation vector. We need to solve for R and $\boldsymbol{v}_0$.

Rotation Component

Suppose $\hat{\boldsymbol{n}}_{m,i}$ is the unit normal, in model coordinates, of face $\boldsymbol{f}_i$, and $\hat{\boldsymbol{n}}_{s,i}$ is the corresponding unit normal in sensor coordinates. Given a two such pairs of model and sensor normals, an estimate of the direction of rotation $\hat{\boldsymbol{r}}_{ij}$ such that a rotation about that direction would take $\hat{\boldsymbol{n}}_{m,i}$ into $\hat{\boldsymbol{n}}_{s,i}$ is given by the unit vector in the direction of

$$(\hat{\boldsymbol{n}}_{m,i} - \hat{\boldsymbol{n}}_{s,i}) \times (\hat{\boldsymbol{n}}_{m,j} - \hat{\boldsymbol{n}}_{s,j}).$$

If there were no error in the sensed normals, we would be done. With error included in the measurements, however, the computed rotation direction $\hat{r}$ could be slightly wrong. One way to reduce the effect of this error is to compute all possible $\hat{r}_{ij}$ as i and j vary over the faces of the pose, and then cluster these computed directions to determine a value for the direction of rotation $\hat{r}$.

Once we have computed a direction of rotation $\hat{r}$, we need to determine the angle θ of rotation about it. This is given by

$$\cos\theta = 1 - \frac{1-(\hat{n}_{s,i}\cdot\hat{n}_{m,i})}{1-(\hat{r}\cdot\hat{n}_{s,i})(\hat{r}\cdot\hat{n}_{m,i})}$$

$$\sin\theta = \frac{(\hat{r}\times\hat{n}_{s,i})\cdot\hat{n}_{m,i}}{1-(\hat{r}\cdot\hat{n}_{s,i})(\hat{r}\cdot\hat{n}_{m,i})}. \qquad (1)$$

Hence, given $\hat{r}$, we can solve for θ. Note that if $\sin\theta$ is zero, there is a singularity in determining θ, which could be either 0 or π. In this case, however, $\hat{r}$ lies in the plane spanned by $\hat{n}_{s,i}$ and $\hat{n}_{m,i}$ and hence, only the $\theta = \pi$ solution is valid.

As before, in the presence of error, we may want to cluster the $\hat{r}$ vectors, and then take the average of the computed values of θ over this cluster.

Finally, given values for both $\hat{r}$ and θ, we can determine the rotation matrix R. Let r_x, r_y, and r_z denote the components $\hat{r}$. Then

$$R = \cos\theta \begin{bmatrix} 1 & 0 & 0 \\ 0 & 1 & 0 \\ 0 & 0 & 1 \end{bmatrix} + (1-\cos\theta)\begin{bmatrix} r_x^2 & r_xr_y & r_xr_z \\ r_yr_x & r_y^2 & r_yr_z \\ r_zr_x & r_zr_y & r_z^2 \end{bmatrix}$$

$$+\sin\theta \begin{bmatrix} 0 & -r_z & r_y \\ r_z & 0 & -r_x \\ -r_y & r_x & 0 \end{bmatrix}.$$

Note that in computing the rotation component of the transformation, we have ignored the ambiguity inherent in the computation. That is, there are two solutions to the problem, $(\hat{r}, \theta)$ and $(-\hat{r}, -\theta)$. We assume that a simple convention concerning the sign of the rotation is used to choose one of the two solutions.

Translation Component

Next, we need to solve for the translation component of the transformation. Suppose we consider three triplets from the pose, $(p_{s,i}, \hat{n}_{s,i}, f_i)$, $(p_{s,j}, \hat{n}_{s,j}, f_j)$, and $(p_{s,k}, \hat{n}_{s,k}, f_k)$ such that the triple product $\hat{n}_{m,i}\cdot(\hat{n}_{m,j}\times\hat{n}_{m,k})$ is non-zero, (i.e. the three face normals are independent). Then, it can be shown that the translation component of the transformation v_0 is given by

$$[\hat{n}_{m,i}\cdot(\hat{n}_{m,j}\times n_{m,k})]v_0 = (\hat{n}_{s,i}\cdot p_{s,i}-d_i)(\hat{n}_{s,j}\times\hat{n}_{s,k})$$
$$+(\hat{n}_{s,j}\cdot p_{s,j}-d_j)(\hat{n}_{s,k}\times\hat{n}_{s,i})$$
$$+(\hat{n}_{s,k}\cdot p_{s,k}-d_k)(\hat{n}_{s,i}\times\hat{n}_{s,j}).$$

As in the case of rotation, if there is no error in the measurements, then we are done. The simplest means of attempting to reduce the effects of error on the computation is to average v_0 over all possible trios of triplets from the pose.

Errors in the Computed Transformation

We now consider possible errors in each of the parameters of the transformation, as a function of error in the sensor measurements. The results are summarized below, more explicit details are given in the Appendix.

Errors in $\hat{r}$: Let $\hat{n}_{m,i}$ be the unit normal of face f_i, in model coordinates, let $\hat{n}'_{m,i}$ be the associated unit normal transformed into sensor coordinates, and let $\hat{n}_{s,i}$ be the actual measured unit normal, in sensor coordinates. Suppose that the sensitivity of the measuring device was ϵ_n, that is

$$\hat{n}'_{m,i}\cdot\hat{n}_{s,i}\geq\epsilon_n.$$

Then an absolute bound on the possible error in the computed value for the direction of rotation $\hat{r}_c$ in relation to the true direction of rotation $\hat{r}_t$ is given by

$$\hat{r}_t\cdot\hat{r}_c \geq \frac{\delta_i\delta_j\sqrt{1-\eta^2}}{\sqrt{1-\{\delta_i\delta_j\eta-\sqrt{1-\delta_i^2}\sqrt{1-\delta_j^2}\}^2}},$$

where

$$\eta = \left(\frac{\hat{n}_{m,i}-\hat{n}'_{m,i}}{|\hat{n}_{m,i}-\hat{n}'_{m,i}|}\right)\cdot\left(\frac{\hat{n}_{m,j}-\hat{n}'_{m,j}}{|\hat{n}_{m,j}-\hat{n}'_{m,j}|}\right)$$

$$\delta_i \geq \sqrt{\frac{\epsilon_n-\hat{n}_{m,i}\cdot\hat{n}'_{m,i}}{1-\hat{n}_{m,i}\cdot\hat{n}'_{m,i}}}.$$

Note that if γ_i is close to one, then the error bound becomes increasingly large. This is to be expected, since in this case, $\hat{n}_{m,i}\approx\hat{n}'_{m,i}$ and thus small errors in the position of $\hat{n}$ can lead to large errors in the position of $\hat{r}$. Similarly, if η is near one, large errors can also result. If we restrict our computation (where possible) to cases where γ_i and η are small, then we have an approximate bound on the error in computing the direction of rotation given by

$$\hat{r}_t\cdot\hat{r}_c\geq\epsilon_n.$$

This bound is supported by the results of the simulations reported in [14].

Errors in θ: We know that the angle of rotation θ is given by

$$\tan\theta = -\frac{\hat{n}'_m\cdot(\hat{r}\times\hat{n}_m)}{(\hat{r}\times\hat{n}'_m)\cdot(\hat{r}\times\hat{n}_m)}$$

where $\hat{n}_m$ is the unit normal of a face in model coordinates; $\hat{n}'_m$ is the corresponding normal transformed into sensor coordinates; and $\hat{r}$ is the direction of rotation.

If we let $\hat{r}_t$ denote the true direction of rotation, $\hat{r}_c$ denote the computed direction of rotation, and $\hat{n}_s$ denote the measured surface normal corresponding to $\hat{n}'_m$, then the constraints on the error in computing the angle θ are that $\hat{r}_t\cdot\hat{r}_c \geq \cos\psi$ and $\hat{n}'_m\cdot\hat{n}_s \geq \epsilon_n = \cos\phi$. In the Appendix we show that the correct value for θ is given by

$$\tan\theta_t = -\frac{1}{\sin\nu}\frac{\cos\alpha}{\cos\beta}$$

where

$$\cos\ \nu = \hat{n}_m \cdot \hat{r}_t$$

$$\cos\ \alpha = \left(\hat{n}'_m \cdot \left(\frac{\hat{r}_t \times \hat{n}_m}{|\hat{r}_t \times \hat{n}_m|}\right)\right)$$

$$\cos\ \beta = \left(\left(\frac{\hat{r}_t \times \hat{n}'_m}{|\hat{r}_t \times \hat{n}'_m|}\right) \cdot \left(\frac{\hat{r}_t \times \hat{n}_m}{|\hat{r}_t \times \hat{n}_m|}\right)\right).$$

Furthermore, we show in the appendix that the worst case for the computed value of θ is given by

$$\tan\ \theta_c = -\frac{1}{\sin\ \omega}\frac{\cos\ (\alpha - [\phi + \xi])}{\cos\ (\beta + [\psi + \gamma])}$$

where

$$\cos\ \omega = \cos\ (\phi + \psi) - (1 - \cos\ \nu)\ \cos\ \phi\ \cos\ \psi$$

$$\cos\ \xi = \cos\ \psi\ \sqrt{\frac{1 - \cos^2\ \nu}{1 - \cos^2\ \psi\ \cos^2\ \nu}}$$

$$\cos\ \gamma = \cos\ \phi\ \cos\ \psi\ \sqrt{\frac{1 - \cos^2\ \nu}{1 - \cos^2\ \omega}}.$$

We could use these expressions to derive bounds on the possible variation in θ as a function of ϕ and ψ, but this is a rather messy task. Instead, we show in the Appendix that if ϕ and ψ are small, then an estimate for $\Delta\theta$ such that

$$\tan\ (\theta_t + \Delta\theta) \approx \tan\ \theta_c$$

is given by

$$|\Delta\theta| \approx |\phi + \psi|.$$

This bound is supported by the results of simulations reported in [14].

Errors in Rv: We have computed expressions for the possible error in $\hat{r}$ and θ. In particular, we will denote the error in θ by $\Delta\theta$ and the vector error in $\hat{r}$ by $\delta\hat{r}$ such that $\hat{r} \cdot \delta\hat{r} = 0$. We now consider the problem of estimating bounds on the possible error in applying the computed rotation matrix to an arbitrary vector $\boldsymbol{v}$. We know that the rotational component of the transformation of $\boldsymbol{v}$ is given by

$$R(\hat{r},\ \theta)\boldsymbol{v} = \cos\ \theta\boldsymbol{v} + (1 - \cos\ \theta)(\hat{r} \cdot \boldsymbol{v})\hat{r} + \sin\ \theta(\hat{r} \times \boldsymbol{v})$$

where $\hat{r}$ and θ are the parameters determining the rotation.

We show in the appendix that if we ignore higher order terms, a Taylor series expansion yields the following bound on errors in the computed value of a rotation:

$$|R(\hat{r} + \delta\hat{r},\ \theta + \Delta\theta)\boldsymbol{v} - R(\hat{r},\ \theta)\boldsymbol{v}| \le (2|\delta\hat{r}| + |\Delta\theta|)|\boldsymbol{v}|.$$

Now, if the errors ϕ and ψ are small, then we know that

$$|\Delta\theta| \le |\phi + \psi|.$$

Furthermore

$$|\delta\hat{r}| = |\sin\ \psi| \approx |\psi|$$

and this implies a bound on variation in $\boldsymbol{v}$ of

$$|3\psi + \phi||\boldsymbol{v}|.$$

Moreover, if we are careful to restrict our computation appropriately, then $\psi \approx \phi$, and thus

$$|R(\hat{r} + \delta\hat{r},\ \theta + \Delta\theta)\boldsymbol{v} - R(\hat{r},\ \theta)| \le |4\phi||\boldsymbol{v}|.$$

Effective Bounds on Rotation Errors

Unfortunately, this is still a fairly weak bound. For example, an error cone of radius $\pi/12$ about the measured surface normals would give rise to potential errors in the computed rotation on the order of the magnitude of the rotated vector. This is obviated to a large extent by the fact that we do not rely on a single measurement in computing the transformation parameters. Rather, we use several sets of measurements, and use the mean value as the result when computing $\hat{r}$ and θ.

To see how this helps reduce the effective bound, consider the following argument. Suppose that the error in computing θ is uniformly distributed over the range $[-2\phi, 2\phi]$. If we take n measurements and average, then the distribution of error about the correct value θ_c should approach a normal distribution, by the Central Limit Theorem. If we assume a uniform distribution for the error in each measurement, then the variance in the error can be shown to equal

$$\frac{4\phi^2}{3}.$$

If there is no systematic error in the measurements, i.e., each measurement error can be considered independent of the others, then the distribution of average error is essentially a zero-mean normal distribution with variance

$$\frac{4\phi^2}{3n}$$

and hence with standard deviation

$$\sqrt{\frac{4\phi^2}{3n}}.$$

Similarly, if the magnitude of the error vector $\delta\hat{r}$ associated with the computation of the direction of rotation $\hat{r}$ is uniformly distributed over its possible range, and the measurements are independent, then the distribution of error in $2|\delta\hat{r}|$ is given by an identical normal distribution, since the maximum error in $|\delta\hat{r}|$ is essentially ϕ. By linearly combining the two distributions, the error in the computation of $R\boldsymbol{v}$ is given by a zero-mean normal distribution with variance $8\phi^2/3n$.

While an absolute bound on the error in computing $R\boldsymbol{v}$ is given by $4|\phi||\boldsymbol{v}|$, tighter, but less certain, bounds are possible. For example, if we impose a 0.95 probability that the error does not exceed the bound, then an expression for this bound is given by the normal distribution error function, and in this particular case, by $(3.92\sqrt{2}/\sqrt{3n})\phi$. As the number of samples n increases, this bound becomes increasingly tighter.

Note that while we have assumed a uniform distribution of

the errors in the individual measurements, this is not a critical assumption. Since we are only seeking estimates for the bounds in computational error, other distributions will give similar results.

In summary, given some lower bound on the number of samples to be used in computing the transformation from model coordinates to sensor coordinates, and given that the assumption of small errors in the measurements of surface orientation holds, then the error in the computed rotation of a vector $\mathbf{v}$ is given by a zero-mean normal distribution, scaled by the magnitude $|\mathbf{v}|$, with standard deviation $(\sqrt{8/3n})\phi$, where n is the number of measurement samples and ϕ is the angle of maximum error in the measurement of surface orientation at each sample point.

Errors in $\mathbf{v}_0$: We know that the translation component of the transformation is given by

$$[\hat{\mathbf{n}}_{m,i}\hat{\mathbf{n}}_{m,j}\hat{\mathbf{n}}_{m,k}]\mathbf{v}_0 = (\hat{\mathbf{n}}'_{m,i} \cdot \mathbf{p}_{s,i} - d_i)(\hat{\mathbf{n}}'_{m,j} \times \hat{\mathbf{n}}'_{m,k}) + (\hat{\mathbf{n}}'_{m,j} \cdot \mathbf{p}_{s,j} - d_j)(\hat{\mathbf{n}}'_{m,k} \times \hat{\mathbf{n}}'_{m,i}) + (\hat{\mathbf{n}}'_{m,k} \cdot \mathbf{p}_{s,k} - d_k)(\hat{\mathbf{n}}'_{m,i} \times \hat{\mathbf{n}}'_{m,j})$$

where $\hat{\mathbf{n}}_{m,i}$ is a face normal in model coordinates, $\hat{\mathbf{n}}'_{m,i}$ is the corresponding face normal transformed into sensor coordinates, $\mathbf{p}_{s,i}$ is the position vector of the contact point in sensor coordinates, and d_i is the constant offset for face i.

If the error in measured surface normals is given by $\epsilon_n = \cos\phi$ such that $\hat{\mathbf{n}}_s \cdot \hat{\mathbf{n}}'_m \geq \epsilon_n$ and the error in measured contact positions is bounded in magnitude by ϵ_d, then the error in each component

$$(\hat{\mathbf{n}}'_{m,k} \cdot \mathbf{p}_{s,k} - d_k)(\hat{\mathbf{n}}'_{m,i} \times \hat{\mathbf{n}}'_{m,j})$$

is bounded in magnitude by

$$\sqrt{[s \sin\zeta - (s+\Delta)\sin(\zeta - 2\phi)]^2 + (s+\Delta)^2 \sin(2\zeta)\sin(4\phi)}$$

where

$$s = \hat{\mathbf{n}}'_{m,k} \cdot \mathbf{p}_{s,k} - d_k$$

$$\Delta \leq \epsilon_d + |\mathbf{p}_{s,k}|\sqrt{2}\sqrt{1-\epsilon_n}$$

$$\cos\zeta = \hat{\mathbf{n}}'_{m,i} \cdot \hat{\mathbf{n}}'_{m,j}.$$

If we restrict our computation to cases in which the faces are nearly orthogonal, then this bound on the components of the translation vector reduces to $|s - (s + \Delta)\cos(2\phi)|$.

Choosing the Parameters

We now consider deriving a formal definition of distinguishability as applied to surface normals and to distances. Consider first the case of distinguishing poses on the basis of measured surface normals. Suppose that α denotes the angle between surface normals associated with two different possible poses. What is the minimum size of α needed to distinguish these poses?

Clearly, the expected normals must differ by an amount that is bigger than the sensitivity of the measurements themselves. Thus, α must at least exceed $2\cos^{-1}\epsilon_n = 2\phi$. Since there is also some error associated with the computed transformations associated with each pose, the angle must also exceed this error. By the previous analysis, if we use a single measurement to determine the rotation matrix R, then this error is bounded by $|4\phi|$ and hence, we have the bound $\alpha \geq 2\phi + 2(4\phi) = 10\phi$. For most values of ϕ, this bound is far too large to be of much use.

If we use several measurements to compute R, however, then more effective bounds can be used. As shown previously, assuming no systematic error implies that the error in the computed surface normal associated with a transformed face is given by a zero-mean normal distribution with standard deviation $(\sqrt{8/3n})\phi$, where n is the number of measurement samples.

This gives us a tighter definition of distinguishable surface normals. In particular, if α denotes the angle between surface normals associated with two distinct poses, those poses are distinguishable if $\alpha \geq 2\phi + 2\rho\phi$. The first term denotes the range of possible error in the measurement of the surface normals, and the second term denotes the range of possible error in the expected values of the surface normals. Here, ρ is a scale factor that is a function of the reliability of the error bound. That is, $\rho(c)$ denotes the point in the normal distribution described above such that c percent of the weight of the distribution lies below the value ρ.

For example, if the cutoff on the reliability of the bound is 0.95, and the number of measurements involved in computing the transformation is at least 10, then $\rho \leq 1.01$. Thus the bound on two surface normals being distinguishable is 4.02ϕ.

We can also derive a formal definition of distinguishability based on position measurements. We first note that if a face is defined by the pair $(\hat{\mathbf{n}}_m, d)$ in model coordinates, such that a point $\mathbf{v}$ lies on the plane of the face if $\mathbf{v}\cdot\hat{\mathbf{n}}_m - d = 0$, then the same face, after transformation, is defined by the pair $(\hat{\mathbf{n}}'_m, d')$, where

$$\hat{\mathbf{n}}'_m = R\hat{\mathbf{n}}_m$$

$$d' = d + (\mathbf{v}_0 \cdot R\hat{\mathbf{n}}_m).$$

We let the error associated with the computed value of $\hat{\mathbf{n}}'_m$ be denoted by $\mathbf{w}$ such that $\mathbf{w}\cdot\hat{\mathbf{n}}'_m = 0$. The magnitude of $\mathbf{w}$ can be bounded above by $|\sin(\rho\phi)|$, by the discussion above. We also let the error in computing $\mathbf{v}_0$ be denoted by $\mathbf{u}$. We now seek a bound on the possible errors in computing the point of intersection of a sensing ray with an object face, due to errors in the computed transformation.

Suppose that the sensing ray is given by $\alpha\hat{\mathbf{s}} + \mathbf{o}$ where $\hat{\mathbf{s}}$ is a specified unit vector, $\mathbf{o}$ is a specified offset vector orthogonal to $\hat{\mathbf{s}}$, and α is the free parameter specifying position along the sensing ray. The correct parameter of intersection of the sensing ray with the transformed face is given by the value of α such that

$$(\alpha\hat{\mathbf{s}} + \mathbf{o}) \cdot \hat{\mathbf{n}}'_m - d' = 0$$

or

$$\alpha_t = \frac{d' - \mathbf{o} \cdot \hat{\mathbf{n}}'_m}{\hat{\mathbf{s}} \cdot \hat{\mathbf{n}}'_m}.$$

On the other hand, if we include the potential error in the computed transformation, then the point of intersection is given by

$$\alpha_m = \frac{[d' + \boldsymbol{u} \cdot (\hat{\boldsymbol{n}}'_m + \boldsymbol{w}) + \boldsymbol{v}_0 \cdot \boldsymbol{w}] - \boldsymbol{o} \cdot (\hat{\boldsymbol{n}}'_m + \boldsymbol{w})}{\hat{\boldsymbol{s}} \cdot (\hat{\boldsymbol{n}}'_m + \boldsymbol{w})}$$

and thus the difference is given by

$$r = \frac{\boldsymbol{u} \cdot \hat{\boldsymbol{n}}'_m + (\boldsymbol{u} + \boldsymbol{v}_0 - \boldsymbol{o}) \cdot \boldsymbol{w}}{\hat{\boldsymbol{s}} \cdot \hat{\boldsymbol{n}}'_m + \hat{\boldsymbol{s}} \cdot \boldsymbol{w}} - \frac{\hat{\boldsymbol{s}} \cdot \boldsymbol{w}}{\hat{\boldsymbol{s}} \cdot \hat{\boldsymbol{n}}'_m + \hat{\boldsymbol{s}} \cdot \boldsymbol{w}} \alpha_t .$$

As a consequence, we can bound the error in the expected intersection point of the sensing ray with the face by

$$r \leq \frac{|\boldsymbol{u}| + (|\boldsymbol{u}| + |\boldsymbol{v}_0 - \boldsymbol{o}|)|\boldsymbol{w}|}{|\hat{\boldsymbol{s}} \cdot \hat{\boldsymbol{n}}'_m| - |\boldsymbol{w}|} + \frac{|\boldsymbol{w}|}{|\hat{\boldsymbol{s}} \cdot \hat{\boldsymbol{n}}'_m| - |\boldsymbol{w}|} \alpha_t ,$$

where

- $\boldsymbol{w}$ is the error in computing $\hat{\boldsymbol{n}}'_m = R\hat{\boldsymbol{n}}_m$;
- α_t is the predicted intersection point;
- $\boldsymbol{u}$ is the error in $\boldsymbol{v}_0$;
- $\hat{\boldsymbol{s}}$ is the sensing direction;
- $\boldsymbol{o}$ is the sensing offset vector;
- $\hat{\boldsymbol{n}}'_m$ is the face normal in transformed coordinates; and
- $\boldsymbol{v}_0$ is the computed translation.

Thus, given this bound, two poses are distinguishable if their expected points of intersection are large enough

$$|\alpha_1 - \alpha_2| \geq 2\epsilon_d + r_1 + r_2 .$$

As in the case of distinguishing on the basis of surface normals, the bounds for $\boldsymbol{w}$ and $\boldsymbol{u}$ may be too large to be practical. We can reduce these bounds by using several measurements to determine a value for $\boldsymbol{v}_0$. As in the previous case, this will lead to a zero-mean normal distribution of expected error, and the effective range of error will be reduced.

Finally, we need to place bounds on the minimum Chebychev values needed to guarantee that perturbations in the computed transformations will not cause the sensing ray to miss the intended face. Let c denote the Chebychev value associated with a particular face, whose transformed unit surface normal is $\hat{\boldsymbol{n}}'_m$, and where the unit sensing ray is given by $\hat{\boldsymbol{s}}$. Then the modified Chebychev value in the sensing plane is given by $c|\hat{\boldsymbol{n}}'_m \cdot \hat{\boldsymbol{s}}|$. At the same time, the variation in the position of a point on the face, as a function of error in the computed transformation, is given by

$$\Delta \boldsymbol{p} = \boldsymbol{u} + \boldsymbol{q} .$$

where, as above, $\boldsymbol{u}$ denotes the error in the computed value of $\boldsymbol{v}_0$ and $\boldsymbol{q}$ is the error in the computed value of $R\boldsymbol{p}$, where $\boldsymbol{p}$ is the Chebychev point in model coordinates. The magnitudes of these error vectors are bounded by the expressions derived above. Since the directions of the error vectors are arbitrary, the condition on the Chebychev value required to ensure contact with the face is

$$c \geq \frac{|\boldsymbol{u}| + |\boldsymbol{q}|}{|\hat{\boldsymbol{n}}'_m \cdot \hat{\boldsymbol{s}}|} .$$

Thus, we have derived conditions on the parameters of the disambiguation algorithm needed to guarantee the performance of the algorithm.

Discussion and Examples

When Do We Compute the Sensing Directions?

We have described a technique for determining optimal sensing directions. We have still to consider, however, how to interface such a technique with the general problem of recognition and localization. The simplest method is to obtain some initial set of sensory data points, apply our recognition technique, and then use the disambiguation process as required, based on the current set of consistent poses. For example, if there are several consistent poses, we could choose the first pair, compute an optimal sensing direction based on that pair and obtain a new data point. Then we could determine which of the set of poses are also consistent with the new data point and iterate. This technique, while applicable to arbitrary sets of objects, has the disadvantage of high computational expense.

In situations in which a large number of objects are possible, we may not be able to do any better than to compute sensing points as needed, based on the current set of feasible poses. In situations involving a single object, however, there may be an alternative method for integrating the computation of sensing positions with the interpretation of the sensory data.

In particular, given the analysis developed here, one can precompute optimal sensing rays as a function of the difference in transformation associated with two poses. Take any pair of poses of an object. There exists a rigid transformation taking one pose into the other, which we can parameterize in some fashion. We compute the optimal sensing direction for this pair of poses, and insert it into a lookup table, whose dimensions are indexed by the parameters of the relative transformation. Since the workspace of the sensory system is bounded, this is a bounded table (that is, the translational degrees-of-freedom are not infinite in extent). The analysis can be used to compute an optimal sensing ray corresponding to each entry of the table, where the parameters of the transformation are quantized to some desired level.

Now, when attempting to disambiguate two possible poses, one simply computes the difference in the transformations, looks up of the precomputed sensing ray in the appropriate slot of the table, transforms that ray by the transformation associated with the first pose, and then senses along that ray to obtain a new data point. That data point is added to the current set of sensory data, and the recognition and localization process is applied. If a unique pose results, the process is stopped; if not, a new sensing ray is obtained and the process continues.

By precomputing the sensing rays, we can avoid the computational expense associated with finding a new sensing position, and at the same time take advantage of the efficiency of the technique is disambiguating multiple poses.

Avoiding False Negatives

We have seen in the previous discussion that the analytic error bounds on the computed transformations for any pose are probably too large to be practical. We argued that one way to reduce these bounds was to use several measurements in the computation of the transformation. This led to a normal distribution of error in each of the components of the transformation, and thus, given a level of desired confidence in the algorithm, tighter bounds on the parameters were possible. In this case, we would expect that in general the algorithm will succeed, and we need only consider alterations to the algorithm to deal with the infrequent case when the errors in the computed transformation do exceed the expected thresholds. There are two situations that can arise in this case. The first is that the perturbation in the transform causes a surface normal to be sensed, that does not agree with any of the expected normals. This is essentially a false negative, since it implies that the poses are not distinguishable. The more damaging case is a false positive, in which the perturbation in the transformation results in a sensor measurement that coincidentally agrees with the wrong pose.

The easiest solution is to use more than one sense point. In this manner, false negatives are easily handled, since the expectation is that not all sensed points will give inconsistent data. This will be especially true if several sensing directions are used, in particular if the sensing directions are orthogonal. As well, it is likely that false positives can also be detected, since the expectation is that the correct pose will be found by most sensor points, again especially if several directions are used, and a simple voting scheme will arrive at the correct answer.

Testing the Algorithm

We have implemented the described technique, and tested it on a number of examples. Because the worst case bounds are so large, we used the approximations described above, with the expectation that on occasion an incorrect decision would be made, but that such errors could be avoided by voting over several additional sensing points.

In particular, we ran the algorithm described in [14] for an object in arbitrary orientation relative to the sensors and with simulated sensing from three orthogonal directions. Whenever there was an ambiguity in interpreting the sensed data, we used the following disambiguation technique. We used the analysis developed above to predict a sensing ray, and for each pose we predicted ranges of expected values for the sensory data along that ray. We then acquired an additional sense point along the chosen sensing ray, and compared the recorded value with the expected ranges to choose a pose.

Using a variety of simulated sensing errors, the disambiguation technique was applied to 1000 ambiguous cases. It was found in 336 of these cases that, due to the large errors inherent in the sensory data, the algorithm could not distinguish reliably between the possible solutions. In all of these cases, the poses differed by the reassignment of one data point from one face to an adjacent face, and this resulted in nearly identical transformations associated with the poses. Relative to the error resolution of the sensing devices, these can be considered to be identical solutions. In 633 of the cases, the disambiguation algorithm was able to determine the correct pose with only a single additional sensory point. In the remaining 31 cases, the algorithm chose an incorrect pose from the set of consistent poses.

We also ran a second version of the disambiguation algorithm on the same set of data. In this case, rather than using predicted range of values to choose a pose, we simply used the technique to generate the next sensing direction, and then ran the RAF recognition algorithm [14], [15] with that sensory point added to the original set of sensory data. In this case, we found that the algorithm identified the correct pose in all 664 cases, with only from one to three additional sensory points required to complete the identification.

Acknowledgment

Tomás Lozano-Pérez was critical to the development of this work, both in terms of the underlying recognition technique and in terms of valuable comments and criticisms of the material presented here. John Hollerbach and Rodney Brooks provided useful comments on earlier drafts.

Appendix

In the appendix, we present a more detailed error analysis of the computation of the transformation from model to sensor coordinates.

Errors in $\hat{r}$

We begin by considering the range of possible errors in the computation of the direction of rotation $\hat{r}$. By the analysis of [14] the rotation direction $\hat{r}$ is computed by taking two pairs $(\hat{n}_m, \hat{n}'_m)$, where $\hat{n}_m$ is the unit normal of a face of the model, and $\hat{n}'_m$ is the same unit normal rotated into sensor coordinates, and letting $\hat{r}$ be the unit vector in the direction of

$$(\hat{n}_{m,i}-\hat{n}'_{m,i})\times(\hat{n}_{m,j}-\hat{n}'_{m,j}).$$

We assume that we are given $\hat{n}_{m,i}$, $\hat{n}'_{m,i}$ and that the sensitivity of the sensor to errors in surface orientation is given by ϵ_n. That is, if $\hat{n}'_{m,i}$ is the correct surface normal transformed into sensor coordinates and $\hat{n}_s$ is the actual measured (or sensed) surface normal, then

$$\hat{n}_s \cdot \hat{n}'_{m,i} \geq \epsilon_n. \tag{4}$$

We will consider two stages in deriving bounds on the error in computing $\hat{r}$. If we let

$$\hat{v}_i=\frac{\hat{n}_{m,i}-\hat{n}'_{m,i}}{|\hat{n}_{m,i}-\hat{n}'_{m,i}|}$$

and

$$\hat{u}_i=\frac{\hat{n}_{m,i}-\hat{n}_{s,i}}{|\hat{n}_{m,i}-\hat{n}_{s,i}|}$$

then the correct value for $\hat{r}$ is given by

$$\hat{r}_t=\frac{\hat{v}_i\times\hat{v}_j}{\sqrt{1-(\hat{v}_i\cdot\hat{v}_j)^2}}$$

and the computed value is given by

$$\hat{r}_c = \frac{\hat{u}_i \times \hat{u}_j}{\sqrt{1-(\hat{u}_i \cdot \hat{u}_j)^2}}.$$

We will first derive bounds on $\hat{v}_i \cdot \hat{u}_i$ and then use the result to bound $\hat{r}_t \cdot \hat{r}_c$.

The vector $\hat{n}_s$ can be represented by the following parameterization

$$\hat{n}_s = \alpha \hat{n}'_m + \beta \hat{n}_m + \delta(\hat{n}_m \times \hat{n}'_m).$$

Then equation (4) produces the inequality

$$\alpha + \beta\gamma \geq \epsilon_n \tag{4'}$$

where $\gamma = \hat{n}_m \cdot \hat{n}'_m$. We will consider the worst case, in which equality holds. Furthermore, the fact that $\hat{n}_s$ is a unit vector yields the following constraint

$$\alpha^2 + 2\alpha\beta\gamma + \beta^2 + \delta^2 = 1. \tag{5}$$

Given $\hat{n}_m$, $\hat{n}'_m$, and $\hat{n}_s$, we first consider the range of possible values for $\hat{n}_m - \hat{n}_s$, relative to $\hat{n}_m - \hat{n}'_m$, that is, we want bounds on the range of possible values for

$$E = \hat{v} \cdot \hat{u} = \frac{(\hat{n}_m - \hat{n}'_m) \cdot (\hat{n}_m - \hat{n}_s)}{|\hat{n}_m - \hat{n}'_m||\hat{n}_m - \hat{n}_s|}.$$

It is straightforward to show that

$$|\hat{n}_m - \hat{n}'_m| = \sqrt{2(1-\gamma)}.$$

Furthermore, using (5), one can show that

$$|\hat{n}_m - \hat{n}_s| = \sqrt{2(1-\beta-\alpha\gamma)}.$$

Finally, expanding out the dot product and substituting yields

$$E = \frac{(1-\beta+\alpha)(1-\gamma)}{2\sqrt{1-\gamma}\sqrt{1-\beta-\alpha\gamma}}.$$

By equation (4′), $\alpha = \epsilon_n - \beta\gamma$, and substitution yields

$$E = \frac{\sqrt{1-\gamma}[1+\epsilon_n - \beta(1+\gamma)]}{2\sqrt{1-\epsilon_n\gamma - \beta(1-\gamma^2)}}.$$

The first problem to consider is what is the minimum value for E as β varies. In particular, we find that

$$\frac{\partial E}{\partial \beta} = \frac{\sqrt{1-\gamma}(1+\gamma)^2}{4} \frac{[\beta(1-\gamma)-(1-\epsilon_n)]}{(1-\epsilon_n\gamma-\beta(1-\gamma^2))^{3/2}}.$$

This is zero when

$$\beta = \frac{1-\epsilon_n}{1-\gamma} \tag{6}$$

and this is a valid value for β provided $\gamma < \epsilon_n$. Taking a second partial derivative of E, we find that the sign of $\partial^2 E/\partial\beta^2$ is given by the sign of

$$\beta(1-\gamma^2) + 3(\epsilon_n - \gamma) + \epsilon_n\gamma - 1.$$

Substituting equation (6), we find that the sign of the second partial derivative is given by the sign of

$$2(\epsilon_n - \gamma)$$

and this is positive, since $\gamma < \epsilon_n$. Hence, E achieves a minimum at the value of β given by (6), and this value is

$$E = \sqrt{\frac{\epsilon_n - \gamma}{1-\gamma}}. \tag{7}$$

If $\gamma > \epsilon_n$, then the minimum value for E occurs for β at the limit of its range, namely $\beta = 1$. In this case, $E \leq 0$ and is minimized when $\gamma = \sqrt{\epsilon_n}$, taking the value

$$E = \frac{-(1-\sqrt{\epsilon_n})}{2}.$$

In general, we will try to restrict the computation of $\hat{r}$ to those cases in which $\gamma < \epsilon_n$, in order to keep the magnitude of the possible error in $\hat{r}$ small. Note that the minimum value for E is monotonic in γ, that is, the minimum E increases as γ decreases towards -1.

Thus we obtain the bound

$$\hat{u}_i \cdot \hat{v}_i \geq \sqrt{\frac{\epsilon_n - \gamma_i}{1-\gamma_i}} = \delta_i$$

where

$$\gamma_i = (\hat{n}_{m,i} \cdot \hat{n}'_{m,i}).$$

We are now interested in obtaining bounds on $\hat{r}_t \cdot \hat{r}_c$. In essence, we have two cones in the Gaussian sphere, centered about $\hat{v}_i$ and $\hat{v}_j$, of radius δ_i and δ_j, respectively. The possible values of $\hat{r}_c$ are given by the normalized cross products of vectors within these cones. Clearly, if the cones overlap, then the computation for $\hat{r}$ is unstable. We avoid this case by requiring that the cones do not overlap.

Note that if all the error in the computation of either $\hat{u}_i$ or $\hat{u}_j$ lies in the plane spanned by $\hat{v}_i$ and $\hat{v}_j$, then the normalization of the cross product will result in the correct value $\hat{r}_t = \hat{r}_c$. The maximum deviation of $\hat{r}_c$ from $\hat{r}_t$ will occur when the error between $\hat{u}_i$ and $\hat{v}_i$ and the error between $\hat{u}_j$ and $\hat{v}_j$ lie maximally separated from this plane. This requires that we check two cases, one in which the errors lie on the same side of the plane and one in which the errors lie on opposite sides of the plane. We now consider the first case.

Let $\eta = \hat{v}_i \cdot \hat{v}_j$. Then

$$\hat{u}_i = \delta_i \hat{v}_i + \sqrt{\frac{1-\delta_i^2}{1-\eta^2}}(\hat{v}_i \times \hat{v}_j)$$

$$\hat{u}_j = \delta_j \hat{v}_j + \sqrt{\frac{1-\delta_j^2}{1-\eta^2}}(\hat{v}_i \times \hat{v}_j)$$

$$\hat{\boldsymbol{u}}_i \times \hat{\boldsymbol{u}}_j = \delta^2(\hat{v}_i \times \hat{v}_j) + \delta_j \sqrt{\frac{1-\delta_i^2}{1-\eta^2}}\,[(\hat{v}_i \times \hat{v}_j) \times \hat{v}_j]$$

$$+\,\delta_i \sqrt{\frac{1-\delta_j^2}{1-\eta^2}}\,[\hat{v}_i \times (\hat{v}_i \times \hat{v}_j)].$$

Thus

$$(\hat{v}_i \times \hat{v}_j) \cdot (\hat{\boldsymbol{u}}_i \times \hat{\boldsymbol{u}}_j) = \delta_i \delta_j [1-\eta^2]$$

and

$$\hat{\boldsymbol{u}}_i \cdot \hat{\boldsymbol{u}}_j = \delta_i \delta_j \eta + \sqrt{1-\delta_i^2}\sqrt{1-\delta_j^2}.$$

Then, by substitution

$$\hat{\boldsymbol{r}}_t \cdot \hat{\boldsymbol{r}}_c \geq \frac{\delta_i \delta_j \sqrt{1-\eta^2}}{\sqrt{1-\{\delta_i \delta_j \eta + \sqrt{1-\delta_i^2}\sqrt{1-\delta_j^2}\}^2}}. \qquad (8)$$

In the second case, we change $\hat{\boldsymbol{u}}_j$ to

$$\hat{\boldsymbol{u}}_j = \delta_j \hat{v}_j - \sqrt{\frac{1-\delta_j^2}{1-\eta^2}}\,(\hat{v}_i \times \hat{v}_j)$$

and thus

$$\hat{\boldsymbol{u}}_i \times \hat{\boldsymbol{u}}_j = \delta_j \delta_i (\hat{v}_i \times \hat{v}_j) + \delta_j \sqrt{\frac{1-\delta_i^2}{1-\eta^2}}\,[(\hat{v}_i \times \hat{v}_j) \times \hat{v}_j]$$

$$-\,\delta_i \sqrt{\frac{1-\delta_j^2}{1-\eta^2}}\,[\hat{v}_i \times (\hat{v}_i \times \hat{v}_j)].$$

Following through the same algebra leads to a bound on the dot product of

$$\hat{\boldsymbol{r}}_t \cdot \hat{\boldsymbol{r}}_c \geq \frac{\delta_i \delta_j \sqrt{1-\eta^2}}{\sqrt{1-\{\delta_i \delta_j \eta - \sqrt{1-\delta_i^2}\sqrt{1-\delta_j^2}\}^2}} \qquad (9)$$

where

$$\eta = \hat{v}_i \cdot \hat{v}_j$$

$$\hat{v}_i = \frac{\hat{\boldsymbol{n}}_{m,i} - \hat{\boldsymbol{n}}'_{m,i}}{|\hat{\boldsymbol{n}}_{m,i} - \hat{\boldsymbol{n}}'_{m,i}|}$$

$$\delta_i \geq \sqrt{\frac{\epsilon_n - \gamma_i}{1-\gamma_i}}$$

$$\gamma_i = \hat{\boldsymbol{n}}_{m,i} \cdot \hat{\boldsymbol{n}}'_{m,i}.$$

It is straightforward to show that the bound in (9) is in fact smaller than the one in (8).

Note that if γ_i is close to one, then the error bound comes increasingly large. This is to be expected, since in this case, $\hat{\boldsymbol{n}}_{m,i} \approx \hat{\boldsymbol{n}}'_{m,i}$, and thus small errors in the position of $\hat{\boldsymbol{n}}'_m$ can lead to large errors in the position of $\hat{\boldsymbol{r}}$. Similarly, if η is near one, large errors can also result. If we restrict our computation (where possible) to cases where γ_i and η are small, then we have an approximate bound on the error in computing the direction of rotation given by $\hat{\boldsymbol{r}}_t \cdot \hat{\boldsymbol{r}}_c \geq \epsilon_n$. This bound is supported by the results of the simulations reported in [14].

Errors in θ: We now want to consider bounds on the possible error in computing the remaining parameter of the rotation component of the transformation, namely, the angle of rotation θ. Given the expressions in (1) for cos θ and sin θ, the value of θ is given by

$$\tan\,\theta = -\frac{\hat{\boldsymbol{n}}'_m \cdot (\hat{\boldsymbol{r}} \times \hat{\boldsymbol{n}}_m)}{(\hat{\boldsymbol{r}} \times \hat{\boldsymbol{n}}'_m) \cdot (\hat{\boldsymbol{r}} \times \hat{\boldsymbol{n}}_m)}$$

where $\hat{\boldsymbol{n}}_m$ is the unit normal of a face in model coordinates and $\hat{\boldsymbol{n}}'_m$ is the corresponding normal transformed into sensor coordinates.

As in the previous section, we let $\hat{\boldsymbol{r}}_t$ denote the true direction of rotation and $\hat{\boldsymbol{r}}_c$ the computed direction of rotation. We will assume that the error in the computed direction of rotation is bounded by

$$\hat{\boldsymbol{r}}_t \cdot \hat{\boldsymbol{r}}_c \geq \delta_r$$

and that the measured value for $\hat{\boldsymbol{n}}'_m$ is given by $\hat{\boldsymbol{n}}_s$ such that

$$\hat{\boldsymbol{n}}'_m \cdot \hat{\boldsymbol{n}}_s \geq \epsilon_n.$$

We shall make use of the following four unit vectors:

$$\hat{\boldsymbol{w}} = \frac{\hat{\boldsymbol{r}}_t \times \hat{\boldsymbol{n}}_m}{|\hat{\boldsymbol{r}}_t \times \hat{\boldsymbol{n}}_m|}$$

$$\hat{\boldsymbol{u}} = \frac{\hat{\boldsymbol{r}}_c \times \hat{\boldsymbol{n}}_m}{|\hat{\boldsymbol{r}}_c \times \hat{\boldsymbol{n}}_m|}$$

$$\hat{\boldsymbol{s}} = \frac{\hat{\boldsymbol{r}}_t \times \hat{\boldsymbol{n}}'_m}{|\hat{\boldsymbol{r}}_t \times \hat{\boldsymbol{n}}'_m|}$$

$$\hat{\boldsymbol{t}} = \frac{\hat{\boldsymbol{r}}_c \times \hat{\boldsymbol{n}}'_m}{|\hat{\boldsymbol{r}}_c \times \hat{\boldsymbol{n}}'_m|}$$

Given these definitions, it is straightforward to show that

$$\tan\,\theta_t = \frac{-1}{|\hat{\boldsymbol{r}}_t \times \hat{\boldsymbol{n}}'_m|}\,\frac{(\hat{\boldsymbol{n}}'_m \cdot \hat{\boldsymbol{w}})}{(\hat{\boldsymbol{s}} \cdot \hat{\boldsymbol{w}})}$$

$$\tan\,\theta_c = \frac{-1}{|\hat{\boldsymbol{r}}_c \times \hat{\boldsymbol{n}}_s|}\,\frac{(\hat{\boldsymbol{n}}_s \cdot \hat{\boldsymbol{u}})}{(\hat{\boldsymbol{t}} \cdot \hat{\boldsymbol{u}})}.$$

Our method in obtaining bounds on the deviation between these two expressions will be to bound $\hat{\boldsymbol{n}}_s \cdot \hat{\boldsymbol{u}}$ as a function of $\hat{\boldsymbol{n}}'_m \cdot \hat{\boldsymbol{w}}$, and to bound $\hat{\boldsymbol{t}} \cdot \hat{\boldsymbol{u}}$ as a function of $\hat{\boldsymbol{s}} \cdot \hat{\boldsymbol{w}}$. Once we have bounds on these expressions, they can be combined to bound the overall expression for tan θ.

First, we consider the range of values for

$$\hat{\boldsymbol{u}} \cdot \hat{\boldsymbol{w}} = \frac{(\hat{\boldsymbol{r}}_c \times \hat{\boldsymbol{n}}_m)}{|\hat{\boldsymbol{r}}_c \times \hat{\boldsymbol{n}}_m|} \cdot \frac{(\hat{\boldsymbol{r}}_t \times \hat{\boldsymbol{n}}_m)}{|\hat{\boldsymbol{r}}_t \times \hat{\boldsymbol{n}}_m|}.$$

In considering the range of values for $\hat{\boldsymbol{u}} \cdot \hat{\boldsymbol{w}}$ we note that because of the normalization of the vectors, any error in $\hat{\boldsymbol{r}}_c$

lying in the $\hat{r}_t - \hat{n}_m$ plane will have no effect on the dot product. Thus the worst case occurs when all of the error lies perpendicular to this plane. Hence, we need only consider the cases where

$$\hat{r}_c = \alpha\hat{r}_t \pm \beta(\hat{r}_t \times \hat{n}_m),$$

where

$$\alpha \geq \delta_r$$
$$1 = \alpha^2 + \beta^2(1 - \cos^2 \nu)$$
$$\cos \nu = \hat{r}_t \cdot \hat{n}_m.$$

Now, the worst case will occur for $\alpha = \delta_r$, in which case

$$\beta^2 = \frac{1 - \delta_r^2}{1 - \cos^2 \nu}$$

so that the worst case will arise for

$$\hat{r}_c = \delta_r \hat{r}_t \pm \sqrt{\frac{1 - \delta_r^2}{1 - \cos^2 \nu}} (\hat{r}_t \times \hat{n}_m).$$

In this case, the following expressions hold:

$$\hat{r}_c \times \hat{n}_m = \delta_r(\hat{r}_t \times \hat{n}_m) \pm \sqrt{\frac{1 - \delta_r^2}{1 - \cos^2 \nu}} (\hat{n}_m \times (\hat{n}_m \times \hat{r}_t))$$
$$(\hat{r}_c \times \hat{n}_m) \cdot (\hat{r}_c \times \hat{n}_m) = 1 - \delta_r^2 \cos^2 \nu$$
$$(\hat{r}_t \times \hat{n}_m) \cdot (\hat{r}_t \times \hat{n}_m) = 1 - \cos^2 \nu$$
$$(\hat{r}_c \times \hat{n}_m) \cdot (\hat{r}_t \times \hat{n}_m) = \delta_r(1 - \cos^2 \nu).$$

Thus we have the bound

$$\hat{u} \cdot \hat{w} \geq \mu$$

where

$$\mu = \delta_r \sqrt{\frac{1 - \cos^2 \nu}{1 - \delta_r^2 \cos^2 \nu}}.$$

At this stage, we have $\hat{n}_s \cdot \hat{n}'_m \geq \epsilon_n$ and $\hat{u} \cdot \hat{w} \geq \mu$. We can visualize this situation by considering two cones in the Gaussian sphere, one centered about $\hat{n}'_m$ with radius ϵ_n, and one centered about $\hat{w}$ with radius μ. We are essentially asking for bounds on the range of dot products between vectors lying within these two different cones. Assuming that the cones do not overlap, the maximum and minimum dot products will occur for the minimum and maximum angles between elements of the cones, respectively, and this clearly occurs for vectors lying in the cones and lying in the $\hat{n}'_m - \hat{w}$ plane.

Suppose we denote

$$\cos \alpha = (\hat{n}'_m \cdot \hat{w})$$
$$\cos \phi = \epsilon_n$$
$$\cos \psi = \delta_r$$
$$\cos \xi = \mu.$$

Clearly, the extremal angles for these two cones are given by

$$\alpha \pm [\phi + \xi].$$

Thus the range of possible values for $\hat{n}_s \cdot \hat{w}$ is bounded by $\cos(\alpha \pm [\phi + \xi])$.

An analogous argument can be made for the dot product $\hat{t} \cdot \hat{u}$. If we let

$$\cos \beta = (\hat{s} \cdot \hat{w})$$
$$\cos \gamma = \rho$$

$$\cos \omega = \delta_r \epsilon_n \cos \nu - \sqrt{1 - \delta_r^2}\sqrt{1 - \epsilon_n^2}$$
$$= \cos(\phi + \psi) - (1 - \cos \nu) \cos \phi \cos \psi$$
$$= (\hat{r}_c \cdot \hat{n}_s)$$

where ρ is the bound

$$\rho = \delta_r \epsilon_n \sqrt{\frac{1 - \cos^2 \nu}{1 - \cos^2 \omega}}$$

then the range of possible values for

$$\hat{t} \cdot \hat{u}$$

is bounded by

$$\cos(\beta \pm [\psi + \gamma]).$$

Finally

$$|\hat{r}_t \times \hat{n}'_m| = \sin \nu$$
$$|\hat{r}_c \times \hat{n}_s| = \sin \omega.$$

Thus by gathering all these expressions together, we obtain the following worst case expressions:

$$\tan \theta_t = \frac{-1}{\sin \nu} \frac{\cos \alpha}{\cos \beta}$$

$$\tan \theta_c = \frac{-1}{\sin \omega} \frac{\cos(\alpha - (\phi + \xi))}{\cos(\beta + (\psi + \gamma))}.$$

We note that this expression for the computed value of $\tan \theta$ has the expected limiting case behavior. In particular, if the error parameters $\phi = \psi = 0$, then the entire expression for $\tan \theta_c$ reduces to that for $\tan \theta_t$.

We now seek an estimate for the error in the computed value of θ, in the special case of ϕ and ψ both small. In particular, we would like an expression for $\Delta\theta$ such that

$$\tan(\theta_t + \Delta\theta) \approx \tan(\theta_c).$$

In this way, we can place a bound on the possible error in the computation of θ.

In the limiting case of ϕ and ψ small, the bound $\mu \approx \delta_r$ so that $\xi \approx \psi$. Furthermore, $\cos \omega \approx \cos \nu$ so that $\cos \gamma \approx \cos \phi \cos \psi \approx \cos(\phi + \psi)$ and hence, $\gamma \approx \phi + \psi$. As a consequence, finding an approximation for the deviation in tan

θ reduces to comparing the worst case of deviation between

$$\tan(\theta_c) = \frac{\cos(\alpha - [\phi + \psi])}{\cos(\beta + [\phi + 2\psi])}$$

and

$$\tan(\theta_t) = \frac{\cos\alpha}{\cos\beta}. \quad (10)$$

By expansion

$$\tan(\theta_t + \Delta\theta) = \frac{\tan\theta_t + \tan\Delta\theta}{1 - \tan\theta_t \tan\Delta\theta}$$

and if we substitute $\Delta\theta \approx \phi + \psi$, and use (10), then this expression can be expanded into

$$\frac{\cos(\alpha - \phi - \psi) + [\cos\beta - \sin\alpha]\sin(\phi + \psi)}{\cos(\beta + \phi + \psi) + [\sin\beta - \cos\alpha]\sin(\phi + \psi)}. \quad (11)$$

Now, if $\cos\beta \approx \sin\alpha$, then the second term in both the numerator and denominator can be ignored, especially since $\sin(\phi + \psi)$ is also small. Requiring this to be true is equivalent to requiring that

$$(\hat{s} \cdot \hat{w})^2 + (\hat{n}'_m \cdot \hat{w})^2 \approx 1$$

that is, that the component of the unit vector $\hat{w}$ in the direction of $\hat{n}'_m \times \hat{s}$ be small. It is straightforward to show that

$$(\hat{w} \cdot (\hat{n}'_m \times \hat{s}))^2 = [\cot\nu(\hat{n}_m \cdot \hat{s})]^2.$$

Since we have already indicated that we will restrict our computation of the transformation parameters to those cases in which $\hat{r} \cdot \hat{n}_m \ll 1$, it follows that $\cot\nu$ is small and the second terms in both the numerator and denominator in (11) can be disregarded.

By dropping these terms, we see that the remaining expression reduces to

$$\tan(\theta_t - [\phi + \psi]) \approx \frac{\cos(\alpha - [\phi + \psi])}{\cos(\beta + [\phi + \psi])}.$$

Thus if ψ is small enough, it follows that the worst case deviation is given by $\theta_c \approx \theta_t + (\phi + \psi)$ and hence that a good approximation to the error $\Delta\theta$ in the computed value of the rotation θ is given by

$$\Delta\theta \approx |\phi + \psi|.$$

Errors in Rv: We have computed expressions for the possible error in $\hat{r}$ and θ. In particular, we will denote the error in θ by $\Delta\theta$ and the vector error in $\hat{r}$ by $\delta\hat{r}$ such that $\hat{r} \cdot \delta\hat{r} = 0$. We now consider the problem of estimating bounds on the possible error in applying the computed rotation matrix to an arbitrary vector $\boldsymbol{v}$. We know that the rotational component of the transformation of $\boldsymbol{v}$ is given by

$$R(\hat{r}, \theta)v = \cos\theta v + (1 - \cos\theta)(\hat{r} \cdot v)\hat{r} + \sin\theta(\hat{r} \times v)$$

where $\hat{r}$ and θ are the parameters determining the rotation.

We first consider the variation of this expression with respect to the angle of rotation. In particular, under the assumption that $\Delta\theta$ is small, the following holds:

$$\begin{aligned} R(\hat{r}, \theta + \Delta\theta)v - R(\hat{r}, \theta)v \\ &= |v|\{(\cos(\theta + \Delta\theta) - \cos\theta)\hat{v} \\ &\quad + (\cos\theta - \cos(\theta + \Delta\theta))(\hat{r} \cdot \hat{v})\hat{r} \\ &\quad + (\sin(\theta + \Delta\theta) - \sin\theta)(\hat{r} \times \hat{v})\} \\ &\approx |v|\{-\Delta\theta \sin\theta\hat{v} + \Delta\theta \sin\theta(\hat{r} \cdot \hat{v})\hat{r} \\ &\quad + \Delta\theta \cos\theta(\hat{r} \times \hat{v})\}. \end{aligned}$$

Straightforward algebraic manipulation shows that the magnitude of this term is given by

$$|v|\Delta\theta\sqrt{1 - (\hat{r} \cdot \hat{v})^2}$$

and this is bounded above by $\Delta\theta|v|$.

Next, we consider the variation with respect to $\hat{r}$, so that

$$\begin{aligned} R(\hat{r} + \delta\hat{r}, \theta) - R(\hat{r}, \theta) = |v|\{(1 - \cos\theta)[(\hat{r} \cdot \hat{v})\delta\hat{r} \\ + (\delta\hat{r} \cdot \hat{v})[\hat{r} + \delta\hat{r}]] \\ + \sin\theta[\delta\hat{r} \times \hat{r}]\}. \end{aligned} \quad (2)$$

We consider the magnitude of the second term in the right hand side of this expression, by taking the dot product of this vector with itself. If we ignore terms in $(\delta\hat{r} \cdot \delta\hat{r})$, since the assumption of $\delta\hat{r}$ small implies such terms are negligible, then the magnitude of the second term in (2) is given by

$$|(1 - \cos\theta)(\delta\hat{r} \cdot \hat{v}) + \sin\theta(\hat{v} \cdot (\hat{r} \times \delta\hat{r}))|. \quad (3)$$

We now consider a bound for this expression. Suppose we let $\hat{k}$ denote the unit vector in the direction of $\delta\hat{r}$, and let $(\hat{k} \cdot \hat{v}) = \cos\zeta$. Since the worst case will occur when $\hat{v}$ lies entirely in the plane spanned by $\delta\hat{r}$ and $\hat{r} \times \delta\hat{r}$, (3) reduces to

$$\begin{aligned} |\delta\hat{r}||(1 - \cos\theta)\cos\zeta + \sin\theta\sin\zeta| \\ = |\delta\hat{r}||\cos\zeta - \cos(\theta + \zeta)|. \end{aligned}$$

It is clear that the worst possible value for this expression is $2|\delta\hat{r}|$. Thus, the maximum value for the magnitude of the second term in (2) is two $|\delta\hat{r}|$ and overall, the maximum deviation due to a variation in $\hat{r}$ is given by

$$2|\hat{v}||\delta\hat{r}|.$$

Finally, we can piece together these two variations. By ignoring higher order terms, it is clear that a Taylor series expansion of $R(\hat{r}, \theta)$ yields the following bound on errors in the computed value of a rotation:

$$|R(\hat{r} + \delta\hat{r}, \theta + \Delta\theta)v - R(\hat{r}, \theta)v| \leq (2|\delta\hat{r}| + |\Delta\theta|)|v|.$$

Now, if the errors ϕ and ψ are small, then we know that

$$|\Delta\theta| \leq |\phi + \psi|.$$

Furthermore

$$|\delta\hat{r}| = |\sin\psi| \approx |\psi|$$

and this implies a bound on variation in v of

$$|3\psi+\phi||v|.$$

Moreover, if we are careful to restrict our computation appropriately, then $\psi \approx \phi$, and thus

$$|R(\hat{r}+\delta\hat{r},\ \theta+\Delta\theta)v - R(\hat{r},\ \theta)| \leq |4\phi||v|.$$

Errors in v_0: We now consider bounds on the error associated with computing the translation component v_0 of the transformation. Recall that the correct form for v_0 is given by

$$\begin{aligned}[\hat{n}_{m,i}\hat{n}_{m,j}\hat{n}_{m,k}]v_0 &= (\hat{n}'_{m,i} \cdot p_{s,i} - d_i)(\hat{n}'_{m,j} \times \hat{n}'_{m,k})\\ &+ (\hat{n}'_{m,j} \cdot p_{s,j} - d_j)(\hat{n}'_{m,k} \times \hat{n}'_{m,i})\\ &+ (\hat{n}'_{m,k} \cdot p_{s,k} - d_k)(\hat{n}'_{m,i} \times \hat{n}'_{m,S5j})\end{aligned}$$

where $\hat{n}_{m,i}$ is a face normal in model coordinates, $\hat{n}'_{m,i}$ is the transformed normal in sensor coordinates, $p_{s,i}$ is the position vector of the contact point in sensor coordinates, and d_i is the constant offset for face i. We will consider error ranges for each of the components

$$(\hat{n}'_{m,k} \cdot p_{s,k} - d_k)(\hat{n}'_{m,i} \times \hat{n}'_{m,j})$$

separately.

We let $s = \hat{n}_{m,k} \cdot p_{s,k} - d_k$ and $v = \hat{n}'_{m,i} \times \hat{n}'_{m,j}$ so that the correct component is simply sv and the computed component is

$$(s+\Delta)(\xi v + \eta \hat{u})$$

where $\hat{u}$ is a unit vector orthogonal to v, and Δ, ξ, and η are values to be determined. We assume that the measured position vector is given by $p_i + \delta p_i$, where δp_i is a vector of magnitude ϵ_d, and the measured normal is given by $\hat{n}_{s,i}$ such that $\hat{n}_{s,i} \cdot \hat{n}'_{m,i} \geq \epsilon_n$.

First, note that the magnitude of the error in computing the component of the translation is given by

$$|sv - (s+\Delta)(\xi v + \eta\hat{u})|$$

$$= \sqrt{[s(1-\xi) - \Delta\xi]^2 (v \cdot v) + \eta^2 (s+\Delta)^2}. \quad (12)$$

Thus we need to find bounds for s, Δ, $(v \cdot v)$, ξ and η^2. We know that s is a given scalar value. If the angle between the face normals is given by $\hat{n}_i \cdot \hat{n}_j = \cos \zeta$, then

$$v \cdot v = 1 - \cos^2 \zeta = \sin^2 \zeta.$$

It is straightforward to show that

$$\begin{aligned}\Delta &= |(\hat{n}_s \cdot (p+\delta p) - d) - (\hat{n}'_m \cdot p - d)|\\ &\leq \epsilon_d + |(\hat{n}_s - \hat{n}'_m) \cdot p|\\ &\leq \epsilon_d + |p|\sqrt{2}\,\sqrt{1-\epsilon_n}.\end{aligned}$$

Next, we consider bounds for ξ, η^2, where

$$\hat{n}_{s,i} \times \hat{n}_{s,j} = \xi(\hat{n}'_{m,i} \times \hat{n}'_{m,j}) + \eta\hat{h} \quad (13)$$

for some unit vector $\hat{h}$ orthogonal to $(\hat{n}'_{m,i} \times \hat{n}'_{m,j})$. Now

$$\begin{aligned}&(\hat{n}_{s,i} \times \hat{n}_{s,j}) \cdot (\hat{n}'_{m,i} \times \hat{n}'_{m,j})\\ &\quad = (\hat{n}_{s,i} \cdot \hat{n}'_{m,i})(\hat{n}_{m,j} \cdot \hat{n}'_{m,j})\\ &\quad\quad - (\hat{n}_{s,i} \cdot \hat{n}'_{m,j})(\hat{n}_{s,j} \cdot \hat{n}'_{m,i})\\ &\quad \geq \epsilon_n^2 - (\hat{n}_{s,i} \cdot \hat{n}'_{m,j})(\hat{n}_{s,j} \cdot \hat{n}'_{m,i}).\end{aligned}$$

Moreover, the worst case (i.e. largest value) for $\hat{n}_{s,i} \cdot \hat{n}'_{m,j}$ occurs at $\cos(\zeta - \cos^{-1}\epsilon_n)$ since this is the smallest angle possible between the cone of radius ϵ_n about $\hat{n}'_{m,i}$ and the vector $\hat{n}'_{m,j}$. (Note that we have assumed that the two cones do not overlap, i.e. $\zeta > 2\cos^{-1}\epsilon_n$). As before, we let $\epsilon_n = \cos \phi$. Then, by substitution and expansion, we get

$$(\hat{n}_{s,i} \times \hat{n}_{s,j}) \cdot (\hat{n}'_{m,i} \times \hat{n}'_{m,j}) \geq \sin \zeta \sin(\zeta - 2\phi).$$

At the same time, from (13)

$$(\hat{n}_{s,i} \times \hat{n}_{s,j}) \cdot (\hat{n}'_{m,i} \times \hat{n}'_{m,j}) = \xi \sin^2 \zeta$$

so that we have the bound

$$\xi \geq \frac{\sin(\zeta - 2\phi)}{\sin \zeta}.$$

Now the length of $(\hat{n}_{s,i} \times \hat{n}_{s,j})$ is given by

$$1 - (\hat{n}_{s,i} \cdot \hat{n}_{s,j})^2$$

and to get a bound on η^2, we want to maximize this expression. As before, the worst case occurs when the $\hat{n}_s$ vectors lie at the limits of their respective cones, and

$$(\hat{n}_{s,i} \cdot \hat{n}_{s,j}) = \cos(\zeta + 2\phi).$$

We also have, however, from (13)

$$\begin{aligned}(\hat{n}_{s,i} \times \hat{n}_{s,j}) \cdot (\hat{n}_{s,i} \times \hat{n}_{s,j}) &= \xi^2 \sin^2 \zeta + \eta^2\\ &\leq 1 - \cos^2(\zeta + 2\phi).\end{aligned}$$

Substitution and expansion yield the following bound

$$\eta^2 \leq \sin(4\phi) \sin(2\zeta).$$

We are now ready to bound the error in computing each component of the translation vector v_0. From (12), the magnitude of the error is given by

$$\sqrt{[s(1-\xi) - \Delta\xi]^2 (v \cdot v) + \eta^2 (s+\Delta)^2}.$$

Substitution of the various bounds yields

$$\sqrt{[s \sin \zeta - (s+\Delta)\sin(\zeta - 2\phi)]^2 + (s+\Delta)^2 \sin(2\zeta)\sin(4\phi)}$$

where

$$s = \hat{n}_{s,k} \cdot p_{s,k} - d_k$$

$$\Delta \leq \epsilon_d + |p_{s,k}|\sqrt{2}\,\sqrt{1-\epsilon_n}$$

$$\cos \zeta = \hat{n}_{m,i} \cdot \hat{n}_{m,j}.$$

Note that as $\phi \to 0$, this bound reduces to $|\Delta \sin \zeta|$. Furthermore, as $\epsilon_d \to 0$, this expression tends to 0, so that the error in the computed translation vanishes as the error in the measurements do.

Typically, we will want to restrict our computations to cases in which the faces are roughly orthogonal, so that $\zeta \approx \pi/2$. In this case, the bound reduces to the simple expression

$$|s-(s+\Delta)\cos(2\phi)|.$$

References

[1] H. H. Baker and T. O. Binford, "Depth from edge and intensity based stereo," *Seventh Int. Joint Conf. Artificial Intelligence*, pp. 631–636, 1981.

[2] D. H. Ballard, "Generalizing the Hough transform to detect arbitrary shapes," *Pattern Recognition*, vol. 13, no. 2, pp. 111–122, 1981.

[3] S. T. Barnard and W. B. Thompson, "Disparity analysis of images," *IEEE Pattern Anal. Mach. Intell.*, PAMI-2, no. 4, pp. 333–340, 1980.

[4] M. Brady, "Smoothed local symmetries and frame propagation," in *Proc. IEEE Pattern Recog. Im. Proc.*, 1982.

[5] R. Brooks, "Symbolic reasoning among 3-dimensional models and 2-dimensional images," *Artificial Intell.*, vol. 17, pp. 285–349, 1981.

[6] P. Brou, "Using the Gaussian image to find orientations of objects," *Int. J. Robotics Res.*, vol. 3, no. 4, pp. 89–125, 1984.

[7] R. C. Bolles and R. A. Cain, "Recognizing and locating partially visible objects: The Local-Feature-Focus method," *Int. J. Robotics Res.*, vol. 1, no. 3, pp. 57–82, 1982.

[8] R. C. Bolles, P. Horaud, and M. J. Hannah, *Robotics Research: The First International Symposium*, M. Brady and R. Paul, Eds. Cambridge, MA: MIT, 1984, pp. 413–424.

[9] O. D. Faugeras, and M. Hebert, "A 3-D recognition and positioning algorithm using geometrical matching between primitive surfaces," in *Proc. Eighth Int. Joint Conf. Artificial Intell.*, Los Altos: William Kaufmann, pp. 996–1002.

[10] P. C. Gaston and T. Lozano-Pérez, "Tactile recognition and localization using object models: The case of polyhedra on a plane." *IEEE Trans. Pattern Anal. Mach. Intell.*, vol. PAMI-6, no. 3, pp. 257–265, 1984.

[11] W. E. L. Grimson, "A computer implementation of a theory of human stereo vision," *Philosophical Trans. Royal Soc. London*, vol. B 292, pp. 217–253, 1981.

[12] ——, "The combinatorics of local constraints in model-based recognition and localization from sparse data," MIT Artificial Intelligence Laboratory, Tech Memo 763, 1984.

[13] ——, "Computational experiments with a feature-based stereo algorithm," *IEEE Trans. Patt. Anal. Mach. Intell.*, vol. 7, pp. 17–34, 1985.

[14] W. E. L. Grimson and T. Lozano-Pérez, "Model-based recognition and localization from sparse range or tactile data," *Int. J. Robotics Res.*, vol. 3, pp. 3–35, 1984.

[15] ——, "Recognition and localization of overlapping parts from sparse data," AIM-841. Massachusetts Institute of Technology, Artificial Intelligence Laboratory, Tech. Memo. AIM-841, 1985.

[16] ——, "Recognition and localization of overlapping parts from sparse data in two and three dimensions," in *Proc. IEEE Intern. Conf. Robotics Automat.*, pp. 61–66, 1985.

[17] L. D. Harmon, "Automated tactile sensing," *Int. J. Robotics Res.*, vol. 1, pp. 3–32, 1982.

[18] W. D. Hillis, "A high-resolution image touch sensor," *Int. J. Robotics Res.*, vol. 1, pp. 33–44, 1982.

[19] S. W. Holland, "A programmable computer vision system based on spatial relationships," General Motors publ. GMR-2078, General Motors, Detroit, MI, 1976.

[20] B. K. P. Horn, "Extended Gaussian images," AIM-740. MIT Artificial Intelligence Laboratory, Tech Memo AIM-740, 1983.

[21] ——, "Picking parts out of a bin," MIT Artificial Intelligence Laboratory, Tech. Memo. AIM-746, 1983.

[22] K. Ikeuchi, "Determining attitude of object from needle map using extended gaussian image," MIT Artificial Intelligence Laboratory, Tech Memo AIM-714, 1983.

[23] K. Ikeuchi and B. K. P. Horn, "An application of photometric stereo," in *Proc. Sixth Int. Joint Conf. Artificial Intell.*, 1979, pp. 413–415.

[24] R. A. Lewis and A. R. Johnston, "A scanning laser range finder for a robotic vehicle," *Fifth Int. Joint Conf. Artificial Intell.*, 1977, pp. 762–768.

[25] D. Marr and H. K. Nishihara, "Representation and recognition of the spatial organization of three-dimensional shapes," in *Proc. R. Soc. Lond.*, vol. B 200, 1978, pp. 269–294.

[26] D. Marr and T. Poggio, "A computational theory of human stereo vision," in *Proc. R. Soc. Lond.*, vol. B 204, 1979, pp. 310–328.

[27] J. E. W. Mayhew and J. P. Frisby, "Psychophysical and computational studies towards a theory of human stereopsis," *Artificial Intell.*, vol. 17, pp. 349–385, 1981.

[28] R. Nevatia, "Structured descriptors of complex curved objects for recognition and visual memory." Ph.D. thesis, Stanford University. Stanford, CA, Tech. Memo. AIM-250, 1974.

[29] R. Nevatia and T. O. Binford, "Description and recognition of curved objects," *Artificial Intell.*, vol. 8, pp. 77–98, 1977.

[30] D. Nitzan, A. E. Brain, and R. O. Duda, "The measurement and use of registered reflectance and range data in scene analysis," *Proc. IEEE*, vol. 65, pp. 206–220, Feb. 1977.

[31] Y. Ohta and T. Kanade, "Stereo by intra- and inter-scanline search using dynamic programming," *IEEE Trans. Patt. Anal. Mach. Intell.*, vol. PAMI-7, no. 2, pp. 139–155, 1985.

[32] K. J. Overton and T. Williams, "Tactile sensation for robots," in *Proc. Seventh Int. Joint Conf. Artificial Intell.*, pp. 791–795, 1981.

[33] W. A. Perkins, "A model-based vision system for industrial parts," *IEEE Trans. Comput.*, vol. C-27, pp. 126–143, 1978.

[34] R. J. Popplestone, C. M. Brown, A. P. Ambler, and G. F. Crawford, "Forming models of plane and cylinder faceted bodies from light stripes," *Fourth Int. Joint Conf. Artificial Intell.*, Tbilisi, Georgia, USSR (September 1975), pp. 664–668.

[35] J. A. Purbrick, "A force transducer employing conductive silicone rubber," *First Int. Conf. Robot Vision Sensory Contr.*, Stratford-upon-Avon, UK, Apr. 1981.

[36] M. H. Raibert and J. E. Tanner, "Design and implementation of a VLSI tactile sensing computer," *Int. J. Robotics Res.*, vol. 1, pp. 3–18, 1982.

[37] J. L. Schneiter, "An optical tactile sensor for robots," S.M. Thesis, Dept. of Mech. Engr., Massachusetts Institute of Technology, Aug. 1982.

[38] Y. Shirai and M. Suwa, "Recognition of polyhedrons with a range finder," *Second Int. Joint Conf. Artificial Intell.*, 1971.

[39] G. Stockman and J. C. Esteva, "Use of geometrical constraints and clustering to determine 3D object pose," Michigan State University Department of Computer Science, East Lansing, MI, Tech. Rep. TR84002, 1984.

[40] K. Sugihara, "Range-data analysis guided by a junction dictionary," *Artificial Intell.*, vol. 12, pp. 41–69, 1979.

[41] S. Tsuji and A. Nakamura, "Recognition of an object in a stack of industrial parts," in *Proc. Fourth Int. Joint Conf. Artificial Intell.*, Los Altos, CA: William Kaufmann, 1975, pp. 811–818.

[42] R. J. Woodham, "Photometric stereo: A reflectance map technique for determining surface orientation from image intensity," *Image Understanding Syst. Industrial Applications*, in *Proc. SPIE*, vol. 155, 1978.

[43] ——, "Photometric method for determining surface orientation from multiple images," *Optical Engineering*, vol. 19, pp. 139–144, 1980.

[44] ——, "Analysing images of curved objects," *Artificial Intelligence*, vol. 17, pp. 117–140, 1981.

Probing Convex Polytopes

D. Dobkin†, *H. Edelsbrunner*‡, *C. K. Yap*§

†Department of Computer Science
Princeton University
Princeton, New Jersey

‡Department of Computer Science
University of Illinois
Urbana-Champaign, Illinois

§Courant Institute of
Mathematical Sciences
New York University
New York, New York

ABSTRACT

We investigate the complexity of determining the shape and presentation (i.e. position with orientation) of convex polytopes in multi-dimensional Euclidean space using a variety of probe models.

1. Introduction

This paper considers the problem of discovering the environment by means of simple sensory equipment. We are motivated by robots that can make 'probes' into their surroundings. The term 'probe' is intended to cover a whole range of crude sensory devices. An example of a probe might be a robot arm moving in a fixed direction until it contacts an obstacle. Such a probe might yield the spatial position of a point on the obstacle. Another example is an ultra-sound device that can detect the proximity (but not the precise location) of other objects. In contrast to such crude devices, one might think of using vision to discover one's environment: the amount of data gathered and processed in vision systems is many orders of magnitude greater than the kind of data gathered by probes. Consequently, the computational issues are rather different. While vision is a much studied topic, the area of processing crude sensory data is relatively new. It is reasonable to ask why should one bother with probes when we can accomplish much more using data-intensive sensory equipment. The answer is that data-intensive sensory equipment (such as a camera) may be uneconomical, too delicate or physically impossible to install. Even if we could gather such data, the processing of such data may be computationally too expensive. Probe devices are usually more robust, cheaper and smaller; the processing of such data is also expected to be relatively cheap. Device engineers have invented many ingenious methods to gather such crude sensory data. The theoretical understanding of processing such data is clearly important for the effective use of such devices. Furthermore, our understanding of the problem enhances our understanding of convex polytopes.

One of the first papers in this area is [CY87] which considers the problem of determining a convex polygon using 'finger' probes. Such probes can be imagined as a point moving from infinity along a straight line until the point contacts an object. The data yielded by the probe is this contact point. (If the probe misses all objects, the contact point is 'at infinity', by convention.) They proved that for a convex n-gon, $3n$ probes are sufficient and $3n-1$ probes are necessary. This may appear a little surprising since their algorithm only assumes that the object is a convex n-gon for some unspecified n. In this paper, we consider probing convex polytopes in d-dimensional Euclidean spaces $\mathcal{E}^d$, $d \geq 2$, and look at several reasonable models of probes:

- a moving point ('finger probe')

---

† NSF Grant MCS-83-03926 and DCR-85-05517

‡ Amoco Foundation Faculty Development in Computer Science

§ NSF Grant DCR-84-01633 and DCR-84-01898

- a moving hyperplane ('hand probe')
- a light-source yielding a silhouette
- a plane yielding a cross-section
- a moving line in $\mathcal{E}^3$
- finger probe with $\varepsilon > 0$ uncertainty

We consider a complexity model that counts the worst case number of probes where the successive probes are adaptive, i.e., can depend of the previous probe outcomes. Note that we do not count the cost of determining successive probes. This assumption, similar to that of counting only comparisons in sorting-related problems, can be justified under various circumstances.

We use the following standard terminology: for any point set $X \subseteq \mathcal{E}^d$, its boundary and interior is denoted ∂X and $int(X)$, respectively. A (possibly non-convex) closed polyhedron P is *star-shaped* if there is a point x in the polyhedron such that the line segment connecting x with any other point of P is also contained in P. The set of all such points x is the *kernel* of P. A *polytope* is a bounded polyhedron. The boundary of a polyhedron $P \subseteq \mathcal{E}^d$ is partitioned in the usual way into i-faces for $i = 0, \ldots, d-1$ where we define an *i-face* to be any relatively open subset of $\mathcal{E}^d$ homeomorphic to $\mathcal{E}^i$. Thus the interior of P is a d-face. Two faces are *incident* if one of the faces is contained in the closure of the other. We use *vertex, edge, facet* and *cell* as synonyms for 0-face, 1-face, $(d-1)$-face and d-face. Let $f_i(P)$ denote the number of i-faces of P. The partition of $\mathcal{E}^d$ into a set of i-faces ($i = 0, \ldots, d$) such that the closure of each face is equal to a union of faces is called a *cell-complex*. In this paper, the cell complex is determined by a finite set of hyperplanes.

The organization of this paper is as follows: section 2 investigates the complexity of finger-probing a convex polytope $P \subseteq \mathcal{E}^d$. We show that $f_0(P) + f_{d-1}(P)$ finger probes are necessary and $f_0(P) + (d+2)f_{d-1}(P)$ finger probes are sufficient. We also show that probing with hyperplanes is dual to finger probing. Section 3 addresses more complex probes such as light-sources which yield silhouettes and intersecting hyperplanes that yield cross-sections. In section 4 we consider line probes in $\mathcal{E}^3$. It is important to realize that in line probes, no information about the points of contact between the line and the polytope is given; that such probes can determine polytopes is not obvious. We give an algorithm using a number of probes that is linear in the number of faces of the polytope. Section 5 examines probes that have uncertainty of $\varepsilon > 0$. The issues here are considerably more subtle. It is not even clear that a meaningful generalization of our other probe models can be posed in this context. Finally, section 6 discusses the results of this paper and indicates directions for future work.

2. Finger Probing

Let $P \subseteq \mathcal{E}^d$ be a convex polytope (so P is closed and bounded) whose interior contains the origin O. Nothing else is known about P. The goal is to infer complete information about P on the basis of probes as we now describe. A *finger probe* F is an unbounded directed line. We imagine a point moving along the directed line F. The *contact point* $C(F, P)$ of F at P is the first point on P that F intersects along its direction. The *probe path* $\pi(F, P)$ of F to P is the directed half-line contained in the line F, with direction consistent with F and terminating at $C(F, P)$. We also write $C(F)$ and $\pi(F)$ for the contact point and path, respectively, when P is understood. If F does not intersect P, we say that the probe *misses* P and the contact point is at infinity, denoted $C(F) = \infty$; the probe path $\pi(F)$ is equal to F in this case. The finger probe model was first studied in [CY87] for the case $d = 2$. We first investigate the case $d = 3$ and then generalize the result to arbitrary d.

2.1. Bounds in $\mathcal{E}^3$

Let $F_1, \ldots, F_m$ be a sequence of probes and let $c_i = C(F_i)$ and $\pi_i = \pi(F_i)$, $i = 1, \ldots, m$, denote the corresponding contact points and probe paths. We next introduce some structures to capture the information gained from these probes. For any set of points X, let $conv(X)$ denote the smallest closed convex set containing X.

$\mathcal{P}_m = \{ c_1, \ldots, c_m \}$ is the (current) set of contact points.

$\mathcal{H}_m = conv(\mathcal{P}_m \cup \{ O \})$ is the convex hull of the set of contact points together with the origin O.

$\mathcal{A}_m$ is the cell complex defined by the planes that support facets of $\mathcal{H}_m$.

Clearly the convexity of P implies

$$\mathcal{H}_m \subseteq P.$$

Next we want to define the smallest set O_m^* that is guaranteed to contain P. Each probe path π_i casts a 'shadow' which consists of all those points x that cannot be contained in P by virtue of the convexity of P and the fact that every point in $\mathcal{P}_m$ lies on the boundary of P. This shadow is precisely defined by the set S_i of those points x such that the probe path π_i would intersect the interior of

$$conv(\mathcal{P}_m \cup \{ O \} \cup \{ x \}).$$

We can define

$$O_m^* = \mathcal{E}^3 - \bigcup_{i=1}^{m} S_i$$

Thus $O_m^* = \mathcal{E}^3$ for $m = 0$. The set O_m^* can be rather complicated to understand, and for the purposes of upper bounds, we prefer to use an alternative set defined as follows. Let v be a contact point in $\mathcal{P}_m$. Let $h_1, \ldots, h_k$ ($k \geq 1$) be the open half-spaces where each h_i does not contain O but is bounded by a plane containing a facet incident to v. Define K_v to be the intersection of $h_1, \ldots, h_k$. If v is a vertex of $\mathcal{H}_m$ then K_v is a cone with vertex v; if v is in the relative interior of an edge (resp. facet) of $\mathcal{H}_m$ then K_v is a wedge (resp. a half-space). Notice that K_v is the set of points x such that v belongs to the interior of $conv(\mathcal{P}_m \cup \{ O, x \})$. It follows that K_v is contained in S_i if $v = c_i$. We now define

$$O_m = \mathcal{E}^3 - \bigcup_{v} K_v$$

where v ranges over $\mathcal{P}_m$. Clearly, for all $m \geq 0$,

$$O_m^* \subseteq O_m$$

and

$$\mathcal{H}_m \subseteq P \subseteq O_m.$$

Lemma 1. O_m is star-shaped with $\mathcal{H}_m$ as its kernel.

Proof. Define $\overline{K}_v$ as the complement of K_v, and define H_v as the kernel of $\overline{K}_v$. Thus, H_v is the intersection of the closed half-spaces that are complements of the half-spaces $h_1, \cdots, h_k$ used in defining K_v. If v is a vertex of $\mathcal{H}_m$ then H_v is the smallest cone with apex v that contains $\mathcal{H}_m$; if v belongs to an edge of $\mathcal{H}_m$ then H_v is a wedge; and if v belongs to a facet of $\mathcal{H}_m$ then H_v is a half-space. In any case, H_v contains $\mathcal{H}_m$, and furthermore, $\mathcal{H}_m$ is the intersection of all H_v, for v ranging over all vertices of $\mathcal{H}_m$. By definition of O_m as the intersection of all $\overline{K}_v$, the kernel of O_m contains the intersection of all H_v. Therefore, $\mathcal{H}_m$ is subset of the kernel of O_m. We conclude that $\mathcal{H}_m$ is exactly the kernel of O_m since every facet of $\overline{K}_v$ contains a facet of O_m, where v is a vertex of $\mathcal{H}_m$. **Q.E.D.**

If O belongs to the interior of $\mathcal{H}_m$, and we only know about the contact points $\mathcal{P}_m$ (i.e. we know nothing about the probe paths) then $\mathcal{H}_m$ and O_m are the strongest possible sets in the sense that $\mathcal{H}_m$ (resp. O_m) cannot be replaced by a larger (resp. smaller) set in the above lemma. Initially, $\mathcal{H}_0$ consists of just the origin and $O_0 = \mathcal{E}^3$.

It is natural to say that P is *determined* when $\mathcal{H}_m = P = O_m$. Now we propose a probing strategy to determine P.

Definition. A facet of $\mathcal{H}_m$ or of O_m is said to be *verified* if the plane defined by that facet contains at least four co-planar points of $\mathcal{P}_m$ such that one of these points is in the relative interior of convex hull formed by the other points; otherwise the facet is *unverified.* The plane containing such a facet is said to be verified or unverified according as the facet is verified or unverified. A probe aimed at the relative interior of an unverified facet is said to be *trying to verify* that facet.

Definition. A vertex of $\mathcal{H}_m$ is said to be *verified* if it is incident to at least three verified facets of $\mathcal{H}_m$; otherwise, it is *unverified.* A vertex of O_m is verified if it is a verified vertex of $\mathcal{H}_m$.

We sometimes call an unverified facet a 'conjectured' facet. An attempted verification of a facet *succeeds* if the contact point lands on the facet. This terminology is justified by the observation that a verified facet of $\mathcal{H}_m$ is necessarily contained in a facet of P. Roughly speaking, our strategy consists of sending probes where each probe tries to verify a conjectured facet. If this objective fails then we get a contact point p which may or may not be co-planar with other facets of $\mathcal{H}_m$; in any case, p is a new vertex of $\mathcal{H}_{m+1}$. If we can somehow guarantee that an unverified facet of $\mathcal{H}_m$ has at most a constant number of vertices, then the pigeonhole principle guarantees that $\mathcal{H}_m$ cannot have too many vertices so that we must eventually succeed in verifying a facet. To guarantee this constant, our probes must avoid verified facets and we need to make sure that we land on at most one unverified facet. To meet the first goal, we use the fact that a verified facet of O_m contains a facet of P and can therefore be used as an upper bound for the facets of P they contain. To meet the second goal, we use the arrangement $\mathcal{A}_m$: we can land on two or more unverified facets only if we land on an edge or a vertex of $\mathcal{A}_m$. The union of edges and vertices of $\mathcal{A}_m$ can be easily avoided, however, since it is only a one-dimensional subcomplex. The constant that we will obtain is four which implies that each verified facet will contain at most five contact points.

Note that P is not completely explored as long as we do not have a contact point at each vertex. For, it is possible that a facet of the polytope (of potentially arbitrarily small area) has been overlooked. The verification of the vertices, however, is automatic: this is because in choosing probes that avoid verified facets of O_m, we are forced to aim at vertices of O_m.

The method just outlined is a generalization of the method of Cole and Yap. It should be noted, however, that the necessity of controlling the number of contact points in a single unverified facet is a new phenomenon in three dimensions. In two dimensions, one of any three collinear points is in the relative interior of the convex hull of the other two. In three dimensions, one can have an arbitrarily large number of co-planar points without having any point in the relative interior of the convex hull of the others. Surprisingly, nothing new arises in dimension four or greater.

The next lemma establishes a bijective correspondence between verified facets of $\mathcal{H}_m$ and verified facets of O_m.

Lemma 2. Each verified facet of $\mathcal{H}_m$ is contained in a verified facet of O_m, and each verified facet of O_m contains a verified facet of $\mathcal{H}_m$.

Proof. Let f be a verified facet of $\mathcal{H}_m$. If v is a point in $\mathcal{P}_m$ in the relative interior of f then K_v is a half-space bounded by the plane through the facet f. This K_v in turn determines a unique verified facet f' of O_m containing f. Conversely, a verified facet of O_m contains four contact points with one in the relative interior of the convex hull of the other three, by definition. These four contact points necessarily lie in a verified facet of $\mathcal{H}_m$. **Q.E.D.**

This lemma allows us to be sloppy when we refer to 'verified facets' without saying whether they are facets of $\mathcal{H}_m$ or of O_m. For the next lemma to make sense, the reader should realize that even when all the facets of O_m are verified, there could be facets of $\mathcal{H}_m$ that are not yet verified.

Lemma 3. The following are equivalent statements: (i) All facets of O_m are verified. (ii) O_m is convex. (iii) each vertex of $\mathcal{H}_m$ is incident to some verified facet of $\mathcal{H}_m$.

Proof. (i) $\Rightarrow$ (ii). Suppose all facets of O_m are verified. If a facet f of O_m is verified then O_m is contained in the half-space containing O determined by the plane of f. Let C denote the intersection of all such (closed) half-spaces. Thus $O_m \subseteq C$. Equality between O_m and C follows since O_m has no facet that does not belong to a facet of C.

(ii) $\Rightarrow$ (iii). Observe that if $v \in \mathcal{H}_m$ is not incident to a verified facet of $\mathcal{H}_m$ then for any ball B centered at v with sufficiently small radius, $B \cap K_v$ does not intersect K_u for all $u \neq v$, $u \in \mathcal{P}_m$. This means that $B - K_v = B \cap O_m$. This shows that O_m is not convex, since $B - K_v$ is not convex.

(iii) $\Rightarrow$ (i). Observe that the proof is complete if we show that any unverified facet of O_m is contained in a facet of K_v for some v that is not incident to any verified facet of $\mathcal{H}_m$. Let f be any unverified facet of O_m. There exists a $v \in \mathcal{P}_m$ such that f is contained in some facet f' of K_v. For the sake of contradiction, assume that some facet of K_v contains a verified facet g of O_m. Let w be a point of $\mathcal{P}_m$ in g (w exists because g is verified). Then $K_v \subseteq K_w$ which implies that one facet of K_v belongs to the plane ∂K_w and that the other facets of K_v are in the interior of K_w. Hence f cannot be a facet of O_m since $O_m \subseteq \mathcal{E}^3 - K_w$. This is a contradiction. **Q.E.D.**

For convenience, we define $\mathcal{V}_m$ to be the intersection of the 'verified' half-spaces where a 'verified' half-space is one containing $\mathcal{H}_m$ and bounded by a verified plane. Initially $\mathcal{V}_m = \mathcal{V}_0$ is the entire space $\mathcal{E}^3$. Clearly $\mathcal{V}_m$ is convex and $O_m \subseteq \mathcal{V}_m$. From the preceding lemma, we see that equality holds precisely when all the facets of O_m are verified. We will use the vertices of $\mathcal{V}_m$ to choose our probe lines.

We shall maintain the following invariant (H) in our probing strategy. This guarantees that each unverified facet of P will contain at most five contact points.

(H) Each unverified facet of $\mathcal{H}_m$ is a triangle, with at most one exception which may be a quadrilateral.

This invariant is easily initialized by forming a tetrahedron $\mathcal{H}_4$ about the origin. Our probing strategy is as follows:

> If all facets of $\mathcal{H}_m$ are verified, we are done. Otherwise, choose any unverified facet f of $\mathcal{H}_m$. If there are any unverified quadrilateral, choose f to be it. Treating the plane of f to be horizontal and lying above the origin, there are two cases: if there is any vertex v of $\mathcal{V}_m$ above the plane of f, then v is unverified and we let F be a probe aimed from v to any point x in the relative interior of f. If v does not exist, pick any F aimed at an interior point x where the path of F right up to x lies entirely in $\mathcal{V}_m$. By a suitable perturbation of x, we can ensure that F satisfies the additional property of not intersecting any vertex or edge of the cell complex $\mathcal{A}_m$.

It is clear that F as specified above exists. For a later application, we made an additional observation about the preceding probing strategy: it is easy to see that we can further assume that the probe we choose is aimed at the origin.

Lemma 4. The invariant (H) is maintained by the outcome of probe F.

Proof. Note that the contact point p of F will lie between v and x; if v does not exist, then p occurs before or at x. If $p = x$ then we have verified the face f. If v exists and $p = v$ then we have verified v; v is now a contact point and it is incident at at least three verified facets of O_{m+1}. Finally, if p is equal to neither x nor v then p is a new vertex of $\mathcal{H}_{m+1}$ and we form some new facets, each incident at p. Then f is no longer a facet. Notice that if a point lies in a cell of $\mathcal{A}_m$ then it is not coplanar with any three contact points that belong to the closure of a common facet of $\mathcal{H}_m$. Furthermore, if a point belongs to a facet of $\mathcal{H}_m$ then it is coplanar with the vertices of only one facet of $\mathcal{H}_m$. Our probing strategy assures that F does not hit a verified facet, and it hits at most one plane defined by an unverified facet of $\mathcal{H}_m$. It follows that F creates at most one new quadrilateral while the only old quadrilateral, if any, disappears. **Q.E.D.**

Using this strategy, we continue until P is determined. The partial correctness of the algorithm is clear. To show termination, note that with each probe, we either increase the number of unverified vertices of $\mathcal{H}_m$ or verify a facet or verify a vertex. By the pigeonhole principle, the number of unverified vertices is at most $3(f_2(P) - k) + 1$ where k is the number of verified facets. Termination is assured since we never verify a facet or a vertex more than once.

There is a contact point at each vertex of $\mathcal{H}_m$, and each facet has at most 5 contact points incident to it (one of these contact points lies in the relative interior of the triangle or quadrilateral formed by the others). This gives an upper bound of $f_0(P) + 5f_2(P)$.

We prove a lower bound by a straightforward adversary argument. First observe that every vertex of P must be probed. Next note that the relative interior of each facet of P must be probed. Combining this with the preceding upper bound, we obtain:

Theorem 5. Let P be a convex polytope in $\mathcal{E}^3$. Let $T_F(P)$ be the worst case number of finger probes necessary to determine P. Then

$$f_0(P) + f_2(P) \leq T_F(P) \leq f_0(P) + 5f_2(P).$$

2.2. Higher-dimensional finger probing

Seeing that probing in 3-dimensions encounters a difficulty that has no analogue in the planar case, it is not immediately clear whether we will encounter yet new difficulties in 4-dimensions and beyond. Fortunately, nothing new arises and we can obtain a fairly straightforward generalization. We merely sketch a proof.

Theorem 6. Let P be a convex polytope in $\mathcal{E}^d$, $d \geq 3$, and $T_F(P)$ be the worst case number of finger probes necessary to determine P. Then

$$f_0(P) + f_{d-1}(P) \leq T_F(P) \leq f_0(P) + (d+2)f_{d-1}(P).$$

Proof. The lower bound argument of the case $d = 3$ clearly generalizes. To obtain the upper bound, we organize our probes so that probes either land on $\mathcal{H}_m$, on a facet of $\mathcal{A}_m$, in a cell of $\mathcal{A}_m$, or at a vertex of O_m. To see this, we can check that all the lemmas for the 3-dimensional algorithm hold. The corresponding invariant is that an unverified facet of $\mathcal{H}_m$ have d contact points, with at most one exception that may have $d + 1$ contact points. **Q.E.D.**

2.3. Hyperplane probes

In contrast to the above, we now define a *probe* to be a moving hyperplane H approaching from infinity in the direction of its normal. The *contact hyperplane* $C(H)$ is the location of H where it first contacts P. Interestingly, this probe model can be reduced to finger probing using a duality transformation. For a point $p \neq O$ in $\mathcal{E}^d$, let its *dual* be the hyperplane $D(p)$ where the vector p is normal to $D(p)$ and the point $\frac{p}{|p|^2}$ belongs to $D(p)$. For any hyperplane h that avoids O, h^+ denotes the closed halfspace bounded by h and containing O. Then define

$$D(P) = \bigcap_{p \in P} D(p)^+$$

as the dual image of P. It is straightforward to verify that $D(p)$ is a hyperplane that touches $D(P)$ in a vertex if and only if p belongs to a facet of P, and that $D(p)$ supports a facet of $D(P)$ if and only if p is a vertex of P. For all computations it is therefore sufficient to consider the dual images of all responses (that is, images under D inverse) and to apply the strategy for finger probing as explained above. It is important to realize that we can apply the finger probing strategy of section 2.1 because, by an earlier remark, we can assume the finger probes of that strategy are directed at the origin. This yields

Theorem 7. Let P be a convex polytope in $\mathcal{E}^d$, $d \geq 3$, and let $L_H(P)$ be the worst case number of hyperplane probes necessary to determine P. Then

$$f_0(P) + f_{d-1}(P) \leq L_H(P) \leq (d + 2)f_0(P) + f_{d-1}(P).$$

3. Cross section and silhouette probing

We consider two models of probing a polytope in 3-dimensions here. In the first case, a probe consists of a direction in which a polytope cross section is to be taken. All cross sections are assumed to pass through the origin so that it suffices to specify the direction via a normal vector. In the second case, we consider probes which are specified by the position of a point source of light. Again, to simplify the results, we assume the point source is at infinity. The result of such a probe is the silhouette of the polytope generated by the light source. Duality shows that these two models are really the same:

Theorem 8. Cross section probing and silhouette probing are duals.

To obtain upper bounds, we reduce finger probing to cross section probing as follows: a finger probe F together with the origin O determine a plane P which we use to define our cross section probe. If F passes through the origin then we can choose P to be any plane containing F. Clearly, the outcome from this cross section probe yields at least as much information as the outcome of F. Thus, it follows from the results in section 2 that for any convex polytope $P \subseteq \mathcal{E}^3$, if $T_C(P)$ is the worse case number of cross section probes through the origin necessary to determine P then

$$T_C(P) \le f_0(P) + 5f_2(P)$$

It is easy to show a linear lower bound. First we observe that for any P, there is a perturbation P' of P such that $T_C(P') \ge \frac{f_0(P')}{2}$. This comes from the fact that any correct algorithm must pass a cross section through each vertex of P. The bound then follows from the fact that we perturb P to P' so that every three vertices of P' define a plane that avoids the origin O. In fact this bound holds for any 'certificate' (non-deterministic algorithm) for P'. *A fortiori* no sublinear algorithm is possible. We may conclude that our above upper bound on $T_C(P)$, which is given by an apparently wasteful reduction of finger-probes to cross-section probes, is at most a constant factor from the optimal.

4. Line probing

We consider probing a polytope in 3-dimensions using line probes. Intuitively, a line probe λ consists of a line sweeping out a plane H_λ. The position of the line at time $t \in \mathbf{R}$ ($\mathbf{R}$ is the set of all real numbers) is λ_t, where all the lines λ_t are parallel. The probe contacts a given bounded closed subset $P \subseteq \mathcal{E}^3$ the first time t_0 such that $\lambda_{t_0} \cap P \ne \emptyset$. It is important to realize that we are told the line λ_{t_0} but get no information about the set $\lambda_{t_0} \cap P$. One motivation for such probes is the IBM RS-1 robot which has LED sensors on two opposing robot fingers. The invisible ray between the two finger corresponds to a sweeping line that is cut off on contact with an object. The paper [CY87] uses the same model to motivate finger probes: here we imagine an object to be a polygonal piece of cardboard somehow supported to stand with one of its edges on a table.

More precisely, a *line probe* λ is a triple

$$(d,\ \hat{\mathbf{n}},\ \hat{\mathbf{v}})$$

where $d \in \mathbf{R}$, $\hat{\mathbf{n}}$ and $\hat{\mathbf{v}}$ are unit vectors that are orthogonal to each other. See figure 1. The probe determines the *probe plane* H_λ that is normal to $\hat{\mathbf{n}}$ at distance d from the origin O. For any $t \in \mathbf{R}$, let λ_t denote the line

$$\{\ s\hat{\mathbf{u}} + (d\hat{\mathbf{n}} + t\hat{\mathbf{v}}) : s \in \mathbf{R}\ \}$$

where $\hat{\mathbf{u}} = \hat{\mathbf{n}} \times \hat{\mathbf{v}}$. Note that λ_t is contained in the plane H_λ and the line moves in the direction of $\hat{\mathbf{v}}$. The *result line* of the probe on a bounded closed subset P is the line λ_{t_0} where

$$t_0 = \min\{\ t \in \mathbf{R} : \lambda_t \cap P \ne \emptyset\ \}.$$

If $\lambda_t \cap P = \emptyset$ for all t then we set $t_0 = \infty$ above and we say that the λ_∞ is undefined. The result of the probe is then said to be *infinite* or *finite* according to whether $t_0 = \infty$ or not.

We call $\hat{\mathbf{u}}$ the *probe orientation* and $\hat{\mathbf{v}}$ the *probe direction*. A probe $\lambda = (d, \hat{\mathbf{n}}, \hat{\mathbf{v}})$ is *centered* if $d = 0$. If $\lambda' = (d', \hat{\mathbf{n}}', \hat{\mathbf{v}}')$ is another probe, then λ and λ' are *opposite* if $H_\lambda = H_{\lambda'}$ and $\hat{\mathbf{v}}' = -\hat{\mathbf{v}}$. Again, λ and λ' are said to be *parallel* if $\hat{\mathbf{u}} = \pm\hat{\mathbf{u}}'$ where as usual $\hat{\mathbf{u}} = \hat{\mathbf{n}} \times \hat{\mathbf{v}}$ is the probe orientation. Note that in this case, either H_λ and $H_{\lambda'}$ are parallel, or else $\hat{\mathbf{u}}$ is parallel to the line $H_\lambda \cap H_{\lambda'}$. More importantly, the result lines (if both finite) of a pair of parallel probes are either identical or they determine a

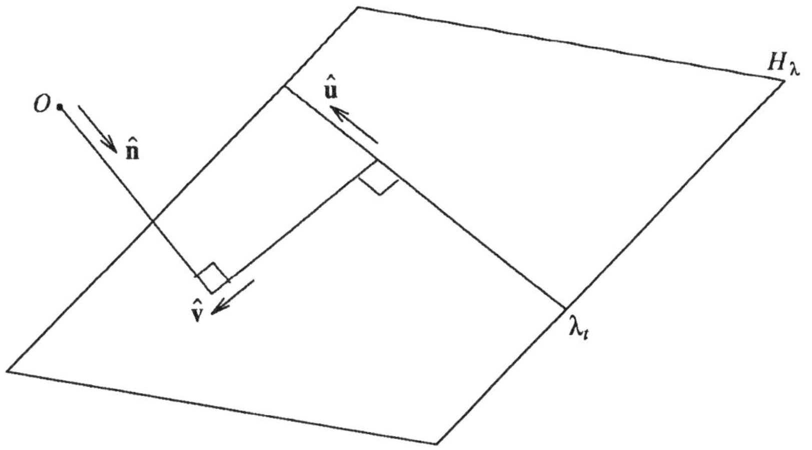

Figure 1.

plane.

Assume we are given a convex polytope P whose interior contains the origin O. In the following, we will say that a point (resp. line, plane) is *verified* at a particular moment if we can deduce from the result lines that the point (resp. line, plane) is a vertex (resp. contains an edge, contains a facet) of P. Before describing the algorithm we describe a basic subroutine.

Half-profile computation. Let H be any closed half-space such that its bounding plane ∂H intersects P. Let $\hat{\mathbf{w}}$ be any direction parallel to the plane ∂H. Let $h(\hat{\mathbf{w}})$ denote the plane normal to $\hat{\mathbf{w}}$ and passing through the origin. Then a $(H, \hat{\mathbf{w}})$*-profile* (of P) is the planar projection of $H \cap P$ along the direction of $\hat{\mathbf{w}}$ onto $h(\hat{\mathbf{w}})$. Let the projection of ∂H in $h(\hat{\mathbf{w}})$ be L. Thus the $(H, \hat{\mathbf{w}})$-profile is a polygon $R \subseteq h(\hat{\mathbf{w}})$ with one side abutting L. Suppose R has m vertices, that is,

$$R = (u_1, u_2, \ldots, u_m).$$

In our application, we can assume that $m \geq 2$ and without loss of generality, assume that the edge $[u_1, u_m]$ abuts L. Again, in our application below, we are told two distinct points on the edge $[u_1, u_m]$. It is not hard to see that a simple modification of the algorithm of Cole and Yap, we can either (a) determine that $m = 2$ using one probe or else (b) determine that $m > 2$ and compute R using $3m - 2$ line probes. The procedure is called a *half-profile computation.* To see why the half-profile computation is essentially a finger probe problem in the plane, observe that if we restrict ourselves to line probes λ whose probe orientations are parallel to $\hat{\mathbf{w}}$, then the behavior of the moving line λ_t is faithfully recorded by the intersection of λ_t with $h(\hat{\mathbf{w}})$. Clearly the intersection of the probe plane H_λ with $h(\hat{\mathbf{w}})$ gives rise to a (planar) finger probe of the $(H, \hat{\mathbf{w}})$-profile.

Remark: Note the similarity between this subcomputation and the shadow probes of the previous section. Since this subcomputation already makes a linear number of probes, and there is a linear lower bound on the number of shadow probes, this suggests a quadratic lower bound in the number of line probes. Thus our linear upper bound on line probing is somewhat surprising.

We are ready to give the overall algorithm. At the beginning of the nth step below ($n \geq 0$), we maintain the following invariant. There is a set $\mathcal{V}_n$ of verified vertices, and $\mathcal{H}_n$ is defined to be the convex hull of $\mathcal{V}_n \cup \{ O \}$. Let $\mathcal{F}_n$ denote the subset of the facets of $\mathcal{H}_n$ each of whose plane is verified.

To initialize, we do a 'full profile' of P along any direction $\hat{\mathbf{w}}$, in exact analogy with the half-profile computation above. That is, we restrict ourselves to line probes whose orientations are parallel to $\hat{\mathbf{w}}$, and hence such line probes are faithfully represented by finger probes in the plane $h(\hat{\mathbf{w}})$, and a direct application of the planar probing problem in [CY87] gives us the desired profile, which is a projection Q

of P onto $h(\hat{w})$.

We then continue as follows. Note that the polygon $Q \subseteq h(\hat{w})$ is such that its vertices are projections of vertices (possibly edges) of P and its edges are projections of edges (and possibly facets) of P. Let e be an edge of Q. Consider the plane $h'(e)$ containing e and perpendicular to $h(\hat{w})$. We know that there is at least one edge of P, and possibly one facet of P, lying in $h'(e)$. Our goal is to verify the vertices of P in $h'(e)$. This is easily reduced to "hyperplane probe" in $h'(e)$, as described in section 2.3. In other words, we restrict our attentions to line probes λ whose probe planes are $h'(e)$. By duality, hyperplane probes are transformed to the planar finger probing problem of Cole and Yap. Note that we already know two parallel hyperplane probe outcomes – namely the two lines parallel to $\hat{w}$ and passing through the endpoints of e. Using at at most $3m$ additional hyperplane probes, we can determine the $m \geq 1$ edges of P lying in $h'(e)$. Although the original formulation of Cole and Yap does not allow $m=1$, this is easily modified to suit our situation, allowing us to charge at most 3 probes to each edge of P found in this way. For reference, we call this probing of $h'(e)$ an *edge verification procedure*. The probes made in the full-profile computation of Q can be charged to the vertices of P that project onto vertices of Q; and hence we charge at most 3 probes to each vertex of P found. We now have our initial set $\mathcal{V}_0$ of verified vertices, and of verified facets $\mathcal{F}_0$. The general nth step is as follows:

> If every facet of $\mathcal{H}_n$ is verified then we halt and $P = \mathcal{H}_n$. Suppose some facet f is still unverified. Let H be the half-space where the plane ∂H supports $\mathcal{H}_n$ at f and H does not contain the interior of $\mathcal{H}_n$. Choose any direction $\hat{w}$ in ∂H such that $\hat{w}$ is not parallel to any edge of f. Construct the $(H, \hat{w})$-profile. If f is actually a facet of P, then one probe would verify this. In this case $\mathcal{H}_{n+1} = \mathcal{H}_n$. Otherwise, we get a non-trivial profile $R = (u_1, \ldots, u_m)$ $(m \geq 3)$ and we next perform an edge verification procedure for each edge $[u_i, u_{i+1}]$ $(i=1, \ldots, u_{m-1})$ as described above.

Complexity analysis: A profile computation that yields a profile of the form $R = (u_1, \ldots, u_m)$ with $m \geq 3$ costs $3m - 2$ probes. Since at least $m-2$ new vertices are revealed in this process, we charge each of the new vertices with at most 7 probes from the profile computation. The edge verification procedure for each of the edges $[u_i, u_{i+1}]$ charges at most 3 probes to each edge that is revealed. This proves a bound of $7f_0(P) + 2f_1(P) + f_2(P)$.

Theorem 9. Let the worst case number of line probes to determine a convex polytope P in $\mathcal{E}^3$ be $T_L(P)$. Then

$$f_0(P) + f_2(P) \leq T_L(P) \leq 7f_0(P) + 2f_1(P) + f_2(P).$$

Proof. We only have to show the lower bound. For each facet f, there is at least one line probe whose result line lies in the plane of f. For each vertex v, there is at least one line probe whose result line passes through v but does not lie in any of the planes defined by a facet incident to v. **Q.E.D.**

It ought to be remarked that the edge verification procedure is a highly unstable numerical procedure. It would be nice to give an alternative algorithm with better stability properties.

5. Probing in the presence of uncertainty

5.1. The model

The issue of probing in the presence of uncertainty is sufficiently intricate that we will restrict ourselves to finger probes in the plane. Let $\varepsilon > 0$ denote a fixed number that quantifies the uncertainty in our model. For any point x let $E_\varepsilon(x)$ denote the closed disk of radius ε centered at x. For any set X of points, define its ε-*expansion* $E_\varepsilon X$ and its ε-*interior* $I_\varepsilon X$ as follows:

$$E_\varepsilon X = \bigcup_{x \in X} E_\varepsilon(x)$$

$$I_\varepsilon X = \{ x : E_\varepsilon(x) \subseteq interior(X) \}$$

$$EI_\varepsilon X = E_\varepsilon X - I_\varepsilon X$$

Clearly $I_\varepsilon X \subseteq X \subseteq E_\varepsilon X$. Let P be a (closed) convex polygonal region. Then $I_\varepsilon P$ is an open set, and $E_\varepsilon P$ and $EI_\varepsilon P$ are closed sets. The set $I_\varepsilon P$, if non-empty, has boundary that forms a convex polygon with at most the same number of sides of P. The boundary of $E_\varepsilon P$ is decomposed into straight line segments and circular arcs as illustrated in figure 2.

We use the above sets to assign meaning to the term 'a probe has ε uncertainty' or 'ε-probes'. Such a probe is again denoted by a directed line. To distinguish probes with uncertainty from our original finger probes, we refer to the latter as 'standard probes' or '0-probes'; we may also write $C_0(F)$ and $\pi_0(F)$ for the contact point and probe path of standard probes. Out of the many possible interpretations of error probes, we choose one with the following semantics: we imagine an ε-probe F as specifying a collection of standard probes F' such that each F' is parallel to F, similarly directed and such that there is a vector $\bar{v}$ of length at most ε where $F' = F + \bar{v}$. Call this collection of standard probes the *ε-bundle* of probes defined by F. Each such F' defines the contact point $C_0(F', P)$, which corresponds to a point $C_0(F', P) - \bar{v}$ on F; this point $C_0(F', P) - \bar{v}$ is called a potential ε-contact point. Note that two different vectors $\bar{v}$ and $\bar{u}$ may give rise to the same standard probe F' but they could be distinguished in that they give rise to distinct ε-contact points. This is captured as follows.

Definition. Let F be an ε-probe and P be any polygon, not necessarily convex. A point p (possibly $p = \infty$) on F is called a *potential ε-contact point* of F at P if there exists a standard probe F' in the ε-bundle defined by F such that either $p = \infty = C_0(F', P)$ or $d(p, C_0(F', P)) \le \varepsilon$.

For any potential ε-contact point x of F, we have a corresponding probe path $\pi(F, x)$ defined analogously as for standard probes: $\pi(F, x)$ is a directed half-line that terminates at x. For simplicity (mainly in upper bound proofs), we prefer to ignore the information provided by the entire probe path $\pi(F, x)$ and simply consider the contact point x. Note that if ∞ is a possible ε-contact point for F then F does not intersect $I_\varepsilon P$; the converse is not true. In any case, let $C_\varepsilon(F, P) \subseteq F$ denote the set of finite potential ε-contact points. If P is understood, we write $C_\varepsilon(F)$ instead of $C_\varepsilon(F, P)$. If P is convex then $C_\varepsilon(F, P)$ is a connected interval of F; if P is non-convex then $C_\varepsilon(F, P)$ need not be connected.

Lemma 10. The set $C_\varepsilon(F, P)$ is equal to $F \cap E_\varepsilon X$ where

$$X = \{\, C_0(F', P) : F' \textit{ is in the } \varepsilon\textit{-bundle of } F \,\}.$$

It follows easily that

(i) For all P, not necessarily convex, $\bigcup_F C_\varepsilon(F, P) \subseteq EI_\varepsilon P$ where the F ranges over all ε-probes.

(ii) If P is *star-shaped* then $\bigcup_F C_\varepsilon(F, P) = EI_\varepsilon P$.

In our model of computation, it is important to realize that the algorithm can only specify ε-probes F but some adversary (relative to P) will specify some ε-contact point in $C_\varepsilon(F)$. There is no assumption that the adversary will even consistently reply with the same contact point for two identical probes.

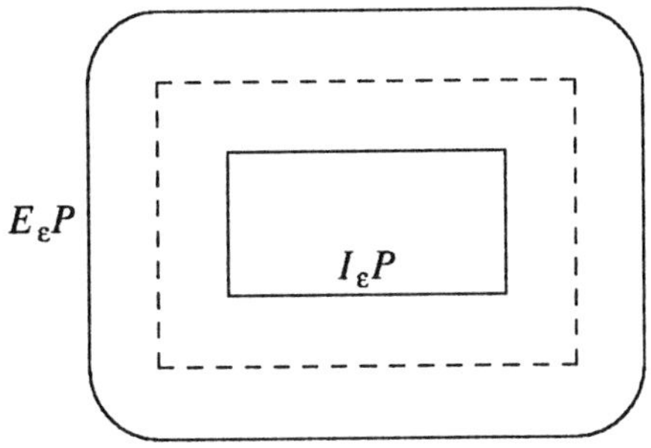

Figure 2.

However it is easy to see that the adversary can do no worse by being consistent. Thus, we may define a P-*adversary* A to be a rule which for each sequence of ε-probes $\bar{F} = (F_1, \ldots, F_k)$, $k \geq 1$, specifies a point $A(\bar{F})$ in $C_\varepsilon(F_k, P)$. We call A a *forgetful* adversary if $A(\bar{F})$ depends only on the last probe F_k. We may similarly define a *probe algorithm* to be a rule B such that for any sequence $\bar{\pi} = (\pi_1, \ldots, \pi_k)$, $k \geq 0$, of probe paths, the rule either decides that it is an 'end-game' or else specifies a probe $B(\bar{\pi})$. We usually consider probe algorithms that are determined only by the contact points on the probe paths π_i $(i = 1, \ldots, k)$.

Let B be any probe algorithm and A be any adversary. Together they determine a unique sequence of probes $F_1, F_2,$. and a unique sequence of probe paths $\pi_1, \pi_2,$. where for each $k \geq 1$,

$$F_k = B(\pi_1, \ldots, \pi_{k-1}), \text{ and } p_k = A(F_1, \ldots, F_k)$$

and $B(\pi_1, \ldots, \pi_k)$ is an 'end-game'. We call the sequence $(F_1, \pi_1, F_2, \ldots, F_k, \pi_k)$ a *game*.

Say a set of points X is *extremal* if

$$X \subseteq \partial(conv(X)).$$

Observe that the set of ε-contact points for P need not be extremal. Despite such difficulties, we will attempt to recover in this new setting concepts from the standard probe case. For example, we can recover the notion of collinearity by saying that three distinct points x_1, x_2, x_3 are ε-*collinear* if there is a line L that has non-empty intersection with each of $E_\varepsilon(x_i)$, $i = 1, 2, 3$.

Definition. Let $\delta > 0$. For any finite set Π of ε-probe paths, let $X(\Pi)$ be the set of finite contact points in Π. A convex polygon Q is said to δ-*fit* Π if Π is consistent with the response of a Q-adversary with δ-error. Let $H_\delta(\Pi)$ denote the set of all convex polygons that δ-fit Π. We define Π to be δ-*consistent* if $H_\delta(\Pi)$ is non-empty. If $\Pi' \subseteq \Pi$ and Π is δ-consistent then we call Π a δ-*consistent extension* of Π'.

A slight clarification may be necessary to our definition of δ-fit since Π is a set of ε-probe paths and yet we consider Q-adversaries with δ-error: this should cause no confusion since an ε-probe path is formally no different than a δ-probe path (both are specified by directed rays). A Q-adversary with δ-error is simply one that treats each ε-probe F as if it were a δ-probe, i.e., it ignores the value of ε but returns a point in $C_\delta(F,Q)$. The key definition in this section is the next idea of a 'certificate'.

Definition. Let $\delta > 0$ and Π a finite set of probe paths. We say that Π δ-*determines* a convex polygon P if for all ε-consistent extensions Π' of Π, P δ-fits Π'. If S is a finite sequence of ε-probes then we say S is a δ-*certificate* for P if the following holds: if Π is any set of probe paths defined by some P-adversary with ε-error in response to S, then Π δ-determines P.

Note that if Π δ-determines P, since Π is a consistent extension of itself, we must have $P \in H_\delta(\Pi)$. The crucial step in this development is to introduce a new parameter $\delta > 0$ that is independent of ε. There is a kind of double-standard in our definition of 'Π δ-determining P': we want P to δ-fit Π' but Π' is an ε-consistent (rather than a δ-consistent) extension of Π. A certificate is essentially a non-deterministic non-adaptive algorithm; it allows one to verify (up to δ precision) whether P has a certain shape. We first show an absolute lower limit on the precision δ of any certificate:

Theorem 11. For any P, if $\delta < 2\varepsilon$ then there does not exist any δ-certificate of P.

Proof. Suppose S is a purported δ-certificate for P. Consider the P-adversary that for any probe gives the response corresponding to the case of no errors. Let Π be the corresponding set of probe paths for S with respect to this adversary. Now choose any edge e of P and let $\bar{v}$ be the vector of length ε that points normal outward from e. Let Q be the polygon obtained by translating P a distance of ε in the direction perpendicularly outward from e, so $Q = P + \bar{v}$. We show that Π could be the response of a Q-adversary (with ε-error) to the probes in S. Let F be a probe in S and let its contact point in Π be x. Let F' be the probe obtained by translating F by the vector $\bar{v}$ (so F' is in the bundle of probes defined by F) First suppose that x is at infinity. Then F misses P implies F' misses Q, so ∞ is an appropriate response of a Q-adversary to F. If x is finite, then $x' = x + \bar{v}$ is the standard contact point of F' with Q. Since $x' \in E_\varepsilon(x)$, again x is an appropriate response of a Q-adversary to F. Finally, to show a ε-consistent extension to Π, let π be the probe path that is aimed normally at the midpoint of edge e with contact point at distance 2ε outside of P. Clearly $\Pi' = \Pi \cup \{ \pi \}$ is a ε-consistent extension of Π. But P does not δ-fit Π, contradiction. **Q.E.D.**

The intuition for the preceding lemma is that there is an ε-uncertainty associated with contact points, but there is an added ε-uncertainty arising from the desire to ensure that 'no further probes' can nullify our present guess about the shape of the polygon.

5.2. Analysis of a certificate

In contrast to the last result, we now show a positive result about certificates. In the context of standard probes, [CY87] noted that every P has a certificate of $2n$ probes, and there are no certificates with fewer probes. We need only a very mild 'largeness' requirement for P, namely:

(∗) $I_{4\varepsilon}P$ is non-empty and each edge of P has length $|e| > 2\varepsilon$.

We show that with only two probes more than in the standard certificate, we can achieve a certificate of precision $\delta = O(\varepsilon)$. Our goal is to analyse the following set $S(P)$ of probes:

(i) For each corner c of P, arbitrarily pick one of the two edges e incident on c and let c' to be the point on e at distance ε away. Send a probe F_c aimed at c' such that F_c is normal to e.

(ii) For each edge e of P, send a probe F_e that is parallel to e and at distance ε+ (ε+ means any value that is infinitesimally greater than ε) from e and such that F_c misses P.

(iii) Let c be a corner with angle less than 60°. Send a probe that misses P but which intersects F_c orthogonally at a distance ε+ from c. Since there are at most two such corners, we use at most two such probes. Call these the *special* probes.

First we prove the following lemma:

Lemma 12. Let Π be any set of probe paths produced by a P-adversary in response to the probes in $S(P)$. Let $\pi = (F, p)$ be a probe path such that if $\Pi \cup \{ \pi \}$ is ε-consistent.

(a) *If F intersects $I_{(2+\sqrt{2})\varepsilon}P$ then $p \neq \infty$.*

(b) *If $p \neq \infty$ then p lies outside of $I_{(2+\sqrt{2})\varepsilon}P$.*

Proof. (a) Since $\Pi \cup \{ \pi \}$ is ε-consistent, let Q be a polygon that ε-fits $\Pi \cup \{ \pi \}$. Assume that F is horizontal and at distance $(2+\sqrt{2})\varepsilon$ below the x-axis. For the sake of contradiction, suppose $p = \infty$ and there is a point $y \in F \cap I_{(2+\sqrt{2})\varepsilon}P$. So there is a standard probe F' in the ε-bundle defined by F such that F' misses Q. Say F' lies above F. Since $E_{(2+\sqrt{2})\varepsilon}(y) \subseteq P$, there is some corner c of P that lies on or above the x-axis. Similarly, there is some corner c' of P that lies on or below the horizontal level $y = -2(2+\sqrt{2})\varepsilon$. Consider the probe path (F_c, x) in Π for some x. It is not hard to see that x is finite (since each edge of P is at least 2ε long. Furthermore, x is at a distance at most $\sqrt{2}\varepsilon$ from c. Hence x is at least 2ε above F. Since x can be the response of a Q-adversary (with ε-error) to F_c, ∂Q has some point q at a distance less than ε from x. Thus q is at least ε distance above F. A similar argument with respect to the corner c' shows that there is a point q' in ∂Q at a distance at least ε below F. This implies that the Q-adversary could not respond with $p = \infty$ for the probe F. This proves that Q does not ε-fit $\Pi \cup \{ \pi \}$, contradiction.

(b) This is similarly shown, and is omitted in this proceedings. **Q.E.D.**

We are now ready to prove the main positive result about certificates.

Theorem 13. Let P be a convex n-gon satisfying the 'largeness requirement' (∗) and $\delta > 2+\sqrt{2}\varepsilon$. Then $S(P)$ as specified above is a δ-certificate for P.

Proof. We have already observed in the above proof that each of the probes F_c gives a finite contact point. Also each F_e and also each of the two special probes yield contact points at infinity. Let Π be any set of probe paths from some P-adversary's response to $S(P)$. For each probe F that has contact point at infinity, we let H_F be the half-space bounded by F which contains P. Let Q be the convex polygon obtained by intersecting the $n+2$ half-spaces H_F. We observe that $Q \subseteq E_{2\varepsilon}P$: it is here that we need the two special probes of $S(P)$.

Now consider any ε-consistent extension Π' of Π. We want to show that P δ-fits Π'. Take any probe path $(F, x) \in \Pi' - \Pi$. If $x = \infty$, then by the last lemma F must miss $I_\delta P$, so (F, x) is consistent with the response of a P-adversary with δ-error. So suppose x is finite. Then by the last lemma again, x is not in $I_\delta P$. It is also easy to see that x is not outside of $E_\varepsilon Q$. This implies that x is inside $E_\delta P$. Therefore $x \in EI_\delta P$. We are not yet done since $x \in EI_\delta P$ is a necessary but not sufficient condition for (F, x) to be a response of a P-adversary with δ-error: we must consider the entire probe path itself. In particular, if the probe path (i.e. the half-line of F terminating at x) intersects an edge e of P then we must be sure that $E_\delta(x) \cap e$ is not empty. This can be checked. **Q.E.D.**

By $2n + O(1)$ probes, we can improve the precision of our certificate $S(P)$ arbitrarily closer to 3ε. To do this, we first generalize the observation that there are at most two angles less than 60°:

Claim: Let $\theta = (1 - \frac{2}{k})\pi$ for some $k \geq 3$. Then every convex polygon P has at most $k-1$ angles that are $< \theta$.

[Proof of claim: If P has k angles which are $< \theta$, form the subpolygon Q with k vertices formed from these vertices of P. Each angle of Q is no larger than the corresponding angle in P. Hence we obtain the contradiction that $k\theta = (k-2)\pi$ is strictly larger than the total interior angles of Q, which is equal to $(k-2)\pi$.] Using this claim, we can now send probes to 'cut off sharp corners' analogous to our two special probes in $S(P)$. The number of 'sharp corners' is a function $k(\delta)$ of δ, and we use $O(k(\delta)) = O(1)$ special probes.

6. Concluding remarks

The study of 'crude' sensory data is relatively new in robotics. In this paper we studied the problem of determining a convex polytope using various models of probing. At first glance, it is surprising that we can even obtain finite bounds on the algorithms without any further information about these polytopes (compare (2) below). Many interesting questions remain.

(1) The section on probes with error offers many interesting prospects for further work. We have isolated some key concepts to show that some of the results on standard probes can be extended in a meaningful way. An open problem is to find efficient algorithms to determine polytopes up to δ-precision. Another intriguing question is the trade-offs between accuracy $\delta = \delta(\varepsilon)$ and the number of probes.

(2) It is natural to try to extend our results on standard probes to non-convex objects. However, *without further assumptions about these objects*, there is no meaningful extension of our results even for star-shaped objects. In other words, there is no algorithm that can determine the objects exactly in a finite number of steps. On the other hand, the problem becomes meaningful if we make additional assumptions on the star-shaped objects similar to (*) in section 5.2. Similarly, the extension to multiple convex polygons can only work with additional information.

(3) The approach to vision known as the 'model-based vision' has been extensively investigated, especially by researchers at Stanford. Abstractly, this is essentially the problem of preprocessing information about the objects or scenes to be identified. Suppose that, given a polygon P, we want to know whether P is a translation and rotation of some convex polygon taken from a given finite set S. The upper and lower bounds of Yap and Cole no longer holds. Indeed, H. Bernstein [Ber86] has observed that $2n + O(1)$ probes suffice (the necessity of this bound is easy).

(4) The probing strategies reported in this paper rely on substantial, although polynomial time, computations which process the results of the past probes and then figures out the next probe. It would be interesting to analyze the complexity of these background computations.

REFERENCES

[Ber86]. Herbert J. Bernstein,"Determining the shape of a convex n-sided polygon by using $2n+k$ tactile probes," *Information Processing Letters*, **22**, pp. 255-260 (1986).

[CY87]. Cole and C. Yap, "Shape from probing", *Journal of Algorithms*, **8**, pp. 19-38 (1987).

6. Motion Planning

A mighty maze! but not without a plan.

Alexander Pope, *An Essay on Man*, Epistle 1 (1733-1734)

The general motion planning problem for a system of autonomous vehicles can be stated as follows: Given (1) an initial state of the vehicles, (2) a desired final state of the vehicles, and (3) any constraints on allowable motions, find a collision-free motion of the vehicles from the initial state to the final state that satisfies the constraints. A state of the vehicles may include a description of such things as the position, orientation, and velocity of each vehicle. Several surveys and overviews of general motion planning exist, for example [4, 22-24].

When it is assumed that the environment is static and the only constraints on the motions of the vehicles are that they do not collide, it suffices to specify collision-free paths for the vehicles to follow. In the majority of the work in motion planning, these conditions are assumed, and such work can be better described as *path planning*.

Consider the case of path planning for a robot vehicle. If the object is allowed to rotate and translate, then the space occupied by the vehicle is uniquely specified by tuples (x,y,θ), where x and y specify the location of some fixed point on the vehicle and θ is the orientation of the vehicle with respect to some fixed coordinate system. That is, configuration space, the space of all such tuples, and hence free space, the space of all tuples that describe collision-free placements of the vehicle, is three dimensional; see Chapter 4. Suppose there are no constraints on the motion of the vehicle other than keeping the vehicle free of collisions. Then the path planning problem can be reduced to to the problem of finding a path in free space. That is, the path planning problem is equivalent to (1) finding a path in free space between the initial and goal configurations or (2) determining that the initial configuration and goal configuration are not in the same path-connected component of free space. An advantage of computing the path connectivity of free space is that this computation can be performed once as a preprocessing step, and then requests for paths between query configurations can be processed without recomputing free space each time. Methods for testing path connectivity in free space for general path planning problems have been developed [5, 21].

Unfortunately, many robot vehicles are nonholonomic systems and hence are not capable of arbitrarily rotating and translating; see Chapter 1. The path planning problem for nonholonomic vehicles becomes complicated, since not all paths in the vehicle's free space correspond to feasible motions. However, Laumond [14] has shown that if two configurations are connected by a path in configuration space that does not correspond to a feasible motion, then there is another path between the configurations that does correspond to a feasible motion. Therefore, the problem of determining if a feasible motion of a nonholonomic vehicle exists is computationally equivalent to determining path connectivity of free space. Unfortunately, the problem of actually computing a feasible motion between two configurations that are known to be path connected in free space has no known complexity bounds. This is due to the fact that Laumond's method [14] of constructing a path that corresponds to a feasible motion may require the vehicle to make a large number of small forward and backward motions. To visualize such a motion, consider a poor driver attempting to parallel park in a small space.

The motion planning problem has received a great deal of attention in the theory community. We shall refer to the theoretical study as *algorithmic motion planning*. In algorithmic motion planning, the goal is to analyze the asymptotic complexity of the problem by gaining a thorough understanding of its underlying structure. For a particular motion planning problem, theoreticians seek exact algorithms that are provably correct, as opposed to approximation schemes. That is, the algorithm will describe a desired motion if one exists or else it will report that no motion exists. The algorithm is accompanied by a proof of its correctness and an analysis of its worst-case asymptotic resource bounds; see [1]. Such algorithms provide an upper bound on the inherent complexity of the problem being studied. Lower bounds on the complexity of motion planning problems are often provided by showing that the problems are NP-complete or PSPACE-complete; see [11]. Hence they are unlikely to have efficient, that is, polynomial time, exact solutions; see [19], for example. The emphasis on developing correct algorithms with provably efficient running times almost always leads to solutions that are far too complex to be implemented. The result obtained by Kedem and Sharir in this chapter is a rare exception. The algorithm they describe for planning the motion of a single vehicle is not only efficient and provably correct but also has been implemented by the authors.

A common assumption in motion planning is that a complete description of the environment in which the vehicle is to move is known a priori. In some situations, it may be unreasonable to expect to be given this complete description of the environment. For example, in undersea or space exploration, the goal is to navigate intelligently in a possibly completely unknown environment. The problem of finding a motion to a target position in an unknown environment is the problem of developing a strategy for solving a maze; see [15], for example. The paper in this chapter by Lumelsky and Stepanov describes several strategies that are guaranteed to find a path if one exists. Of course, the path found is not likely to be the shortest path possible although the information gathered as the vehicle wanders about could be used later to find more efficient motions.

Having a complete knowledge of the environment often makes it possible to find globally optimal motions such as motions that follow a shortest path to the goal. The book by Akman [2] is a good survey of shortest path problems. Mitchell and Papadimitriou [16] describe a method for finding least cost paths in an environment consisting of regions with different costs associated with moving through each region. This models the problem of moving a vehicle outdoors, where it may be more costly to travel through regions of mud or tall grass rather than a region of hard soil. Another useful type of motion is one that keeps the vehicle as far away as possible from obstacles in the environment. Such a motion must follow the edges of the geometric structure known as the *Voronoi diagram* [9]. A simplified version of this structure is described in the reprinted paper by Canny and Donald in Chapter 4 [6]. A point on an edge of the Voronoi diagram is equidistant from two obstacles and these are the closest obstacles to the point. Moving off the edge to one side or the other brings the vehicle closer to one of the obstacles than if it stayed on the edge.

In addition to the fact that many robot vehicles are nonholonomic systems, many also have a lower bound on their turning radius, due to dynamic constraints, for example. The path that such a vehicle can follow has a bound on its curvature. The study of such bounded curvature paths has a fairly long history. Dubins [8] described a method for computing the shortest bounded curvature path in the absence of obstacles when the vehicle cannot go in reverse. Reeds and Shepp [18] provided a similar result in the case in which reversals are allowed. Laumond [14] discussed the problem of computing a bounded curvature path in the presence of obstacles when reversals are allowed. Results [10] indicate that the general problem of path planning under the bounded turning radius constraint is difficult when reversals are not allowed and so compromises are necessary in order to find efficient algorithms. The paper in this chapter by Wilfong describes an algorithm for a restricted version of the bounded turning radius problem.

Although much work has been concentrated on path planning, more general motion planning problems have recently been considered. An initial study of planning motions when the moving object has an acceleration bound [17] describes a method for finding a collision-free, acceleration-bounded motion of a point on a line moving within oscillating barriers. Several algorithmic motion planning results [3, 7, 20], have been described for problems in which a speed bound has been imposed on an object moving in an environment containing moving obstacles. A heuristic approach [12] to motion planning in the presence of moving obstacles proceeds by first finding a collision-free path in the environment in which the obstacles are fixed in their initial positions and then attempting to compute a motion of the object along the path so that the object avoids collisions with the moving obstacles while obeying a speed bound. Given speed bounds, acceleration bounds, and an error bound $\varepsilon > 0$, it is possible [6] to compute a motion that takes a time no more than $1 + \varepsilon$ times the true optimal time solution efficiently.

An interesting approach that combines the problems of path planning with those of control is the *potential field method*. This method places artificial repelling forces on the obstacles and an attracting force at the goal, and the resulting field is used to guide the vehicle towards the goal while avoiding obstacles. This method has been demonstrated to work well in practice when the environment is not densely occupied by obstacles. The paper by Khatib in this chapter introduces this method. For an excellent overview of this method, the reader is urged to consult [13].

Although the theoretical approach has the drawback that the algorithms produced tend to be far too complicated to implement, the heuristic approach suffers from the problem that there is no guarantee that the methods will work in all cases. However, the hope is that the theoretical results can reveal the fundamental structure of the problem, thus providing a framework to assist in making decisions as to what compromises should be made when developing a heuristic approach. Also, the success of a heuristic approach should perhaps indicate to theoreticians that the model they have been considering is incorrect in that worst-case situations seldom, if ever, occur in practice.

References

1. Aho, A., Hopcroft, J., and Ullman, J. *The Design and Analysis of Computer Algorithms*. Addison-Wesley, Reading, Mass., 1974.

2. Akman, V. *Unobstructed Shortest Paths in Polyhedral Environments*. Vol. 251. Springer-Verlag, New York, 1987.

3. Aronov, B., Fortune, S., and Wilfong, G. Minimum Speed Motions. Robotics Research Technical Report 163, NYU-Courant Institute, New York, July 1988.

4. Brady, Hollerbach, Johnson, Lozano-Pérez, and Mason *Robot Motion: Planning and Control*. MIT Press, Cambridge, Mass., 1982.

5. Canny, J. *The Complexity of Robot Motion Planning*. MIT Press, Cambridge, Mass., 1988.

6. Canny, J., Donald, B., Reif, J., and Xavier, P. On the Complexity of Kinodynamic Planning. In *Proceedings of the 29th IEEE Symposium on Foundations of Computer Science*, IEEE, New York, 1988, pp. 306-316.

7. Canny, J., and Reif, J. New Lower Bound Techniques for Robot Motion Planning Problems. In *Proceedings of the 28th IEEE Symposium on Foundations of Computer Science*, IEEE, New York, 1987, pp. 49-60.

8. Dubins, L. On Curves of Minimal Length with a Constraint on Average Curvature, and with Prescribed Initial and Terminal Positions and Tangents. *Am. J. Math. 79* (1957), 497-516.

9. Fortune, S. A Sweepline Algorithm for Voronoi Diagrams. *Algorithmica 2* (1987), 153-174.

10. Fortune, S., and Wilfong, G. Planning Constrained Motion. In *Proceedings of the 20th Annual ACM Symposium on Theory of Computing*, ACM, New York, 1988, pp. 445-457.

11. Garey, M., and Johnson, D. *Computers and Intractability: A Guide to the Theory of NP-Completeness*. W.H. Freeman and Co., San Francisco, Calif., 1979.

12. Kant, K., and Zucker, S. Toward Efficient Trajectory Planning: The Path-Velocity Decomposition. *Intl. J. Robotics Res. 5*, 3 (1986), 72-89.

13. Koditschek, D.E. Robot Planning and Control via Potential Functions. In *The Robotics Review 1*, O. Khatib, J. Craig, and T. Lozano-Pérez, Eds. MIT Press, Cambridge, Mass., 1989, pp. 349-368.

14. Laumond, J. Feasible Trajectories for Mobile Robots with Kinematic and Environment Constraints. In *Intelligent Autonomous Systems*, 1986, pp. 346-354.

15. Lumelsky, V., and Skewis, T. A Paradigm for Incorporating Vision in the Robot Navigation Function. In *IEEE International Conference on Robotics and Automation*, IEEE, New York, 1988, pp. 734-739.

16. Mitchell, J., and Papadimitriou, C. The Weighted Region Problem. In *Proceedings of the 3rd Annual Symposium on Computational Geometry*, ACM, New York, 1987, pp. 30-38.

17. Ó'Dúnlaing, C. Motion Planning with Inertial Constraints. *Algorithmica 2*, 4 (1987), 431-475.

18. Reeds, J.A., and Shepp, L.A. Optimal Paths for a Car That Goes Both Forward and Backwards. *Pacific J. Math.* (to be published).

19. Reif, J. Complexity of the Mover's Problem and Generalizations. In *Proceedings of the 20th IEEE Symposium on Foundations of Computer Science*, IEEE, New York, 1979, pp. 421-427.

20. Reif, J. and Sharir, M. Motion Planning in the Presence of Moving Obstacles. In *Proceedings of the 26th IEEE Symposium on Foundations of Computer Science* , IEEE, New York, 1985, pp. 144-154.

21. Schwartz, J., and Sharir, M. On the Piano Movers' Problem: II. General Techniques for Computing Topological Properties of Real Algebraic Manifolds. *Adv. Appl. Math. 4* (1983), 298-351.

22. Schwartz, J., and Sharir, M. Motion Planning and Related Geometric Algorithms in Robotics. In *Proc. Intl. Cong. Math.*, 2, (1986), 1594-1611.

23. Sharir, M. Algorithmic Motion Planning in Robotics. *Computer 22*, 3 (1989), 9-20.

24. Yap, C. Algorithmic Motion Planning. In *Advances in Robotics, Volume 1: Algorithmic and Geometric Aspects*, J. Schwartz and C. Yap, Eds. Lawrence Erlbaum Associates., Hillsdale, N.J., 1986.

An Automatic Motion Planning System for a Convex Polygonal Mobile Robot in 2-Dimensional Polygonal Space

K. Kedem[1] *and M. Sharir*[1,2]

[1] Computer Science Department
School of Mathematical Sciences
Tel Aviv University
and
[2] Courant Institute of Mathematical Sciences
New York University

Abstract

We present an automatic system for planning the (translational and rotational) collision-free motion of a convex polygonal body B in two-dimensional space bounded by a collection of polygonal obstacles. The system consists of a (combinatorial, non-heuristic) motion planning algorithm, based on sophisticated algorithmic and combinatorial techniques in computational geometry, and is implemented on a Cartesian robot system equipped with a 2-D vision system. Our algorithm runs in the worst-case in time $O(kn\lambda_6(kn)\log kn)$, where k is the number of sides of B, n is the total number of obstacle edges, and $\lambda_6(r)$ is the (nearly-linear) maximum length of an $(r,6)$ Davenport Schinzel sequence. Our implemented system provides an "intelligent" robot that, using its attached vision system, can acquire a geometric description of the robot and its polygonal environment, and then, given a high-level motion command from the user, can plan a collision-free path (if one exists), and then go ahead and execute that motion.

1. Introduction

A major motivation of research in robotics is the development of "intelligent" robots, namely machines that can gather information about their environment, analyze it, and plan their tasks accordingly. In other words, the robot should be told what its assignment is, in a very high-level and "gross" style, and the actual task analysis and planning should then be performed by the system in a completely automatic manner. Availability of robot systems fully capable of achieving these goals is still a dream for the future, but the strive to develop such systems is of crucial significance in robotics research, since they will make future robot systems much more flexible, versatile and autonomous, and substantially easier to program.

One of these intelligent tasks is automatic collision-free motion planning. In a typical setting, the system is expected to possess (or to acquire) information about the geometry of itself and its environment, and is then given a command to move from its current position to a desired target position, so that during that motion it avoids collision with any obstacle (or collision between two subparts of the moving system). There are two basic approaches in this area. One is path planning with incomplete information about the obstacles (cf. [Lu], [LuS]), where sensors of various types gather information about the obstacle space as the robot is moving, thus changing dynamically the planned path. The other approach, which is the one used and presented here, is path planning with complete prior information about the geometry and positions of the robot and the obstacles. We also follow the (by now classical) approach which seeks to obtain combinatorial non-heuristic and worst-case efficient algorithms, which (if the geometric data is accurate) are guaranteed to yield a path when one exists.

There is a variety of other, more heuristic techniques used in the design of motion planning systems, ranging from slicing up the configuration space of the moving system into a number of slices and exploring each slice in an approximate fashion (see [BLP]), to producing collision-free motions induced by a potential-field associated with the obstacles. Although considerable progress has been made using more heuristic approaches (see e.g.

Work on this paper has been supported by Office of Naval Research Grant N00014-82-K-0381, National Science Foundation Grant No. NSF-DCR-83-20085, and by grants from the Digital Equipment Corporation, and the IBM Corporation.

[LP]), these alternative methods do not achieve fully satisfactory performance, and the limitations of their applicability is perhaps one of the reasons why the computational geometry based approach to motion planning has gained so much popularity recently. Another motivation for studying combinatorial algorithmic approaches to motion planning is the realization that the problem has a very rich mathematical content, and that one needs to fully explore the mathematical and geometric structure of the problem before beginning to negotiate approximating or heuristic techniques. These algorithmic investigations have so far been very successful, have led to many insightful developments in the field, and have resulted in rich cross-fertilization between computational geometry and motion planning (as well as other areas in theoretical robotics). See [SS2], [Ya], [HSS] for recent summaries and compendium of results in the area. However, this algorithmic approach has been so far mostly theoretical, and applications of the numerous algorithms and techniques developed to date are still slow to emerge.

In this paper we study one of these motion planning problems, namely that of planning the (translational and rotational) motion of a convex polygonal mobile robot in a 2-D polygonal space, and obtain an asymptotically efficient algorithm for it. We then go further and provide an implemented system which carries out this algorithm on a Cartesian robot arm equipped with a 2-D vision system. This work is thus one of the initial steps towards interfacing between the rapidly growing theory of motion planning and the slower-to-follow pragmatic applications.

In more detail, we study here the following problem. Let B be a convex polygonal object having k vertices and edges, free to move (translate and rotate) in a two-dimensional space V bounded by a collection of polygonal obstacles ("walls") having altogether n corners. We want to automatically plan and perform a continuous obstacle-avoiding motion of B between any two specified initial and final placements (see Fig. 1.1).

This specific motion planning problem has been considered by Schwartz and Sharir [SS], who present an $O((kn)^5)$ algorithm for its solution (which applies to non-convex moving objects as well). Since then this classical "piano-movers" problem had been studied extensively, and several efficient algorithms had since been developed for certain special cases of it [OY], [OSY1], [OSY2], [LS1], [KS], [LS2], [SiS].

The algorithm developed in this paper is based on a generalization and combination of the algorithms developed by Leven and Sharir [LS1] and by Sifrony and Sharir [SiS] for the case in which B is a line segment. The Leven-Sharir algorithm partitions the (3-dimensional) space FP of free placements of B (also known as the *free configuration space* of B) into simple disjoint and connected cells, and then determines the adjacency between these cells. This yields an abstract representation of FP by a *connectivity graph* whose nodes are these cells and whose edges connect pairs of adjacent cells. Once the cells containing the specified initial and final placements of B are determined, the motion-planning problem is then reduced to a simple graph searching.

The second algorithm [SiS] also reduces the problem to a combinatorial graph searching, but uses a different graph, called the *vertex graph*, whose nodes are the corners of FP, and whose edges connect pairs of corners that are adjacent along edges of FP, or along some additional auxiliary arcs in FP.

Our algorithm constructs an "intermediate" kind of graph, which we call an *edge graph;* its nodes are edges of FP and its edges connect pairs of adjacent FP-edges (in a sense to be defined more precisely below). Our algorithm begins by restricting the motion of B to be purely translational at some fixed orientation θ. This motion has only two degrees of freedom, so that it is easier to calculate its associated restricted 2-D space of free placements FP_θ (a task which has already been carried out in [KS], [KLPS]; cf. also [LS2]), and to represent it as the union of 2-D polygonal regions having simple shape. Each such region can be given a discrete combinatorial labeling that does not depend continuously on θ. Roughly speaking, we represent FP_θ by a graph VG_θ, whose nodes are the corners of FP_θ, and whose edges connect pairs of adjacent corners. The nodes of VG_θ are given discrete combinatorial labels (that do not depend continuously on θ).

Next we observe that this combinatorial description of FP_θ will not change as θ varies slightly, unless θ is one of finitely many *critical orientations*, at which some critical condition, which affects the combinatorial structure of FP_θ, occurs.

As it turns out, the most complex of these critical orientations are those at which the object B makes three simultaneous contacts with the obstacles, without penetrating into any obstacle. If B is a line segment (a "ladder"), then it is shown in [LS1] that the total number of such critical placements of B is $O(n^2)$, which consequently leads to an $O(n^2 \log n)$ algorithm for the desired motion planning. If B is a convex k-gon which is free only to translate in V but not to rotate, then the motion planning problem becomes simpler and can be accomplished in time $O(kn \log kn)$ [KS], [LS2]. This follows from the property, which is proved in [KS] and will be used below, that the number of placements of B (all having the same given orientation) at which it simultaneously touches two obstacles, without penetrating into any obstacle, is $O(n)$ (provided that B and the obstacles are in "general position"; cf. [KS] and below). If B is also allowed to rotate, then the corresponding critical orientations are much harder to analyze. Since each contact of B with the walls is a contact either of a corner of B with a wall edge or of an edge of B with a wall corner, a crude and straightforward upper bound on the number of these critical placements of triple contact of B is $O((kn)^3)$. Moreover, if B is nonconvex, then there are cases where the number of these critical placements of B is indeed $\Omega((kn)^3)$. However, a recent result of Leven and Sharir [LS3] shows that if B is convex, then the number of these critical orientations is only $O(kn\lambda_6(kn))$, where $\lambda_s(q)$ is the maximum length of a (q,s) *Davenport Schinzel sequence* as defined in [HS], and is known to be an almost linear function of n for any fixed s. The known bounds are

$\lambda_1(n) = n$; $\lambda_2(n) = 2n-1$.

$\lambda_3(n) = \Theta(n\alpha(n))$ [HS], where $\alpha(n)$ is the functional inverse of Ackermann's function, and thus grows extremely slowly (it is ≤ 5 for all practical values of n).

$\lambda_4(n) = \Theta(n \cdot 2^{\alpha(n)})$ [ASS].

$\lambda_{2s}(n) = O(n \cdot 2^{O(\alpha(n)^{s-1})})$, for $s>2$ [ASS].

$\lambda_{2s+1}(n) = O(n \cdot \alpha(n)^{O(\alpha(n)^{s-1})})$, for $s \geq 2$ [ASS].

$\lambda_{2s}(n) = \Omega(n \cdot 2^{\Omega(\alpha(n)^{s-1})})$, for $s>2$ [ASS].

Using these bounds together with the techniques of [LS1], [SiS] and [KS], we next extend each node in VG_θ, by varying θ, into a node which represents an edge of FP, and then construct an *edge graph* EG which represents adjacency of such edges along the boundary of FP. There are several technical issues involved in the analysis and construction of EG. The first issue is topological, and has to do with ensuring that EG properly captures the connectivity of FP, in the sense that to each connected component of FP there corresponds a unique connected component of EG within it. To preserve connectivity in this manner, one has to add enough connections between edges of FP, and we show below how to achieve this in a rather simple manner. This, combined with a procedure that moves B from any given placement in a canonical manner to a placement that lies on an edge of FP, gives us a method for transforming any motion planning instance to path searching in the graph EG.

The second issue involves the efficiency of our algorithm. We show that since the number of critical orientations is only nearly quadratic in kn, and since FP_θ generally undergoes only a constant number of combinatorial changes at each critical orientation, it follows that the overall complexity of EG is only nearly quadratic. We then develop an efficient (albeit complicated) procedure for calculating all critical orientations, and combine it with an efficient incremental technique for constructing EG by stepping through the list of critical orientations, and updating EG with new features as they materialize at the critical orientations. All this finally yields a motion planning algorithm for a convex polygonal object B which runs in time $O(kn\lambda_6(kn)\log kn)$.

Section 2 describes the recursive discrete representation of FP by its associated edge-graph EG, as outlined above, and Section 3 provides more algorithmic details involved in the construction of EG, and in its use for actual motion planning.

We have implemented this algorithm into a motion planning system that uses an IBM RS/II Cartesian arm (as a device to move our 2-D "robot"), equipped with a "home-made" 2-D computer vision system developed at the Robotics Lab of New York University. As is typical in motion planning algorithms, our technique functions best as a preprocessing step, performed once and for all in a static environment, so that any subsequent motion command can be processed and executed efficiently (by a path searching in EG). Nevertheless, since the worst-case complexity of EG, and the time needed to compute it is, relatively speaking, small, our aim has been to test our algorithm also in dynamic situations when the environment, or even the moving system itself, keep changing, so the algorithm may have to be run from scratch many times. A typical set-up for such a situation, which is the one that we have used in our experiments, is having a collection of convex flat objects scattered on a floor (in placements known to the planner), so that each motion command asks for moving any one of them to a new placement. Thus both the environment, and also the moving system, can change between successive motion commands (but in a controlled manner known to the planning system). Our implementation and some experimental results with it are reported in Section 4.

For lack of space, many technical details of our analysis and implementation have been omitted in this version. More details can be found in [Ke].

2. Discrete Recursive Representation of the Free Configuration Space

Let B be a bounded convex k-sided polygonal body, whose interior is nonempty. B is free to move (translate and rotate) in a bounded two dimensional open region V having a polygonal boundary with n corners altogether (we refer to edges of V as "walls").

Let P be an arbitrary fixed reference point in the interior of B, and let Q be an arbitrary corner of B. Each placement Z of B in the plane can be represented by the parameters (X, θ), where X is the position of P, and where θ is the orientation of the vector PQ.

As in [SS], we define a *free placement* (X,θ) of B to be a placement at which B is fully contained within V; a *semi free placement* (X,θ) of B is defined to be a placement at which B may touch some walls, but not penetrate into the interior of the wall region V^c. The set FP of all free placements of B is an open three dimensional manifold, and the set SFP of all semi free placements is closed.

We assume that B and V lie in "general positions" so as to avoid degeneracies in the resulting configuration space. Roughly speaking, this means that (i) there does not exist a placement of B in which it meets four independent constraints involving contacts with obstacles; and (ii) there do not exist two placements of B with the same orientation such that B meets at each of them three independent constraints involving contacts with obstacles (cf. [LS3] for more detail). We also assume for simplicity that no wall edge is horizontal; this can always be enforced by an appropriate rotation of V.

Since the motion of B has three degrees of freedom, we first analyze, as in [LS1], only purely translational motion of B (involving just two degrees of freedom), and only then treat the case of general motion of B, including rotation. This will enable us to obtain recursively a combinatorial representation of each cross-section of FP at a fixed θ, from which we will then construct a certain discrete graph which represents the entire space FP in a "connectivity preserving" manner, and which allows us to reduce the motion planning problem to a discrete problem of path searching through that graph.

2.1. The Case of Translational Motion of B

Definition 2.1: (a) ([LS3]) A (potential) *contact pair* O is a pair (W,S) such that either W is a wall edge and S is a corner of B, or W is a wall corner and S is a side of B. In the first case we call the pair a contact pair of type I and in the second case a contact pair of type II.

(b) A contact pair of type III is a pair $O=(W,S)$ where W is a wall corner and S is a corner of B.

(c) An actual *obstacle contact* (i.e. a contact of B with an obstacle) is said to *involve* the contact pair $O=(W,S)$ if this contact is of a point on S against a point of W, and furthermore if this contact is *locally free,* i.e. B and the wall region are openly disjoint in the neighborhood of this contact point.

(d) (cf. [KS]) Let A be one of the convex polygonal obstacles into which V^c is decomposed. The *expanded obstacle* A_θ^* associated with A for a given orientation θ of B is the pointwise vector difference $A - B_\theta$, where B_θ is the standard placement of the moving object B in which P lies at the origin, rotated by θ. A_θ^* is also a convex polygonal region whose sides are vector (Minkowski) differences of the form $W - S$ where (W,S) is a contact pair of type I or II, and whose vertices have a similar representation for contact pairs (W,S) of type III.

It follows from the results of [KS] that the restricted free configuration space FP_θ, which is the space of all free placements of B having orientation θ, is the complement of the union

$$K_\theta = \bigcup_{i=1}^{m} (A_i)_\theta^* = \bigcup_{i=1}^{m} (A_i - B_\theta)$$

where $A_1, \ldots, A_m$ are the convex polygonal regions into which V^c is decomposed. See Fig. 2.1 for an illustration of K_θ and of FP_θ. The boundary of K_θ (and also of FP_θ) thus consists of a collection of polygonal curves having finitely many corners. An edge on the boundary of K_θ is a connected portion of an edge of an expanded obstacle induced by a type I or a type II contact pair, and each vertex of K_θ is either a (convex) corner of an expanded obstacle induced by a type III contact pair, or a (non-convex) intersection point of two sides of different expanded obstacles, each induced by a contact pair of type I or of type II.

It is shown in [KS] (cf. also [LS2]) that, for each fixed orientation θ of B, the number of non-convex corners of K_θ (which correspond to semifree placements of B in which it makes two distinct obstacle contacts simultaneously) is only $O(n)$. A corresponding $O(kn \log^2 kn)$ algorithm for the calculation of FP_θ, which takes advantage of this fact in a divide and conquer approach, is presented in [KS] (see also [LS2], [BZ]).

We extend this algorithm to obtain a representation of FP_θ by a *vertex graph* VG_θ, whose set of nodes include all corners of K_θ, plus auxiliary corners u^*, one per each *vertically maximal* corner u of K_θ (i.e. a convex corner u of K_θ which has the largest y-coordinate among all points of K_θ in a sufficiently small neighborhood of u), so that the segment uu^* is vertical (with u^* lies above u), and its interior does not meet K_θ. The edges of VG_θ are either portions of the boundary of FP_θ connecting pairs of adjacent corners, or vertical segments uu^* as above; see Fig. 2.2.

It is crucial for the analysis of general motion of B, as given below, that VG_θ will not depend continuously on θ, but rather remain invariant except at certain finitely many *critical orientations,* to be defined below. This is achieved by assigning discrete labelings to each node of VG_θ, which do not depend continuously on θ, and which describe each node by the pattern of obstacle contacts giving rise to it. Thus each convex corner of K_θ is labeled by the type III contact pair that induces it; each non-convex corner is labeled by the two contact pairs (of type I or II) that induce it, etc.

The following theorem completes the analysis of the translational case.

Theorem 2.1: Two nodes u, v of VG_θ belong to the same connected component of VG_θ if and only if the corners they represent lie on the boundary of the same connected component of FP_θ.

Sketch of Proof: If the edge (u,v) belongs to VG_θ then by definition u and v lie on the (boundary of the) same connected component of FP_θ. Thus the "only if" part of the Theorem follows by transitive closure. As to the "if" part, it is clearly true in the case in which u and v lie on the same connected component of the boundary of a connected component Q of FP_θ. If Q has more than one boundary component, then we let A_u (resp. A_v) denote the set of all nodes in VG_θ reachable from u (resp. from v) by a path in VG_θ. We claim that both A_u and A_v contain corners lying on the (unique) exterior boundary E of Q (intuitively, we cannot get "stuck" at an interior component of VG_θ, because we can always shoot from a vertically maximal corner upwards to a "higher" component, so that u and v lie in the same connected component of VG_θ. □

2.2. The case of a general motion of B

We now turn to the general case in which B can both translate and rotate. To do so we first consider how the combinatorial characterization of FP_θ, provided by the graph VG_θ, changes as θ varies.

Definition 2.2: An orientation θ of B is called a *critical orientation*, if one of the following conditions occurs:

(i) There exists a semi-free placement of B at orientation θ at which either it makes simultaneously three distinct obstacle contacts involving contact pairs of types I or II, or it makes simultaneously two obstacle contacts, one involving a contact pair of type III and another involving a contact pair of type I or II. In other words, either three edges of expanded obstacles meet at the same non-convex corner of K_θ, or a convex corner of K_θ meets another edge of K_θ (see Fig. 2.3(a,b)).

(ii) There exists a vertically maximal convex corner u of K_θ whose associated auxiliary corner u^* coincides with a (convex or non-convex) corner of K_θ (see Fig. 2.3(c)).

(iii) Two adjacent edges of K_θ become collinear (see Fig. 2.3(d)), or an edge of K_θ becomes horizontal (see Fig. 2.3(e)).

Lemma 2.2: The vertex graph VG_θ does not change as θ varies in a sufficiently small neighborhood of any non-critical orientation. Furthermore, for each such sufficiently small neighborhood, the connected component $Q(A,\theta)$ of FP_θ, corresponding to a fixed connected component A of VG_θ, varies continuously (in the Hausdorff topology of sets) with θ.

Sketch of Proof: First observe that as long as condition (iii) does not arise, each edge or convex corner of an expanded obstacle, corresponding to some contact pair O, continues to appear on the boundary of that obstacle, and varies continuously with θ. Moreover, as long as conditions (i) and (iii) do not arise, each intersection between any two such edges is transversal, and thus also varies continuously with θ. Extending these observations, we show that as θ varies slightly, no connected subcomponent of FP_θ (or even no component of the boundary of FP_θ) shrinks to a point and disappears, no such component newly appears, no two such components merge into a single component, nor does such a component split into two subcomponents. Furthermore, each connected component of the boundary of FP_θ retains the same combinatorial representation as a circular sequence of convex and non-convex corners induced by the same combinations of obstacle contacts, and each of these corners varies continuously with θ. This also implies that each connected component of FP_θ varies continuously with θ. Similar arguments show that the absence of condition (ii) ensures that the relative positions of all auxiliary corners, and thus finally VG_θ itself, remains constant as θ varies through non-critical orientations. □

Critical orientations of type (i) have been analyzed in [LS3], where it has been shown that there are at most $O(kn\lambda_6(kn))$ such orientations. We describe below an algorithm for the calculation of these orientations, which runs in time $O(kn\lambda_6(kn)\log kn)$.

Critical orientations of type (iii) are trivial to analyze. In fact, there are only $O(kn)$ such orientations, whose calculation is straightforward.

Finally, critical orientations of type (ii) require some analysis, similar to, but somewhat simpler than that in [LS3]. We show below that the number of such critical orientations is also at most $O(kn\lambda_6(kn))$, and that they can also be calculated in time $O(kn\lambda_6(kn)\log kn)$.

Let θ_0 be a noncritical orientation, and let u be a corner in VG_{θ_0}. We will associate with u a *life-span* $L_u = (\theta_1,\theta_2)$, which is the maximal (open) interval containing θ_0 such that $u \in VG_\theta$ for all $\theta \in (\theta_1,\theta_2)$, and such that u does not become coincident with any other corner of VG_θ for any such θ. We also associate a similar life-span interval $L_e \subset L_u \cap L_v$ with each edge $e = (u,v)$ of VG_{θ_0}, defined in a similar manner. We call the pair (u,L_u) an *extended* corner, and the pair (e,L_e) an *extended* edge.

We next construct an *(extended) edge graph* EG, whose nodes are all these extended corners, and each of whose edges connects an extended corner (u,L_u) to another extended corner (v,L_v), if (u,v) is an edge of some VG_θ (so that the life span of the edge (u,v) is non-empty).

The graph EG is used to represent FP as follows. Each node (u,L_u) of EG represents an edge of the boundary of FP which consists of all placements $Z = (X,\theta)$ of B such that $\theta \in L_u$ and such that at placement Z, either B makes the double contact inducing u in case u is a non-convex corner of K_θ, or B makes the type III contact inducing u if u is a convex corner of K_θ; if u is an auxiliary corner, then (u,L_u) represents an arc along a face of FP, consisting of placements (X,θ) at which B makes a single contact of type I or II, such that X lies directly above the convex corner of K_θ inducing u.

One can show that the converse statement also holds, namely each edge e of FP arises as the intersection of two constraint boundary surfaces, and thus must correspond to some node of EG defined in terms of some convex or non-convex corner of VG_θ (for θ in the life-span of e), induced by the pair of contact pairs that describe the boundary surfaces intersecting at e (see the full version [Ke] for more details). The only exceptions are at degenerate situations in which all placements along e have the same orientation θ; these can arise in certain degenerate situations (cf. Proposition 2.1 of [LS3]), e.g. when e arises from two contact pairs (W_1,S_1), (W_2,S_2), where W_1, W_2 are both wall corners and $S_1 = S_2$, or in some similarly degenerate cases. Nevertheless, we show that the edges of FP that are represented as nodes of EG are sufficient for adequate representation of FP.

The edges of EG also have natural interpretation in terms of FP. Let $\xi = [(u,L_u),(v,L_v)]$ be an edge of EG, and let e_u, e_v denote the two edges of FP corresponding to (u,L_u) and (v,L_v) respectively. Then ξ represents adjacency of e_u and e_v along a face f of FP; more precisely, for each $\theta \in L_{(u,v)}$, the cross-section of f at θ contains a straight segment whose endpoints are (labeled by) u and v. An exception is when $v = u^*$; then ξ represents in a similar manner adjacency of e_u and e_{u^*} along an auxiliary "vertical" surface inside FP.

We claim that EG captures the connectivity of FP in the following sense:

Proposition 2.3: (A) For each connected component A of EG there corresponds a unique connected component Q of FP, such that the nodes in A represent precisely those edges of FP which bound Q (and which do not lie in a single θ-cross section of FP).

(B) Each connected component of FP has at least one bounding edge which is represented by a node of EG.

Sketch of Proof: The proof of (B) is straightforward: Let Q be a connected component of FP, and let $Z = (X,\theta) \in Q$. The cross section $Q_\theta = Q \cap FP_\theta$ of Q contains X and has thus nonempty interior. Since V is assumed to be bounded, so must be Q_θ, which therefore must contain at least one non-convex corner u of K_θ (e.g. that with the largest y coordinate). Then (u,L_u) is a node of EG which represents an edge of Q.

The proof of (A) is more complicated, and consists of two parts. In the first part we argue that if (u,L_u), (v,L_v) are two adjacent nodes of EG, then they represent edges of FP which bound the same connected component. This claim follows immediately from definition of EG. Thus it follows by transitive closure that all the nodes in a connected component of EG represent edges of FP bounding the same connected component of FP.

The second part proves the converse statement, namely that any pair of edges of FP which bound the same connected component of FP, and neither of which is contained in a single θ-cross section of FP, are represented by nodes of EG which belong to the same connected component of this graph. The proof of this statement consists of the following steps.

The set of all critical orientations partitions the angular space into disjoint open *non-critical intervals*. Define a decomposition of FP into disjoint connected *cells* as follows. Let I be a non-critical interval, and let A be a connected component of VG_θ for any (hence all) $\theta \in I$. The cell $c = c(I,A)$ of FP associated with I and A is defined as

$$c(I,A) = \{(X,\theta) : \theta \in I,\ X \in Q(A,\theta)\},$$

where, as above, $Q(A,\theta)$ is the (unique) connected component of FP_θ whose boundary corners appear in A.

It is clear that the cells defined in this manner are open, connected, and pairwise disjoint, and that the union of their closures covers the entire space FP.

Two cells $c = c(I,A)$, $c' = c(I',A')$ are called *adjacent* if the intervals I, I' have a common endpoint θ^* and there exists a free placement $(X^*,\theta^*) \in FP$ lying in the closures of both cells c, c'.

It then follows that the transitive closure of this adjacency relationship defines a graph on these cells, so that two cells c, c' lie in the same connected component of the graph if and only if they are contained in the same connected component of FP.

To complete the proof of (A), it suffices to show that

(a) The boundary of each cell c of FP contains at least one edge which is contained in an edge e_u represented by some node (u,L_u) of EG.

(b) If two cells $c = c(I,A)$, $c' = c(I',A')$ of FP are adjacent, and if (u,L_u), (v,L_v) are two nodes of EG whose corresponding edges e_u, e_v intersect the boundaries of c, c' respectively, then they lie in the same connected component of EG. Moreover, the same property also holds if $c = c'$.

See the full version for more details. □

Using the above observations, we can also define a map $\Phi(Z)$ which maps each $Z \in FP$ to a node of EG representing an edge of FP which bounds the connected component of FP containing Z. Specifically, given any free placement $Z = (X,\theta)$ of B, we can translate B upwards from Z until it reaches a new placement $Z^* = (X^*,\theta)$ at which it makes contact with an obstacle. If this contact involves a contact pair of type III, or consists of two simultaneous contacts involving pairs of types I or II, then X^* is a corner of FP_θ, and we let $\Phi(Z)$ be the node of EG induced by that corner. Otherwise, we continue to translate B from Z^* leftwards, maintaining the obstacle contact that involves the same contact pair, until we reach a placement $Z^{**} = (X^{**},\theta)$ for which X^{**} is a corner of FP_θ, and then continue as above. Again, the fact that FP_θ is bounded implies that Z^*, Z^{**}, and thus also

$\Phi(Z)$, are always well defined.

We can then obtain the following main theorem of this section.

Theorem 2.4: Let Z, $Z' \in FP$ be two free placements of B. Then B can move in FP from Z to Z' if and only if the nodes of EG which correspond to the two edges $\Phi(Z)$, $\Phi(Z')$ of FP, belong to the same connected component of EG.

3. Algorithmic Details

We next present a brief overview of an efficient implementation of our solution. It consists of the following stages

(I) Calculating all critical orientations. This stage calculates critical orientations of each of the three types. The most difficult orientations to calculate are those of type (i), and we will describe their calculation in more detail. Our algorithm is based on the following property, proved in [LS3]. Let $O_1=(W_1,S_1)$, $O_2=(W_2,S_2)$ be two contact pairs inducing a simultaneous obstacle contact at some placement Z of B at orientation θ. Each of the two contacts O_1, O_2 is between a point and a segment. Extend the two corresponding segments to full lines, and let q be their intersection point. We say that O_2 *bounds* O_1 at θ, if $B^* = conv\ (S_1 \cup S_2)$ always intersects W_2 as we translate B from Z along the line containing the contact segment of O_1 in the direction of the intersection point q of the two extended lines, until the last placement at which S_1 still touches W_1.

We will further say that at orientation θ O_2 bounds O_1 towards an endpoint E of the edge induced by O_1, if the translation of B from Z as above terminates at the contact E (which is a corner-corner contact by definition). See Figure 3.1 for an illustration.

Proposition 3.1 [LS3]: If at some placement $Z=(X,\theta)$, B makes simultaneously two obstacle contacts involving the contact pairs O_1, O_2, then either O_1 bounds O_2 at θ, or O_2 bounds O_1 at θ, except in the degenerate case where the two extended lines are coincident or parallel.

Let O_1 be any contact pair and let E be one of its endpoints. Consider all contact pairs that bound O_1 (at any orientation θ) towards E. For each such contact pair O_2 we define a (partial) *bounding function* $F_{O_1O_2}(\theta)$, to be the distance of the endpoint of W_1 at which the contact E is made from the contact point of S_1 and W_1 (at the placement of simultaneous (O_1,O_2) contact), if O_1 is a contact pair of type I, or the distance of the endpoint of S_1 involved in E from the contact point of S_1 and W_1, if O_1 is a contact pair of type II. See Figure 3.2.

We partition the collection of bounding functions $\{F_{O_1O_i}\}$ of O_1 into two classes $\mathbf{A}_L(O_1)$, $\mathbf{A}_R(O_1)$ so that for all functions $F_{O_1O_i}$ in $\mathbf{A}_L$, O_i bounds O_1 towards the left endpoint W_L of W_1 (or S_L of S_1) whereas for all functions in $\mathbf{A}_R$, O_i bounds O_1 towards the other right endpoint of W_1 (or of S_1). Note that there may exist a contact pair O_i for which a bounding function $F_{O_1O_i}$ appears in both collections $\mathbf{A}_L$, $\mathbf{A}_R$, but then these two functions will have openly-disjoint domains of definition [LS3]. Also the domain of definition of each $F_{O_1O_i}$ consists of at most five connected intervals [LS3]. If the domain of such a $F_{O_1O_i}$ is not connected, we regard this function as several (≤ 5) distinct partially defined functions having connected (and pairwise openly-disjoint) domains (cf. [LS3]).

The usefulness of the notion of bounding is that, for each θ, the minimum of all functions $F_{O_1O}(\theta)$ in, say $\mathbf{A}_L(O_1)$, gives a "barrier" along O_1 beyond which (i.e. between it and the left endpoint of O_1) no placement of B with contact O_1 is (semi)free. Similarly, the minimum of all functions in $\mathbf{A}_R(O_1)$ constitutes an opposite barrier, so that all semifree placements of B with contact O_1 at orientation θ must lie between these two minima.

To formalize this observation, define, for each contact pair O, the lower envelope

$$\Psi_{L;O}(\theta) = \min_i \{ F_{OO_i}(\theta) \mid F_{OO_i} \in \mathbf{A}_L(O) \} ,$$

and the corresponding envelope $\Psi_{R;O}$ for the collection $\mathbf{A}_R(O)$.

By the preceding observation, one of the following two cases can arise at a type (i) critical orientation θ, involving a simultaneous triple contact induced by the contact pairs O_1, O_2, O_3 (up to permuting these three contact pairs).

(1) θ is the orientation of an intersection point of two functions $F_{O_1O_2}$ and $F_{O_1O_3}$ lying along the same lower envelope $\Psi_{E;O_1}$.

(2) θ is the orientation of an intersection point of a function $F_{O_1O_2}$ on, say $\Psi_{L;O_1}$ and a (reflected and shifted transform of a) function $F_{O_1O_3}$ lying along $\Psi_{R;O_1}$ (The second function is shifted and reflected so that both functions measure distance from the same endpoint of O_1).

In other words, either θ represents a "breakpoint" on some lower envelope, or θ is an intersection of the two lower envelopes associated with the same contact O_1.

Unfortunately, there is also a third, and most complex subtype of critical orientations of type (i), namely it can also be the case that

(3) At orientation θ, O_1 bounds O_2 (towards an endpoint E_2), O_2 bounds O_3 (towards an endpoint E_3) and O_3 bounds O_1 (towards an endpoint E_1), and the envelopes $\Psi_{E_1;O_1}$, $\Psi_{E_2;O_2}$, $\Psi_{E_3;O_3}$ are indeed attained at θ by the functions $F_{O_3O_1}$, $F_{O_1O_2}$, $F_{O_2O_3}$

respectively.

See Figure 3.3 for an illustration of this criticality. The problem with the third subtype of orientations is that their presence can be detected only by looking at the three corresponding envelopes simultaneously.

We thus calculate the desired critical orientations as follows.

(a) We construct all bounding functions, and distribute them among the various collections $A_L(O)$, $A_R(O)$. Each function is broken into (at most 5) pieces, each of which is defined over a connected interval. By definition, each collection consists of $O(kn)$ partial functions.

(b) We next compute the individual lower envelopes of each collection separately. It was shown in [LS3] that any pair of bounding functions $F_{O_1O_2}$, $F_{O_1O_3}$ intersect in at most 4 points. It follows from the results of [HS], [At] that the combinatorial complexity of each lower envelope is at most $O(\lambda_6(kn))$, and that it can be calculated, using a straightforward divide-and-conquer technique, in time $O(\lambda_6(kn)\log kn)$. Repeating this step $O(kn)$ times, we obtain all lower envelopes in total time $O(kn\lambda_6(kn)\log kn)$. This step already produces the critical orientations of subtype (1).

(c) We next compute critical orientations of the second subtype, by taking each pair of matching envelopes $\Psi_{L;O}$, $\Psi_{R;O}$, merging their breakpoints into a single sorted list, and finding the intersections between these envelopes in each interval delimited by a pair of adjacent breakpoints in constant time. This step takes time of the same order of magnitude as the preceding one.

(d) Finally, we compute orientations of the third subtype. Roughly speaking, we sort all breakpoints in all envelopes into one big list, and then process these breakpoints in order. For every interval I between two successive breakpoints, each envelope is attained by a fixed function. By examining the envelopes over I, it is easy to find all candidate triples (O_1,O_2,O_3) of contact pairs that can yield an orientation of the desired kind. We store all these potential orientations in a priority queue; those that already fall inside I are output, and the others will be processed when we reach the interval containing them. The crucial observation is that when we pass from I to the next interval I', most envelopes do not change, which implies that only $O(1)$ new candidate triples can arise. It thus takes only a constant number of steps (each of a logarithmic cost) to process each new interval, so this step can also be implemented in time $O(kn\lambda_6(kn)\log kn)$. More details are given in the full paper.

Calculation of critical orientations of type (ii) uses a similar but considerably simpler procedure. In particular, we show that these orientations all arise as breakpoints along lower envelopes of certain collections of functions of θ, each of which measures the vertical distance between a (vertically maximal) corner-corner contact and some expanded obstacle edge lying above it, at any given θ. In the full version we provide an analysis on the structure of these functions, show that the domain of definition of each such function consists of at most 5 intervals, and that any pair of functions in the same collection intersect at most four times. Thus, proceeding as in (a) above, we conclude that the number of type (ii) critical orientations is $O(kn\lambda_6(kn))$, and that they can be calculated in time $O(kn\lambda_6(kn)\log kn)$. Calculation of type (iii) orientations is straightforward.

Once all these orientations are calculated, we merge them into a single list $0<\theta_1<\theta_2<\cdots<\theta_w<2\pi$ of length $O(kn\lambda_6(kn))$.

(II) Constructing VG_θ and EG. As in Section 2, we represent each node ξ of the graph EG by a pair (u,L_u), where $L_u=(\theta_1,\theta_2)$ is the angular life span of ξ, and where u is the (discrete labeling of the) corner of VG_θ, for $\theta\in L_u$, that lies on the edge of FP represented by ξ.

The calculation of EG will be accomplished in an incremental manner, similar to that of [LS1]. That is, we process critical orientations in increasing order, maintaining the "cross-section" graph VG_θ and use it to update EG at each critical orientation. At each such orientation θ we determine those nodes of EG whose life-span terminates at θ (these are nodes whose corresponding corners have to be deleted from VG_θ), and the new nodes whose life span starts at θ. Nodes of the first kind will already have been stored in EG, and we update their life span by adding θ as its terminal orientation. Nodes of the second kind are added to EG, with θ as the initial orientation of their life span, and with the corresponding terminal orientation being presently left undefined. EG is also augmented by edges connecting the nodes just inserted with nodes already present in EG, as described in Subsection 2.2 (note that these edges correspond to new edges in $VG_{\theta'}$, for θ' slightly larger than θ).

To initialize EG, we first calculate the graph $VG_{\theta=0}$. For each corner u of this graph we create a corresponding initial node in EG whose label contains u, but with both initial and terminal life-span orientations left undefined. Then we step through the list of critical orientations in the manner described above.

At the end of this procedure, the nodes and edges of EG whose terminal life-span orientation is still undefined are in 1-1 correspondence with the initial nodes and edges of EG, as created from the initial graph $VG_{\theta=0}$. We match these nodes in pairs, thus obtaining the actual life span of each, and thereby completing the construction of EG.

Remark. The construction of critical orientations does not guarantee that the criticality that such an orientation represents actually occurs at a semi-free placement of B. We thus need to check each critical orientation for "authenticity", and discard it if it is spurious. More details about this are given in the full version.

We can now state our main result. Its proof is omitted in this version.

Theorem 3.2. The graph EG has at most $O(kn\lambda_6(kn))$ nodes and edges. The algorithm just described computes EG in $O(kn\lambda_6(kn)\log kn)$ time.

Remarks. (1) When $m \ll n$ the complexity of VG_θ for any fixed θ can be shown to be only $O(km+n)$ which can be significantly smaller than $O(kn)$. We exploit this fact to show that a better bound on the complexity of our algorithm is only $O((km+n)\lambda_6(kn)\log kn)$.

(2) The example given in [LS3] can be used to show that the combinatorial complexity of FP can be $\Omega(k^2n^2)$ in the worst case, so that our algorithm is close to being optimal among algorithms that calculate the entire space FP. See also a related result by O'Rourke [OR] for the case of a moving line segment.

(III) Actual motion planning. Once EG is available, actual motion-planning between any two given placements of B can be easily accomplished, as follows. Let $Z_1 = (X_1,\theta_1)$, $Z_2 = (X_2,\theta_2)$ be two given free placements of B. We first calculate the nodes $\Phi(Z_1)$, $\Phi(Z_2)$ of EG (where Φ is the "pushing" map defined in section 2). It is easy to calculate these nodes in $O(kn)$ time in a straightforward manner. Next determine by simple graph searching whether $\Phi(Z_1)$ and $\Phi(Z_2)$ belong to the same connected component of EG (this can be accomplished in time $O(kn\lambda_6(kn))$). If not, then by Theorem 2.4 no collision-free motion of B between placements Z_1 and Z_2 is possible. Otherwise, let $\pi = (\xi_1 = \Phi(Z_1), \xi_2, \ldots, \xi_t = \Phi(Z_2))$ be a path in EG connecting $\Phi(Z_1)$ to $\Phi(Z_2)$. We show that π can be transformed into a continuous semi-free motion of B from Z_1 to Z_2. This motion has an initial and final portions which respectively translate B from Z_1 to $\Phi(Z_1)$, and from $\Phi(Z_2)$ back to Z_2 as in the definition of Φ. The remainder of the motion is a sequence of elementary submotions, alternating between the two following kinds.

(a) A "gliding" motion of B along the FP-edge corresponding to some EG-node ξ_i (here B moves so as to maintain a double contact induced by two contact pairs of type I or II, or B rotates while maintaining a contact induced by a type III pair); this motion proceeds until B reaches an orientation θ_i at which it can "cross" to the next FP-edge ξ_{i+1};

(b) A translation of B at orientation θ_i from a position along the FP-edge corresponding to ξ_i to a position along the FP-edge corresponding to ξ_{i+1}. This translation is along the straight segment connecting these two positions in VG_{θ_i}.

4. The Implementation

In this section we describe our implementation of a motion planning system based on our algorithm. As mentioned in the introduction, there have not been many attempts so far to implement a system of this kind in an actual robotic environment, by applying combinatorial and non-heuristic techniques such as those used in computational geometry (except for very simple special cases, such as that of a mobile circular robot, or a single moving point, etc.) We believe that it is important to implement and experiment with algorithms of this kind, so as to fully explore their pragmatic potential, understand their weaknesses (and overcome them), and learn about the problems that arise when such a system confronts the real world, such as handling the acquisition of information about the robot's environment, analyzing and controlling various algorithmic issues pertaining to the geometry, kinematics, dynamics, and real-time control of mechanical and other systems (not all these issues were addressed in our own project, though), and coping with the relevant problems in an algorithmically efficient manner. Moreover, it is hoped that actual experimentation would help to uncover new issues in the design of such a system, lead to a reformulation of certain portions of the problem to facilitate easier solutions thereof, and gain better understanding of optimal ways to make heuristic shortcuts and approximations, when and if needed, that would improve the actual performance of such a system. Another issue that arises in implementation is that of software development, which can be expected to be quite complex (it was in our case), given the sophisticated nature of the problem and of its algorithmic solutions. Finally, the issue of numerical precision has to be addressed.

Given the broad range of issues listed above, and considering the complex structure of the algorithm that we have implemented, our system addresses only some of these issues, and concentrates mainly on demonstrating the pragmatic potential of our algorithm. Since we expect our algorithm to be useful in many practical situations involving a mobile robot operating in a static environment (e.g. on a factory floor, in the office or at home), and even in a dynamic environment which changes in a manner known to the robot (as in the discussion in the introduction), we regard our system as a prototype implementation that can be further specialized and improved for actual pragmatic applications.

4.1. System Overview

The implementation and experimentation with our algorithm has taken place in the Robotics Laboratory of the Courant Institute, New York university.

We have created several flat polygonal objects to represent the obstacles. Some of the objects were convex and had handles, so that one of them could be designated as the moving robot B and its planned motion be executed by an actual robot (an IBM RS/II Cartesian arm) which would hold the handle of B and move it on a horizontal work-table amidst the other objects. We have employed (and slightly fine-tuned) a vision system that includes a video camera connected to a SUN workstation, and uses a 2-D shape analysis program [KSSS], developed at the Robotics Lab. This system is capable of obtaining a smoothed polygonal contour of each object (with a number of sides that is roughly the same as the number of actual sides of the object), as well as its present position on the work-table. Having a graphical display of this acquired information available, the user can then select the object B to be moved, and also specify (graphically) its target placement. The system then runs the algorithm described above. If a collision-free motion has been found, the system converts it to a sequence of basic moves of the grasping robot, and sends this sequence to the RS/II as a sequence of AML commands, which are then executed by the robot, who grasps B by its handle and acts as a "motor" of B (i.e. moves B on the plane of the work-table and turns it when necessary).

A main controller (see the system-flow block diagram in Fig. 4.1) was written in C on a VAX/785 computer under the UNIX operating system (which is a trademark of AT&T Bell Labs). Its task was to run in series the vision system on the SUN, to activate the motion planning system on the VAX after shape contour data has been acquired, and finally to communicate between the VAX and the robot's IBM computer, sending motion commands to the robot and getting acknowledgements from the robot about the success (or failure) of the completion of each command.

The motion planning routines were written in Fortran and C and the code length is about 12000 lines. Our system has been written as a prototype model, where the main goal was to demonstrate the feasibility and usefulness of our algorithm in an actual planning system. We have thus not invested in making the code ultra-efficient; for example, the system is written in a highly modular style, and its subsystems spend significant portion of their time on I/O communication. We expect that careful re-coding of our system would result in considerable speed-up, which would enable it to respond in real-time to planning tasks of moderately large size (the system runs reasonably efficiently even in its current form).

4.2. Performance and Experimentation

In implementing our system we had two goals in mind. The most important one was the actual programming of our algorithm, and the demonstration of its pragmatic applicability on real (or simulated) data. The second goal was to examine its performance as part of an integrated robotic system. We tested the implementations according to these two objectives.

One type of data that is of interest concerns statistical characteristics of the performance of our algorithm, such as its average run time and storage requirements. In other words, we are interested in "calibrating the constants" in our asymptotic bound, so that we can demonstrate that, for a typical (or even "difficult") instance of the problem, the number of critical orientations and the size of the graph EG are of reasonable size (perhaps even subquadratic on the average). Our initial experiments support this fact. (See a detailed analysis in [Ke].)

As it turns out, running the whole system from start (arranging objects on the work-table) to actual motion of the robot, takes approximately 3-4 minutes, out of which the vision program that we use consumes about 2 minutes, and I/O time dominates the rest.

Even though our implementation was not meant to be as robust as possible, we still had to address several issues of numerical precision. We list them below.

The first issue is the precision of the program variables. For example, in our algorithm we sort the critical orientations and build EG incrementally updating it at each critical orientation in turn. While testing we found that in some cases, orientations that were very close in value were processed in an opposite, physically inconsistent order.

Another numerical precision problem arises in processing pictures acquired by the camera. Some of them are standard vision problems like scaling and distortion of information. The others, more crucial to our implementation, involve the quality of the output of the shapes program, which tends to produce a polygonal representation involving more corners than the objects really have. This, in turn, results (i) in having more obstacle corners and thus worse time performance by the program, and (ii) having many similar (almost parallel) slopes of the sides of the objects leading to numerical instability and other problems.

Finally, since our algorithm dictates motion of the robot in which it often touches the obstacles, we were concerned that the robot might push aside some obstacles as it moves. In [Ke] we give the description of the techniques we used to overcome these difficulties.

Examples

As an example, consider the layout shown in Figure 4.2. B has $k=3$ corners, and the obstacles have $n=13$ corners. Figure 4.2 gives a graphical output of our system, showing the motion of B from the initial to the final placement, as planned by our algorithm. Figure 4.3 shows an overlay of all the graphs VG_θ drawn around each critical orientation for some range of θ. Some typical cross-sections VG_θ are shown individually in Figure 4.4. In Figure 4.4 the obstacles are shaded, and the expanded obstacles are drawn around them. For example, Figure 4.4(a) shows a critical orientation θ in which (within the dotted placement of B) a new component of FP_θ newly appears (shown as a tiny triangle inside the dotted placement). In Figure 4.4(c) this little component has grown, and is about to merge together with the outside component of FP_θ (the dotted placement of B shows the criticality, through which B can barely translate from the "inside" free region to the outside region. An intermediate critical orientation is shown in Figure 4.4(b).

Our program (without the vision portion) runs on this example in less than a minute. In this example we have 425 critical orientations, many of them redundant, and EG has only 120 nodes and 220 edges. These numbers compare very favorably with our upper bounds $O(kn\lambda_6(kn))$ for these quantities.

In the second example (Fig. 4.5) B has $k = 5$ corners and the obstacles have $n = 35$ corners. Run-time of this example is two minutes, the number of critical orientations is 555, and EG has 990 nodes and 2000 edges.

A third example is shown in Figures 4.6 through 4.8, including a motion of B between a pair of initial and final placements (Figure 4.6), an overlay graphical representation of the cross-sections of EG (Figure 4.7), and one individual cross section VG_θ (Figure 4.8), where a criticality involving contact between a convex corner of K_θ and another edge of K_θ is shown in the middle.

References

[ASS] P. Agarwal, M. Sharir and P. Shor, Sharp upper and lower bounds for the length of general Davenport Schinzel sequences, to appear in *J. Combin. Theory, Ser. A.*

[At] M. Atallah, Some dynamic computational geometry problems, *Comp. Math. Appl.* 11 (1975), pp. 1171-1181.

[BZ] B.K. Bhattacharya and J. Zorbas, Solving the two-dimensional findpath problem using a line-triangle representation of the robot, *J. Algorithms*, 9 (1988), pp. 449-469.

[BLP] R.A. Brooks and T. Lozano-Perez, A subdivision algorithm in configuration space for findpath with rotation, *Proc. IJCAI-83*, Karlsruhe, 1983.

[HS] S. Hart and M. Sharir, Nonlinearity of Davenport-Schinzel sequences and of generalized path compression schemes, *Combinatorica* 6 (1986), pp. 151-177.

[HSS] J. Hopcroft, J. Schwartz and M. Sharir (Eds.), *Planning, Geometry, and Complexity of Robot Motion*, Ablex Pub. Co., Norwood, NJ 1987.

[Ke] K. Kedem, Problems in planning collision-free motion for a rigid robot system in the plane: theory and application, Ph.D. thesis, Comp. Sci. Dept, Tel-Aviv University, July 1988.

[KLPS] K. Kedem, R. Livne, J. Pach and M. Sharir, On the union of Jordan regions and collision-free translational motion amidst polygonal obstacles, *Discrete Comput. Geom.* 1 (1986), pp. 59-71.

[KS] K. Kedem and M. Sharir, An efficient algorithm for planning collision-free translational motion of a convex polygonal object in 2-dimensional space amidst polygonal obstacles, *Proc. ACM Symp. on Computational Geometry* 1985, pp. 75-80.

[KSSS] A. Kalvin, E. Schonberg, J.T. Schwartz and M. Sharir, Two dimensional model based boundary matching using footprints, *Int. J. Robotics Research* 5 (4) (1986), pp. 38-55.

[LS1] D. Leven and M. Sharir, An efficient and simple motion planning algorithm for a ladder moving in two-dimensional space amidst polygonal barriers, *J. Algorithms* 8 (1987), pp. 192-215.

[LS2] D. Leven and M. Sharir, Planning a purely translational motion for a convex object in two-dimensional space using generalized Voronoi diagrams, *Discrete Comput. Geom.* 2 (1987), pp. 9-31.

[LS3] D. Leven and M. Sharir, On the number of critical free contacts of a convex polygonal object moving in 2-D polygonal space, *Discrete Comput. Geom.* 2 (1987), pp. 255-270.

[LP] T. Lozano-Perez, A simple motion planning algorithm for general robot manipulators, *IEEE J. Robotics and Automation*, 3 (3) (1987), pp. 224-238.

[Lu] V. J. Lumelsky, Dynamic path planning for planar articulated robot arm moving amidst moving obstacles, *Automatica*, 23 (5) (1987), pp. 551-570.

[LuS] V. J. Lumelsky and A. A. Stepanov, Path planning strategies for a point mobile automaton moving amidst unknown obstacles of arbitrary shape, *Algorithmica*, 2 (4) (1987), pp. 403-430.

[OY] O'Dunlaing, C. and Yap, C., A 'retraction' method for planning the motion of a disc, *J. Algorithms* 6 (1985), pp. 104-111.

[OSY1] O'Dunlaing, C., Sharir, M. and Yap, C., Generalized Voronoi diagrams for a ladder: I. Topological analysis, *Comm. Pure Appl. Math.* 39 (1986), pp. 423-483.

[OSY2] O'Dunlaing, C., Sharir, M. and Yap, C., Generalized Voronoi diagrams for a ladder: II.

Efficient construction of the diagram, *Algorithmica* 2 (1987), pp. 27-59.

[OR] J. O'Rourke, A lower bound for moving a ladder, Tech. Rept. JHU/EECS-85/20, The Johns Hopkins University, 1985.

[SS] J.T. Schwartz and M. Sharir, On the Piano Movers' problem: I. The case of a two-dimensional rigid polygonal body moving amidst polygonal barriers, *Comm. Pure Appl. Math.* 36 (1983), pp. 345-398.

[SS2] J.T. Schwartz and M. Sharir, Motion planning and related geometric algorithms in robotics, *Proc. International Congress of Mathematicians,* Berkeley, August 1986, Vol. 2, pp. 1594-1611.

[SiS] S. Sifrony and M. Sharir, An efficient motion planning aLgorithm for a rod moving in two-dimensional polygonal space, *Algorithmica* 2 (1987), pp. 367-402.

[Ya] C.K. Yap, Algorithmic motion planning, in *Advances in Robotics,* Vol. 1 (J.T. Schwartz and C.K. Yap, Eds.), Lawrence Erlbaum, Hillsdale, NJ 1987.

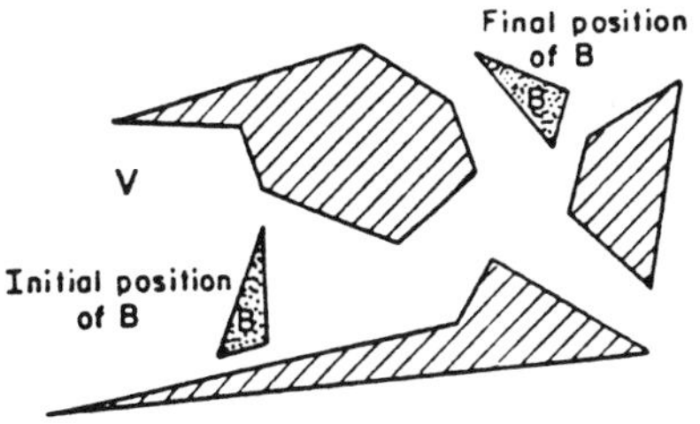

Figure 1.1.

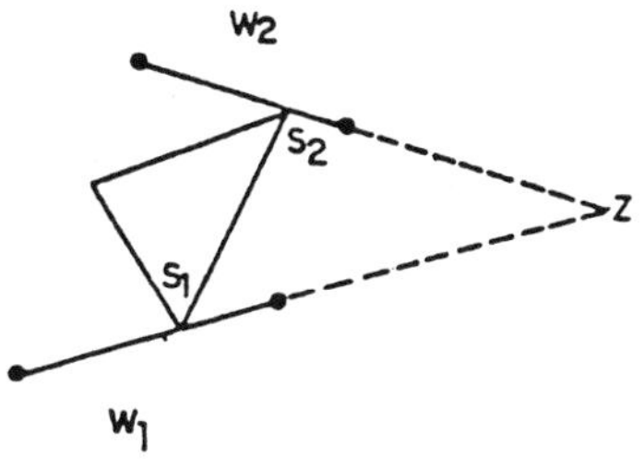

Figure 3.1.

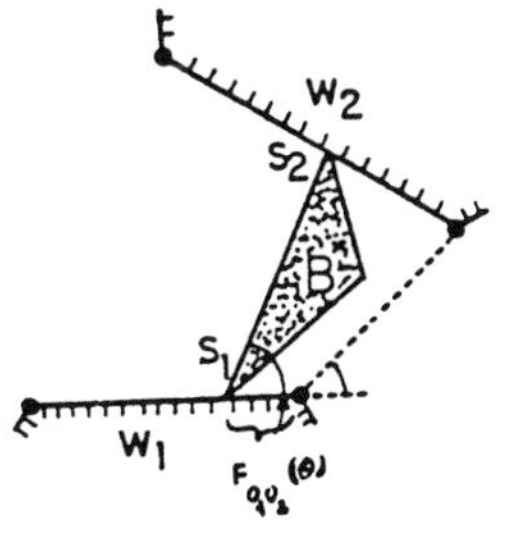

Figure 3.2.

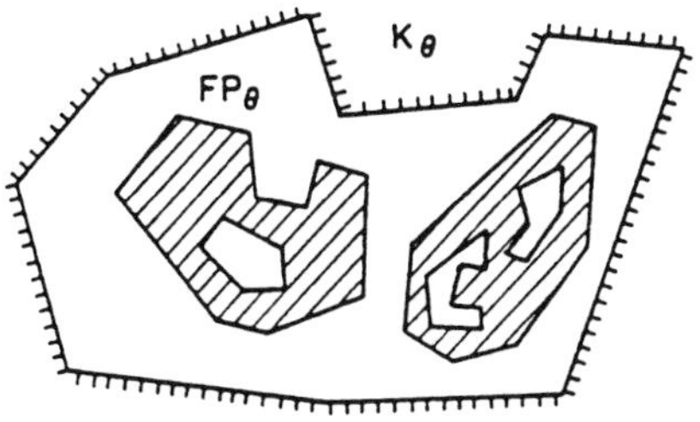

Figure 2.1.

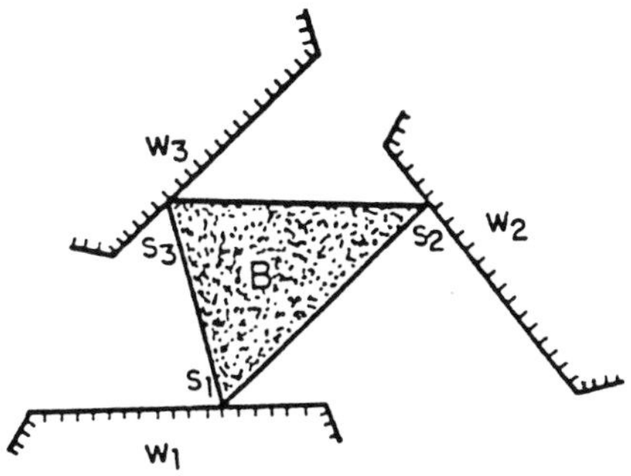

Figure 3.3.

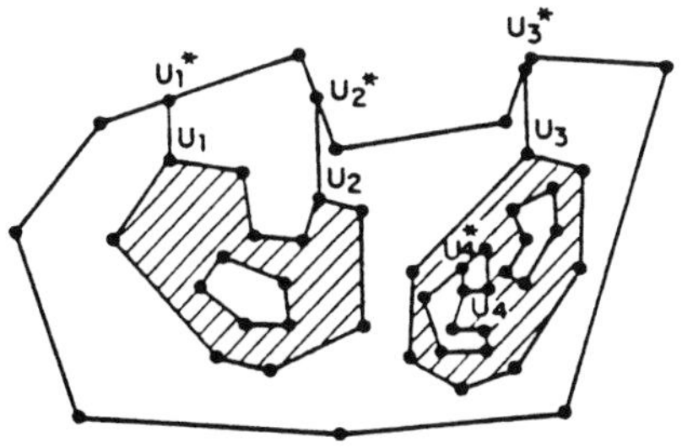

Figure 2.2.

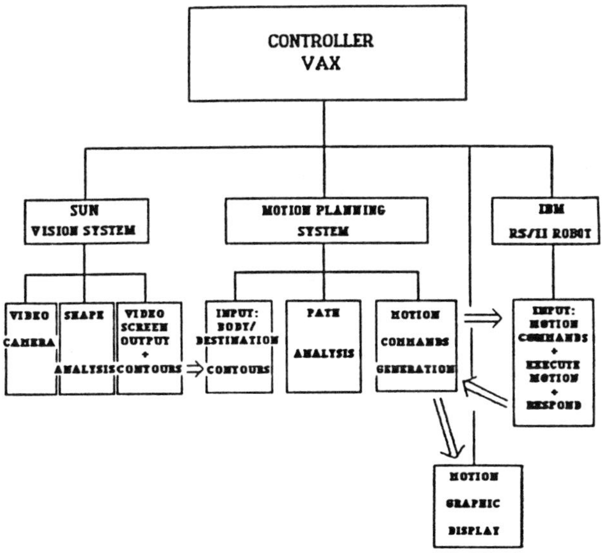

Figure 4.1.

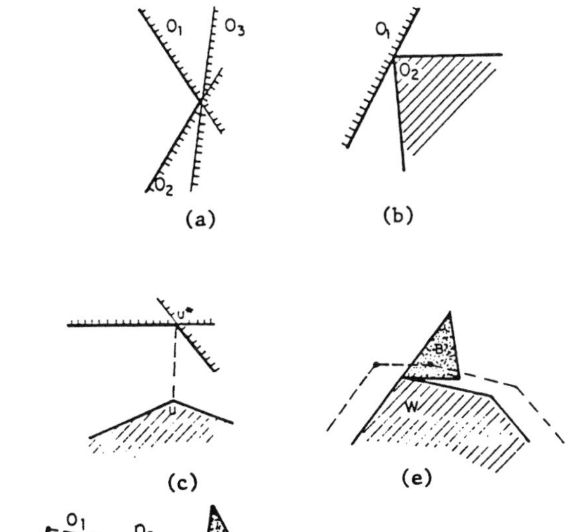

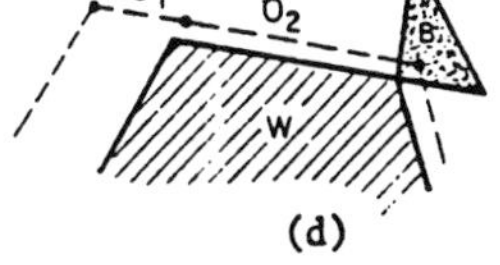

Figure 2.3.

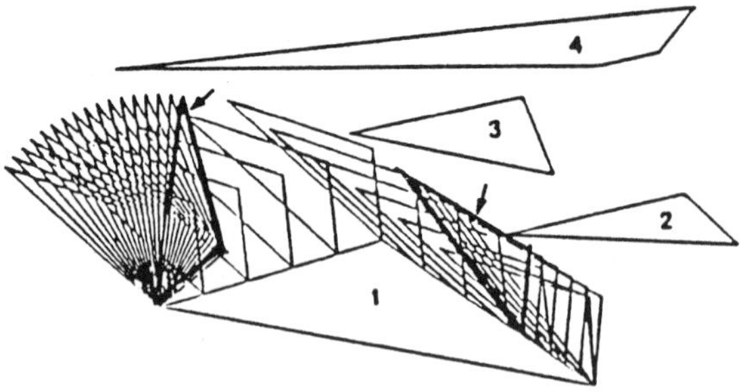

Figure 4.2.

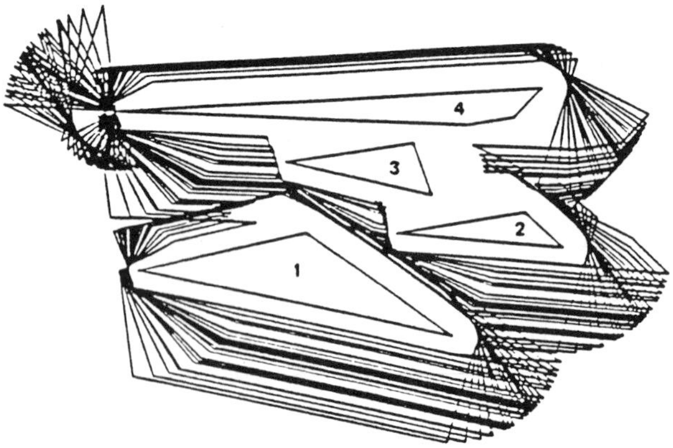

Figure 4.3.

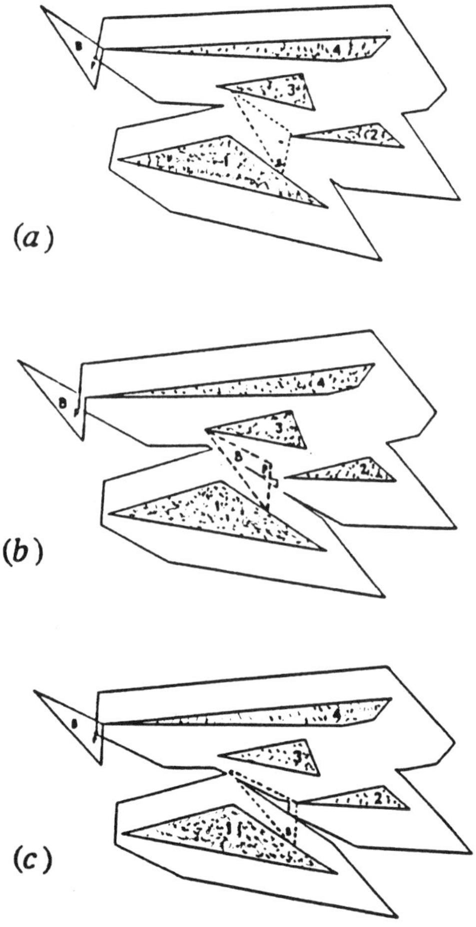

Figure 4.4.

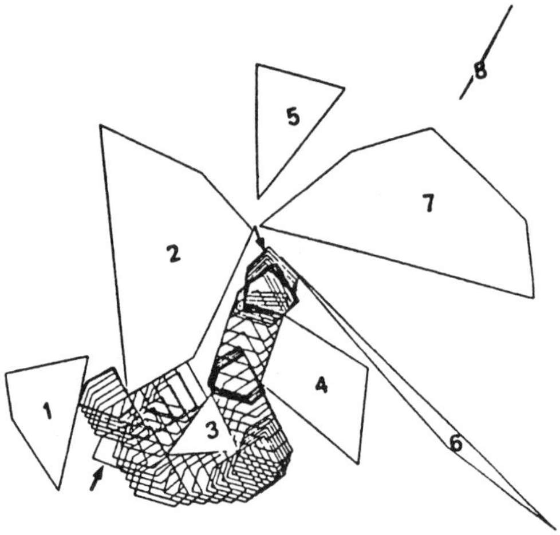

Figure 4.5.

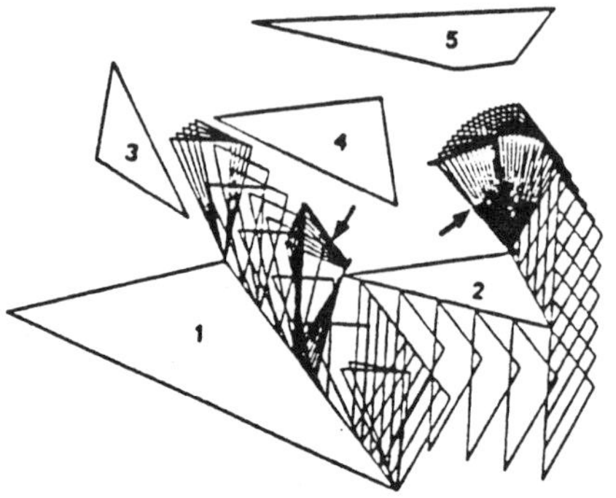

Figure 4.6.

Figure 4.7.

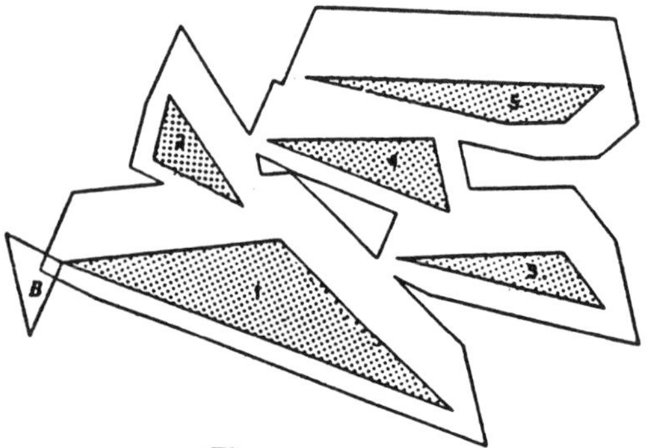

Figure 4.8.

Path-Planning Strategies for a Point Mobile Automaton Moving Amidst Unknown Obstacles of Arbitrary Shape[1]

Vladimir J. Lumelsky[2] and Alexander A. Stepanov[3]

Abstract. The problem of path planning for an automaton moving in a two-dimensional scene filled with unknown obstacles is considered. The automaton is presented as a point; obstacles can be of an arbitrary shape, with continuous boundaries and of finite size; no restriction on the size of the scene is imposed. The information available to the automaton is limited to its own current coordinates and those of the target position. Also, when the automaton hits an obstacle, this fact is detected by the automaton's "tactile sensor." This information is shown to be sufficient for reaching the target or concluding in finite time that the target cannot be reached. A worst-case lower bound on the length of paths generated by any algorithm operating within the framework of the accepted model is developed; the bound is expressed in terms of the perimeters of the obstacles met by the automaton in the scene. Algorithms that guarantee reaching the target (if the target is reachable), and tests for target reachability are presented. The efficiency of the algorithms is studied, and worst-case upper bounds on the length of generated paths are produced.

Key Words. Robotics, Robot motion planning, Collision avoidance algorithms, Planning with uncertainty, Provable algorithms.

1. Introduction. To plan a path for a mobile automaton (a mobile robot) means finding a continuous trajectory leading from the initial position of the automaton to its target position. In this paper, we consider a case where the automaton is a point, and the environment (the scene) in which the automaton travels is defined in a two-dimensional plane. The scene can be filled with unknown obstacles of an arbitrary shape and size. The information about the obstacles comes from a simple sensor whose capability is limited to detecting an obstacle only when the automaton hits it. The main question being asked is whether, under such a model, a provable path-planning algorithm can be designed.

The current research on robot path planning can be classified into two large categories depending on which of the two following basic models is being used. In the first model, called *path planning with complete information* (another popular term is the *piano movers problem*), perfect information about the obstacles is assumed. In the second model, called *path planning with incomplete information*, an element of uncertainty about the environment is present. Another important distinction can be made between the provable (other terms—exact, algorithmic) and heuristic approaches. In these terms, this paper addresses the problem of designing provable path-planning algorithms in the framework of the model with incomplete information.

---

[1] Supported in part by the National Science Foundation Grant DMC-8519542.
[2] Yale University, New Haven, CT 06520, USA.
[3] Polytechnic University, Brooklyn, NY 11201, USA.

Received November, 1985; revised December 1, 1986. Communicated by Chee-keng Yap.

A brief review of the state of affairs in both categories of works is in order here; the review is by no means inclusive. The first model—the piano movers problem—is formulated as follows.[4] Given a solid object, in two- or three-dimensional space (2D or 3D), of known size and shape, its initial and target position and orientation, and a set of obstacles whose shapes, positions, and orientations in space are fully described, the task is to find a continuous path for the object from the initial position to the target position while avoiding collisions with obstacles along the way. It is assumed that the surfaces of the moving object and of the obstacles are algebraic; in most works, a stricter requirement of planar surfaces is imposed.

Because full information is assumed, the whole operation of path planning is a one-time, off-line operation. The main difficulty is not in proving that an algorithm that would guarantee a solution exists, but in obtaining a computationally efficient scheme. By a solution, we mean either reaching the target or concluding in finite time that the target cannot be reached. Conceptually, cases of arbitrary complexity can be considered, given the fact that a solution is always feasible. Another apparent advantage of dealing with complete information is that any optimization criteria (finding the shortest path, or the minimum-time path, or the safest path, etc.) can be easily introduced.

The computational complexity of the problem was first realized when Reif [6] showed that the general piano movers problem is PSPACE-hard; he also sketched a possible solution for moving a solid object in polynomial time, by direct computation of the "forbidden" volumes in spaces of higher dimensions.[5] Schwartz and Sharir [1] presented a polynomial-time algorithm for a two-dimensional piano movers problem with convex polygon obstacles. In a number of works (e.g., Lozano-Perez [2]), the solid object is viewed as shrinking to a point while the obstacles are viewed as expanding accordingly, to compensate for the shrinking object. The resulting *configuration space* has higher dimensionality compared with the original *workspace*—one extra dimension per each degree of rotational freedom. In general, the obstacles in the configuration space have nonplanar walls—even if the original obstacles are polyhedral. In order to keep the problem manageable, various constraints are typically imposed.

Moravec [3] considers a path-planning algorithm in two dimensions with the object presented as a circle. Brooks [4], in his treatment of a two-dimensional path-planning problem with a convex polygon object and convex polygon obstacles, uses a generalized cylinders presentation [5] to reduce the problem to a graph search. A generalized cylinder is formed by a volume swept by a cross-section (in general, of varying shape and size) moving along the cylinder axis (in general, a spine curve).

A version of the piano movers problem where the moving object is allowed to consist of a number of free-hinged links is more difficult. On a heuristic level, this version was started by Pieper [7] and then investigated by Paul [8] because

[4] A good survey of the work on provable algorithms for the problem can be found in [17].

[5] Higher dimensions d appear when one takes into account the orientation of the moving object along its way; $d = 3$ for the two-dimensional case, and $d = 6$ for the three-dimensional case.

of its obvious relation to path generation and coordinate transformation problems of multiple-degrees-of-freedom robot arms. Recently, new approaches for this version were considered in [9] and [10]. The most general algorithm (although very expensive computationally) for moving a free-hinged body was given by Schwartz and Sharir [9]; the technique is based on the general method of cell decomposition; the moving object and the obstacles are assumed to be limited by algebraic surfaces.

From the application standpoint, unless there is a reason to believe that the obstacles in the scene are polyhedral or, if the algorithm in question allows it, at least algebraic, the algorithms above cannot be applied directly. Then, an appropriate approximation has to be performed, which can introduce problems of its own. For example, given a prescribed accuracy of the approximation, the operation of approximating nonlinear surfaces with linear constraints itself requires exponential time [21]. Also, the space of possible approximations with constraints of a given order is not continuous in the approximation accuracy: in other words, a slight change in the specified accuracy of the approximation can cause a dramatic change in the approximated surfaces and eventually in the generated paths. On the other hand, the approximation itself depends on considerations that are secondary to the path-planning problem: these are, for example, the accuracy of the presentation of actual obstacles by polygons, or—a conflicting criterion—computational costs of processing the resulting connectivity graph.

The attractiveness of the model of *path planning with incomplete information* for robotics lies in the possibility of naturally introducing a powerful notion of feedback control, and thus transforming the operation of path planning into a continuous on-line process. In turn, using sensory feedback in the control scheme allows one to ease the requirements on the shape and location of the obstacles in the scene, and even lift the requirement that the obstacles be stationary. An important component of the model is the fact that the information about the environment has a local character—this is because at any given moment the robot sensors can provide information only about its immediate surroundings. This component changes the problem rather significantly. To start with, it is not clear whether an algorithm exists which would guarantee reaching a global goal (here, the robot target position) based on local means (the sensor information).

The question of reaching a global goal with local means presents a fundamental problem, various formulations of which have been studied in a number of areas: game theory (differential games and macroeconomics, e.g., [18]; collective behavior, e.g., [19]), computer science (maze search [14]), and studies in geometry [16]. The difficult question of relationship between uncertainty and the algorithm complexity has been studied in [20].

In the context of robot path planning, works related to the model with incomplete information have come primarily from studies on autonomous vehicle navigation; so far, they have been limited to various heuristics. In [11]-[13] a two-dimensional navigation problem is considered. Typically, obstacles are approximated by polygons; produced paths lie along the edges of the *connectivity graph* formed by the straight line segments connecting the obstacle vertices, the start point, and the target point, with a constraint on nonintersection of the graph edges with the obstacles. Path planning is limited to the automaton's immediate surroundings for which information on the scene is available—for example, from a vision module. Within this limited area, the problem is actually treated as one with complete information.

One problem of working with incomplete information is that, because of the dynamic character of the incoming (sensor) information, the path cannot be preplanned, and so its global optimality is ruled out. Instead, one can judge the algorithm performance based on how it compares with other existing or theoretically feasible algorithms, or how optimal they are locally, or how "reasonable" they look from the human traveler standpoint.

Another inherent difficulty in designing algorithms for the model with incomplete information is that the problem dimensionality cannot be made arbitrarily high. Because the information about the obstacles is unknown, a natural way to limit the number of options available to the automaton at each step is to impose a constraint on the problem dimensionality. Otherwise, when an object meets an obstacle in the three-dimensional space, it has an infinite number of possibilities for passing around the obstacle. Hence the significance of the requirement introduced in this paper that the environment be limited to the two-dimensional plane (actually, our algorithms can be used on any surface homeomorphic to a plane). With this constraint, every time the automaton encounters an obstacle it can turn only left or right along the obstacle boundary. Essentially, the algorithms described below are exploiting the Jordan Curve Theorem, which states that any closed curve homeomorphic to a circle drawn around and in the vicinity of a given point on an orientable surface divides the surface into two separate domains, for which the curve is their common boundary [22]. Similar ideas can be used for designing algorithms for robot arm manipulators [23], [24].

In terms of the available information, the model considered in this paper can be viewed as being diametrically opposite to the piano movers model: instead of the full information about the obstacles assumed in the latter model, no information about the obstacles is given to the automaton in our model. The only available input information includes the automaton's own coordinates and those of the target. The automaton's capability for learning about an obstacle is limited to the "ultra-local" information provided by the automaton's "tactile sensor." In other words, the automaton learns about the presence of an obstacle only when it hits it.

Under the proposed algorithms, the automaton is continuously analyzing the incoming information about its current surroundings and is continuously planning its path. This is somewhat similar to the approach utilized in [16] for treating geometric phenomena based on local information. No approximation of the obstacles is done, and, consequently, no connectivity graphs or other intermediate computational structures are used. Since no reduction to a discrete space takes place, all points of the scene are available for the purpose of path planning. Because the model is continuous, the criteria typically used for evaluating algorithm performance—such as computational complexity as a function of the number of vertices of the obstacles, or the time or memory required—are not applicable. Instead, a new performance criterion based on the length of the generated paths as a function of the obstacle perimeters is introduced.

One point should be mentioned. The main goal of this work is to investigate how productive the model with incomplete information is for designing provable robot path-planning algorithms. Specifically, given the local nature of the incoming sensor information about the environment, is a global solution feasible? Furthermore, what are the minimum resources—the minimum knowledge and the minimum memory—that the automaton's model must assume in order to guarantee a solution? As a result of such narrow problem formulation, a number of important issues are left out. No learning of any kind is considered. Although the results we obtain are applicable to various types of sensor feedback, the presentation is limited to tactile sensing. The automaton's size and shape, which have to be taken into consideration in applications, are simply ignored in this work; a point automaton is considered.

In Section 2 the model of the environment and of the automaton is formulated. Assuming this model, a worst-case lower bound is produced in Section 3 for the general path-planning problem. In Sections 4 and 5 two *basic algorithms* for path planning are described and their convergence properties are analyzed. Each algorithm has quite different characteristics, and, depending on the scene, one can produce a shorter path than the other. For these algorithms, the worst-case upper bounds on the length of the generated paths are established, and tests for target reachability are formulated. This is followed, in Section 6, by an improved version of the path-planning algorithm which, while guaranteeing termination, combines good features of both basic algorithms without sacrificing much of their clarity. Finally, in Section 7, all three algorithms are compared, and an additional insight into the algorithms' mechanisms is provided by showing their relevance to the maze search problem.

2. Model. The model includes two parts—one related to the geometry of the scene, and the other related to the characteristics and capabilities of the point mobile automaton (MA).

Environment. The scene is a plane with a set of obstacles and the points Start (S) and Target (T) in it. Each obstacle is a simple closed curve of finite length such that a straight line will cross it only in finitely many points; a case when the straight line coincides with a finite segment of the obstacle boundary is not a "crossing." (An equivalent term used in the text for a simple closed curve is the *obstacle boundary*.) Obstacles do not touch each other; that is, a point on an obstacle belongs to one, and only one, obstacle. A scene can contain only a locally finite number of obstacles; this means that any disc of finite radius intersects a finite set of obstacles. Note that the model does not require that the set of obstacles is finite.

Automaton. MA is a point; this means that an opening of any size between two distinct obstacles is considered to be passable. The only information MA is provided with by its sensors is (1) its current coordinates, and (2) the fact of contacting an obstacle. MA is also given the coordinates of the Target. Thus, it can always calculate its direction toward and its distance from the Target. The memory available for storing data or intermediate results is limited to a few computer words. The motion capabilities of MA include three possible actions: move toward the Target on a straight line; move along the obstacle boundary; stop.

DEFINITION 1. A *local direction* is a once and for all decided direction for passing around an obstacle. For the two-dimensional problem, it can be either left or right.

Because of the uncertainty involved, every time MA meets an obstacle, there is no information or criteria which could help it decide whether it should go around the obstacle from the left or from the right. For the sake of clarity, and without losing generality, assume that the local direction of MA is always *left* (as in Figure 4). Unless stated otherwise, MA will be assumed to follow the local direction while walking around obstacles.

DEFINITION 2. MA is said to *define a hit point H* on the obstacle, when, while moving along a straight line toward the Target, MA contacts the obstacle at the point H. It *defines a leave point L* on the obstacle, when it leaves the obstacle at the point L in order to continue its straight line walk toward the Target. (See, for example, Figure 4.)

If MA moves along a straight line toward the Target and the line touches some obstacle tangentially then there is no need to invoke the procedure for walking around the obstacle—MA just continues its straight line walk toward the Target. In other words, no H or L points will be defined in this case. Because of that, no point of an obstacle can be defined as both an H and an L point. In order to define an H or an L point, the corresponding straight line has to produce a "real" crossing of the obstacle; that is, in the vicinity of the crossing, a finite segment of the line should lie inside the obstacle, and a finite segment of it should lie outside the obstacle.

Throughout, the following notation is used:

D is the (Euclidean) distance from the Start to the Target;

$d(A, B)$ is the distance between points A and B of the scene; thus, $d(\text{Start}, \text{Target}) = D$;

$d(A)$ is used as a shorthand notation for $d(A, \text{Target})$;

$d(A_i)$ signifies the fact that the point A is located on the boundary of the ith obstacle met by MA on its way to the Target;

P is the total length of the path generated by MA on its way from the Start to the Target;

p_i is the perimeter of the ith obstacle met by MA.

The performance of the path-planning algorithms will be evaluated based on the quantity $\sum_i p_i$, the sum of perimeters of obstacles met by MA on its way to the Target, or of obstacles contained in a specific area of the scene. This quantity will allow us to compare various path-planning procedures in terms of the length of the paths they produce.

3. The Lower Bound for the Path-Planning Problem. This lower bound, formulated in Theorem 1 below, determines what performance can be expected in the worst case from any path-planning algorithm operating within the framework of our model. The bound is formulated in terms of the length of the path generated by the automaton on its way from the point Start (S) to the point Target (T). The bound is a powerful means for measuring performance of various path-planning procedures.

THEOREM 1. *For any path-planning algorithm satisfying the assumptions of our model, any (however large) $P>0$, any (however small) $D>0$, and any (however small) $\delta>0$, there exists a scene for which the algorithm will generate a path of length P, and*

$$P \geq D + \sum_i p_i - \delta, \tag{1}$$

where D is the distance between the points Start and Target, and p_i are perimeters of the obstacles intersecting the disc of radius D centered at the Target.

PROOF. We want to prove that for any given algorithm a scene can be designed for which the length of the path generated by this unknown algorithm—say, Algorithm X—will satisfy (1). Algorithm X can be of any type: it can be deterministic or random; its intermediate steps may or may not depend on intermediate results; etc. The only information known about Algorithm X is that it operates within the framework of our model of the automaton MA and the environment (Section 2). The proof consists of designing a special scene (a set of obstacles) and then proving that the scene will force Algorithm X to generate a path not shorter than P in (1).

The following scheme, consisting of two stages, is suggested for designing the required scene (called the *resultant scene*). At the first stage, a *virtual obstacle* is introduced; this is an obstacle parts or all of which, but not more, will eventually produce, once the second stage is completed, the *actual obstacle(s)* of the resultant scene.

Consider a virtual obstacle shown in Figure 1(a). It presents a corridor, of finite width $2W>\delta$, and of finite length L. One end of the corridor is closed. The corridor is positioned such that the point S is located at the middle point of the closed end; the corridor opens in the direction opposite to the line (S, T). The thickness of the corridor walls is negligible compared with δ. Still in the first stage, MA is let walk from S to T along the path prescribed by Algorithm X. On its way, MA may or may not touch the virtual obstacle.

When the path is complete, the second stage starts. A segment of the virtual obstacle is said to be *actualized* if all points of the inside wall of the segment have been touched by MA. If MA touched the inside wall of the virtual obstacle at some length l, then the actualized segment is exactly of length l. If MA was continuously touching the virtual obstacle at a point and then bouncing back, the corresponding actualized area is considered to be a segment of length δ

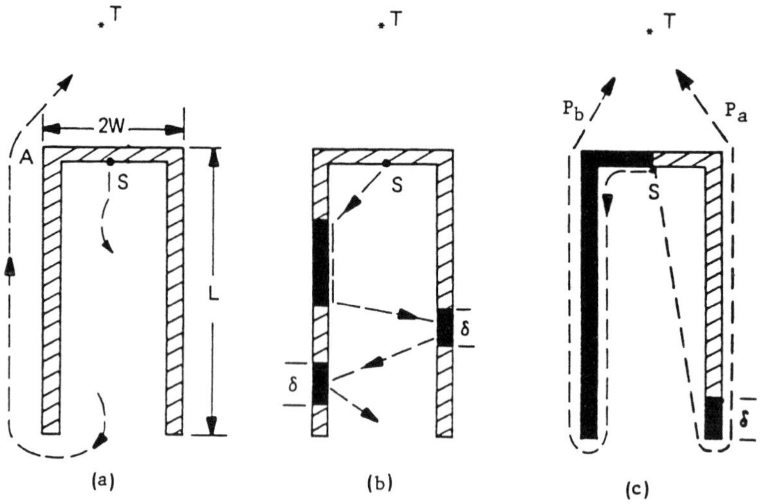

Fig. 1. Illustration for Theorem 1. Actualized segments of the maximum obstacle are shown as solid (S, Start point, T, Target point).

around the point of contact. If two segments of MA's path along the virtual obstacle are separated by an area of the virtual obstacle which MA did not touch, then MA is said to have actualized two separate segments of the virtual obstacle.

We produce the resultant scene by designating as actual obstacles only those areas of the virtual obstacle which have been actualized. Thus, if an actualized segment is of length l, then the perimeter of the corresponding actual obstacle is equal to $2l$; this takes into account the inside and the outside walls of the segment, and also the fact that the thickness of the wall is negligible.

This method for producing the resultant scene can be justified by the fact that, under the accepted model, the behavior of MA is affected only by those obstacles which it touches along its way. Indeed, the produced path could have appeared, under Algorithm X, in two different scenes: in the scene with the virtual obstacle, and in the resultant scene. One can argue, therefore, that the areas of the virtual obstacle, which MA did not touch along its way, might never have existed, and Algorithm X produced its path not in the scene with the virtual obstacle but in the resultant scene. This means that the performance of MA in the resultant scene can be judged against (1). This completes the design of the scene. Note that depending on MA's behavior under Algorithm X, zero, one, or more actual obstacles can be created in the scene of Figure 1.

Next, we have to prove that MA's path in the resultant scene satisfies (1). Since MA starts at a distance $D = d(S, T)$ from T, it obviously cannot avoid the term D in (1). We concentrate, then, on the second term in (1). One can see by now that the main idea behind the described process of designing the resultant scene is to force MA to generate, for each actual obstacle, a segment of the path at least as long as the total length of the boundary of that obstacle. Note that this characteristic of the path is independent of Algorithm X.

MA's path in the scene can be divided into two parts, $P1$ and $P2$; $P1$ corresponds to MA's traveling inside the corridor, and $P2$ corresponds to its traveling outside the corridor; the same notation is used to indicate the length of the corresponding part. Both parts can become intermixed since, after having left the corridor, MA can temporarily return into it. Since part $P2$ starts at the exit point of the corridor then

$$P2 \geq L + C, \tag{2}$$

where $C = \sqrt{D^2 + W^2}$ is the hypotenuse AT of the triangle ATS (Figure 1(a)). As for part $P1$ of the path inside the corridor, it can be, depending on Algorithm X, any curve. Observe that in order to defeat the bound (1), Algorithm X has to decrease the "path per obstacle" ratio as much as possible. What is important for the proof is that, from the "path per obstacle" standpoint, every segment of $P1$ that does not result in the creation of the equivalent segment of an actualized obstacle makes the path worse. All possible alternatives for $P1$ can be clustered to three groups. These groups are discussed separately below.

1. Part $P1$ of the path never touches the walls of the virtual obstacle (Figure 1(a)). As a result, no actual obstacles will be created, $\sum_i p_i = 0$; but, as one can see, the resulting path is $P > D$. Therefore, for this kind of Algorithm X the theorem holds. Moreover, at the final evaluation, where only actual obstacles count, this MA strategy will not be judged as being very efficient: it creates an additional path component at least equal to $(2 \cdot L + (C - D))$—in a scene with no obstacles!
2. MA touches one or both inside walls of the virtual obstacle more than once (Figure 1(b)). In other words, between the consecutive touchings MA is temporarily "out of touch" with the virtual obstacle. As a result, part $P1$ of the path will produce a number of disconnected actual obstacles; the smallest of these, of length δ, correspond to point touchings. Observe that, in terms of the "path per obstacle" assessment, this kind of strategy is not very wise either. First, for each actual obstacle, a segment of the path at least as long as the obstacle perimeter is created; besides, additional segments of $P1$, due to traveling between the actual obstacles, are produced. Each of these additional segments is at least not smaller than $2W$, if the two consecutive touchings correspond to the opposite walls of the virtual obstacle, or at least not smaller than the distance between two sequentially visited disconnected actual obstacles on the same wall. Thus, the length P of the path exceeds the right side in (1), and the theorem holds.
3. MA touches the inside walls of the virtual obstacle at most once. This case includes various possibilities, from a point touching, which creates a single actual obstacle of length δ, to the case when MA closely follows the inside wall of the virtual obstacle. As one can see in Figure 1(c), this case contains the most interesting paths. The shortest possible path would be created if MA went directly from S to the furthest point of the virtual obstacle and then directly to T (path P_a, Figure 1(c)). (Given the fact that MA knows nothing

about the obstacles, this kind of a path can be produced only by accident.) The total perimeter of the obstacle(s) here is 2δ, and the theorem clearly holds.

Finally, the most efficient path, from the "path per obstacle" standpoint, is produced if MA closely follows the inside wall of the virtual obstacle, and then goes directly to the point T (path P_b, Figure 1(c)). Here MA is doing its best in trying to compensate each segment of the path with an equivalent segment of the actual obstacle. In this case, the generated path P is equal to

$$P = \sum_i p_i + \sqrt{D^2 + W^2} - W \tag{3}$$

(in our case, there is only one term in $\sum_i p_i$). Since no constraints have been imposed on the choice of the lengths D and W, take them such that

$$\delta \geq D + W - \sqrt{D^2 + W^2} \tag{4}$$

which is always possible because the right-hand side in (4) is nonnegative for any D and W. Reverse the sign (and the inequality) in (4), and add $(D + \sum_i p_i)$ to both its sides; this produces

$$\sum_i p_i + \sqrt{D^2 + W^2} - W \geq D + \sum_i p_i - \delta. \tag{5}$$

Comparing (3) with (5), observe that (1) is satisfied. This exhausts all possible cases of path generation by Algorithm X. □

We conclude this section with two remarks. First, by appropriately selecting multiple virtual obstacles, Theorem 1 can be extended to an arbitrary number of obstacles. Second, for the lower bound to hold, the constraints on the information available to MA can be relaxed significantly. Namely, the only required constraint is that MA does not have complete information about the scene at any time.

In the following sections, three path-planning algorithms are introduced, their performances analyzed, and the upper bounds on the length of their generated paths derived.

4. First Basic Algorithm: Bug1

4.1. Procedure. The procedure Bug1 is to be executed at any point of a continuous path. Figure 2 demonstrates the behaviour of MA. The goal is to generate a path from the Start to the Target. When meeting an ith obstacle, MA defines a hit point H_i, $i = 1, 2, \ldots$. When leaving the ith obstacle, to continue its travel toward the Target, MA defines a leave point L_i; initially, $i = 1$; $L_0 = \text{Start}$. The procedure uses three registers, R_1, R_2, R_3, to store intermediate information; all three are reset to zero when a new hit point, H_i, is defined. Specifically, R_1 is used to store the coordinates of the current point, Q_m, of the minimum distance

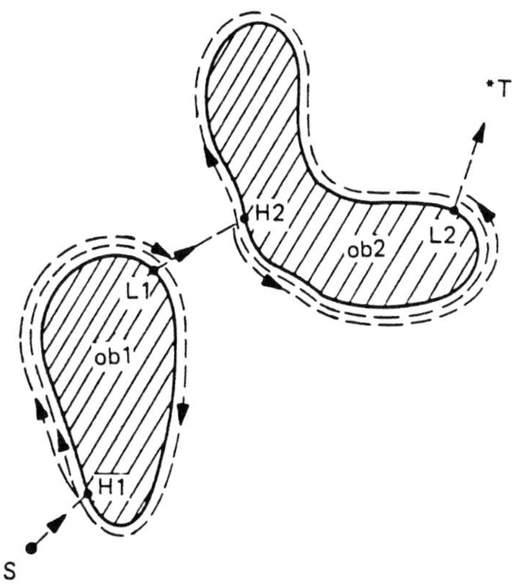

Fig. 2. Automaton's path (dotted lines), Algorithm Bug1 (*ob*1, *ob*2, obstacles; *H*1, *H*2, hit points; *L*1, *L*2, leave points).

between the obstacle boundary and the Target (this takes one comparison at each path point); R_2 integrates the length of the obstacle boundary starting at H_i; and R_3 integrates the length of the obstacle boundary starting at Q_m. (In case of many choices for Q_m, any one of them can be taken.) The test for target reachability mentioned in step 3 of the procedure is explained in Section 4.3. The procedure consists of the following steps:

1. From the point L_{i-1}, move toward the Target along a straight line until one of the following occurs:
   (a) The Target is reached. The procedure stops.
   (b) An obstacle is encountered and a hit point, H_i, is defined. Go to step 2.
2. Using the local direction, follow the obstacle boundary. If the Target is reached, stop. After having traversed the whole boundary and having returned to H_i, define a new leave point $L_i = Q_m$. Go to step 3.
3. Apply the test for target reachability. If the Target is not reachable, the procedure stops. Otherwise, using the content of the registers R_2 and R_3, determine the shorter way along the boundary to L_i, and use it to get to L_i; set $i = i+1$ and go tô step 1.

*4.2. Characteristics of Bug*1. In this section the characteristics and performance of the algorithm are analyzed.

LEMMA 1. *When, under Bug*1, MA *leaves a leave point of an obstacle in order to continue its way toward the Target, it never returns to this obstacle again.*

PROOF. Assume that on its way from the Start to the Target MA does meet some obstacles. We number those obstacles in the order in which MA meets them. Then, the following sequence of distances appears:

$$D, d(H_1), d(L_1), d(H_2), d(L_2), d(H_3), d(L_3), \ldots.$$

If the point Start happened to be on the boundary of an obstacle and the line (Start, Target) crosses that obstacle then $D = d(H_1)$.

It has been mentioned in Section 2 that if MA's path touches an obstacle tangentially then there is no need to invoke the procedure for walking around an obstacle—MA just continues its straight line walk toward the Target. In all other cases of meeting the ith obstacle, unless the Target lies on the boundary of the obstacle, a relation holds: $d(H_i) > d(L_i)$. This is because, on the one hand, according to the model, any straight line (except a line that touches the obstacle tangentially) crosses the obstacle at least in two distinct points (finite "thickness" of obstacles), and, on the other hand, according to Algorithm Bug1, the point L_i is the closest point from the obstacle to the Target. Starting from L_i, MA walks straight to the Target until it meets the $(i+1)$th obstacle. Since, according to the model, obstacles do not touch one another, then $d(L_i) > d(H_{i+1})$. Therefore, our sequence of distances satisfies the relation,

$$d(H_1) > d(L_1) > d(H_2) > d(L_2) > d(H_3) > d(L_3) > \cdots, \tag{6}$$

where $d(H_1)$ is or is not equal to D. Since $d(L_i)$ is the shortest distance from the ith obstacle to the Target, and since (6) guarantees that Algorithm Bug1 monotonically decreases the distances $d(H_i)$ and $d(L_i)$ to the Target, Lemma 1 follows. □

The lemma thus guarantees that the strategy will never create cycles.

COROLLARY. *Under Bug1, independent of the geometry of an obstacle,* MA *defines not more than one hit and not more than one leave point on it.*

To produce an upper bound on the length of the paths generated by Bug1, an assurance is needed that on its way to the Target MA always encounters only a finite number of obstacles. This is not obvious since, while following Algorithm Bug1, MA can "look" at the Target not only from different distances but also from different directions; that is, besides moving toward the Target, it may also rotate around the Target (see Figure 3). Hence the following lemma.

LEMMA 2. *Under Bug1, on its way to the Target* MA *can meet only a finite number of obstacles.*

PROOF. Although, while walking around an obstacle, MA can, at some moments, be at distances much larger than D from the Target (see Figure 3), the straight line segments of its path toward the Target are always within the same circle of radius D centered at the Target; this is guaranteed by inequality (6). Since, according to our model, any disc of finite radius can intersect with only a finite number of obstacles, the lemma follows. □

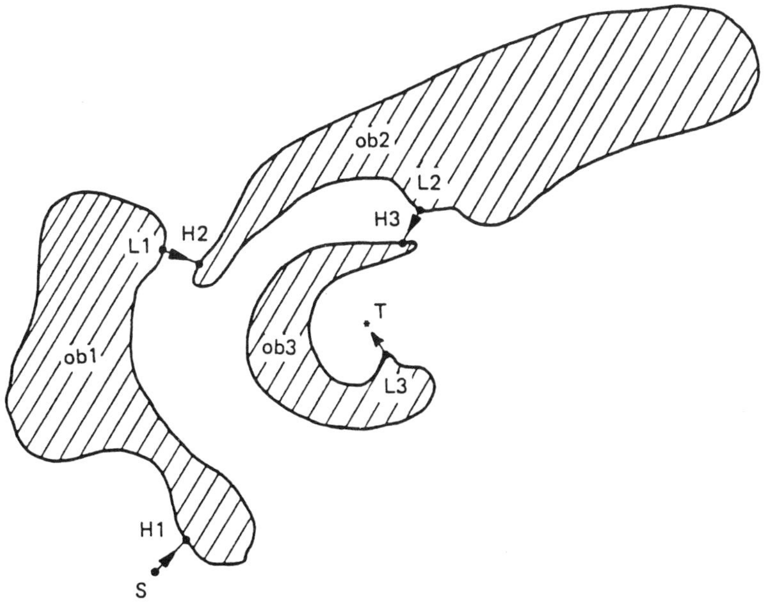

Fig. 3. Algorithm Bug1. Arrow lines indicate straight line segments of the automaton's path. Segments around obstacles are not shown; these are similar to the ones shown in Figure 2.

COROLLARY. *The only obstacles that can be met by* MA (*under Algorithm Bug*1) *are those which intersect the disc of radius D centered at the Target.*

Together, Lemma 1, Lemma 2, and the corollary guarantee convergence of Algorithm Bug1.

At this point we can establish the performance of the procedure Bug1, in terms of the length of the paths it generates. The following theorem gives an upper bound on the path lengths produced by Bug1.

THEOREM 2. *The length of the path produced by the procedure Bug*1 *will never exceed the limit*

$$P = D + 1.5 \cdot \sum_i p_i, \tag{7}$$

where $\sum_i p_i$ *refer to the perimeters of the obstacles intersecting the disc of radius D centered at the Target.*

PROOF. Any path can be looked at as consisting of two parts: straight line segments of the path between the obstacles, and the path segments related to walking around the obstacles. Due to inequality (6), the sum of the straight line segments will never exceed D. As to the path segments around the obstacles, Algorithm Bug1 requires that, in order to define a leave point on the ith obstacle, MA has to make a full circle around it; this produces a path segment equal to one perimeter, p_i, of the ith obstacle. By the time MA is prepared to walk from the hit to the leave point, in order to depart for the Target, it knows the direction (go left or go right) of the shorter path to the leave point. Thus, its path segment between the hit and the leave points along the boundary of the ith obstacle will not exceed $0.5 \cdot p_i$. Summing up the partial estimates for the straight line segments of the path and for the segments around the obstacles met by MA on its way to the Target, we obtain (7). □

Analysis of the procedure Bug1 shows that the requirement that MA has to know its own coordinates at any instance can be eased. It suffices if MA is capable of positioning itself at the circle of a given radius centered at the Target. In other words, what is important for Bug1 is not the actual position of MA but its direction toward and its distance from the Target. Assume that instead of the coordinates of the current point Q_m of the minimum distance between the obstacle and the Target, we store in the register R_1 the minimum distance itself. Then, in step 3 of Bug1, MA can reach the point Q_m by comparing its current distance from the Target with the content of the register R_1. If more than one point of the current obstacle lie at the minimum distance from the Target, any one of them can be used as the leave point, without affecting the convergence of the procedure.

4.3. Test for Target Reachability. Every time MA completes the exploration of a new obstacle i, it defines on it a point L_i. Then, MA leaves the ith obstacle (according to Lemma 1, it will never return to it) and starts moving from L_i, to the Target along the straight line segment (L_i, Target). Since the point L_i is by definition the closest point of the ith obstacle to the Target, then normally there should be no points of the ith obstacle between L_i and the Target. Because of the model assumption that the obstacles do not touch each other, the point L_i cannot belong to any other obstacle but i. Therefore, if MA, after having arrived at L_i in step 3 of the algorithm, discovers that the straight line (L_i, Target) crosses some obstacle at point L_i, this can only mean that the crossed obstacle is i and that the Target is not reachable—either the Start or the Target point is *trapped* inside the ith obstacle.

To show that this is true, let O be a simple closed curve, X be a point in the scene not belonging to O, L the point on O closest to X, and (L, X) the straight line segment connecting L and X; all of these are defined in the plane. The segment (L, X) is said to be *directed outward* if a finite part of it in the vicinity of the point L is located outside of the curve O; otherwise, if (L, X) penetrates inside the curve O in the vicinity of L, it is said to be *directed inward.*

The following statement holds: if the segment (L, X) is directed inward then X is inside O. The condition is necessary because if X were outside the curve O then some other point of O would appear in the intersection of (L, X) and O which would be closer to X than L; by definition of the point L, this is impossible. The condition is also sufficient because if (L, X) is directed inward and L is the point of the curve O closest to X then (L, X) cannot cross any other point of O and, therefore, X must lie inside O. This fact is used in the following test.

TEST FOR TARGET REACHABILITY. If, while using Algorithm Bug1, after having defined a point L on an obstacle, MA discovers that the straight line segment (L, Target) crosses the obstacle at the point L, then the Target is not reachable.

5. Second Basic Algorithm: Bug2

5.1. Procedure. The procedure Bug2 is executed at any point of a continuous path. Again, the goal is to generate a path from the Start to the Target. As will be seen, in its travel under Bug2, on the one hand, MA can meet the same obstacle i more than once, but, on the other hand, the algorithm has no way of distinguishing between different obstacles. Because of that, the subscript i will be used only when referring to more than one obstacle; in addition, the superscript j will be used to indicate the jth occurrence of the hit or leave points on the same or on a different obstacle. Initially, $j = 1$; $L^0 = \text{Start}$. The test for target reachability built into steps 2(b) and 2(c) of the procedure is explained in Section 5.3. One can follow the procedure using the example shown in Figure 4. The algorithm consists of the following steps:

1. From the point L^{j-1}, move along the straight line (Start, Target) until one of the following occurs:
   (a) The Target is reached. The procedure stops.
   (b) An obstacle is encountered and a hit point, H^j, is defined. Go to step 2.
2. Using the accepted local direction, follow the obstacle boundary until one of the following occurs:
   (a) The Target is reached. The procedure stops.

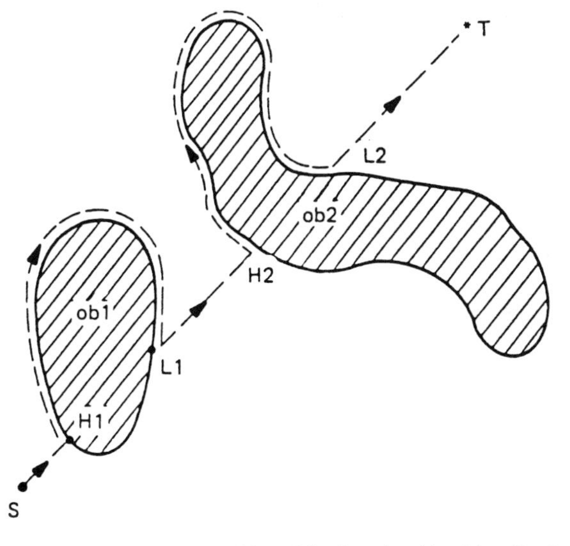

Fig. 4. Automaton's path (dotted line) under Algorithm Bug2.

(b) The line (Start, Target) is met at a point Q such that the distance $d(Q) < d(H^j)$, and the line (Q, Target) does not cross the current obstacle at the point Q. Define the leave point $L^j = Q$. Set $j = j + 1$. Go to step 1.
(c) The automaton returns to H^j and thus completes a closed curve (the obstacle boundary) without having defined the next hit point, H^{j+1}. The target is trapped and cannot be reached. The procedure stops.

Unlike Algorithm Bug1, more than one hit and more than one leave point can be generated on a single obstacle (see, for example, Figure 5). Also, note that the relationship between the perimeters of the obstacles and the length of the paths generated by Bug2 is not as clear as in the case of Bug1. In Bug1 the perimeter of an obstacle met by MA is covered at least once, and never more than 1.5 times. In Bug2, however, more options appear. A path segment around an obstacle generated by MA is sometimes shorter than the obstacle perimeter (compare Figures 2 and 4). In some other cases, when a straight line segment of

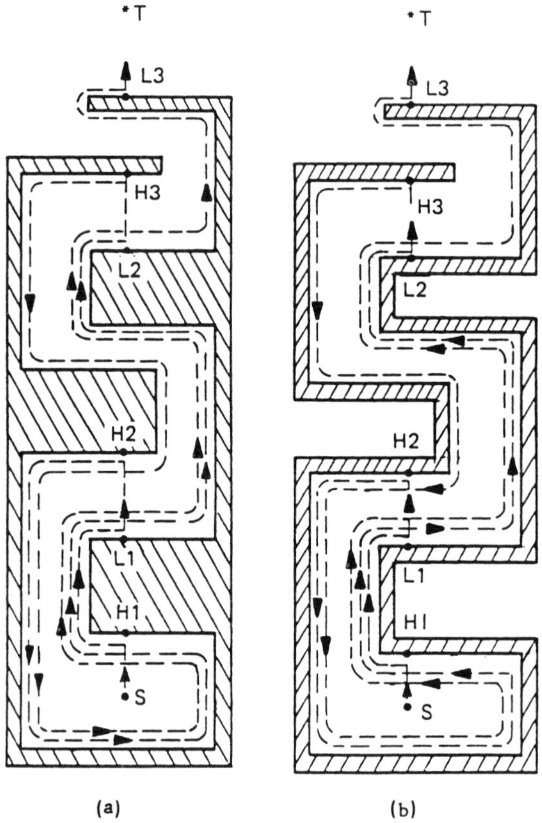

Fig. 5. Automaton's path around a maze-like obstacle (in-position case) under Algorithm Bug2. In terms of path complexity, both obstacles (a) and (b) are the same, whereas for (a) the straight line (S, T) crosses the obstacle 10 times, $n_i = 10$, and for (b), $n_i = 16$. At most, the path passes one segment (here ($H1$, $L1$)) three times; that is, there are at most two local cycles.

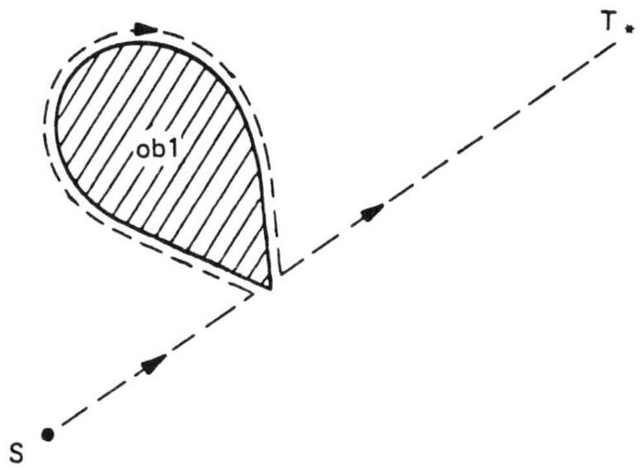

Fig. 6. A case when, under Algorithm Bug2, the automaton will have to make almost a full circle around a convex obstacle.

the path meets the obstacle almost tangentially and MA goes around the obstacle in a "wrong" direction, the path can actually be equal to the obstacle's full perimeter (Figure 6). And finally, as Figures 5 and 7 demonstrate, the situation can get even worse, and MA may have to pass along some segments of a maze-like obstacle more than once (see more on this case in the next section).

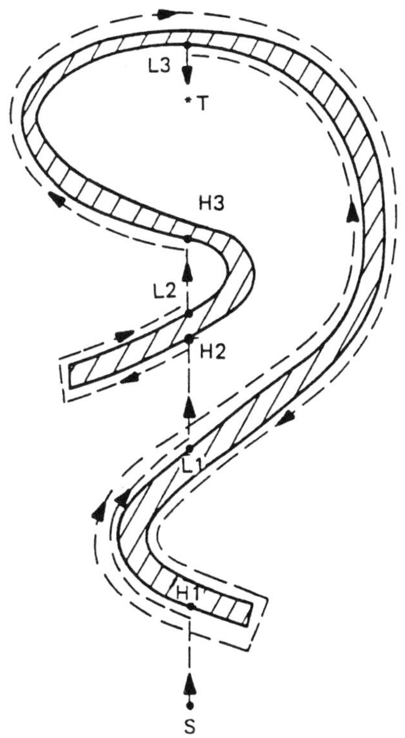

Fig. 7. Automaton's path in case of an in-position scene; here S is outside the obstacle, and T is inside.

5.2. Characteristics of Bug2

LEMMA 3. *Under Bug2, on its way to the Target* MA *can meet only a finite number of obstacles.*

PROOF. Although, while walking around an obstacle, MA can, at some moments, be at distances much larger than D from the Target, its straight line segments toward the Target are always within the same circle of radius D centered at the Target. This is guaranteed by the algorithm condition that $d(L^j, \text{Target}) > d(H^j, \text{Target})$ (see step 2 of Algorithm Bug2). Since, according to our model, any disc of finite radius can intersect with only a finite number of obstacles, the lemma follows. □

COROLLARY. *The only obstacles that can be met by* MA *under Algorithm Bug*1 *are those which intersect the disc of radius D centered at the Target. Moreover, the only obstacles that can be met by* MA *are those that intersect the straight line (Start, Target).*

DEFINITION 3. For a given local direction, a *local cycle* is created when MA has to pass some point of its path more than once.

In the example in Figure 4, no cycles are created; in Figures 5 and 7, there are some local cycles.

DEFINITION 4. A term *in-position* refers to such a mutual position of the pair of points (Start, Target) and a given obstacle where (1) the straight line segment (Start, Target) crosses the obstacle boundary at least once, and (2) either the Start or the Target lie inside the convex hull of the obstacle. A term *out-position* refers to such a mutual position of the pair (Start, Target) and the obstacle in which both points Start and Target lie outside the convex hull of the obstacle. A given scene is referred to as an in-position case if at least one obstacle in the scene, together with the Start and the Target points, creates an in-position condition; otherwise, the scene present an out-position case.

For example, the scene in Figure 3 presents an in-position case; without the obstacle *ob*3, it would have presented an out-position case. Below, n_i is the number of intersections between the straight line (Start, Target) and the ith obstacle; thus, n_i is a characteristic of the set (scene, Start, Target) and not of a specific algorithm. Obviously, for any convex obstacle $n_i = 2$.

If an obstacle is not convex but still $n_i = 2$, the path generated by Bug2 can be as simple as that for a convex obstacle (Figure 4, obstacle *ob*2). It can become more complicated if $n_i > 2$. In Figure 5(a) and (b), the segment of the boundary from $H1$ to $L1$, $(H1, L1)$, will be passed three times; segments $(L1, L2)$ and $(H2, H1)$, twice each; and segments $(L2, L3)$ and $(H3, H2)$, once each.

LEMMA 4. *Under Bug*2, MA *will pass any point of the* ith *obstacle boundary at most* $n_i/2$ *times.*

Proof. As one can see, the procedure Bug2 does not distinguish whether two consecutive obstacle crossings by the straight line (Start, Target) correspond to the same or to different obstacles. Without loss of generality, assume that only one obstacle is present; then, the index i can be dropped. For each hit point, H^j, the procedure will make MA walk around the obstacle until it reaches the corresponding leave point, L^j; therefore, all H and L points appear in pairs, (H^j, L^j). Because, under the accepted model, obstacles are of finite "thickness," for each pair (H^j, L^j) an inequality holds: $d(H^j) > d(L^j)$. After leaving L^j, MA walks along a straight line to the next hit point, H^{j+1}. Since, according to the model, the distance between two crossings of the obstacle by a straight line is finite then $d(L^j) > d(H^{j+1})$. This produces an inequality for all the H and L points,

$$d(H^1) > d(L^1) > d(H^2) > d(L^2) > d(H^3) > d(L^3) > \cdots . \tag{8}$$

Therefore, although any H or L point may be passed more than once, it will be defined as an H (correspondingly, L) point only once; thus, it can generate only one new passing of the same segment of the obstacle perimeter. In other words, each pair (H^j, L^j) can give rise to only one passing of a segment of the obstacle boundary. □

The lemma guarantees that the procedure terminates, and gives a limit on the number of generated local cycles. Using the lemma, an upper bound on the length of the paths generated by Bug2 can be produced.

Theorem 3. *The length of a path generated by the procedure Bug2 never exceeds the limit*

$$P = D + \sum_i \frac{n_i p_i}{2}, \tag{9}$$

where p_i refer to the perimeters of the obstacles intersecting the straight line segment (Start, Target).

Proof. Any path can be looked at as consisting of two parts: straight line segments of the line (Start, Target) (between the obstacles intersecting the line) and the path segments related to walking around the obstacle boundaries. Because of inequality (8), the sum of the straight line segments will never exceed D. As to the path segments around the obstacles, there is an upper bound guaranteed by Lemma 4 for each obstacle met by MA: that is, not more than $n_i/2$ passings along the same segment of the obstacle boundary will take place. Because of Lemma 3 (see the proof of the lemma), only those obstacles that intersect the straight line (Start, Target) should be counted. Summing up the straight line segments and those corresponding to walking around the obstacles, we obtain (9). □

Theorem 3 suggests that in some special scenes, while under the procedure Bug2, MA may have to go around the obstacles any (large albeit finite) number of times. An important question, then, is how typical such scenes are, and, in particular, what characteristics of the scene influence the length of the path. Theorem 4 and its corollary below address this question. They suggest that the mutual position of the Start point, the Target point, and the obstacles can affect the path length rather dramatically. Together, they significantly improve the upper bound on the length of the paths generated by Bug2—in out-position scenes in general and in scenes with convex obstacles in particular.

THEOREM 4. *Under the procedure Bug2, in the case of an out-position scene* MA *will pass any point of the obstacle boundary at most once.*

In other words, if the mutual position of the obstacle and of the points Start and Target satisfies the out-position requirements, the estimate on the length of the path for the procedure Bug2 reaches its lower bound (1).

PROOF. Figure 8 illustrates the proof. Shaded areas in the figure correspond to one or many obstacles. The boundaries of these areas of the obstacles can be of any shape which is indicated by dotted lines.

Consider an obstacle met by MA on its way to the Target, and consider an arbitrary point Q on the obstacle boundary; assume that Q is not a hit point. Because the obstacle boundary is a simple closed curve, the only way that MA can reach point Q is to come to Q from a previously defined hit point. Move from Q along the already generated part of the path in the direction opposite to

Fig. 8. Illustration for Theorem 4.

the accepted local direction, until the closest hit point along the path is encountered—say, this is H^j. We are interested only in those cases where Q is involved in at least one local cycle, that is, when MA passes the point Q more than once. For this event to occur, MA has to pass H^j at least as many times. In other words, if MA does not pass H^j more than once, it cannot pass Q more than once.

According to the procedure Bug2, the first time MA reaches the point H^j is along the straight line (Start, Target), or, more precisely, along the straight line segment (L^{j-1}, Target). Then, MA turns left and starts walking around the obstacle. To form a local cycle on this path segment, MA has to return to the point H^j again. Since a point can be defined as a hit point only once (see the proof for Lemma 4), the next time MA returns to the point H^j it must approach it from the right (see Figure 8), along the obstacle boundary. Therefore, after having defined H^j, in order to reach it again, this time from the right, MA somehow must cross the straight line (Start, Target) and enter its right semiplane. This can take place in one of only two ways—outside or inside the interval (Start, Target). Consider both cases.

1. The crossing occurs outside the interval (Start, Target). This case can correspond only to an in-position configuration (see Definition 4). Theorem 4, therefore, does not apply.
2. The crossing occurs inside the interval (Start, Target). We now want to prove that such a crossing of the path with the interval (Start, Target) cannot produce local cycles. Notice that the crossing cannot occur anywhere within the interval (Start, H^j) because otherwise at least a part of the straight line segment (L^{j-1}, H^j) would be included inside the obstacle. This is impossible because MA is known to have walked along the whole segment (L^{j-1}, H^j). If the crossing occurs within the interval (H^j, Target) then at the crossing point MA would define the corresponding leave point, L^j, and start moving along the line (Start, Target) toward the Target until it defined the next hit point, H^{j+1}, or reached the Target. Therefore, between H^j and L^j, MA could not have reached into the right semiplane of the line (Start, Target) (see Figure 8).

Since the above argument holds for any Q and the corresponding H^j, we conclude that in an out-position case MA will never cross the interval (Start, Target) into the right semiplane, which prevents it from producing local cycles. □

So far, no constraints on the shape of the obstacles have been imposed. In a special case when all the obstacles in the scene are convex, no in-position configurations can appear, and the upper bound on the length of the path can be improved as follows.

COROLLARY. *If all the obstacles in the scene are convex then, in the worst case, the length of the path produced by the procedure Bug2 is*

$$P = D + \sum_i p_i \qquad (10)$$

and, on the average,

(11) $$P = D + 0.5 \cdot \sum_i p_i,$$

where p_i refers to the perimeters of the obstacles intersecting the straight line segment (Start, Target).

Consider a statistically representative number of scenes with a random distribution of convex obstacles over each scene, a random distribution of points Start and Target over the set of scenes, and a fixed local direction as defined above. Then, the straight line (Start, Target) will cross all the obstacles it meets in such a way that for some obstacles MA will have to walk around them so as to cover the bigger part of their perimeters (as in case of the obstacle *ob*1, Figure 4), and, for some other obstacles, MA will cover only a smaller part of their perimeters (as in case of the obstacle *ob*2, Figure 4). On the average, one would expect a path that satisfies (11). As for (10), Figure 6 presents an example of such a "bad" scene. The corollary thus assures that for a wide range of scenes the length of paths generated by Algorithm Bug2 will not exceed the universal lower bound (1).

5.3. Test for Target Reachability. As Lemma 4 suggests, under Bug2 MA may pass the same point H^j of a given obstacle more than once, thus producing a finite number p of local cycles, $p = 0, 1, 2, \ldots$. The proof to the lemma indicates that, after having defined a point H^j, MA will never define this point again as an H or an L point. Therefore, on each of the subsequent local cycles (if any), the point H^j will be passed not along the straight line (Start, Target) but along the obstacle boundary. Every time after leaving the point H^j MA can expect one of the following to occur:

MA will never return again to H^j; this happens, for example, if it leaves the current obstacle altogether, or reaches the Target,

MA will define at least the first pair of the points $(L^j, H^{j+1}), \ldots$ and then return to the point H^j, to start a new local cycle,

MA will come back to the point H^j without having defined on the previous cycle a point L^j. In other words, MA could find no other intersection point Q of the line $(H^j$, Target) with the current obstacle such that Q would be closer to the Target than H^j, and the line (Q, Target) would not cross the current obstacle at Q. This can happen only if either MA or the Target are trapped inside the current obstacle (see Figure 9). The condition is both necessary and sufficient, which can be shown similarly to the proof in the target reachability test for the procedure Bug1, Section 4.3.

Based on this observation, a test for Target reachability for the procedure Bug1 can be formulated as follows.

TEST FOR TARGET REACHABILITY. If, on the pth local cycle, $p = 0, 1, \ldots$, after having defined a point H^j, MA returns to this point before it defines at least the

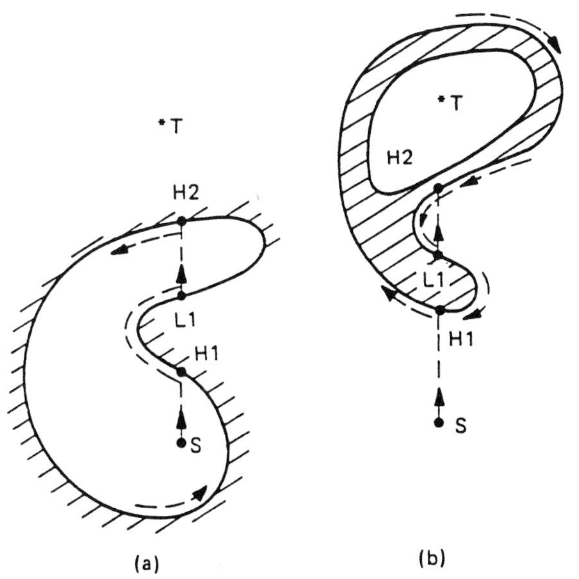

Fig. 9. Examples of traps. The path (dotted line) is executed under Algorithm Bug2. After having defined the point $H2$, the automaton returns to it before it defines any new L point. Therefore, the Target is not reachable.

first two out of the possible set of points L^j, $H^{j+1}, \ldots, H^k$, it means that MA has been *trapped* and hence the Target is not reachable.

6. Improving the Performance of the Basic Algorithms. Each of the algorithms Bug1 and Bug2 has a clear and simple underlying idea; each has its pluses and minuses. Namely, Bug1 never creates any local cycles, but it tends to be "over-cautious" and never covers less than the full perimeter of an obstacle. The procedure Bug2, on the other hand, is more "human" in that it takes advantage of the simple situations, but may become quite inefficient in more difficult cases. The better features of both procedures are combined in the following procedure, called BugM1 (for "modified"). The procedure BugM1 combines the efficiency of the procedure Bug2 in simpler scenes (where MA will pass only portions, instead of the full perimeters, of the obstacles, see Figure 4) with the more conservative strategy of the procedure Bug1 (which limits the corresponding segment of the path around an obstacle to 1.5 of its perimeter, see the bound (7)). In BugM1, for a given point on the path, the number of local cycles containing this point is never larger than two; in other words, MA will never pass the same point of the obstacle boundary more than three times. Although the flow of action in BugM1 is not as "clean" as in the basic algorithms, their termination properties are retained.

The procedure BugM1 is executed at any point of the continuous path. Instead of using the fixed straight line (Start, Target), as in Bug2, BugM1 uses a straight line (L_i^j, Target), with a changing point L_i^j; here, L_i^j indicates the jth leave point on an obstacle. The procedure uses three registers, R_1, R_2, R_3, to store intermediate information; all three are reset to zero when a new hit point, H_i^j, is defined. Specifically, R_1 is used to store the coordinates of the current point, Q_m, of the minimum distance between the obstacle boundary and the Target; R_2 integrates the length of the obstacle boundary starting at H_i^j; and R_3 integrates the length of the obstacle boundary starting at Q_m. (In case of many choices for Q_m, any one of them can be taken.) The test for target reachability mentioned in step 2(d) is explained in Section 5.3. Initially, $i=1, j=1$; $L_1^0 =$ Start. The procedure consists of the following steps:

1. From the point L_i^{j-1}, move along the line (L_i^{j-1}, Target) toward the Target until one of the following occurs:
   (a) The Target is reached. The procedure stops.
   (b) An obstacle is encountered and a hit point, H_i^j, is defined. Go to step 2.
2. Using the accepted local direction, follow the obstacle boundary until one of the following occurs:
   (a) The Target is reached. The procedure stops.
   (b) The line (L_i^{j-1}, Target) is met inside the interval (L_i^{j-1}, Target), at a point Q such that the distance $d(Q) < d(H_i^j)$, and the line (Q, Target) does not cross the current obstacle at the point Q. Define the leave point $L_i^j = Q$. Set $j=j+1$. Go to step 1.
   (c) The line (L_i^{j-1}, Target) is met outside the interval (L_i^{j-1}, Target). Go to step 3.
   (d) The automaton returns to H_i^j and thus completes a closed curve (the obstacle boundary) without having defined the next hit point. The Target cannot be reached. The procedure stops.
3. Continue following the obstacle boundary. If the Target is reached, stop. Otherwise, after having traversed the whole boundary and having returned to H_i^j, define a new leave point $L_i^j = Q_m$. Go to step 4.
4. Using the content of R_2 and R_3, determine the shorter way along the obstacle boundary to L_i^j, and use it to get to L_i^j. Apply the test for target reachability (as in Section 4.3). If the Target is not reachable, the procedure stops. Otherwise, designate $L_i^0 = L_i^j$, set $i=i+1$, $j=1$, and go to step 1.

Why does the procedure BugM1 converge? Depending on the scene, the flow of the algorithm fits one of the following two cases:

1. For a given scene, if the condition in step 2(c) of the procedure is never satisfied then the actual flow of the algorithm is that of Bug2, for which convergence has already been shown. In this case, the straight lines (L_i^j, Target) always coincide with the straight line (Start, Target), and no local cycles appear.
2. If, on the other hand, the scene presents an in-position case then the condition in step 2(c) is satisfied at least once; that is, MA crosses the straight line (L_i^{j-1}, Target) outside the interval (L_i^{j-1}, Target). This indicates that there is a danger of multiple local cycles. At this point, MA switches to a more conservative

approach offered by Algorithm Bug1, instead of risking an uncertain number of local cycles it might now expect from the procedure Bug2 (see Lemma 4). MA does this by executing steps 3 and 4 of BugM1 which are identical to steps 2 and 3 of the procedure Bug1.

After one execution of steps 3 and 4, the last leave point on the obstacle is defined, L_i^j, which is guaranteed to be closer to the Target than the corresponding hit point H_i^j (see inequality (7), Lemma 1). Then, MA leaves the ith obstacle and never returns to it again (Lemma 1). From now on, the algorithm (specifically, its steps 1 and 2) will be using the straight line (L_i^0, Target) as the "leading thread." (Note that, in general, the line (L_i^0, Target) does not coincide with the straight lines (L_{i-1}^0, Target) or (Start, Target).) One execution of the sequence of steps 3 and 4 of BugM1 is equivalent to one execution of steps 2 and 3 of Bug1, which guarantees the reduction by one of the number of obstacles that MA will meet on its way. Therefore, as in Bug1, the convergence of this case is guaranteed by Lemma 1, Lemma 2, and its corollary. Since cases 1 and 2 are independent and they exhaust all possible cases, the procedure BugM1 converges.

7. Concluding Remarks. The two bounds given by (1) and (7) indicate a gap in the path length estimates for the general problem of path planning among unknown obstacles, between at least $\sum p_i$ given by the lower bound (1), and at most $1.5 \cdot \sum p_i$ assured by the upper bound (7) of Algorithm Bug1. This poses an interesting problem of narrowing the gap either by finding a higher lower bound, or by introducing a better path-planning algorithm and lowering the upper bound.

The only work that these authors are aware of on convergence of motion-planning algorithms with uncertainty is the Pledge algorithm [16], which addresses a problem different from ours—namely, how to escape from a maze (and not how to find a given point inside or outside the maze). Very little is known about the performance of such algorithms; there are no performance estimates for the Pledge algorithm. In this paper we limit the performance of such algorithms by the worst-case lower bound (1), the worst-case upper bounds (7), (9), and (10), and by the average estimate (11) above.

On the theoretical level, therefore, the question "How reasonable can the paths generated by the algorithms for robot path planning among unknown obstacles be?" has a simple answer: Algorithm Bug1 is the best that can be offered today. Another advantage to Bug1 is that it does not require knowledge of the robot's current coordinates; it is sufficient if the robot can measure its distance from and its direction toward the target.

On a more practical level, Bug1 is a rather "conservative" algorithm, whose thoroughness in investigating each obstacle may or may not fit our notion of "being reasonable." Algorithm Bug2, on the other hand, is more "aggressive" and more efficient in many cases. Its behavior seems more reasonable, more "human." And, as happens sometimes with humans when they desperately try to reach their target and instead get lost in the woods, Bug2 pays a high price on those rare occasions when the set (obstacles, Start, Target) presents an in-position arrangement (see Section 5.2). In this sense, Algorithm BugM1 provides a compromise between the pluses and minuses of Algorithms Bug1 and Bug2.

One gets an additional insight into the operation and the "area of expertise" of the basic algorithms by trying them in maze search problems. The problem of search in an unknown or a known maze can be formulated in a number of ways. In one version (see, e.g., [14]), the automaton, after starting at an arbitrary cell of an unknown maze, must eventually visit every single cell of the maze, without passing through any barriers. Any pair of cells in the maze is assumed to be connected via other cells. Note that in this version there is no notion of a specific target cell, and no sense of direction is present. Because of that, neither of the basic algorithms can be used.

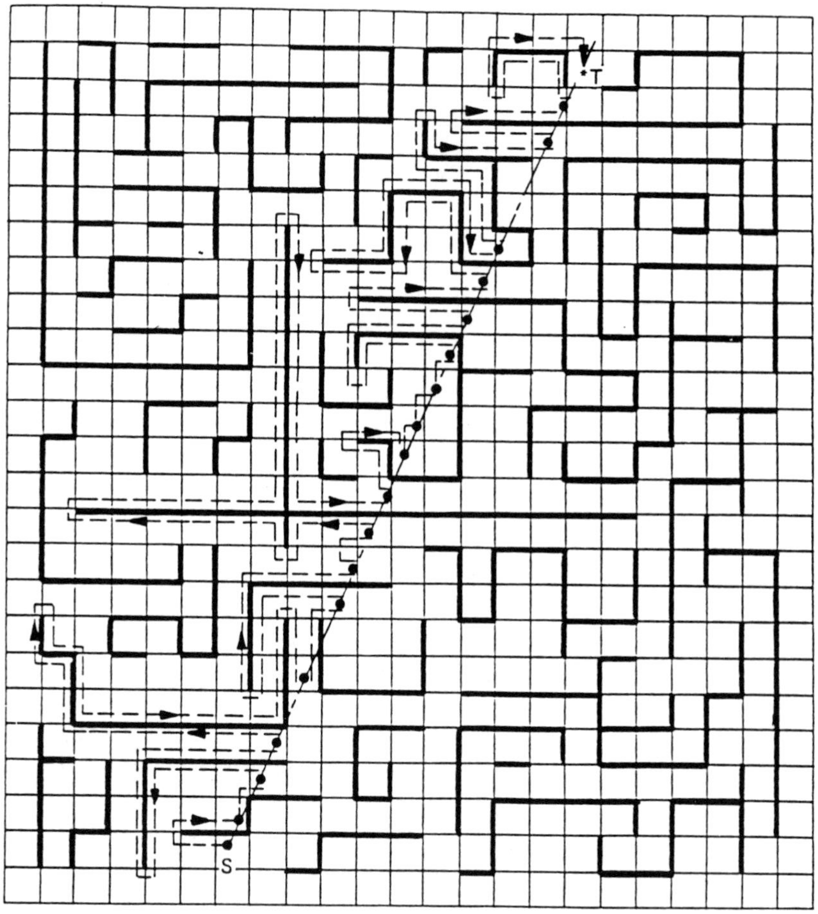

Fig. 10. Example of a walk in a maze using Algorithm Bug2; S = Start, T = Target. Points in which the automaton's path (dotted line) crosses the imaginary straight line (S, T) are indicated by dots. Maze barriers are shown as heavy lines.

In another version of the maze search problem, given a starting cell, the automaton has to find an exit from an unknown maze; the coordinates of the exit are not known. Although no target is explicitly presented here, either of our algorithms can be used. Namely, the automaton can choose any point somewhere in infinity as its target position, and then use the basic algorithms as usual. Then, if the exit exists, it is guaranteed to be found.

In still another version of the maze search problem [15], the barriers of the maze are formed by a set of horizontal and vertical line segments; unlike the two previous versions, full information about the maze is assumed. Given the coordinates of two points (cells) in a maze, the problem is to find a route from one to the other. Assuming that the barriers have nonzero thickness, our work shows that full information about the maze is redundant; local information (from any mechanism simulating a "tactile sensor") is sufficient to solve the problem. It is not clear whether this observation can have interesting consequences for applications—for example, for the routing problem in VLSI design.

Since the performance of Algorithm Bug2 can differ remarkably, depending on whether it deals with an in-position or an out-position case (see the estimates (9)-(11)), its behavior in a maze depends largely on the local topology of the barriers and, specifically, on whether the barriers in the maze form connected or disconnected patterns. This is demonstrated in an example shown in Figure 10. (To fit a typical convention of the maze literature, we present a discrete version of the continuous path-planning problem: the maze is a rectangular cell structure, with each cell being a little square; any cell crossed by the straight line (S, T) is considered to be lying on the line.) At first, one might think that the automaton is dealing here with an in-position scene and, therefore, Algorithm Bug2 is likely to produce an inefficient path with local cycles. Observe that this is not the case: given the fact that the automaton knows nothing about the maze, it performs quite well.

References

[1] J. T. Schwartz and M. Sharir, On the "piano movers" problem: I. The case of a two-dimensional rigid polygonal body moving amidst polygonal barriers, *Comm. Pure Appl. Math.*, **36** (1983), 345-398.

[2] T. Lozano-Perez and M. Wesley, An algorithm for planning collision-free paths among polyhedral obstacles, *Comm. ACM*, **22** (1979), 560-570.

[3] H. Moravec, The Stanford cart and the CMU rover, *Proc. IEEE*, **71** (1983), 872-874.

[4] R. A. Brooks, Solving the find-path problem by good representation of free space, *IEEE Trans. Systems Man Cybernet.*, **13** (1983).

[5] T. O. Binford, Visual perception by computer, *Proceeding of the IEEE Systems Science and Cybernetics Conference*, Miami, FL, 1971.

[6] J. Reif, Complexity of the mover's problem and generalizations, *Proceedings of the 20th Symposium on the Foundations of Computer Science*, 1979.

[7] D. L. Pieper, The kinematics of manipulators under computer control, Ph.D. Thesis, Stanford University, 1968.

[8] R. Paul, Modeling trajectory calculation and servoing of a computer controlled arm, Ph.D. Thesis, Stanford University, 1972.

[9] J. T. Schwartz and M. Sharir, On the "piano movers" problem: II. General techniques for computing topological properties of real algebraic manifolds, *Adv. in Appl. Math.*, **4** (1983), 298–351.

[10] J. Hopcroft, D. Joseph, and S. Whitesides, On the movement of robot arms in 2-dimensional bounded regions, *Proceedings of the IEEE Foundations of Computer Science Conference*, Chicago, IL, 1982.

[11] B. Bullock, D. Keirsey, J. Mitchell, T. Nussmeier, and D. Tseng, Autonomous vehicle control: an overview of the hughes project, *Proceedings of the IEEE Computer Society Conference "Trends and Applications, 1983: Automating Intelligent Behavior"*, Gaithersburg, MD, 1983.

[12] A. M. Thompson, The navigation system of the JPL robot, *Proceedings of the Fifth Joint International Conference on Artificial Intelligence*, Cambridge, MA, 1977.

[13] D. M. Kersey, E. Koch, J. McKisson, A. M. Meystel, and J. S. B. Mitchell, Algorithm of navigation for a mobile robot, *Proceedings of the International Conference on Robotics*, Atlanta, GA, 1984.

[14] M. Blum and D. Kozen, On the power of the compass (or, why mazes are easier to search than graphs), *Proceedings of the 19th Annual Symposium on Foundation of Computer Science*, Ann Arbor, MI, 1978.

[15] W. Lipski and F. Preparata, Segments, rectangles, contours, *J. Algorithms*, **2** (1981), 63–76.

[16] H. Abelson and A. diSessa, *Turtle Geometry*, MIT Press, Cambridge, MA, 1980, pp. 176–199.

[17] C. Yap, Algorithmic motion planning, in *Advances in Robotics*, Vol. 1 (J. Schwartz and C. K. Yap, eds.), Lawrence Erlbaum, NJ, 1986.

[18] L. Meijdam and A. de Zeeuw, On expectations, information, and dynamic game equilibria, in *Dynamic Games and Applications in Economics* (T. Basar, ed.), Springer-Verlag, New York, 1986.

[19] E. Moore, The firing squad synchronization problem, in *Sequential Machines* (E. Moore, ed.), Reading, MA, 1964.

[20] J. Traub, G. Wasilkowski, and H. Wozniakowski, *Information, Uncertainty, Complexity*, Addison-Wesley, Reading, MA, 1983.

[21] J. Reif, A survey on advances in the theory of computational robotics, in *Adaptive and Learning Systems* (K. Narendra, ed.), Plenum, New York, 1986.

[22] W. S. Massey, *Algebraic Topology*, Harcourt, Brace, & World, New York, 1967.

[23] V. Lumelsky, Effect of robot kinematics on motion planning in unknown environment, *Proceedings of the 24th IEEE Conference on Decision and Control,* Fort Lauderdale, FL, 1985.

[24] V. Lumelsky, Continuous motion planning in unknown environment for a 3D cartesian robot arm, *Proceedings of the IEEE International Conference on Robotics and Automation*, San Francisco, 1986.

MOTION PLANNING FOR AN AUTONOMOUS VEHICLE

Gordon T. Wilfong

AT&T Bell Laboratories
Murray Hill, NJ 07974-2070

ABSTRACT

The use of autonomous vehicles for moving materials between workstations is an important consideration in the overall design of a flexible automated factory. An algorithm for computing a collision-free motion for a vehicle with limited steering range is presented and its running time is analyzed. When given m "lanes" on which the vehicle is allowed to move in a polygonal environment of complexity n the algorithm produces a motion with the minimum number of turns between two query placements of the vehicle in time $O(m^2)$ after $O(m^2(n^2 + \log m))$ preprocessing. A restricted version of the algorithm has been implemented.

Introduction

Theoretical results providing upper and lower bounds for a wide variety of motion planning problems have been developed. (See [4] for a good overview.) Most of these algorithms are too complicated to be of practical value. However the lower bounds as well as the detailed analysis of these algorithms should be studied by those who wish to develop practical algorithms, to determine where the inherent complexities of a particular problem lie. An understanding of these complexities should guide one to a choice of a special case (or subproblem) that can be solved with a practical, yet provably efficient and correct, algorithm. In this paper we consider such a special case of a seemingly difficult motion planning problem. An algorithm is described for the special case and a proof of correctness and complexity analysis of the algorithm are provided. While our current implementation does not contain the full generality of the algorithm it is a first step in that direction and further generalization of the implementation should not be difficult.

The problem of computing a collision-free motion for a vehicle (see Figure 1) with a limited steering range in an environment containing various obstacles is studied. A collision-free motion is defined to be a motion throughout which the vehicle does not intersect the interior of any obstacle. The vehicle has two passive wheels near the rear and the center of rotation of the vehicle (denoted by z) is the midpoint of the rear axle. The front wheel w is used to drive and steer the vehicle. For a more complete description see the work by Cox [1] and Nelson and Cox [3]. The vehicle is not permitted to move in reverse and so the (directed) path p traced out by z during a motion has the property that if a line L was drawn lengthwise along the center of the vehicle then L would always be tangent to p. Thus a directed path p completely defines a motion of the vehicle. Therefore the goal is to compute a path of z so that the resulting motion of the vehicle is collision-free.

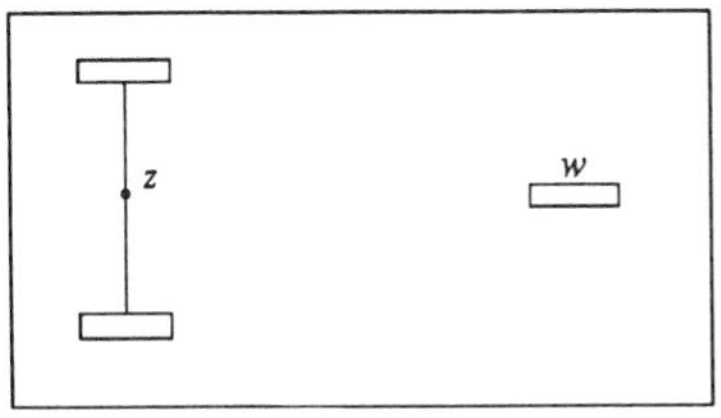

Figure 1. Vehicle model

1. Problem definition

The vehicle is to move in an environment containing polygonal obstacles. The maximal closed line segments in the boundary of an obstacle are called *obstacle walls* and the points where two obstacle walls meet are called *obstacle corners*. Since the vehicle has a limit on its steering range a path that z can follow has a limit on its curvature. Let $r_{\min}$ be the lower bound on the radius of curvature along an allowable path. The analysis found in [2] dealing with the problem of computing a path of bounded curvature for a point in such an environment seems to indicate that even if the geometry of the vehicle is ignored the problem remains difficult. Since, in this paper, the geometry of the vehicle will be accounted for, the problem is restricted to the case where there are *lanes* (line segments in the environment) given such that if L, the line down the middle of the vehicle, is collinear with a lane c and z is on c then the vehicle is not colliding with any obstacle. The motion of the vehicle is restricted so that z remains on some lane except to move between lanes in the following manner. If z is moving in direction d_1 along lane c_1 approaching the intersection I of c_1 with lane c_2 then z is permitted to follow a circular arc path of radius $r \geq r_{\min}$ from c_1 to c_2 so that z now moves along c_2 away from I in direction d_2. Such an arc is denoted by $Arc_t(r)$ where $t = (c_1,c_2,d_1,d_2)$. It is said that t is a *turn onto* c_2 *from* c_1. Lanes c_1 and c_2 are referred to as the *entrance lane* and *exit lane* respectively, of the turn t. The directions d_1 and d_2 are called the *entrance direction* and *exit direction* of t, respectively. If $Arc_t(r)$ defines a collision-free motion of the vehicle then r is a *free radius for turn* t. Notice that for each intersection I of two lanes there are eight types of turns since there are two lanes that could be the entrance lane (the remaining one being the exit lane) and there are two choices for each of the entrance and exit directions.

When changing lanes it is clearly desirable to get on the new lane as close to the intersection as possible so as to "overshoot" as few other intersections along the new lane (since the vehicle is not permitted to reverse). However the choice of radius for the turn is restricted by $r_{\min}$ and the obstacles. Therefore a goal will be to determine the minimum free radius for each turn.

The restriction to motion along designated lanes makes sense in a factory environment where it may be desirable (for safety reasons) to restrict the motion of the vehicle to certain "lanes of traffic" rather than allowing the vehicle to move arbitrarily about the factory floor. The restriction on the manner in which the vehicle is permitted to change lanes was made since allowing arbitrary paths for moving from one lane to another becomes the general problem before lanes were introduced.

An algorithm is described that, after preprocessing the given polygonal environment and collection of lanes, quickly responds to queries of the form: "For initial direction d_o and position z_o on lane c_o and final direction d_f and position z_f on lane c_f, determine a collision-free motion that consists of motions along lanes separated by single arc motions (of radius at least $r_{\min}$) between intersecting lanes so that the vehicle moves from z_o in direction d_o to where the vehicle is moving in direction d_f at z_f and the motion has the minimum number of turns if such a motion exists (or report "failure" otherwise)."

2. Motion planning algorithm

An overview of the algorithm is presented and then a more detailed analysis is given. The worst-case running time of the algorithm will be analyzed in terms of n, the number of obstacle corners and m, the number of lanes.

2.1 Outline of the algorithm

Let t be a turn and $r_t \geq r_{\min}$ be the minimum free radius for t. (If there is no free radius for t then t is removed from the collection of turns considered.) If t is a turn from lane c_1 onto lane c_2 then the *starting position* of t, denoted by $S(t)$, is the position along c_1 tangent with $Arc_t(r_t)$ and the *final position* of t, denoted by $F(t)$, is the position along c_2 tangent with $Arc_t(r_t)$. See Figure 2 for an example of $S(t)$ and $F(t)$.

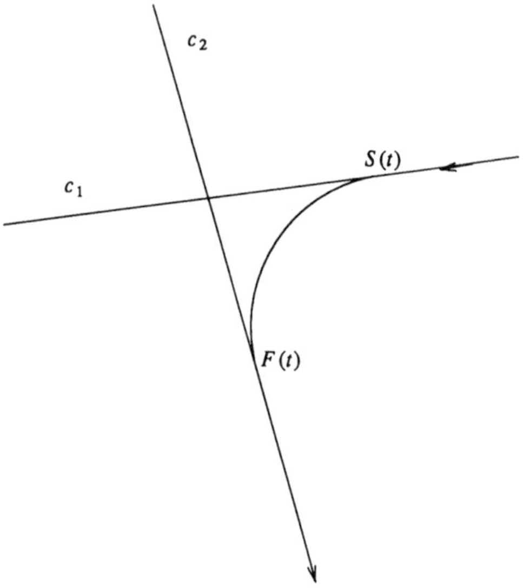

Figure 2. Starting and final positions of turn t

The following steps are the preprocessing steps of the algorithm. The steps are performed once to build the data structures used for quick responses to queries.

i) For each turn t, compute r_t.
ii) Build the digraph $D=(V,E)$ where for each turn t there is a vertex in V labeled t and there is a directed edge in E from the vertex labeled t_i to the vertex labeled t_j if the exit lane of t_i is the same as the entrance lane of t_j, the exit direction d of t_i is the entrance direction of t_j and $S(t_j)$ is in direction d from $F(t_i)$.

Throughout the paper it is assumed that the query to be processed requests a motion starting in direction d_o from position z_o on lane c_o and ending moving in direction d_f at position z_f on lane c_f. The algorithm performs the following steps for such a query.

i) Compute T_o, the set of all turns t from c_o with entrance direction d_o such that $S(t)$ is in direction d_o from z_o.
ii) Compute T_f, the set of all turns t onto c_f with exit direction d_f where z_f is in direction d_f from $F(t)$.
iii) Search D for a directed path from any vertex labeled by a turn in T_o to any vertex labeled by a turn in T_f.

2.2 Calculating minimum free radius for a turn

Let $Sweep_t(r)$ denote the region of the plane swept out by the vehicle as it follows $Arc_t(r)$. Figure 3 shows $Sweep_t(r)$ where t is a turn from c_1 onto c_2. The vehicle starts the turn at position z_1 and ends at position z_2.

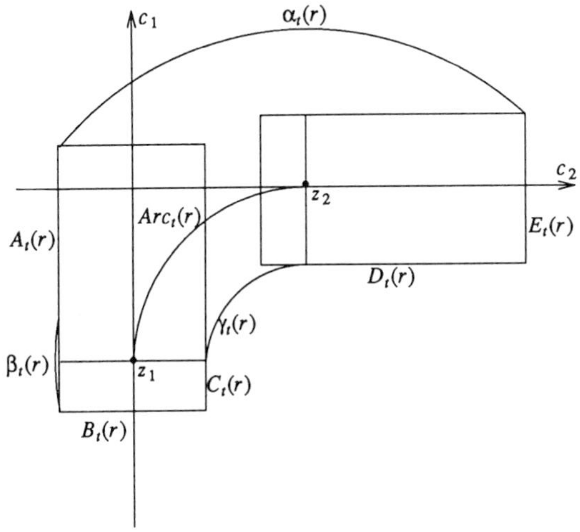

Figure 3. Swept-out region at a turn

For fixed t, $Sweep_t(r)$ changes shape continuously as a function of r. In order to find the minimum free radius for t we compute a set of "critical radii" for t so that for every radius r in the open interval between two given successive critical radii for t the statement

''$Sweep_t(r)$ intersects the interior of no obstacle''

(denoted by $Q_t(r)$) is either always true or always false. (For a given turn t, two critical radii r and r' where $r < r'$ are called *successive* if there are no critical radii r'' for t such that $r < r'' < r'$.) Notice that $Q_t(r)$ being true means r is a free radius for t. Continuity arguments show that if $Q_t(r)$ is true for any (and hence all) r in the open interval between two successive critical radii for t then it is also true at the critical radii bounding the interval. Thus the goal is to compute, for each turn t, the critical radii for t and then the minimum free critical radius for t is the minimum free radius r_t.

2.2.1 Critical radii

Intuitively, r is a critical radius for turn t if some part of the boundary of $Sweep_t(r)$ touches (without crossing over) some obstacle wall. Thus as r is increased continuously from $r_{\min}$, the only possible values for r where $Q_t(r)$ changes from true to false (or false to true) are at the critical radii. The notation used below to refer to the various segments of the boundary of $Sweep_t(r)$ can be found in Figure 3. The *starting vertices* of $Sweep_t(r)$ are the points on the boundary of $Sweep_t(r)$ where $A_t(r)$ meets $\alpha_t(r)$, $\beta_t(r)$ meets $B_t(r)$ or $B_t(r)$ meets $C_t(r)$. The *final vertices* of $Sweep_t(r)$ are the points on the boundary of $Sweep_t(r)$ where $\alpha_t(r)$ meets $E_t(r)$ or $E_t(r)$ meets $D_t(r)$.

Suppose $t = (c_1,c_2,d_1,d_2)$ and I is the intersection of c_1 and c_2. Let e_1 be the endpoint of c_1 such that I is in direction d_1 from e_1 (it may be that $e_1 = I$) and define $R_1(t)$ so that $Arc_t(R_1(t))$ is tangent to c_1 at e_1. Let e_2 be the endpoint of c_2 such that e_2 is in direction d_2 from I and define $R_2(t)$ so that $Arc_t(R_2(t))$ is tangent to c_2 at e_2. If $R_1(t) < r_{\min}$ or $R_2(t) < r_{\min}$ then remove t from consideration because the vehicle can not make such a turn. Otherwise define $r_{\max}(t)$ to be the minimum of $R_1(t)$ and $R_2(t)$.

A radius r is said to be a *critical radius* for turn t if $r_{\min} \le r \le r_{\max}(t)$ and at least one of the following holds.

(1) $r = r_{\min}$
(2) $r = r_{\max}(t)$
(3) $\alpha_t(r)$ is tangent to an obstacle wall
(4) $\beta_t(r)$ is tangent to an obstacle wall
(5) an obstacle corner is on $\alpha_t(r)$
(6) an obstacle corner is on $\beta_t(r)$
(7) an obstacle corner is on $\gamma_t(r)$
(8) $B_t(r)$ (not including its endpoints) hits an obstacle corner
(9) $B_t(r)$ is collinear with obstacle wall W if W is perpendicular to the entrance lane of turn t
(10) a starting vertex of $Sweep_t(r)$ hits an obstacle wall W if W is not parallel to the entrance lane of turn t
(11) $E_t(r)$ (not including its endpoints) hits an obstacle corner
(12) $E_t(r)$ is collinear with obstacle wall W if W is perpendicular to the exit lane of turn t
(13) a final vertex of $Sweep_t(r)$ hits an obstacle wall W if W is not parallel to the exit lane of turn t

Lemma 1: There are $O(n)$ critical radii for each turn and they can be computed in $O(n)$ time.

Let R_1 and R_2 be successive critical radii for t. Then $J = (R_1,R_2)$ is called a *critical interval for t*. The next result shows that the collection of all free radii for t consists of a subset of the critical radii for t and a collection of the critical intervals for t.

Lemma 2: Let R_1 and R_2 be successive critical radii for turn t and let $J = (R_1,R_2)$. Suppose $r_1,r_2 \in J$. Then $Q_t(r_1)$ if and only if $Q_t(r_2)$.

Proof. Follows by continuity and a straight forward geometric argument. □

The above facts allow us to conclude that if r is the minimum free radius for a turn t then r must be a critical radius for t.

2.2.2 Testing for free radii

The previous section showed that in order to compute the minimum free radius for a turn t it is sufficient to find the minimum free critical radius for t. This section describes how to test if a critical radius is free.

If radius r is not free for turn t then either some obstacle is completely contained in $Sweep_t(r)$ or there is some obstacle O such that there are interior points of O in the interior of $Sweep_t(r)$ and other interior points of O outside of $Sweep_t(r)$. (By the definition of lanes, $Sweep_t(r)$ cannot be completely contained in any obstacle.) Thus to decide if a radius r is free for t, first test to see if some obstacle is completely contained in $Sweep_t(r)$ by choosing an interior point from each obstacle and checking if any of these points are in the interior of $Sweep_t(r)$. If at least one of these points is in $Sweep_t(r)$ then r is not free for t. Otherwise no obstacle is completely contained in $Sweep_t(r)$ and so test if some obstacle partially overlaps $Sweep_t(r)$. This can be done by checking if any obstacle corner is in the interior of $Sweep_t(r)$ or any obstacle wall passes through the interior of $Sweep_t(r)$. This takes $O(1)$ time per corner or wall.

The above discussion shows that testing to see if a radius is free for a given turn can be done in $O(n)$ time.

Lemma 3: Computing the least free radius for each turn can be done in $O(m^2n^2)$ total time.

Proof. By Lemma 1, the $O(n)$ critical radii for a turn can be computed in $O(n)$ time. Sorting them takes $O(n\log n)$ time. In increasing order starting at the smallest critical radii test if the radius is free for the turn. This takes $O(n)$ time for each critical radii. Therefore computing the minimum free radius for a turn (or determining that there is no free radius for a turn) can be done in $O(n^2)$ time for each of $O(m^2)$ turns. □

2.3 Finding a motion

We describe how a motion is obtained once the minimum free radius for each turn has been computed. The directed graph $D = (V,E)$ is constructed as follows. There is a vertex labeled t in V for each turn t such that there is some free radius r ($r_{\min} \le r \le r_{\max}(t)$). There is a directed edge from the vertex labeled t_1 to the vertex labeled t_2 if the exit lane c of t_1 is the entrance lane of t_2, the exit direction d of t_1 is the same as the entrance direction of t_2 and $S(t_2)$ is at least as far along c in direction d as $F(t_1)$.

Lemma 4: Let d and d' be the entrance direction of t and the exit direction of t' respectively. There is a directed path in D from the vertex labeled t to the vertex labeled t' if and only if there is a motion (of the restricted kind under consideration) from $S(t)$ moving in direction d to position $F(t')$ moving in direction d'.

Proof. Follows from the definition of D and the observation that there is a (restricted) motion if and only if there is one where each lane change is the arc whose radius is the minimum free radius for that turn. □

The above result says that finding motions of the vehicle (of the restricted kind being considered) is equivalent to searching the directed graph D for a path. We now consider the complexity of searching for a path in D.

Lemma 5: D has size $O(m^3)$ but it can be implicitly constructed in time $O(m^2\log m)$ time (after the minimum free radius for each turn has been computed) and a breadth-first search of D can subsequently be performed in $O(m^2)$ time.

Proof. There are $O(m^2)$ intersections of lanes and so there are $O(m^2)$ vertices in V and all can be computed in $O(m^2)$ time. There are $O(m)$ turns involving any given lane. Thus there are $O(m^2)$ directed edges in E from vertices labeled by turns onto a given lane. Therefore there are a total of $O(m^3)$ edges in E.

We wish to do a breadth-first search of D without explicitly computing and exploring all $O(m^3)$ edges. For each lane c and each direction d along c, a list $L(c,d)$ is maintained of turns from c (along with their starting positions on c) that have d as their entrance direction. Each $L(c,d)$ is sorted by order of starting positions. Each such list can be constructed in $O(m\log m)$ time. This $O(m)$ sized list implicitly contains the adjacency lists of all $O(m)$ turns onto c with exit direction d. (Vertex v is on the adjacency list of vertex u if there is a directed edge from u to v.)

As a vertex is visited during the breadth-first search, it is removed from the list $L(c,d)$ that it appears in and so only one edge into each vertex is ever considered. Therefore the search can be done in $O(|V|) = O(m^2)$ time after $O(m^2\log m)$ time to compute the $L(c,d)$'s. □

In order to find a motion from some initial position and direction to a final position and direction, first compute the set T_o of all the turns t from c_o with entrance direction d_o such that $S(t)$ is further along c_o in direction d_o than z_o. Also compute all turns in T_f the set of turns t onto c_f with exit direction d_f such that $F(t)$ is before z_f in direction d_f. Clearly these computations can be done in time $O(m)$ once the minimum free radii have been calculated. Compute D and perform a breadth-first search of D as indicated above starting with T_o as the set of visited vertices and ending when some vertex in the goal set T_f has been reached. The resulting path will have the minimum number of turns in it by definition of breadth-first search. Therefore we have the following result.

Theorem 6: Using $O(m^2(n^2 + \log m))$ preprocessing time, queries requesting a motion with a minimum number of turns from an initial position and direction to a final position and direction can be answered in time $O(m^2)$.

Proof. Computing the minimum free radii can be done in $O(m^2n^2)$ by Lemma 3. A representation of the directed graph D can be constructed in $O(m^2\log m)$ time by Lemma 5. For a given query, T_o and T_f can be computed in $O(m)$ time and D can be searched for a directed path from a vertex in T_o to a vertex in T_f in $O(m^2)$ time by Lemma 5. □

3. Current implementation and future work

We have implemented a simplified version of the algorithm for a restricted case. The case considered is where all lanes are parallel to the x-axis or the y-axis and the obstacles are all rectangles whose sides are also parallel to the x-axis or the y-axis. The entire environment may be enclosed within a bounding rectangle. An example environment is shown in Figure 4. The dashed lines represent the lanes and the rectangles are the obstacles.

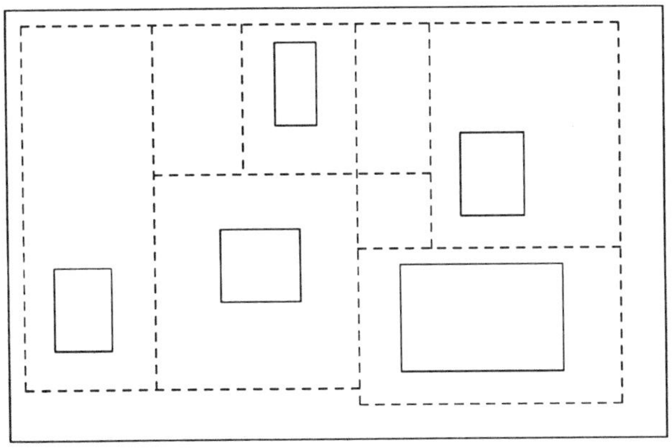

Figure 4. Typical environment

For practical reasons, following the minimum free radius arc for each turn may not be the most desirable tactic because this usually results in the vehicle brushing against some obstacle. Because of errors in positioning of the vehicle this strategy can cause unexpected collisions. Control of the vehicle when it is at or near its turning limit may also be difficult. Therefore in the implementation, all intervals of free radii are computed so that the user can choose from a wider range of free radii. For instance, one may wish to choose radii midway within a wide interval of free radii to ensure a large distance between the vehicle and obstacles. Of course, such a heuristic cannot guarantee that a path will be found even if one exists.

There are many areas of possible improvement for the current implementation. For example, the implementation could be generalized to allow lanes in arbitrary directions and arbitrary polygonal obstacles. Another improvement that would be simple to achieve would be to find a shortest length path amongst all paths with the minimum number of turns. Also the current implementation does not take advantage of the "implicit" adjacency lists described in the previous section.

Lanes are assumed to be given as part of the input to the problem. An interesting question to ask is "Given a description of the obstacles, what is a "good" collection of lanes and how are they to be computed?"

Acknowledgments

A graphics package written by E.N. Pinson for allowing graphic input and the simulation of the motion resulting from the implemented algorithm was extremely useful in the debugging process as well as for viewing the computed motion. Discussions concerning the nature of a vehicle's motion with I.J. Cox and W.L. Nelson and discussions about bounded curvature paths with S.J. Fortune were extremely useful in the development of this work.

REFERENCES

1. Cox, I.J. Blanche: An Autonomous Robot Vehicle for Structured Environments. In *Proc. IEEE Int. Conf. Robotics and Automation*, 1988.
2. Fortune, S.J. and Wilfong, G.T. Constrained Motion Planning. In *Proc. ACM Symp. Theory of Comput.*, 1988.
3. Nelson, W.L. and Cox, I.J. Local Path Control for an Autonomous Vehicle. In *Proc. IEEE Int. Conf. Robotics and Automation*, 1988.
4. Yap, C.K. Algorithmic Motion Planning. In *Advances in Robotics, Volume 1: Algorithmic and Geometric Aspects*, C.K. Yap and J.T. Schwartz, Eds. Lawrence Erlbaum Assoc., Hillsdale NJ, 1986.

Oussama Khatib
Artificial Intelligence Laboratory
Stanford University
Stanford, California 94305

Real-Time Obstacle Avoidance for Manipulators and Mobile Robots

Abstract

This paper presents a unique real-time obstacle avoidance approach for manipulators and mobile robots based on the artificial potential field concept. Collision avoidance, traditionally considered a high level planning problem, can be effectively distributed between different levels of control, allowing real-time robot operations in a complex environment. This method has been extended to moving obstacles by using a time-varying artificial potential field. We have applied this obstacle avoidance scheme to robot arm mechanisms and have used a new approach to the general problem of real-time manipulator control. We reformulated the manipulator control problem as direct control of manipulator motion in operational space—the space in which the task is originally described—rather than as control of the task's corresponding joint space motion obtained only after geometric and kinematic transformation. Outside the obstacles' regions of influence, we caused the end effector to move in a straight line with an upper speed limit. The artificial potential field approach has been extended to collision avoidance for all manipulator links. In addition, a joint space artificial potential field is used to satisfy the manipulator internal joint constraints. This method has been implemented in the COSMOS system for a PUMA 560 robot. Real-time collision avoidance demonstrations on moving obstacles have been performed by using visual sensing.

1. Introduction

In previous research, robot collision avoidance has been a component of higher levels of control in hierarchical robot control systems. Collision avoidance has been treated as a planning problem, and research in this area has focused on the development of collision-free path planning algorithms (Lozano-Perez 1980; Moravec 1980; Chatila 1981; Brooks 1983). These algorithms aim at providing the low level control with a path that will enable the robot to accomplish its assigned task free from any risk of collision.

From this perspective, the function of low level control is limited to the execution of elementary operations for which the paths have been precisely specified. The robot's interaction with its environment is then paced by the time cycle of high level control, which is generally several orders of magnitude slower than the response time of a typical robot. This places limits on the robot's real-time capabilities for precise, fast, and highly interactive operations in a cluttered and evolving environment. We will show that it is possible to extend greatly the function of low level control and to carry out more complex operations by coupling environment sensing feedback with the lowest level of control.

Increasing the capability of low level control has been the impetus for the work on real-time obstacle avoidance that we discuss here. Collision avoidance at the low level of control is not intended to replace high level functions or to solve planning problems. The purpose is to make better use of low level control capabilities in performing real-time operations. At this low level of control, the degree or *level of competence* (Brooks 1984) will remain less than that of higher level control.

The *operational space formulation* is the basis for the application of the potential field approach to robot manipulators. This formulation has its roots in the work on end-effector motion control and obstacle avoidance (Khatib and Le Maitre 1978) that has been implemented for an MA23 manipulator at the Laboratoire d'Automatique de Montpellier in 1978. The operational space approach has been formalized by constructing its basic tool, the equations of motion in the operational space of the manipulator end effector.

The International Journal of Robotics Research,
Vol. 5, No. 1, Spring 1986

Details of this work have been published elsewhere (Khatib 1980; Khatib 1983; Khatib 1985). In this paper, we review the fundamentals of the operational space formulation and the artificial potential field concept. We present the integration of this collision avoidance approach into the operational space control system and its real-time implementation. The extension of this work to link collision avoidance is also developed.

2. Operational Space Formulation

An *operational coordinate system* is a set x of m_0 *independent* parameters describing the manipulator end-effector position and orientation in a frame of reference R_0. For a nonredundant manipulator, these parameters form a set of configuration parameters in a domain of the operational space and constitute a system of generalized coordinates. The kinetic energy of the holonomic articulated mechanism is a quadratic form of the generalized velocities,

$$T(\mathrm{x}, \dot{\mathrm{x}}) = \tfrac{1}{2}\dot{\mathrm{x}}^T \Lambda(\mathrm{x})\dot{\mathrm{x}}, \tag{1}$$

where $\Lambda(\mathrm{x})$ designates the symmetric matrix of the quadratic form, *i.e.*, the kinetic energy matrix. Using the Lagrangian formalism, the end-effector equations of motion are given by

$$\frac{d}{dt}\left(\frac{\partial L}{\partial \dot{\mathrm{x}}}\right) - \frac{\partial L}{\partial \mathrm{x}} = \mathrm{F}, \tag{2}$$

where the Lagrangian $L(\mathrm{x}, \dot{\mathrm{x}})$ is

$$L(\mathrm{x}, \dot{\mathrm{x}}) = T(\mathrm{x}, \dot{\mathrm{x}}) - U(\mathrm{x}), \tag{3}$$

and $U(\mathrm{x})$ represents the potential energy of the gravity. The symbol F is the operational force vector. These equations can be developed (Khatib 1980; Khatib 1983) and written in the form,

$$\Lambda(\mathrm{x})\ddot{\mathrm{x}} + \mu(\mathrm{x}, \dot{\mathrm{x}}) + \mathrm{p}(\mathrm{x}) = \mathrm{F}, \tag{4}$$

where $\mu(\mathrm{x}, \dot{\mathrm{x}})$ represents the centrifugal and Coriolis forces and $\mathrm{p}(\mathrm{x})$ the gravity forces.

The control of manipulators in operational space is based on the selection of F as a command vector. In order to produce this command, specific forces Γ must be applied with joint-based actuators. The relationship between F and the joint forces Γ is given by

$$\Gamma = J^T(\mathrm{q})\,\mathrm{F}, \tag{5}$$

where q is the vector of the n joint coordinates and $J(\mathrm{q})$ the Jacobian matrix.

The decoupling of the end-effector motion in operational space is achieved by using the following structure of control,

$$\mathrm{F} = \Lambda(\mathrm{x})\mathrm{F}^* + \mu(\mathrm{x}, \dot{\mathrm{x}}) + \mathrm{p}(\mathrm{x}), \tag{6}$$

where F^* represents the command vector of the decoupled end effector that becomes equivalent to a *single unit mass.*

The extension of the operational space approach to redundant manipulator systems is discussed in Khatib (1980); Khatib (1983). The integration of active force control for assembly operations is presented in Khatib (1985).

3. The Artificial Potential Field Approach

We present this method in the context of manipulator collision avoidance. Its application to mobile robots is straightforward. The philosophy of the artificial potential field approach can be schematically described as follows:

> *The manipulator moves in a field of forces. The position to be reached is an attractive pole for the end effector and obstacles are repulsive surfaces for the manipulator parts.*

We first consider the collision avoidance problem of a manipulator end effector with a single obstacle O. If x_d designates the goal position, the control of the manipulator end effector with respect to the obstacle O can be achieved by subjecting it to the artificial potential field,

$$U_{art}(\mathrm{x}) = U_{\mathrm{x}_d}(\mathrm{x}) + U_O(\mathrm{x}). \tag{7}$$

This leads to the following expression of the potential

energy in the Lagrangian (Eq. 3),

$$U(\mathrm{x}) = U_{art}(\mathrm{x}) + U_g(\mathrm{x}), \tag{8}$$

where $U_g(\mathrm{x})$ represents the gravity potential energy. Using Lagrange's (Eq. 2) and taking into account the end-effector dynamic decoupling (Eq. 6), the command vector F^*_{art} of the decoupled end effector that corresponds to applying the artificial potential field U_{art} (Eq. 7) can be written as

$$\mathrm{F}^*_{art} = \mathrm{F}^*_{\mathrm{x}_d} + \mathrm{F}^*_{O}, \tag{9}$$

with

$$\begin{aligned} \mathrm{F}^*_{\mathrm{x}_d} &= -\mathrm{grad}[U_{\mathrm{x}_d}(\mathrm{x})], \\ \mathrm{F}^*_{O} &= -\mathrm{grad}[U_O(\mathrm{x})]. \end{aligned} \tag{10}$$

The symbol $\mathrm{F}^*_{\mathrm{x}_d}$ is an attractive force allowing the point x of the end effector to reach the goal position x_d, and F^*_O represents a *Force Inducing an Artificial Repulsion from the Surface* (FIRAS, from the French) of the obstacle created by the potential field $U_O(\mathrm{x})$. The symbol $\mathrm{F}^*_{\mathrm{x}_d}$ corresponds to the proportional term, *i.e.* $-k_p(\mathrm{x} - \mathrm{x}_d)$, in a conventional PD servo where k_p is the position gain. The attractive potential field $U_{\mathrm{x}_d}(\mathrm{x})$ is simply

$$U_{\mathrm{x}_d}(\mathrm{x}) = \tfrac{1}{2} k_p(\mathrm{x} - \mathrm{x}_d)^2. \tag{11}$$

$U_O(x)$ is selected such that the artificial potential field $U_{art}(x)$ is a positive continuous and differentiable function which attains its zero minimum when $\mathrm{x} = \mathrm{x}_d$. The articulated mechanical system subjected to $U_{art}(\mathrm{x})$ is stable. Asymptotic stabilization of the system is achieved by adding dissipative forces proportional to $\dot{\mathrm{x}}$. Let k_v be the velocity gain; the forces contributing to the end-effector motion and stabilization are of the form,

$$\mathrm{F}^*_m = -k_p(\mathrm{x} - \mathrm{x}_d) - k_v\dot{\mathrm{x}}. \tag{12}$$

This command vector is inadequate to control the manipulator for tasks that involve large end-effector motion toward a goal position without path specification. For such a task, it is better for the end effector to move in a straight line with an upper speed limit.

Rewriting Eq. (12) leads to the following expression, which can be interpreted as specifying a desired velocity vector in a pure velocity servo-control,

$$\dot{\mathrm{x}}_d = \frac{k_p}{k_v}(\mathrm{x}_d - \mathrm{x}). \tag{13}$$

Let V_{max} designate the assigned speed limit. The limitation of the end-effector velocity magnitude can then be obtained (Khatib, Llibre, and Mampey 1978),

$$\mathrm{F}^*_m = -k_v(\dot{\mathrm{x}} - \nu\dot{\mathrm{x}}_d), \tag{14}$$

where

$$\nu = \min\left(1, \frac{V_{\max}}{\sqrt{\dot{\mathrm{x}}_d^T\dot{\mathrm{x}}_d}}\right). \tag{15}$$

With this scheme, the velocity vector $\dot{\mathrm{x}}$ is controlled to be pointed toward the goal position while its magnitude is limited to $V_{\max}$. The end effector will then travel at that speed in a straight line, except during the acceleration and deceleration segments or when it is inside the repulsive potential field regions of influence.

4. FIRAS Function

The artificial potential field $U_O(\mathrm{x})$ should be designed to meet the manipulator stability condition and to create at each point on the obstacle's surface a potential barrier which becomes negligible beyond that surface. Specifically, $U_O(\mathrm{x})$ should be a nonnegative continuous and differentiable function whose value tends to infinity as the end effector approaches the obstacle's surface. In order to avoid undesirable perturbing forces beyond the obstacle's vicinity, the influence of this potential field must be limited to a given region surrounding the obstacle.

Using analytic equations $f(\mathrm{x}) = 0$ for obstacle description, the first artificial potential field function used (Khatib and Le Maitre 1978) was based on the values of the function $f(\mathrm{x})$,

$$U_O(\mathrm{x}) = \begin{cases} \dfrac{1}{2}\eta\left(\dfrac{1}{f(\mathrm{x})} - \dfrac{1}{f(\mathrm{x}_0)}\right)^2 & \text{if } f(\mathrm{x}) \leq f(\mathrm{x}_0), \\ 0 & \text{if } f(\mathrm{x}) > f(\mathrm{x}_0). \end{cases} \tag{16}$$

Fig. 1. An n-ellipsoid with $n = 4$.

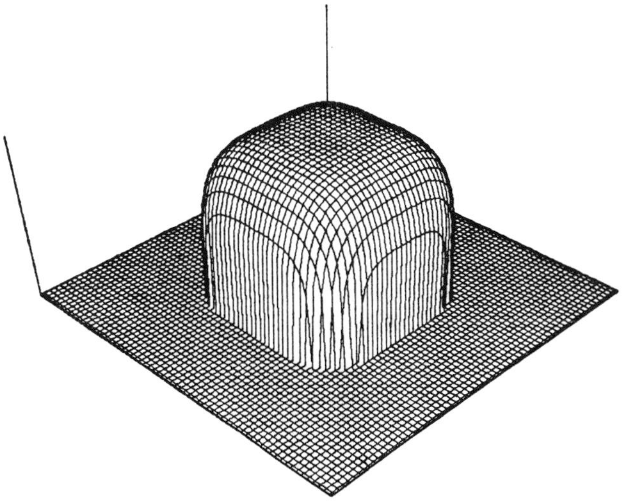

The region of influence of this potential field is bounded by the surfaces $f(\mathrm{x}) = 0$ and $f(\mathrm{x}) = f(\mathrm{x}_0)$, where x_0 is a given point in the vicinity of the obstacle and η a constant gain. This potential function can be obtained very simply in real time since it does not require any distance calculations. However, this potential is difficult to use for asymmetric obstacles where the separation between an obstacle's surface and equipotential surfaces can vary widely.

Using the shortest distance to an obstacle O, we have proposed (Khatib 1980) the following artificial potential field;

$$U_O(\mathrm{x}) = \begin{cases} \frac{1}{2}\eta\left(\frac{1}{\rho} - \frac{1}{\rho_0}\right)^2 & \text{if } \rho \leq \rho_0, \\ 0 & \text{if } \rho > \rho_0, \end{cases} \tag{17}$$

where ρ_0 represents the limit distance of the potential field influence and ρ the shortest distance to the obstacle O. The selection of the distance ρ_0 will depend on the end effector operating speed V_{max} and on its deceleration ability. End-effector acceleration characteristics are discussed in Khatib and Burdick (1985).

Any point of the robot can be subjected to the artificial potential field. The control of a *Point Subjected to the Potential* (PSP) with respect to an obstacle O is achieved using the FIRAS function,

$$\mathrm{F}^*_{(O,psp)} = \begin{cases} \eta\left(\frac{1}{\rho} - \frac{1}{\rho_0}\right)\frac{1}{\rho^2}\frac{\partial\rho}{\partial\mathrm{x}} & \text{if } \rho \leq \rho_0, \\ 0 & \text{if } \rho > \rho_0, \end{cases} \tag{18}$$

where $\frac{\partial\rho}{\partial\mathrm{x}}$ denotes the partial derivative vector of the distance from the PSP to the obstacle,

$$\frac{\partial\rho}{\partial\mathrm{x}} = \left[\frac{\partial\rho}{\partial x}\,\frac{\partial\rho}{\partial y}\,\frac{\partial\rho}{\partial z}\right]^T. \tag{19}$$

The joint forces corresponding to $\mathrm{F}^*_{(O,psp)}$ are obtained using the Jacobian matrix associated with this PSP. Observing Eqs. (6) and (9), these forces are given by

$$\Gamma_{(O,psp)} = J^T_{psp}(\mathrm{q})\Lambda(\mathrm{x})\mathrm{F}^*_{(O,psp)}. \tag{20}$$

5. Obstacle Geometric Modelling

Obstacles are described by the composition of *primitives.* A typical geometric model base includes primitives such as a point, line, plane, ellipsoid, parallelepiped, cone, and cylinder. The first artificial potential field (Eq. 16) requires analytic equations for the description of obstacles. We have developed analytic equations representing envelopes which best approximate the shapes of primitives such as a *parallelepiped, finite cylinder,* and *cone.*

The surfaee, termed an *n-ellipsoid,* is represented by the equation,

$$\left(\frac{x}{a}\right)^{2n} + \left(\frac{y}{b}\right)^{2n} + \left(\frac{z}{c}\right)^{2n} = 1, \tag{21}$$

and tends to a parallelepiped of dimensions (2a, 2b, 2c) as n tends to infinity. A good approximation is obtained with $n = 4$, as shown in Fig. 1. A cylinder of elliptical cross section ($2a$, $2b$) and of length $2c$ can be approximated by the so-called *n-cylinder* equation,

$$\left(\frac{x}{a}\right)^{2} + \left(\frac{y}{b}\right)^{2} + \left(\frac{z}{c}\right)^{2n} = 1. \tag{22}$$

The analytic description of primitives is not necessary for the artificial potential field (Eq. 17) since the continuity and differentiability requirement is on the shortest distance to the obstacle. The primitives above, and more generally all convex primitives, comply with this requirement.

Determining the orthogonal distance to an n-ellipsoid or to an n-cylinder requires the solution of a complicated system of equations. A variational procedure for the distance evaluation has been developed that avoids this costly computation. The distance expressions for other primitives are presented in Appendices I through III.

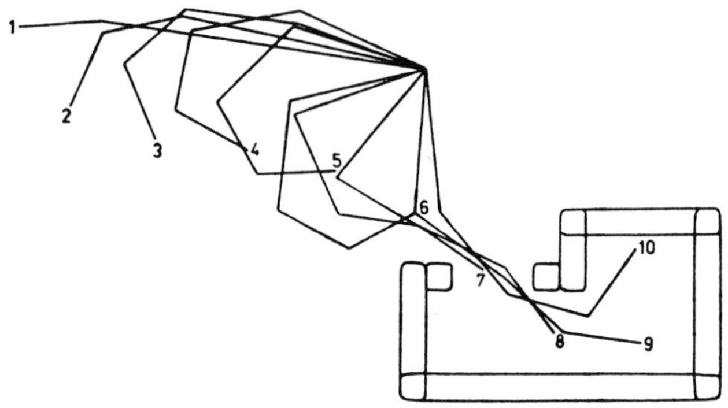

Fig. 2. Displacement of a 4 dof manipulator inside an enclosure.

6. Robot Obstacle Avoidance

An obstacle O_i is described by a set of primitives $\{P_p\}$. The superposition property (additivity) of potential fields enables the control of a given point of the manipulator with respect to this obstacle by using the sum of the relevant gradients,

$$\mathbf{F}^*_{O_i,psp} = \sum_p \mathbf{F}^*_{(P_p,psp)}. \tag{23}$$

Control of this point for several obstacles is obtained using

$$\mathbf{F}^*_{psp} = \sum_i \mathbf{F}^*_{(O_i,psp)}. \tag{24}$$

It is also feasible to have different points on the manipulator controlled with respect to different obstacles. The resulting joint force vector is given by

$$\Gamma_{obstacles} = \sum_j J^T_{psp_j}(\mathbf{q})\Lambda(\mathbf{x})\mathbf{F}^*_{psp_j}. \tag{25}$$

Specifying an adequate number of PSPs enables the protection of all of the manipulator's parts. An example of a dynamic simulation for a redundant 4 *dof* manipulator operating in the plane is shown in the display of Fig. 2.

The artificial potential field approach can be extended to *moving obstacles* since stability of the mechanism persists with a continuously time-varying potential field.

The manipulator obstacle avoidance problem has been formulated in terms of *collision avoidance of links* rather than points. Link collision avoidance is achieved by continuously controlling the link's closest point to the obstacle. At most, n PSPs then have to be considered. Additional links can be artificially introduced or the length of the last link can be extended to account for the manipulator tool or load. In an articulated chain, a link can be represented as the line segment defined by the Cartesian positions of its two neighboring joints. In a frame of reference R, a point $m(x, y, z)$ of the link bounded by $m_1(x_1, y_1, z_1)$ and $m_2(x_2, y_2, z_2)$ is described by the parametric equations,

$$\begin{aligned} x &= x_1 + \lambda(x_2 - x_1), \\ y &= y_1 + \lambda(y_2 - y_1), \\ z &= z_1 + \lambda(z_2 - z_1). \end{aligned} \tag{26}$$

The problem of obtaining the link's shortest distance to a parallelepiped can be reduced to that of finding the link's closest point to a vertex, edge, or face. The analytic expressions of the link's closest point, the distance, and its partial derivatives are given in Appendix I. In Appendices II and III these expressions are given for a cylinder and a cone, respectively.

7. Joint Limit Avoidance

The potential field approach can be used to satisfy the manipulator internal joint constraints. Let $\underline{q}_i$ and $\overline{q}_i$ be respectively the minimal and maximal bounds of the ith joint coordinate q_i. These boundaries can contain q_i by creating barriers of potential at each of the hyperplanes $(q_i = \underline{q}_i)$ and $(q_i = \overline{q}_i)$. The corresponding joint forces are

$$\gamma_{\underline{q}_i} = \begin{cases} \eta \left(\dfrac{1}{\underline{\rho}_i} - \dfrac{1}{\underline{\rho}_{i(0)}} \right) \dfrac{1}{\underline{\rho}_i^2} & \text{if } \underline{\rho}_i \leq \underline{\rho}_{i(0)}, \\ 0 & \text{if } \underline{\rho}_i > \underline{\rho}_{i(0)}, \end{cases} \tag{27}$$

and

$$\gamma_{\bar{q}_i} = \begin{cases} -\eta \left(\dfrac{1}{\bar{\rho}_i} - \dfrac{1}{\bar{\rho}_{i(0)}} \right) \dfrac{1}{\bar{\rho}_i^2} & \text{if } \bar{\rho}_i \leq \bar{\rho}_{i(0)}, \\ 0 & \text{if } \bar{\rho}_i > \bar{\rho}_{i(0)}, \end{cases} \tag{28}$$

where $\underline{\rho}_{i(0)}$ and $\bar{\rho}_{i(0)}$ represent the distance limit of the potential field influence. The distances $\underline{\rho}_i$ and $\bar{\rho}_i$ are defined by

$$\begin{aligned} \underline{\rho}_i &= q_i - \underline{q}_i, \\ \bar{\rho}_i &= \bar{q}_i - q_i. \end{aligned} \tag{29}$$

8. Level of Competence

The potential field concept is indeed an attractive approach to the collision avoidance problem and much research has recently been focused on its applications to robot control (Kuntze and Schill 1982; Hogan 1984; Krogh 1984). However, the complexity of tasks that can be implemented with this approach is limited. In a cluttered environment, local minima can occur in the resultant potential field. This can lead to a stable positioning of the robot before reaching its goal. While local procedures can be designed to exit from such configurations, limitations for complex tasks will remain. This is because the approach has a *local* perspective of the robot environment.

Nevertheless, the resulting potential field does provide the global information necessary and a collision-free path, if attainable, can be found by linking the absolute minima of the potential. Linking these minima requires a computationally expensive exploration of the potential field. This goes beyond the real-time control we are concerned with here but can be considered as an integrated part of higher level control. Work on high level collision-free path planning based on the potential field concept has been investigated in Buckley (1985).

9. Real-Time Implementation

Finally, the global control system integrating the potential field concept with the operational space approach has the following structure:

$$\Gamma = \Gamma_{motion} + \Gamma_{obstacles} + \Gamma_{joint-limit}, \tag{30}$$

where Γ_{motion} can be developed (Khatib 1983) in the form,

$$\begin{aligned} \Gamma_{motion} = {} & J^T(\mathrm{q})\Lambda(\mathrm{q})\mathrm{F}_m^* + \tilde{B}(\mathrm{q})[\dot{\mathrm{q}}\dot{\mathrm{q}}] \\ & + \tilde{C}(\mathrm{q})[\dot{\mathrm{q}}^2] + \mathrm{g}(\mathrm{q}), \end{aligned} \tag{31}$$

where $\tilde{B}(\mathrm{q})$ and $\tilde{C}(\mathrm{q})$ are the $n \times n(n-1)/2$ and $n \times n$ matrices of the joint forces under the mapping into joint space of the end-effector Coriolis and centrifugal forces. The symbol $\mathrm{g}(\mathrm{q})$ is the gravity force vector, and the symbolic notations $[\dot{\mathrm{q}}\dot{\mathrm{q}}]$ and $[\dot{\mathrm{q}}^2]$ are for the $n(n-1)/2 \times 1$ and $n \times 1$ column matrices,

$$\begin{aligned} [\dot{\mathrm{q}}\dot{\mathrm{q}}] &= [\dot{q}_1\dot{q}_2\ \dot{q}_1\dot{q}_3\ \ldots\ \dot{q}_{n-1}\ \dot{q}_n]^T, \\ [\dot{\mathrm{q}}^2] &= [\dot{q}_1^2\ \dot{q}_2^2\ \ldots\ \dot{q}_n^2]^T. \end{aligned} \tag{32}$$

In this control structure, dynamic decoupling of the end effector is obtained using the end-effector dynamic parameters (EEDP) $\Lambda(\mathrm{q})$, $\tilde{B}(\mathrm{q})$, $\tilde{C}(\mathrm{q})$ and $\mathrm{g}(\mathrm{q})$, which are configuration dependent. In real time, these parameters can be computed at a lower rate than that of the servo control. In addition, the integration of an operational position and velocity estimator allows a reduction in the rate of end-effector position computation, which involves evaluations of the manipulator geometric model. This leads to a two-level control system architecture (Khatib 1985):

- A low rate *parameter evaluation level* that updates the end-effector dynamic coefficients, the Jacobian matrix, and the geometric model.
- A high rate *servo control level* that computes the command vector using the estimator and the updated dynamic coefficients.

The control system architecture is shown in Fig. 3 where *np* represents the number of PSPs. The Jacobian matrices $J_{psp_j}^T(\mathrm{q})$ have common factors with the end-effector Jacobian matrix $J^T(\mathrm{q})$. Thus, their evaluation does not require significant additional computation.

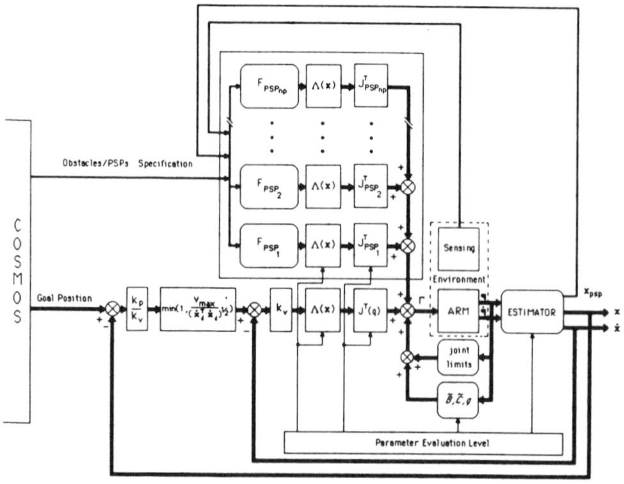

Fig. 3. Control system architecture.

10. Applications

This approach has been implemented in an experimental manipulator programming system *Control in Operational Space of a Manipulator-with-Obstacles System* (COSMOS). Demonstration of real-time collision avoidance with links and moving obstacles (Khatib et al. 1984) have been performed using a PUMA 560 and a Machine Intelligence Corporation vision module.

We have also demonstrated real-time end-effector motion and active force control operations with the COSMOS system using wrist and finger sensing. These include contact, slide, insertion, and compliance operations (Khatib, Burdick, and Armstrong 1985).

In the current multiprocessor implementation (PDP 11/45 and PDP 11/60), the rate of the servo control level is 225 Hz while the coefficient evaluation level runs at 100 Hz.

11. Summary and Discussion

We have described the formulation and implementation of a real-time obstacle avoidance approach based on the artificial potential field concept. Collision avoidance, generally treated as high level planning, has been demonstrated to be an effective component of low level real-time control in this approach. Further, we have briefly presented our operational space formulation of manipulator control that provides the basis for this obstacle avoidance approach, and have described the two-level architecture designed to increase the real-time performance of the control system.

The integration of this low level control approach with a high level planning system seems to be one of the more promising solutions to the obstacle avoidance problem in robot control. With this approach, the problem may be treated in two stages:

- at high level control, generating a global strategy for the manipulator's path in terms of intermediate goals (rather than finding an accurate collision-free path);
- at the low level, producing the appropriate commands to attain each of these goals, taking into account the detailed geometry and motion of manipulator and obstacle, and making use of real-time obstacle sensing (low level vision and proximity sensors).

By extending low level control capabilities and reducing the high level path planning burden, the integration of this collision avoidance approach into a multi-level robot control structure will improve the real-time performance of the overall robot control system. Potential applications of this control approach include moving obstacle avoidance, collision avoidance in grasping operations, and obstacle avoidance problems involving multimanipulators or multifingered hands.

Appendix I: Link Distance to a Parallelepiped

The axes of the frame of reference R are chosen to be the parallelepiped axes of symmetry. The link's length is l and the dot product is $(\cdot)$.

Distance to a Vertex

The closest point m of the line (Eq. 26) to the vertex v is such that

$$\lambda = \frac{(\mathrm{vm}_1) \cdot (\mathrm{m}_1\mathrm{m}_2)}{l^2}. \qquad \text{(A1-1)}$$

The link's closest point m is identical to m_1 if $\lambda \leq 0$; it is identical to m_2 if $\lambda \geq 1$ and it is given by Eq. (26) otherwise. The shortest distance is therefore,

$$\rho = \begin{cases} [\rho_1^2 - \lambda^2 l^2]^{1/2}, & \text{if } 0 \leq \lambda \leq 1, \\ \rho_1, & \text{if } \lambda < 0, \\ \rho_2 & \text{if } \lambda > 1, \end{cases} \quad \text{(A1-2)}$$

where ρ_1 and ρ_2 are the distance to the vertex from m_1 and m_2, respectively. The distance partial derivatives are

$$\frac{\partial \rho}{\partial \mathrm{x}} = \left[\frac{x}{\rho}\ \frac{y}{\rho}\ \frac{z}{\rho}\right]^T. \quad \text{(A1-3)}$$

Distance to an Edge

By a projection in the plane perpendicular to the considered edge (xoy, yoz, or zox), this problem can be reduced to that of finding the distance to a vertex in the plane. This leads to expressions similar to those of (A1-1)–(A1-3) with a zero partial derivative of the distance *w.r.t.* the axis parallel to the edge.

Distance to a Face

In this case, the distance can be directly obtained by comparing the absolute values of the coordinates of m_1 and m_2 along the axis perpendicular to the face. The partial derivative vector is identical to the unit normal vector of this face.

Appendix II: Link Distance to a Cylinder

The frame of reference R is chosen such that its z-axis is the cylinder axis of symmetry and its origin is the cylinder center of mass. The cylinder radius and height are designated by r and h, respectively.

Distance to the Circular Surface

The closest point of the link (Eq. 27) to the circular surface of the cylinder can be deduced from the distance to a vertex considered in the xoy plane and by allowing for the radius r.

Distance to the Circular Edges

The closest distance to the cylinder circular edge can be obtained from that of the circular surface by taking into account the relative z-coordinate of m to the circular edge, *i.e.*, $(z + h/2)$ for the base and $(z - h/2)$ for the top. The distance partial derivative vector results from the torus equation,

$$[x^2 + y^2 + (z \pm h/2)^2 - r^2 - \rho^2]^2 = 4r^2[\rho^2 - (z \pm h/2)^2]. \quad \text{(A2-1)}$$

This vector is

$$\frac{\partial \rho}{\partial \mathrm{x}} = \left[\zeta\frac{x}{\rho}\ \zeta\frac{y}{\rho}\ \frac{z \pm h/2}{\rho}\right]^T, \quad \text{(A2-2)}$$

with

$$\zeta = \frac{x^2 + y^2 + (z \pm h/2)^2 - r^2 - \rho^2}{x^2 + y^2 + (z \pm h/2)^2 + r^2 - \rho^2}. \quad \text{(A2-3)}$$

The distance to the planar surfaces is straightforward and can be simply obtained as in Appendix I.

Appendix III: Link Distance to a Cone

In this case, the frame of reference R is chosen such that its z-axis is the cone axis of symmetry and its origin is the center of the cone circular base. The cone base radius, height and half angle are represented respectively by r, h, and β.

Distance to the Cone-Shaped Surface

The problem of locating $m(x, y, z)$ is identical to that for the cylinder case. The distance can be written as

$$\rho = z \sin(\beta) + (\sqrt{x^2 + y^2} - r) \cos(\beta). \quad \text{(A3-1)}$$

The partial derivatives come from the equation,

$$x^2 + y^2 = r_z^2, \tag{A3-2}$$

where

$$r_z = \tan(\beta)[h + \rho \sin(\beta) - z]. \tag{A3-3}$$

They are

$$\frac{\partial \rho}{\partial x} = \left[\frac{x}{r_z \tan(\beta)} \; \frac{y}{r_z \tan(\beta)} \; \frac{1}{\sin(\beta)}\right]^T. \tag{A3-4}$$

The problem of the distance to the cone circular edge is identical to that of the cylinder circular edge in Appendix II. The distance to the cone vertex is solved as in Appendix I.

Acknowledgments

Tom Binford and Bernie Roth have encouraged and given support for the continuation of this research at Stanford University. I thank also Harlyn Baker, Peter Blicher, and Jeff Kerr for their help in the preparation of the original manuscript.

References

Brooks, R. 1983. Solving the find-path problem by good representation of free space. *IEEE Sys., Man Cyber.* SMC-13:190–197.

Brooks, R. 1984 Aug. 20–23, Kyoto, Japan. Aspects of mobile robot visual map making. *2nd Int. Symp. Robotics Res.*

Buckley, C. 1985. *The application of continuum methods to path planning.* Ph.D. Thesis (in progress). Stanford University, Department of Mechanical Engineering.

Chatila, R. 1981. *Système de navigation pour un robot mobile autonome: modélisation et processus décisionnels.* Thèse de Docteur-Ingénieur. Université Paul Sabatier. Toulouse, France.

Hogan, N. 1984 June 6–8, San Diego, California. Impedance control: an approach to manipulation. 1984 *Amer. Control Conf.*

Khatib, O., Llibre, M. and Mampey, R. 1978. Fonction decision-commande d'un robot manipulateur. *Rapport No. 2/7156.* DERA/CERT. Toulouse, France.

Khatib, O. and Le Maitre, J. F. 1978 September 12–15, Udine, Italy. Dynamic control of manipulators operating in a complex environment. *Proc. 3rd CISM-IFToMM Symp. Theory Practice Robots Manipulators,* 267-282. Elsevier. 1979.

Khatib, O. 1980. *Commande dynamique dans l'espace opérationnel des robots manipulateurs en présence d'obstacles.* These de Docteur-Ingénieur. École Nationale Supérieure de l'Aéronautique et de l'Espace (ENSAE). Toulouse, France.

Khatib, O. 1983 December 15–20, New Delhi. Dynamic control of manipulators in operational space. *6th CISM-IFToMM Congress Theory Machines Mechanisms,* 1128–1131.

Khatib, O., et al. 1984 June, *Robotics in three acts* (Film). Stanford University. Artificial Intelligence Laboratory.

Khatib, O. 1985 September 11–13, Tokyo. The operational space formulation in robot manipulators control. *15th Int. Symp. Indust. Robots.*

Khatib, O. and Burdick, J. 1985 November, Miami, Florida. Dynamic optimization in manipulator design: the operational space formulation. *ASME Winter Annual Meeting.*

Khatib, O., Burdick, J., and Armstrong, B. 1985. *Robotics in three acts–Part II* (Film). Stanford University, Artificial Intelligence Laboratory.

Krogh, B. 1984 August, Bethlehem, Pennsylvania. A generalized potential field approach to obstacle avoidance control. *SME Conf. Proc. Robotics Research: The Next Five Years and Beyond.*

Kuntze, H. B., and Schill, W. 1982 June 9–11, Paris. Methods for collision avoidance in computer controlled industrial robots. *12th ISIR.*

Lozano-Perez, T. 1980. Spatial planning: a configuration space approach. *AI Memo 605.* Cambridge, Mass. MIT Artificial Intelligence Laboratory.

Moravec, H. P. 1980. *Obstacle avoidance and navigation in the real world by a seeing robot rover.* Ph.D. Thesis. Stanford University, Artificial Intelligence Laboratory.

7. Systems

Things that we have to learn to do we learn by doing them.

Aristotle, *Nicomachean Ethics*, Bk. ii, Ch. 1 (335 BC)

The difficulties encountered when one is actually building an autonomous robot vehicle are twofold. Not only must all the technical issues discussed in the preceding chapters be resolved, but problems associated with any complex systems engineering project must also be dealt with. These include problems of reliability, self-test, fault-tolerance, and software engineering, particularly maintainability and expandability. It is not surprising then that many demonstrations consist of a video tape! We believe it is important to build systems. Often, the interaction between the different subsystems cannot be predicted or simulated beforehand. The final chapter then is a description of four autonomous robot vehicle projects.

Some of the earliest work on autonomous mobile robots was performed by Nilsson [2]. We begin with a review by Moravec of work on mobile robots beginning in the early 1960s at Stanford University and continuing at Carnegie Mellon University in the early 1980s. Follow-up work by Moravec is reported in [1]. This is followed by a detailed description of many of the components that make up the HILARE mobile robot project. Next, Wilcox and Gennery provide a description of the JPL Rover, an autonomous robot vehicle designed for interplanetary exploration. Last, but by no means least, we return to Stanford University and a paper by Kriegman et al. on their current autonomous vehicle project.

This is just a small sample of the many systems that have been developed. Descriptions of other systems can be found in a diverse set of literature, including [3–9] to name but a few.

References

1. Moravec, H.P. Three Degrees for a Mobile Robot. In *Proceedings of the ASME on Advanced Automation: 1984 and Beyond*, Vol. 1, ASME, 1984, pp. 274-278.
2. Nilsson, N.J. A Mobile Automaton: An Application of Artificial Intelligence Techniques. In *International Joint Conference on Artificial Intelligence*, 1969, pp. 509-520.
3. *IEEE Trans. Robotics and Automation.*
4. *International J. Robotics Research.*
5. *J. Robotic Systems.*

6. *Proceedings of the IEEE International Conference on Robotics and Automation.*

7. *Proceedings of the IEEE/RSJE International Workshop on Intelligent Robots and Systems.*

8. *Proceedings of the International Joint Conference on Artificial Intelligence.*

9. *Proceedings of the SPIE Conference on Mobile Robots.*

The Stanford Cart and the CMU Rover

HANS P. MORAVEC, MEMBER, IEEE

Invited Paper

Abstract—The Stanford Cart was a remotely controlled TV-equipped mobile robot. A computer program was written which drove the Cart through cluttered spaces, gaining its knowledge of the world entirely from images broadcast by an on-board TV system. The CMU Rover is a more capable, and nearly operational, robot being built to develop and extend the Stanford work and to explore new directions.

The Cart used several kinds of stereopsis to locate objects around it in three dimensions and to deduce its own motion. It planned an obstacle-avoiding path to a desired destination on the basis of a model built with this information. The plan changed as the Cart perceived new obstacles on its journey.

The system was reliable for short runs, but slow. The Cart moved 1 m every 10 to 15 min, in lurches. After rolling a meter it stopped, took some pictures, and thought about them for a long time. Then it planned a new path, executed a little of it, and paused again. It successfully drove the Cart through several 20-m courses (each taking about 5 h) complex enough to necessitate three or four avoiding swerves; it failed in other trials in revealing ways.

The Rover system has been designed with maximum mechanical and control system flexibility to support a wide range of research in perception and control. It features an omnidirectional steering system, a dozen on-board processors for essential real-time tasks, and a large remote computer to be helped by a high-speed digitizing/data playback unit and a high-performance array processor. Distributed high-level control software similar in organization to the Hearsay II speech-understanding system and the beginnings of a vision library are being readied.

By analogy with the evolution of natural intelligence, we believe that incrementally solving the control and perception problems of an autonomous mobile mechanism is one of the best ways of arriving at general artificial intelligence.

Introduction

Experience with the Stanford Cart [8], [9], [11], a minimal computer-controlled mobile camera platform, suggested to me that, while maintaining such a complex piece of hardware was a demanding task, the effort could be worthwhile from the point of view of artificial intelligence and computer vision research. A roving robot is a source of copious and varying visual and other sensory data which force the development of general techniques if the controlling programs are to be even minimally successful. By contrast, the (also important) work with disembodied data and fixed robot systems often focuses on relatively restricted stimuli and small image sets, and improvements tend to be in the direction of specialization. Drawing an analogy with the natural world, I believe it is no mere coincidence that in all cases imaging eyes and large brains evolved in animals that first developed high mobility.

Manuscript received December 17, 1982; revised February 23, 1983. The Stanford Cart work conducted at the Stanford University Artificial Intelligence Laboratory, was supported over the years 1973–1980 by the Defense Advanced Research Projects Agency, the National Science Foundation, and the National Aeronautics and Space Administration. The CMU Rover has been supported at the Carnegie-Mellon University Robotics Institute since 1981 by the Office of Naval Research under Contract N00014-81-0503.

The author is with the Robotics Institute, Carnegie-Mellon University, Pittsburgh, PA 15213.

Fig. 1. The Stanford Cart.

Fig. 2. The Cart on an obstacle course.

The Stanford Cart

The Cart [10] was a minimal remotely controlled TV-equipped mobile robot (Fig. 1) which lived at the Stanford Artificial Intelligence Laboratory (SAIL). A computer program was written which drove the Cart through cluttered spaces, gaining its knowledge of the world entirely from images broadcast by the on-board TV system (Fig. 2).

The Cart used several kinds of stereo vision to locate objects around it in three dimensions (3D) and to deduce its own motion. It planned an obstacle-avoiding path to a desired destination on the basis of a model built with this information. The plan changed as the Cart perceived new obstacles on its journey.

The system was reliable for short runs, but slow. The Cart moved 1 m every 10 to 15 min, in lurches. After rolling a meter it stopped, took some pictures, and thought about them for a long time. Then it planned a new path, executed a little of it, and paused again.

It successfully drove the Cart through several 20-m courses (each taking about 5 h) complex enough to necessitate three or four avoiding swerves. Some weaknesses and possible im-

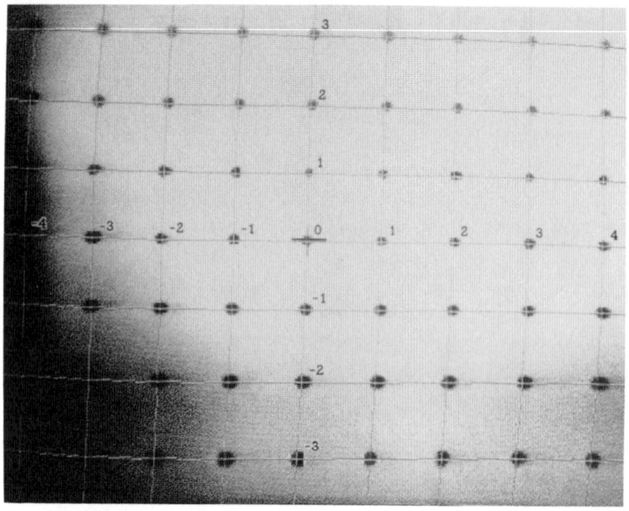

Fig. 3. A Cart's eye view of the calibration grid. The Cart camera's focal length and distortions were determined by parking the Cart a precise distance in front of, and aiming its camera at, a carefully painted array of spots pattern, and running a calibration program. The program located the spots in the image and fitted a 2D, third-degree polynomial which converted actual positions in the image to coordinates in an ideal unity focal length camera. The picture presented here was obtained by running a corresponding grid in the unity focal length frame through the inverse function of the polynomial so obtained and superimposing it on the raw spot image.

provements were suggested by these and other, less successful, runs.

A Cart Run

A run began with a calibration of the Cart's camera. The Cart was parked in a standard position in front of a wall of carefully painted spots. A calibration program noted the disparity in position of the spots in the image seen by the camera with their position predicted from an idealized model of the situation. It calculated a distortion correction polynomial which related these positions, and which was used in subsequent ranging calculations (Fig. 3).

The Cart was then manually driven to its obstacle course (littered with large and small debris) and the obstacle-avoiding program was started. It began by asking for the Cart's destination, relative to its current position and heading. After being told, say, 50 m forward and 20 to the right, it began its maneuvers. It activated a mechanism which moved the TV camera, and digitized nine pictures as the camera slid in precise steps from one side to the other along a 50-cm track.

A subroutine called the *Interest Operator* was applied to one of these pictures. It picked out 30 or so particularly distinctive regions (features) in this picture. Another routine called the *Correlator* looked for these same regions in the other frames (Fig. 4). A program called the *Camera Solver* determined the 3D position of the features with respect to the Cart from their apparent movement from image to image (Fig. 5).

The *Navigator* planned a path to the destination which avoided all the perceived features by a large safety margin. The program then sent steering and drive commands to the Cart to move it about a meter along the planned path. The Cart's response to such commands was not very precise. The camera was then operated as before, and nine new images were acquired. The control program used a version of the Correlator to find as many of the features from the previous location as possible in the new pictures, and applied the camera solver. The program

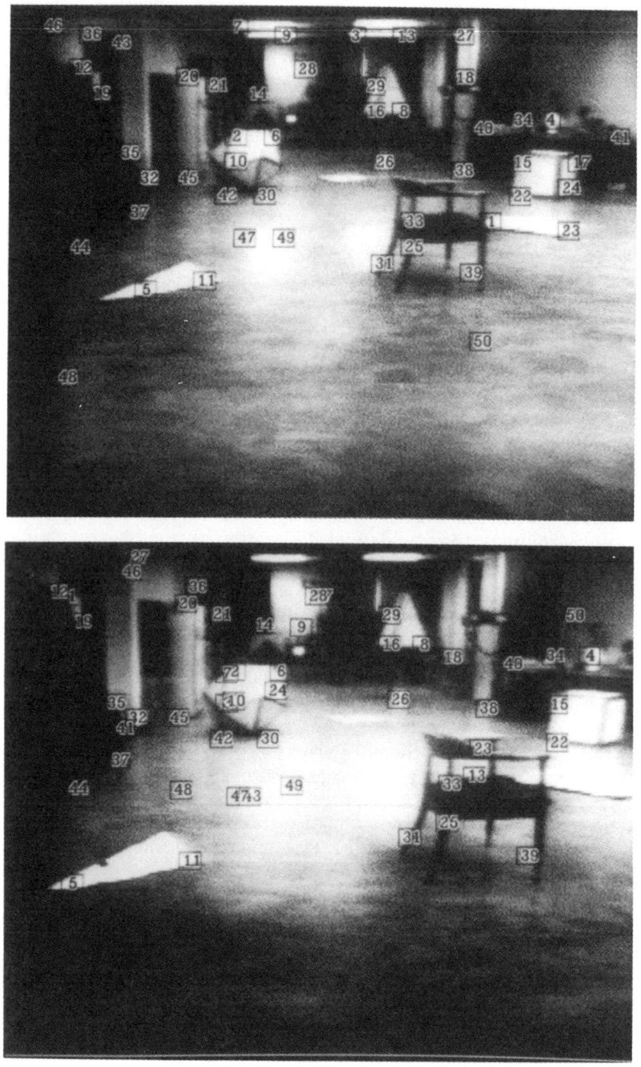

Fig. 4. Interest Operator and Correlator results. The upper picture shows points picked out by an application of the Interest Operator. The lower picture shows the Correlator's attempt to find the same points in an image of the same scene taken from a different point of view.

then deduced the Cart's actual motion during the step from the apparent 3D shift of these features. Some of the features were pruned during this process, and the Interest Operator was invoked to add new ones.

This repeated until the Cart arrived at its destination or until some disaster terminated the program. Figs. 6 and 7 document the Cart's internal world model at two points during a sample run.

Some Details

The Cart's vision code made extensive use of a reductions of each acquired image. Every digitized image was stored as the original picture accompanied by a pyramid of smaller versions of the image reduced in linear size by powers of two, each successive reduction obtained from the last by averaging four pixels into one.

Camera Calibration

The camera's focal length and geometric distortion were determined by parking the Cart a precise distance in front of a wall of many spots and one cross. A program digitized an

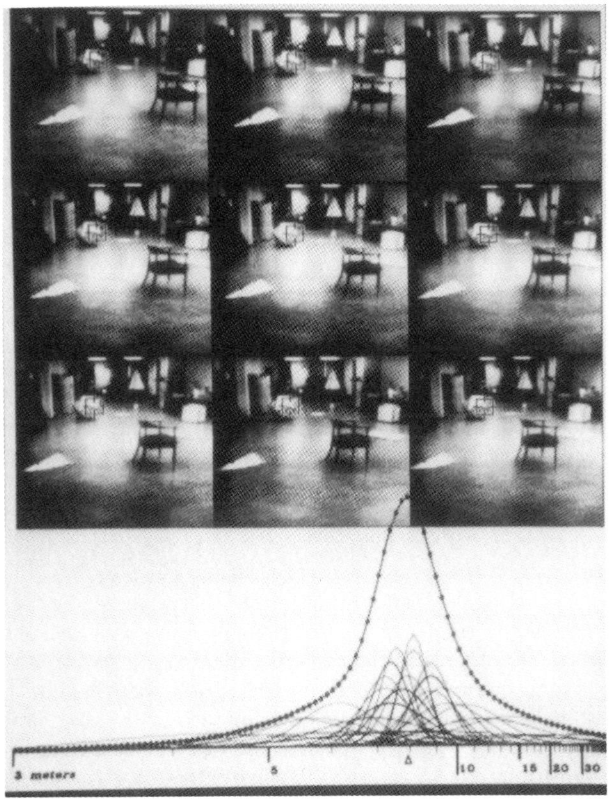

Fig. 5. Slider stereo. A typical ranging. The nine pictures are from a slider scan. The Interest Operator chose the marked feature in the central image, and the Correlator found it in the other eight. The small curves at bottom are distance measurements of the feature made from pairs of the images. The large beaded curve is the sum of the measurements over all 36 pairings. The horizontal scale is linear in inverse distance.

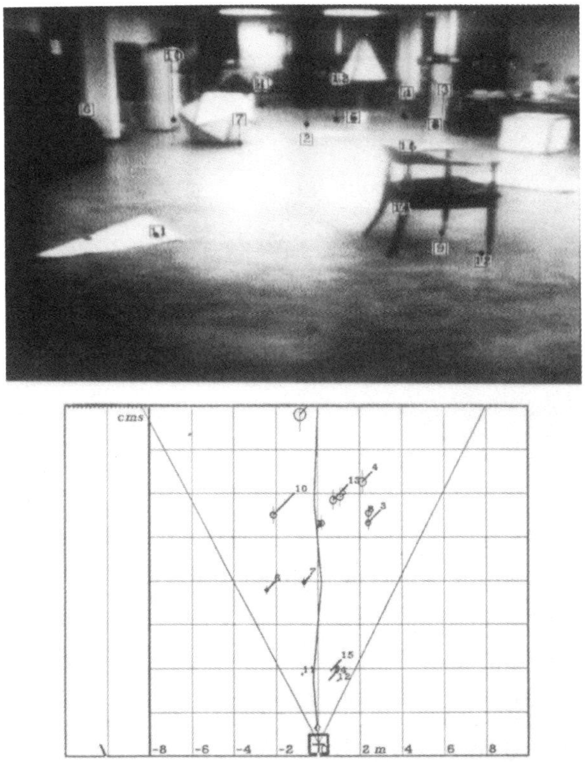

Fig. 6. A Cart obstacle run. This and the following diagram are plan views of the Cart's internal world model during a run of the obstacle-avoiding program. The grid cells are 2 m^2, conceptually on the floor. The Cart's own position is indicated by the small heavy square, and by the graph, indicating height, calibrated in centimeters, to the left of the grid. Since the Cart never actually leaves or penetrates the floor, this graph provides an indication of the overall accuracy. The irregular, tick marked, line behind the Cart's position is the past itinerary of the Cart as deduced by the program. Each tick mark represents a stopping place. The picture at top of the diagrams is the view seen by the TV camera. The two rays projecting forward from the Cart position show the horizontal boundaries of the camera's field of view (as deduced by the camera calibration program). The numbered circles in the plan view are features located and tracked by the program. The centers of the circles are the vertical projections of the feature positions onto the ground. The size of each circle is the uncertainty (caused by finite camera resolution) in the feature's position. The length of the 45° line projecting to the upper right, and terminated by an identifying number, is the height of the feature above the ground, to the same scale as the floor grid. The features are also marked in the camera view, as numbered boxes. The thin line projecting from each box to a lower blob is a stalk which just reaches the ground, in the spirit of the 45° lines in the plan view. The irregular line radiating forwards from the Cart is the planned future path. This changes from stop to stop, as the Cart fails to obey instructions properly, and as new obstacles are detected. The small ellipse a short distance ahead of the Cart along the planned path is the planned position of the next stop.

image of the spot array, located the spots and the cross, and constructed a two-dimensional (2D) polynomial that related the position of the spots in the image to their position in an ideal unity focal length camera, and another polynomial that converted points from the ideal camera to points in the image. These polynomials were used to correct the positions of perceived objects in later scenes (Fig. 3).

The algorithm began by determining the array's approximate spacing and orientation. It reduced by averaging and trimmed the picture to 64 by 64, calculated the Fourier transform of the reduced image, and took its power spectrum, arriving at a 2D transform symmetric about the origin, and having strong peaks at frequencies corresponding to the horizontal and vertical as well as half-diagonal spacings, with weaker peaks at the harmonics. It multiplied each point $[i, j]$ in this transform by point $[-j, i]$ and points $[j - i, j + i]$ and $[i + j, j - i]$, effectively folding the primary peaks onto one another. The strongest peak in the 90° wedge around the y axis gave the spacing and orientation information needed by the next part of the process.

The Interest Operator described later was applied to roughly locate a spot near the center of the image. A special operator examined a window surrounding this position, generated a histogram of intensity values within the window, decided a threshold for separating the black spot from the white background, and calculated the centroid and first and second moment of the spot. This operator was again applied at a displacement from the first centroid indicated by the orientation and spacing of the grid, and so on, the region of found spots growing outward from the seed.

A binary template for the expected appearance of the cross in the middle of the array was constructed from the orientation/spacing data from the Fourier transform. The area around each of the found spots was thresholded on the basis of the expected cross area, and the resulting two-valued pattern was convolved with the cross template. The closest match in the central portion of the picture was declared to be the origin.

Two least squares polynomials (one for X and one for Y) of third (or sometimes fourth) degree in two variables, relating the actual positions of the spots to the ideal positions in a unity focal length camera, were then generated and written into a file. The polynomials were used in the obstacle avoider to cor-

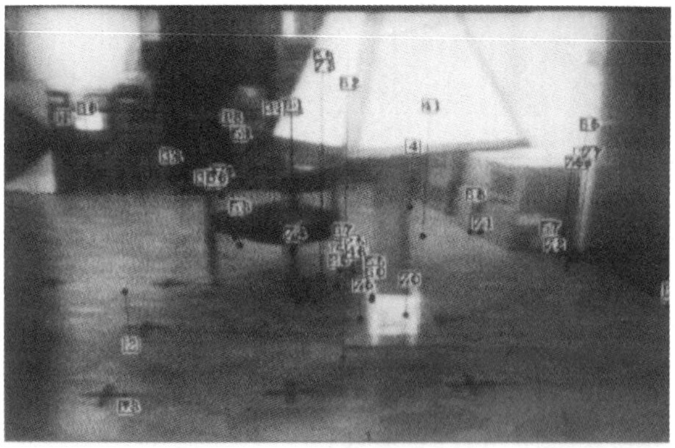

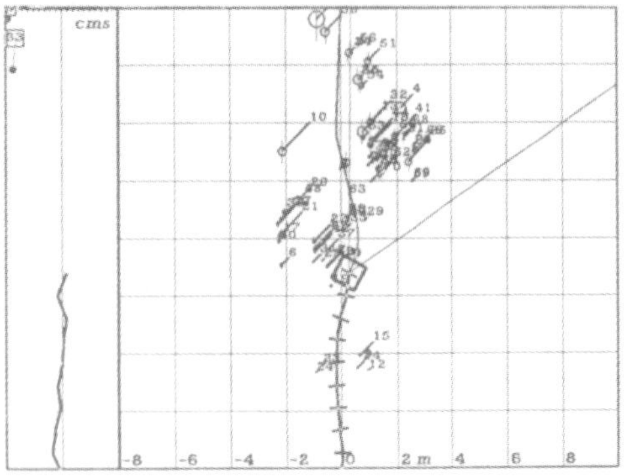

Fig. 7. After the eleventh lurch the Cart has rounded the chair, the icosahedron, and is working on the cardboard tree. The world model has suffered some accumulated drift error, and the oldest acquired features are considerably misplaced.

rect for camera roll, tilt, lens focal length, and long-term variations in the vidicon geometry.

Interest Operator

The Cart vision code dealt with localized image patches called features. A feature is conceptually a point in the 3D world, but it was found by examining localities larger than points in pictures. A feature was good if it could be located unambiguously in different views of a scene. A uniformly colored region or a simple edge is not good because its parts are indistinguishable. Regions, such as corners, with high contrast in orthogonal directions are best.

New features in images were picked by a subroutine called the *Interest Operator*, which returned regions that were local maxima of a directional variance measure, defined below. The idea was to select a relatively uniform scattering of good features over the image, so that a few would likely be picked on every visible object while textureless areas and simple edges were avoided.

Directional variance was measured over small square windows. Sums of squares of differences of pixels adjacent in each of four directions (horizontal, vertical, and two diagonals) over each window were calculated, and the window's interest measure was the minimum of these four sums. Features were chosen where the interest measure had local maxima. The chosen features were stored in an array, sorted in order of decreasing interest measure (Fig. 4–top).

Once a feature was chosen, its appearance was recorded as series of excerpts from the reduced image sequence. A 6 by 6 window was excised around the feature's location from each of the variously reduced pictures. Only a tiny fraction of the area of the original (unreduced) image was extracted. Four times as much area (but the same number of pixels) of the X2 reduced image was stored, sixteen times as much of the X4 reduction, and so on, until at some level we had the whole image. The final result was a series of 6 by 6 pictures, beginning with a very blurry rendition of the whole picture, gradually zooming in linear expansions of two, to a sharp closeup of the feature.

Correlation

Deducing the 3D location of features from their projections in 2D images requires that we know their position in two or more such images. The *Correlator* was a subroutine that, given a feature description produced by the interest operator from one image, found the best match in a different, but similar, image. Its search area could be the entire new picture, or a rectangular subwindow.

The search used a coarse to fine strategy that began in reduced versions of the pictures. Typically, the first step took place at the X16 (linear) reduction level. The 6 by 6 window at that level in the feature description, that covered about one seventh of the total area of the original picture, was convolved with the search area in the correspondingly reduced version of the second picture. The 6 by 6 description patch was moved pixel by pixel over the approximately 15 by 16 destination picture, and a correlation coefficient was calculated for each trial position. The position with the best match was recorded. The 6 by 6 area it occupied in the second picture was mapped to the X8 reduction level, where the corresponding region was 12 pixels by 12. The 6 by 6 window in the X8 reduced level of the feature description was then convolved with this 12 by 12 area, and the position of best match was recorded and used as a search area for the X4 level. The process continued, matching smaller and smaller, but more and more detailed windows, until a 6 by 6 area was selected in the unreduced picture (Fig. 4–bottom).

This "divide and conquer" strategy was, in general, able to search an entire picture for a match more quickly (because most of the searching was done at high reduction levels) and more reliably (because context up to and including the entire picture guided the search) than a straightforward convolution over even a very restricted search area.

Slider Stereo

At each pause on its computer-controlled itinerary, the Cart slid its camera from left to right on a 52-cm track, taking nine pictures at precise 6.5-cm intervals. Points were chosen in the fifth (middle) of these nine images, either by the Correlator to match features from previous positions, or by the Interest Operator. The camera slid parallel to the horizontal axis of the (distortion corrected) camera coordinate system, so the parallax-induced displacement of features in the nine pictures was purely horizontal.

The Correlator was applied eight times to look for the points chosen in the central image in each of other eight pictures. The search was restricted to a narrow horizontal band. This had little effect on the computation time, but it reduced the probability of incorrect matches. In the case of correct matches, the distance to the feature was inversely proportional to its displacement from one image to another. The uncertainty in such a measurement is the difference in distance a shift one pixel in the image would make. The uncertainty varies inversely with the physical separation of the camera positions where the pictures were taken (the stereo baseline). Long baselines give more accurate distance measurements.

After the correlation step the program knew a feature's position in nine images. It considered each of the 36 (9 values taken 2 at a time) possible image pairings as a stereo baseline, and recorded the estimated (inverse) distance of the feature in a histogram. Each measurement added a little normal curve to the histogram, with mean at the estimated distance, and standard deviation inversely proportional to the baseline, reflecting the uncertainty. The area under each curve was made proportional to the product of the goodness of the matches in the two images (in the central image this quantity is taken as unity), reflecting the confidence that the correlations were correct. The distance to the feature was indicated by the largest peak in the resulting histogram, if this peak was above a certain threshold. If below, the feature was forgotten (Fig. 5).

The Correlator sometimes matched features incorrectly. The distance measurements from incorrect matches in different pictures were not consistent. When the normal curves from 36 pictures pairs are added up, the correct matches agree with each other, and build up a large peak in the histogram, while incorrect matches spread themselves more thinly. Two or three correct correlations out of the eight usually built a peak sufficient to offset a larger number of errors. In this way, eight applications of a mildly reliable Operator interacted to make a very reliable distance measurement.

Motion Stereo

After having determined the 3D location of objects at one position, the computer drove the Cart about a meter forward. At the new position, it slid the camera and took nine pictures. The Correlator was applied in an attempt to find all the features successfully located at the previous position. Feature descriptions extracted from the central image at the last position were searched for in the central image at the new stopping place.

Slider Stereo then determined the distance of the features so found from the Cart's new position. The program now knew the 3D position of the features relative to its camera at the old and the new locations. Its own movement was deduced from 3D coordinate transform that related the two.

The program first eliminated mismatches in the correlations between the central images at the two positions. Although it did not yet have the coordinate transform between the old and new camera systems, the program knew the distance between pairs of feature positions should be the same in both. It made a matrix in which element $[i, j]$ is the absolute value of the difference in distances between points i and j in the first and second coordinate systems divided by the expected error (based on the one pixel uncertainty of the ranging). Each row of this matrix was summed, giving an indication of how much each point disagreed with the other points. The idea is that while points in error disagree with virtually all points, correct positions agree with all the other correct ones, and disagree only with the bad ones. The worst point was deleted, and its effect removed from the remaining points in the row sums. This pruning was repeated until the worst error was within the error expected from the ranging uncertainty.

After the pruning, the program had a number of points, typically 10 to 20, whose position error was small and pretty well known. The program trusted these, and recorded them in its world model, unless it had already done so at a previous position. The pruned points were forgotten forevermore.

The 3D rotation and translation that related the old and new Cart position was then calculated by a Newton's method iteration that minimized the sum of the squares of the distances between the transformed first coordinates and the raw coordinates of the corresponding points at the second position, with each term divided by the square of the expected uncertainty in the 3D position of the points involved.

Path Planning

The Cart vision system modeled objects as simple clouds of features. If enough features are found on each nearby object, this model is adequate for planning a noncolliding path to a destination. The features in the Cart's 3D world model can be thought of as fuzzy ellipsoids, whose dimensions reflect the program's uncertainty of their position. Repeated applications of the Interest Operator as the Cart moves caused virtually all visible objects to be become modeled as clusters of overlapping ellipsoids.

To simplify the problem, the ellipsoids were approximated by spheres. Those spheres sufficiently above the floor and below the Cart's maximum height were projected on the floor as circles. The 1-m^2 Cart itself was modeled as a 3-m circle. The path-finding problem then became one of maneuvering the Cart's 3-m circle between the (usually smaller) circles of the potential obstacles to a desired location. It is convenient (and equivalent) to conceptually shrink the Cart to a point, and add its radius to each and every obstacle. An optimum path in this environment will consist of either a straight run between start and finish, or a series of tangential segments between the circles and contacting arcs (imagine loosely laying a string from start to finish between the circles, then pulling it tight).

The program converted the problem to a shortest path in graph search. There are four possible paths between each pair of obstacles because each tangent can approach clockwise or counterclockwise. Each tangent point became a vertex in the graph, and the distance matrix of the graph (which had an entry for each vertex pair) contained sums of tangential and arc paths, with infinities for blocked or impossible routes. The shortest distance in this space can be found with an algorithm whose running time is $O(n^3)$ in the number of vertices, and the Cart program was occasionally run using this exact procedure. It was run more often with a faster approximation that made each obstacle into only two vertices (one for each direction of circumnavigation).

A few other considerations were essential in path planning. The charted routes consisted of straight lines connected by tangent arcs, and were thus plausible paths for the Cart, which steered like an automobile. This plausibility was not necessarily true of the start of the planned route, which, as presented thus far, did not take the initial heading of the Cart into account. The plan could, for instance, include an initial segment going off 90° from the direction in which the Cart pointed, and thus be impossible to execute. This was handled by including a pair of "phantom" obstacles along with the real perceived ones. The phantom obstacles had a radius equal to the Cart's minimum steering radius, and were placed, in the planning process, on either side of the Cart at such a distance that after their radius was augmented by the Cart's radius (as happened for all the obstacles), they just touched the Cart's centroid, and each other, with their common tangents being parallel to the direction of the Cart's heading. They effectively blocked the area made inaccessible to the Cart by its maneuverability limitations (Fig. 6).

Lozano-Pérez and Wesley [7] describe an independently developed, but very similar, approach to finding paths around polygonal obstacles.

Path Execution

After the path to the destination had been chosen, a portion of it had to be implemented as steering and motor commands and transmitted to the Cart. The control system was primitive. The drive motor and steering motors could be turned on and off at any time, but there existed no means to accurately determine just how fast or how far they had gone. The program made the best of this bad situation by incorporating a model of the Cart that mimicked, as accurately as possible, the Cart's actual behavior. Under good conditions, as accurately as possible means about 20 percent; the Cart was not very repeatable, and was affected by ground slope and texture, battery voltage, and other less obvious externals.

The path executing routine began by excising the first 0.75 m of the planned path. This distance was chosen as a compromise between average Cart velocity, and continuity between picture sets. If the Cart moved too far between picture-digitizing sessions, the picture would change too much for reliable correlations. This is especially true if the Cart turns (steers) as it moves. The image seen by the camera then pans across the field of view. The Cart had a wide angle lens that covers 60° horizontally. The 0.75 m, combined with the turning radius limit (5 m) of the Cart resulted in a maximum shift in the field of view of 15°, one quarter of the entire image.

The program examined the Cart's position and orientation at the end of the desired 0.75-m lurch, relative to the starting position and orientation. The displacement was characterized by three parameters; displacement forward, displacement to the right, and change in heading. In closed form, the program computed a path that accomplished this movement in two arcs of equal radius, but different lengths. The resulting trajectory had a general "S" shape. Rough motor timings were derived from these parameters. The program then used a simulation that took into account steering and drive motor response to iteratively refine the solution.

Cart Experiments

The system described above only incompletely fulfills some of the hopes I had when the work began many years ago.

One of the most serious limitations was the excruciating slowness of the program. In spite of my best efforts, and many compromises in the interest of speed, it took 10 to 15 min of real time to acquire and consider the images at each meter long lurch, on a lightly loaded DEC KL-10. This translated to an effective Cart velocity of 3 to 5 m an hour. Interesting obstacle courses (two or three major obstacles, spaced far enough apart to permit passage within the limits of the Cart's size and maneuverability) were about 20 m long, so interesting Cart runs took 5 h.

The reliability of individual moves was high, as it had to be for a 20-lurch sequence to have any chance of succeeding, but the demanding nature of each full run and the limited amount of time available for testing (discussion of which is beyond the scope of this paper) after the bulk of the program was debugged, ensured that many potential improvements were left untried. Three full (about 20-m) runs were digitally recorded and filmed, two indoors and one outdoors. Two indoor false starts, aborted by failure of the program to perceive an obstacle, were also recorded. The two long indoor runs were nearly perfect.

In the first long indoor run, the Cart successfully slalomed its way around a chair, a large cardboard icosahedron, and a cardboard tree then, at a distance of about 16 m, encountered a cluttered wall and backed up several times trying to find a way around it (Figs. 6 and 7 are snapshots from this run).

The second long indoor run involved a more complicated set of obstacles, arranged primarily into two overlapping rows blocking the goal. I had set up the course hoping the Cart would take a long, picturesque (the runs were being filmed) "S" shaped path around the ends of the rows. To my chagrin, it instead tried for a tricky shortcut. The Cart backed up twice to negotiate the tight turn required to go around the first row, then executed several tedious steer forward/backup moves, lining itself up to go through a gap barely wide enough in the second row. This run had to be terminated, sadly, before the Cart had gone through the gap because of declining battery charge and increasing system load.

The outdoor run was less successful. It began well; in the first few moves the program correctly perceived a chair directly in front of the camera, and a number of more distant cardboard obstacles and sundry debris. Unfortunately, the program's idea of the Cart's own position became increasingly wrong. At almost every lurch, the position solver deduced a Cart motion considerably smaller than the actual move. By the time the Cart had rounded the foreground chair, its position model was so far off that the distant obstacles were replicated in different positions in the Cart's confused world model, because they had been seen early in the run and again later, to the point where the program thought an actually existing distant clear path was blocked. I restarted the program to clear out the world model when the planned path became too silly. At that time the Cart was 4 m in front of a cardboard icosahedron, and its planned path lead straight through it. The newly reincarnated program failed to notice the obstacle, and the Cart collided with it. I manually moved the icosahedron out of the way, and allowed the run to continue. It did so uneventfully, though there were continued occasional slight errors in the self-position deductions. The Cart encountered a large cardboard tree towards the end of this journey and detected a portion of it only just in time to squeak by without colliding (Fig. 2 was a photograph taken during this run).

The two short abortive indoor runs involved setups nearly identical to the two-row successful long run described one paragraph ago. The first row, about 3 m in front of the Cart's starting position contained a chair, a real tree (a small cypress in a planting pot), and a polygonal cardboard tree. The Cart saw the chair instantly and the real tree after the second move, but failed to see the cardboard tree ever. Its planned path around the two obstacles it did see put it on a collision course with the unseen one. Placing a chair just ahead of the cardboard tree fixed the problem, and resulted in a successful run. The finished program never had trouble with chairs.

Problems

These tests revealed some weaknesses in the program. The system did not see simple polygonal (bland and featureless) objects reliably, and its visual navigation was fragile under certain conditions. Examination of the program's internal workings suggested some causes and possible solutions.

The program sometimes failed to see obstacles lacking sufficient high contrast detail within their outlines. In this regard, the polygonal tree and rock obstacles I whimsically constructed to match diagrams from a 3D drawing program, were a terrible mistake. In none of the test runs did the programs ever fail to see a chair placed in front of the Cart, but half the time they did fail to see a pyramidal tree or an icosahedral rock made of clean white cardboard. These contrived obstacles were picked up reliably at a distance of 10 to 15 m, silhouetted against a relatively unmoving (over slider travel and Cart lurches) background, but were only rarely and sparsely seen at closer range, when their outlines were confused by a rapidly shifting background, and their bland interiors provided no purchase for the interest operator or correlator. Even when the artificial obstacles were correctly perceived, it was by virtue of only two to four features. In contrast, the program usually tracked five to ten features on nearby chairs.

In the brightly sunlit outdoor run, the artificial obstacles had another problem. Their white coloration turned out to be much brighter than any "naturally" occurring extended object. These super bright, glaring, surfaces severely taxed the very limited dynamic range of the Cart's vidicon/digitizer combination. When the icosahedron occupied 10 percent of the camera's field of view, the automatic target voltage circuit in the electronics turned down the gain to a point where the background behind the icosahedron appeared nearly solid black.

The second major problem exposed by the runs was glitches in the Cart's self-position model. This model was updated after a lurch by finding the 3D translation and rotation that best related the 3D position of the set of tracked features before and after the lurch. In spite of the extensive pruning that preceded this step (and partly because of it, as is discussed later), small errors in the measured feature positions sometimes caused the solver to converge to the wrong transform, giving a position error beyond the expected uncertainty. Features placed into the world model before and after such a glitch were not in the correct relative positions. Often an object seen before was seen again after, now displaced, with the combination of old and new positions combining to block a path that was in actuality open.

This problem showed up mainly in the outdoor run. I had observed it indoors in the past, in simple mapping runs, before the entire obstacle avoider was assembled. There appear to be two major causes for it, and a wide range of supporting factors.

Poor seeing, resulting in too few correct correlations between the pictures before and after a lurch, was one culprit. The highly redundant nine-eyed stereo ranging was very reliable, and caused few problems, but the nonredundant correlation necessary to relate the position of features before and after a lurch was error prone. Sometimes the mutual-distance invariance pruning that followed was overly aggressive, and left too few points for a stable least squares coordinate fit.

The outdoor runs encountered another problem. The program ran so slowly that shadows moved significantly (up to a half meter) between lurches. Their high contrast boundaries were favorite points for tracking, enhancing the program's confusion.

Quick Fixes

Though elaborate (and thus far untried in our context) methods such as edge matching may greatly improve the quality of automatic vision in the future, subsequent experiments with the program revealed some modest incremental improvements that would have solved most of the problems in the test runs.

The issue of unseen cardboard obstacles turns out to be partly one of overconservatism on the program's part. In all cases where the Cart collided with an obstacle it had correctly ranged a few features on the obstacle in the prior nine-eyed scan. The problem was that the much more fragile correlation between vehicle forward moves failed, and the points were rejected in the mutual distance test. Overall, the nine-eyed stereo produced very few errors. If the path planning stage had used the prepruning features (still without incorporating them permanently into the world model) the runs would have proceeded much more smoothly. All of the most vexing false negatives, in which the program failed to spot a real obstacle, would have been eliminated. There would have been a very few false positives, in which nonexistent ghost obstacles would have been perceived. One or two of these might have caused an unnecessary swerve or backup, but such ghosts would not pass the pruning stage, and the run would have recovered after the initial, noncatastrophic, glitch.

The self-position confusion problem is related, and in retrospect may be considered a trivial bug. When the Path Planner computed a route for the Cart, another subroutine took a portion of this plan and implemented it as a sequence of commands to be transmitted to the Cart's steering and drive motors. During this process, it ran a simulation that modeled the Cart acceleration, rate of turning, and so on, and which provided a prediction of the Cart's position after the move. With the old hardware, the accuracy of this prediction was not great, but it nevertheless provided much *a priori* information about the Cart's new position. This information was used, appropriately weighted, in the least squares coordinate system solver that deduced the Cart's movement from the apparent motion in 3D of tracked features. It was not used, however, in the mutual distance pruning step that preceeded this solving. When the majority of features had been correctly tracked, failure to use this information did not hurt the pruning. But when the seeing was poor, it could make the difference between choosing a spuriously agreeing set of mistracked features and the small correctly matched set. Incorporating the prediction into pruning, by means of a heavily weighted point that the program treats like another tracked feature, removed almost all the positioning glitches when the program was fed the pictures from the outdoor run.

More detail on all these areas can be found in [10].

The CMU Rover

The major impediments to serious extensions of the Cart work were limits to available computation, resulting in debilitatingly long experimental times, and the very minimal nature of the robot hardware, which precluded inexpensive solutions for even most basic functions (like "roll a meter forward").

We are addressing these problems at CMU in an ambitious new effort centered around a new, small but sophisticated mobile robot dubbed the CMU Rover. The project so far has been focused on developing a smoothly functional and highly capable vehicle and associated support system which will serve a wide variety of future research.

The shape, size, steering arrangements, and on-board as well as external processing capabilities of the Rover system were chosen to maximize the flexibility of the system (naturally limited by present-day techniques).

The robot is cylindrical, about a meter tall and 55 cm in diameter (Fig. 8) and has three individually steerable wheel assemblies which give it a full three degrees of freedom of mobility in the plane (Figs. 9 and 10). Initially it will carry a

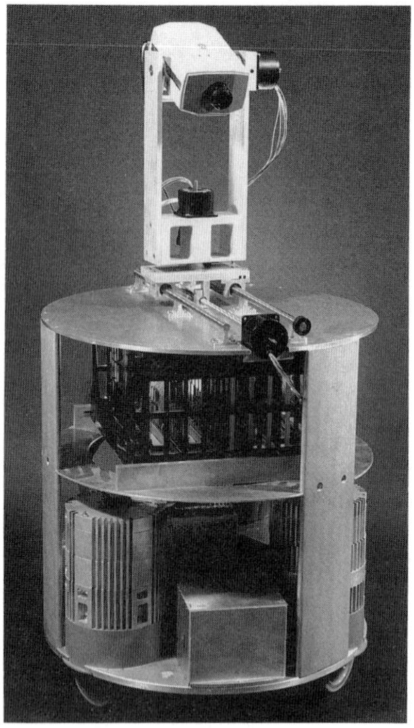

Fig. 8. The CMU Rover.

Fig. 9. The Rover wheelbase. The steering angle and drive of each wheel pair is individually controlled. The rover's trajectory will be an arc about any point in the floor plane if lines through the axles of all three wheels intersect at the center of that arc.

Fig. 10. The Rover wheel assembly. The steering motor is shown attached to the wheel assembly, part of the drive motor is shown detached.

TV camera on a pan/tilt/slide mount, several short-range infrared and long-range sonar proximity detectors, and contact switches. Our design calls for about a dozen on-board processors (at least half of them powerful 16-bit MC68000's) for high-speed local decision making, servo control, and communication (Fig. 13).

Serious processing power, primarily for vision, is to be provided at the other end of a remote-control link by a combination of a host computer VAX 11/780 an ST-100 array processor (a new machine from a new company, Star Technologies Inc., which provides 100 million floating-point operations per second) and a specially designed high-performance analog data acquisition and generation device. The Stanford Cart used 15 min of computer time to move a meter. With this new CMU hardware, and some improved algorithms, we hope to duplicate (and improve on) this performance in a system that runs at least ten times as fast, leaving room for future extensions.

The 15 min for each meter-long move in the Cart's obstacle-avoiding program came in three approximately equal chunks. The first 5 min were devoted to digitizing the nine pictures from a slider scan. Though the SAIL computer had a flash digitizer which sent its data through a disk channel into main memory at high speed, it was limited by a poor sync detector. Often the stored picture was missing scanlines, or had lost vertical sync, i.e., had rolled vertically. In addition, the image was quantized to only 4 bits per sample; 16 grey levels. To make one good picture the program digitized 30 raw images in rapid succession, intercompared them to find the largest subset of nearly alike pictures (on the theory that the nonspoiled ones would be similar, but the spoiled ones would differ even from each other) and averaged this "good" set to obtain a less noisy image with 6 bits per pixel. The new digitizing hardware, which can sample a raw analog waveform, and depends on software in the array processor to do sync detection, should cut the total time to under 1 s per picture. The next 5 min was spent doing the low-level vision; reducing the images, sometimes filtering them, applying the interest operator, especially the correlator, and statistical pruning of the results. The array processor should be able to do all this nearly 100 times faster. The last 5 min were devoted to higher level tasks; maintaining the world model, path planning, and generating graphical documentation of the program's thinking. Some steps in this section may be suitable for the array processor, but in any case we have found faster algorithms for much of it; for instance, a shortest path in graph algorithm which makes maximum use of the sparsity of the distance matrix produced during the path planning.

We hope eventually to provide a manipulator on the Rover's topside, but there is no active work on this now. We chose the high steering flexibility of the current design partly to ease the requirements on a future arm. The weight and power needed can be reduced by using the mobility of the Rover to substitute for the shoulder joint of the arm. Such a strategy works best if the Rover body is given a full three degrees of freedom (X,

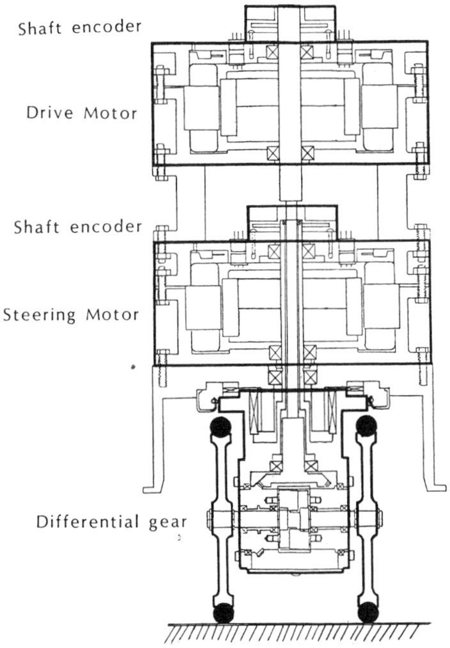

Fig. 11. Diagram of the Rover wheel assembly.

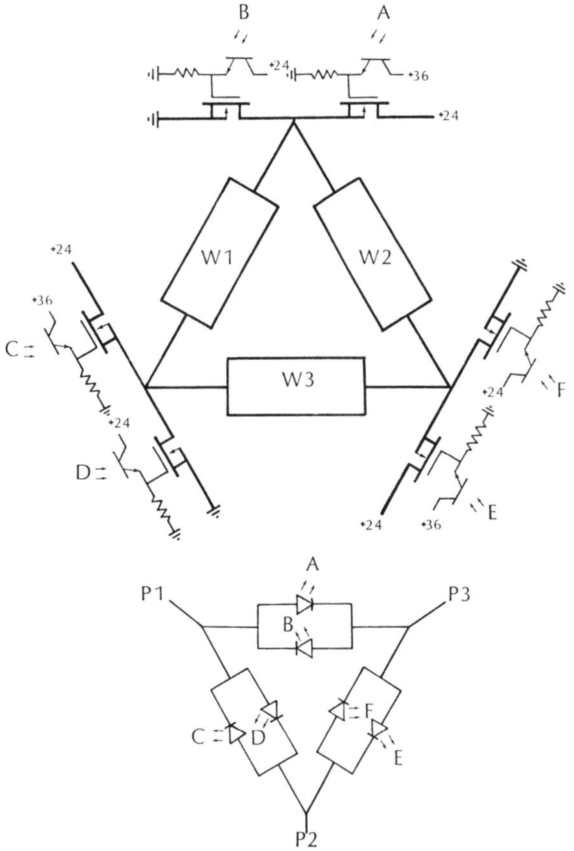

Fig. 12. Rover brushless motor drive circuitry. Each samarium-cobalt brushless motor in the wheel assemblies is sequenced and servoed by its own microprocessor. A processor generates three bipolar logic signals *P1*, *P2*, and *P3* which control the currents through the three phase motor windings *W1*, *W2*, and *W3* through the IR light-emitting diode (LED) phototransistor optical links *A* through *F* shown. Each phototransistor controls a power field-effect transistor which switches power to the windings. The circuitry in the upper, optically isolated, portion of the diagram is contained within its motor's housing, on an annular circuit board, using the housing as heat sink. The LED's poke through holes in the cover. Motor direction is controlled by sequencing order; torque is adjusted by pulsewidth modulation.

Y, and angle) in the plane of the floor. Conventional steering arrangements, as in cars, give only two degrees at any instant.

Rover Details

Three degrees of freedom of mobility are achieved by mounting the chassis on three independently steerable wheel assemblies (Figs. 9–12). The control algorithm for this arrangement at every instant orients the wheels so that lines through their axles meet at a common point. Properly orchestrated, this design permits unconstrained motion in any (2D) direction, and simultaneous independent control of the robot's rotation about its own vertical axis. An unexpected benefit of this agility is the availability of a "reducing gear" effect. By turning about the vertical axis while moving forward the robot derives a mechanical advantage for its motors. For a given motor speed, the faster the Rover spins, the slower it travels forward, and the steeper the slope it can climb. (Visualization of this effect is left as an exercise for the reader.)

To permit low-friction steering while the robot is stationary, each assembly has two parallel wheels connected by a differential gear. The drive shaft of the differential goes straight up into the body of the robot. A concentric hollow shaft around this one connects to the housing of the differential (Fig. 11). Turning the inner shaft causes the wheels to roll forwards or backwards, turning the outer one steers the assembly, with the two wheels rolling in a little circle. The assemblies were manufactured for us by Summit Gear Corp.

Each shaft is connected to a motor and a 4000-count/revolution optical shaft encoder (Datametrics K3). The two motors and two encoders are stacked pancake fashion on the wheel assembly, speared by the shafts. There are no gears except for the ones in the differential. (Fig. 11 shows a schematic cross section of a complete motor/wheel-assembly structure, Fig. 10 a partially assembled stack in the flesh.)

The motors are brushless with samarium–cobalt permanent-magnet rotors and three-phase windings (Inland Motors BM-3201). With the high-energy magnet material, this design has better performance when the coils are properly sequenced than a conventional rotating coil motor. The coils for each are energized by six power MOSFET's (Motorola MTP1224) mounted in the motor casing and switched by six optoisolators (to protect the controlling computers from switching noise) whose LED's are connected in bidirectional pairs in a delta configuration, and lit by three logic signals connected to the vertices of the delta (Fig. 12).

The motor sequencing signals come directly from on-board microprocessors, one for each motor. These are CMOS (Motorola MC146805 with Hitachi HM6116 RAM's) to keep power consumption low. Each processor pulsewidth modulates and phases its motor's windings, and observes its shaft encoder, to servo the motor to a desired motion (supplied by yet another processor, a Motorola 68000, the *Conductor* as a time parameterized function). Though the servo loop works in its present form, several approximations were necessary in this real-time task because of the limited arithmetic capability of the 6805. We will be replacing the 6805 with the forthcoming MC68008, a compact 8-bit bus version of the 68000.

The shaft encoder outputs and the torques from all the motors, as estimated by the motor processors, are monitored by another processor, the *Simulator*, a Motorola MC68000 (with all CMOS support circuitry the power requirement for our

32K 68000 is under 1 W. The new high-performance 74HC series CMOS allows operation at full 10-MHz speed.), which maintains a dead-reckoned model of the robot's position from instant to instant. The results of this simulation (which represents the robot's best position estimate) are compared with the desired position, produced by another 68000, the *Controller*, in the previously introduced the *Conductor*, which orchestrates the individual motor processors. The Conductor adjusts the rates and positions of the individual motors in an attempt to bring the Simulator in line with requests from the Controller, in what amounts to a highly nonlinear feedback loop.

Other on-board processors are as follows:

Communication — A 68000 which maintains an error corrected and checked packet infrared link with a large controlling computer (a VAX 11/780 helped out by an ST-100 array processor and a custom high-speed digitizer) which will do the heavy thinking. Programs run in the Controller are obtained over this link.

Sonar — A 6805 which controls a number of Polaroid sonar ranging devices around the body of the Rover. These will be used to maintain a rough navigation and bump-avoidance model. All measurements and control functions of this processor and the following ones are available (on request over a serial link) to the Controller.

Camera — A 6805 which controls the pan, tilt, and slide motors of the onboard TV camera. The compact camera broadcasts its image on a small UHF or microwave transmitter. The signal is received remotely and the video signal captured by a high-bandwidth digitizer system and then read by the remote VAX. There are tentative plans for a minimal vision system using a 68000 with about 256K of extra memory on-board the Rover, for small vision tasks when the Rover is out of communication with the base system.

Proximity — A 6805 which monitors several short-range modulated infrared proximity detectors which serve as a last line of defense against collision, and which sense any drop off in the floor, and contact switches.

Utility — A 6805 which senses conditions such as battery voltage and motor temperature, and which controls the power to nonessential but power-hungry systems like the TV camera and transmitter.

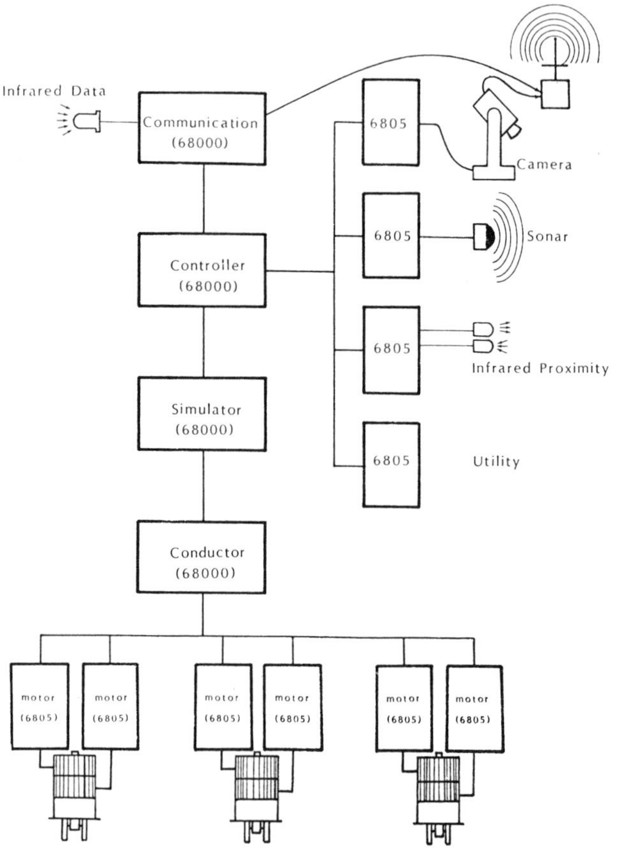

Fig. 13. The Rover's on-board processors.

Communication between processors is serial, via Harris CMOS UART's, at a maximum speed of 256 kBd. The Conductor talks with the motor processors on a shared serial line and the Controller communicates with the Sonar, Camera, Proximity, Utility, and any other peripheral processors by a similar method.

The processors live in a rack on the second storey of the robot structure (Figs. 8 and 13), between the motor and battery assembly (first floor) and the camera plane (penthouse). Fig. 13 shows the initial interconnection.

The Rover is powered by six sealed lead-acid batteries (Globe gel-cell 12230) with a total capacity of 60 A · h at 24 V. The motors are powered directly from these, the rest of the circuitry derives its power indirectly from them through switching dc/dc converters (Kepco RMD-24-A-24 and Semiconductor Circuits U717262). Each 6805 processor draws about one eighth of a watt, each 68000 board only 1 W.

Physically, the robot is a meter tall and 55 cm in diameter. It weighs 90 kg. The maximum acceleration is one quarter *g*, and the top speed is 10 km/h. With appropriate on-board programming the motions should be very smooth. The great steering flexibility will permit simulation of other steering systems such as those of cars, tanks, and boats and other robots by changes in programming.

Progress

As of this writing (January 1983) the robot's major mechanical and electronic structures are complete, mostly tested, and being assembled. An instability in the servo control algorithms which had held us up for several months has finally been solved, and we expect baby's first "steps" early in 1983. The digitizer unit which will receive and generate video and other data for the robot is under construction. Its specifications include four channels each with 2 Mbytes of memory and two ports able to transfer data at 100 Mbytes/s.

Promises

The high-level control system has become very interesting. Initially, we had imagined a language in which to write scripts for the on-board Controller similar to the **AL** manipulator language developed at Stanford [6], from which the commercial languages **VAL** at Unimation [15] and the more sophisticated **AML** [14] at IBM were derived. Paper attempts at defining the structures and primitives required for the mobile application revealed that the essentially linear control structure of these state-of-the-art arm languages was inadequate for a rover. The essential difference is that a rover, in its wanderings, is regularly "surprised" by events it cannot anticipate, but with which it must deal. This requires that routines for responding to various situations can be activated in arbitrary order, and run concurrently.

We briefly examined a production system, as used in many "expert systems," as the framework for the requisite real-time concurrency, but have now settled on a structure similar to that developed for the CMU **Hearsay II** speech understanding project [5]. Independent processes will communicate via messages posted on a commonly accessible data structure we call a *Blackboard*. The individual processes, some of which will run under control of a spare real-time operating system on one or more of the onboard 68000's, others of which will exist at the other end of a radio link on the VAX, change their relative priority as a consequence of relevant messages on the blackboard. For instance, a note from several touch sensors signaling a collision is taken as a cue by the avoidance routine to increase its running rate, and to post messages which trigger the motor coordinating routines to begin evasive actions. We plan to implement the multiple processes required for this task on each of several of the on-board 68000's with the aid of a compact (4K), efficient real-time operating system kernel called VRTX available from Hunter & Ready. A more detailed description of the state of this work may be found in [4].

Other interesting preliminary thinking has resulted in a scheme by which a very simple arm with only three actuators will enable the robot, making heavy use of its great steering flexibility, to enter and leave through a closed standard office door (Fig. 14).

Stepping into a more speculative realm, we are considering approaches to model-based vision [2] which would permit recognition of certain classes of objects seen by the robot. Discussions with the distributed sensor net crew here at CMU [13] has raised the possibility of equipping the robot with ears, so it could passively localize sound, and thus perhaps respond correctly, both semantically and geometrically, to a spoken command like "Come here!" (using, in addition, speech understanding technology also developed at CMU [16]).

We are also toying with the idea of a phased array sonar with about 100 transducers operating at 50 kHz which, in conjunction with the high-speed analog conversion device mentioned above and the array processor, would be able to produce a modest resolution depth map (and additional information) of a full hemisphere in about 1 s, by sending out a single spherical pulse, then digitally combining the returned echoes from the individual transducers with different delay patterns to synthesize narrow receiving beams.

Philosophy

It is my view that developing a responsive mobile entity is the surest way to approach the problem of general intelligence in machines.

Though computers have been programmed to do creditable jobs in many *intellectual* domains, competent performance in *instinctive* domains like perception and common sense reasoning is still elusive. I think this is because the instinctive skills are fundamentally much harder. While human beings learned most of the intellectual skills over a few thousand years, the instinctive skills were genetically honed for hundreds of millions of years, and are associated with large, apparently efficiently organized, fractions of our brain; vision, for example, is done by a specialized 10 percent of our neurons. Many animals share our instinctive skills, and their evolutionary record provides clues about the conditions that foster development of such skills. A universal feature that most impresses me in this context is that *all animals that evolved perceptual and behavioral competence comparable to that of humans first adopted a mobile way of life.*

This is perhaps a moot point in the case of the vertebrates, which share so much of human evolutionary history, but it is dramatically confirmed among the invertebrates. Most molluscs are sessile shellfish whose behavior is governed by a nervous system of a few hundred neurons. Octopus and squid are molluscs that abandoned life in the shell for one of mobility; as a consequence, they developed imaging eyes, a large (annular!) brain, dexterous manipulators, and an unmatched million-channel color display on their surfaces. By contrast, no sessile animal nor any plant shows any evidence of being even remotely this near to the human behavioral competence.

My conclusion is that *solving the day to day problems of developing a mobile organism steers one in the direction of general intelligence, while working on the problems of a fixed entity is more likely to result in very specialized solutions.*

I believe our experience with the control language for the Rover vis a vis the languages adequate for a fixed arm, is a case in point. My experiences with computer vision during the Cart work reinforce this opinion; constantly testing a program against fresh real-world data is nothing at all like optimizing a program to work well with a limited set of stored data. The variable and unpredictable world encountered by a rover applies much more selection pressure for generality and robustness than the much narrower and more repetitive stimuli experienced by a fixed machine. Mobile robotics may or may not be the fastest way to arrive at general human competence in machines, but I believe it is one of the surest roads.

Related Work

Other groups have come to similar conclusions, and have done sophisticated mobile robot work in past [12], [17]. The robotics group at Stanford has acquired a new, experimental,

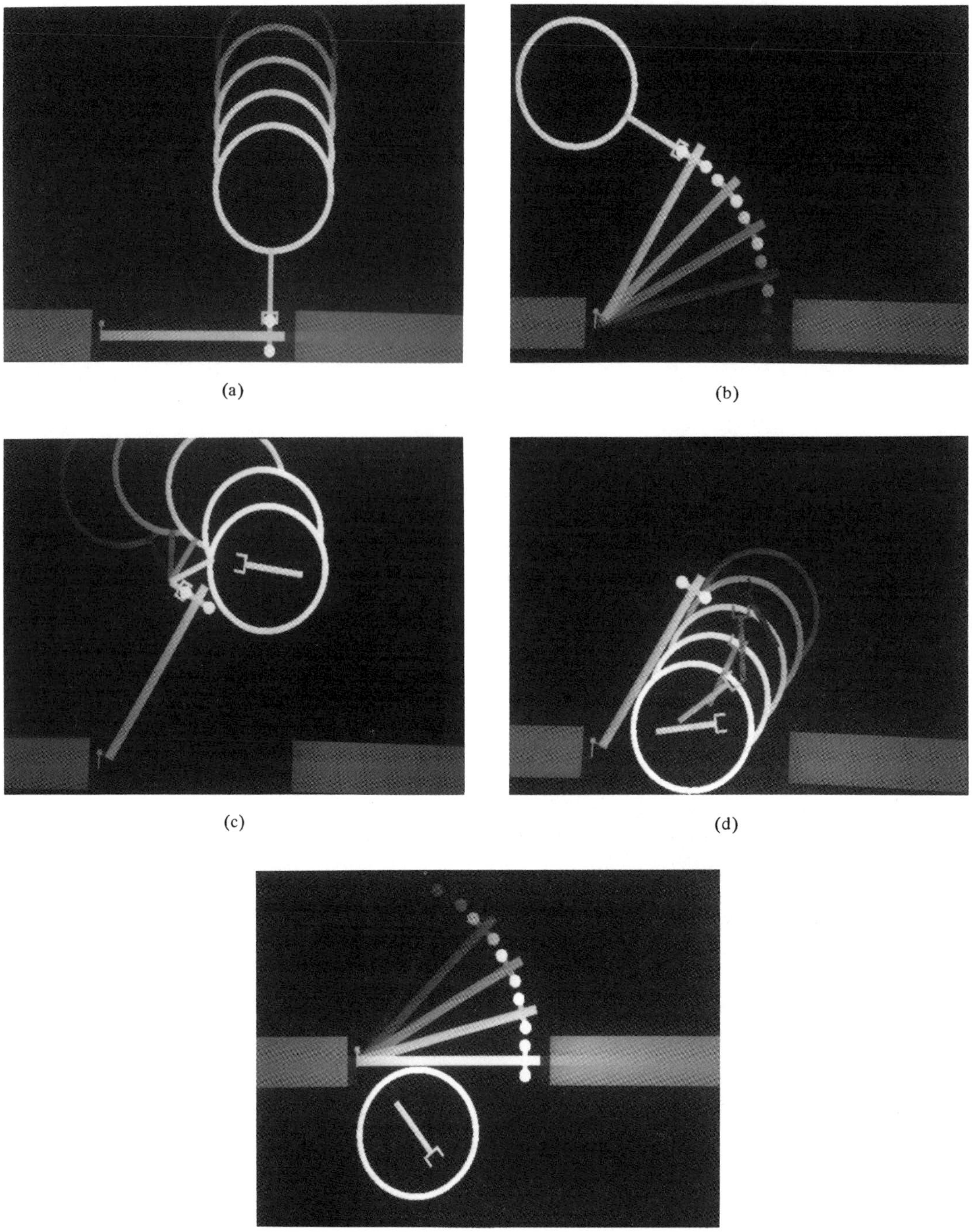

Fig. 14. The Rover goes through a closed door. Using a simple arm with only three powered actuators and two passive hinges, greatly helped by the wheel flexibility, the Rover deals with a self-closing office door. The door and knob are visually located, the Rover extends its arm, and approaches the door (a). The arm grasps the knob, twists it open, and the Rover backs up in an arc, partially opening the door (b). The Rover rolls around the door edge, while retaining its grasp on the knob; passive hinges in the arm bend in response (c). The Rover body now props open the door; the arm releases and retracts, and the Rover rolls along the front of the door (d). The Rover moves in an arc outward, allowing door to close behind it (e).

mobile robot from Unimation Inc., and plans research similar to ours [1]. This new Unimation rover [3] is very similar in size, shape, and mechanical capabilities to the machine we are building. It achieves a full three degrees of freedom of floor-plane mobility by use of three novel "omnidirectional" wheels which, by virtue of rollers in place of tires, can freely move in a direction broadside to the wheel plane as well as performing the usual wheel motion under motor control.

References

[1] T. O. Binford, "The Stanford mobile robot," personal communication, Stanford Univ. Comput. Sci. Dept., Stanford, CA, Oct. 1982.

[2] R. A. Brooks, "Symbolic reasoning among 3-D models and 2-D images," Ph.D. dissertation, Stanford Univ., Stanford, CA, June 1981.

[3] B. Carlisle and B. Shimano, "The Unimation mobile robot," personal communication, Unimation Inc., Mountain View, CA, Aug. 1981.

[4] A. Elfes and S. N. Talukdar, "A distributed control system for the CMU Rover," 8th Int. Joint Conf. on Artificial Intelligence, Karlsruhe, West Germany (IJCAI), Aug. 1983.

[5] L. D. Erman and V. R. Lesser, "The HEARSAY-II speech-understanding system: Integrating knowledge to resolve uncertainty," *Commun. ACM*, vol. 23, no. 6, June 1980.

[6] R. Goldman and S. Mujtaba, *AL User's Manual.* 3rd ed., Computer Science STAN-CS-81-889 (Rep. AIM-344), Stanford Univ., Dec. 1981.

[7] T. Lozano-Pérez and M. A. Wesley, "An algorithm for planning collision-free paths among polyhedral obstacles," *Commun. ACM*, vol. 22, no. 10, pp. 560–570, Oct. 1979.

[8] H. P. Moravec, "Towards automatic visual obstacle avoidance," in *Proc. 5th Int. Joint Conf. on Artificial Intelligence* (MIT, Cambridge, MA), pp. 584 (IJCAI), Aug. 1977.

[9] ——, "Visual mapping by a robot rover," in *Proc. 6th Int. Joint Conf. on Artificial Intelligence* (Tokyo, Japan), pp. 599–601 (IJCAI), Aug. 1979.

[10] ——, "Obstacle avoidance and navigation in the real world by a seeing robot rover," Ph.D. dissertation, Stanford Univ., Sept. 1980 (published as *Robot Rover Visual Navigation.* Ann Arbor, MI: UMI Research Press, 1981.

[11] ——, "Rover visual obstacle avoidance," in *Proc. 7th Int. Joint Conf. on Artificial Intelligence* (Vancouver, B.C., Canada), pp. 785–790 (IJCAI), Aug. 1981.

[12] B. Raphael, *The Thinking Computer.* San Francisco, CA: W. H. Freeman and Company, 1976.

[13] R. F. Rashid and G. G. Robertson, "Accent, a communication oriented operating system kernel," Carnegie-Mellon Univ., Pittsburgh, PA, Tech. Rep. CS-81-123, Oct. 1981.

[14] R. H. Taylor, P. D. Summers, and J. M. Meyer, "AML: A manufacturing language," IBM Thomas J. Watson Research Center, Yorktown Heights, NY, Res. Rep. RC-9389, Apr. 1982.

[15] Unimation, Inc., "User's guide to VAL, a robot programming and control system, version II," Unimation Inc., Tech. Rep., Feb. 1979.

[16] A. Waibel and B. Yegnanarayana, "Comparative study of nonlinear time warping techniques for speech understanding," Carnegie-Mellon Univ., Comput. Sci. Dep., Pittsburgh, PA, Tech. Rep. 125, 1981.

[17] Y. Yakimovsky and R. Cunningham, "A system for extracting three-dimensional measurements from a stereo pair of TV cameras," *Comput. Graphics Image Proces.*, vol. 7, pp. 195–210, 1978.

An Integrated Navigation and Motion Control System for Autonomous Multisensory Mobile Robots

Georges Giralt, Raja Chatila, and Marc Vaisset

An essential task for a mobile robot system is navigation and motion control. The characteristics of perception required by environment modeling or motion control are very different. This may be basically obtained using several sensors. The described NMC system integrates the elementary data acquisition, modeling, planning, and motion control subsystems. A set of rules determines the dynamic structure and the behavior of the system and provides a man/machine and system to system interface.

1 Introduction

Research on mobile robots began in the late sixties with the Stanford Research Institute's pioneering work. Two versions of SHAKEY, an autonomous mobile robot, were built in 1968 and 1971. The main purpose of this project was "to study processes for the real-time control of a robot system that interacts with a complex environment" ⟨NIL 69⟩. Indeed, mobile robots were and still are a very convenient and powerful support for research on artificial intelligence oriented robotics. They possess the capacity to provide a variety of problems at different levels of generality and difficulty in a large domain including perception, decision making, communication, etc., which all have to be considered within the scope of the specific constraints of robotics: on-line computing, cost considerations, operating ability, and reliability.

A second and quite different trend of research began around the same period. It was aimed at solving the problem of robot vehicle locomotion over rough terrain. The work focus was the design and the study of the kinematics and dynamics of multilegged robots ⟨McG 79⟩.

During the seventies various reasons, such as too remote real-world applications and lack of efficient on-board instrumentation (computers, sensors, etc.), slowed the research thrust in the field and even lead to important funding cuts. Meanwhile the so-called industrial robots, i.e., manipulator robots, became the main body of a fast expanding field of robotics.

The present renewal of interest in mobile robots started in the late seventies fostered by powerful on-board signal and data processing capacities offered by microprocessor technology.

Today, in 1983, the scientific reasons for using mobile robots as a support for concep-

The authors would like to thank Malik Ghallab for his suggestions and contribution. Indeed, this work was only made possible thanks to the group efforts of the entire project.

tual and experimental work in advanced robotics hold more than ever. Furthermore a number of real-world applications can now be realistically envisionned, some for the near future. These applications range from intervention robots operating in hostile or extremely dangerous environments to day-to-day machines in highly automated factories using flexible manufactoring systems (FMS) technology.

In this paper we focus on aspects of the HILARE project's current research that we believe are the key to autonomous mobile robots development: system integration, multisensory driven navigation, and motion control.

2 Overview of HILARE, A Mobile Robot

The HILARE project started by the end of 1977 at LAAS ⟨GIR 79⟩. The project's goal is to perform general research in robotics and robot perception and planning. A mobile robot was constructed to serve as an experimental means.

The environment domain considered is a world of a flat or near flat smooth floor with walls which include rooms, hallways, corridors, various portable objects, and mobile or fixed obstacles.

2.1 The Physical Infrastructure

The vehicle has three wheels as shown in figure 1. The two rear wheels are powered by stepping motors and the front wheel is free. This structure is simple but allows the robot to perform such trajectories as straight lines, circles and clothoids.

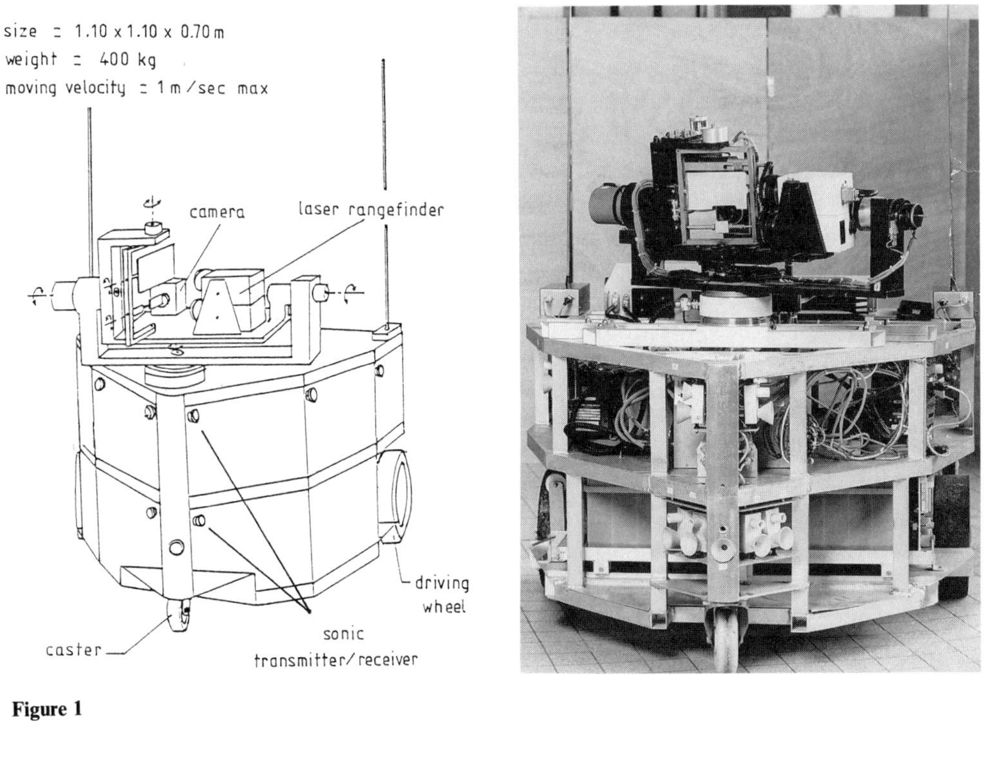

Figure 1

The perception system is composed of two separate subsystems serving different purposes: an ultrasonic system and a vision-based system. To improve odometry path-control, two optical encoders are also used. A manipulator is to be put on the robot in the future. The computer system supporting the various robot functions has a distributed multilevel architecture (figure 2): several (currently six) robot-borne microprocessors are radio-linked to a 32-bit computer accessing one or more other larger or similar processors.

2.2 The Perception System

A multisensory system provides the robot with the information it needs about its environment. The various sensors are used independently or in concert.

Ultrasonic Perception

A set of 14 ultrasonic emitter-receivers distributed on the vehicle provides the range data up to 2 m. The system has two functions:

1. An alarm function that warns the robot of the near vicinity of some object. The reaction of the robot is usually to come to a full stop if moving, but in some circumstances it will try to avoid the detected object.
2. A closed-loop local obstacle avoidance function. In this mode the robot uses the range data to move along an object, maneuvering to stay at a fixed distance from its surface.

Vision

A camera and a laser range-finder are the main perception system. They are mounted on a pan and tilt platform. The laser can be used in scanning mode, or it can measure ranges within the camera's field using a retractable mirror. This provides the robot with 3-D data about its environment ⟨FER 82⟩.

Position Referencing

HILARE's position can be obtained either relatively to objects and specific environment patterns or in a constructed frame of reference. To do this, HILARE is equipped with an infrared triangulation system which operates in areas where fixed beacons are installed.

2.3 Decision Making and Execution Monitoring

One important question that arises in the research area of decision-making system organization is the extent of decision distribution in the system, the degree of decomposition of the system in independent modules, and their level of abstraction. The issues of synchronization and communication between modules have to be addressed to answer this question.

The robot decision-making system is composed of several specialized decision modules (SDMs) some of which are implemented on different processors. Several architectures are being examined wherein either distribution or centralization is enhanced. One ap-

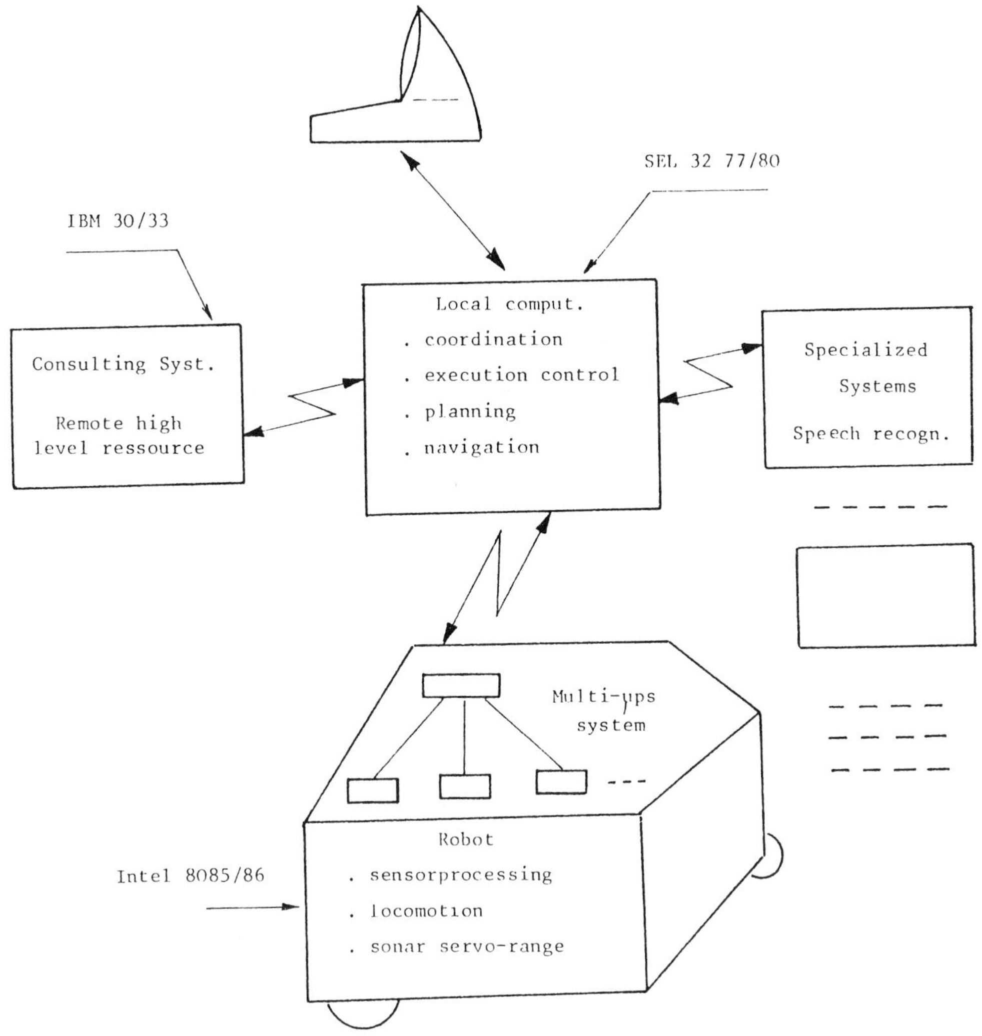
SEL 32 77/80
IBM 30/33
Local comput.
. coordination
. execution control
. planning
. navigation
Consulting Syst.
Remote high
level ressource
Specialized
Systems
Speech recogn.
Multi-μps
system
Robot
. sensorprocessing
. locomotion
. sonar servo-range
Intel 8085/86

Figure 2

proach is to consider a large number of SDMs organized in a fixed hierarchical structure corresponding to a predetermined ordering. The hierarchy expresses the fact that a lower level module is a primitive of the next one in the structure. On top is a general planner and a plan execution monitor. The execution monitor is a central system through which modules interact. It also controls the robot's interaction with the outer world, i.e., sensing and operator-machine communication.

Another architecture wherein distribution is enhanced is currently being considered. In this structure, the robot decision-making system is composed of a small number of specialized decision modules ⟨GIR 79, CHA 82⟩. Scene analysis, planning, and navigation are three such modules. Each SDM has the necessary expertise in its domain and performs the part of the overall plan that is in its domain. An SDM is composed of a rule-based system and a communication interface. SDMs make use of a hierarchy of specialized processing modules (SPMs) that perform various computations at different levels of abstraction. At the lowest level are the robot's sensors and effectors.

The SDMs that will compose the decision-making system of HILARE include general planner (GPL), navigation and motion control system (NMCS), object modeling and scene comprehension, natural language communication, and manipulator control.

The general planner produces abstract plans that may include parallel nodes. The plan nodes are subgoals to be accomplished by the various SDMs that will generate their own plans to achieve them. Some of the SDMs plans are based on the results of the SPMs and may call for sensing or physical action. The GPL "feeds" the other SDMs with new subgoals depending on the results of the previous steps.

In this approach, the SDMs communicate with each other through a common database. This approach is somewhat similar to HEARSAY II ⟨ERM 80⟩ and to the Hayes-Roths' "opportunistic planning" ⟨HAY 79⟩ The common database is actually partitionned into two components: an announcement database (ADB) and an information database (IDB). ADB and IDB are small and are permanently accessed by the SDMs and are their communication means. The SDMs put in the ADB the subproblems that are not in their domain, and these are considered by the other SDMs as requests or goals to be achieved. The SDMs put in the IDB the results of their plans or any new knowledge they have that is of general interest. Each SDM controls the execution of the part of the plan in its own domain.

3 Navigation and Motion Control System

3.1 System Overview

Environment dependent sensory driven navigation is at present the key issue in mobile robots research. And this is still more the case when we consider real-world applications.

The navigation and motion control system (NMCS) is one of HILARE's specialized decision modules. Its domain is all that concerns the robot's mobility activity. The basic procedures it makes use of are routing, navigation, low-level vision, locomotion, position finding, and local obstacle avoidance. The sensors and effector are camera, laser and

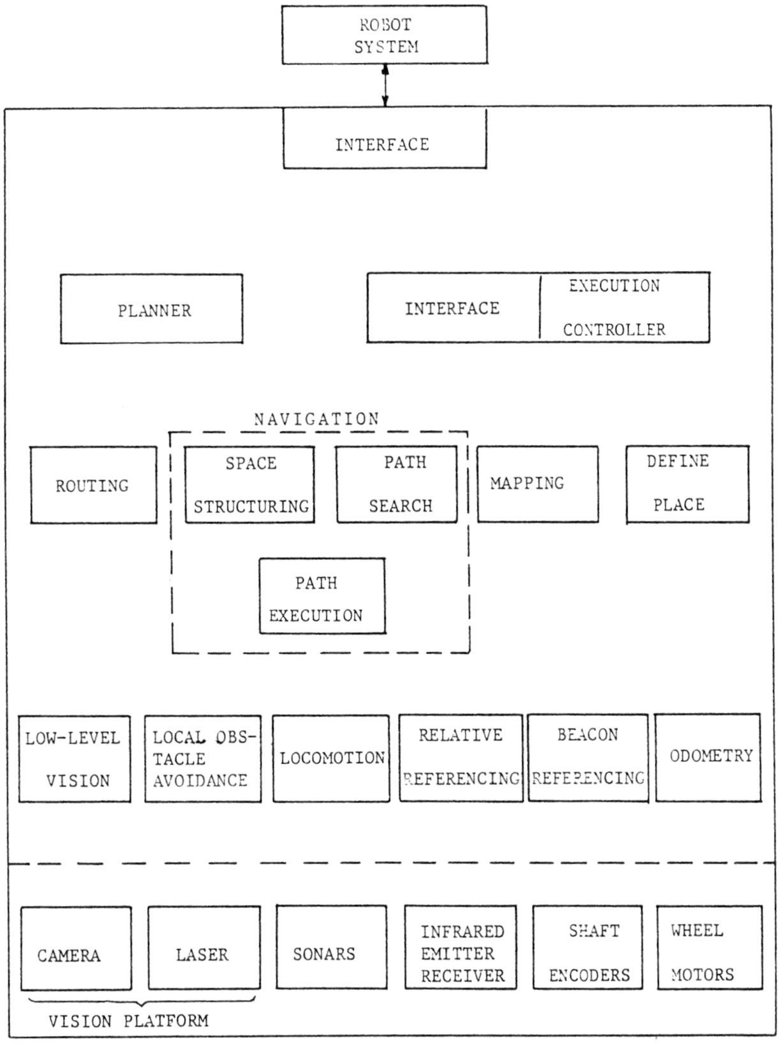

Figure 3
Navigation and motion control system.

vision platform positionning, ultrasonic transducers, infrared transmitters, shaft encoders, and wheel motors (figure 3).

NMCS will use all these resources to produce a general plan of action in the navigation domain and to control its validity and its execution using the sensory information.

NMCS is composed of a planner, an execution controller (expressed in two separate rule systems), and an interface. The rules are the operational specialized knowledge that the system possesses in its domain, and they are fired according to the factual knowledge in its database. The various procedures and sensors/effectors are used in the action part of some of the rules. The planner produces the sequence of actions to be performed by the system's SPMs and sends it step by step to the controller. This last actually has to call the SPMs to accomplish the plan steps, and to monitor their interactions. The SPMs are activated by the controller but they exchange their data directly. The controller is able to handle some situations where a plan step cannot be executed and to perform the necessary mending. It acknowledges the planner after the completion of each step, and in case of failure, the robot current state and the reason of failure are sent back to the planner. The robot state is a global variable expressing its location and some other parameters (see the example).

The rule formalism has some interesting features (flexibility, event driven), but it could be rather inefficient when compared with procedural formalisms. In order to keep the power of a rule system and increase its efficiency, we compile the rules into a decision tree. When the rule set is extended, a new compilation is required, but this is not a frequent event. The interface is the part of NMCS that handles all the exchanges with the other decisional modules or—as is currently the case—with the user.

In this section we first briefly describe the space modeling and path finding techniques investigated in HILARE's project. Then we focus on the SPMs integration and on the execution control aspects.

3.2 Space Models

The knowledge necessary for navigation and motion control concerns mainly the space configuration and the robot's characteristic parameters and location.

HILARE's world is a human-made environment composed of rooms and work stations and containing several objects to be avoided (obstacles) or manipulated. Objects are defined, at this level, by their geometric features (shape and dimensions).

Space is composed of places, e.g., rooms, work areas, corridors, hallways, and connectors between places (e.g., doors and gateways).

Space is represented at two levels:

1. the topological level where places are nodes, and connectors are arcs in a connectivity graph (figure 4) and

2. the geometric level, which assigns dimensions to the elements of the connectivity graph: width to the connectors and boundary dimensions to the places.

Every place has an explicit operator provided or an implicit frame of reference. The latter is set by the robot while learning about its world.

Places can also contain specific locations, such as landmarks and work stations, which

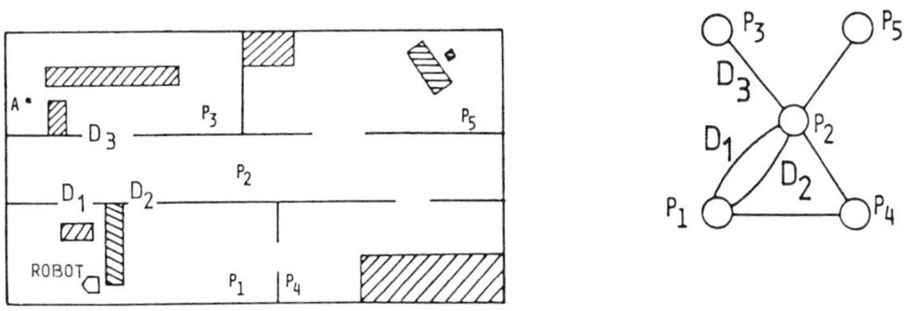

Figure 4
Topological representation of space: a connectivity graph whose nodes are places (rooms, . . .) and arcs are traversible boundaries (doors, . . .) between places.

can be used in similar fashion to the frame of reference to define at a relational level the robot position.

The space structure is further refined considering the internal topology of places. These may contain a variety of static or movable objects (e.g., obstacles and elements of the internal space structure). An internal topological model is created by the definition of polygonal cells:

1. polygonal cells O_i which are floor polygonal projections of the objects and the walls, and
2. convex polygonal cells C_i, which represent empty space.

C_i cells are bounded by segments of O_i and segments of other $C_j, j \neq i$. Segments of space within a place which are either partially undefined or nonconvex are noted C_i^* cells.

Hence, a connectivity graph between cells is created whose nodes are cells and arcs are the connectors i.e., the traversible frontiers between cells (figure 5).

3.3 Path Determination

Let us consider the case, which can be shown to be general, where the robot possesses information on the general setting of its environment (e.g., connectivity graph between places) and no information on the internal structure of places (e.g., obstacles). Now a goal as "Move to place P_3 location $A(X, Y)$" (figure 4) is first interpreted using ROUTING a specialized processing module (SPM).

Routing

Based on the connectivity graph of figure 4, this expert module transforms "Move to P_3 location $A(X, Y)$" into a sequence of elementary point-to-point moves.

For instance,

"Move to door D_1 (X, Y),"
"Move to D_3 (X, Y),"
"Move to A (X, Y)."

The routing expert database contains only topological knowledge and the graph search would consequently provide the best path in terms of search depth only (number

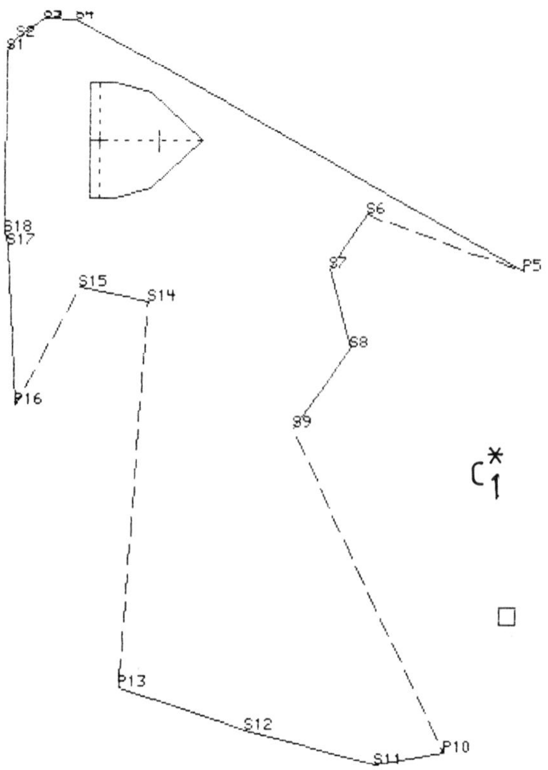

5a

Figure 5 (pp. 199–204)
Example of robot navigation: (a) first vision; (b) first cell structuring and graph; (c) the robot moves to the best frontier of the unknown space; (d) second vision; (e) second cell structuring and modified graph; (f) goal is reached.

of crossed places between start and goal nodes). In order to avoid this limitation, a cost function has to be defined taking into account other information of the robot's world.

The concept of place allows for a nested decomposition of space since a place *j* can be itself described as a set of places ^i_j. Hence in HILARE's decision-making structure a ROUTING expert can be included both at the NMCS and at the GPL levels.

Everyone of the elementary moves is now to be executed within a place whose internal structure is unknown. This task, at the very core of the navigation system, is performed by two cooperating subsystems: space structuring and path search.

Space Structuring

Let us consider now the general problem facing HILARE: the task of going from an initial location *R* to the goal *G*. Nothing else but the reference frame is given to the robot.

The first action to be carried out is a perception of the environment (e.g., using the laser system), which provides the robot with an outline of the perceived space.

In figure 5 an example of environment perception is shown and the S_i and P_i types of vertices are defined.

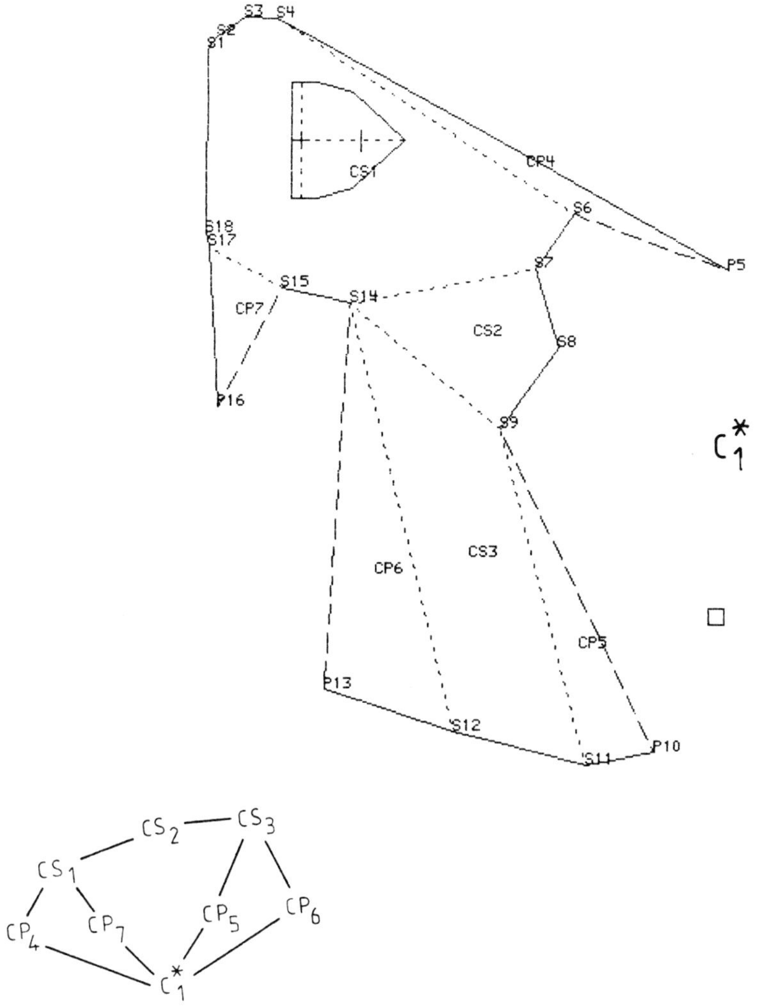
S1
S2
S3
S4
CS1
CP4
S6
P5
S18
S17
S7
S15
S14
CP7
CS2
S8
P16
S9
C_1^*
CS3
CP6
CP5
P13
S12
S11
P10
CS_2
CS_3
CS_1
CP_4
CP_7
CP_5
CP_6
C_1^*

5b

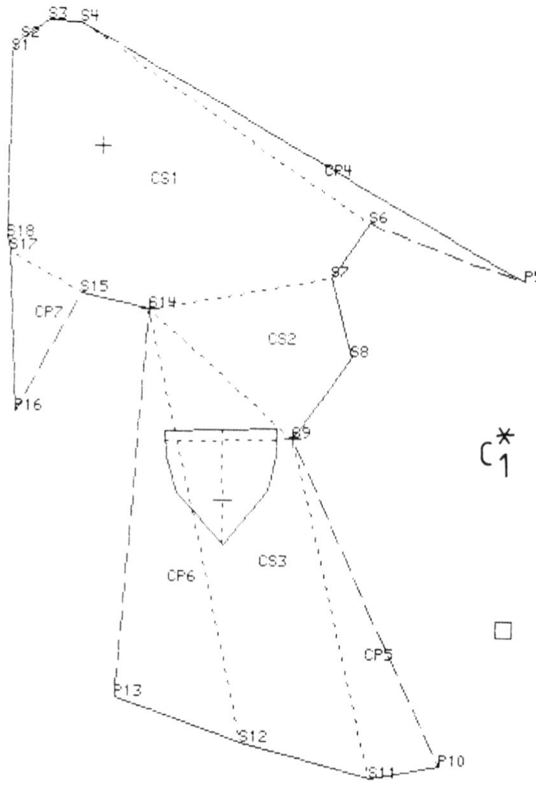

5c

The set of CS_i cells represents the empty space elements which are completely determined. The segments S_jP_i or P_jS_i are used as a basis for further expansion of the CS_i set. This is accomplished in a step-by-step manner while the robot navigates with the aim either of reaching a given location or of exploring its unknown surroundings. This differs significantly from the path-finding approach proposed in ⟨LOZ 81, BRO 82⟩.

Every action, and mainly the actual elementary movements of the robot, is determined by a decision rule based on the current status of the connectivity graph of C_i and C_i^* cells, and a heuristic cost function.

Arcs are labeled. Information includes traversability whenever the frontier segment between the related cells is larger than the robot's width.

Path Search

Together with the connectivity graph previously defined a second graph is constructed whose nodes and arcs are, respectively, the frontier segments and cells.

The cost function used has the following general form:

$$F = \sum_i (F_i^1 + F_i^2);$$

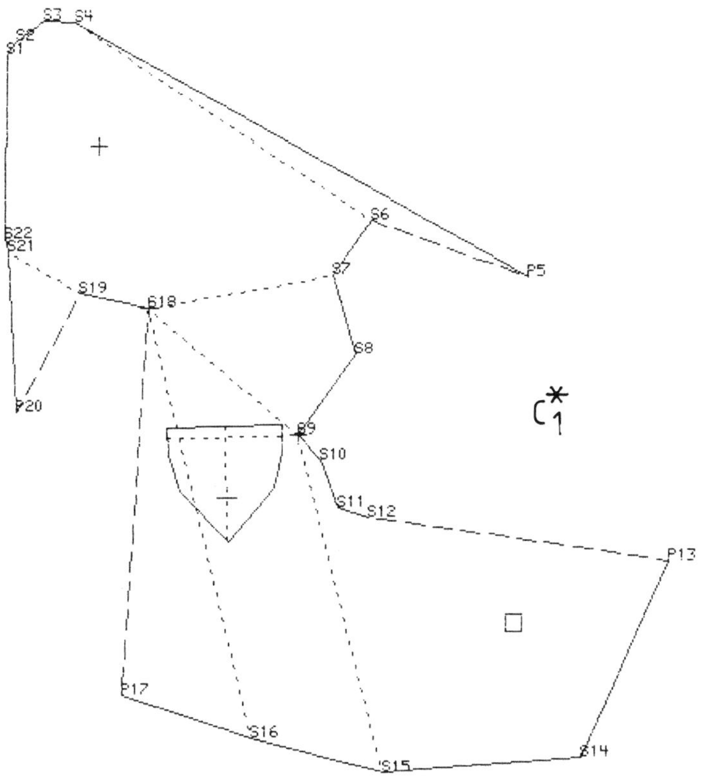

5d

F_i^1 measures every cost associated with determined trajectory travel (CS_i and CP_i cells); F_i^2 measures uncertainty and is evaluated over C_i^* cells.

After giving either the robot-centered or absolute coordinates of the goal to the robot, the path search is performed on the segment graph. We use an A^* algorithm ⟨HAR 68, NIL 80⟩, guided by F. For the evaluation of a minimum distance path, a segment is currently represented by its center. The robot and goal, represented by points, are added to the graph after determination of the cells in which they are.

The robot is obviously always within a CS-cell. If the goal is inside a CS or CP cell all the path is in the known part of the robot's world and the heuristic function reduces to F^1. Otherwise, we introduce in F^2 an evaluation of the information that may be gathered on the unknown world at the selected frontier segment and an evaluation of obstacle density in the region of the unknown world to be traversed, in addition to a lower bound of the distance to be covered ⟨CHA 81, 82⟩. The search procedure is performed using the whole graph at each stage so that the best frontier segment with the C^* containing the goal is selected. Furthermore, if we assume that obstacles do not change place, the cell structuring is definitive and enables the robot to select another frontier segment when it reaches a dead end, for example, without any need for new perceptions and structurings.

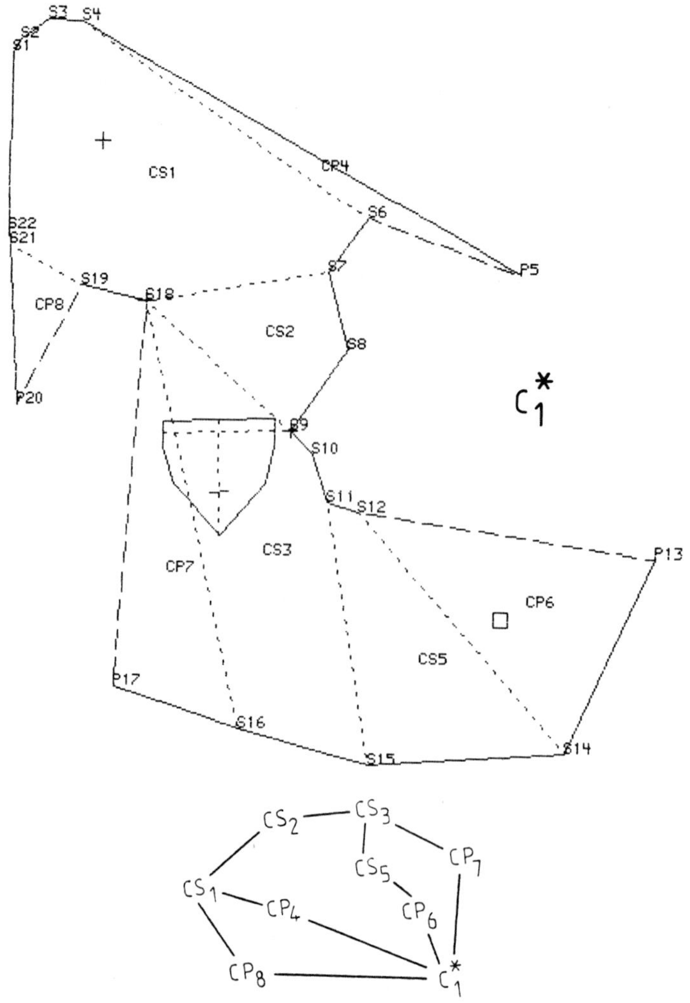
S3 S4
S2
S1
CS1
CP4
S22
S21
S6
P5
S7
S19
CP8
S18
CS2
S8
C*1
P20
S9
S10
S11 S12
P13
CS3
CP7
CP6
CS5
P17
S16
S15
S14
CS2
CS3
CP7
CS5
CS1
CP4
CP6
CP8
C*1

5e

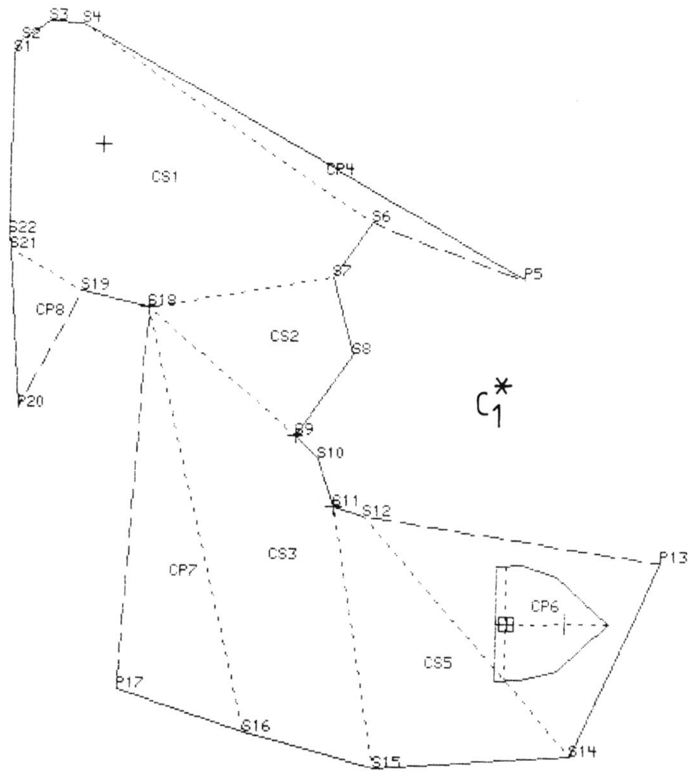

5f

The example given in figure 5 shows how robot movements and space structuring actually interact.

Path execution is accomplished by a lower module which has to take into account the robot maneuvering capabilities and a smoothed near-optimal time trajectory is finally obtained composed of straight lines, clothoids, and circles.

3.4 Automatic Environment Modeling

To date, two aspects which concern automatic environment modeling have been investigated: (1) environment mapping and (2) environment structure acquisition.

Environment Mapping

In the previous section, when HILARE had completed its task, it had obtained a partial map of its environment. This map could become complete after execution of several random displacements. To systematize this mapping function, a specific operator, MAP SPACE (X_0, Y_0), has been defined which associates a decision rule based on segments of the type S_iP_j or S_iS_j to a cost function Fm similar to F. Such an operator produces a map of a place in an optimized way (figure 6). This map will either be referred to the previously given frame of reference (X_0, Y_0) or to a coordinate system automatically defined by the robot and transmitted to the operator.

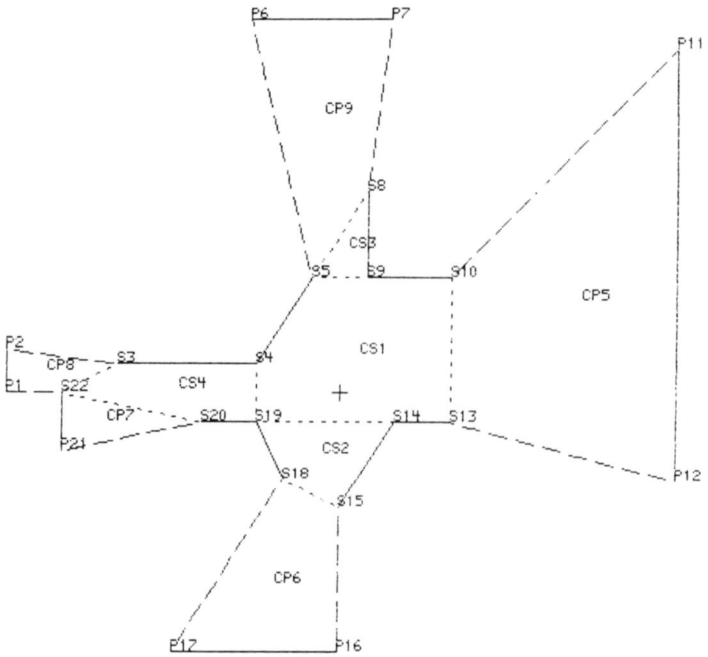

Figure 6a
Space mapping: first view and cell structure. The robot will explore until no more *SP* segments (or *P* vertices) remain.

Environment Structure Acquisition
Notice that automatic mapping can be generalized when the environment explored by the robot is not a simple place but rather a more complex arrangement of places whose structure is not known to the robot. The mapping obtained either in a systematical way or through random moves will produce eventually, together with a map of objects O_i, a set of cells representing empty space. Over this topological model of the robot environment, an automatic partition into places can be performed by use of graph decomposition techniques (figure 7).

Current work at LAAS solves the problem whenever the global connectivity graph can be decomposed in a tree of biconnected graphs ⟨LAU 83⟩.

On this basis a DEFINE PLACE operator can be defined which allows the robot automatically to acquire the overall structure and through subgraph labeling to produce an internal coding for the places.

3.5 The Specialized Processing Modules
In sections 3.3 and 3.4 several specialized processing modules have been introduced, i.e., routing, space structuring, path search, map space, and define place.

All of them concern HILARE's high level understanding of space and mainly control, at a high level, the robot navigation. More specialized processing modules are needed actually to control robot motion.

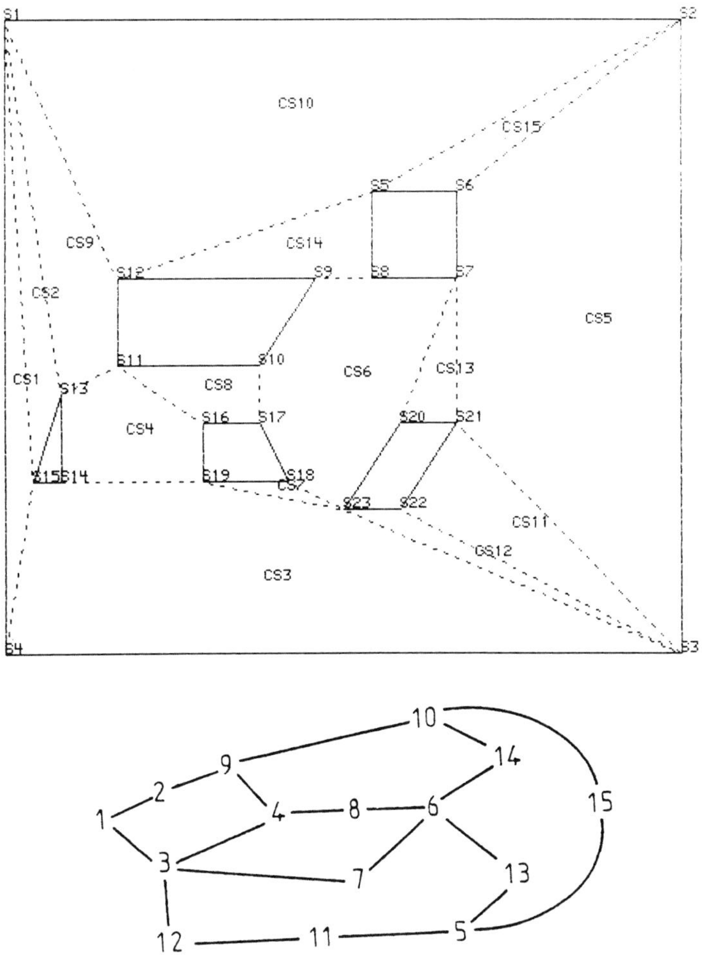

Figure 6b
Space mapping: final cell structure and graph.

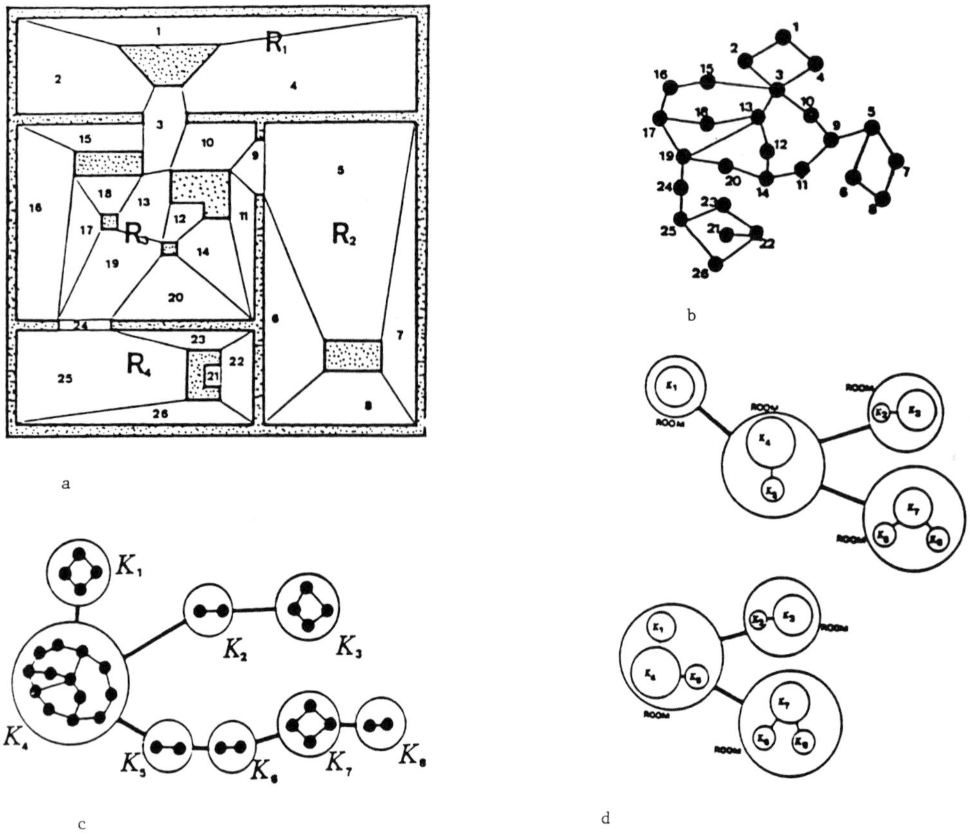

Figure 7
Environment structure acquisition: (a) environment structuring; (b) the cell graph; (c) the decomposition tree; (d) two possible labelings of the decomposition tree.

Low Level Vision

In the navigation domain a representation of the environment is needed wherein all objects are considered as obstacles, and a low-level vision providing a geometrical model is sufficient for this purpose. This system is based on a polyhedral modeling of obstacles and detects the object planar surfaces by a process performing a contour extraction, and after a filtering to eliminate too small or noise generated contours, the laser range-finder is used to determine the planar surfaces slope ⟨FER 82⟩. Objects ground projection or trace is computed from this information and used to build or update the geometrical world model.

Local Obstacle Avoidance and Maneuvering

This module uses the ultrasonic range data and provides for local maneuvering and obstacle avoidance. It is used, for example, when an obstacle is detected on the robot's path whereas none was in that location in the current space model.

Three categories of obstacles are defined: fixed (e.g., walls and internal structure elements), movable, and mobile. Movable objects can, in some conditions defined by

GPL, be considered as occasional obstacles and be disregarded for space structuring purposes. This means they can be avoided by the robot in a way analogous to when it encounters a mobile obstacle.

Thus if the obstacle is identified as occasional it will be merely avoided, taking into account the local space configuration. The rule system contains the conditions under which this may be accomplished without questioning and modifying the world model by a new costly 3-D perception. Another situation where this module may be used is when the robot moves in a corridor: no obstacles are to be avoided and it has only to move along a wall. Proximity navigation and sonar servoranging are also useful in the approach phase of docking maneuvers.

This processing module is also activated by two operators—Move Along, $(Z; v; C_1, C_2)$; Dock, $(\text{Frame}; Z; C_1, C_2)$—where Z is the list of servoing parameters, v is the robot speed, and C_1, C_2 are two predicates which define stop conditions for success and failure.

Position Finding

Position finding decomposes into three modules:

1. Absolute position measurement using external beacons that determine a frame reference. The beacons are infrared diodes and the camera is used to detect them.
2. Position computing using the shaft encoders to integrate the robot's movements and deduce its current position knowing it initial state (odometry).
3. Relative position measurement using some environment or characteristic object features.

Two important implemented cases are docking, where specific proximity sensors are used (e.g., sonar), and fixed obstacle edge referencing, using a laser.

Locomotion

This low-level module operates the wheel motors to produce the robot's moves. It computes and controls the accelerations and rotation speeds of each wheel to execute trajectories (straight lines, circles, clothoids) consistent with the elementary move orders it receives.

3.6 The Execution Control Rule System

Plan execution is mainly monitored through sensory information. The multisensory approach is very rich in the execution control context since some sensors, such as the laser or the ultrasonic range-finders, have a good real-time response and can thus be used to check rapidly the robot's world model, whereas the camera is used to build a new one. On the other hand, one sensor, such as the laser, may be used differently in the acquisition phase, where what is important is to provide a complete perception, and in execution monitoring, where quick measurements are often sufficient.

In the execution control mode, the system has to take into account the various inaccuracies induced by its sensors and effectors and the uncertainties in its world model.

This important problem in robotics has been addressed by several authors since the birth of the field (e.g., ⟨MUN 71⟩ and ⟨BRO 82⟩), and its solution is not straightforward. Our approach to this problem is based on the use of the multisensory data to check the information of one sensor by another and of the system's ability to react rapidly to changes in its factual knowledge. Thus some rules contain the necessary actions to be performed in specified failure conditions in the current plan step.

Sensor Integration

To date HILARE's perception system consists of five independent sensing devices: 1 TV camera, 1 laser range-finder, 14 ultrasonic emitter-receivers, 1 infrared emitter-receiver, and 2 optical shaft encoders.

Through the SPM structure and the execution control rule system the sensing devices operate in three modes:

1. *Cooperative* That is, for instance, the case of the TV camera and the laser range-finder jointly used to obtain 3-D scene analysis.
2. *Monitoring* In this very important mode one of the sensors monitors the operation of a module involving other independent sensing devices. This is, for instance, the case of the Move Along operator, which involves the sonars and can be monitored by odometry or laser position referencing.
3. *Cross Checking* In several configurations data can be obtained from two or more independent measures. For instance, after a "Move to" step both odometry and triangulation using laser telemetry information will allow one to compute the current robot location.

Given accuracy information on the different sensors, the cross checking mode provides

i. a dynamic ranking between measures and
ii. a consistency test (both measures match within accuracy ranges).

Whenever the latter fails, validation and, if necessary, higher level intervention will be requested by the system.

The Rule System

It is necessary to coordinate and organize the interactions of the various system's modules, and this can be achieved in several ways ⟨ELF 83⟩. We have chosen to investigate the rule approach, which appears to be very suitable to describing the actions that a system has to undertake in a set of situations that might occur in a specific domain ⟨VAI 83⟩.

The controller's rules are expressed in a LISP-based system. Each rule has a condition field and an action field. The condition field has 3 parts: ⟨pattern in database⟩, ⟨pattern not in database⟩, and ⟨evaluable expression⟩.

The action field directly modifies the database or is an explicit call of one or more of the system's procedures whose results will in turn change the robot's state description.

Rule examples

R_1: ⟨MONITORP-Si MOVEP⟩NIL NIL → STOP

Actions are monitored through sensors. This function is expressed by the predicate MONITORP-Si which concerns in this rule the sonar sensors. MOVEP is a predicate that expresses the motion state of the robot. The action STOP actually stops the robot. Hence it modifies the state parameters and deletes MOVEP.

R2: ⟨MONITORP-Si AVOIDP⟩⟨MOVEP⟩NIL
→ MOVE ALONG ($Z; v; C_1, C_2$) MONITOR S_i, C_1)

The move orders from the planner contain the predicate AVOIDP that enables the local avoidance of occasional obstacles.

The MOVE ALONG order is executed using the sonar system with a termination success condition.

In general, AVOIDP is added by the planner in places known to be usually empty and where the robot's trajectories are straight lines. Occasional obstacles are stated to be rather small (boxes). Hence the termination success condition is that the robot aims again toward the goal. The execution of the MOVE ALONG order is thus monitored through odometry and if necessary absolute position measurement.

R3: ⟨MONITORP-Si NIL ⟨POSITION-UNCERTAINTY.⟩.Z⟩
→ MEASURE-POSITION

R4: ⟨MONITORP-Si NIL ⟨POSITION-UNCERTAINTY.≤.Z⟩
→ MOVE TO (X, Y) MONITOR (S_i, C_i)

MONITORP-Si is added to the database after the avoidance successful completion. POSITION-UNCERTAINTY is compared with a preset value Z and a measure of the absolute position is made whenever it is necessary (R3). Otherwise the next MOVE order can be executed monitored by the ultrasonic system (R4).

The controller is an asynchronous sequential process that reacts to interruptions coming through its interface and according to its database contents, which includes the current state of the system. An interruption corresponds to a stimulus that has to be processed. A stimulus can be an explicit order to be executed or a particular event occurrence reported by a sensor which has been requested to do so earlier. This sequential interruption-driven process allows for parallelism in the execution of some specified orders if they do not share the same primitive processing modules (considered as ressources) and the sensors and effectors they operate.

The system has to react in real time to rapid changes in its environment. Hence, efficiency aspects become proeminent, and inefficiency is one of the most important disadvantages of rule system ⟨DAV 77⟩. To increase efficiency, we compile the rules into a decision tree. This is possible only if some constraints are met in the rule: at each cycle, there must be at most one rule that matches the context. The tree nodes are the conditions in the left-hand side of the rules. A path from the root to a leaf in the tree corresponds to one rule to be executed. For the rule compilation, we use algorithms developed by Ghallab ⟨GHA 82⟩.

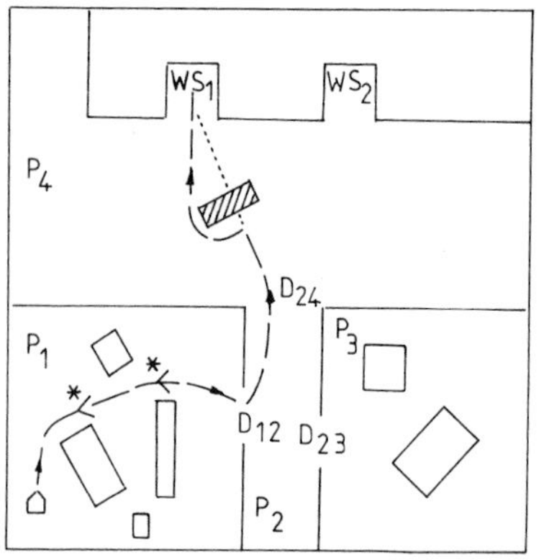

Figure 8
Example of robot task: "MOVE TO WS1." The general layout is known. Place 1 is initially partially known. Place 2 is a corridor. WS1 and WS2 are known work stations. * denotes that 3-D vision is performed.

3.7 The Execution Controller Interface

The interface deals with the input to the controller from the planner and the various other modules. Three different types of orders can reach the interface:

1. Imperative orders that must be processed immediately by the controller. This type of command can be produced by the planner or by the user and is transmitted directly in this case to the controller's interface. It supersedes any other order or sequence of orders in the interface and is considered in the next controller cycle, interrupting any current action.

2. Sequenced commands that have to be executed in a specified order. This sequence is usually a plan or a part of a plan. A given step of the sequence cannot be executed before the successful completion of the preceding one.

3. Commands that can be executed whenever the system's state enables their execution, without any specified ordering.

4 Example

Let us consider the situation (figure 8) in which the robot is initially in a room (place 1) and receives the order to go to work station 1:

"MOVE TO WS1."

In this initial state, the robot's world model contains a partial knowledge of the obstacle configuration in place 1 and the complete layout of the other places. Place 2 is known to be a corridor, and place 4 is an empty room with work station of known location and docking features.

This order is to be processed by the robot decision system and, at the NMCS level, a sequence of actions to be executed will eventually be produced. In this sequence, the robot will first navigate in place 1, avoiding the obstacles it gradually discovers, to reach the door D12 to place 2, where the next order MOVE ALONG is processed. This is an efficient step since place 2 is a corridor where there is no need for costly 3-D vision. The MOVE ALONG order is achieved on a success termination condition when the robot's position measured by a monitoring sensor (in this case odometry), reaches the set value of the entrance D24 of place 4. At this stage, two sequenced orders remain to be processed:

(MOVE TO (X, Y) (DOCK (frame (WS1); Z; C_1, C_2))

where X, Y are coordinates in the access area of WS1 where the robot has to dock. Place 4 is empty of obstacles, but occasional movable objects may be present in it. Thus the MOVE TO order will authorize to avoid such objects without any need to modify the current world model and plan.

Let us examine the system's behavior when the event of the existence of an unpredicted object on the robot straight line path is reported by the sonar system.

The database contents is

DB1: (MONITORP-Si MOVEP AVOIDP)

Rule R1 is valid and executed producing

DB2: (MONITORP-Si AVOIDP)

which tiggers rule R2. The MOVE ALONG order parameters are computed from the robot's state description.

The execution of MOVE ALONG deletes MONITORP-Si and adds MOVEP and AVOIDANCE:

DB3: (MOVEP AVOIDP AVOIDANCE)

After completion of this order we have

DB4: (MONITORP-Si AVOIDP)

Then either rule R3 or R4 is valid depending on the evaluation of position uncertainty. In any case, R4 will be eventually fired. The execution of R4 deletes MONITORP-Si and adds MOVEP,

DB5: (MOVEP AVODIP)

The completion of this order deletes MOVEP and AVOIDP since the goal is reached:

DB6: NIL

The next order, DOCK, is processed according to the robot's state:

DB7: (DOCK MOVEP)

The execution proceeds till the final completion of the sequence.

5 Conclusion

Research on mobile robots is today an important and fast growing field ⟨GIR 83⟩. Autonomous mobility is, we believe, the central problem at hand.

Current research on HILARE emphasizes this aspect. The navigation and motion control system which has been presented in this paper is now partially integrated and *in situ* experimentation is being carried out. Complete integration is in progress ⟨VAI 83⟩ and extensive experimentation to test fully the execution control rule system will then be achieved.

Among the other subjects which are investigated in the HILARE project we devote also an important effort to high level decision making, i.e., the general planner and the distributed decision structure and learning.

Learning is indeed among the most important research topics in robotics and very much a wide-open field. In the HILARE project, learning plays a central role, although we intend to keep ambitions and claims on the subject at a precisely defined and tractable level. In this paper learning was considered in two aspects related to automatic environment modeling.

References

⟨BRO 82⟩ R. A. Brooks, "Symbolic error analysis and robot planning," Intern. Jal of Robotics Research, vol. 1, Winter 1982.

⟨BRO 83⟩ R. A. Brooks and T. Lozano-Perez, "A subdivision algorithm in configuration space for findpath with rotation," Proc. of the 8th IJCAI Karlsruhe, W. Germany, 1983.

⟨CHA 81⟩ R. Chatila, "Système de navigation pour un robot mobile autonome: modélisation et processus décisionnels," Thesis Dissertation, U. P. S. Toulouse, France, July 1981.

⟨CHA 82⟩ R. Chatila, "Path planning and environment learning in a mobile robot system," Proc. of the European Conference on Artificial Intelligence, Orsay, France, July 1982.

⟨DAV 77⟩ R. Davis and J. King, "An overview of production systems," in Machine Intelligence, vol. 8, Chichester: Ellis Horwood, 1977.

⟨ELF 83⟩ A. Elfes and S. Talukdar, "A distributed control system for the CMU Rover," Proc. of the 8th IJCAI, Karlsruhe, W. Germany, 1983.

⟨ERM 80⟩ L. D. Erman et al., "The Hearsay-II speech-understanding system: integrating knowledge to resolve uncertainty," ACM Computing Surveys, vol. 12, June 1980.

⟨FER 82⟩ M. Ferrer, "Système multisenseur de perception 3D pour le robot mobile HILARE," Thesis Dissertation, U.P.S., Toulouse, France, December 1982.

⟨GHA 82⟩ M. Ghallab, "Optimisation de processus décisionnels pour la robotique," Thèse d'Etat, U.P.S., Toulouse, France, October 1982.

⟨GIR 79⟩ G. Giralt, R. P. Sobek, and R. Chatila, "A multi-level planning and navigation system for a mobile robot; a first approach to HILARE," Proc. 6th IJCAI, Tokyo, Japan, August 1979.

⟨GIR 83⟩ G. Giralt, "Mobile Robots," NATO Advanced Study Institute on Robotics and Artificial Intelligence, Castelvecchio Pascoli (BARGA), Italy, 26 June–8 July 1983.

⟨HAR 68⟩ P. E. Hart, N. J. Nilsson, and B. Raphael, "A formal basis for the heuristic determination of minimum cost paths," IEEE Trans. on System Science and Cybernetics, vol. 4, 1968.

⟨HAY 79⟩ B. Hayes-Roth and F. Hayes-Roth, "A cognitive model of planning," Cognitive Science, vol. 3, 1979.

⟨LAU 83⟩ J. P. Laumond, "Model structuring and concept recognition: two aspects of learning for a mobile robot," Proc. 8th IJCAI, Karlsruhe, W. Germany, 1983.

⟨LOZ 81⟩ T. Lozano-Perez, "Automatic planning of manipulator transfer movements," IEEE Trans. on Syst. Man and Cyb., vol. SMC-11, 1981.

⟨McG 79⟩ R. B. McGhee et al., "Adaptive Locomotion of a multilegged robot over rough terrain," IEEE Trans. Syst. Man and Cyber., vol. SMC 9, 1979.

⟨MUN 71⟩ J. H. Munson, "Execution, and monitoring in an uncertain environment," 2nd IJCAI, London, September 1971.

⟨NIL 69⟩ N. J. Nilsson, "A mobile automaton: an application of artificial intelligence techniques," Proc. of the 1st IJCAI, Washington, D.C., May 1969.

⟨NIL 80⟩ N. J. Nilsson, "Principles of artificial intelligence," Tioga Pub., Palo Alto, California, 1980.

⟨VAI 83⟩ M. Vaisset, "Intégration des structures décisionnelle et informatique pour le robot mobile HILARE," Thesis Dissertation, December 1983.

A MARS ROVER FOR THE 1990's

BRIAN H. WILCOX and DONALD B. GENNERY
Robotics and Teleoperators Research Group, Jet Propulsion Laboratory, California Institute of Technology, 4800 Oak Grove Drive, Pasadena, California, 91109, USA.

Some technical issues concerning a Mars rover launched in the 1990's are discussed. Two particular modes of controlling the travelling of the vehicle are described. In one mode, most of the control is from Earth, by human operators viewing stereo pictures sent from the rover and designating short routes to follow. In the other mode, computer vision is used in order to make the rover more autonomous, but reliability is aided by the use of orbital imagery and approximate long routes sent from Earth. In the latter case, it is concluded that average travel rates of around 10 km/day are feasible.

1. INTRODUCTION

In the 1970's, there was a Mars rover research program at the Jet Propulsion Laboratory (JPL) sponsored by the National Aeronautics and Space Administration (NASA), and results of that research were reported at the time [1-4]. However, the programme ended in 1979. Recently, interest in planetary rovers has revived. A JPL study [5] and a workshop [6] have helped to define the characteristics that a Mars rover launched in the 1990's should have. A small-scale internal research project at JPL is developing a few of the needed techniques. (Our current experimental rover is shown in Figure 1). Current NASA planning is for a launch of a Mars rover and sample return mission perhaps in 1998, if funding is forthcoming. This paper discusses some of the technical issues involved.

2. VISUAL NAVIGATION AND HAZARD AVOIDANCE

Because of the long signal time to Mars (anywhere from 6 minutes to 45 minutes for a round trip at the speed of light), it is impractical to have a Mars rover that is teleoperated from Earth (that is, one in which every individual movement would be controlled by a human being). Therefore, some autonomy on the rover is needed. On the other hand, a highly autonomous rover (which could travel safely over long distances for many days in unfamiliar territory without guidance form Earth and obtain samples on its own) is beyond the present state of the art of artificial intelligence, and thus can be ruled out for a rover launched before the year 2000. In between the two extremes just mentioned, various degrees of autonomy are possible. Two in par-

Fig. 1. JPL experimental rover.

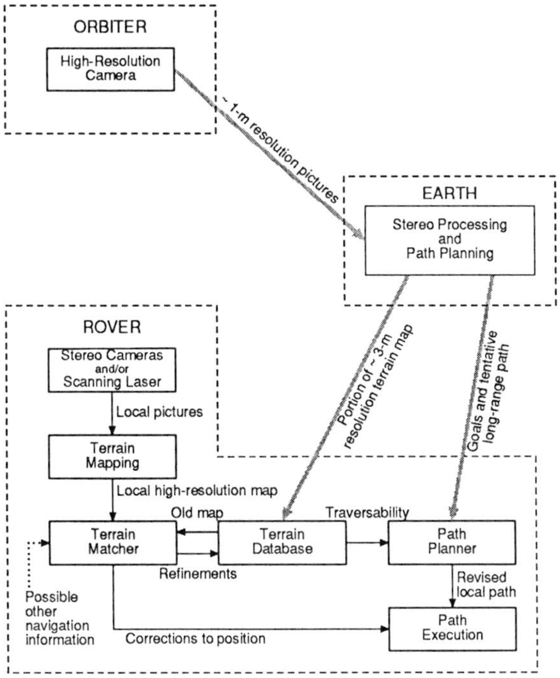

Fig. 2. Semiautonomous operation of Mars rover.

ticular that have been studied at JPL are called Computer Aided Remote Driving (CARD) and Semiautonomous Mobility (SAM).

With CARD, stereo pictures from the rover are sent to Earth where they are viewed by a human operator using a stereo display. He designates a path for the vehicle to follow as far ahead as he can see a safe path accurately in three dimensions. A ground-based computer computes the turn angles and path segment distances that define the path. This information is sent to the rover, which executes the path by dead reckoning, perhaps aided by computer vision. A new stereo pair of pictures is taken from the new position, and the whole process repeats. Depending on the terrain, the rover might travel about 20 meters on each of these iterations, each of which would take typically about 30 minutes. This results in an average speed of roughly one centimeter per second. (The CARD concept originally was developed by Wilcox for another project [7]).

The CARD method might be suitable for controlling the rover in an area where many scientific experiments need to be done, so that the operator would want to command the rover to drive only a short distance before it needs to stop to perform some experiment or to obtain a sample. Also, in particulary difficult areas, it might be used for safety. However, when a long-distance traverse is needed, a faster means of travel is highly desirable.

In the SAM method, local routes are planned autonomosly using images obtained on the vehicle, but they are guided by global routes planned less frequently by human beings using a topographic map, which is obtained from images produced by a vehicle orbiting Mars. The orbiter could be a precursor mission which would map a large area of Mars in advance, or it could be part of the same mission and map areas only as they are needed. (The SAM concept was developed by Gennery, based on a suggestion from others at JPL that, since orbital images probably will be available for landing safety certification, a way should be found to use them for rover navigation. It formed a part of two Mars rover studies [5,6]).

The sequence of operations in the portion of SAM involving Earth is as follows. As commanded from Earth, the orbiter takes a stereo pair of pictures (by taking the two pictures at different points in the orbit) of an area to be traversed (if this area is not already mapped). These pictures might have a resolution of about one meter, although poorer resolution could be used. The pictures are sent to Earth, where they are used by a human operator (perhaps with computer assistance) to designate an approximate path for the vehicle to follow designed to avoid large obstacles, dangerous areas, and dead-end paths. This path and a topographic map for the surrounding area are sent from Earth to the rover. This process repeats as needed, perhaps once for each traverse between sites where experiments are to be done, or perhaps once per day or so on long traverses.

The sequence of operations in the portion of SAM taking place on Mars is as follows. The rover views the local scene and, by means discussed below, computes a local topographic map. This map is matched to the local portion of the global map sent from Earth, as constrained by knowledge of the rover's current position from other navigation devices or previous positions, in order to determine the accurate rover position and to register the local map to the global map. The local map (from the rover's sensors) and the global map (from the Earth) are then combined to form a revised map that has high resolution in the vicinity of the rover. This map is analysed by computation on the rover to determine the safe areas over which to drive. A new path then is computed, revising the approximate path sent from Earth, since with the local high resolution map small obstacles can be seen which might have been missed in the low-resolution pictures used on Earth. Using the revised path, the rover then drives ahead a short distance (perhaps ten meters), and the process repeats.

With the computing power that it will be practical to put on a Mars rover in the 1990's, the computations needed to process a stereo pair of images and perform the other calculations needed may require roughly 60 seconds. If these are needed every 10 meters and it takes the rover 30 seconds to drive 10 meters, the resulting average rate of travel is 10 meters every 90 seconds, which is 11 cm/sec or 10 km/day. (Likely ranges for these values are discussed in Section 4). This is much faster than the CARD method allows. The SAM method also does not require continuous communication with Earth, as does CARD. If a ten-kilometer path is designated from Earth each time, only one communication per day is needed, and the rover could continue to drive all night, using strobe lights for illumination. On the other hand, the SAM method is more reliable than fully autonomous operation, because of the human guidance and the overview that the orbital data provides. However, for backup in case of communication failure, the ability for more autonomous operation is present, since the rover can plan its own path based on the area that it can see, provided that some global navigation system is available. Missions further in the future, after more sophisticated algorithms have been developed and more confidence in their abilities has been gained through experience, may be allowed to operate autonomously most of the time.

The steps in the SAM method are illustrated in Fig. 2. Portions of the steps involved now will be discussed in more detail.

On Earth, an elevation map must be produced. This can be done by human operators using a stereo display, by a computer stereo program, or by a combination of these. Probably, the best combination is to generate the elevation map by computer, and then to have human operators correct its mistakes. (The resolution of the elevation map probably would be slightly lower than that of the pictures from which it was produced). A large amount of computing will be needed for this task, but sufficiently powerful computers already exist, and since the computation is done on Earth there should be no problem. A portion of this elevation map (or some representation extracted from it, as described below for the rover) is sent to the rover.

The other task done on Earth is the planning of the global path for the rover. This can be done in several ways. Probably the best way is for a human operator to view in a stereo display the stereo pictures from the orbiter and to designate the path manually. Another possibility is to use the elevation map discussed above, possibly in conjunction with the brightness data in the images, for path planning either by a human operator, a computer path planning algorithm, or a combination of both.

There are several types of computations that need to be done on the rover (or on a Mars orbiter in constant communication with the rover). These include the computation of a depth map, the computation of a topographic map, the matching of this map to the global data base and merging with it, analyzing the traversability of the area, planning a path, and monitoring the execution of the path. Some of these now will be discussed in more detail.

The first step in the processing on the rover is the production of the depth map (the distances to densely packed points over the field of view of the sensing device). One way of obtaining this is with a scanning laser range finder, which produces the depth map directly. Another way is to use two or more cameras for stereo vision. By the usual stereo process of matching and triangulation the depth map is computed. Other computer vision techniques, such as shape from shading and texture analysis, can aid in this process. Each approach has advantages and disadvantages. Stereo vision usually is more accurate at close ranges but less accurate at long ranges than laser range finders. On the other hand, laser range finders are limited in the maximum range at which they are effective. Laser range finders tend to make fewer errors than stereo vision, but each can fail to produce results under different conditions. Stereo vision and other computer vision techniques are attrative since stereo cameras will be available for scientific sample designation or core locating purposes in any event. Also, solid state cameras are small, have no moving parts, and take less power than laser scanners. Stereo matching is a computation-intensive process which is benefiting greatly from recent advances in microelectronic fabrication, while scanning mirrors remain prone to mechanical problems. Most likely, a rover should use both a scanning laser range finder and stereo cameras, to produce the best results by combining their measurements and to provide reliability in case of failure.

The depth map is transformed into an elevation map (altitudes for densely but unequally spaced horizontal positions). An important issue is whether to keep the data in the iconic form of the elevation map, in which case the topographic map sent from Earth also would be in this form, or to reduce the data to a more symbolic form.

In the iconic case, the elevation map from one view is merged with the elevation map in the data base by a process of correlation and averaging, which also produces the best estimate of vehicle position as that which produces the best correlation. (Information other than elevations, such as reflectance, could be used also). However, this computation is more complicated than ordinary correlation bacause the points are not equally spaced, there may be significant uncertainties in their horizontal positions, and there may be occasional mistakes in the stereo data. This approach is currently being studied at JPL, and we hope to demonstrate a simple version of it in 1987 with our experimental rover.

In the symbolic case, some description of objects in the scene would be extracted, for example ellipsoidal approximations of rocks [8] together with descriptions of ground slope. The same type of description would be developed from the orbital images on Earth and sent to the rover, and the matching and merging process would use these symbolic descriptions [9]. Such symbolic methods may be studied in the future.

With either kind of description, the local data are merged with the global data base to produce an updated data base. In some cases, each new view would be merged immediately with the global data base. However, in some cases, the matching process may not be able to correlate accurately with the global data base because of the lack of prominent features, but there may be enough smaller features to correlate with the high-resolution views seen previously from nearby locations. Therefore, a local data base could be built up by merging several local views. Then when suffiently prominent features are encountered to match well to the global data base, the local data base would be merged with it. In general, there could be a hierarchy of data bases produced in this manner.

The traversability can be determined by analysing the data base to determine the slope and roughness of the ground at each horizontal position. This can be done by local least-square fits of planar or other surfaces and analysis of the residuals. A way of doing this for the iconic representation will be tried in the current JPL project. (If the data base is in symbolic form, the information may already be there in the form needed).

3. GLOBAL NAVIGATION

In addition to the local visual navigation, the rover should have at least one, and preferably several, ways of determining its absolute position on Mars.

The rover probably will have an inertial navigator, whose elements consist of gyroscopes and accelerometers. Mechanical gyroscopes usually need to be fairly massive in order to achieve high accuracy. However, JPL is developing for other space applications a Fiber Optic Rotation Sensor (FORS) [10,11], which in effect is a laser gyroscope with no moving parts. Drift rates of 0.001 degree per hour can be achieved in a three-axis device consuming 10 watts, with a mass of 5 kg. Expected improvements will allow a slight reduction in power and a 10-fold reduction in volume by the mid 1990's. This will allow the rover to have a true-north reference good to better than a minute of arc, an absolute latitude determination good to better than a kilometer, and a longitude drift of about one km/day. The longitude can be corrected by occasional celestial observations, for which the rover's vision cameras can be used. (The celestial fixes also can be used to calibrate the inertial components). In addition to being a position-measuring device, the inertial navigator's accurate orientation information, together with a wheel-revolution counter, makes possible accurate dead-reckoning for both CARD and SAM operation.

Various radio navigation systems are possible. For example, radio beacons could be deposited on Mars and detected by the rover. At least, the sample return vehicle should have a beacon to aid the rover in finding it when it needs to deliver samples that it has gathered. Also, differential very-long-baseline interferometry ($\triangle$VLBI) allows the measurement from Earth of the difference in positions of two transmitters in the vicinity of Mars (such as the position of the rover relative to a beacon on the surface) with an accuracy of about 200 meters.

If the orbiter is still active during the rover's operation and if the resolution of its pictures is good enough, it is possible that the rover could be seen in the orbiter's pictures. Operators on Earth then could determine the rover's position.

4. COMPUTATION REQUIREMENTS

One of the conculsions of the recent workshop [6] is that three numbers are key to the design of the Mars Rover: the power requirements of computing in watts per MIPS (million instructions per second), the power requirements of mobility in watts per kilogram of rover mass per m/sec of forward travel, and the number of instructions needed to sense the terrain, to perceive hazards and obstacles, to model the terrain/vehicle interaction, and to plan a safe course for the vehicle, expressed in million instructions per meter of forward travel.

These three numbers, plus the vehicle mass and power budget, can be combined using very simple algebra to give the vehicle speed, the needed capacity of the on-board computer (in MIPS), and the distribution of power between the computing and mobility subsystems. It was estimated that a radiation-hardened, flight qualified, general purpose multiprocessor could be configured in the mid 1990's with 3-5 MIPS performance for 20-25 watts power consumption, complete with its necessary I/O and memory. This translates into 4 to 8 watts per MIPS. It also was estimated that the mobility power requirements will range from 0.6 to 8 w/kg per m/sec (corresponding to an effective frictional coefficient of 0.15 to 2). Lastly, it was estimated that some 50 to 500 million computer instructions will need to be performed for each meter of safe travel by a rover with any significant degree of autonomy.

Table 1. (From [6]): 1000 kg Rover with 500 watt power supply.

CASE	AVERAGE ROVER SPEED	POWER DISTRIBUTION	COMPUTER PERFORMANCE
Worst case	4 cm/sec	67% Mobility 33% Computation	20 MIPS
Moderate case	17 cm/sec	69% Mobility 31% Computation	25 MIPS
Best case	62 cm/sec	75% Mobility 25% Computation	30 MIPS
Nominal Mobility Worst Computing Worst Software	8 cm/sec	33% Mobility 67% Computation	40 MIPS

These three sets of estimates for the three key numbers give the results in Table 1, where the geometric mean for the high and low estimates is used for the moderate value, and where it is assumed that the rover has a mass of 1000 kg and a 500 watt power supply.

It is encouraging that, even in the worst case for all parameters, the average rover speed will be some four times faster than can be achieved by Earth-based path designation (CARD). The last case in Table 1 is based on worst-case estimates for computer and software performance but using the nominal estimate for the power requirements for mobility. This seems historically justified based on the relatively slow progress in making high performance, deep-space qualified electronics and also the slow progress in autonomous system software development. Mobility, it is assumed, is the best understood of the three. In this case, two thirds of the power goes to computing, and a computer with up to 40 MIPS performance would be used. This is at least 100 times the performance of any computer that has heretofore been used on a planetary spacecraft. These figures lead to the conclusion that custom VLSI or special computational hardware (such as a single-instruction-stream multiple-data-stream processor array or a pipelined image processor) would be very attractive in performing some of the necessary computations while using significantly less power than the estimates leading to Table 1.

5. MANIPULATOR CONTROL

A manipulator arm and associated control system must be provided for sampling operations on Mars. The sampling subsystem must allow the Earth-bound operator to designate a desired sample or coring loction, and for the rover to retrieve that sample or core without detailed low-level instruction or interaction. Additional tasks such as righting an overturned rover or simple instrument repair conceivably could be required. These two extremes lead one to think in terms of a "strong arm" and a "smart arm". However, it is not necessarily true that the strong arm cannot be precise, even delicate, in its actions. Also, limited-degree-of-freedom programmable jigs may allow even a single arm to perform all necessary functions.

A Mars rover manipulator requires a combination of weight, stength, power, and reach which is very different from that of commercial robot manipulators. Even light-duty commercial electric manipulators generally weigh 20 times their payload rating, and consume a kilowatt or more of power. They offer only position-servoing, which means that forces on the environment are not sensed or controlled. Present manipulator command and control systems program the robots by moving them to the desired points with a "teach pendant" and recording the positions, a technique not suitable for planetary exploration. Multivariable position and force control techniques, which are - particularly necessary for precision manipulators, are not commercially available. Three-dimensional spatial calibration with a camera system also is not available. Some of this capability is being developed under the NASA Telerobotics program; much work specific to rovers needs to be done.

A manipulator control subsystem for a Mars rover consists of two parts: a supervisory command-language interpreter to translate human-generated commands and cursor designations into a series of actuation commands, and a manipulator control system to execute the actuation commands to control both the position and the contact forces and torques of the manipulator relative to the environment. If the general multi-processor is to perform all the servo-level, coordination-level, and process-level control of the arm(s) then clearly appropriate I/O channels must emanate from the various processors to allow high-rate control. The Earth-based control station computer would translate human three-dimensional designations and symbolic references (e.g. "pick up the rock at this location" after appropriate postioning of the 3-D cursor) into rover-centered coordinates and plan a nominal collision-free trajectory for the manipulator. The on-board computer would servo to this nominal position-force-time history and compensate for disturbances, check for acceptable error bounds, and halt or recover upon error conditions, notifying the operator of the problem with as much dianostic information as possible on the first transmission. The supervisory command language should be a compromise between a completely autonomous manipulator control system and direct teleoperation. The supervisory language, together with the control system, will allow the rover to conduct rather complex operations such as complete sampling, coring, or other manipulation operations within each speed-of-light round trip.

6. COMMUNICATION BANDWIDTH

It is anticipated that the rover will carry stereo cameras with about one-milliradian resolution. A Computer-Aided Remote Driving command cycle would require that the forward quadrant of panorama be transmitted to Earth for use by the human operator in planning the local path. This means that about a million pixels from each camera need to be transmitted (giving a 1.5 x 0.7 radian field of view) in a time short compared to the 30-minute typical speed-of-light round trip to Mars, say 5 minutes. For 8-bit pixels compressed to 4 bits (a modest compression which should allow perfect image reconstruction even with large scene activity), this would be 8 megabits in 300 seconds, or just under 30 kbit/sec. This communication data rate for the downlink will permit a fallback to entirely Earth-based control, and will allow the scientific team to select sampling or drilling locations. (The transmission of the designated path from Earth to Mars involves very little data).

For the semi-autonomous control approach, the downlink requirements for the rover are much less severe (being mostly for the science package), but the requirements on the uplink to the rover and the downlink from the imaging orbiter are more severe. If the orbiter is to image a 10-km square area each day at meter-scale resolution then (again in stereo and again compressed to 4 bits per pixel) some 800 million bits would have to be transmitted form the orbiter to Earth, which at 100 kbit/sec would take about 2 hours. (In some cases a strip much narrower than 10 km might suffice, with a considerable reduction in the

Fig. 3. A rover collects samples on the surface of Mars in this depiction by artist Ken Hodges. In the background is the ascent vehicle and an orbiter (not to scale) flies overhead. A Mars Rover Sample Return mission is now being investigated in a joint study conducted by the Jet Propulsion Laboratory and the Lyndon B. Johnson Space Center. *NASA*

transmission requirement). This imagery would be processed into an elevation map (perhaps with 3-meter resolution compressed to 4 bits per pixel) and a strip perhaps 400 meters wide by 10 km long centered on the nominal route plan would be uplinked to the rover for use in local path refinement. This would require about 1.8 megabits to be uplinked to the rover, which would take 15 minutes at 2 kbit/sec. If higher resolution images are available and no communication orbiter is available, it would be preferable to increase this 2 kbit/sec uplink rate to allow the daily map to be sent to the rover quickly (especially if the rover must be stopped during communication with Earth). On the other hand, if a symbolic rather than an iconic representation is used, even less transmission from Earth to the rover might be needed.

7. SCALING CONSIDERATIONS

Although it is usually assumed that a single rover/lander system is to be launched from a single Space Shuttle without any in-orbit assembly or other complexities, obviously other scenarios are possible. It is natural to ask whether larger or smaller rovers would be superior to the few-meter-scale rover usually considered. Some of the scaling consideration are discussed below.

In the case of surface mobility, the fractal nature [12] of the surface determines whether larger or smaller rovers are more capable. If the fractual dimension increases with scale (say over the range of 0.1 to 10 meters), then smaller rovers are better (they fit between the obstacles that the larger rover can't get over), whereas if the fractal dimension decreases with scale, then larger rovers are better (they can drive over the obstacles that the smaller rover can't fit between. If the fractal dimension is constant, then size doesn't matter. Note that at the largest scales (above 10 km) the dimension decreases to 2 (since the surface is effectively two-dimensional at the scale of the planet). However, it is unknown what the trends of fractal dimensions are over the scale of interest in the areas on the Martian surface likely to be visited. Future orbital missions may supply the missing data.

Since rolling friction requires approximately constant energy per unit distance per unit weight, small rovers can be assumed to move as far and as fast as larger rovers if their power supply scales with their mass. This may be counter-intuitive, but it assumes that wind resistance and other dynamic effects are negligible, since rover motion will almost certainly be quasi-static. However, solar power goes with square of the scale, while rolling friction goes with the cube; thus solar power is attractive for small rovers. At some scale solar power will provide more energy than radioisotope power sources, fuel cells, or other alternative power supplies. However, the absence of solar power at night and the substantial reduction that will be observed during dust storms and the frequent morning ground fog found by the Viking landers limits the attractiveness of solar power even for small rovers. Also, the waste heat of a radioisotope helps to keep the rover warm (temperatures drop to 160 K at night), overcoming a problem for small rovers.

The computation power per image needed for computer vision is independent of the scale, if the angular resolution and size of the images are constant. However, the stereo baseline (separation of the stereo cameras) is proportional to the size of the rover. Since all the linear dimensions involved can be scaled proportionally, the distance at which the rover can reliably see obstacles having a certain proportion to its size is proportional to its size. Therefore, if the rover is limited by its computational ability (assumed to be constant), its average speed is proportional to its size, for autonomous or semiautomous operation. However, the computational power may be proportional to the cube of size (because the mass of the computer and the mass of the power supply may be the limiting factors), in which case the average speed would be proportional to the fourth power of size.

The 30 kbit/sec data rate needed for the Computer-Aided Remote Driving concept requires fixed power and antenna surface area of some 100 watts and 0.2 m^2 respectively (at K_a band) to communicate with the existing Deep Space Network. If lower data rates are used (i.e. for semiautonomy) then perhaps 1-10 kbit/sec could be used, allowing power and surface reqirements as low as a few watts and 10 cm^2 (again at K_a band). Optical communication might have much lower power, aperature, and weight requirements. Note that CARD is an essential fall-back for rover control and for scientific sampling operations.

Most of the above points seem to make larger rovers more desirable. However, many small rovers may be much more reliable than a single large rover, although presumably the large rover has individual components which are more robust.

8. CONCLUSIONS

It is practical for a Mars rover launched in the 1990's to travel over Mars at approximately 10 km/day with partial autonomy, and to collect samples as directed from Earth. The safe guidance and control of the rover on Mars at such speeds will require substantial on-board sensing and computing resources. The availability of a high-resolution terrain data base will greatly reduce the risks of semiautonomous operation of the rover.

REFERENCES

1. M.D. Levine, D.A. O'Handley, and G. M. Yagi, 'Computer Determination of Depth Maps', *Computer Graphics and Image Processing,* **2**, 131-150 (1973).
2. D.A. O'Handley, 'Scene Analysis in Support of a Mars Rover', *Computer Graphics and Image Processing*, **2**, 281-297 (1973).
3. R.A. Lewis and A.R. Johnston, 'A Scanning Laser Rangefinder for a Robotic Vehicle', Proc. Fifth International Joint Conference on Artificial Intelligence, Cambridge, Massachusetts, 762-768 (1977).
4. A.M. Thompson, 'The Navigation System of the JPL Robot', Proc. Fifth International Joint Conference on Artificial Intelligence, Cambridge, Massachusetts, 749-757 (1977).
5. J.E. Randolph (ed.), 'Mars Rover 1996 Mission Concept', JPL D-3922, Jet Propulsion Laboratory, Pasadena, California (1986).
6. J.C. Mankins (ed.), Proc. Technology Planning Workshop for the Mars Rover, Jet Propulsion Laboratory, Pasadena, California (1987).
7. K.G. Holmes, B.H. Wilcox, J.M. Cameron, B.K. Cooper, and R.A. Salo, 'Robotic Vehicle Computer Aided Remote Driving', Vol. **1**, JPL D-3282, Jet Propulsion Laboratory, Pasadena, California (1986).
8. D.B. Gennery, 'Object Detection and Measurement Using Stereo Vision', Proc. Sixth International Joint Conference on Artificial Intelligence, Tokyo, Japan, 320-327 (1979).
9. D.B. Gennery, 'A Feature-Based Scene Matcher', Proc. Seventh International Joint Conference on Artificial Intelligence, Vancouver, British Columbia, 667-673 (1981).
10. B.R. Youmans, W.C. Gross, R.K. Bartman, and N.M. Nerheim, '1.3μm All-Fiber Passive Optical Rotation Sensor', Proc. Optical Testing and Metrology, *SPIE* **661**, Quebec City, Canada, 177-179 (1986).
11. R.K. Bartman, B.R. Youmans, and N.M. Nerheim, 'Integrated Optics Implementation of a Fiber Optic Rotation Sensor: Analysis and Development', Fiber Optic Gyros: tenth Anniversary Conference, *SPIE* **719**, Cambridge, Massachusetts, 122-134 (1986).
12. B.B. Mandelbrot, 'Fractals: Form, Chance and Dimension', W.H. Freeman and Co., San Francisco, 1977.

ACKNOWLEDGEMENT

The research described in this publication was carried out by the Jet Propulsion Laboratory, California Institute of Technology, under contract with the National Aeronautics and Space Administration.

A MOBILE ROBOT: SENSING, PLANNING AND LOCOMOTION *

David J. Kriegman
Ernst Triendl
Thomas O. Binford

Robotics Laboratory
Department of Computer Science
Stanford University, California 94305

Abstract

A mobile robot architecture must include sensing, planning, and locomotion which are tied together by a model or map of the world based on sensor information, a priori knowledge and generic models. The architecture of a Stanford's autonomous mobile robot is described including its distributed computing system, locomotion, and sensing. Additionally, some of the issues in the representation of a world model are explored. Sensor models are used to update the world model in a uniform manner, and uncertainty reduction is discussed

1 Introduction

Stanford's autonomous mobile robot project is aimed at developing a vehicle that navigates within unstructured man made environments such as the inside of a building. To accomplish this goal, the robot must be able to sense its environment as well as move about within it.

The vehicle at Stanford is a testbed for a variety of sensors each with its own possibilities and limitations. Generally, one finds a tradeoff between resolution, accuracy, repeatability, and speed in sensing and processing; so no single form is adequate for all situations. The most flexible yet most costly sense, in terms of processing complexity, is computer vision. Vision affords a very detailed description of the environment. On a broader and coarser scale, ultrasonics are used to determine the presense of obstacles that the vision system may have overlooked. For closer proximity and protection, there is tactile sensing. The characteristics of these sensors must be modeled and their information must be intelligently integrated into a map or world model that is used for navigation and in the future, task planning.

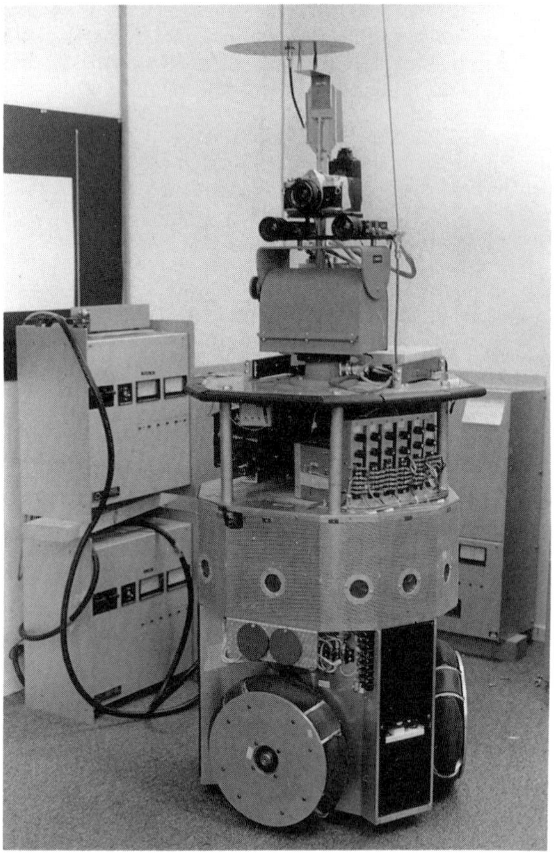

Figure 1: The robot

*Support for this work was provided by the Air Force Office of Scientific Research under contract F33615-85-C-5106, Image Understanding contract N00039-84-c-0211 and Autonomous Land Vehicle contract AIDS-1085S-1. David Kriegman was supported by a fellowship from the Fannie and John Hertz Foundation.

2 System Architecture

The Stanford Mobile Robot (figure 1), known as the Mobile Autonomous Robot Stanford (MARS) or more affectionately, Mobi, is an omnidirectional vehicle (Three Degrees of freedom) with a novel three wheel configuration. The robot is essentially cylindrical with a diameter of 65cm and height of 170cm. The wheels, forming the edges of an equilateral triangle, have six contoured passive rollers along their chords, allowing the wheel to move sideways while driven. With a combination of driving the wheels about their major axis, and passive turning of the rollers, the vehicle can translate while simultaneously rotating. With this form of locomotion, the vehicle is well suited for indoor travel and maneuvering about tight obstacles. However, because of the small roller diameter, the vehicle is confined to traveling over relatively smooth terrain.

Each of the wheels' motor contains a shaft encoder that is used for odometry and trajectory following. Trajectories are specified by two systems: B-splines connect knot points to form a path with smoothly changing velocity. A second system, called Smooth Driver, connects straight line segments with clothoid curves yielding trajectories that have continuous curvature and smooth acceleration along the path [6].

The robot has four sensing modalities: vision, acoustics, tactile, and odometry. Stereo vision, using two onboard cameras, returns the 3 dimensional location of vertical lines within its field of view. and the details of this system can be found in [10]. The cameras are mounted on a pan/tilt head, giving them two degrees of freedom. The acoustic system is composed of twelve polaroid sensors equally spaced about the circumference of the robot. These sensors return a distance that is proportional to the time of flight of an echoed acoustic chirp. To a first approximation, the acoustics find the nearest object within a thirty degree cone. Though the acoustic system has very low angular resolution, it measures depth quite accurately making it useful for guarded moves. The information can also be incorporated into the world model as well as checking consistency of the existing model. As a last line of defense, if the vision and acoustic systems miss an object, there is a "tactile" sensing system, composed of twelve bumpers with internal tape switches. Besides emergencies, the information can augment the world model. The bumpers are also useful for navigating through tight areas such as doorways where the edges of the door are within the minimum ultrasonic range and not within the region of stereo vision. Finally, odometry determines the robot's position.

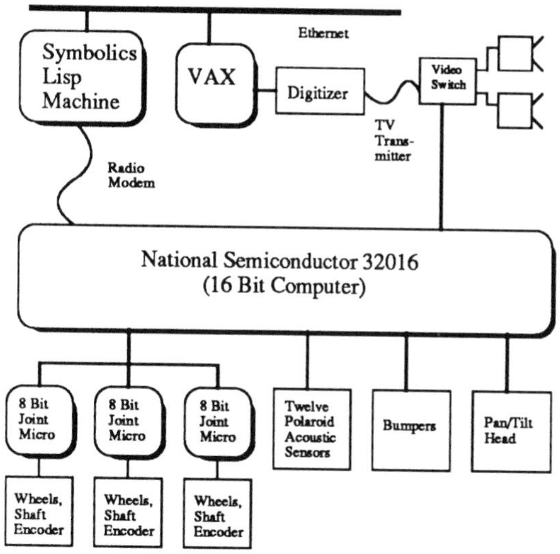

Figure 2: The robot's architecture

The robot's computational system is distributed onboard and offboard, and figure 2 presents an overview of the architecture and subsystems. Onboard, a sixteen bit computer (National Semiconductor 32016) is responsible for trajectory planning, sensor data acquisition and communication. Because of the unnecessary burden of onboard program development, most of the intelligence is offboard, and so the onboard computer executes simple commands. Additionally, it handles real time emergencies such as stopping when a bumper is unexpectedly pressed (a crash).

The robot communicates, via a digital radio link, with a (necessarily offboard) Symbolics Lisp Machine, which is responsible for planning operations such as sensor interpretation (except vision), sensor fusion, uncertainty reduction, map making, and path planning. The lisp machine also graphically displays the robot's status, including position and sensor values. For safety, planning algorithms are developed and tested on a simulator. A TV transmitter sends the video signal from the cameras to the digitizer. From there the image is transferred, via DMA link, to a VAX which does the image processing and computes stereo correspondence points that are sent to the Lisp Machine over an ethernet.

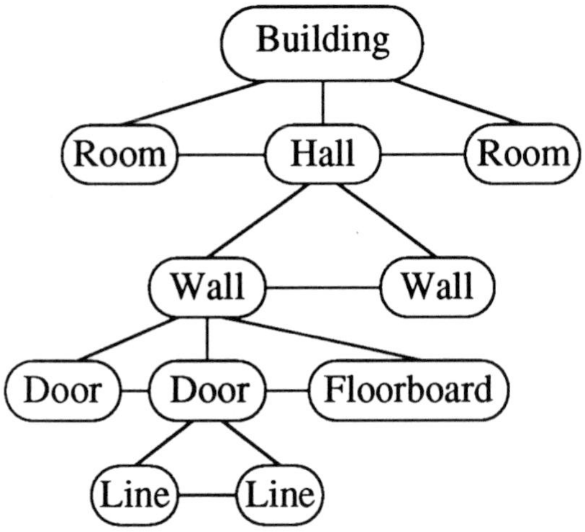

Figure 3: Hierarchical model

3 World Model

A critical design issue is the representation of the map or world model [3, 2, 7] The map must richly describe the world without burdensome details. Consider the purpose of a map. Generally, a mobile robot must reach a location at which to perform a task (e.g. getting my coffee cup from my desk). Without a map the robot could search all of the offices in my building and check for coffee cups using a straight forward search algorithm. Instead, the map can be searched for my room and then the robot can go there and look for the coffee cup.

Since the cost of a graph search is O(n), keeping the map small will speed search. An alternative to a small map, is a hierarchical representation so the search will be over a smaller map at each level of the hierarchy.

To make robots more useful, the robot, rather than a person, should build the maps, not the person. Thus, the map is built bottom up, and the lowest level of information is closest to the actual sensor measurement while higher levels of the hierarchy become more abstract and symbolic. Figure 3 shows the hierarchy of objects that might describe a hallway in a building. At the lowest level are the points and lines found from the sensor information. This information is fit to a model of generic objects such as walls, doors, and windows. Two parallel walls that bound an elongated region of free space could be a hall, and hallways are found in buildings. So, when searching for a route between rooms in a building, we would search at the building level for route and then

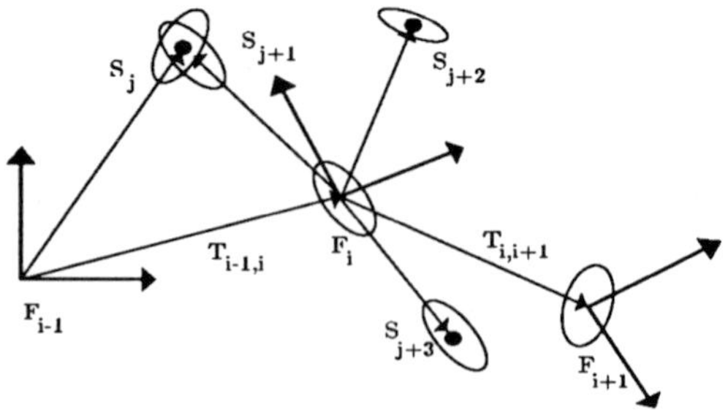

Figure 4: Relationship of frames and sensor readings

find paths along successive levels of the map.

Another issue is whether to represent the world as 2 or 3 dimensional. Since our robot is confined to the indoors and only travels on fairly flat level surfaces, a 2-D coordinate frame represents robot location. However, the robot observes objects that are not confined to lie on the floor, and their location is in 3 dimensions.

Furthermore, all sensing modalities posses some degree of uncertainty in measurement. Because the inferred location of an object sensed from two distinct locations would be different without considering uncertainty, this must be represented, and so the map cannot be a grid based on an absolute coordinate system but must express the relationship of points to one another [2]. So, sensor data and their interpretation are represented relative to the location of the robot and the map becomes stretchy. Since the map cannot be based on a grid, hierarchical grid systems such as quad-trees are inappropriate.

A final issue is uncertainty representation. Uncertainty manifolds have been proposed where the actual value of a measurement is within a bounded region [2] When determining the uncertainty of a sequence of motions, for example, the bounding manifold of the convolution of the individual manifolds is used. This has the unfortunate property of being overly conservative. Instead of using the bounding region, we describe measurements by a normal distribution (Gaussian), retaining the first two moments of the measurement probability density function (i.e mean vector, $\vec{\mu}$ and covariance matrix, Σ). All sensor data is gathered at discrete locations, and for each sensing, we create a Frame, F_i which is a local coordinate system with the robot at the origin. This frame is related to the previous frame, F_{i-1} by a transformation $\vec{T}_{i-1,i} = (x_i y_i \phi_i)^T$ which has a mean vector and covariance matrix. Additionally, each sensor reading S_j is related to the frame where the reading was taken. An example of these relationships is shown in figure 4. Locations are drawn as ellipses because ellipsoids are the equi-probability density contours of a multivariate normal distribution. The larger the ellipse, the greater the uncertainty.

Uncertainty can be propagated or compounded [9]. For example if we know $\vec{T}_{i-1,i}$ and $\vec{T}_{i,i+1}$ we can determine the mean and covariance matrix of $\vec{T}_{i-1,i+1}$. Additionally, the uncertainty of the location of a point can be reduced if the point was sensed from two different frames. In figure 4, the point described by S_j and S_{j+1} are the same, and we can get tighter bounds on this location as well as the relationship between F_{i-1} and F_i.

4 Sensor Models

The mobile robot receives information about its environment from its shaft encoders, contact bumpers, ultrasonics, and stereo vision. This information has some common characteristics (e.g. spatial location of objects) and some unique characteristics (e.g. vision can give the color of an object). The fundamental measurement that all of these sensors yield is the position of an object s_i which is described by a multivariate normal probability density function with mean vector $\vec{\mu}_{s_i} = (\mu_{x_i} \mu_{y_i} \mu_{z_i})^T$ and covariance matrix

$$\boldsymbol{\Sigma}_{s_i} = \begin{pmatrix} \sigma_{x_i x_i} & \sigma_{x_i y_i} & \sigma_{x_i z_i} \\ \sigma_{x_i y_i} & \sigma_{y_i y_i} & \sigma_{y_i z_i} \\ \sigma_{x_i z_i} & \sigma_{y_i z_i} & \sigma_{z_i z_i} \end{pmatrix} \quad (1)$$

and so the probability that a sensed point s_i is at $\vec{x}$ is:

$$f_{s_i}(\vec{x}) = \frac{e^{-\frac{1}{2}(\vec{x}-\vec{\mu})^T \Sigma^{-1} (\vec{x}-\vec{\mu})}}{2\pi\sqrt{|\Sigma_i|}} \quad (2)$$

4.1 Robot Motion

As the robot moves, shaft encoder readings $\vec{\varphi}$ are mapped through the kinematics K to determine robot velocity and integrated to calculate vehicle location. So the motion is

$$\vec{p} = \int_{t_0}^{t} K(\frac{\partial \vec{\varphi}}{\partial t}) dt \quad (3)$$

After the robot has moved far enough a frame is created and related to the previous one by a transformation described by a normal random vector, $\vec{T}_i = (x_i y_i \phi_i)^T$.

When our robot moves, it is generally accurate in distance traveled but its orientation can be off by a few degrees. This leads to fairly large Cartesian errors in the orthogonal direction to the travel direction. Analogously, when a flagpole moves in the wind, its height will not change drastically, but the top of the pole may move parallel to the ground by many feet. Thus for a commanded straight line motion of:

$$\Delta\vec{P} = \begin{pmatrix} \Delta x \\ \Delta y \\ \Delta\phi \end{pmatrix} = \begin{pmatrix} \Delta r cos \Delta\theta \\ \Delta r sin \Delta\theta \\ \Delta\phi \end{pmatrix} \quad (4)$$

the mean, μ_p is computed by equation 3, and in the direction of motion, the covariance matrix of the uncertainty in position is

$$\Sigma_r = F(\Delta r, \Delta\phi) \quad (5)$$

Σ_r can be rotated by θ yielding Σ_p. Thus, the transformation between two successive frames is described by μ_p and Σ_p.

4.2 Bumpers

The mobile robot has twelve bumpers with internal contact switches along the edges of a nonregular Dodecagon. In addition to protection when the robot accidently crashes, the bumpers provide very definite information about the presence and location of an object; this information can be added to the map. If we assume that we only contact one object (i.e. either a corner or wall) at a time, then the geometry of the bumpers allows us to make the following interpretations.

a. If two adjacent bumpers are contacted as in figure 5a, then the contact was in the corner and we can localize the contact point to that corner with a fairly high degree of certainty. Additionally, if that point is part of a wall, we have a bounds on the wall's direction relative to the robot.

b. If only one bumper is contacted, (figure 5b), then the point of contact can be considered uniformly distribed along the length and depth of the bumper.

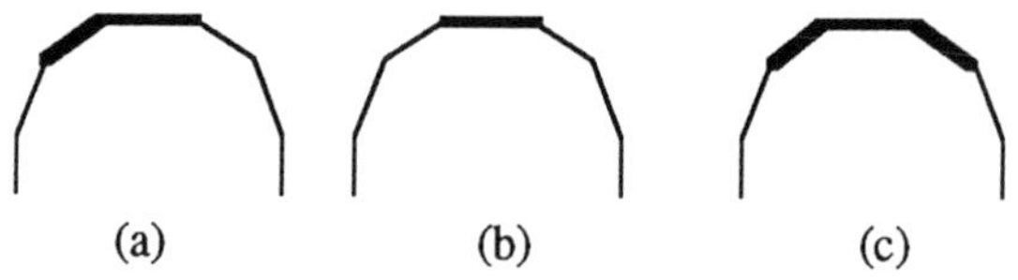

Figure 5: Bumper Contacts: (a) corner of robot contacts a corner or plane, (b) bumper contacts a corner, (c) bumper contacts a plane

c. Finally, if the contact is with three adjacent bumpers (figure 5c) then the contact is planar, and the direction of contact is that of the central bumper with an uncertainty based on the depth of the bumpers.

These interpretations have been used for passing through doorways where there is only one contact at any time.

4.3 Ultrasonics

Currently, twelve Polaroid ultrasonic sensors [8] surround the robot, and provide direct range information at 10 readings per second. The sensing rate is due to the speed of sound and time required for a previous chirp to be attenuated by air. From experimentation, which has been confirmed by others[7, 4], we have found that the system is sensative over approximately 30°. This rather wide spread is one of the complications of using ultrasonics because it tends to blur edges and corners. The other major drawback is specularities due to the long acoustic wavelength ($\sim .6cm$). Many objects have surfaces that are smooth compared to the acoustic wavelength causing the transmitted sound to be specularly reflected, and so the nearest object may not cause the first echo.

Figure 6 shows an overhead view of a portion of a hallway and figure 7 presents a rotational scan of the hallway at five degree increments with five readings taken at each position. The lines represent the sensor's central axis, and the distance from the inner end-point of the lines to the circles on that line are the distances of the each sonar reading. Some of the interesting observations about this scan are noted below and are marked by letters in the figure and text.

First there is small amount of noise on each sensor reading (a) which leads to a rather tight distribution of distances especially when the echo is strong such as when the axis of the sensor is close to the normal of

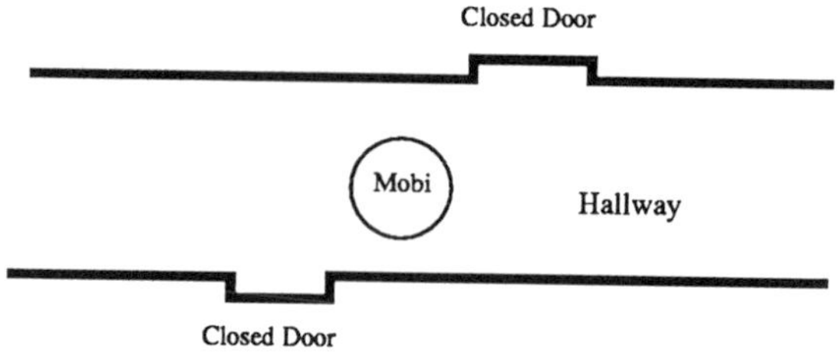

Figure 6: Overhead view of the hallway

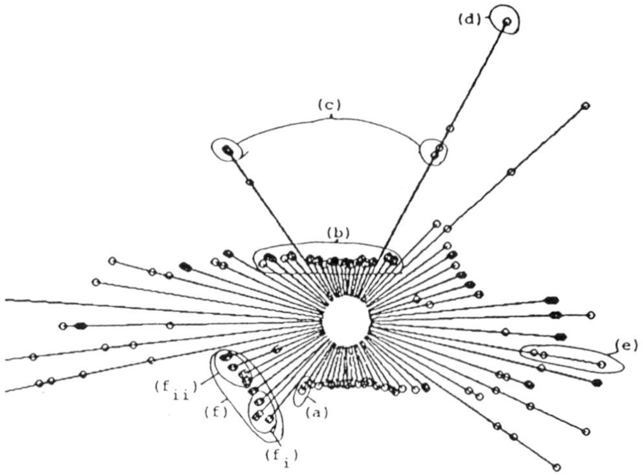

Figure 7: Sonar scan of the hallway in figure 6

the surface. This uncertainty can be easily modeled as a Gaussian stochastic process. Straight lines can be readily extracted to determine the position and direction of the wall as in (b) using a least mean square best line fit. This has been applied to a wall following algorithm. Also specularities are seen at some angles due to the smoothness of the walls at larger grazing angles as point (c) indicates. Note that a specular reflection off of the near wall and then a good back reflection by the far wall causes this reading. Point (d) shows a reading where there were actually reflections off both walls. In the direction down the hall (e) shows that larger variation in the sensor readings due to the larger distances and angles of incidence with all surfaces.

A final interesting point to note is (f) which is caused by the corner of a closed door and its molding. The doors are very smooth and quite specular so even at moderate incidence angles, there are reflections. However, the door edge is slightly recessed (7 cm), and this forms a dihedral where any ultrasonic energy that is close enough to the corner will

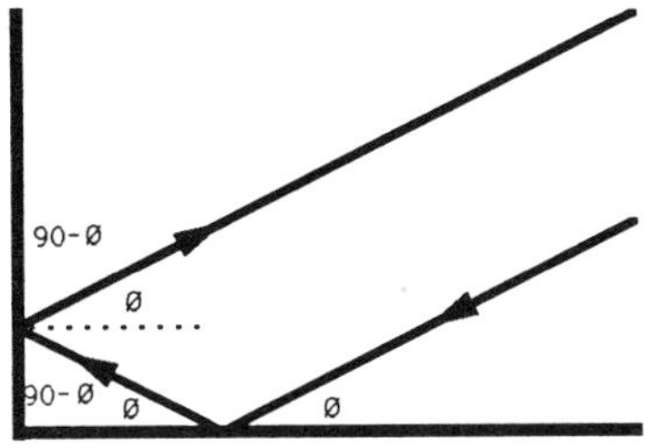

Figure 8: Acoustic reflections in a dihedral

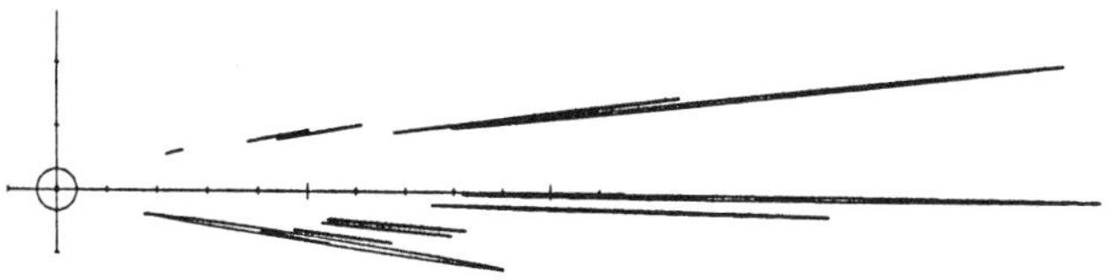

Figure 9: Stereo correspondance points in a hallway from a single position

be specularly reflected off one edge and then reflected by the second edge, leaving in the same direction as the incoming chirp as shown in figure 8. Because the reflections occur over the entire length and width of the dihedral, a very strong echo is received. Given that there are no objects closer than the Dihedral, the width of the apparent dihedral will be on the order of the sensor beam width as shown by the points highlighted in (f_{ii}). Note that a corner such as a post will also yield the same width reading however the echo will not be nearly as strong. The readings in (f_i) are interesting because the acoustic chirp's first point of contact is the door, but this is specularly reflected, and the dihedral causes the first returned echo.

After a scan such as figure 7 we can extract straight line segments such as were highlighted by (b) If its length is on the order of a beam width such as the one extracted by (f), then there are two possible interpretations, either a straight line or a corner (dihedral). Additionally, these straight line readings inform us that the region between the robot and the segment is free space. The consistency of these features with the map can be ascertained, and the map can be updated. Additionally, based on the extracted straight lines and the world model, specularities such as points (c) and (d) can be found. Finally, those readings that can not be modeled as straight lines or corner points can be added to the map as representing a surface patch that could lie anywhere along 30° arc. Because the surface patch causing the echo is more likely near the axis of the sensor[7], and to make the representation more uniform with other point locations, a chord approximates the arc, and the location is represented as a normal distribution. The patch's mean is along the sensor axis, and there is a large variance along the chord and a small variance in the direction of the sensor axis.

4.4 Vision

The stereo vision system described in [10] locates vertical edges in space. These are incorporated into the map as extending down to the ground plane and upward to the ceiling, and so only the (x,y) location is of concern. As with the other sensor measurements, there is uncertainty in the location of the vertical which we describe by a normal distribution with a mean vector of $\vec{\mu} = (\mu_x \mu_y)^T$ and a covariance matrix:

$$\Sigma = \frac{4d^2}{(\mu_{x_r} - \mu_{x_l})^4}\sigma^2 \begin{pmatrix} \mu_{x_l}^2 + \mu_{x_r}^2 & -f(\mu_{x_r} + \mu_{x_l}) \\ -f(\mu_{x_l} + \mu_{x_r}) & 2f^2 \end{pmatrix} \tag{6}$$

where μ_{x_l} and μ_{x_r} are the locations of the edges in the left and right images. σ^2 is the variance in the location of image vertical, f is the focal length of the lenses and $2d$ is the stereo baseline. This has been derived in Appendix A. Figure 9 shows the resulting error ellipses of a stereo image looking down the hallway. The ellipses grow quickly in length with respect to the mean distance, becoming long and narrow. Also, notice the one large nearby ellipse caused by poor localization of a low resolution edge.

Stereo correspondence points are generally the most accurate form of sensed measurement available. Of course, one must consider that stereo has a high computational cost and only covers a rather narrow field of view. An interesting point is that at even moderate distances, the uncertainty in distance measurement from stereo becomes larger than angular uncertainty which is complementary to the acoustics that have broad angular resolution and good depth accuracy.

5 Reducing Uncertainty

Generally, it is important to be able to recognize that two observed points are either the same or part of the same structure. If they are the same, then they will have similar characteristics (e.g. color, grey level gradient) as well as proximity as measured from two different frames. Additionally, points must be aggregated to instantiate a model. This can be more easily accomplished if uncertainty is reduced. From experience, we have found the most influential source of uncertainty when fusing information from different scenes is the motion transform between them. Furthermore, uncertainty is compounded across multiple transforms, so aggregating points between views that are a large distance apart becomes more difficult; thus, uncertainty in odometry must be reduced.

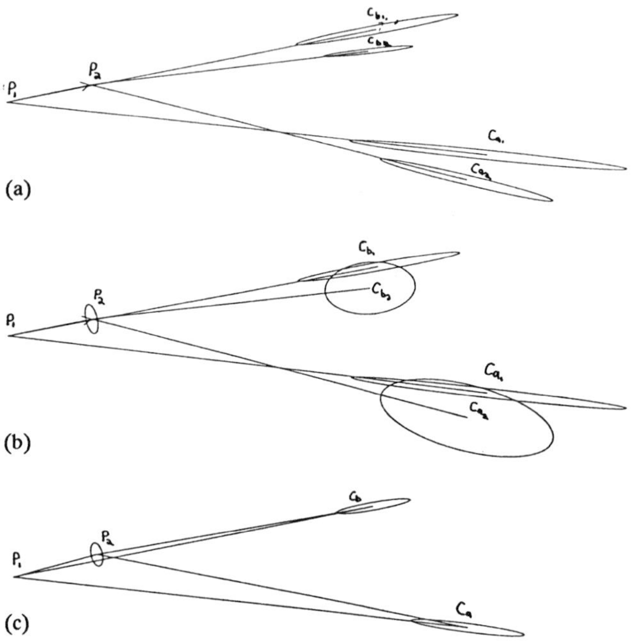

Figure 10: Uncertainty reduction, merging two views: (a) Two points seen from different locations, (b) Uncertainty with respect to P_1, (c) Reduced uncertainty

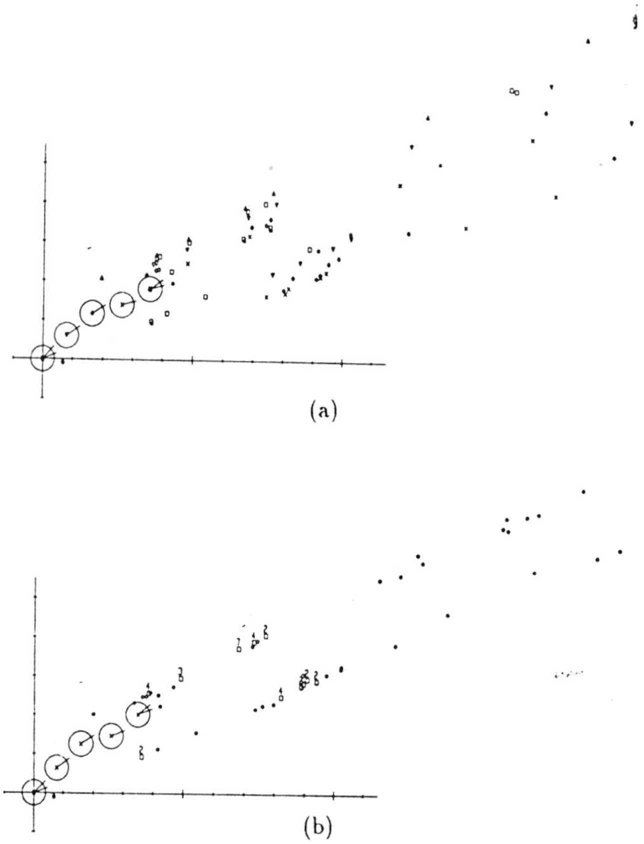

Figure 11: Motion sequence: (a) Location of correspondence points with odometry determining the robot's position, (b) Location of the same points after kalman filtering

Uncertainty can reduced by applying a Kalman filter to the sensor information and this is explained through an example using stereo vision as the sensor in Figure 10. Consider, (in figure 10a) the uncertainty ellipses of two points (C_a and C_b), determined from equation 6, that were viewed from two different locations, (P_1 and P_2). This figure shows that there was an error in motion transform, $T_{1,2}$ as determined by odometry because correspondence points C_{a_1} and C_{a_2} should be coincident as should C_{b_1} and C_{b_2}. Given the uncertainty in robot location, P_2, which is determined from equations 4 and 5, and represented by the ellipse about its mean in figure 10b. The uncertainty in the location of the stereo points C_{a_2} and C_{b_2} with respect to location P_1 are found by compounding [9], also shown in figure 10B. Correspondence between vertical edges from two successive stereo views can be determined by using the edge characteristics. In this case $C_{a_1} \equiv C_{a_2}$ and $C_{b_1} \equiv C_{b_2}$, so we can apply a Kalman filter to reduce the uncertainty in the stereo correspondences and motion transform. The transform and the correspondence points as seen from P_1 become our initial state vector along with their covariances. The stereo measurements as well as the transform are assumed to be independent. The correspondence points as seen from P_2 becomes the measurement vector, and the Kalman filter [5] is applied to obtain a better estimate of the state as shown in figure 10c. Note that the motion uncertainty has been reduced as has the location of the correspondence points as viewed from both positions.

This technique can be used to integrate motion sequences and build models from multiple views. Figure 11a shows a motion sequence composed of correspondence points found from eight positions. The scale is 1 meter, and the shape of the mark of the correspondence and the mark at the center of the robot indicates the robot's position when sensing that point. In this figure, the robot's location was determined purely by odometry. Between each successive location, motion correspondences were determined along with the error models of odometry and stereo. The Kalman filter was applied, and the points in figure 11b form the two walls of a hallway, demonstrating that motion uncertainty was greatly reduced. The circles were seen only once while the squares depict merged points. The observant reader will notice that one of the correspondence points is touching mobi's perimeter; Mobi crashed in this run.

6 Conclusions

We have presented the architecture of Stanford mobile robot and described the modeling of information from the four sensor sources, stereo vision, acoustics, bumpers and odometry. This information is used to update a world model whose hierarchical representation ranges from symbolic information to sensor data. All locations in the world model are relative to other points and measurement uncertainty is explicitly represented by multivariate normal distributions. Uncertainty can be reduced by applying a Kalman filter to the data and has been applied to reducing motion uncertainty.

Appendix: Derivation of Uncertainty in Stereo

Given a stereo vision system of two cameras with parallel image planes and horizontal epipolar lines, the uncertainty in the location of stereo correspondence point can be found (equation 7). Consider the overhead view of the stereo arrangement in figure 12 during this derivation. Only the projection onto the ground plane is considered. Given the focal length of the cameras, f, the baseline distance, $2d$ between

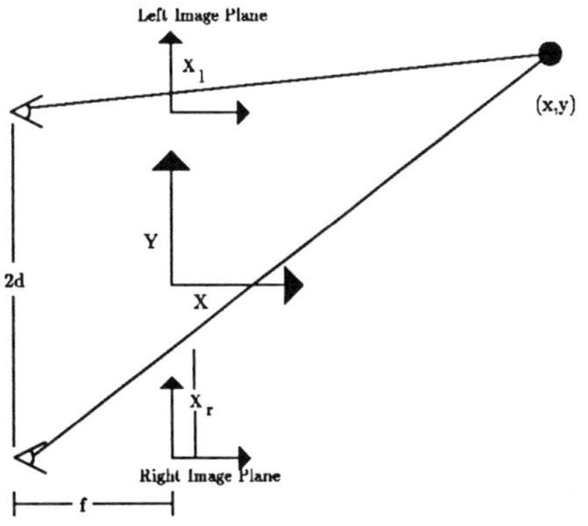

Figure 12: The stereo setup

cameras, and the location of the points in the image plane, x_l and x_r, which are described by a normal distribution. Thus, the location of a point in the image is given by a mean μ_{x_i} and variance $\sigma_{x_i x_i}$. From [1] we know that the location of the point in space is:

$$\vec{x} = \begin{pmatrix} x \\ y \end{pmatrix} = \begin{pmatrix} F(x_l, x_r) \\ G(x_l, x_r) \end{pmatrix} = \begin{pmatrix} \frac{d(x_r + x_l)}{x_r - x_l} \\ f - \frac{2df}{x_r - x_l} \end{pmatrix} \tag{7}$$

Since a linear function of normal random variables is normal, nonlinear F and G are linearized about the mean of x_l and x_r with a Taylor Series expansion:

$$\begin{aligned} x \approx\ & F(\mu_{x_l}, \mu_{x_r}) + \left.\frac{\partial F(x_l, x_r)}{\partial x_l}\right|_{\substack{x_l = \mu_{x_l} \\ x_r = \mu_{x_r}}} (x_l - \mu_{x_l}) \\ & + \left.\frac{\partial F(x_l, x_r)}{\partial x_r}\right|_{\substack{x_l = \mu_{x_l} \\ x_r = \mu_{x_r}}} (x_r - \mu_{x_r}) \\ =\ & d\frac{\mu_{x_r} + \mu_{x_l}}{\mu_{x_r} - \mu_{x_l}} + \frac{2d\mu_{x_r}}{(\mu_{x_r} - \mu_{x_l})^2}(x_l - \mu_{x_l}) \\ & - \frac{2d\mu_{x_l}}{(\mu_{x_r} - \mu_{x_l})^2}(x_r - \mu_{x_r}) \end{aligned} \tag{8}$$

and similarly.

$$y \approx f - \frac{2df}{\mu_{x_r} - \mu_{x_l}} - \frac{2df}{(\mu_{x_r} - \mu_{x_l})^2}(x_l - \mu_{x_l}) + \frac{2df}{(\mu_{x_r} - \mu_{x_l})^2}(x_r - \mu_{x_r}) \tag{9}$$

Now, the Jacobian, J, of the transform is:

$$J = \begin{pmatrix} \frac{\partial x}{\partial x_l} & \frac{\partial x}{\partial x_r} \\ \frac{\partial y}{\partial x_l} & \frac{\partial y}{\partial x_r} \end{pmatrix} = \begin{pmatrix} \frac{2d\mu_{x_r}}{(\mu_{x_r} - \mu_{x_l})^2} & \frac{-2d\mu_{x_l}}{(\mu_{x_r} - \mu_{x_l})^2} \\ \frac{-2df}{(\mu_{x_r} - \mu_{x_l})^2} & \frac{2df}{(\mu_{x_r} - \mu_{x_l})^2} \end{pmatrix} \tag{10}$$

so the mean of the location of the point in space is determined from equation 7:

$$\vec{\mu} = \begin{pmatrix} \mu_x \\ \mu_y \end{pmatrix} = \begin{pmatrix} F(\mu_{x_l}, \mu_{x_r}) \\ G(\mu_{x_l}, \mu_{x_r}) \end{pmatrix} \tag{11}$$

and the covariance is:

$$\Sigma = \begin{pmatrix} \sigma_{xx} & \sigma_{xy} \\ \sigma_{xy} & \sigma_{yy} \end{pmatrix} = J \begin{pmatrix} \sigma_{x_l}^2 & 0 \\ 0 & \sigma_{x_r}^2 \end{pmatrix} J^T \tag{12}$$

and assuming that $\sigma_{x_l}^2 = \sigma_{x_r}^2 = \sigma^2$ then the covariance matrix of a stereo correspondence point is:

$$\Sigma = \frac{4d^2}{(\mu_{x_r} - \mu_{x_l})^4}\sigma^2 \begin{pmatrix} \mu_{x_l}^2 + \mu_{x_r}^2 & -f(\mu_{x_r} + \mu_{x_l}) \\ -f(\mu_{x_l} + \mu_{x_r}) & 2f^2 \end{pmatrix} \tag{13}$$

Acknowledgements

We'd like to thank Soon Yau Kong, Ron Fearing, Giora Gorali, Shaul Fishman and Joel Burdick, for all their help and suggestions.

References

[1] D.H. Ballard and C.M. Brown. *Computer Vision*. Prentice-Hall, Englewood Cliffs, NJ, 1982.

[2] R.A. Brooks. Visual map making for a mobile robot. In *IEEE International Conference on Robotics and Automation*, 1985.

[3] R. Chatila and J.P. Laumond. Position referencing and consistent world modeling for mobile robots. In *IEEE International Conference on Robotics and Automation*, 1985.

[4] Anita M. Flynn. Redundant sensors for mobile robot navigation. Artificial Intelligence Laboratory AI-TR-859, MIT, October 1985. MS Thesis.

[5] T. Kailath. *Lectures on Wiener and Kalman Filtering*. Springer-Verlag, 1981.

[6] Y. Kanayama and N. Miyake. Trajectory generation for mobile robots. In *International Symposium on Robotics Research*, 1985.

[7] Hans P. Moravec and Alberto Elfes. High resolution maps from wide-angle sonar. In *IEEE International Conference on Robotics and Automation*, pages 1151–6, April 1986.

[8] Ultrasonic ranging system. Commericial Battery Division, Polaroid Corp., Cambridge, Mass., 1984.

[9] R Smith and P. Cheeseman. Estimating uncertain spatial relationships in robotics. In *AAAI Workshop on Uncertainty in Artificial Intelligence*, August 1986.

[10] E. Triendl and D.J. Kriegman. Stereo vision and navigation within buildings. In *IEEE International Conference on Robotics and Automation*, 1987.

Author Index

Subject Index

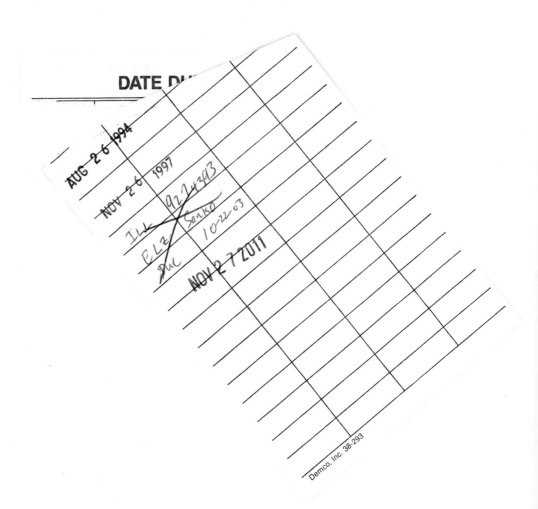